Modern Chemical Enhanced Oil Recovery

Modern Chemical Enhanced Oil Recovery

Theory and Practice

James J. Sheng, Ph. D.

ELSEVIER

AMSTERDAM • BOSTON • HEIDELBERG • LONDON
NEW YORK • OXFORD • PARIS • SAN DIEGO
SAN FRANCISCO • SINGAPORE • SYDNEY • TOKYO
Gulf Professional Publishing is an imprint of Elsevier

G P P

Gulf Professional Publishing is an imprint of Elsevier
30 Corporate Drive, Suite 400
Burlington, MA 01803, USA

The Boulevard, Langford Lane
Kidlington, Oxford, OX5 1GB, UK

Library of Congress Cataloging-in-Publication Data

Sheng, James J.
Modern chemical enhanced oil recovery : theory and practice / James J. Sheng.
p. cm.
ISBN 978-1-85617-745-0
1. Enhanced oil recovery. 2. Oil reservoir engineering. 3. Oil fields—Production methods. I. Title.
TN871.S516 2010
622'.33827–dc22 2010026763

British Library Cataloguing-in-Publication Data

A catalogue record for this book is available from the British Library.

For information on all Gulf Professional Publishing publications visit our Web site at *www.elsevierdirect.com*

Printed in the United States
10 11 12 13 14 10 9 8 7 6 5 4 3 2 1

Contents

Preface

With growing global energy demand and depleting reserves, enhanced oil recovery (EOR) from existing or brown fields has become more and more important. Among the various enhanced oil recovery methods, chemical EOR has been labeled an expensive method, and field applications have been almost completely stopped during the past two decades worldwide except China, although limited university research was continued. Because we are facing the difficulty of replacing depleting reserves with "cheap" oil and rising oil price, chemical EOR has drawn increasing interest from oil companies, especially national oil companies.

In particular, technical references on chemical EOR are needed because petroleum professionals have not been trained in this area in the past 20 years. Except for some chapters in a few books that discuss general EOR, a comprehensive and systematic chemical EOR book has not been published. The purpose of this book is to complement the current literature on EOR. More important, it summarizes the results of research, pilot tests, and field applications in China because oil companies and research organizations there have continually made the effort to develop and apply chemical EOR technology during the past three decades.

This book is written mainly for petroleum professionals. Because overwhelming parameters are needed to describe a chemical EOR process, it is not practical to measure every one of them; therefore an effort has been made to collect, synthesize, and summarize available data, especially Chinese information that is inaccessible in Western literature. An effort has also been made to cover comprehensively the fundamental theories and practices related to alkaline (A), surfactant (S), and polymer (P) flooding processes, especially alkaline-surfactant-polymer (ASP) flooding that has barely been discussed in any enhanced oil recovery book in English.

Many pilot studies and field cases have been summarized; an effort has been made to select these studies and cases so that each one addresses unique issues. This book also proposes some new concepts and ideas or hypotheses. Several of them need to be validated by further research, and some may stimulate other research interests. From this standpoint, this book could be useful to researchers. The basic theories and sample calculations should help students and professionals who are less experienced in this area. This book also may be used by environmental engineering professionals who work on cleaning up wastes and nonaqueous phase liquids (NAPL). In addition, an effort has been made to strike an ideal balance between theory and practice; in addition, extensive references are provided.

The flow of logic of this book is as follows:

- Chapter 1 introduces general EOR and how this book is organized.
- Chapter 2 discusses the fundamentals of chemical transport and fractional flow analysis.
- Chapter 3 reviews salinity and ion exchange, and the effect of salinity on waterflooding.
- Chapter 4 proposes a new mobility control requirement for enhanced oil recovery processes.
- Chapter 5 presents fundamentals and field practices of polymer flooding.
- Chapter 6 reviews polymer viscoelastic behavior.
- Chapter 7 discusses the fundamentals, concepts, and issues related to surfactant flooding.
- Chapter 8 proposes new concepts of optimum salinity type and optimum salinity profile in surfactant flooding.
- Chapter 9 discusses surfactant-polymer interactions.
- Chapter 10 presents the fundamentals and modeling of alkaline flooding.
- Chapter 11 discusses alkaline-polymer interactions.
- Chapter 12 discusses alkaline-surfactant synergy.
- Chapter 13 focuses on emulsion and ASP field applications.

Acknowledgments

While doing research on chemical EOR, I had opportunities to discuss the subject with several gurus and many experts. I sincerely appreciate those opportunities. Most of all, I greatly appreciate the insights gained from Dr. Gary Pope, professor at the University of Texas at Austin, through his explanation and stimulating discussion of some controversial and sometimes confusing issues. Such appreciation is extended to Dr. Larry Lake, professor at the University of Texas at Austin, and Dr. George Hirasaki, professor at Rice University. I am grateful to Dr. Mojdeh Delshad, professor at the University of Texas at Austin, for her help in using UTCHEM, a chemical simulator developed at the University of Texas at Austin.

I also appreciate the stimulating and valuable discussions I have had with these chemical EOR experts: Larry N. Britton, University of Texas at Austin; Maurice Bourrel, TOTAL Petrochemical; Danielle Morel, Pascal Gauer, and Gilles Bourdarot, TOTAL E&P; Ramon Bentsen, University of Alberta; Brij Maini, University of Calgary; Michael Prats and Scott Wellington, Shell E&P; Tor Austad, University of Stavanger; Shunhua Liu, Occidental Oil and Gas; and Harry L. Chang, Chemor Tech International. I am also grateful to Wilson Chin of StrataMagnetic Software for his advice to write a book. My thanks also go to Kenneth P. McCombs, Irene Hosey, Marilyn Rash, and the staff at Elsevier for their support, which made this book's writing more enjoyable.

My biggest thanks must go to my wife, Ying Zhang, for her patience, understanding, and support. I owe much to my daughters, Emily and Selena, for the time I should have shared with them as their father. I am also thankful to my greater family for their continuing support. Finally, I am indebted to the authors and publishers who have given me the permission to use their results and copyrighted materials.

Nomenclature

A	(1) area, L^2, m^2, or (2) molecular interaction in R ratio, or (3) pre-exponential factor in Eq. 10.40, or (4) alkaline flooding
$\mathbf{A^-}$	in situ generated anionic surfactant (soap)
$\mathbf{a_0}$	cross-sectional area occupied by the hydrophilic group at the micelle surface, L^2
$\mathbf{A_H}$	empirical parameter in the Hand equation
$\mathbf{A_{H0}}$	A_H at $C_{seD} = 0$ (very low salinity)
$\mathbf{A_{H1}}$	A_H at $C_{seD} = 1$ (optimum salinity)
$\mathbf{A_{H2}}$	A_H at $C_{seD} = 2$ (two times optimum salinity)
$\mathbf{a_L}$	empirical constant in the Langmuir equation
$\mathbf{a_i}$	empirical constant in the Langmuir equation for component i adsorption (i = 4 or p for polymer, 3 for surfactant)
$\mathbf{a_{i1}}$	empirical constants to define a_i
$\mathbf{a_{i2}}$	empirical constant to define a_i, unit of the inverse of salinity
AN	acid number, mg KOH/g oil
$\mathbf{A_p}$	frequency factor in the polymer viscosity equation in Eq. 5.6
$\mathbf{A_{pi}}$	fitting constants in the polymer viscosity correlation, i = 1, 2, 3
$\mathbf{B_H}$	empirical parameter in the Hand equation
$\mathbf{b_{kr}}$	empirical parameter to define F_{kr}, unit of the inverse of C_p
$\mathbf{b_L}$	empirical constant in the Langmuir equation, unit of the inverse of concentration
$\mathbf{b_i}$	empirical constant in the Langmuir equation for component i adsorption (i = 4 or p for polymer, 3 for surfactant), unit of the inverse of C_p
$\mathbf{B_w}$	water formatin volume factor, L^3/L^3
C	(1) concentration, m/L^3, meq/mL, or (2) constant defined in Eq. 4.16
C′	(1) constant defined in Eq. 4.18, or (2) constant to define the elongation pressure drop in Eq. 6.22
$\mathbf{C_0}$	initial concentration at t = 0, m/L^3
$\mathbf{C_{33}\vert_{CseDm}}$	C_{33} at C_{seDm} (m = 0, 1, 2 for C_{seD} = 0, 1, and 2, respectively)
$\mathbf{C_3^m}$	surfactant concentration in the microemulsion phase, meq/mL water
$\mathbf{C_{33maxm}}$	parameters related to C_{33} and calculated using Eq. 7.14 at C_{seD} = 0, 1, 2
$\mathbf{C_{33max,km}}$	parameters related to C_{33} in the presence of alcohol at C_{seD} = 0, 1, 2

$C_{33max,tm}$	parameters related to C_{33} in the presence of two alcohols at $C_{seD} = 0, 1, 2$
C^*_{51}	anion concentration in the absence of alcohol or divalents, meq/mL
C^*_{51op}	optimum anion concentration in the absence of alcohol or divalents, meq/mL
C^s_6	divalent concentration bounded to surfactant micelles, meq/mL water
C_c	empirical constant to define the elongational viscosity in Eq. 6.10
CEC	(1) cation exchange capacity, various units, or (2) critical electrolyte concentration (Chapter 9)
C_{el}	constant to define t_c
C_H	empirical constant in the Huh IFT equation
C_{Ho}	empirical constant in the Huh IFT equation for σ_{mo}
C_{Hw}	empirical constant in the Huh IFT equation for σ_{mw}
C_i	concentration of fluid species i, m/L^3, meq/mL, mol/L
$\bar{C}_i$	concentration of matrix-adsorbed solute i, meq/mL, mole/L
$\hat{C}_i$	adsorbed solute i concentration, various units
c_{kr}	empirical parameter to define F_{kr}
$\bar{\bar{C}}_m$	concentration of micelle-associated cation m, m/L^3, mole/L water
cP	unit of viscosity, centipoise
C_p	polymer concentration, m/L^3, wt.% or mg/L, or alkaline injection concentration (Chapter 10)
$\hat{C}_{ps}$	polymer adsorption on unit surface area m/L^2, mol/m^2
C_{se}	effective salinity, m/L^3, meq/mL
C_{seD}	C_{se}/C_{seop}
C_{sel}	lower effective salinity limit of a type III microemulsion, m/L^3, meq/mL
C_{seop}	optimum effective salinity of a microemulsion, m/L^3, meq/mL
C_{sep}	effective salinity for polymer, m/L^3, meq/mL
C_{seu}	upper effective salinity limit of a type III microemulsion, m/L^3, meq/mL
C^t_n	total concentration of component n, m/L^3, mole/L
d	(1) unit of time, day, or (2) unit of permeability, darcy
D	depth, L, ft
D_0	diffusion coefficient in a bulk liquid or gas phase, L^2/t, m^2/s
d_{10}	grain diameter at 10% cumulative fraction, L
D_c	convective dispersion components, L^2/t, m^2/s
D_i	ratio of retained chemical (i) concentration to the injected, m/m
D_L	longitudinal dispersion coefficient, L^2/t, m^2/s
d_p	diameter of particles of a sand pack, L^2
D_T	transverse dispersion coefficient, L^2/t, m^2/s

D_τ	diffusion coefficient in the porous medium, L^2/t, m^2/s
E	energy, mL^2/t^2
E_a	activity energy of polymer solution
E_H	empirical parameter in the Hand equation
f	dilution factor
F	(1) unit of force (Nomenclature), or (2) flux, $m/t/L^2$
F_c	capillary force, F
F_H	empirical parameter in the Hand equation
F_I	inhomogeneity factor for the porous medium
F_{kr}	permeability reduction factor for porous media during polymer flow
F_{krr}	residual permeability reduction factor for porous media after polymer flow
FADS	UTCHEM input parameter to adjust surfactant adsorption due to polymer adsorption
F_r	resistance factor for porous media during polymer flooding
F_R	formation electrical resistivity factor
FR	filtration ratio
f_6^s	fraction of the total divalent cations bound to surfactant micelles
f_k^s	volume fraction of alcohol (k = 7, 8) in the total volume of surfactant and alcohol, L^3/L^3
F_{SP}	empirical parameter to adjust surfactant adsorption due to polymer adsorption
F_v	viscous force, F
f_w	water cut, L^3/L^3, %, fraction
f_{w3}	water cut at the displacement front in surfactant flooding, L^3/L^3, %, fraction
f_{we}	water cut at the effluent, L^3/L^3, %, fraction
f'_{we}	$(\partial f_w/\partial S_w)_{S_{we}}$ at the effluent
f_{wf}	water cut at the waterflood front, L^3/L^3, %, fraction
f'_{wf}	$(\partial f_w/\partial S_w)_{S_{wf}}$ at the waterflood front (specific velocity)
f_{wi}	water cut at the initial water saturation, L^3/L^3, %, fraction
f_{wp}	water cut at the displacement front in polymer flooding, L^3/L^3, %, fraction
g	acceleration of gravity, L/t^2
G'	elastic or storage modules, m/Lt^2
G''	viscous or loss modulus, m/Lt^2
G^*	complex dynamic modulus
G_{en}	contribution to the elastic modulus due to polymer chain entanglement
HA_o	pseudo-acid component in oil, m/L^3, mol/L
HA_w	pseudo-acid component in water, m/L^3, mol/L
HBNC70	UTCHEM input parameter $C_{seD} = 0$, vol.%
HBNC71	height of binodal curve at optimum $C_{seD} = 1$ in UCHEM, vol.%

HBNC72	UTCHEM input parameter at $C_{seD} = 2$, vol.%
HEC	hydrogen exchange capacity, m/L^2, meq/m^2
$\mathbf{H_{pi}}$	fitting parameters in the Healy et al. IFT correlation (p = 1 for σ_{mw}, 2 for σ_{mo}; i = 1, 2, 3)
I	ionic strength of a solution, unit of solubility
IFT	interfacial tension, dyne/cm, mN/m
k	permeability, L^2, md
$\bar{\mathbf{k}}$	average permeability, L^2, md
K	(1) constant fitting the power-law equation to describe the bulk viscosity, mt^{n-1}/L, or (2) Boltzmann constant (Chapter 6), or (3) partition coefficient (Chapter 7), or (4) equilibrium constant (Chapter 10)
$\mathbf{K_A}$	acid dissociation constant
$\mathbf{K_{A-B}}$	equilibrium constant for the ion exchange between *solute* ion A and ion B
$\mathbf{K^{G}_{A-B}}$	equilibrium constant for the ion exchange in the Gapon convention
$\mathbf{K^{G-T}_{A-B}}$	equilibrium constant for the ion exchange in the Gaines–Thomas convention
$\mathbf{K^{V}_{A-B}}$	equilibrium constant for the ion exchange in the Vaneslow convention
$\mathbf{K_D}$	partition coefficient of the molecular acid
$\mathbf{K_e}$	ion-exchange equilibrium constant, unit of the inverse of concentration
$\mathbf{K_F}$	empirical constant in the Freundlich isotherm
$\mathbf{k_H}$	horizontal permeability, L^2, md
$\mathbf{k_r}$	relative permeability, fraction
$\mathbf{k_{ro}}$	oil relative permeability, fraction
$\mathbf{k_{ro3}}$	oil relative permeability during polymer displacement, fraction
$\mathbf{k_{ro1}}$	oil relative permeability before polymer contact, S_w increasing, fraction
$\mathbf{k_{rop}}$	oil relative permeability after polymer contact, S_w increasing, fraction
$\mathbf{k_{rw}}$	water relative permeability, fraction
$\mathbf{k_{rwp}}$	water relative permeability after polymer contact, fraction
$\mathbf{K_s}$	surfactant partition coefficient, C_{32}/C_{31}
$\mathbf{k_v}$	vertical permeability, L^2, md
$\mathbf{k_{wr}}$	water relative permeability at the residual oil saturation, fraction
L	(1) unit of length (Nomenclature), or (2) unit of liter, or (3) outlet distance from the inlet, L
$\mathbf{L_c}$	length of the hydrophobic group
m	(1) unit of mass (Nomenclature), or (2) unit of length, m, or (3) unit of solubility, moles/kg solvent, or (4) order of reaction (Chapter 10)

M	(1) unit of solubility, moles/L solution, or (2) molecular weight, Dalton (Da)
$\mathbf{m_c}$	empirical constant to define the elongational viscosity in Eq. 6.10
md	millidarcy, unit of permeability
me or **ME**	microemulsion phase
$\mathbf{m_H}$	slope of the Hall plot, psi/STB/d
$\mathbf{m_{km}}$	slope of $C_{33max,km}$ vs. f_k^s (k = 7, 8; m = 0, 1, 2)
mPa·s	milliPascal·s, unit of viscosity,
$\mathbf{M_r}$	mobility ratio
$\mathbf{M_{roc}}$	ratio of displacing fluid mobility to oil mobility in an assumed oil channel, defined in Eq. 4.14
MW	molecular weight, Dalton (Da)
n	(1) exponent (Chapters 2, 3, and 7), or (2) distance in the direction normal to the oil/water interface, L (Chapter 4), or (3) polymer-specific empirical constant in the Carreau equation or in a power-law viscosity equation (Chapters 5 and 6), or (4) number density of crosslinkers (Chapter 6)
N	(1) number of molecules or moles (Chapter 2), or (2) solubility unit, meq/mL
$\mathbf{n_2}$	empirical constant to define the elongational viscosity in Eq. 6.11
$\mathbf{N_B}$	bond number
$\mathbf{N_C}$	(1) capillary number, or (2) number of components
$\mathbf{(N_C)_c}$	critical capillary number
$\mathbf{(N_C)_{max}}$	maximum desaturation capillary number
$\mathbf{(N_C)_t}$	total desaturation capillary number
$\mathbf{N_{Da}}$	Damköhler number
$\mathbf{N_{De}}$	Deborah number
$\mathbf{N_F}$	number of freedom
$\mathbf{N_p}$	cumulative oil recovered in subsurface pore volume, L^3/L^3
$\mathbf{N_P}$	number of phases
$\mathbf{N_{Pe}}$	Peclet number
$\mathbf{N_T}$	trapping number
n(x)	normal density function
OOIP	original oil in place, bbl
p	pressure, m/Lt^2, Pa, MPa, or psi
P	polymer flooding
$\mathbf{p_\alpha}$	empirical or fitting parameter for polymer viscosity
$\mathbf{p_e}$	formation pressure, psi
PL	left plait point
$\mathbf{p_{of}}$	pressure at the front x_{of}, m/Lt^2
PR	right plait point
$\mathbf{p_{tf}}$	well tubing head pressure, psi

PV	pore volume, fraction or % of PV
$\overline{\mathbf{PV}}$	normalized injection PV in Eq. 1.1
$\mathbf{PV_0}$	pore volume when a chemical flood is started, PV
$\mathbf{PV_1}$	total injection pore volume of waterflood or chemical food at the final cutoff, PV
$\mathbf{PV_{1c}}$	total injection pore volume of chemical food at the final cutoff, PV
$\mathbf{PV_{1w}}$	total injection pore volume of waterflood at the final cutoff, PV
q, Q	flow or injection rate, L^3/t, cm^3/s, STB/d, m^3/d, ton/d
$\mathbf{Q_v}$	cation exchange capacity (CEC), meq/mL PV
r	(1) distance in the radial direction, L, m, or (2) reaction term, $m/t/L^3$
R	(1) gas constant (8.314 J/°K/mol), or (2) radius of a capillary or a pore, L, or (3) R-ratio (Chapter 7), or (4) solubilization ratio
$\mathbf{r_b}$	diameter of glass beads, L, cm
$\mathbf{R_{Ci}}$	retardation factor of concentration C_i
$\mathbf{R_D}$	alkaline net loss rate due to dissolution, m/tL^3
$\mathbf{R_E}$	(1) alkaline net loss rate due to ion exchange, m/tL^3, or (2) retardation factor due to ion exchange
RF	(1) oil recovery factor, fraction or %, or (2) resistance factor
$\mathbf{R_{i3s}}$	solubilization ratios based on surfactant volume only in the microemulsion phase (i = 1, 2)
$\mathbf{R_{i3t}}$	solubilization ratios based on total volume of soap and surfactant in the microemulsion phase (i = 1, 2)
RRF	residual resistance factor defined in the literature, which is the same as residual permeability reduction factor defined in this book, F_{krr}
s	unit of time, second
S	(1) saturation, L^3/L^3, fraction, or (2) surfactant flooding
SG(–)	negative salinity gradient
SG(+)	positive salinity gradient
$\mathbf{S_{mro}}$	residual microemulsion saturation in microemulsion–oil conjugates, L^3/L^3, fraction
$\mathbf{S_{mrw}}$	residual microemulsion saturation in microemulsion–water conjugates, L^3/L^3, fraction
$\overline{\mathbf{S}}_\mathbf{o}$	normalized movable oil saturation in , L^3/L^3, fraction
$\mathbf{S_{oi}}$	initial oil saturation, L^3/L^3, fraction
$\mathbf{S_{orc}}$	residual oil saturation in surfactant flooding, L^3/L^3, fraction
$\mathbf{S_{orm}}$	residual oil saturation in oil–microemulsion conjugates, L^3/L^3, fraction
$\mathbf{S_{orw}}$	residual oil saturation in oil–water conjugates, L^3/L^3, fraction
$\mathbf{S_p}$	slope of $\dfrac{\mu_p^0 - \mu_w}{\mu_w}$ vs. C_{sep} on a log–log plot

SPI surfactant-polymer interaction or incompatibility
ΣS_{pr} sum of residual saturations of all the phases except the phase p, L^3/L^3, fraction or %
S_r pore surface area, L^2/m, m^2/g rock
$(SR)_s$ solubilization ratio when only the surfactant volume is used to define the ratio
$(SR)_{total}$ solubilization ratio when the total volume of surfactant and soap is used to define the ratio
$\bar{S}_w$ average water saturation, L^3/L^3, fraction
S_{w1} average water saturation from chemical denuded front to waterflood, L^3/L^3, fraction
$\bar{S}_{w1}$ water saturation in the chemical denuded zone, L^3/L^3, fraction
S_{w3} water saturation at the chemical (surfactant) front, L^3/L^3, fraction
S_{wb} water saturation at the boundary between injected water and initial water, L^3/L^3, fraction
S_{wc} connate (interstitial) water saturation, L^3/L^3, fraction
S_{we} water saturation at the effluent, L^3/L^3, fraction
S_{wf} water saturation at the waterflood front, L^3/L^3, fraction
S_{wi} initial water saturation, L^3/L^3, fraction
S_{wp} water saturation at the displacement front in polymer flooding, L^3/L^3, fraction
$\bar{S}_{wp}$ average water saturation in the polymer zone, L^3/L^3, fraction
S_{wrm} residual water saturation in water–microemulsion conjugates, L^3/L^3, fraction
S_{wro} residual water saturation in water–oil conjugates, L^3/L^3
t (1) unit of time (Nomenclature), or (2) time, t, s, or days
T (1) unit of temperature (Nomenclature), or (2) absolute temperature in °K or °C
t_c time scale of observation (characteristic time), t
Tc cloud point, T, °C
TDS total dissolved solids, ppm or %
TEC total exchangeable cations, mmol/kg rock
T_p parameter in a capillary desaturation curve equation for the phase p
t_r relaxation time, t, s
t_{re} residence time, t, s
u Darcy velocity, L/t, m/s, ft/d
U $\mathbf{U = (1 - V_i(t)/V_p)/\sqrt{V_i(t)/V_p}}$
v interstitial velocity, L/t, m/s, ft/d
$\bar{v}$ average velocity in a capillary or a pore, L/t, m/s
v_{C_i} velocity of concentration C_i, L/t, m/s
$v_{\Delta S_w}$ velocity at the water saturation shock, L/t, m/s

V	volume, L^3
$\mathbf{V_i(t)}$	injection pore volume at time t, L^3
$\mathbf{V_l}$	liquid molar volume of the substance, cm^3/mol
$\mathbf{V_p}$	pore volume, L^3, m^3, cm^3
We	Weissenberg number
$\mathbf{W_i}$	cumulative water injection, STB
WOR	water/oil ratio
x	distance, L, m, cm
X	(1) exchange site on the solid material (clay) (Chapter 3), or (2) mole fraction
$\mathbf{x_{D1}}$	front of chemical denuded zone
$\mathbf{x_{Df}}$	water front
$\mathbf{x_{Dp}}$	polymer concentration front
$\mathbf{x_{w3}}$	surfactant at the chemical (surfactant) front, L
$\mathbf{x_{of}}$	location of water front in the oil channel, L
$\mathbf{z_i}$	charge of ion i

Greek Symbols

α	(1) polymer-specific empirical constant in the Carreau equation (Eq. 5.5), or (2) dipping angle or angle formed by $\vec{x}$ and a vector
α_1–α_5	microemulsion phase viscosity parameters in Eq. 7.80
α_L	longitudinal dispersivity, L, m
α_T	transverse dispersivity, L, m
β	$= 1 - R_n/R_p$
β_6	effective salinity parameter for divalents (calcium)
β_6^s	slope parameter for divalents (calcium) in a surfactant system
β_7	slope parameter for alcohol dilution in a surfactant system
β_I	equivalent exchangeable fraction for ion I
β_I^M	molar exchangeable fraction for ion I
β_p	effective salinity parameter for divalents (calcium) to calculate C_{sep}
β_T	temperature coefficient in defining effective salinity, 1/T
ε	small perturbation, L
$\dot{\varepsilon}$	stretch rate, 1/t
γ	ratio of the displacement (strain) to its original length
$\dot{\gamma}$	shear rate, 1/t, 1/s
$\dot{\gamma}_{1/2}$	shear rate at which viscosity is the average of μ_p^0 and μ_w, 1/t, 1/s
$\dot{\gamma}_c$	empirical parameter to calculate polymer viscosity in porous media defined in Eq. 5.26
$\dot{\gamma}_{eq}$	equivalent shear rate in the porous medium, 1/t, 1/s
$\dot{\gamma}_w$	shear rate at the wall of a capillary or a pore, 1/t, 1/s
Δ	operator that refers to a discrete change

ΔH_r	reaction enthalpy, J/mol
ΔH_r^0	reaction enthalpy at 25°C, J/mol
ϕ	porosity, fraction or %
ϕ_{IPV}	inaccessible pore volume, L^3/L^3, fraction
Φ	packing factor
Φ_p	potential of displacing fluid, m/Lt^2
ρ	density, m/ L^3, g/cm^3
λ	(1) mobility, L^3t/m (Chapter 4), or (2) polymer-specific empirical constant in the Carreau equation (Eq. 5.5)
λ_2	empirical constant to define elongational viscosity
λ_r	relative mobility, L^3t/m
μ	viscosity, m/Lt, mPa·s (cP)
μ'	dynamic viscosity defined in Eq. 6.6, m/Lt
μ''	dynamic viscosity defined in Eq. 6.7, m/Lt
μ^*	complex viscosity
μ_∞	polymer viscosity at infinite shear rate (solvent viscosity), m/Lt
μ_{app}	(1) apparent polymer viscosity in porous media, m/Lt, or (2) apparent viscosity of viscoelastic polymer solution, m/Lt
μ_{el}	elongational viscosity, m/Lt
μm	micrometer, unit of length
μ_{max}	empirical constant to define maximum elongational viscosity
μ_p	polymer viscosity, m/Lt, mPa·s
μ_p^0	polymer viscosity at zero shear rate, m/Lt, mPa·s
μ_{sh}	shear-thinning viscosity, m/Lt, mPa·s
μ_w	water viscosity, m/Lt, mPa·s
σ	(1) normal stress, m/Lt^2, or (2) interfacial tension, m/t^2, mN/m
σ^2	variance
τ	tortuosity of the porous medium
τ_{12}	shear stress, m/Lt^2
$\tau_{11} - \tau_{22}$	first normal stress difference, m/Lt^2, Pa
τ_r	shear stress at r in the pipeline, m/Lt^2
τ_{rr}	normal stress, m/Lt^2
θ	(1) phase shift, or (2) contact angle
ω	(1) angular frequency, 1/t, or (2) interpolation parameter to define k_{rm}

Superscripts

–	(1) adsorption associated with matrix through ion exchange (Chapter 10), or (2) average, or (3) normalized
^	adsorbed
=	adsorption associated with micelles
0	initial
′	(1) derivative, or (2) transformed coordinate system (Chapter 7)

*****	(1) limiting cases for the left and right plait points, or (2) the salinity in the absence of alcohol or divalents
e	end point (maximum saturation)
eq	equilibrium, used in an equilibrium constant
ex	ion exchange on matrix, used in an ion exchange constant
exm	ion exchange on micelles, used in an ion exchange constant
n	time step
opt	optimum
s	bounded to surfactant micelles
sp	solubility product in a solubility product constant
w	exponent of concentration (coefficient of concentration in the reaction equation)

Subscripts

0	at $c_{seD} = 0$
1	(1) water component, or (2) at $c_{seD} = 1$
2	(1) oil component, or (2) at $c_{seD} = 2$
3	surfactant
a	(1) ahead of the displacing front, or (2) air
(a)	aqueous
A	advancing
b	(1) behind the displacing front, or (2) bead
bt	breakthrough
C	amphiphilic membrane
d	downstream
D	(1) dimensionless, or (2) rock dissolution by alkali
el	(1) elongational, or (2) ion exchange
f	flowing
h	surfactant head
i	(1) initial, or (2) inlet, or (3) species index (first position on composition variables)

- **1** water
- **2** oil
- **3** surfactant
- **4** polymer
- **5** anion
- **6** divalents
- **7** cosolvent 1 (alcohol 1)
- **8** cosolvent 2 (alcohol 2)
- **o** oil
- **p** polymer
- **s** surfactant

j	phase index (second subscript in composition variables) **1** water-rich phase **2** oil-rich phase **3** microemulsion **w** water, or water-rich phase **o** oil-rich phase **m** microemulsion
inj	injection
l	surfactant tail
M	invariant point
m	(1) mixture, or (2) microemulsion phase
n	pore throat (neck)
nw	non-wetting phase
o	(1) outlet, or (2) oil
ob	oil bank
of	oil front
op	optimum
p	(1) polymer, or (2) pore body, or (3) phase or displacing phase
p′	displaced phase
P	plait point
PL	left plait point
PR	right plait point
r	(1) rock, or (2) residual
R	receding
ref	reference
s	surfactant
(s)	solid
se	effective salinity
sh	shear-thinning
si	silicate
t	total
u	upstream
w	(1) water, or (2) wet phase, or (3) well
wb	water boundary between the chemical denuded zone and the initial water zone
wc	interstitial connate water
wf	(1) water front, or (2) well flowing
x, y, z	in x, y, z direction

Chapter 1

Introduction

1.1 ENHANCED OIL RECOVERY'S POTENTIAL

Today fossil fuels supply more than 85% of the world's energy. Currently, we are producing roughly 87 million barrels per day—32 billion barrels per year in the world. That means every year the industry has to find twice the remaining volume of oil in the North Sea just to meet the target to replace the depleted reserves. Of the 32 billion barrels produced each year, almost 22 billion come out of sandstone reservoirs. The reserves and production ratios in sandstone fields have around 20 years of production time left. The proven and probable reserves in carbonate fields have around 80 years of production time left (Montaron, 2008). With global energy demand and consumption forecast to grow rapidly during the next 20 years, a more realistic solution to meet this need lies in sustaining production from existing fields for several reasons:

- The industry cannot guarantee new discoveries.
- New discoveries are most likely to lie in offshore, deep offshore, or difficult-to-produce areas.
- Producing unconventional resources would be more expensive than producing from existing brown fields by enhanced oil recovery (EOR) methods.

Figure 1.1 shows the US oil volume distribution in 1993 (Green and Willhite, 1998). The total oil discovered up to 1993 was 536 billion barrels, with the total produced being 162 billion barrels (30% of the total discovered) and the reserves being 23 billion barrels (4% of the total discovered). This is the number that could be produced economically using conventional methods. The remaining oil in the reservoirs was 351 billion barrels, or 66% of the total discovered. If EOR can recover half of the remaining (i.e., 176 billion barrels), then we could double the currently projected recoverable reserves. Similarly, we could have additional reserves of 2 trillion barrels worldwide.

1.2 DEFINITIONS OF EOR AND IOR

Depending on the producing life of a reservoir, oil recovery can be defined in three phases: primary, secondary, and tertiary. *Primary recovery* is recovery by natural drive energy initially available in the reservoir. It does not require

Modern Chemical Enhanced Oil Recovery. DOI: 10.1016/B978-1-85617-745-0.00001-2

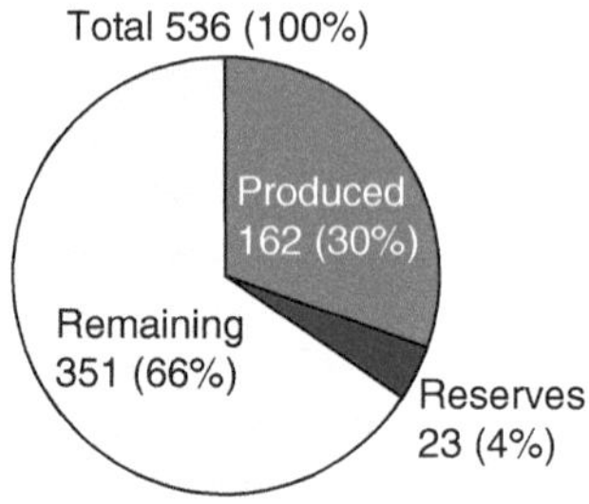

FIGURE 1.1 US oil volume distribution in 1993.

injection of any external fluids or heat as a driving energy. The natural energy sources include rock and fluid expansion, solution gas, water influx, gas cap, and gravity drainage. *Secondary recovery* is recovery by injection of external fluids, such as water and/or gas, mainly for the purpose of pressure maintenance and volumetric sweep efficiency. *Tertiary recovery* refers to the recovery after secondary recovery. It is characterized by injection of special fluids such as chemicals, miscible gases, and/or the injection of thermal energy.

Enhanced oil recovery is oil recovery by injection of gases or chemicals and/or thermal energy into the reservoir. It is not restricted to a particular phase, as defined previously, in the producing life of the reservoir. Another term, *improved oil recovery* (IOR), is also used in the petroleum industry. The terms EOR and IOR have been used loosely and interchangeably at times. Some feel that the two terms are synonymous; others feel that IOR covers just about anything, including infill drilling and reservoir characterization.

Workable definitions of EOR and IOR are necessary not just for improved communication, but also to recoverable reserves booking, contract negotiations, government incentives, taxation, and regulatory authorities when looking at fiscal issues (Stosur et al., 2003). The following sections summarize the existing definitions used in the petroleum industry and then propose this book's definitions of EOR and IOR.

1.2.1 Existing Definitions

Apparently, it has been agreed among petroleum professionals that IOR is a general term that implies improving oil recovery by any means; EOR is more specific in concept and can be considered a subset of IOR. According to Taber et al. (1997a), EOR simply means that something other than plain water or brine is injected into the oil reservoir, whereas IOR is a term used more broadly. According to Green and Willhite (1998), the term EOR is used to replace tertiary recovery because the chronological term does not describe some actual operation such as thermal recovery in a viscous oil reservoir. In this case, thermal recovery might be the only way to be able to recover significant oil. EOR results principally from the injection of gases and chemicals and/or the use of thermal energy. IOR includes EOR but also encompasses a broader range of activities—for example, reservoir characterization, improved reservoir management, infill drilling, horizontal well drilling, and sweep efficiency improvement. Selamat et al. (2008) included workover, step-out drilling, and infill drilling into IOR. Jørgenvåg and Sagli (2008) included any activity into IOR programs that may improve oil rate and recovery, whereas EOR refers to reservoir processes that recover oil not produced by secondary processes (Stosur et al., 2003).

High-pressure nitrogen injection is considered an EOR process (Manrique et al., 2007). Some authors (e.g., Moritis, 2000) classify immiscible gas injection as EOR too. In those cases, processes other than pressure maintenance are involved, and the processes result in more oil recovered.

Thomas (2008) defined EOR as a process to reduce oil saturation below the residual oil saturation (S_{or}). Recovery of oils retained due to capillary forces (after waterflooding in light oil reservoirs) and oils that are immobile or nearly immobile due to high viscosity (heavy oils and tar sands) can be achieved only by lowering the oil saturation below S_{or}. Such a case needs a definition of residual oil saturation different from the conventional one.

1.2.2 Proposed Definitions

The terms EOR and IOR should refer to reservoir processes. Any practices that are independent of the recovery process itself should not be grouped into either EOR or IOR. Such practices include reservoir characterization, reservoir simulation, use of hardware and equipment (pumps, down-hole separators, etc.), use of special well types (horizontal wells, multilaterals, smart wells, etc.), improved reservoir management, infill drilling, and so on. Oil here means hydrocarbon, including oil and natural gas.

Improved oil recovery refers to any reservoir process to improve oil recovery. Virtually, this term comprises all but the primary processes (Stosur et al., 2003). The following is an incomplete list:

- EOR processes
- Near wellbore conformance control (cement plug/gel treatment for water and gas shutoff)
- Immiscible gas injection (dry gas, CO_2, nitrogen, alternating or co-injection with water)
- Water injection, cyclic water injection
- Well stimulation (acidizing and fracturing)

Enhanced oil recovery refers to any reservoir process to change the existing rock/oil/brine interactions in the reservoir. Here is an incomplete list:

- Thermal recovery: in situ combustion—forward: dry, wet, Toe-to-Heel Air Injection (THAI), and CAPRI (i.e., variation of THAI with a catalyst for in situ upgrading); reverse, high-pressure air injection; steam soak and cyclic huff-and-puff steam flood; SAGD, VAPEX (solvent gas VAPor EXtraction), Expanding Solvent VAPEX (ES-VAPEX) or ES-SAGD; Steam And Gas Push (SAGP); hot water drive; electromagnetic

- Miscible flooding: CO_2, nitrogen, flue gas, hydrocarbon, solvent
- Chemical flooding: polymer, deep-formation profile control using gels, surfactant, alkaline, emulsion, foam, and their combinations
- Microbial

The classification could never be satisfactory because several processes can be combined. For example, chemicals are added to thermal and miscible EOR processes. Conformance control near wellbore zones, such as cement plug/gel treatment for water and gas shutoff, and acidizing and fracturing well stimulation are grouped into IOR because these processes are not in the reservoir scale even though there are some interactions near the wellbore.

1.3 GENERAL DESCRIPTION OF CHEMICAL EOR PROCESSES

This book focuses on chemical EOR processes, including alkaline (A), surfactant (S), polymer (P), and any combination of these processes. We discuss emulsion whenever it relates to any chemical processes. In addition, we briefly describe foam when presenting an application of ASP with foam. Emulsion and foam are more related to mobility control. These two processes are not discussed in detail because they are thermodynamically unstable processes quite different from the stable processes we deal with here. Rather, we discuss the general mobility control requirement in EOR processes in Chapter 4.

The mobility control process is based primarily on maintaining a favorable mobility ratio to improve sweep efficiency. Figure 1.2 provides an example of macroscopic displacement efficiency improvement by polymer flooding over waterflooding. The mobility control process is closely coupled with every chemical process. There is hardly any chemical application without injecting a mobility controlling agent. Another fact that further justifies using a mobility control agent (e.g., polymer) is that water phase relative permeability is increased in surfactant flooding. Therefore, this book first discusses the general concept of mobility control, followed by polymer flooding.

A fundamental chemical process is surfactant flooding in which the key mechanism is to reduce interfacial tension (IFT) between oil and the displacing fluid. The mechanism, because of the reduced IFT, is associated with the increased capillary number, which is a dimensionless ratio of viscous-to-local capillary forces. Experimental data show that as the capillary number increases, the residual oil saturation decreases (Lake, 1989). Therefore, as IFT is reduced through the addition of surfactants, the ultimate oil recovery is increased. In alkaline flooding, the surfactant required to reduce IFT is generated in situ by the chemical reaction between injected alkali and naphthenic acids in the

FIGURE 1.2 Schematic of macroscopic displacement efficiency improvement by polymer flooding (b) over waterflooding (a). *Source: Courtesy of Surtek, a chemical EOR service company in Golden, Colorado.*

crude oil. However, more important mechanisms in alkaline injection are its synergy with surfactant and its function to reduce surfactant (even polymer) adsorption.

The synergy makes the alkaline-surfactant process more robust and results in a wider range of application conditions. In modern chemical EOR, the most important processes are to reduce the amount of injected chemicals and to fully explore the synergy of different processes. This effort has resulted in the alkaline-surfactant-polymer (ASP) process. Laboratory studies, pilot tests, and field applications have demonstrated the greatest potential for enhancing oil recovery. However, some problems, such as scaling and emulsion, have also emerged in practical applications. Although ASP has the greatest potential, the practical problems lead operators to consider chemical processes without alkaline injection. Other factors challenging chemical EOR processes include expensive water treatment such as filtering, softening, and post-filtering; disposal of produced chemical solution; initial capital investment for facilities and equipment; and so on.

1.4 PERFORMANCE EVALUATION OF EOR PROCESSES

A common measure of the success of an EOR process is the incremental oil recovery factor. Figure 1.3 shows the schematic of incremental oil recovery from an EOR process. The oil production rates from B to C are extrapolated rates, and the cumulative oil at D is the predicted ultimate oil recovery had the EOR process not been initiated at B. The time from B to C is required to

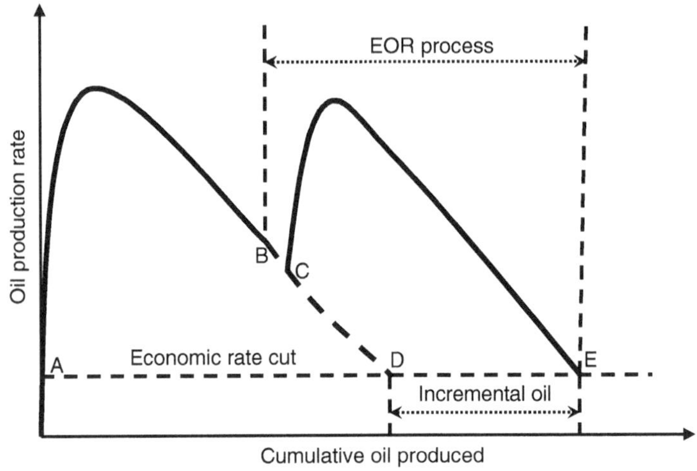

FIGURE 1.3 Incremental oil recovery from an EOR process.

respond to an EOR process. The cumulative oil at E is the ultimate oil recovery at the end of the EOR process. Consequently, the difference of cumulative oil between E and D is the incremental EOR oil recovery. For a chemical EOR process, the EOR oil is generally the incremental over waterflooding. Incremental enhanced oil recovery is commonly represented by the incremental oil recovery factor, which is the incremental oil recovered divided by the original oil in place (OOIP). Be aware that instead of using OOIP, we sometimes use the remaining oil after waterflooding to calculate the incremental oil recovery factor.

Another measure of the success of chemical EOR is the amount of chemical injected in pounds per barrel of incremental oil produced (lb/bbl), or tons of oil produced per ton of chemical injected, a figure often used in China to represent polymer flooding efficiency. Chang et al. (2006) reported that incremental oil recovery factors of up to 14% of the OOIP have been obtained in polymer flooding good-quality reservoirs, and incremental oil recovery factors of up to 25% of OOIP have been reported in ASP pilot areas.

To estimate ultimate oil recovery, we have to extrapolate the production rate to an economic cutoff at which the production wells are shut-in. In waterflooding and chemical flooding, the economic cutoff is generally 98% water cut. The ultimate oil recovery will be the cumulative oil production by the cutoff.

If the economic cutoff of 98% water cut is used for both waterflooding and chemical flooding, then the total injection pore volumes (PVs) from these two processes could be different. Generally, the total injection PV in waterflooding

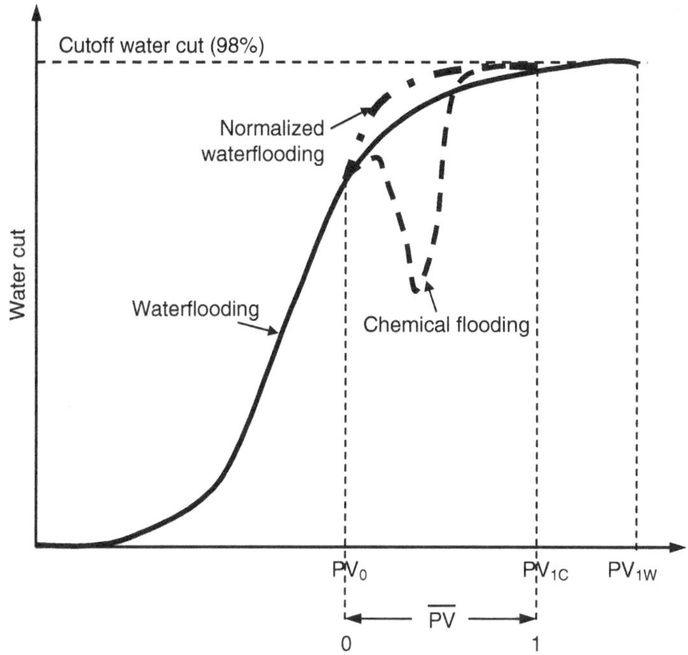

FIGURE 1.4 Schematic of water-cut curves of waterflooding, chemical flooding, and normalized waterflooding.

is larger than that in chemical flooding. To compare the performance of the two processes at any time t, we should normalize the injection pore volume using the equation

$$\overline{PV} = \frac{PV(t) - PV_0}{PV_1 - PV_0}, \tag{1.1}$$

where $\overline{PV}$ is the normalized injection PV, PV(t) is the injection PV of waterflood or chemical flood at any time, PV_0 is the start time of chemical flood, and PV_1 is the total injection PV of waterflood or chemical food at the final cutoff.

The schematic interpretation of the idea is shown in Figure 1.4. If we take the water-cut curve of a chemical flood as the base ($\overline{PV} = 0$ and 1 corresponding to PV_0 and PV_{1C}, respectively, for it), the normalized water-cut curve for waterflooding is shown in the figure. We can see that the normalized water-cut curve for waterflooding is above the original curve. It is implied that waterflood performance is actually worse (a higher water cut) if we take into account that more water is injected compared with the chemical flood.

1.5 SCREENING CRITERIA FOR CHEMICAL EOR PROCESSES

A publication that specifically focuses on the screening criteria for chemical processes has not been seen in the literature. Screening criteria for broader EOR processes have been discussed by several researchers—for example, Taber et al. (1997a, 1997b), Al-Bahar et al. (2004), and Dickson et al. (2010). This section briefly summarizes several critical parameters regarding chemical EOR application conditions. Many parameters could affect chemical EOR processes; however, the most critical parameters should be reservoir temperature, formation salinity and divalent contents, clay contents, and oil viscosity. For polymer flooding, permeability is another critical parameter.

1.5.1 Formation

Almost all chemical EOR applications have been in sandstone reservoirs, except a few stimulation projects and a few that have not been published have been in carbonate reservoirs. One reason for fewer applications in carbonate reservoirs is that anionic surfactants have high adsorption in carbonates. Another reason is that anhydrite often exists in carbonates, which causes precipitation and high alkaline consumption. Clays also cause high surfactant and polymer adsorption and high alkaline consumption. Therefore, clay contents must be low for a chemical EOR application to be effective.

1.5.2 Oil Composition and Oil Viscosity

Oil composition is very important to alkaline-surfactant flooding because different surfactants must be selected for different oils, but it is not critical to polymer flooding. According to Taber et al. (1997a, 1997b), oil viscosity should be less than 35 mPa·s for A/S projects. For polymer flooding, oil viscosity could be 10 to 150 mPa·s. Sorbie (1991) defined 30 mPa·s as the upper limit of oil viscosity for polymer flooding, and 70 mPa·s as the maximum. In Chinese ASP projects, the oil viscosity is around 10 mPa·s, whereas for polymer projects, the median viscosity is about 20 mPa·s with the maximum viscosity being about 90 mPa·s. Recently, there has been an increasing research interest in chemical EOR for oils with higher viscosities.

1.5.3 Formation Water Salinity and Divalents

Formation water salinity and divalents are critical to chemical EOR processes for both surfactants and polymers. Although chemical suppliers claim their products can be tolerant to high salinity, most of the chemical EOR processes have been applied in low-salinity reservoirs. For most of the Chinese EOR projects, the formation water salinity is below 10,000 ppm, and fresh water is injected. The criterion Al-Bahar et al. (2004) discussed is 50,000 ppm salinity and 1000 ppm hardness. This 1000 ppm hardness is probably too high or needs extra chelating agents. It must be emphasized that the salinity and divalent limits depend on the type of polymer used. Biopolymer xanthan is much more salinity or hardness tolerant than HPAM.

1.5.4 Reservoir Temperature

According to Taber et al. (1997a, 1997b), the reservoir temperature should be lower than 93°C for A/S/P projects, but the average temperature for the actual A/S field projects they reported was 27°C. The average temperature for their reviewed 171 polymer projects was 49°C for the projects (Taber et al., 1997b); however, some chemical suppliers state that polymer can be applied up to 120°C. Daqing reservoir temperature is about 45°C. The maximum temperature for a few Chinese projects was in the order of 80°C. The criterion Al-Bahar et al. (2004) used is 70°C, which is on the lower side. Sorbie's (1991) upper limit for polymer is 80°C, and the maximum is 95°C.

1.5.5 Formation Permeability

High permeability is favorable to chemical flooding, and it is critical to polymer flooding. Low-permeability formation will have injectivity and excess retention problems. Interestingly, Taber et al. (1997a) showed that although the criterion for chemical projects is greater than 10 md, the average permeabilities in their reviewed actual projects were 450 md for A/S and 800 md for polymer flooding. In Chinese chemical EOR projects, the permeability is 100 to 1000s md.

The data provided here can serve as a reference for potential projects. Among the parameters discussed in the preceding paragraphs, reservoir temperature and water salinity are the most critical parameters. However, as chemical products are improved, the criteria will be changed. From the current chemical EOR technology, extensive laboratory measurements still are needed for every project. Simulation work is needed to analyze laboratory data and upscale to a field model for potential prediction. The chemical EOR application in fields of high temperature and high salinity is still a challenging task.

1.6 NAMING CONVENTIONS AND UNITS

The chapters in this book are organized based on individual processes and combinations of individual processes. The individual processes are mainly polymer flooding (P), surfactant flooding (S), and alkaline flooding (A). A mixed process can be any combination of these individual processes—for example, surfactant-polymer flooding (SP) and alkaline-surfactant-polymer flooding (ASP). Here, the hyphen (-) between individual processes represents a combination. In the literature, a forward slash (/) is used more often—for example, alkaline/surfactant/polymer (A/S/P). We propose that if chemicals are mixed and injected in a single slug, a hyphen should be used; if chemicals are injected in sequential slugs, a forward slash should be used. For example, if surfactant and polymer are mixed and injected in a single slug, the term surfactant-polymer (SP or S-P) should be used. If surfactant and polymer are injected in sequentially separate slugs, the term surfactant/polymer (S/P) should be used. In this book, the numerical and alphabetical notations for phases are as shown in Table 1.1. The numerical and alphabetical notations for components (species) are as shown in Table 1.2.

A parameter V_{ij} means Parameter V, Component i in Phase j; for example, V_{13} means the water volume in the microemulsion phase. For this example, V_{wm} is the alphabetical form. The numerical and alphabetical notations for total concentrations of components are as shown in Table 1.3.

Sometimes the fluid concentration is expressed in the units M, m, and N. M is an abbreviation of the solubility unit, molarity, which is the number of moles (or gram formula weights) of solute in one liter (L) of solution. The abbreviation m is another unit, molality, which is the number of moles (or gram formula weights) of solute in one kilogram of solvent. A molar concentration is labeled with a square bracket. The unit of N is meq/mL. The unit of

TABLE 1.1 Numerical and Alphabetical Phases Notations

Phase	Numerical (j)	Alphabetical
Aqueous	1	W
Oleic	2	O
Microemulsion	3	M

TABLE 1.2 Numerical and Alphabetical Components Notations

Component (species)	Numerical (i)	Alphabetical	Units
Water	1	W	volume fraction
Oil	2	O	volume fraction
Surfactant	3	S	volume fraction
Polymer	4	P	wt.%, mg/L (ppm)
Anion	5		meq/mL, mg/L (ppm), wt.%
Divalents	6		meq/mL, mg/L (ppm), wt.%
Cosolvent 1 (alcohol 1)	7		volume fraction
Cosolvent 2 (alcohol 2)	8		volume fraction

TABLE 1.3 Numerical and Alphabetical Total Concentrations of Components Notations

Total Component	Numerical	Alphabetical
Water	C_1	C_w
Oil	C_2	C_o
Surfactant	C_3	C_s

meq/mL is equivalent mole/L, or the equivalent milimole/mL. The general units of parameters are listed in the Nomenclature section at the beginning of this book. Some of the units used frequently in this book are listed there as well. For some formulas or equations, if the units are not specified, use of consistent units or the SI units should be assumed.

1.7 ORGANIZATION OF THIS BOOK

The basic chemical processes are polymer flooding (Chapter 5), surfactant flooding (Chapter 7), and alkaline flooding (Chapter 10). The fundamentals of these processes are detailed in their respective chapters. There are a few combinations of these basic processes. The important aspects of each combination are their interactions and their synergies. Therefore, we have dedicated specific chapters to these interactions and synergies: Chapter 9 for the surfactant-polymer interaction and compatibility, Chapter 11 for the alkali-polymer interaction and synergy, and Chapter 12 for the alkali-surfactant synergy. Chapter 13 describes alkaline-surfactant-polymer flooding; it focuses on the practical issues of the ASP process and discusses pilot tests and field applications.

Understanding the mechanisms of chemical flow helps us design chemical flooding. The transport of chemicals and fractional flow theories are summarized in Chapter 2. Salinity plays an important role in chemical flooding, so we discuss the salinity effects and ion exchange in Chapter 3 and the optimum salinity profile in Chapter 8. Mobility control is compulsory for a chemical flooding process. Consequently, we discuss the general mobility control requirement in EOR processes in Chapter 4. It has been observed in Daqing that polymer viscoelastic properties can reduce residual oil saturation and further improve polymer performance; therefore, this issue is reviewed in Chapter 6.

Chapter 2

Transport of Chemicals and Fractional Flow Curve Analysis

2.1 INTRODUCTION

Diffusion and dispersion are important mechanisms for the transport of chemicals. This chapter first addresses diffusion and dispersion in the single phase flow. Then it discusses the fractional flow curve analysis in the water/oil two-phase flow. Fractional flow curve analysis may not provide an accurate estimate of actual field flood performance, but it is a good tool for mechanism analysis.

2.2 DIFFUSION

This section discusses diffusion coefficients in a bulk phase and a porous medium. It also briefly introduces a statistical representation of diffusion. Diffusion is less significant in reservoir flow than dispersion and their mechanisms are different, but the discussion of diffusion provides an analog to the formulation of dispersion.

2.2.1 Diffusion in a Bulk Liquid or Gas Phase

The basic concept of diffusion refers to the net transport of material within a single phase in the absence of mixing (by mechanical means or by convection). Both experiment and theory have shown that diffusion can result from pressure gradients (pressure diffusion), temperature gradients (thermal diffusion), external force fields (forced diffusion), and concentration gradients. Only the last type is considered in this book; that is, the discussion is limited to diffusion caused by the concentration difference between two points in a stagnant solution. This process, called molecular diffusion, is described by Fick's laws. His first law relates the flux of a chemical to the concentration gradient:

$$F = -D_0 \frac{\partial C}{\partial x}, \tag{2.1}$$

where F is the flux (mol/s/cm^2), C is the concentration (mol/cm^3), and D_0 is the diffusion coefficient (cm^2/s) in a bulk liquid or gas phase. The values of

Modern Chemical Enhanced Oil Recovery. DOI: 10.1016/B978-1-85617-745-0.00002-4

diffusion coefficients of ions in water are reported in Appelo and Postma (2007). For multicomponent solutions, an average diffusion coefficient of 1.3 $\times 10^{-5}$ cm^2/s may be used. According to these authors, the values at any temperature can be obtained using Eq. 2.2,

$$D_{0,T} = \frac{D_{0,298} T \mu_{298}}{298 \mu_T}, \tag{2.2}$$

where T is the absolute temperature in degrees Kelvin (°K), and μ is the water viscosity in mPa·s. In the equation, the reference temperature is 298 °K. For neutral organic molecules, the diffusion coefficient can be estimated from Eq. 2.3 (Lyman et al., 1990; Schwartzenbach et al., 1993),

$$D_{0,298} = 2.8 \times 10^{-5} V_l^{-0.71}, \tag{2.3}$$

where V_l is the liquid molar volume of the substance in cm^3/mol.

2.2.2 Diffusion in a Tortuous Pore

Equation 2.1 defines the flux in a bulk liquid or gas phase (with unit porosity or in a straight capillary). The effect of tortuous paths has to be considered in a porous medium. We use the effective diffusion coefficient, D_τ, to replace D_0 in Eq. 2.1 to consider the effect of tortuosity. The relationship between D_τ and D_0 may be defined (Childs, 1969) using Eq. 2.4,

$$D_\tau = \frac{D_0}{\tau^2}, \tag{2.4}$$

where τ is the tortuosity of the porous medium, which is defined as the length of the actual travel path taken by a solute divided by the straight line distance. And the tortuosity is related to the formation electrical resistivity factor, F_R, defined as the electrical resistivity of a porous medium with a liquid that conducts electricity divided by the electrical resistivity of the liquid in the porous medium,

$$\tau^2 = F_R \phi, \tag{2.5}$$

where ϕ is the porosity in fraction. The empirical relationship between the formation electrical resistivity factor and porosity takes the form of Archie's law,

$$F_R = \phi^{-n}, \tag{2.6}$$

where the exponent n varies from 1.4 to 2.0 (McNeil, 1980). When n is taken to be 2, we have

$$D_\tau = D_0 \phi. \tag{2.7}$$

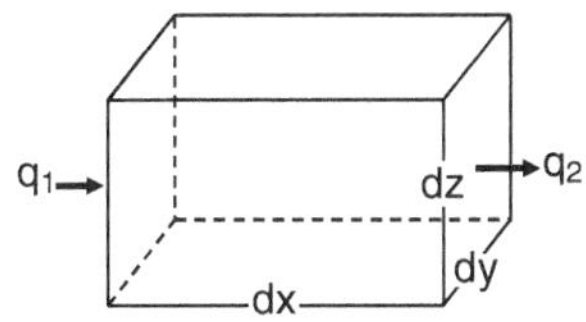

FIGURE 2.1 Concentration changes in a small volume due to diffusion.

For a small volume as shown in Figure 2.1, the mass that enters from the left side is

$$q_1 = Fdydz\phi = -D_\tau \frac{\partial C}{\partial x} dydz\phi, \tag{2.8}$$

and the mass that leaves from the right side is

$$q_2 = \left(F + \frac{\partial F}{\partial x}dx\right)dydz\phi = -D_\tau\left[\frac{\partial C}{\partial x} + \frac{\partial}{\partial x}\left(\frac{\partial C}{\partial x}\right)dx\right]dydz\phi. \tag{2.9}$$

The mass balance for the small volume is

$$\frac{\partial C}{\partial t}(dxdydz\phi) = q_1 - q_2 = D_\tau \frac{\partial^2 C}{\partial x^2}(dxdydz\phi). \tag{2.10}$$

Thus, we have

$$\frac{\partial C}{\partial t} = D_\tau \frac{\partial^2 C}{\partial x^2}. \tag{2.11}$$

Equation 2.11 is known as Fick's second law of diffusion.

2.2.3 Statistical Representation of Diffusion

The diffusion process is molecular in nature. It results from the random Brownian motion of molecules in solution. Appelo and Postma (2007) showed that the solution of Eq. 2.11 can be related to the normal density function. Consider the initial condition where no chemical is present at time $t < 0$, N moles are injected at the origin, $x = 0$. This is known as a single shot input or Dirac delta function. As $t \to 0$, $C = 0$ everywhere except at the origin where $C \to \infty$. The solution of Eq. 2.11 for the initial conditions stated is

$$C(x, t) = \frac{N}{\sqrt{4\pi D_0 t}} \exp\left(\frac{-x^2}{4D_0 t}\right), \tag{2.12}$$

where N is the input mass (moles) at time $t = 0$ at $x = 0$. $C(x, t)$ is expressed in mol/cm for convenience because we consider only one dimension.

Fundamentally, Eq. 2.12 is analogous to the normal density function (the Gaussian curve),

$$n(x) = \frac{N}{\sqrt{2\pi\sigma^2}} \exp\left(\frac{-(x - x_0)^2}{2\sigma^2}\right), \tag{2.13}$$

where x_0 is the average location (in Eq. 2.12, $x_0 = 0$), and σ^2 is the variance of the distribution. Diffusion can therefore be treated as a statistical process. Because Eqs. 2.12 and 2.13 are fundamentally the same, the variance σ^2 is related to the diffusion coefficient by

$$\sigma^2 = 2D_0t, \tag{2.14}$$

where σ has the dimension of length. According to the statistics, $2\sigma = 2\sqrt{2D_0t}$ represents the distance comprising 68% of the original mass. This simple formula, first derived by Einstein, provides a rapid estimate of the mean diffusion length. For the diffusion in three dimensions, the squared distances should be additive. Thus, when $\sigma_x^2 = \sigma_y^2 = \sigma_z^2$, the sphere where 68% of a point source is located has a radius $r = \sigma_{xyz}$:

$$\sigma_{xyz} = \sqrt{\sigma_x^2 + \sigma_y^2 + \sigma_z^2} = \sqrt{6D_0t}. \tag{2.15}$$

2.3 DISPERSION

Dispersion is an important issue in chemical processes in porous media, but we are really challenged to quantify this parameter because of its scale dependency. This section presents the empirical correlations to estimate dispersion coefficients in the laboratory scale and discusses the methods to estimate dispersivities in the large field scale.

2.3.1 Concept of Dispersion

The previous section discussed diffusion in the absence of gross fluid movement. If fluids are flowing through a porous medium, some additional mixing may be taking place. This increased mixing caused by uneven fluid flow or caused by concentration gradients resulting from fluid flow will be designated *dispersion*. In other words, dispersion is caused by variations (heterogeneity) in the velocity, whereas molecular diffusion is caused by the concentration gradient. There are two types of dispersion. One is the dispersion in the longitudinal direction or in the direction of gross fluid movement, and it is represented by D_L (K_ℓ in the petroleum literature), the longitudinal dispersion coefficient. The other one is the dispersion transverse to the direction of gross fluid movement, and it is represented by D_T (K_t in the petroleum literature), the transverse dispersion coefficient.

Bear (1972) suggested hydrodynamic dispersion is the macroscopic outcome of the actual movements of the individual tracer particles through the pores and various physical and chemical phenomena that take place within the pores. This movement can arise from a variety of causes. Dispersion is the mixing of two miscible fluids caused by diffusion, local velocity gradients (as between a pore wall and a pore center), locally heterogeneous streamline lengths, and mechanical mixing in pore bodies, according to Lake (1989). The physical process behind dispersion is different from diffusion, which will be more evident in the subsequent discussion; however, we still use the *form* of Fick's law (Eq. 2.1) to quantify dispersion:

$$F = -D_L \frac{\partial C}{\partial x}. \tag{2.16}$$

Here, we have substituted D_L, the longitudinal dispersion coefficient, in Eq. 2.16 for D_τ, the diffusion coefficient, in Eq. 2.1.

2.3.2 Estimate Longitudinal Dispersion Coefficient from Experimental Data

When we derived the diffusion equation (Eq. 2.11), there was no bulk fluid flow. Referring to Figure 2.1, in the case of bulk flow, the mass that enters from the left side is the sum of the dispersion component and the flow component:

$$q_1 = F dydz\phi + v dydz\phi = \left(-D_L \frac{\partial C}{\partial x} + vC\right) dydz\phi. \tag{2.17}$$

Here, the longitudinal dispersion coefficient D_L is used, and v is the interstitial velocity equal to the Darcy velocity, u, divided by the porosity, ϕ. Similarly, the mass that leaves from the right side is

$$\begin{aligned} q_2 &= \left(F + \frac{\partial F}{\partial x} dx\right) dydz\phi + \left(vC + v\frac{\partial C}{\partial x} dx\right) dydz\phi \\ &= \left\{-D_L\left[\frac{\partial C}{\partial x} + \frac{\partial}{\partial x}\left(\frac{\partial C}{\partial x}\right) dx\right] + \left(vC + v\frac{\partial C}{\partial x} dx\right)\right\} dydz\phi. \end{aligned} \tag{2.18}$$

The mass balance for the small volume is

$$\frac{\partial C}{\partial t}(dxdydz\phi) = q_1 - q_2 = \left(D_L \frac{\partial^2 C}{\partial x^2} - v\frac{\partial C}{\partial x}\right)(dxdydz\phi). \tag{2.19}$$

Thus, we have

$$\frac{\partial C}{\partial t} + v\frac{\partial C}{\partial x} - D_L \frac{\partial^2 C}{\partial x^2} = 0. \tag{2.20}$$

Equation 2.20 is the advection-dispersion (AD) equation. In the petroleum literature, the term convection-diffusion (CD) equation is used, or simply diffusion equation (Brigham, 1974). When a reaction term is included, the term advection-reaction-dispersion (ARD) equation is used elsewhere. When the adsorption term is expressed as a reaction term, the ARD equation is as discussed later in Section 2.4. Several solutions of Eq. 2.20 have been presented in the literature, depending on the boundary conditions imposed. In general, they are various combinations of the error function. When the porous medium is long compared with the length of the mixed zone, they all give virtually identical results.

For a core flood, the initial and boundary conditions are

$$C(x,0) = C_0, \quad x \geq 0, \tag{2.21}$$

$$C(x \to +\infty, t) = C_0, \quad t \geq 0, \tag{2.22}$$

$$C(0,t) = C_{inj}, \quad t \geq 0, \tag{2.23}$$

where C_0 and C_{inj} are the concentrations at $t = 0$ and the injection concentration, respectively. The dimensionless forms of Eqs. 2.20 through 2.23 are

$$\frac{\partial C_D}{\partial t_D} + \frac{\partial C_D}{\partial x_D} - \frac{1}{N_{Pe}} \frac{\partial^2 C_D}{\partial {x_D}^2} = 0, \tag{2.24}$$

$$C_D(x_D, 0) = 0, \quad x_D \geq 0, \tag{2.25}$$

$$C_D(x_D \to +\infty, t_D) = 0, \quad t_D \geq 0, \tag{2.26}$$

$$C_D(0, t_D) = 1, \quad t_D \geq 0, \tag{2.27}$$

where

$$C_D = \frac{C - C_0}{C_{inj} - C_0}, \tag{2.28}$$

$$N_{Pe} = \frac{vL}{D_L}, \tag{2.29}$$

$$t_D = \frac{vt}{L}, \tag{2.30}$$

$$x_D = \frac{x}{L}. \tag{2.31}$$

L is the dimension parallel to bulk flow (length), and N_{Pe} is the Peclet number. According to Naiki (1979), the solution of Eq. 2.24 under the initial and boundary conditions (Eqs. 2.25–2.27) is

$$C_D = \frac{1}{2}\operatorname{erfc}\left(\frac{x_D - t_D}{2\sqrt{t_D/N_{Pe}}}\right) + \frac{1}{2}\exp(x_D N_{Pe})\operatorname{erfc}\left(\frac{x_D + t_D}{2\sqrt{t_D/N_{Pe}}}\right), \quad (2.32)$$

where

$$\operatorname{erfc}(x) = \frac{2}{\sqrt{\pi}}\int_x^{+\infty} \exp(-u^2)\,du, \quad (2.33)$$

is the complementary error function. The values of the error function and the complementary error function are presented in Table 2.1 and Figure 2.2. When t_D and/or N_{Pe} is large, or when the inlet boundary appears as if it were a long distance from the displacing front for most of the flood, the second term is omitted, and the solution, Eq. 2.32, becomes

$$C_D = \frac{1}{2}\operatorname{erfc}\left(\frac{x_D - t_D}{2\sqrt{t_D/N_{Pe}}}\right). \quad (2.34)$$

The solution, Eq. 2.34, corresponds to the solution of Eq. 2.24 with the boundary condition, Eq. 2.27, changed to Eq. 2.35:

$$C_D(-\infty, t_D) = 1, \quad t_D \geq 0. \quad (2.35)$$

The dimensional form of Eq. 2.34 is

$$C_D = \frac{1}{2}\left[1 - \operatorname{erf}\left(\frac{x - vt}{2\sqrt{D_L t}}\right)\right], \quad (2.36)$$

but the dimensionless C_D is still used for the convenience of plotting on probability paper. Here, one of the error-function properties has been used:

$$\operatorname{erf}(x) = 1 - \operatorname{erfc}(x) = \frac{2}{\sqrt{\pi}}\int_0^x \exp(-u^2)\,du. \quad (2.37)$$

If the initial condition, Eq. 2.25, and the boundary condition, Eq. 2.27, are changed to

$$C_D(x_D, 0) = 1, \quad x_D \geq 0, \quad (2.38)$$

$$C_D(0, t_D) = 0, \quad t_D \geq 0, \quad (2.39)$$

respectively, the solution shown in Eq. 2.36 becomes the one in Perkins and Johnston (1963):

$$C_D = \frac{1}{2}\left[1 + \operatorname{erf}\left(\frac{x - vt}{2\sqrt{D_L t}}\right)\right]. \quad (2.40)$$

TABLE 2.1 Values of Error Function and Complementary Error Function

x	erf(x)	erfc(x)	x	erf(x)	erfc(x)
0.00	0.0000000	1.0000000	1.30	0.9340079	0.0659921
0.05	0.0563720	0.9436280	1.40	0.9522851	0.0477149
0.10	0.1124629	0.8875371	1.50	0.9661051	0.0338949
0.15	0.1679960	0.8320040	1.60	0.9763484	0.0236516
0.20	0.2227026	0.7772974	1.70	0.9837905	0.0162095
0.25	0.2763264	0.7236736	1.80	0.9890905	0.0109095
0.30	0.3286268	0.6713732	1.90	0.9927904	0.0072096
0.35	0.3793821	0.6206179	2.00	0.9953223	0.0046777
0.40	0.4283924	0.5716076	2.10	0.9970205	0.0029795
0.45	0.4754817	0.5245183	2.20	0.9981372	0.0018628
0.50	0.5204999	0.4795001	2.30	0.9988568	0.0011432
0.55	0.5633234	0.4366766	2.40	0.9993115	0.0006885
0.60	0.6038561	0.3961439	2.50	0.9995930	0.0004070
0.65	0.6420293	0.3579707	2.60	0.9997640	0.0002360
0.70	0.6778012	0.3221988	2.70	0.9998657	0.0001343
0.75	0.7111556	0.2888444	2.80	0.9999250	0.0000750
0.80	0.7421010	0.2578990	2.90	0.9999589	0.0000411
0.85	0.7706681	0.2293319	3.00	0.9999779	0.0000221
0.90	0.7969082	0.2030918	3.10	0.9999884	0.0000116
0.95	0.8208908	0.1791092	3.20	0.9999940	0.0000060
1.00	0.8427008	0.1572992	3.30	0.9999969	0.0000031
1.10	0.8802051	0.1197949	3.40	0.9999985	0.0000015
1.20	0.9103140	0.0896860	3.50	0.9999993	0.0000007

It is not possible to predict the dispersion coefficient for a given system from fundamental principles; however, we can estimate D_L by conducting an experimental miscible flood and empirically fitting concentration data to the appropriate solution. According to Eq. 2.36, a plot of C_D versus $(x - vt)/\sqrt{t}$ will yield a straight line on an arithmetic-probability paper. Thus, we can estimate D_L. It is not convenient, however, to measure concentration at an arbitrary location x. We usually measure the concentration at the exit end of the core or tube. By setting $x = L$, we have Eq. 2.41.

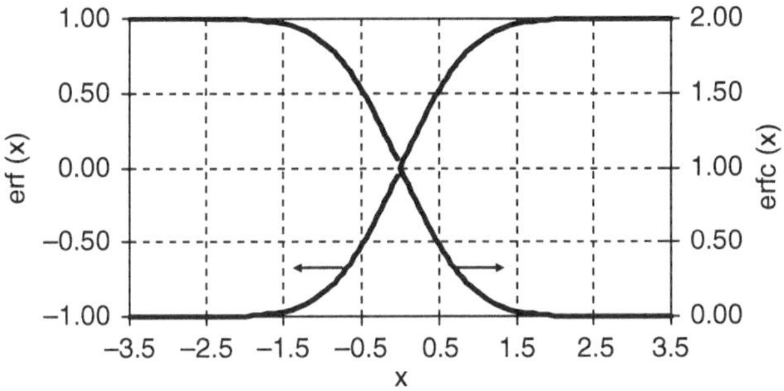

FIGURE 2.2 The error function and its complementary error function.

$$\frac{x - vt}{\sqrt{t}} = \frac{L - vt}{\sqrt{t}} = \frac{A\phi L - A\phi vt}{\sqrt{A\phi vt}} \frac{\sqrt{vL}}{\sqrt{A\phi L}} = \frac{V_p - V_i(t)}{\sqrt{V_i(t)}} \frac{\sqrt{vL}}{\sqrt{V_p}} = \frac{1 - V_i(t)/V_p}{\sqrt{V_i(t)/V_p}} \sqrt{vL}, \tag{2.41}$$

where L is usually the core length, A is the cross-sectional area, V_p is the pore volume, and $V_i(t)$ is the injection pore volume at time t. Then Eq. 2.36 becomes

$$C_D = \frac{1}{2}\left[1 - \text{erf}\left(\frac{U\sqrt{vL}}{2\sqrt{D_L}}\right)\right], \tag{2.42}$$

where $U = (1 - V_i(t)/V_p)/\sqrt{V_i(t)/V_p}$. Now U, a time-dependent variable, is the only parameter in the error-function argument that varies with C_D. U versus C_D should generate a straight line on the probability paper.

At $C_D = 0.9$, from Eq. 2.42 we have

$$0.9 = \frac{1}{2}\left[1 - \text{erf}\left(\frac{U_{90}\sqrt{vL}}{2\sqrt{D_L}}\right)\right], \tag{2.43}$$

where U_{90} is the value of U read from the straight line at $C_D = 0.9$ (90% on the probability paper). When we use one of the error-function properties

$$\text{erf}(-x) = -\text{erf}(x), \tag{2.44}$$

Equation 2.42 becomes

$$0.8 = \text{erf}\left(-\frac{U_{90}\sqrt{vL}}{2\sqrt{D_L}}\right). \tag{2.45}$$

Looking in Table 2.1, or Figure 2.2, at erf(x) = 0.8, by interpolation we have x = 0.90622. Thus, Eq. 2.45 becomes

$$0.90622 = -\frac{U_{90}\sqrt{vL}}{2\sqrt{D_L}}. \tag{2.46}$$

At $C_D = 0.1$, $U = U_{10}$. According to Eq. 2.42, we have

$$0.90622 = \frac{U_{10}\sqrt{vL}}{2\sqrt{D_L}}. \tag{2.47}$$

Adding Eq. 2.46 and Eq. 2.47 gives

$$D_L = \left(\frac{U_{10} - U_{90}}{3.625}\right)^2 vL. \tag{2.48}$$

Thus, D_L may be calculated from the readings U10 and U90 on the straight line on the probability paper. Similarly, we may calculate D_L from the other readings; for example:

$$D_L = \left(\frac{U_{20} - U_{80}}{2.38}\right)^2 vL, \tag{2.49}$$

$$D_L = \left(\frac{U_5 - U_{95}}{4.65}\right)^2 vL. \tag{2.50}$$

2.3.3 Empirical Correlations for the Longitudinal Dispersion Coefficient

Empirical correlations for the longitudinal dispersion coefficient are based on the premise that D_L can be represented as the sum of molecular diffusion (D_τ) and convective dispersion components (D_c):

$$D_L = D_\tau + D_c. \tag{2.51}$$

At relatively low flow rates, the convective component is negligible, and the diffusion component is dominant. As shown in Figure 2.3, at high flow rates, the diffusion component is negligible, and the convective component is dominant. Between these extremes, both components contribute to the overall dispersion process, and this is the regime commonly encountered in reservoir flow processes. Note that the dimensionless Peclet number is defined in Figure 2.3 as

$$N_{Pe} = \frac{vd_p}{D_0}, \tag{2.52}$$

where d_p is the diameter of particles of a sand pack.

Perkins and Johnston (1963) presented the following correlation for the longitudinal dispersion coefficient,

$$\frac{D_L}{D_0} = \frac{1}{F_R\phi} + 0.5\frac{vF_I d_p}{D_0}, \tag{2.53}$$

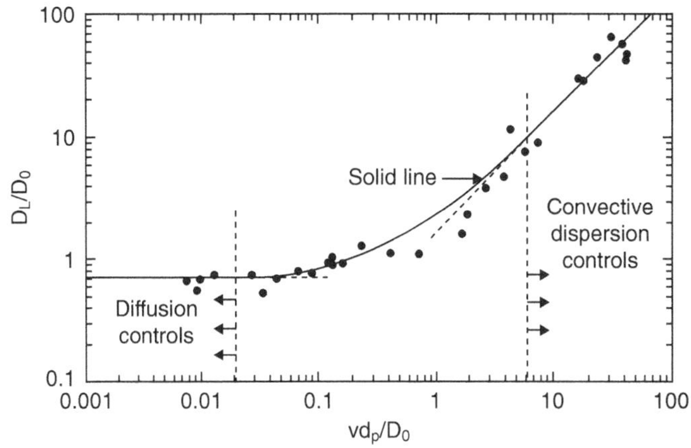

FIGURE 2.3 A plot of longitudinal dispersion coefficients for unconsolidated, random packs of uniform-size sand or beads. *Source: Perkins and Johnston (1963).*

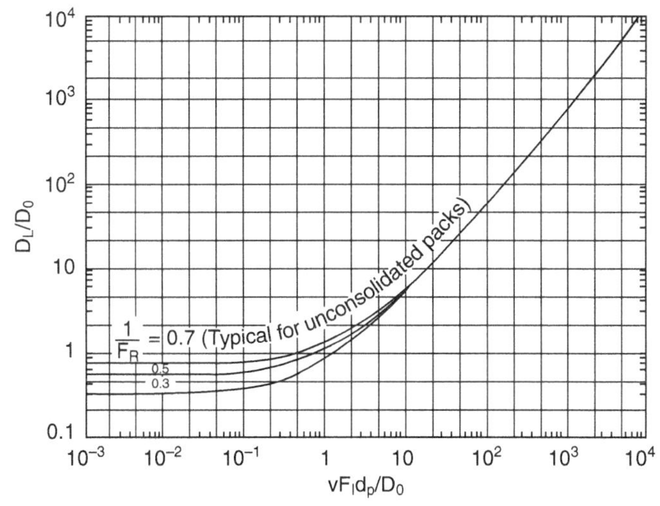

FIGURE 2.4 A plot of longitudinal dispersion coefficients for porous media. *Source: Perkins and Johnston (1963).*

for $vF_Id_p/D_0 < 50$, where v is the interstitial velocity, d_p is the average grain particle diameter, D_0 is the molecular diffusion coefficient in a bulk liquid or gas phase, and F_I is the inhomogeneity factor for the porous medium. Equation 2.53 is in dimensionless form, and any consistent set of units is applicable; it is plotted in Figure 2.4, along with the relationship for $vF_Id_p/D_0 > 50$. For a typical random pack, F_I is 3.5. Literature data suggest that packing of large beads is usually better than for small beads. Hence, we should expect the value

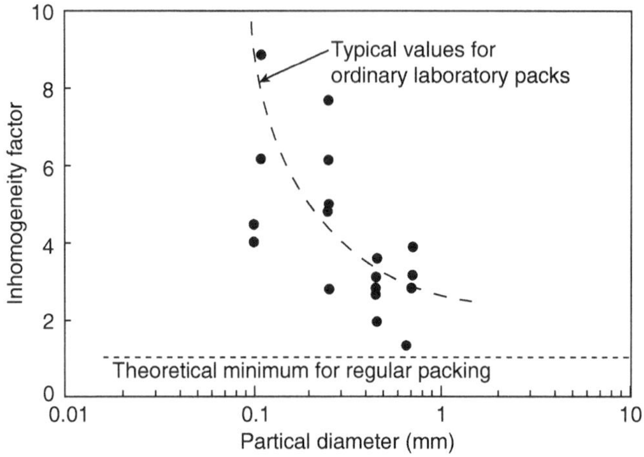

FIGURE 2.5 Inhomogeneity factor for random packs of spheres. *Source: Perkins and Johnston (1963).*

of F_I to be a bit smaller for random packs of large beads, as shown by Figure 2.5 (but F_I should never be less than unity). Also, F_I may be larger for poorly packed beads. For consolidated porous media, F_I cannot easily be separated from d_p, and the product of $F_I d_p$ is often used. Green and Willhite (1998) reported $F_I d_p$ values for several outcrop sandstones. The average value is 0.36 cm.

Within the range of applicability of Eq. 2.53, the convective component of the dispersion coefficient, $0.5vF_I d_p$, is proportional to the first power of the velocity, if composition is equalized in pore spaces by diffusion. At higher velocities for $vF_I d_p/D_0 > 50$, there is insufficient time for diffusion to equalize concentration within each pore. Most data indicate that D_L/D_0 varies with $(vF_I d_p/D_0)^{1.2}$ (Perkins and Johnston, 1963). Salter and Mohanty (1982) found that dispersion coefficients increase roughly linearly with velocity, indicating dispersion, not diffusion, governs the flow within the flowing wetting phase. They also found that dispersion coefficients in multiphase flow are higher than that in a single phase flow by up to one order of magnitude.

2.3.4 Empirical Correlations for the Transverse Dispersion Coefficient

The total transverse dispersion coefficient is the sum of the diffusion coefficient in the porous medium (which is the same as in the longitudinal direction unless the porous medium is anisotropic) and the transverse convective dispersion coefficient. Perkins and Johnston (1963) presented the correlation for the transverse dispersion coefficient in Eq. 2.54.

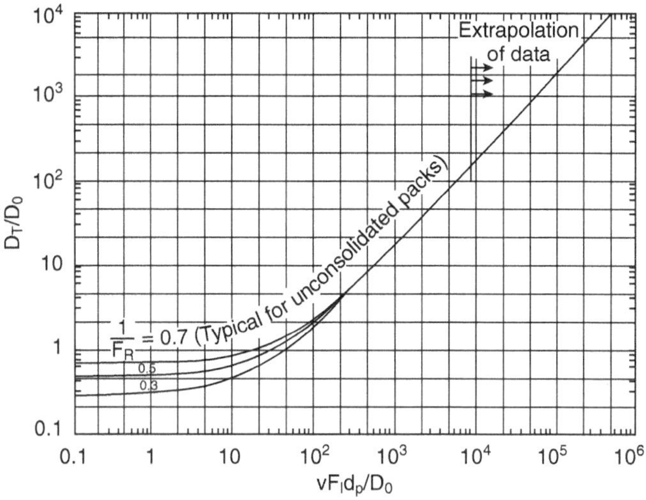

FIGURE 2.6 Transverse dispersion coefficients for porous media. *Source: Perkins and Johnston (1963).*

$$\frac{D_T}{D_0} = \frac{1}{F_R \phi} + 0.0157 \frac{vF_I d_p}{D_0}, \tag{2.54}$$

for $vF_Id_p/D_0 < 10^4$. The correlation is shown in Figure 2.6. The inhomogeneity, F_I, is assumed to have the same value in correlations for both D_L and D_T. Comparing Eq. 2.54 with Eq. 2.53, we can see that the convective component of the transverse dispersion component, $0.0157vF_Id_p$, is an order of magnitude smaller than the corresponding component of longitudinal dispersion.

2.3.5 Evaluation of the Contributions of Diffusion, Convection, and Dispersion to the Front Spread

Diffusion, convection, and dispersion all contribute to the spread of a front. Let us see how much each mechanism contributes to the spread. First, let us see when the diffusion transport is important as compared to the convective transport. We use $\sqrt{2D_0t}$ to calculate the spreading distance from a point source; 68% of the injected source is within this distance. Table 2.2 shows the results for different time periods compared with the traveled distances during the same time periods by a convective flow of 1 m/day. A typical flow rate in petroleum reservoirs is 1 m/day (interstitial velocity). A typical value of diffusion coefficient of 4×10^{-10} m^2/s in a porous medium is used. In the first 5 seconds, the diffusive transport is more important than the convective transport. Soon after, the convective flow becomes the dominant mechanism.

TABLE 2.2 Spreading of a Point Source through Diffusion Compared with Convective Transport

Time t (s)		Diffusion (cm) $\sqrt{2D_0t}$	Convective Distance (cm) vt
1	0.017 min	0.003	0.001
5	0.08 min	0.006	0.006
30	0.5 min	0.015	0.035
60	1 min	0.022	0.069
3600	1 hr	0.170	4.167
21600	6 hr	0.416	25.000
86400	1 day	0.831	100.000
31536000	1 year	15.884	36500.000

Now we compare the values of diffusion coefficient and convective dispersion coefficient. For a typical value of $F_Id_p = 0.36$ cm, the ratio of convective term to diffusion term is $vF_Id_p/D_0 = (1\text{ m }/86400\text{ s})(0.0036\text{ m})/(4 \times 10^{-10}\text{ m}^2/\text{s}) = 105$. Referring to Figure 2.3, we can see that the mechanism of transport in typical reservoir flow is convection dominated.

Finally, we compare the values of longitudinal and transverse dispersion coefficients at typical reservoir flow conditions. Using Eqs. 2.53 and 2.54, we have

$$\frac{D_L}{D_0} = \frac{1}{F_R\phi} + 0.5\frac{vF_Id_p}{D_0} = \frac{1}{(0.3^{-2})(0.3)} + 0.5\frac{(1/86400)(0.0036)}{4\times10^{-10}} = 52.38,$$

$$\frac{D_T}{D_0} = \frac{1}{F_R\phi} + 0.0157\frac{vF_Id_p}{D_0} = \frac{1}{(0.3^{-2})(0.3)} + 0.0157\frac{(1/86400)(0.0036)}{4\times10^{-10}} = 1.95.$$

Here, we use $F_R = \phi^{-2}$. Now we have $D_L/D_T = 27$.

2.3.6 Dispersivity

Equation 2.53 shows that when the convective term, $0.5vF_Id_p$, is high, the dispersion coefficient is proportional to the velocity, if F_I, d_p, and D_0 in the porous medium are assumed to be unchanged. Then if we define another parameter, α_L,

$$\alpha_L = \frac{D_L}{v}, \tag{2.55}$$

this parameter will be a better characteristic of the porous medium because it is independent of flow velocity. It is called the longitudinal dispersivity. For

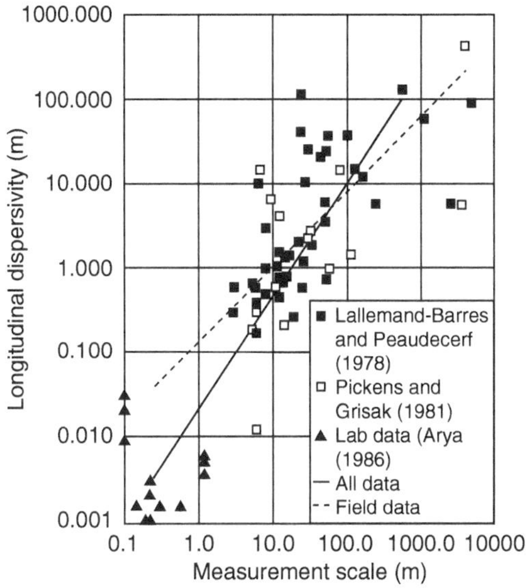

FIGURE 2.7 Field and laboratory dispersivity data. *Source: Arya et al. (1988).*

sand packs, Perkins and Johnston (1963) reported that the effective particle size for the log-normal distributions is the particle size corresponding to the 10% cumulative fraction (d_{10}). Thus, we have

$$\alpha_L = 0.5F_I d_{10} = 1.75 d_{10}, \tag{2.56}$$

if the diffusion term is negligible. F_I is assumed to be 3.5 in Eq. 2.56.

Arya et al. (1988) reported some of published experimental and field data of the longitudinal dispersivity, α_L, as shown in Figure 2.7. Their log–log least-squares fits of the data are

$$\alpha_L = 0.229L^{0.755} \tag{2.57}$$

for field data and

$$\alpha_L = 0.044L^{1.13} \tag{2.58}$$

for all the experimental and field data. Here, L is the measured length scale, and both L and α_L are in meters. The laboratory data themselves do *not* show a good trend.

Equations 2.57 and 2.58 and Figure 2.7 show that the dispersivity is a scale-dependent property, and it increases with the length scale. A similar trend has been reported elsewhere (Appelo and Postma, 2007). The general trend is that the longitudinal dispersivity is about one tenth of the measurement scale. The scale-dependent property makes the simulation of dispersion process difficult.

Both heterogeneity and dispersion cause mixing in a reservoir, and the appropriate longitudinal and transverse dispersivities depend on how the field flow model is set up. When a detailed heterogeneous model is used with a fine grid system, smaller dispersivities are used because heterogeneity is considered using the fine grid system. When a coarse grid system is used, large effective dispersivities have to be used.

Similarly, we have the transverse dispersivity, α_T:

$$\alpha_T = \frac{D_T}{v}. \tag{2.59}$$

Transverse dispersivity has been less studied, but it is smaller than the longitudinal dispersivity. Measurements from tracer studies indicate that the transverse horizontal dispersivity is about 10% of the longitudinal dispersivity in the bedding plane, and the transverse vertical dispersivity is about 1% (Gelhar, 1997). Klenk and Grathwohl (2002) found that the transverse vertical dispersivity was determined mostly by diffusion.

2.4 RETARDATION OF CHEMICALS IN SINGLE-PHASE FLOW

The general advection-reaction-dispersion equation is

$$\frac{\partial C_i}{\partial t} = -v\left(\frac{\partial C_i}{\partial x}\right) - \frac{\partial \hat{C}_i}{\partial t} + D_L\left(\frac{\partial^2 C_i}{\partial^2 x^2}\right), \tag{2.60}$$

where C_i is the solute i concentration—for example, mass/PV (pore volume), $\hat{C}_i$ is the adsorbed concentration with the unit (mass/PV), D_L is the longitudinal dispersion (m^2/s), v is the solution (e.g., water) interstitial velocity (m/s) equal to u/ϕ, and u is Darcy velocity. Three terms appear on the right side of Eq. 2.60. The first represents Advective flow; the second, adsorption (chemical Reactions); and the third, Dispersion. Therefore, it is commonly called the ARD equation. This sequence of the three terms may be the order of their relative importance.

The preceding equation is for 1D isothermal single phase flow. The fluid is incompressible. Gravity and capillary forces are not included. When dispersion is also neglected, $D_L = 0$, and Eq. 2.60 becomes

$$\frac{\partial C_i}{\partial t} = -v\left(\frac{\partial C_i}{\partial x}\right) - \frac{\partial \hat{C}_i}{\partial t}. \tag{2.61}$$

For a constant concentration C_i, we have

$$dC_i = 0 = \left(\frac{\partial C_i}{\partial t}\right)_x \cdot dt + \left(\frac{\partial C_i}{\partial x}\right)_t \cdot dx, \tag{2.62}$$

this gives

$$-\left(\frac{dx}{dt}\right)_{C_i} = \left(\frac{\partial C_i}{\partial t}\right)_x \Big/ \left(\frac{\partial C_i}{\partial x}\right)_t . \tag{2.63}$$

If we combine Eqs. 2.61 and 2.63, we have

$$\left(\frac{dx}{dt}\right)_{C_i} \equiv v_{C_i} = \frac{v}{\left(1 + \frac{d\hat{C}_i}{dC_i}\right)} . \tag{2.64}$$

The retardation factor is defined as

$$R_{C_i} = 1 + \frac{d\hat{C}_i}{dC_i}, \tag{2.65}$$

and the retardation equation is defined as

$$v_{C_i} = \frac{v}{R_{C_i}} . \tag{2.66}$$

Note that Eq. 2.66 is derived from the interstitial velocities v and v_{C_i}. Obviously, Eq. 2.66 also holds for Darcy velocities.

2.5 TYPES OF FRONTS

A chemical concentration or fluid saturation varies in time and location. When its variation is presented in the plot of concentration or saturation versus location (distance) at one time snapshot, as shown in Figure 2.8a, it is called a profile. When its variation is presented in the plot of concentration or saturation versus time at one fixed location (distance), as shown in Figure 2.8b, it is called a history. The history at the production end (well) is a production history or an

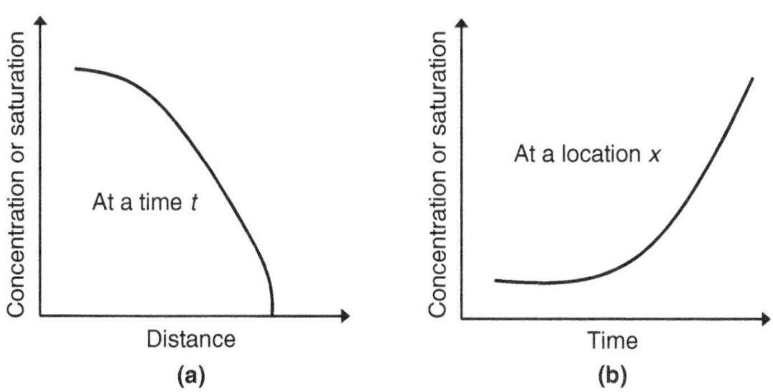

FIGURE 2.8 Concentration or saturation profile and history: (a) profile and (b) history.

effluent history. A front is a distinct concentration or saturation profile that travels through a porous medium in the direction of fluid flow. Alternative terms used for the front are wave, boundary, and transition.

Equation 2.66 shows that the velocity of a chemical solute concentration v_{C_i} is slower than the solution (water) velocity v by a retardation factor R_{C_i}. The retardation is caused by the adsorption of the chemical on the solid. Adsorption can be defined using the Freundlich isotherm or Langmuir isotherm. The general form of the Freundlich isotherm is

$$\hat{C}_i = K_F C_i^n, \tag{2.67}$$

where C_i is the equilibrium concentration in the system, and K_F and n are empirical constants obtained by fitting experimental data. The units of these variables must be consistent. To avoid any mistake caused by the units, it is suggested that the units used in fitting experimental data be the same as those used in a prediction model.

The general form of the Langmuir isotherm is

$$\hat{C}_i = \frac{a_L C_i}{1 + b_L C_i}, \tag{2.68}$$

where a_L and b_L are empirical constants. The unit of b_L is the reciprocal of the unit of C_i. a_L is dimensionless. Note that C_i and $\hat{C}_i$ should be in the same unit. In Eq. 2.68, when b_L is zero, it will become Eq. 2.67 with $K_F = a_L$ and $n = 1$. Figure 2.9 shows the two Langmuir-type isotherms with a_L and b_L marked inside the figure.

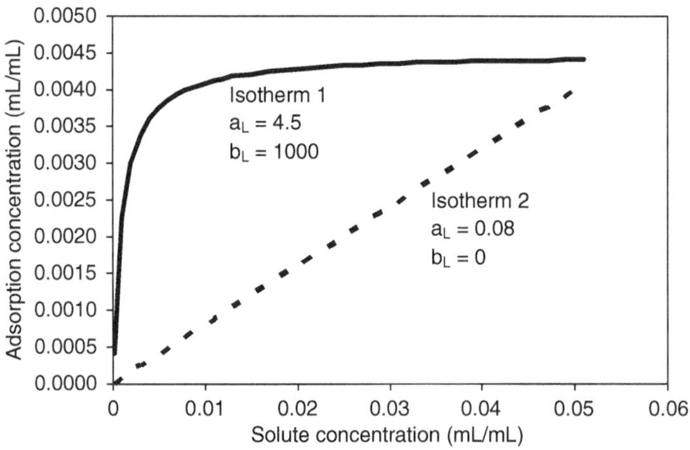

FIGURE 2.9 Langmuir isotherm (Isotherm 1) and Freundlich isotherm (Isotherm 2).

The Langmuir isotherm is commonly used in describing chemical adsorption, such as polymer and surfactant adsorption. Therefore, in the following examples, we will use the Langmuir isotherm to discuss the three different types of fronts.

From Eq. 2.68, we have

$$\frac{d\hat{C}_i}{dC_i} = \frac{a_L}{(1+b_L C_i)^2}. \tag{2.69}$$

The retardation equation (Eq. 2.66) may be written in terms of distances traveled by water solution and the chemical:

$$x_{C_i} = \frac{x_w}{R_{C_i}}. \tag{2.70}$$

The preceding equation may be further written in terms of pore volume (PV),

$$(PV)_{C_i} = \frac{(PV)_w}{R_{C_i}}, \tag{2.71}$$

where $(PV)_w$ and $(PV)_{Ci}$ are the total pore volume of the water injected and the pore volume that the concentration C_i has traveled from the injection point, respectively.

2.5.1 Spreading Front

Assume the initial chemical concentration C_i is 0.05 mL/mL solution volume in the system. It is to be flushed by a 0.005 mL/mL water volume. The Langmuir isotherm 1 in Figure 2.9 is used in this situation, where a_L and b_L are 4.5 and 1000, respectively. If we inject one PV water solution with $C_i = 0.005$ mL/mL solution volume, let us see how far different concentrations have traveled. Such concentration distribution is called a profile, as introduced previously. The calculation of concentrations is presented in Table 2.3. The locations of different solute concentrations after one PV solution injection are shown in Figure 2.10. Table 2.3 and Figure 2.10 show that a higher concentration has a lower retardation factor and travels faster than a lower concentration, and the concentrations in between travel at a velocity in between, resulting in a spreading front (broadening front). A spreading front occurs when the downstream initial concentration travels faster than the upstream injection concentration, as in this case.

2.5.2 Indifferent Front

Now it is important to keep everything in the preceding spreading-front situation unchanged, except that Isotherm 1 is replaced by Isotherm 2 shown in

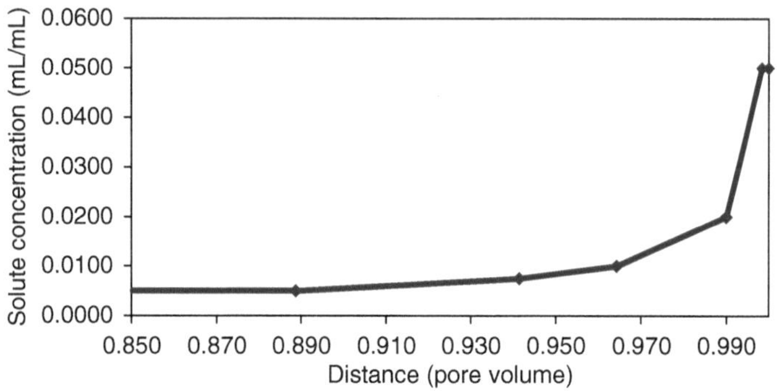

FIGURE 2.10 Spreading front when the downstream initial concentration travels faster than the upstream injection concentration.

TABLE 2.3 Spreading Front after One PV Injection

C_i	$\hat{C}_i$ Eq. 2.68	$d\hat{C}_i/dC_i$ Eq. 2.69	R_{C_i} Eq. 2.65	$(PV)_{C_i}$ Eq. 2.71
0.0500				1.000
0.0500	0.00441	0.00173	1.002	0.998
0.0200	0.00429	0.01020	1.010	0.990
0.0100	0.00409	0.03719	1.037	0.964
0.0075	0.00397	0.06228	1.062	0.941
0.0050	0.00375	0.12500	1.125	0.889
0.0050				0.000

Figure 2.9. For Isotherm 2, which is linear, $a_L = 0.08$ and $b_L = 0$, or $\hat{C}_i = 0.08C_i$. The calculation is presented in Table 2.4, and the locations of different solute concentrations after one PV solution injection are shown in Figure 2.11. For the linear adsorption, the concentration is delayed, while the shape of the concentration front remains unchanged. Such a front is called an indifferent front, which occurs when the slope of the adsorption isotherm is independent of concentration (constant), or the velocities of different concentrations are the same.

2.5.3 Sharpening Front

The situation of a sharpening front to be discussed is the same as that of the preceding spreading front, except that the initial concentration and the injection

TABLE 2.4 Indifferent Front after One PV Injection

C_i	$\hat{C}_i$ Eq. 2.68	$d\hat{C}_i/dC_i$ Eq. 2.69	R_{C_i} Eq. 2.65	$(PV)_{C_i}$ Eq. 2.71
0.0500				1.000
0.0500	0.0040	0.08	1.08	0.926
0.0200	0.0016	0.08	1.08	0.926
0.0100	0.0008	0.08	1.08	0.926
0.0075	0.0006	0.08	1.08	0.926
0.0050	0.0004	0.08	1.08	0.926
0.0050				0.000

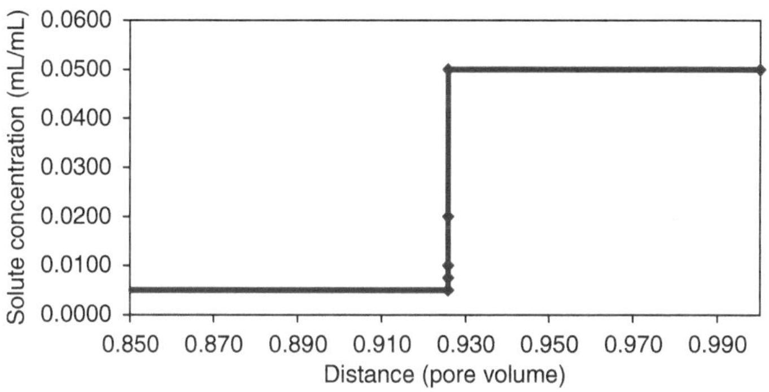

FIGURE 2.11 Indifferent front when the adsorption is linear.

concentration are exchanged; that is, the initial chemical concentration is 0.005 mL/mL solution volume, while the injection concentration is 0.05 mL/mL solution volume. If the same calculation is performed as in the situation of spreading front (i.e., using Eqs. 2.65, 2.68, 2.69, and 2.71), the results are as presented in Table 2.5 and by the dotted line in Figure 2.12. The dotted line shows that a higher concentration has overtaken a lower concentration. It is impossible for this situation to happen. Take a look at the concentration of 0.05 mL/mL. When it tries to overtake the low concentration of 0.005 mL/mL ahead of it, several things happen.

According to the Langmuir isotherm, the higher the concentration, the higher the adsorption is. Therefore, more adsorption will occur for the overtaking high concentration. Thus, the high concentration has to be reduced and retarded to meet the adsorption requirement by the high concentration. Then

TABLE 2.5 Calculated Results Using Eqs. 2.65, 2.68, 2.69, and 2.71

C_i	$\hat{C}_i$ Eq. 2.68	$d\hat{C}_i/dC_i$ Eq. 2.69	R_{C_i} Eq. 2.65	$(PV)_{C_i}$ Eq. 2.71
0.0050				1.000
0.0050	0.0038	0.1250	1.125	0.889
0.0075	0.0040	0.0623	1.062	0.941
0.0100	0.0041	0.0372	1.037	0.964
0.0200	0.0043	0.0102	1.010	0.990
0.0500	0.0044	0.0017	1.002	0.998
0.0500				0.000

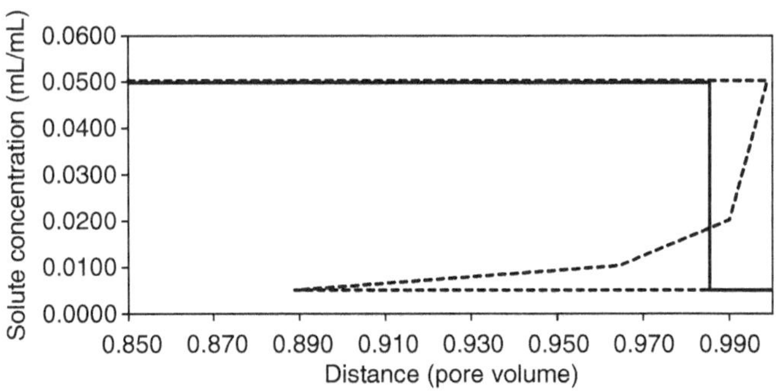

FIGURE 2.12 Sharpening front in the solid line when the upstream injection concentration travels faster than the downstream initial concentration; the front in the dotted line is calculated using Eqs. 2.65, 2.68, 2.69, and 2.71.

the subsequent high concentration of 0.05 mL/mL comes to the front to displace the denuded fluid; by this time the adsorption requirement is met, and the subsequent high concentration raises the lower concentration at the front to the high concentration of 0.05 mL/mL. As a result, the high concentration of 0.05 mL/mL cannot overtake the low concentration of 0.005 mL/mL. Instead, it raises the low concentration of 0.005 mL/mL at the front to the high concentration value (0.05 mL/mL), after meeting the higher adsorption requirement.

Here, it is assumed that the injection solution velocity is slow enough so that the system's equilibrium is reached. Thus, a sharpening front is formed where the concentration is jumped from 0.005 to 0.05 mL/mL, as shown by the solid line in Figure 2.12. Over the sharpening front, the

TABLE 2.6 Sharpening Front after One PV Injection

C_i	$\hat{C}_i$ Eq. 2.68	$\Delta\hat{C}_i/\Delta C_i$	R_{C_i} Eq. 2.72	$(PV)_{C_i}$ Eq. 2.71
0.0050				1.000
0.0050	0.0038	0.0147	1.015	0.986
0.0500	0.0044	0.0147	1.015	0.986
0.0500				0.000

intermediate concentrations between the high concentration 0.05 mL/mL and the low concentration 0.005 mL/mL do not exist. Therefore, all the calculations in Table 2.3 for those intermediate concentrations are not valid. The calculation can be done only for the two end concentration points. For the step change in concentration, the retardation equation, Eq. 2.65, should be changed to Eq. 2.72 for a sharpening front. The calculation based on Eq. 2.72 for the sharpening front is presented in Table 2.6.

$$R_{C_i} = 1 + \frac{\Delta\hat{C}_i}{\Delta C_i}. \tag{2.72}$$

To discern whether a front is sharpening, we compare the velocities at the initial concentration and final concentration. If the velocity at the final concentration is higher than that at the initial concentration, the front will be a sharpening front, or shock. In the opposite situation, in which the velocity at the final concentration is lower than that at the initial concentration, the front will be a spreading front or broadening front. If the velocity is independent of the concentration, the front will be an indifferent front. To discern the types of fronts, some (e.g., Pope, 1980) compare the upstream and downstream velocities that correspond to the velocities at the *final* and *initial concentrations*, the terms used here.

Others compare the slopes ($d\hat{C}_i/dC_i$) of the sorption isotherm at the final and initial concentrations. If the slope, $d\hat{C}_i/dC_i$, is smaller for the final concentration than that for the initial solution, a jumplike concentration change, a sharpening front, will form. If the slope is greater for the final concentration, however, a broadening front or spreading front will form. When the slope is constant (linear isotherm), the front is not affected by concentration-dependent retardation, and we have an indifferent front. A smaller slope will result in a higher velocity because the smaller slope represents less adsorption so that the solute can travel faster. Therefore, comparing the slope is equivalent to comparing velocity.

2.6 FRACTIONAL FLOW CURVE ANALYSIS OF TWO-PHASE FLOW

This section introduces the concept of saturation shock and discusses the fractional flow curve analysis of different processes.

2.6.1 Saturation Shock

Before discussing fractional flow analysis, we first need to derive the moving velocity of a saturation discontinuity or shock. Figure 2.13 shows a saturation shock from S_{w2} to S_{w1}. S_{w2} moves from x_1 to x_2 during the time interval $\Delta t = t_2 - t_1$. The total injection rate, q_t, is constant, but the water cut changes from f_{w1} to f_{w2}, which corresponds to S_{w1} and S_{w2}, respectively. Therefore, during the time interval, Δt, the total incremental water injected into the block from x_1 to x_2 is $(q_t)(\Delta t)(f_{w2}-f_{w1})$. Meanwhile, this incremental water injected results in the increase in saturation from S_{w1} to S_{w2}. The material balance of water gives

$$A\phi(x_2 - x_1)(S_{w2} - S_{w1}) = q_t \Delta t (f_{w2} - f_{w1}). \tag{2.73}$$

Then the velocity $v_{\Delta S_w}$ at which saturation shock exists is

$$v_{\Delta S_w} = \left(\frac{dx}{dt}\right)_{\Delta S_w} = \frac{q_t}{A\phi}\left(\frac{f_{w2} - f_{w1}}{S_{w2} - S_{w1}}\right). \tag{2.74}$$

According to the Buckley–Leverett (1942) theory, the velocity of the saturation S_{w2} is

$$v_{S_{w2}} = \left(\frac{dx}{dt}\right)_{S_{w2}} = \frac{q_t}{A\phi}\left(\frac{\partial f_w}{\partial S_w}\right)_{S_{w2}}. \tag{2.75}$$

At the contact between the shock and continuous saturation distribution, these velocities must be equal. From Eqs. 2.74 and 2.75, we have what is shown in Eq. (2.76).

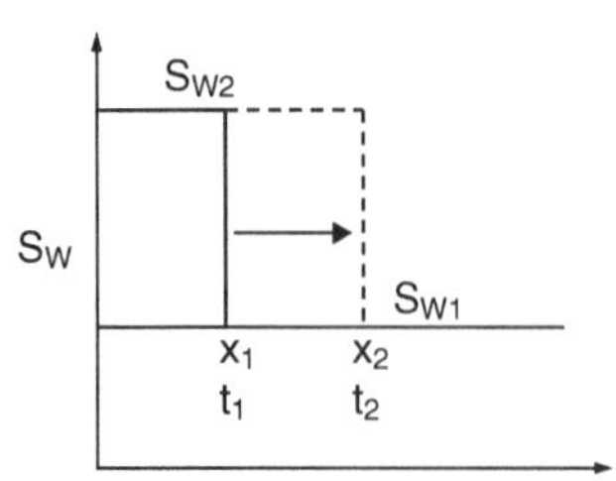

FIGURE 2.13 Schematic of saturation shock.

$$\left(\frac{\partial f_w}{\partial S_w}\right)_{S_{w2}} = \frac{f_{w2} - f_{w1}}{S_{w2} - S_{w1}}. \tag{2.76}$$

Similar to the concentration sharpening front (shock), the saturation shock front discussed previously may not always form. A shock front may form if the saturation velocities upstream are greater than those downstream. This is true for most oil/water fractional flow curves between certain limits of saturation, depending on the curvature of the fractional flow curve (Pope, 1980). If we cannot draw a tangent to the fractional flow curve, then a good flood front will not form (Craig, 1971).

2.6.2 Fractional Flow Equation

The advection-reaction-dispersion equation defined by Eq. 2.60 is for an isothermal single phase flow in one dimension. The fluid is incompressible. Gravity and capillary forces are not included. For multiphase flow, because chemicals are usually injected in the water phase, the advection term in the previous equation should be multiplied by water fraction f_w, and the left side should be multiplied by water saturation S_w. When dispersion is also neglected, $D_L = 0$. Equation 2.60 therefore becomes

$$\frac{\partial (S_w C_i)}{\partial t} = -v\left(\frac{\partial (f_w C_i)}{\partial x}\right) - \frac{\partial \hat{C}_i}{\partial t}. \tag{2.77}$$

Note that in Eq. 2.60, both C_i and $\hat{C}_i$ are in mass/(PV). C_i is always expressed in solution volume, generally water volume. Therefore, C_i is in mass/(PV water), but $\hat{C}_i$ is in *mass/PV* in Eq. 2.77. Their units are *different* now.

Equation 2.77 should also be applied to the water component—that is, $C_i = C_w$. Because the chemical solute composition is small, C_w can be assumed to be constant. And water retention is negligible. Then for the water component, Eq. 2.77 becomes

$$\frac{\partial S_w}{\partial t} = -v\frac{\partial f_w}{\partial x}. \tag{2.78}$$

Expanding Eq. 2.77 and combining it and Eq. 2.78, we have

$$S_w\left(\frac{\partial C_i}{\partial t}\right) = -vf_w\left(\frac{\partial C_i}{\partial x}\right) - \frac{\partial \hat{C}_i}{\partial t}. \tag{2.79}$$

In the preceding equation, $v = q_t/(A\phi)$, where q_t is the total injection rate, A is the flow area, and ϕ is the porosity. When we define $x_D = x/L$ and $t_D = tq_t/(AL\phi)$, Eq. 2.79 becomes

$$S_w\left(\frac{\partial C_i}{\partial t_D}\right) = -f_w\left(\frac{\partial C_i}{\partial x_D}\right) - \frac{\partial \hat{C}_i}{\partial t_D}, \tag{2.80}$$

Because $\hat{C}_i$ is a function of C_i, we have

$$\frac{\partial \hat{C}_i}{\partial t_D} = \left(\frac{d\hat{C}_i}{dC_i}\right)\left(\frac{\partial C_i}{\partial t_D}\right) = D_i\left(\frac{\partial C_i}{\partial t_D}\right). \tag{2.81}$$

In the preceding equation, we define the frontal advance lag for concentration C_i:

$$D_i = \frac{d\hat{C}_i}{dC_i}. \tag{2.82}$$

Combining Eqs. 2.80 and 2.81, we have

$$(S_w + D_i)\left(\frac{\partial C_i}{\partial t_D}\right) + f_w\left(\frac{\partial C_i}{\partial x_D}\right) = 0. \tag{2.83}$$

For a constant concentration C_i, we have

$$dC_i = 0 = dt_D\left(\frac{\partial C_i}{\partial t_D}\right)_{x_D} + dx_D\left(\frac{\partial C_i}{\partial x_D}\right)_{t_D}, \tag{2.84}$$

this gives

$$-\left(\frac{dx_D}{dt_D}\right)_{C_i} = \left(\frac{\partial C_i}{\partial t_D}\right)_{x_D} \Big/ \left(\frac{\partial C_i}{\partial x_D}\right)_{t_D}. \tag{2.85}$$

If we combine Eqs. 2.83 and 2.85, we have

$$v_{C_i} = \left(\frac{dx_D}{dt_D}\right)_{C_i} = \frac{f_w}{(S_w + D_i)}. \tag{2.86}$$

Note that the preceding velocity is the interstitial injection velocity normalized by $q_t/(A\phi)$, and that it is dimensionless. Lake (1989) and Green and Willhite (1998) used the term *specific* velocity for the dimensionless velocity. In this book, we follow their terminology. Corresponding to the front of the component C_i, we assume the water saturation is S_{w3}. According to the Buckley–Leverett theory (1942), the specific velocity of S_{w3} is

$$v_{S_{w3}} = \left(\frac{dx_D}{dt_D}\right)_{S_{w3}} = \left(\frac{\partial f_w}{\partial S_w}\right)_{S_{w3}}. \tag{2.87}$$

Because S_{w3} is the water saturation at the chemical front of C_i, their specific velocities must be the same, resulting in

$$\left(\frac{\partial f_w}{\partial S_w}\right)_{S_{w3}} = \left(\frac{f_w}{S_w + D_i}\right)_{S_{w3}, C_i}. \tag{2.88}$$

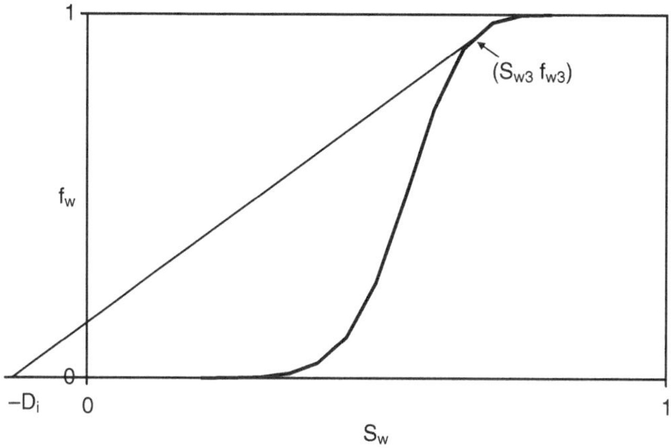

FIGURE 2.14 Construction of tangent to find S_{w3}.

Equation 2.88 shows that S_{w3} can be found by drawing a tangent to the f_w versus S_w curve for the injected water solution with the chemical component i from the point $(S_w, f_w) = (-D_i, 0)$, as shown in Figure 2.14.

Next, we discuss the application of Eq. 2.88 in different chemical flood processes.

2.6.3 Retardation of Chemicals in Two-Phase Flow

When a chemical solution is injected into a reservoir at interstitial water saturation, due to chemical retention, a denuded water zone is formed at the injection front, which causes a chemical shock at x_{w3}, as shown in Figure 2.15. This chemical shock causes the saturation shock from S_{w3} to S_{w1} at x_{w3}. The denuded water displaces the interstitial water. There is a boundary between the denuded water and the displaced interstitial water at x_{wb}.

The velocity of this boundary and its relation with the chemical shock velocity can be determined by making a material balance on the retaining chemical: the amount of chemical with concentration C_i in the injected solution in the denuded water zone from x_{w3} to x_{wb} must equal the amount of chemical retained behind the front x_{w3}. Mathematically,

$$C_i A\phi (x_{wb} - x_{w3}) S_{w1} = \hat{C}_i A\phi x_{w3}. \tag{2.89}$$

Thus, we have

$$x_{wb} = x_{w3}\left(1 + \frac{\hat{C}_i/C_i}{S_{w1}}\right) = x_{w3}\left(1 + \frac{D_i}{S_{w1}}\right), \tag{2.90}$$

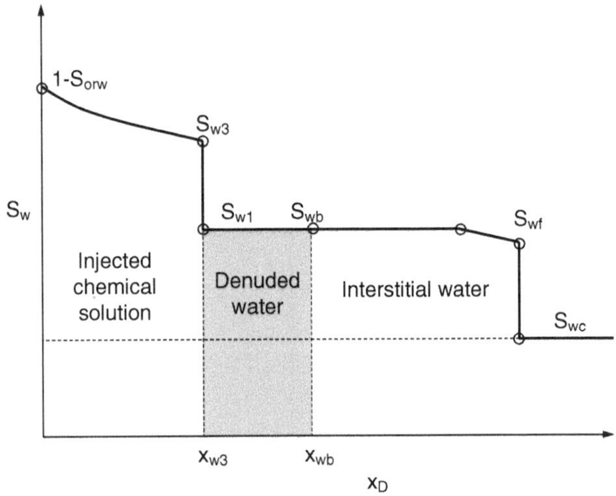

FIGURE 2.15 Saturation profile when a chemical solution is injected in a reservoir at interstitial water saturation.

where

$$D_i = \frac{\hat{C}_i}{C_i}, \tag{2.91}$$

with the injection concentration C_i jumping to zero to satisfy the adsorption $\hat{C}_i$. Equation 2.90 can be expressed in specific velocities as

$$v_{Dw3} = \frac{v_{Dwb}}{1 + \dfrac{D_i}{S_{w1}}}, \tag{2.92}$$

where v_{Dwb} is the specific velocity of the solution if no chemical retention exists, while v_{Dw3} is the specific moving velocity of the injected chemical C_i with retention $\hat{C}_i$. Equation 2.92 shows that the chemical moving velocity is retarded by a factor

$$R_{C_i} = 1 + \frac{D_i}{S_{w1}}. \tag{2.93}$$

Compared with Eq. 2.65, Eq. 2.93 has an extra term S_{w1}. As mentioned earlier, C_i and $\hat{C}_i$ have different units in the definition of D_i in Eq. 2.91. The units of C_i and $\hat{C}_i$ are mass/PV water (PV_{water}) and mass/PV, respectively. If $\hat{C}_i$ is made to have the same unit as C_i in mass/PV water, then we have

$$D_i\left[\frac{PV_{water}}{PV}\right] = D_i\left[\frac{PV_{water}}{PV_{water}}\right] S_{w1}. \tag{2.94}$$

If we multiply D_i by the water saturation S_{w1}, then the retardation defined in Eq. 2.93 would be the same as Eq. 2.65. In other words, the retardation factor is simply $R_{C_i} = 1 + d\hat{C}_i / dC_i$ with Ci and $\hat{C}_i$ in the same unit.

To determine the boundary velocity, we must consider the fact that the velocities at x_{w3} must travel at the same velocity. According to Eqs. 2.87 and 2.88, we have

$$v_{Dw3} = \frac{f_{w3}}{S_{w3} + D_i} = \frac{f_{w3} - f_{w1}}{S_{w3} - S_{w1}} = \frac{f_{w1}}{S_{w1} + D_i}. \quad (2.95)$$

Substituting v_{Dw3} in Eq. 2.95 for v_{Dw3} in Eq. 2.92 yields

$$v_{Dwb} = \left(\frac{f_{w1}}{S_{w1} + D_i} \right) \left(1 + \frac{D_i}{S_{w1}} \right) = \frac{f_{w1}}{S_{w1}} = \frac{f_{wb}}{S_{wb}}. \quad (2.96)$$

2.6.4 Fractional Flow Curve Analysis of Waterflooding

During a waterflood, the injected water displaces the interstitial water as well as oil. There are two fluid boundaries: one between the injected water and the interstitial water, and the other between the displaced interstitial water and the oil ahead. We want to find the water saturation at the boundary between the injected water and interstitial water. Because a nonadsorbing chemical travels at the same velocity as the water front, we can use Eq. 2.88 to find the water saturation at the front (the boundary between the injected and interstitial water), S_{wb}. From Eq. 2.88, if D_i is zero, we have

$$v_{S_{wb}} = \left(\frac{\partial f_w}{\partial S_w} \right)_{S_{wb}} = \left(\frac{f_w}{S_w} \right)_{S_{wb}}. \quad (2.97)$$

Equation 2.97 shows that S_{wb} can be determined by drawing the tangent to the f_w versus S_w curve from the origin (0, 0), as shown in Figure 2.16.

The waterflood front is given by the classical Buckley–Leverett theory by drawing the tangent to the f_w versus S_w curve from (S_{wc}, 0), as shown in Figure 2.16. The corresponding equation is

$$\left(\frac{\partial f_w}{\partial S_w} \right)_{S_{wf}} = \frac{f_{wf}}{S_{wf} - S_{wc}}, \quad (2.98)$$

where S_{wc} is the connate (interstitial) water saturation. The saturation profile in Figure 2.17 shows that the injected water displaces the original interstitial water ahead of the water boundary x_{wb}. The front is a sharpening front from S_{wf} to S_{wc}. From $1–S_{orw}$ to S_{wf}, it is a spreading wave because there is no chemical shock that causes a saturation chock (*cf* . Figure 2.15). In Figure 2.17, the flow behavior of the injected water is assumed to be the same as that of the interstitial water. When we do not consider the displacement of interstitial water

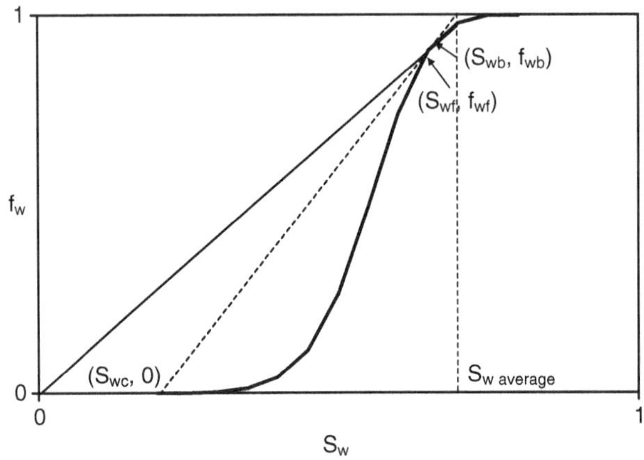

FIGURE 2.16 Construction of tangent to find S_{wb}, the saturation at the boundary between the injected water and interstitial water.

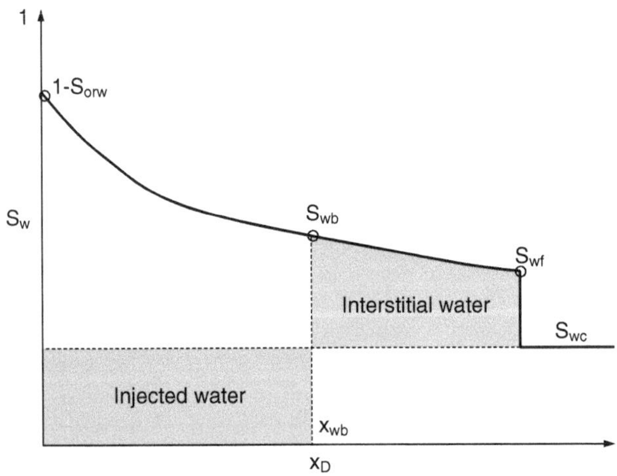

FIGURE 2.17 Water saturation profile showing interstitial water displaced by injected water.

by the injected water, S_{wb} or x_{wb} is not relevant. Only the displacing front S_{wf} exists.

The recovered oil, N_p, in subsurface pore volume is described by the average water saturation change

$$N_p = \overline{S}_w - S_{wc}, \tag{2.99}$$

where $\overline{S}_w$ is the average water saturation in the entire oil zone. Because S_{wf} moves at the specific velocity of $f'_{wf} = (\partial f_w / \partial S_w)_{S_{wf}}$, then the breakthrough time, t_{Dbt}, at $x_D = 1$ is

$$t_{Dbt} = \frac{1}{f'_{wf}}. \quad (2.100)$$

Before water breakthrough, only oil is produced, and the volume of oil produced is equal to the water injected. Therefore, the oil recovered at any time t_D *before* water breakthrough is

$$N_p = \bar{S}_w - S_{wc} = \frac{t_D}{f'_{wf}}. \quad (2.101)$$

The average water saturation at any time after breakthrough is computed from the Welge (1952) equation,

$$\bar{S}_w = S_{we} + \frac{1 - f_{we}}{f'_{we}}, \quad (2.102)$$

where the subscript e means at the effluent end ($x_D = 1$), and $f'_{we} = (\partial f_w / \partial S_w)_{S_{we}}$.

2.6.5 Fractional Flow Curve Analysis of Polymer Flooding

In the case of polymer flooding with a sharpening front, polymer concentration jumps from zero (its initial value) to its injection concentration C_{inj}. D_i in Eq. 2.88 becomes D_p. In this case, the high polymer concentration solution flushes the initial zero polymer concentration solution. As discussed in Section 2.5 on types of fronts, there is a concentration shock. Corresponding to this concentration shock, there is a saturation shock from S_{wp} to S_{w1}. The specific velocity of this saturation shock, $(v_D)_{\Delta S_w}$, is

$$(v_D)_{\Delta S_w} = \frac{f_{wp} - f_{w1}}{S_{wp} - S_1}, \quad (2.103)$$

where f_{wp} and f_{w1} are the water cuts corresponding to S_{wp} and S_{w1}, respectively. On the other hand, according to the Buckley–Leverett theory, the specific velocity of S_{wp} is

$$(v_D)_{S_{wp}} = \left(\frac{\partial f_w}{\partial S_w}\right)_{S_{wp}} = \frac{f_{wp}}{S_{wp} + D_p}. \quad (2.104)$$

These two specific velocities must travel at the same specific velocity. Thus,

$$\frac{f_{wp} - f_{w1}}{S_{wp} - S_1} = \frac{f_{wp}}{S_{wp} + D_p}. \quad (2.105)$$

Equation 2.105 shows that S_{wp} can be found by drawing a tangent to the f_w versus S_w curve for the polymer solution from the point $(S_w, f_w) = (-D_p, 0)$, as shown in Figure 2.18. The water saturation profile is shown in Figure 2.19. The average water saturation is given by Eq. 2.106.

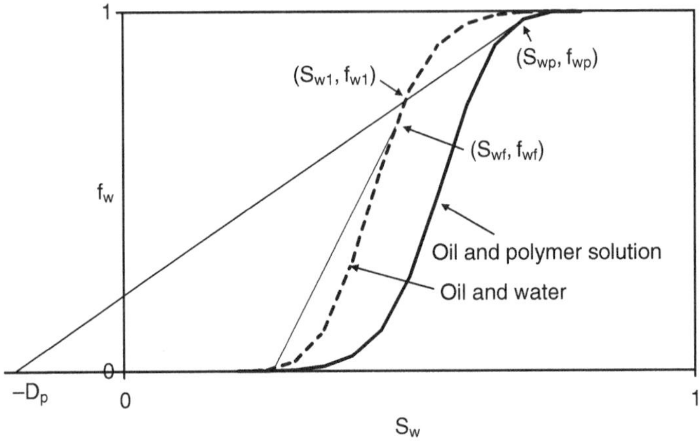

FIGURE 2.18 Construction of tangent to find S_{wp}, S_{w1}, and S_{wf}.

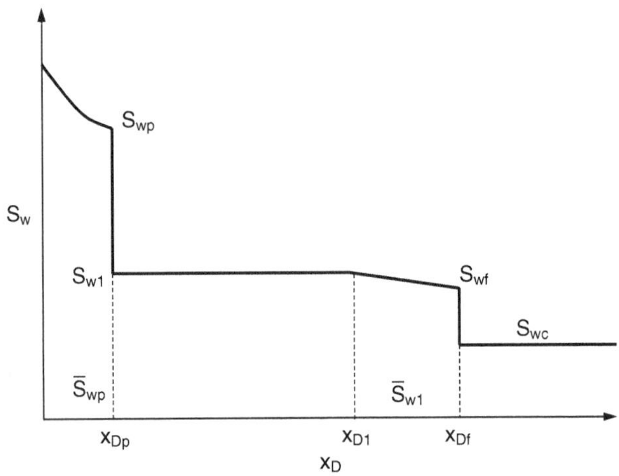

FIGURE 2.19 Saturation profile for polymer flood started at interstitial water saturation when $S_{w1} > S_{wf}$.

$$\begin{aligned}\bar{S}_w &= \int_0^{x_{Dp}} S_w dx_D + \int_{x_{Dp}}^{x_{D1}} S_{w1} dx_D + \int_{x_{D1}}^{x_{Df}} S_w dx_D + \int_{x_{Df}}^{1} S_{wc} dx_D \\ &= \bar{S}_{wp} x_{Dp} + S_{w1}(x_{D1} - x_{Dp}) + \bar{S}_{w1}(x_{Df} - x_{D1}) + S_{wc}(1 - x_{Df}),\end{aligned} \tag{2.106}$$

where $\bar{S}_{wp}$ and $\bar{S}_{w1}$ are the average water saturations in the respective polymer and water front regions. When the polymer solution begins at time zero, $\bar{S}_{wp}$ is calculated from the expanded Welge equation,

$$\overline{S}_{wp} = S_{wp} + t_D \frac{1 - f_{wp}}{x_{Dp}}, \quad (2.107)$$

and $\overline{S}_{w1}$ is given by (Willhite, 1986)

$$\overline{S}_{w1} = \frac{x_{Df} S_{wf} - x_{D1} S_{w1}}{x_{Df} - x_{D1}} - t_D \frac{f_{wf} - f_{w1}}{x_{Df} - x_{D1}}. \quad (2.108)$$

When the injection starts at $t_D = 0$, the locations of saturations are given by the Buckley–Leverett theory:

$$x_{Df} = f'_{wf} t_D, \quad (2.109)$$

$$x_{D1} = f'_{w1} t_D, \quad (2.110)$$

and

$$x_{Dp} = f'_{wp} t_D. \quad (2.111)$$

Before water breakthrough, the oil recovered at any time t_D is given by Eq. 2.101.

Between water breakthrough and arrival of the oil bank, x_{D1},

$$\overline{S}_w = \overline{S}_{wp} x_{Dp} + S_{w1}(x_{D1} - x_{Dp}) + \overline{S}_{w1}(1 - x_{D1}). \quad (2.112)$$

When Eqs. 2.107 and 2.108 are substituted for $\overline{S}_{wp}$ and $\overline{S}_{w1}$, respectively, and Eq. 2.111 is used for x_{Dp} and $f'_{wp} = (f_{wp} - f_{w1})/(S_{wp} - S_{w1})$, Eq. 2.112 is simplified to become

$$\overline{S}_w = S_{we} + t_D(1 - f_{we}). \quad (2.113)$$

When we derive Eq. 2.113, $x_{Df} = 1$, and S_{wf} and f_{wf} become S_{we} and f_{we} at the effluent end, respectively. During the time $t_{Df} \le t_D \le t_{D1}$, S_{we} increases from S_{wf} to S_{w1}. When the oil bank arrives at the end of the system (x_{D1} =1), from Eq. 2.112, the average water saturation is given by

$$\overline{S}_w = \overline{S}_{wp} x_{Dp} + S_{w1}(1 - x_{Dp}). \quad (2.114)$$

When we substitute Eqs. 2.107 and 2.111 for $\overline{S}_{wp}$ and x_{Dp}, respectively, and use $f'_{wp} = (f_{wp} - f_{w1})/(S_{wp} - S_{w1})$, Eq. 2.114 becomes

$$\overline{S}_w = S_{w1} + t_D(1 - f_{w1}). \quad (2.115)$$

When S_{wp} arrives at the end, $x_{D3} = 1$, and $t_D = 1/f'_{wp}$. Therefore, for $t_D \ge 1/f'_{wp}$,

$$\overline{S}_w = S_{we} + t_D(1 - f_{we}). \quad (2.116)$$

Note that Eqs. 2.113, 2.115, and 2.116 follow the form of the Welge equation.

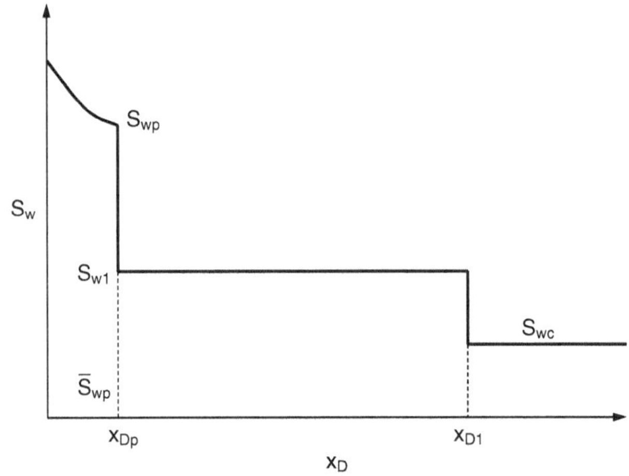

FIGURE 2.20 Saturation profile for polymer flood started at interstitial water saturation when $S_{w1} < S_{wf}$.

When S_{w1} is less than S_{wf}, the oil bank forms immediately and overtakes S_{wf}. Then the uniform water saturation, S_{w1}, is formed (Green and Willhite, 1998). Figure 2.20 shows the saturation profile in this situation. Water breaks through at

$$t_{D1} = \frac{S_{w1} - S_{wc}}{f_{w1}}, \tag{2.117}$$

and S_{wp} breaks through at

$$t_{Dp} = \frac{S_{wp} - S_{w1}}{f_{wp} - f_{w1}}. \tag{2.118}$$

Polymer inaccessible pore volume results in a faster polymer velocity that is opposite to the polymer adsorption effect. The polymer inaccessible pore volume effect can be included in the D_p term. Lake (1989) explicitly added the term $-\phi_{IPV}$ in Eq. 2.104 to include this effect. The unit of ϕ_{IPV} is fraction of porosity.

Polymer floods, like any other chemical floods, will be more efficient if they are started at low initial water saturations. Due to practical feasibilities, however, they are more often started at high initial water saturations. One reason is that we need some waterflood history to better understand the reservoir so that we can design a proper polymer flood program. Figure 2.21 shows the water saturation profile when a polymer flood is started at a high initial water saturation. Corresponding to Figure 2.21, Figure 2.22 shows the fractional flow curves. The individual specific velocities are also marked in these figures, and they are defined next.

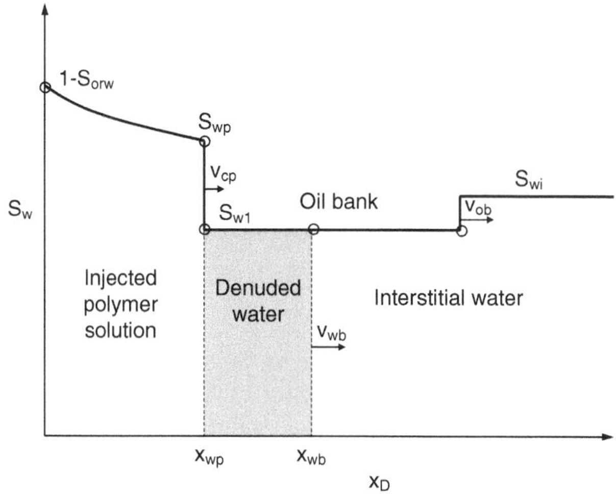

FIGURE 2.21 Water saturation profile when a polymer flood is started at a high initial water saturation.

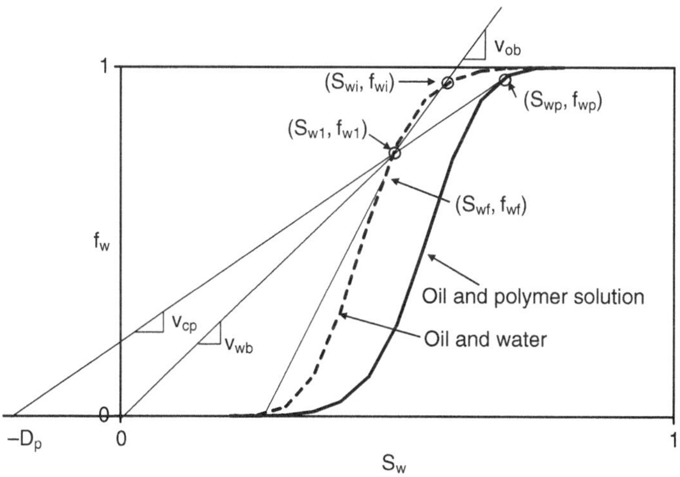

FIGURE 2.22 Graphical construction of polymer flood fractional curves.

The polymer concentration shock, corresponding to the saturation shock from S_{wp} to S_{w1}, moves at v_{cp}:

$$v_{cp} = \frac{f_{wp} - f_{w1}}{S_{wp} - S_{w1}}. \tag{2.119}$$

The boundary between denuded water and the initial water moves at v_{wb}:

$$v_{wb} = \frac{f_{w1}}{S_{w1}}. \tag{2.120}$$

The front of the oil bank moves at v_{ob}:

$$v_{ob} = \frac{f_{wi} - f_{w1}}{S_{wi} - S_{w1}}. \tag{2.121}$$

2.6.6 Fractional Flow Curve Analysis of Surfactant Flooding

When surfactant solution is injected in a reservoir, it contacts with oil to form three types of microemulsion, depending on the local salinity. Here, we discuss only the fractional curve analysis of Winsor I microemulsion. For a discussion of fractional flow of Winsor II without retention, see Lake (1989). Fractional flow treatment for three-phase microemulsion flood (Winsor III) has not been extensively investigated (Giordano and Salter, 1984).

In a Winsor I system, the surfactant is in the water phase, and some oil is solubilized in the water phase as well. Thus, the aqueous phase viscosity is higher than that of the originally existing water. In most cases, polymer is added in the surfactant solution to increase solution viscosity. Therefore, a typical fractional flow curve of surfactant solution/oil shifts to the right of the water/oil fractional curve, as shown in Figure 2.23, which shows a fractional flow diagram of a Winsor I microemulsion flood. Note that the immobile water saturation, S_{wc}, for the oil/surfactant fractional curve is smaller than for the oil/water fractional curve. Figure 2.23 shows that S_{w1} is less than S_{wf}. The water saturation profile is shown in Figure 2.24. The specific velocity at the shock front is shown in Eq. 2.122.

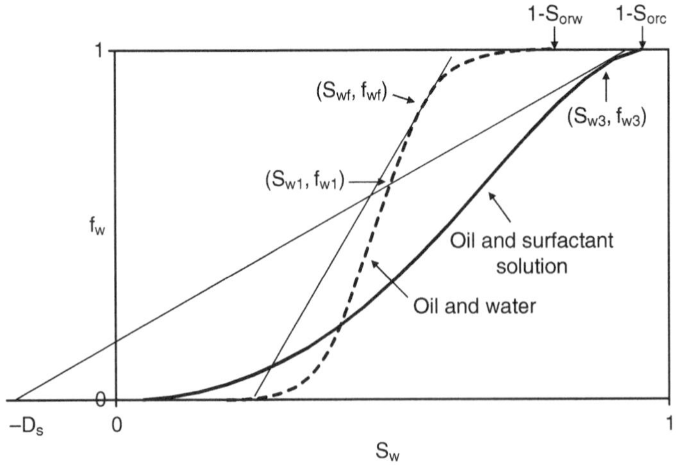

FIGURE 2.23 Fractional flow diagram of a Winsor I microemulsion flood.

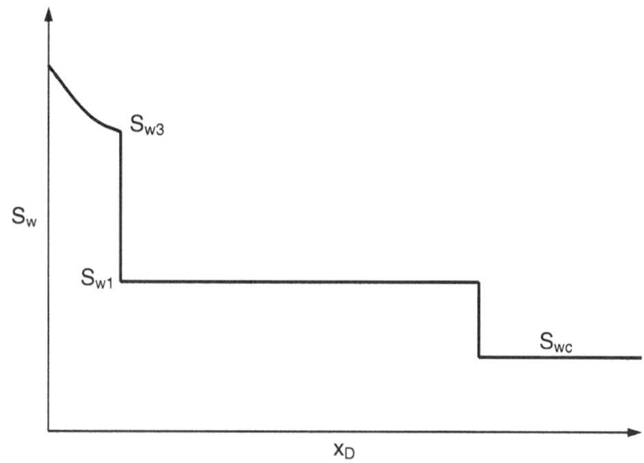

FIGURE 2.24 Saturation profile for a Winsor I microemulsion flood started at interstitial water saturation when $S_{w1} < S_{wf}$.

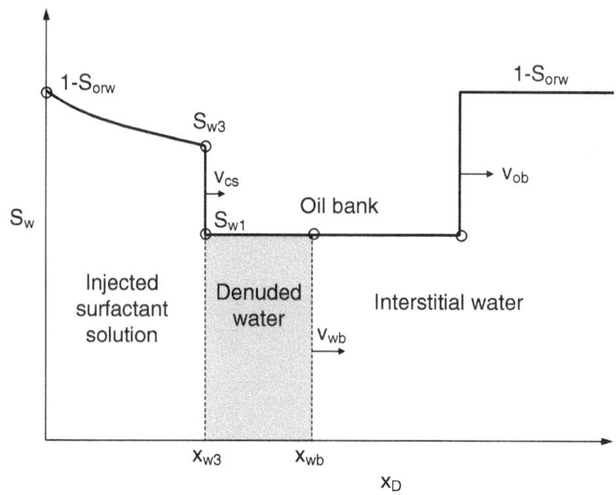

FIGURE 2.25 Saturation profile for a Winsor I microemulsion flood started at waterflood residual oil saturation, S_{orw}.

$$\frac{dx_D}{dt_D} = \frac{f_{w1}}{S_{w1} + D_s} = \frac{f_{w3} - f_{w1}}{S_{w3} - S_{w1}} = \frac{f_{w3}}{S_{w3} + D_s}. \tag{2.122}$$

A surfactant flood can recover the oil left from a waterflood. Sometimes, a surfactant flood is applied at the waterflood residual oil saturation, S_{orw}. When a surfactant flood is started, S_{orw}, the water saturation profile is as shown in Figure 2.25. Corresponding to Figure 2.25, Figure 2.26 shows the fractional

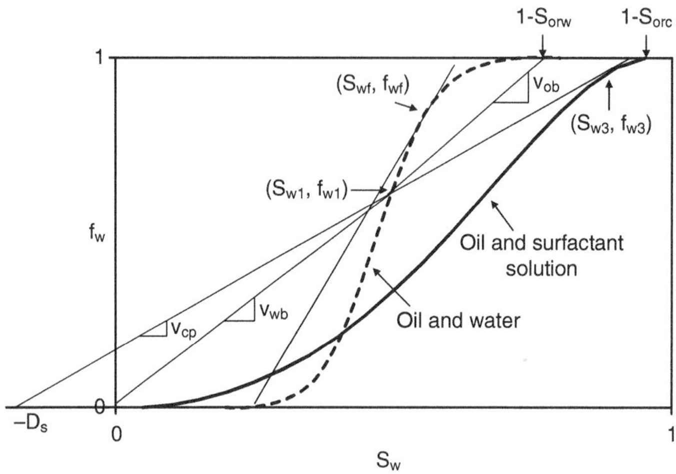

FIGURE 2.26 Fractional flow diagram of Winsor I microemulsion flood at waterflood residual oil saturation, S_{orw}.

flow curves. The individual specific velocities are also marked in these figures, and they are defined next.

The surfactant concentration shock, corresponding to the saturation shock from S_{w3} to S_{w1}, moves at v_{cs}:

$$v_{cs} = \frac{f_{w3} - f_{w1}}{S_{w3} - S_{w1}}. \tag{2.123}$$

The boundary between the denuded water and the initial water moves at v_{wb} defined in Eq. 2.120. The front of the oil bank moves at v_{ob}:

$$v_{ob} = \frac{1 - f_{w1}}{1 - S_{orw} - S_{w1}}. \tag{2.124}$$

Chapter 3

Salinity Effect and Ion Exchange

3.1 INTRODUCTION

Salinity is essential for all chemical processes. It directly affects polymer viscosity, and it determines the type of microemulsion a surfactant can form. Salinity effects in waterflooding, in both sandstone and carbonate reservoirs, have recently drawn research interest. This chapter briefly discusses salinity and ion exchange. At the end of this chapter, the salinity effects on waterflooding in sandstone and carbonate reservoirs are summarized.

3.2 SALINITY

Salinity can be represented in several ways. One of the simplest ways to quantify salinity is to use total dissolved solids (often abbreviated TDS). TDS is the total mass content of dissolved ions and molecules or suspended microgranules in a liquid medium. Generally, the operational definition is that the solids must be small enough to survive filtration through a sieve size of two micrometers. Because sodium chloride (NaCl) is the main salt in saline water, we commonly use the mass of sodium chloride as the salinity. The common units are ppm or wt.%. Of course, we can almost always use meq/mL.

Because of electrical charge neutrality, the total mass of negative ions (anions) should equal the total mass of positive ions (cations) in a system. Sometimes, we use only the total amount of anions to represent salinity. The unit used in this case is meq/mL. Because sodium chloride is the main salt in saline water, we may simply use the total amount of Cl^- to represent salinity.

Because the effects of monovalents, divalents, or multivalents could be significantly different, we generally separate the ions into two groups: monovalents represented by the sodium ion Na^+ and divalents and multivalents represented by the calcium ion Ca^{2+}. Sometimes, we use the terms salinity to represent the total amount of anions and hardness to represent multivalents. Because the amount of all anions should be equal to the amount of all cations in meq/mL, the amount of monovalents is equal to the total anions minus the amount of multivalents. Here, we use the unit meq/mL. This is done in UTCHEM-9.0 (2000), a chemical simulator that we will use extensively to generate results in this book. According to its user manual, the input parameters are the

Modern Chemical Enhanced Oil Recovery. DOI: 10.1016/B978-1-85617-745-0.00003-6

concentrations of all the anions and the divalent cations of brine in equivalents if gel reactions or alkaline reactions are not included.

The preceding discussion shows that salinity is represented in different ways, but it is most important that the input values for the initial formation water and the injected brines are consistent with the laboratory data. Note that C50 and C60 are the UTCHEM input parameters for the concentrations of all the anions and the divalent cations, respectively, in the initial formation brine. C(M,5,1) and C(M,6,1) are the UTCHEM input parameters for the concentrations of all the anions and the divalents in the injection water, respectively. Inside the parentheses, M represents the injection well number, 5 represents all anions (mainly Cl^-), 6 represents divalents or multivalents (mainly Ca^{2+}), and 1 represents water phase. These concentrations are explained in Example 3.1 later in this chapter.

To combine the effects of monovalents and divalents, UTCHEM uses the concept of effective salinity, which is defined as

$$C_{se} = \frac{C_{51} + (\beta_6 - 1)C_{61}}{C_{11}}, \tag{3.1}$$

where C_{51}, C_{61}, and C_{11} are the anion, calcium, and water concentrations in the aqueous phase, respectively, and β_6 is the effective salinity parameter for calcium measured in the laboratory and is an input parameter in UTCHEM.

To consider the effect of divalent cations bound to micelles, the effective salinity is (Hirasaki, 1982a)

$$C_{se} = \frac{C_{51}}{1 - \beta_6^s f_6^s}, \tag{3.2}$$

where β_6^s is the slope parameter for divalent in a surfactant system, a positive constant, and f_6^s is the fraction of the total divalent cations bound to surfactant micelles as $f_6^s = C_6^s / C_3^m$. Here, C_6^s is the total divalent cations bound to surfactant micelles, and $C_3^m = 1000C_3/(C_1M_3)$ is the surfactant concentration in *meq/mL water*. Here, M_3 is the equivalent weight of the surfactant, and C_1 and C_3 are volume fractions of water and microemulsion phases, respectively.

If we want to consider the effect of temperature, the effective salinity is

$$C_{se} = \frac{C_{51}}{1 + \beta_T(T - T_{ref})}, \tag{3.3}$$

where β_T is the temperature coefficient, and T and T_{ref} are the temperature and reference temperature, respectively. The products of β_T and T should be dimensionless. The effective salinity decreases with the temperature for anionic surfactants but increases with the temperature for nonionic surfactants.

To consider the effect of alcohol on the effective salinity, we extend the Hirasaki (1982a) model. The effective salinity is

$$C_{se} = \frac{C_{51}}{1+\beta_7 f_7^s}, \tag{3.4}$$

where β_7 is the slope parameter for alcohol dilution determined by matching an experimental salinity requirement diagram, and f_7^s is defined as the volume fraction of alcohol in the total volume of surfactant and alcohol:

$$f_7^s = \frac{V_7}{V_7 + V_3} = \frac{1}{1+(C_{73})^{-1}}. \tag{3.5}$$

So far we have summarized different ways to present salinity. Strictly speaking, the effect of an ion may be different from the effect of any other one, so we should take into account every ion separately. We have to use a kind of pseudo-ion concept, however, to present salinity because (1) we do not fully understand the effect of each ion, (2) it is difficult and tedious to present the effects of so many ions and their reactions, and (3) the effect of an ion could be different in different processes.

In alkaline-related chemical processes, the addition of alkalis such as sodium carbonate increases ionic strength (salinity). As alkaline concentration is increased, the optimum salinity is decreased. The order to decrease the optimum salinity is (Martin and Oxley, 1985)

$$NH_4OH < Na_2O(SiO_2)_{3.2} < Na_2CO_3 < NaOH < KOH.$$

We must, however, take into account some chemical reactions in alkaline flooding. Therefore, alkalis are modeled differently (see Chapter 10 on alkaline flooding).

Ionic strength is related to salinity. The ionic strength of a solution is a measure of the concentration of ions in that solution. The ionic strength, I, of a solution is a function of the concentrations of *all* the ions present in that solution (Atkins and de Paula, 2006)

$$I = \frac{1}{2}\sum_{i=1}^{n} C_i z_i^2, \tag{3.6}$$

where C_i is the molar concentration of ion i (M = mol/L), z_i is the charge number of that ion, and the sum is taken over all ions in the solution. For a 1 : 1 electrolyte such as NaCl, ionic strength is equal to the concentration, but for $CaCl_2$ the ionic strength is four times higher. Generally, multivalent ions contribute strongly to the ionic strength.

For example, the ionic strength of a mixed 0.050 M $CaCl_2$ and 0.020 M NaCl solution is

$$\begin{aligned} I &= 1/2\big(0.050\times1\times(+2)^2 + 0.050\times2\times(-1)^2 \\ &\quad + 0.020\times1\times(+1)^2 + 0.020\times1\times(-1)^2\big) \\ &= 0.17\ \text{M} \end{aligned}$$

Because volumes are no longer strictly additive in nonideal solutions, it is often preferable to work with molality, m_i (mol/kg H_2O), rather than concentration (mol/L). In that case, ionic strength is defined as

$$I = \frac{1}{2}\sum_{i=1}^{n} m_i z_i^2. \tag{3.7}$$

Example 3.1 Calculate Salt Contents of Formation Water and Injection Water

Table 3.1 is a water analysis report. Find the input values of salinities for UTCHEM simulation and discuss the salinities in the table.

Solution

For the formation water, Table 3.1 shows that the amount of total cations is 58279.3 mg/L, which is 36884.8 mg/L less than that of total anions, 95164.1 mg/L. The difference of the total anions and the total cations in meq/L, however, is only −0.4. This is the measurement or calculation error because the difference should be zero. Similarly, for the injection water, the difference of the total anions and the total cations is 0.2 in meq/L. These data show that the unit of meq/L or meq/mL should be used to present the ion concentrations.

In this example, the TDS of formation water is 153443.4 mg/L, or 5358.8 meq/L. The TDS of injection water is 39615.0 mg/L, or 1379.0 meq/L. The UTCHEM input parameters, C50 and C60 for the total anions and the total divalents, are 2.6792 meq/mL and 0.7497 meq/mL, respectively. The UTCHEM input parameters, C(M,5,1) and C(M,6,1) for the total anions and total divalents in the injection water, are 0.6896 meq/mL and 0.1651 meq/mL, respectively.

Table 3.1 also shows that Cl^- is the dominant ion in both formation water and injection water. Sometimes, we may use the amount of Cl^- to represent the salinity. Salinity effects are very important in alkaline flooding. This concept could be better understood after a discussion of alkaline flooding (see Example 10.4).

3.3 ION EXCHANGE

Formation rocks contain materials like clay minerals, organic matter, and metal oxy-hydroxides that can sorb chemicals. The general term sorption is used to describe the various processes that include adsorption, absorption, and ion exchange. The term adsorption refers to the adherence of a chemical to the solid surface, absorption suggests that the chemical is taken up into the solid, and ion exchange involves replacement of one chemical for another at the solid surface (Appelo and Postma, 2007). In chemical EOR processes, we deal with adsorption and ion exchange. Adsorption is addressed in Chapters 5 and 7 on polymer flooding and surfactant flooding, respectively. This section discusses ion exchange. Usually, the ion exchange processes in EOR processes are cation exchange. Therefore, quite often, we just use the term cation exchange. Pope

TABLE 3.1 Water Analysis Report

		Formation Water		Injection Water	
		A *mg/L*	*B = A/MW* × |*charge*| *meq/L*	*C* *mg/L*	*D = C/MW* × |*charge*| *meq/L*
I1	Na^+ (23, +1)	44388.0	1929.9	12060.0	524.3
I2	Ca^{2+} (40, +2)	12238.0	611.9	502.0	25.1
I3	Mg^{2+} (24, +2)	1653.3	137.8	1680.0	140.0
I4	Cl^- (35.5, –1)	94976.7	2675.4	22040.0	620.8
I5	HCO_3^- (61, –1)	40.7	0.7	163.0	2.7
I6	SO_4^{2-} (96, –2)	146.7	3.1	3170.0	66.0
Total cations (I1 + I2 + I3)		58279.3	2679.6	14242.0	689.4
Total anions (I4 + I5 + I6)		95164.1	2679.2	25373.0	689.6
TDS (sum of I1 to I6)		153443.4	5358.8	39615.0	1379.0
Total anions – Total cations		36884.8	–0.4	11131.0	0.2
Total monovalents (I1)		44388.0	1929.9	12060.0	524.3
Total divalents (I2 + I3)		13891.3	749.7	2182.0	165.1

et al. (1978) provided the basic theory without dispersion about cation exchange in chemical flooding.

3.3.1 Ion Exchange Equations

An ion is an atom with an electric charge due to gain or loss of electrons. Ion exchange equilibria have been described by empirically and theoretically derived equations. They follow the form of the general law of mass action, which is

$$aA + bB \leftrightarrow cC + dD. \tag{3.8}$$

The distribution at equilibrium of the species at the left and right sides of the reaction is given by

$$K = \frac{[C]^c[D]^d}{[A]^a[B]^b}, \tag{3.9}$$

where K is the equilibrium constant, and the bracketed quantities denote activities or effective concentration. The law of mass action is applicable to any type of reaction, the dissolution of minerals, the formation of complexes between dissolved species, the dissolution of gases in water, and so on. For example, the cation exchange between Ca^{2+} and Na^+ is

$$Ca^{2+} + 2(Na\text{-}X) \leftrightarrow Ca\text{-}X_2 + 2Na^+, \tag{3.10}$$

where X is the exchange site on the solid material (clay), and the exchange (equilibrium) constant is

$$K_{Ca-Na} = \frac{[Ca\text{-}X_2][Na^+]^2}{[Na\text{-}X]^2[Ca^{2+}]}. \tag{3.11}$$

The ions that are in the subscript of the exchange constant are written in the order in which they appear as solute ions in the reaction. For example, in $K_{Ca\text{-}Na}$, Ca^{2+} appears in the solution first as a solute and has ion exchange with Na^+ attached to the solid surface. After ion exchange, the Na^+ attached before ion exchange appears in the solution as a solute. The magnitude of $K_{Ca\text{-}Na}$ indicates the relative tendency of the two ions to react with the sites on the clay, or the affinities of the two ions for the solid. The larger the value of $K_{Ca\text{-}Na}$, the greater the tendency of Ca^{2+} to attach to the clay compared with the tendency of Na^+. The value of $K_{A\text{-}B}$ for any pair of ions (A, B) is a function of the type of ion and nature of the solid. The order of affinity of several ions for the clay sites is

$$Li^+ < Na^+ < K^+ < Rb^+ < Cs^+ < Mg^{2+} < Ca^{2+} < Sr^{2+} < Ba^{2+} < H^+. \tag{3.12}$$

Species that have high charge densities (multivalents or small ionic radii) are more tightly held by the anionic sites. This observation suggests a possible explanation for the permeability-reducing behavior of Na^+. The large Na^+ cations disrupt the clay particles when they intrude into the structure. But only a small amount of another cation is sufficient to prevent this situation because most other naturally occurring cations are more tightly bound than Na^+ (Lake, 1989).

The name of the exchange equilibrium constant defined in Eq. 3.11 is used in UTCHEM and in Tan (1982). Lake (1989) used selectivity or selectivity coefficient, reflecting the tendency of an ion to attach to the clay compared with the tendency of another ion. Appelo and Postma (2007) used exchange coefficient because the values depend on the type of exchanger present in the soil and on the solution composition.

The capacity of cation exchange for a given rock is expressed in terms of the cation exchange capacity (CEC), usually given in the unit of milliequivalent per kilogram of rock (meq/kg). The capacity can also be expressed in terms of unit pore volume (PV). The unit conversion is

$$\mathrm{CEC}\left[\frac{\mathrm{meq}}{\mathrm{kg\ rock}}\right]=\mathrm{CEC}\left[\frac{\mathrm{meq}}{\mathrm{L\ PV}}\right]\left[\frac{\phi\mathrm{L\ PV}}{\mathrm{L\ bulk\ rock}}\right]\left[\frac{\mathrm{L\ bulk\ rock}}{(1-\phi)\mathrm{L\ rock}}\right]\left[\frac{\mathrm{L\ rock}}{\rho_r\ \mathrm{kg\ rock}}\right]$$
$$=\mathrm{CEC}\left[\frac{\mathrm{meq}}{\mathrm{L\ PV}}\right]\frac{\phi}{\rho_r(1-\phi)}, \qquad (3.13)$$

where ρ_r is in g/mL, and ϕ is in fraction. The preceding conversion also applies from CEC in meq/g rock to CEC in meq/mL PV. Note that the unit milliequivalent is used for CEC to consider the exchange of ions with different charges. In converting one unit system to another, we use an equivalent relationship; for example, one unit of bulk volume is equivalent to ϕ unit PV. Table 3.2 gives the cation exchange capacities of several rocks. The data that are plotted in Figure 3.1. This figure shows that if the first data point for Bandera is ignored, the cation exchange capacities demonstrate a good linear relationship with the surface area per gram of rock. The linear relationship is

$$\mathrm{CEC}=10.846S_r-5.4448, \qquad (3.14)$$

TABLE 3.2 Cation Exchange Capacities of Selected Rocks

Sandstone	Surface Area (m²/g)	CEC (meq/kg)
Bandera	5.5	11.99
Berea	0.93	5.28
Coffeyville	2.85	23.92
Cottage Grove	2.30	17.96
Noxie	1.43	10.01
Torpedo	2.97	29.27

Source: *Crocker et al., 1983.*

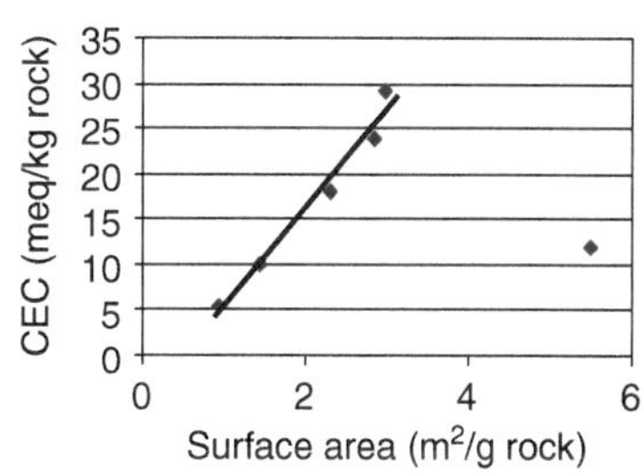

FIGURE 3.1 Cation exchange capacity versus surface area for sandstone rocks.

TABLE 3.3 Calcium Cation Exchange Capacities of California Oil Sands

Sands	Montmorillonite (wt.%)	CEC (meq/kg)
Wilmington		
Composite	1.2	20 ± 9
B110 at 4,913 ft	1.5	24 ± 13
B110 at 4,914 ft	0.27	9 ± 8
Huntington		
4,894 ft	0.85	13 ± 5
4,897 ft	0.52	13 ± 5
Coalinga	0.45	9 ± 5

Source: *Somerton and Radke, 1983.*

where CEC is in meq/kg rock, and the surface area S_r is in m^2/g (10^3 m^2/kg) rock.

Table 3.3 gives the calcium CECs for the six California oil sands at ambient temperature (Somerton and Radke, 1983). Also shown are the weight percentages of montmorillonite in these sands. These data are obtained for the grain size fraction < 43 μm. Note that the exchange capacities generally parallel the montmorillonite contents of the sands.

Table 3.4 gives the cation exchange capacities of common soil and sediment materials. An empirical formula that relates the CEC to the percentages of clay (< 2 μm) and organic carbon at near neutral pH is (e.g., Breeuwsma et al., 1986)

$$\text{CEC}[\text{meq/kg}] = 7 \times (\%\ \text{clay}) + 35 \times (\%\ \text{C}). \tag{3.15}$$

The equations of ion exchanges need further discussion. Equations 3.10 and 3.11 follow the general law of mass action. Strictly speaking, the use of activities instead of concentrations is required. For adsorbed cations, however, there is no unifying theory to calculate activity coefficients, and different conventions are in use (Appelo and Postma, 2007). Hill and Lake (1978) chose to express all concentrations in milliequivalents per milliliter pore volume. The activity of each exchangeable ion is expressed as a fraction of a total number, either as a molar fraction or as an equivalent fraction. The total number can be based on the number of exchange sites or on the number of exchangeable cations. Depending on which total number is used and whether a molar fraction or an equivalent fraction is used, different conventions to define the exchange coefficient have been used. Before we present these conventions, we define the equivalent exchangeable fraction and molar fraction.

TABLE 3.4 Cation Exchange Capacities (meq/kg) of Common Rocks and Clays

	From Appelo and Postma (2007)	From Holm and Robertson (1981)
Kaolinite	30–150	122
Halloysite	50–100	
Montmorillonite	800–1200	1170
Vermiculite	1000–2000	
Glauconite	50–400	
Illite	200–500	250
Halloysite		150
Aquagel (bentonite)		800–1500
Chlorite	100–400	
Allophane	up to 1000	
Goethite and hematite	up to 1000 (pH > 8.3)	
Organic matter (C)	1500–4000 (at pH = 8)	
or, accounting for pH-dependence:	510 × pH − 590 = CEC per kg organic carbon	

For ion I^{i+} the equivalent exchangeable fraction β_I is calculated as

$$\beta_I = \frac{\text{meq I-X}_i \text{ per kg sediment}}{\text{CEC}} = \frac{\text{meq I-X}_i}{\sum_{I,J,K,\cdots} \text{meq I-X}_i}, \tag{3.16}$$

where I, J, K, … are the exchangeable cations, with charges i, j, and k. A molar fraction β_I^M is likewise defined as

$$\beta_I^M = \frac{\text{mmol I-X}_i \text{ per kg sediment}}{\text{TEC}} = \frac{(\text{meq I-X}_i)/i}{\sum_{I,J,K,\cdots} (\text{meq I-X}_i)/i}, \tag{3.17}$$

where TEC denotes total exchangeable cations in mmol/kg rock.

As an example, for the exchange of Na^+ and Ca^{2+}, the number of exchangeable cations is used in the Gaines–Thomas (1953) convention. According to the Gaines–Thomas convention, Eqs. 3.10 and 3.11 are rewritten as

$$\tfrac{1}{2}Ca^{2+} + Na{-}X \leftrightarrow \tfrac{1}{2}(Ca{-}X_2) + Na^+, \tag{3.18}$$

with

$$K^{GT}_{Ca-Na} = \frac{[Ca\text{-}X_2]^{0.5}[Na^+]}{[Na\text{-}X][Ca^{2+}]^{0.5}} = \frac{\beta^{0.5}_{Ca}[Na^+]}{\beta_{Na}[Ca^{2+}]^{0.5}}. \quad (3.19)$$

The use of the molar fractions in Eq. 3.19 leads to the Vanselow (1932) convention:

$$K^{V}_{Ca-Na} = \frac{[Ca\text{-}X_2]^{0.5}[Na^+]}{[Na\text{-}X][Ca^{2+}]^{0.5}} = \frac{(\beta^{0.5}_{Ca})^{M}[Na^+]}{(\beta_{Na})^{M}[Ca^{2+}]^{0.5}}. \quad (3.20)$$

If the activities of the adsorbed ions are expressed as a fraction of the number of exchange sites (-X), then Eq. 3.18 becomes

$$\tfrac{1}{2}Ca^{2+} + Na\text{-}X \leftrightarrow Ca_{0.5}\text{-}X + Na^+, \quad (3.21)$$

with

$$K^{G}_{Ca-Na} = \frac{[Ca_{0.5}\text{-}X][Na^+]}{[Na\text{-}X][Ca^{2+}]^{0.5}} = \frac{\beta_{Ca}[Na^+]}{\beta_{Na}[Ca^{2+}]^{0.5}}. \quad (3.22)$$

Equations 3.21 and 3.22 are the Gapon (1933) convention. In this case, the molar and equivalent exchangeable fractions are the same because both are based on a single exchange site with the subscript of X, i, in Eq. 3.17 equal to 1.

The CEC (in meq/kg) of a rock is most likely constant, whereas TEC (in mmol/kg) of a heterovalent system varies with the relative amount of cations with different charges that neutralize the constant CEC. In most situations the activities of exchangeable cations are therefore calculated more conveniently as exchangeable fractions with respect to a fixed CEC.

Note that activities or equivalent molar concentrations are used in Eqs. 3.10 and 3.11. UTCHEM follows this convention. When the activity is calculated with respect to the number of exchangeable cations, which is indicated as $[I\text{-}X_i]$ and used in the Gaines–Thomas convention, or with respect to the number of exchangeable sites as $[I_{1/i}\text{-}X]$ and used in the Gapon convention, the use of fractions for the activities of exchangeable ions always satisfies $\sum\beta = 1$. Most important, you should be aware that the values of exchange coefficients could be different for the different conventions and units used. In the case of homovalent exchange, the coefficients for the Gapon and Vanselow conventions are identical to the Gaines–Thomas values. For heterovalent exchange, it is possible to derive the coefficients for the binary case, as shown next.

Starting from the exchange equation

$$Na^+ + \tfrac{1}{i}(I\text{-}X_i) \leftrightarrow Na\text{-}X + \tfrac{1}{i}(I^{i+}), \quad (3.23)$$

we have the coefficient for the Gaines–Thomas convention

$$K^{GT}_{Na-I} = \frac{[Na\text{-}X][I^{i+}]^{1/i}}{[I\text{-}X_i]^{1/i}[Na^+]} = \frac{\beta_{Na}[I^{i+}]^{1/i}}{\beta_I^{1/i}[Na^+]}, \quad (3.24)$$

where $[I^{i+}]$ and $[Na^+]$ are in molar units. Including the relation between β_I and β_I^M into the previous K^{GT}_{Na-I} leads to

$$\begin{aligned} K^{GT}_{Na-I} &= \frac{\beta_{Na}[I^{i+}]^{1/i}}{\beta_I^{1/i}[Na^+]} = \frac{\beta^M_{Na}}{(\beta^M_I)^{1/i} i^{1/i}} \left\{ \frac{[Na\text{-}X]+[I\text{-}X_i]/i}{[Na\text{-}X]+[I\text{-}X_i]} \right\}^{1-1/i} \frac{[I^{i+}]^{1/i}}{[Na^+]} \\ &= K^V_{Na-I}[1+(i-1)\beta_{Na}]^{1-1/i}(i^{-1}). \end{aligned} \quad (3.25)$$

Equation 3.25 is derived using Eqs. 3.16 and 3.17, and adding the term $[Na\text{-}X] + [I\text{-}X_i]$ which is $\beta_{Na} + \beta_I = 1$. For a homovalent exchange, $i = 1$, the preceding equation results in $K^{GT}_{Na-I} = K^V_{Na-I}$.

For the coefficient based on the Gapon convention, Eq. 3.23 becomes

$$Na^+ + (I_{1/i}\text{-}X) \leftrightarrow Na\text{-}X + \tfrac{1}{i}(I^{i+}), \quad (3.26)$$

with

$$\begin{aligned} K^G_{Na-I} &= \frac{[Na\text{-}X][I^{i+}]^{1/i}}{[I_{1/i}\text{-}X][Na^+]} = \frac{\beta_{Na}[I^{i+}]^{1/i}}{\beta_I[Na^+]} = \frac{\beta_{Na}[I^{i+}]^{1/i}}{(\beta_I)^{1/i}[Na^+]} \frac{1}{(\beta_I)^{1-1/i}} \\ &= \frac{K^{GT}_{Na-I}}{(\beta_I)^{1-1/i}} = \frac{K^{GT}_{Na-I}}{(1-\beta_{Na})^{1-1/i}}. \end{aligned} \quad (3.27)$$

In Eq. 3.27, the fact that the molar and equivalent exchangeable fractions are the same if the Gapon convention is used. In other words, $[I_{1/i}\text{-}X] = \beta_I$, and $[Na\text{-}X] = \beta_{Na}$. For a homovalent exchange, $i = 1$, the preceding equation results in $K^{GT}_{Na-I} = K^G_{Na-I}$.

The cation exchange coefficients relative to Na^+ for various ions following the Gaines–Thomas convention (Eq. 3.24) are reported in Appelo and Postma (2007), based partly on a compilation by Bruggenwert and Kamphorst (1982). The given ranges represent many measurements from different soils and for different clay minerals.

3.3.2 Values of Other Exchange Coefficients

When the exchange coefficients for some reactions are known, exchange coefficients among other cation pairs can be obtained by combining the known reactions. For example, the exchange relation for Al^{3+} and Ca^{2+} can be as follows:

$$Na^+ + \tfrac{1}{2}(Ca\text{-}X_2) \leftrightarrow Na\text{-}X + \tfrac{1}{2}(Ca^{2+}), \quad K_{Na-Ca} = 0.4;$$

and

$$Na^+ + \tfrac{1}{3}(Al\text{-}X_3) \leftrightarrow Na\text{-}X + \tfrac{1}{3}Al^{3+}, \quad K_{Na-Al} = 0.7.$$

When we subtract the two reactions and divide the two exchange coefficients, we get

$$\tfrac{1}{3}Al^{3+} + \tfrac{1}{2}(Ca\text{-}X_2) \leftrightarrow \tfrac{1}{3}(Al\text{-}X_3) + \tfrac{1}{2}(Ca^{2+}), \quad K_{Al-Ca} = 0.4/0.7 = 0.6.$$

3.3.3 Calculation of Exchange Composition

For any pair of cations, we have two equivalent exchangeable fractions—β_1 and β_2. These two parameters are related through their exchange coefficient, $K_{1\text{-}2}$. They must also satisfy the condition $\beta_1 + \beta_2 = 1$. Then, from these two conditions, β_1 and β_2 are determined. If the CEC is known, their exchange compositions can be calculated. This principle can be extended to more than two cations.

Example 3.2 Calculate Cation Exchange Compositions

Calculate the exchangeable cations in a sandstone core with a cation exchange capacity of 0.13822 meq/mL, in equilibrium with the formation water with Na^+ = 0.05 meq/mL, Ca^{2+} = 0.01 meq/mL. The value of $K_{Ca\text{-}Na}$ is 3.0.

Solution

We use Eq. 3.19, which follows the Gaines–Thomas convention:

$$3.0 = \frac{\beta_{Ca}^{0.5}[Na^+]}{\beta_{Na}[Ca^{2+}]^{0.5}} = \frac{\beta_{Ca}^{0.5}0.05}{\beta_{Na}0.005}.$$

We also have

$$\beta_{Na} + \beta_{Ca} = 1.$$

Solving the two equations simultaneously yields both β_{Na} = 0.923 and β_{Ca} = 0.073. The exchange compositions are [Na-X] = 0.1276 meq/mL and [Ca-X_2] = 0.01062 meq/mL.

Note that in the preceding calculation, the cation concentrations in the water are in mole/mL. If they are in meq/mL, the exchangeable fractions become β_{Na} = 0.78 and β_{Ca} = 0.22. This example shows that even if the same convention is used, the calculated fractions are quite different by simply using different units of cation concentrations in the water.

The previous calculations are based on the Gaines–Thomas convention. If they are calculated based on the law of mass action (Eq. 3.11), and the unit of meq/mL is used, the exchangeable fractions become β_{Na} = 0.25 and β_{Ca} = 0.75! Now these calculated exchangeable fractions from the law of mass action are significantly different from those from the Gaines–Thomas convention.

3.3.4 Calculation of Mass Action Constant at Different Temperatures

Changes of equilibrium constants with temperature are usually described with the van't Hoff equation (Atkins and de Paula, 2006).

$$\frac{d\ln K}{dT}=\frac{\Delta H_r}{RT^2}, \tag{3.28}$$

where ΔH_r is the reaction enthalpy, or the heat lost or gained by the chemical system. For exothermal reactions, ΔH_r is negative (the system loses energy and heats up), and for endothermal reactions, ΔH_r is positive (the system cools). R is the gas constant (8.314 J/K/mol). At 25°C, the value of the reaction enthalpy, ΔH_r^0, is calculated from the formation enthalpies.

This equation shows that K increases with temperature for positive ΔH_r and decreases with temperature for negative ΔH_r. Usually, ΔH_r^0 is constant within the range of a few tenths of degrees. Therefore, the preceding equation can be integrated to give two temperatures:

$$\log K_{T_1}-\log K_{T_2}=\frac{-\Delta H_r^0}{2.303R}\left(\frac{1}{T_1}-\frac{1}{T_2}\right). \tag{3.29}$$

Some values of ΔH_r^0 are provided by Dria et al. (1988).

3.3.5 Effect of Diluting an Equilibrium Solution

When fresh water displaces saline water, dilution occurs. During dilution, divalent ions are preferentially adsorbed in comparison to a monovalent ion. For example, dilution of the solution with distilled water is accompanied by an increase of the monovalent ions (Na^+) relative to the divalent cations (Ca^{2+}) when the equilibrium with the exchanger is maintained. This effect follows from Eq. 3.19:

$$\frac{[Na^+]}{\sqrt{[Ca^{2+}]}}=K_{Ca-Na}^{GT}\frac{[Na\text{-}X]}{\sqrt{[Ca\text{-}X_2]}}. \tag{3.30}$$

When the right side of this equation is constant, a 10-fold dilution of the Na^+ concentration is accompanied by a 100-fold dilution of the Ca^{2+} concentration. Similarly, for Al^{3+}-Na^+ exchange, if Na^+ is diluted 10 times, Al^{3+} must be diluted 1000 times. For ions with the same charge, the dilution is the same. The dilution factor, f, can be used to calculate the composition behind the displacing front from that ahead of the front (Appelo and Postma, 2007),

$$\frac{[Na^+]_a}{f}+\frac{2\{[Mg^{2+}]_a+[Ca^{2+}]_a\}}{f^2}=\sum\{i[I^{i+}]_b\}=\sum\{i[I^{i-}]_b\}, \tag{3.31}$$

or in general,

$$\sum\left\{\frac{i[I^{i+}]_a}{f^i}\right\}=\sum\{i[I^{i+}]_b\}=\sum\{i[I^{i-}]_b\}, \tag{3.32}$$

where the subscripts a and b indicate ahead of and behind the displacing front, $[I^{i+}]$ represents the ion *molar* concentration with charge i, and $\sum\{i[I^{i+}]_b\}$ means the sum (total) of *equivalent* molar concentrations of the cations. Because anions are nonsorbed species, the total of equivalent molar concentrations of the cations equals the total equivalent molar concentrations of the anions, $\sum\{i[I^{i-}]_b\}$; see Example 3.3.

From Eq. 3.30, we see that if we can keep the ratio of $\frac{[Na^+]}{\sqrt{[Ca^{2+}]}}$ unchanged in neighboring slugs, the change in the cation compositions due to cation exchange will be minimized. This has some merits in practical application. For example, if we design the cation compositions in the chemical slug and chase water in such a way that the concentration ratio of the predominant monovalent to the predominant divalent is substantially the same as that of the existing formation water, the cation compositions in the chemical slug will not be changed by cation exchange so that the optimum salinity and hardness can be maintained. Hill et al. (1978) filed a patent about this idea. However, it is difficult to achieve that effect because of dispersion and diffusion, multicomponent ion exchange, surfactant complexation, and so on.

Example 3.3 Dilution Calculation

The original formation water concentrations are $[Na^+]$ = 86.5 mmol/L, $[Mg^{2+}]$ = 18.2 mmol/L, and $[Ca^{2+}]$ = 11.1 mmol/L. The anions in the injected water total 14.7 meq/L. Calculate the concentrations of Na^+, Ca^{2+}, and Mg^{2+} in mg/L.

Solutions

The total equivalent molar concentration of the cations is the same as that of the anions, which is 14.7 meq/L. According to Eq. 3.31,

$$\frac{86.5}{f}+\frac{2(18.2+11.1)}{f^2}=14.7,$$

we obtain the dilution factor f = 6.5. Then in the injected water,

$$[Na^+]_b=[Na^+]_a/f=86.5/6.5=13.3\text{ meq/L}=306.1\text{ mg/L},$$

$$[Mg^{2+}]_b=[Mg^{2+}]_a/f^2=2(18.2)/(6.5^2)=0.86\text{ meq/L}=10.5\text{ mg/L},$$

$$[Ca^{2+}]_b=[Ca^{2+}]_a/f^2=2(11.1)/(6.5^2)=0.535\text{ meq/L}=10.5\text{ mg/L}.$$

3.3.6 Chromatography

Initially, formation water and oil are at equilibrium with reservoir rocks. When a new fluid whose ion compositions are different from the formation is injected, a new equilibrium will be established after ion exchange. Strongly selected cations will displace other ions from the exchanger and be transported at a relatively low velocity. The reason is that the cations in the injected solution are sorbed more strongly than the existing cations. The cation with the lowest selectivity comes to the producer first, then the next favored, and so on. We use Example 3.4 to illustrate this chromatography of cation exchange. To do that, we need to review the retardation equation introduced in Chapter 2. The retardation factor is

$$R_C = 1 + \frac{d\hat{C}}{dC} = 1 + \left|\frac{\Delta\hat{C}}{\Delta C}\right|, \tag{2.65}$$

and the retardation equation is defined as

$$u_C = \frac{u}{R_C}. \tag{2.66}$$

Example 3.4 Multicomponent Chromatography of Cation Exchange

A reservoir initially is filled with water (for the simplicity of illustration, no oil is assumed in the reservoir). The CEC is 1.1 meq/L PV. The initial water is 1 mM $NaNO_3$ and 0.2 mM KNO_3. The reservoir is flooded with 0.6 mM $CaCl_2$ water. The exchange coefficient $K_{K\text{-}Na}$ is 5. Estimate the concentration histories of these ions at the production well.

Solution

We know the order of affinity of these ions is $Na^+ < K^+ < Ca^{2+}$. Without calculation, we expect that Na^+ will be flushed out earlier than K^+, followed by Ca^{2+}. We assume that the displacement is piston like, and the diffusion and dispersion are not included. Initially, Na^+ and K^+ "stick" to the rock based on their exchangeable fractions. The injected Ca^{2+} will replace Na^+ and K^+. We want to find out how long it takes to flush out Na^+ and K^+. To do that, we need to know how much Na^+ and K^+ stick to the rock initially.

The cation exchange between K^+ and Na^+ is

$$K^+ + \text{Na-X} \leftrightarrow \text{K-X} + Na^+, \tag{3.33}$$

with

$$K_{K-Na} = \frac{[\text{K-X}][Na^+]}{[\text{Na-X}][K^+]}. \tag{3.34}$$

From Eq. 3.34, we have

$$K_{K-Na} = \frac{[\text{K-X}][Na^+]}{[\text{Na-X}][K^+]} = \frac{[\text{K-X}](1)}{[\text{Na-X}](0.2)} = 5.$$

We also have [K-X] + [Na-X] = CEC = 1.1 meq/(L PV). Thus, we have [K-X] = [Na-X] = 0.55 meq/(L PV).

Continued

Example 3.4 Multicomponent Chromatography of Cation Exchange—*Continued*

Because we are not using any specific software to do the detailed calculation, we just analyze several points. From the start to one pore volume of injection, the solution in its initial composition, Na^+, K^+, and NO_3^-, is produced. From one pore volume onward, NO_3^- has been completely produced, and Cl^- of 1.2 mM is produced. For the whole process, the toal cations must equal the total anion, which is 1.2 mM. After one pore volume, the produced cations must be the cations initially at the exchanger (rock). Because Na^+ is the least selected cation, it must be preferably displaced by Ca^{2+}. In other words, the exchange is emptied of Na^+ before K^+. The elution of Na^+ ends with the retardation, $R_C = 1+ (0.55 - 0)/(1.0 - 0) = 1.55$ PV, based on the retardation equation.

Next, the K^+ concentration in the solution increases to 1.2 mM to compensate for the anion charge. At this time, Ca^{2+} is retarded and has not arrived at the production end yet. Therefore, K^+ is the only cation in the solution. Meanwhile, K^+ is the only cation on the exchanger, and [K-X] increases to the CEC of 1.1 mM. K^+ is depleted with the retardation $R_C = 1+ (1.1 - 0)/(1.2 - 0) = 1.917$ PV. Afterward, Ca^{2+} arrives. Figure 3.2 shows the effluent concentration histories.

Note that in this example many assumptions have been made. The objective is to explain the chromatographic separation of cations due to their selectivity on the exchanger. In reality, because of the dispersion and the diffusion, the initial concentration profiles will be changed. Therefore, selectivity should be based on changed concentrations. Such calculations must be performed using software such as PHREEQC, which can be downloaded free from the web at *www.brr.cr.usgs.gov/projects/GWC_coupled/phreeqc/* or *www.xs4all.nl/~appt/downl.html*.

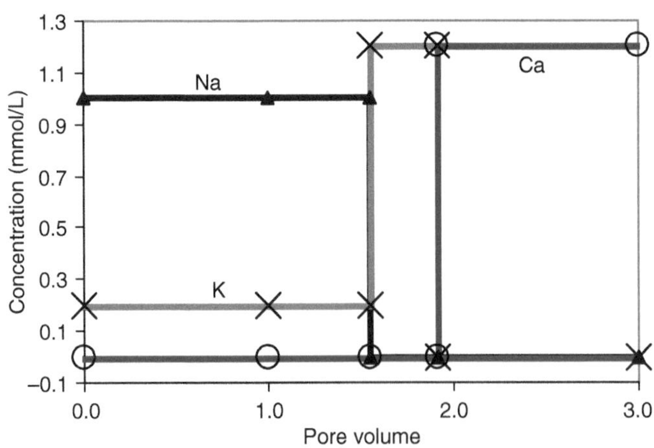

FIGURE 3.2 Approximate histories of Na^+, K^+, and Ca^{2+} at the effluent end.

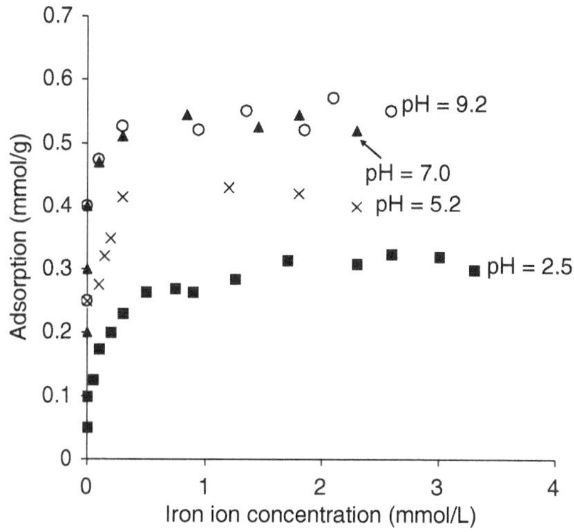

FIGURE 3.3 Cation exchange between iron ion and sodium-montmorillonite at 25°C. *Source: Yang et al. (2002a).*

3.3.7 Effect of pH

Cation exchange is affected by pH. Figure 3.3 shows an example of pH effect. This figure shows the cation exchange between iron ion and sodium-montmorillonite at different values of pH and iron concentrations. The exchange follows the Langmuir-type adsorption (20–65°C). At a given temperature, the cation exchange capacity increases with pH.

3.4 LOW-SALINITY WATERFLOODING IN SANDSTONE RESERVOIRS

The EOR potential of low-salinity waterflooding was not recognized until Morrow and his coworkers started to work on the effect of brine composition on oil recovery; this effort showed that changes in injection brine composition improved recovery (Jadhunandan and Morrow, 1991, 1995; Yildiz and Morrow, 1996). Since then, Tang and Morrow (1997, 1999) advanced the research on the impact of brine salinity on oil recovery, followed by active research by the oil company British Petroleum, or BP (Webb et al., 2004, 2005; McGuire et al., 2005). BP's work includes numerous core floods at ambient and reservoir conditions with live oils in both secondary and tertiary modes, single-well tracer tests, and log-inject-log tests. The company's work led to the registration of the LoSal™—EOR process trademark. Meanwhile, the researchers from several oil companies and universities worked on this topic as well. Several mechanisms of low-salinity waterflooding have been proposed in the literature;

however, there is no consensus about the primary mechanisms. This section briefly summarizes the status of the subject.

3.4.1 Observations of Low-Salinity Waterflooding Effect

BP laboratory results (Lager et al., 2006) showed an average benefit of 14% with low-salinity brine, and a large scatter of results from +4 to +40% was observed. Such a wide spread of results was also observed by Morrow and coworkers. In some core floods, no incremental oil recovery was observed.

Single-well chemical tracer tests performed by BP Alaska resulted in reduction in remaining oil saturation of 6 to 12% OOIP (McGuire et al., 2005). In a log-inject-log test, typically 0.1 to 0.15 pore volumes of high-salinity brine were injected first into the volume of interest to obtain the baseline residual oil saturation. This was followed by sequences of more dilute brine followed by high-salinity brine. Multiple log passes were conducted during each brine injection. At least three further passes were run to ensure that a stable saturation value had been established after injection of each type of brine. The results showed 0.25 to 0.5 reduction in residual oil saturation when waterflooding with low-salinity brine (Webb et al., 2004).

Preflush using low-salinity water before surfactant-polymer flooding was carried to condition the reservoir in the early days. Such preflush water should bring incremental oil if low-salinity waterflooding worked. However, apparently, no oil rate increase was observed during the fresh water preflush in the North Burbank Unit surfactant-polymer pilot in Osage County, Oklahoma (Trantham et al., 1978) and Loudon surfactant pilot (Pursley et al., 1973). Other observations include the following:

- If no connate water saturation was present, no benefit was seen (Tang and Morrow, 1999; Sharma and Filoco, 2000; Zhang and Morrow, 2006).
- Refined (unpolarized) oils had no benefit (Tang and Morrow, 1999).
- Generally, the salinity of injected water must be significantly low to see the benefit—for example, 1500 ppm (Zhang et al., 2007b), <4000 ppm TDS (Web et al., 2005).
- The recovery benefit appeared to increase with clay content, especially kaolinite content (Seccombe et al., 2008).

3.4.2 Proposed Mechanisms

This section summarizes and discusses several mechanisms proposed in the literature regarding low-salinity waterflooding.

Fine Migration or Permeability Reduction

In principle, clay tends to hydrate and swell when contacting with fresh water—that is, water containing salts in amounts insufficient to prevent swelling and hydration of the clay. A less-saline solution affects the dispersion of clay and silt in the formation. The clay and silt, upon dispersion, become mobile and follow the paths taken by the greatest proportion of the flowing water. These paths are the domains of high permeability, and the mobile clay and silt become lodged in the smaller pore spaces of these domains and reduce the flow of water through these pore spaces. The permeability of the domains where clay and silt lodge is accordingly reduced, and the water is forced to take other flow paths.

As a result, the permeabilities in these domains within the formation become more uniform. Reduction in permeability in the more permeable domains improves the mobility ratio of waterflood. Premature breakthrough is thus reduced, and the efficiency of the waterflood is improved (Boston et al., 1969). Poorly cemented clay particles, such as kaolinite and illite, can become detached during aqueous flow, especially when flowing brines become fresher.

Martin (1959) and Bernard (1967) observed that clay swelling and/or dispersion accompanied by increased pressure drop resulted in incremental oil recovery. Tang and Morrow (1999) concluded that fine mobilization (mainly kaolinite) increased recovery based on their observations: (1) fired/acidized Berea core showed insensitivity of salinity on oil recovery, whereas unfired Berea core did show sensitivity; and (2) for clean sandstones, the increase in oil recovery with the decrease in salinity was less than that for the clay sands. Figure 3.4 shows some of their results. In the tests, the reservoir CS core was used. The reservoir brine, CS RB, was used as connate brine for the entire CS core tests.

Cyclic waterfloods were run in sequence and are indicated by a test number (T#) and flood number (F#). Stages of injection that involved changes in brine composition are indicated as a, b, and c. The cyclic waterflood was to make cyclic use of the same core to run several repeated floods, and before each flood, S_{wi} was reestablished and aging with oil was repeated. From Figure 3.4, we can see the following:

1. As the injection brine salinity was reduced to 0.1 CS RB, more oil was produced. When Ca^{2+}/Na^{+} ratio was increased, however, no more oil was produced.
2. During the stage the injection brine salinity was reduced to 0.1 CS RB, a higher-pressure drop was observed, indicating permeability reduction caused by fine migration. This observation was also made by Zhang et al. (2007b).
3. The maximum pH value was about 9.
4. The observed increase in waterflood recovery indicated that rock surface wettability might be changed as a result of cyclic waterflooding—from Flood 1 (T1F1) to Flood 3 (T1F3).

Lager et al. (2006) and Zhang et al. (2007b) reported that no clays were produced at the effluent stream. However, that does not mean there was no fine

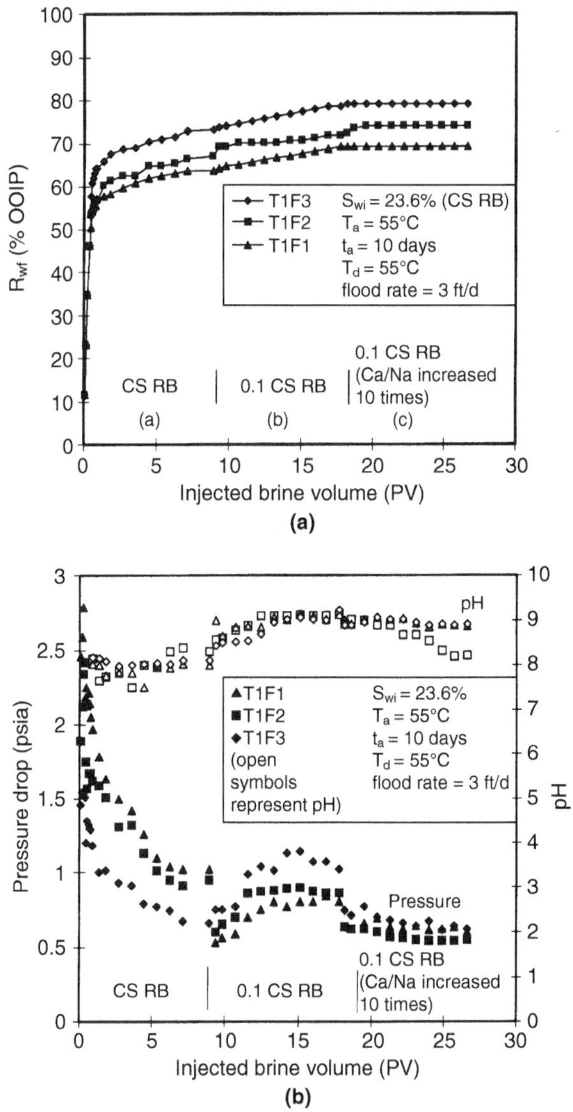

FIGURE 3.4 Effects of changes in injection brine composition on the recovery of CS crude oil during cyclic waterflooding of a CS core: (a) cyclic waterflooding of CS sandstone, and (b) pH of effluent brine and pressure drop. *Source: Tang and Morrow (1999).*

migration. Kia et al. (1987) reported that freshwater flooding of sandstones previously exposed to sodium salt solutions resulted in the release of clay particles and a drastic reduction in permeability. The permeability reduction was lessened, however, when calcium ions were also present in the salt solution. Formation damage was virtually eliminated when the solution composition was adjusted to give calcium surface coverage greater than a critical value of 75%, or when a solution Ca^{2+} fraction is greater than 20 to 30%.

Khilar and Fogler (1984) results showed a 30% reduction in permeability when the pretreatment was carried out with cesium-salt solutions, a reduction of more than 95% with a sodium-salt pretreatment, and virtually no reduction when the divalent cation existed in the solution. In terms of oil recovery, however, Yildiz and Morrow (1996) observed that for Moutray crude oil and Berea cores, the oil recovery factor using 2% $CaCl_2$ was 5.5% higher than that using 4% NaCl plus 0.5% $CaCl_2$. The fine migration in the former case should be lower than in the latter case. Yildiz and Morrow also found the effect of brine composition was highly specific to the crude oil and aging conditions. With Alaskan crude oil, the results are opposite that of Moutray oil.

pH Effect

Valdya and Fogler (1992) studies showed that dispersion of clays is minimized at low pH. Salinity reduction induces a pH increase, which amplifies the release of fines and leads to a drastic reduction in permeability. They reported little change in permeability when fluids with increasing pH were injected until an injection pH of 9 was reached. At a pH > 11, a rapid and drastic decrease in the permeability was observed; however, at typical low-salinity flooding, pH is lower than 9, as shown in Figure 3.4. In alkaline flooding, pH is usually 11 to 13. Zhang et al. (2007b) reported that after low-salinity brine injection, a slight rise and drop in pH were observed. There is no clear relationship between effluent pH and recovery. High pH may induce IFT reduction or emulsification in alkaline flooding and fine migration. Low pH in low-salinity waterflooding raises a question about the pH effect proposed by McGuire et al. (2005).

Multicomponent Ion Exchange

Owing to the different affinities of ions on rock surfaces, the result of multicomponent ion exchange (MIE) is to have multivalents or divalents such as Ca^{2+} and Mg^{2+} strongly adsorbed on rock surfaces until the rock is fully saturated. Multivalent cations at clay surfaces are bonded to polar compounds present in the oil phase (resin and asphaltene) forming organo-metallic complexes and promoting oil-wetness on rock surfaces. Meanwhile, some organic polar compounds are adsorbed directly to the mineral surface, displacing the most labile cations present at the clay surface and enhancing the oil-wetness of the clay surface. During the injection of low-salinity brine, MIE will take place, removing organic polar compounds and organo-metallic complexes from the surface and replacing them with uncomplexed cations (Lager et al., 2006). In theory, desorption of polar compounds from the clay surface should lead to a more water-wet surface, resulting in an increase in oil recovery.

Lager et al. (2006) reported that their experimental results matched the prediction from their hypothesis. First, the North Slope core sample was prepared to the representative initial water saturation and aged in the dead crude oil. The initial screening experiments were conducted at 25°C. A conventional high-salinity waterflood gave a recovery of 42% OOIP, and a tertiary low-salinity flood resulted in a total recovery of 48% OOIP (i.e., an additional 6% OOIP). A second suite of experiments was conducted at the reservoir temperature (102°C). A conventional high-salinity waterflood resulted in a recovery of 35% OOIP. The core was flushed with the brine containing only high-salinity NaCl until Ca^{2+} and Mg^{2+} was effectively eluted from the pore surface. The initial water saturation was reestablished, and the sample was aged in the crude oil. A high-salinity waterflood consisting of NaCl (no Ca^{2+} and Mg^{2+}) resulted in a recovery of 48% OOIP. A tertiary low-salinity flood was then conducted (again no Ca^{2+} and Mg^{2+}), and no additional recovery observed.

This sequence indicated that high-salinity connate brine containing Ca^{2+} and Mg^{2+} resulted in the low recovery factors (42% and 35%). Removing Ca^{2+} and Mg^{2+} from the rock surface before waterflooding led to a higher recovery factor (48%) irrespective of salinity. They noted that no improvement in oil recovery was observed when low salinity is injected into a clastic reservoir where the mineral structure has been preserved. Apparently, their proposed MIE explains why low-salinity waterflooding did not work when a core was acidized and fired. The reason is the cation exchange capacity of the clay minerals was destroyed. This explains why low-salinity water injection has little effect on mineral oil, as reported by Zhang et al. (2007b), because no polar compounds are present to strongly interact with the clay minerals. Their proposed mechanism of multicomponention exchange (MIE) is supported by the pore-scale model proposed by Sorbie and Collins (2010). However, Zhang et al. (2007b) reported that additional recovery was obtained when adding divalent ions to the injection brine.

Further Discussion

Apparently, mechanisms of low-salinity waterflooding are related to the DLVO theory, which is named after Derjaguin, Landau, Verwey, and Overbeek. The theory describes the force between charged surfaces interacting through a liquid medium. It combines the effects of the van der Waals attraction and the electrostatic repulsion due to the so-called double layer of counter ions.

Fine migration occurs if the ionic strength of injected brine is less than the critical flocculation concentration, which is strongly dependent on the relative concentration of divalent cations. Divalent cations have been known to stabilize clay by lowering the Zeta potential, resulting in lowering the repulsive force. In addition, the adsorption of divalents at the oil/water and water/sand interfaces changes the water-wet to oil-wet (Liu et al., 2007). Kia et al. (1987) have reported that in the presence of Na^+, the surface of kaolinite carries a negative charge, and the electrical charge present on the edge surface is a strong function of the solution pH. Most of the reported values show that edges of kaolinite particles are negatively charged when pH is higher than 6 to 8. For some brine compositions, both oil/brine and brine/solid interfaces have the same charge (Buckley et al., 1998). Thus, there is electrostatic repulsion between these interfaces. When low-salinity brine is injected, the repulsion is increased. When high-salinity brine is injected, due to screening of the surface charges (Adamson, 1997), the repulsion is decreased.

For the exact same mechanism that the electrostatic repulsion is increased when low-salinity brine is injected, the water film between the oil/brine and brine/particle will be more stable. The rock surfaces will be more water-wet. Thus, oil recovery will be higher.

Sharma and Filoco (2000) reported that the salinity of connate water was found to be the primary factor controlling oil recovery. They attributed this dependence to alteration of the wettability to mixed-wet conditions from water-wet conditions. They did not observe that salinity of injection brine affected the oil recovery factor. Their results clearly showed that oil recovery was higher for lower connate brine salinities. This suggests that changes in connate water salinity may cause changes in wettability (from water-wet toward mixed-wet) of the pore space. Their experiments showed that brine film was more stable at high salinities (*inconsistent* with the DLVO theory—more stable film probably by invoking hydrophobic interaction). The higher stability of brine films at higher salinity suggests that low connate water salinity causes cores to become mixed-wet (more oil-wet). Mixed-wet cores show lower residual oil saturations than strongly water-wet or oil-wet cores; that is, the oil recovery is higher.

The challenge to define low-salinity waterflooding mechanisms is that counterexamples can always be found to negate the mechanism proposed. Another puzzle about low-salinity waterflooding is that the incremental oil recovery is relatively high compared to other chemical flooding processes, such as surfactant flooding. Is the low-salinity waterflooding so powerful? In Daqing chemical EOR floods, fresh water (< 1000 ppm) was injected into reservoirs of about 7000 ppm. Then how much incremental recovery is due to the freshwater flooding? After we identify the real mechanisms of low-salinity waterflooding, the result should definitely help us design chemical flooding.

3.5 SALINITY EFFECT ON WATERFLOODING IN CARBONATE RESERVOIRS

Chalk is the dominant oil-containing carbonate formation in the North Sea. Because of the soft nature of the biogenic sediment, the reservoirs are usually naturally fractured. The permeability of the matrix blocks is low, approximately 2 md, and the porosity can be very high, nearly 0.5. The reservoir temperatures are high, in the range of 90 to 130°C. During the primary production phase, purely by pressure depletion of the Ekofisk field, compaction and subsidence occurred, which contributed to 40% of the drive mechanism. Water injection in the Ekofisk field started in 1987 in order to give pressure support and prevent compaction. Injection of seawater was a great success, and the oil recovery is estimated to be approximately 50%. Apparently, seawater improved the water wetness of chalk, which increases the oil recovery by spontaneous imbibition and viscous displacement.

It was also observed that the compaction did not stop in the waterflooded areas, even though the reservoir was repressurized to the initial condition. Thus, seawater appeared to have a special interaction with chalk at high temperatures, which has an impact on oil recovery and rock mechanics (Austad et al., 2008). Austad and his coworkers started to work on the issues related to seawater flooding in carbonate reservoirs in 1990s. In the next section, the salinity effect on oil recovery is briefly summarized.

3.5.1 Wettability Alteration by Seawater Injection into Chalk Formation

Figure 3.5 shows a series of imbibition tests (Zhang et al., 2007a). In the first series, the experiment was designed to study the interplay between the different potential determining ions (Mg^{2+}, Ca^{2+}, and SO_4^{2-}) present in seawater. The cores were prepared using the oil with a high acid number of 2.07 mg KOH/g oil and the brine with no potential determining ions present. The initial water saturation for the four cores was quite similar, 22 to 23%. The cores were treated and aged, and the imbibition tests were run with different fluids at successively higher temperatures: 70, 100, and 130°C. The imbibition at 70°C was performed using modified seawater without Ca^{2+} and Mg^{2+}, but with different amounts of SO_4^{2-} present.

The imbibing fluids were named SW0×0S, SW0 (×1S), SW0×2S, and SW0×4S (SW0×iS denotes the seawater with i times the SO_4^{2-} concentration of the seawater), and the ionic strength was kept constant and was similar to that of the seawater by adjusting the amount of NaCl. In all cases, the oil

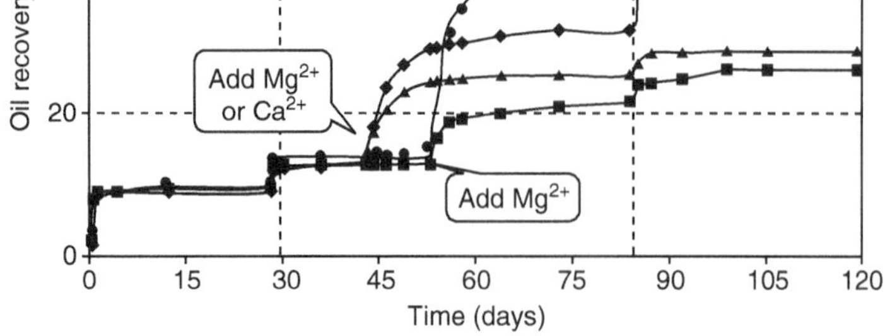

FIGURE 3.5 Spontaneous tests at different Mg^{2+}, Ca^{2+}, and SO_4^{2-} concentrations and at different temperatures. *Source: Zhang et al. (2007a).*

recovery was low, about 10%, which was interpreted to be caused by fluid expansion and some heterogeneities in the wetting conditions.

Then the temperature was increased to 100°C. A small increase in oil recovery was noticed, which could be related to fluid expansion. Thus, SO_4^{2-} alone as a potential determining ion was not able to increase spontaneous imbibition of water in chalk by wettability alteration. Then Mg^{2+} and Ca^{2+} were added to the respective imbibing fluids (CM-1 and CM-3 at the 43rd day, and CM-2 and CM-4 at the 53rd day). The Mg^{2+} and Ca^{2+} concentrations were similar to those in the seawater—that is, $[Mg^{2+}] = 0.045$ mol/L and $[Ca^{2+}] = 0.013$ mol/L. In all cases, a sudden increase in oil recovery was noticed. For the tests in which Mg^{2+} was added (CM-2, CM-1, and CM-4), the oil recovery increased to about 20, 32, and 42% as the concentration of SO_4^{2-} in the imbibing fluid was increased by 0×, 1×, and 4× the concentration present in the seawater, respectively.

This result showed that the oil recovery was strongly related to the concentration of SO_4^{2-} present. Ca^{2+} was added to the solution (CM-3) containing two times the concentration of SO_4^{2-} in the seawater. Even though the concentration of Ca^{2+} was about four times lower than the concentration of Mg^{2+}, the oil recovery increased to about 25% (CM-3). However, the recovery in the case of adding Mg^{2+} (CM-1) is higher than that in the case of adding Ca^{2+} (CM-3).

Finally, the temperature was increased to 130°C. The oil recoveries from the tests with 1× and 4× SO_4^{2-} (CM-1 and CM-4, respectively) increased to 50 and 60%, respectively, compared with the recovery of about 25% for the test with 0× SO_4^{2-} (CM-2). Thus, the efficiency of Mg^{2+} with SO_4^{2-} as wettability modifiers increased drastically as the temperature was increased. The core exposed to Mg^{2+} without SO_4^{2-} present (CM-2) resulted in marginal extra oil recovery, which might be related to fluid expansion. It seems that Mg^{2+} acted as a wettability modifier only when SO_4^{2-} was present and the temperature was high.

The following may be summarized about wettability alteration from what is shown in Figure 3.5:

- There must be Ca^{2+} and SO_4^{2-} or Mg^{2+} and SO_4^{2-}.
- Mg^{2+} is better than Ca^{2+}.
- The higher the reservoir temperature, the better the wettability alteration.
- Mg^{2+} reactivity increases with temperature.

Based on these experimental results, a chemical mechanism for the wettability modification was suggested, as illustrated by Figure 3.6. At low and high temperatures, SO_4^{2-} adsorbs onto the positively charged chalk surface. Ca^{2+} may react with the adsorbed carboxylic group to form a complex and release it from the surface, as shown in part a of the figure. At high temperature, Mg^{2+} may displace the Ca^{2+}–carboxylate complex (as shown in part b of the figure).

This suggests that the small and strongly solvated Mg^{2+} is able to substitute Ca^{2+} in a Ca^{2+}–carboxylate complex, although the Ca^{2+}–carboxylate bond is

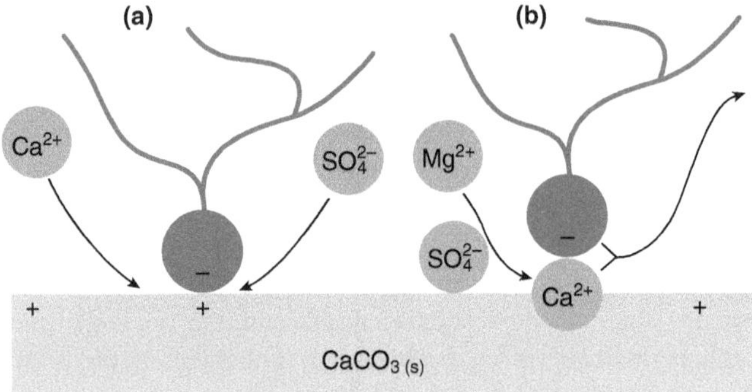

FIGURE 3.6 Schematic model of the suggested mechanism for the wettability alteration induced by seawater. (a) represents the mechanism when Ca^{2+} and SO_4^{2-} are active at lower temperature; and (b) represents the mechanism when Mg^{2+} and SO_4^{2-} are active at a higher temperature. *Source: Zhang et al. (2007a).*

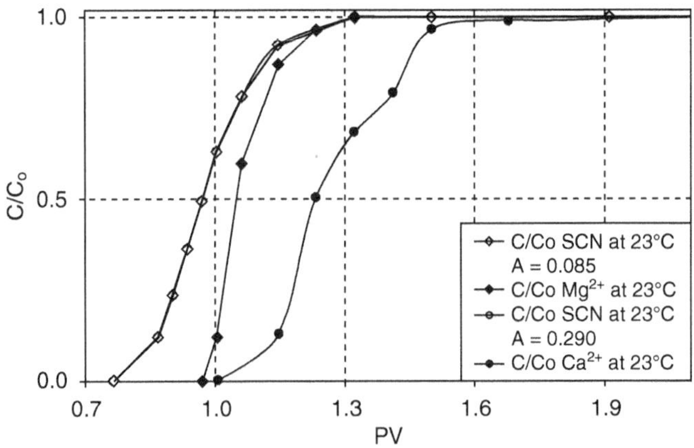

FIGURE 3.7 Comparison of affinities of Ca^{2+} and Mg^{2+} to chalk at 23°C. *Source: Zhang et al. (2007a).*

normally stronger than the Mg^{2+}–carboxylate bond. As SO_4^{2-} adsorbs on the chalk surface, more divalents can be adsorbed on the surface due to less electrostatic repulsion (Zhang et al., 2007a). As the complexes are displaced from the chalk surfaces, the surfaces become more water-wet. The suggested mechanism illustrated in Figure 3.6 is based on the access of injected water to the bonding between the carboxylic group and the chalk surface. The active potential determining ions in seawater can be active only through the aqueous phase. The carboxylic group, –COO–, is of course a strong hydrophilic group, which can create some water saturation close to the bonding sites at the chalk surface.

The preceding suggested mechanism is supported by the fact that the wettability modification using Mg^{2+} and SO_4^{2-} is active only at high temperatures.

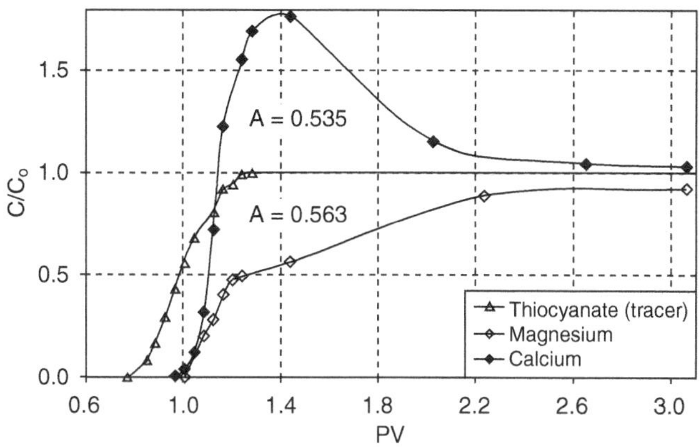

FIGURE 3.8 Comparison of affinities of Ca^{2+} and Mg^{2+} to chalk at 130°C. *Source: Zhang et al. (2007a).*

The affinity of Ca^{2+} and Mg^{2+} to chalk can be verified by the experimental data shown in Figures 3.7 and 3.8. At the low temperature (23°C, Figure 3.7), Ca^{2+} has a higher affinity to the chalk than Mg^{2+}. At the higher temperature (130°C, Figure 3.8), Ca^{2+} has less affinity to the chalk than Mg^{2+} because it breaks through a little bit earlier than Mg^{2+}, and more importantly, its concentration is higher than the injected concentration, which means Mg^{2+} displaced Ca^{2+} to the front. The effluent concentrations of the nonadsorbing tracer SCN^- work as the reference line.

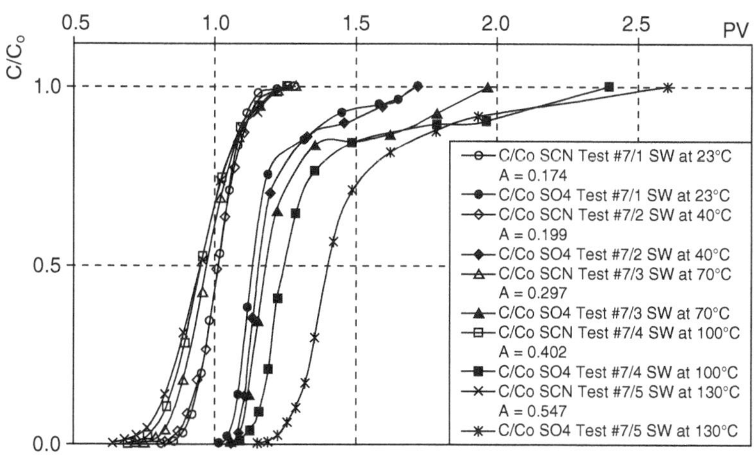

FIGURE 3.9 Retention of SO_4^{2-} in chalk cores at different temperatures. *Source: Strand et al. (2006).*

The striking difference in salinity between the Ekofisk formation water and the injected seawater is SO_4^{2-}, which is abundant in the seawater but almost zero in the formation water. The preceding wettability modification mechanism is also supported by the fact that SO_4^{2-} has strong affinity onto the chalk surfaces. Figure 3.9 compares SO_4^{2-} affinity with that of the nonadsorbing tracer SCN^- toward the chalk surface at different temperatures. SO_4^{2-} retarded more and more relative to SCN^- as the temperatures increased, and a large increase in adsorption occurred between 100 and 130°C (Strand et al., 2006).

Chapter 4

Mobility Control Requirement in EOR Processes

4.1 INTRODUCTION

Mobility control is one of the most important concepts in any enhanced oil recovery process. It can be achieved through injection of chemicals to change displacing fluid viscosity or to preferentially reduce specific fluid relative permeability through injection of foams, or even through injection of chemicals, to modify wettability. This chapter does not address a specific mobility control process. Instead, it discusses the general concept of the mobility control requirement in enhanced oil recovery (EOR).

The existing concept of mobility control is that the displacing fluid mobility should be equal to or less than the (minimum) total mobility of displaced multiphase fluids. This chapter first uses a simulation approach to demonstrate that the existing concept is invalid; the simulation results suggest that the displacing fluid mobility should be related to the displaced oil phase mobility, rather than the total mobility of the displaced fluids. From a stability point of view, a new criterion regarding the mobility control requirement is derived when one fluid displaces two mobile oil and water fluids. The chapter presents numerical verification and analyzes some published experimental data to justify the proposed criterion.

4.2 BACKGROUND

For the convenience of discussion, we first define relative oil, water, and total mobility. The mobility is defined as the effective permeability (k) divided by the viscosity (μ) of the phase:

$$\lambda = \frac{k}{\mu}. \tag{4.1}$$

In the preceding equation, if k is replaced by relative permeability, k_r, we have relative mobility, λ_r:

$$\lambda_{rj} = \frac{k_{rj}}{\mu_j}. \tag{4.2}$$

Modern Chemical Enhanced Oil Recovery. DOI: 10.1016/B978-1-85617-745-0.00004-8

In the preceding equation, the subscript j represents the phase j; j = w, o, t for water phase, oil phase, and total relative mobility, respectively. The unit of relative mobility is the inverse of the viscosity unit, for example, $(\text{mPa·s})^{-1}$ or $(\text{cP})^{-1}$. An example of water and oil relative permeability curves is shown in Figure 4.1. The corresponding water, oil, and total relative mobilities are shown in Figure 4.2, with the water and oil viscosities being 1 and 10 mPa·s, respectively. Figure 4.2 also shows the minimum total relative mobility, the water mobility, oil mobility, and total mobility at a given saturation. The total mobility is the sum of water and oil mobilities.

When discussing viscous fingering, generally we deal with the case of displacing *one mobile* fluid (e.g., oil) by another fluid (e.g., water). The concept is that the displacing fluid mobility in the upstream (λ_u) should be equal to or less than the displaced fluid mobility in the downstream (λ_d):

$$\lambda_u \leq \lambda_d. \tag{4.3}$$

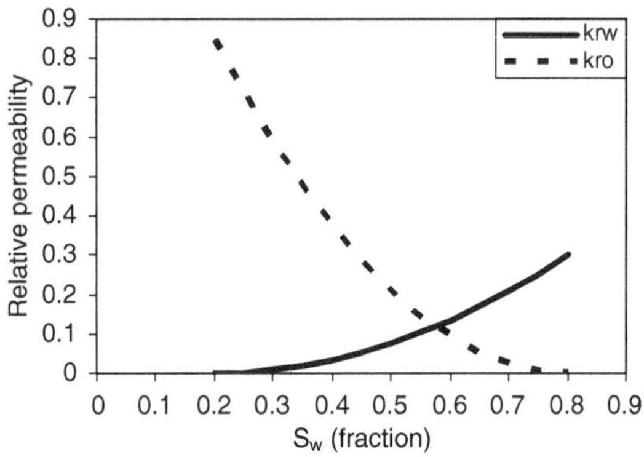

FIGURE 4.1 Water and oil relative permeabilities.

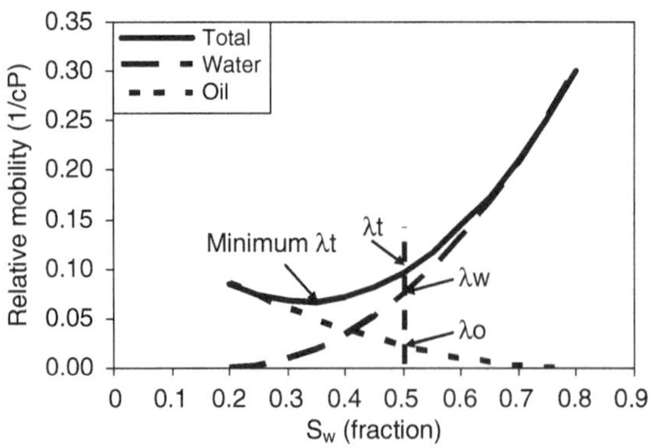

FIGURE 4.2 Water, oil, and total relative mobilities.

We often use the term mobility ratio (M_r), which is defined as the ratio of the displacing phase mobility to the displaced phase mobility:

$$M_r = \frac{\lambda_u}{\lambda_d}. \tag{4.4}$$

Here, the subscripts u and d represent upstream and downstream, respectively. Conventionally, a mobility ratio equal to or less than one ($M_r \leq 1$) is favorable, and $M_r > 1$ is unfavorable (Craig, 1971).

In the waterflooding process, Craig et al. (1955) found that if the water mobility was defined at the average water saturation behind the displacement flood front—that is, $\lambda_u = \lambda(\overline{S}_w)$—the data on areal-sweep versus mobility ratio would match those data obtained by Slobod and Caudle (1952) and Dyes et al. (1954) using miscible fluids in which there was no saturation gradient behind the front. Although the displacing fluid mobility should include the mobility of the movable oil behind the flood front, the oil mobility was considered to be insignificant compared with the water mobility. When they discussed this subject, *only* oil was assumed to be movable ahead of the front. In other words, the oil saturation ahead of the front is the initial oil saturation ($S_{oi} = 1 - S_{wc}$). Here, S_{wc} is the immobile connate water saturation; however, there is no theoretical justification for using this method of calculating mobility ratio.

In enhanced oil recovery processes, such as polymer flooding, one fluid (or even several fluids) displaces several mobile fluids (e.g., water and oil). According to the conventional concept, when one or several fluids displace several mobile fluids ahead, the *total* mobility of displacing fluids should be equal to or less than the *total* mobility of the several displaced fluids (Dyes et al., 1954; Lake, 1989):

$$M_r = \frac{\sum(\lambda_{rj})_u}{\sum(\lambda_{rj})_d} \leq 1. \tag{4.5}$$

In a case in which several mobile fluid saturations are not known, the total mobility of these fluids cannot be calculated because it is a function of saturations, and these saturations are generally unknown. Gogarty (1969) and Gogarty et al. (1970) chose to use the minimum total mobility to avoid this problem:

$$M_r = \frac{[\sum(\lambda_{rj})_u]_{minimum}}{\sum(\lambda_{rj})_d} \leq 1. \tag{4.6}$$

Prats (1982) stated that a parameter commonly used in reservoir engineering as a measure of the stability of a displacement front in the absence of capillary and gravity forces is the ratio of the pressure gradient, $\partial p/\partial n$, normal to and on the downstream side of the displacement front to that on the upstream side. There is no proof for this statement, however (Michael Prats, personal communication, May 12, 2008). From Darcy's law, this pressure gradient ratio can be expressed as

$$\frac{\left(\frac{\partial p}{\partial n}\right)_d}{\left(\frac{\partial p}{\partial n}\right)_u} = \frac{\left(\frac{u}{\lambda}\right)_d}{\left(\frac{u}{\lambda}\right)_u}. \tag{4.7}$$

Here, u is the volumetric velocity normal to the front. In waterflooding, if we assume that *only* water flows upstream and *only* oil downstream, it follows that $u_d = u_u$. Then the previous equation becomes

$$\frac{\left(\frac{\partial p}{\partial n}\right)_d}{\left(\frac{\partial p}{\partial n}\right)_u} = \frac{\lambda_u}{\lambda_d}. \tag{4.8}$$

The preceding equation shows that the ratio of pressure gradient is equivalent to the mobility ratio in this case, being the same as the well-known concept of mobility ratio.

None of the previous discussions take into account one important factor: the difference in phase velocities. In any real EOR process, several phases flow at different velocities before and behind the displacement front. This chapter uses a simulation approach to demonstrate that the conventional concept (the inequality 4.5) is invalid to define favorable or unfavorable displacement. A new concept is proposed instead.

4.3 SETUP OF SIMULATION MODEL

Our first task is to evaluate the validity of the conventional concept about the mobility control requirement using a simulation approach. This model uses the UTCHEM-9.0 simulator (2000). The dimensions of the two-dimensional XZ cross-section model are 300 ft × 1 ft × 10 ft. One injection well and one production well are at the two extreme ends in the X direction, and they are fully penetrated. The injection velocity is 1 ft/day; the initial water saturation and oil saturation are 0.5. The displacing fluid is a polymer solution. The purpose of using the polymer solutuion in the model is to change the viscosity of the displacing fluid. Therefore, polymer adsorption, shear dilution effect, and so on are not included in the model. To simplify the problem, it is assumed that the oil and water densities are the same; that the capillary pressure is not included; that the relative permeabilities of water and oil are straight lines with the connate water saturation and residual oil saturation equal to 0; and that the water and oil viscosity is 1 mPa·s. Under these assumptions and conditions, we can know the fluid mobilities at any saturation. The model uses an isotropic permeability of 10 mD.

The grid blocks tested are listed in Table 4.1. The recovery factors (RF) from each grid shown in the table are all greater than 99.48% at one pore volume (PV) of injection. That means, at least from the recovery factor point of view, all these models provide reasonably accurate results (close to theoretical RF of 100% for the built base model with the mobilities of displacing and displaced fluids being equal).

Figure 4.3 shows the recovery factors (RF) and water cuts (f_w) for the six cases of different grids. The recovery factors versus injection pore volumes all fall on almost the same curve. We can, however, see some difference in the water-cut curves after 0.97 PV. Grid02 and Grid04 have the same number of blocks (120) in the X direction, but different in the Z direction. Grid02 has 10 blocks, whereas Grid04 has 1 block in the Z direction. If we look closely at the figure and data file (not shown here), we can see that the water cuts and recovery

TABLE 4.1 Grid Sensitivity

Case ID	Grids	RF (%)
Grid01	1D, 360 × 1 × 1	99.83
Grid02	2D, 120 × 1 × 10	99.48
Grid03	2D, 240 × 1 × 10	99.74
Grid04	1D, 120 × 1 × 1	99.48
Grid05	1D, 240 × 1 × 1	99.74
Grid06	2D, 300 × 1 × 10	99.80

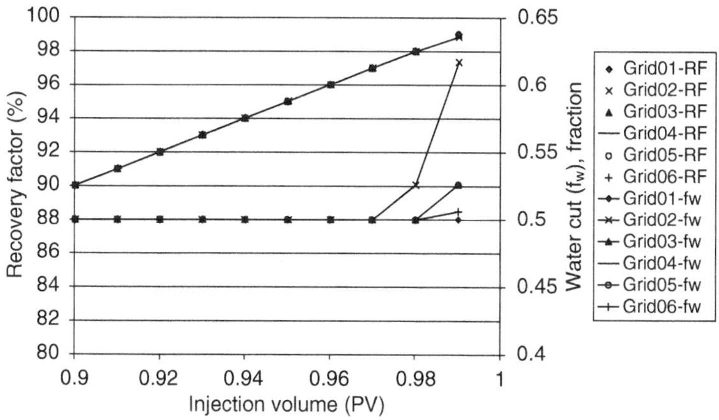

FIGURE 4.3 Recovery factors and water cuts for the six cases of different grids.

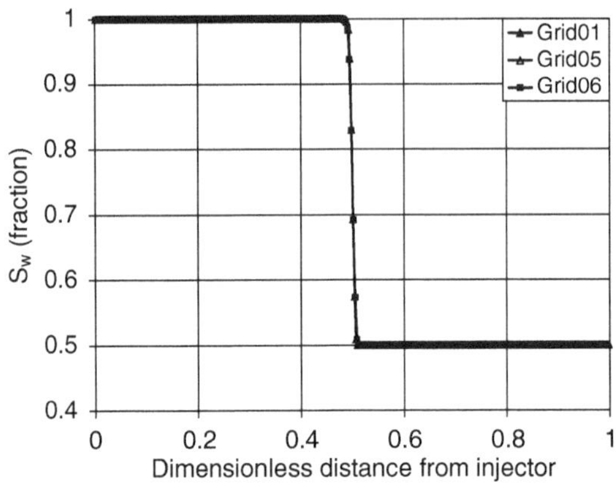

FIGURE 4.4 Water saturation profiles at 0.5 PV injection for Grid01, Grid05, and Grid06.

factors for these two cases are exactly the same at the same injection PV. In other words, flow behavior is the same at different vertical layers.

The same observation applies to Grid03 and Grid05. We have to compare only the grid sensitivities for cases Grid04, Grid05, Grid06, and Grid01, with their numbers of blocks in the X direction being 120, 240, 300, and 360, respectively. Figure 4.3 shows that the water-cut curve in Grid06 is very close to that in Grid01, with the water cut difference at one PV injection being 1%. The water saturation profiles in Grid01, Grid05, and Grid06 at 0.5 PV injection are compared in Figure 4.4, which shows the saturation profiles almost overlap each other at the front. Therefore, Grid06 should be fine enough, and it is taken as the base grid.

4.4 DISCUSSION OF THE CONCEPT OF THE MOBILITY CONTROL REQUIREMENT

As mentioned earlier, when one fluid displaces several mobile fluids ahead, it is assumed that the displacing fluid mobility should be equal to or less than the *total* mobility of the several mobile fluids ahead, according to the literature (Gogarty, 1969; Gogarty et al., 1970; Lake, 1989). This section discusses the validity of this statement. The displacing fluid is a polymer solution (water phase). The viscosity of a polymer solution is changed to a target viscosity by varying polymer concentration in the solution.

We start with Case visc01, which is the same as the base model Grid06, a homogeneous model with permeability of 10 mD. In visc01, the polymer solution viscosity behind the displacing front, the oil viscosity, and the water viscosity in the displaced zone (ahead of the displacing front) are the same (1 mPa·s). Therefore, the mobility of the polymer solution in the displacing zone is the same as the total mobility of water and oil in the displaced zone in which the initial water saturation and oil saturation are the same (0.5) and their relative permeability is the same (0.5). In mathematical formula, this mobility is expressed by

$$\frac{k_{rw}(S_w=1)}{\mu_p}=\frac{1}{1}=\frac{k_{rw}(S_{wi}=0.5)}{\mu_w}+\frac{k_{ro}(S_{wi}=0.5)}{\mu_o}=\frac{0.5}{1}+\frac{0.5}{1}.$$

In this case, $\mu_p = 1$ mPa·s.

Case visc02 is the same as Case visc01, except that the oil viscosity is increased to 100 mPa·s, and the polymer concentration is adjusted using Eq. 4.5 so that the polymer mobility is equal to the total mobility of oil and water phases ahead of the displacing front. In this case, $\mu_p = 1.98$ mPa·s.

Case visc03 is the same as Case visc02, except that the polymer concentration is adjusted so that the polymer mobility is the same as the oil mobility only (not total mobility). Note that in Case visc03, as well as in Cases visc01 and visc02, the initial oil saturation is 0.5. In this situation, the cross-section area available for polymer to displace the oil phase is half the whole cross-section area. The other half cross-section area is used for polymer to displace the water phase ahead. In other words, the polymer mobility to displace the oil is reduced by half. Mathematically, we should determine the polymer viscosity required using the following equation:

$$\frac{k_{rw}(S_w=1)}{\mu_p}\times(1-S_{wi})=\frac{k_{ro}(S_{wi})}{\mu_o}. \tag{4.9}$$

The preceding equation is derived later in Section 4.5. From this equation, we have $\mu_p = 100$ mPa·s.

Now we have the three cases: visc01, visc02, and visc03. The recovery factors at 1 PV injection and the main conditions are presented in Table 4.2.

TABLE 4.2 Recovery Factors at Different Mobilities

Case ID	RF (%)	μ_o, mPa·s	Mobility (λ), $(\text{mPa·s})^{-1}$
visc01	99.78	1	$\lambda_{inj} = \lambda_t$
visc02	3.00	100	$\lambda_{inj} = \lambda_t$
visc03	98.34	100	$\lambda_{inj} = \lambda_o$
visc04	28.20	10	$\lambda_{inj} = \lambda_t$
visc05	98.34	10	$\lambda_{inj} = \lambda_o$
visc08*	99.79	10	$\lambda_{inj} = \lambda_t$

**Same as visco04 except that the initial water saturation is changed from 0.5 to 0.0.*

Interestingly, although in the two cases visc01 and visc02, the injection fluid mobility is the same as the total mobility of oil and water ahead in their respective cases, the recovery factors at 1 PV injection are extremely different (98.78% in visc01 versus 3% in visc02). According to the conventional theories, however, the recovery factors in visc01 and visc02 should be similar because in both of these cases the ratio of the displacing fluid mobility to the total mobility of the displaced oil and water is 1.

In Case visc03, even though the oil viscosity is 100 mPa·s, the same as that in visc02, when the injection fluid mobility is adjusted to be the same as the oil mobility only (not the total mobility) based on Eq. 4.9, the recovery factor is 98.34%, almost the same as that in Case visc01 (only 1% different). Based on these results, we can see that to satisfy the mobility control requirement for a high oil recovery factor (favorable displacement condition), the injection mobility should be equal to or less than the oil mobility corrected by the initial oil saturation by Eq. 4.9, not the total mobility of fluids ahead of the displacing front.

Figure 4.5 shows the recovery factors and water cuts for visc01, visc02, and visc03. For visc01, water and oil viscosities are the same. For the whole injection period, oil is produced, and the water cut is maintained at 50%. For visc02, although the displacing fluid mobility is the same as the total mobility of the displaced oil and water, because of the relatively high mobility of water phase, the water (polymer solution and initial mobile water) bypasses the high viscous oil. Therefore, the water cut is very high (> 98%) during the entire injection period. For visc03, the displacing fluid mobility is the same as the oil mobility corrected by initial oil saturation. Before 0.5 PV injection, because of the low viscosity (1 mPa·s) of initial mobile water (compared with 100 mPa·s of the oil), the water cut is close to 1.0. Meanwhile, the high viscous displacing fluid

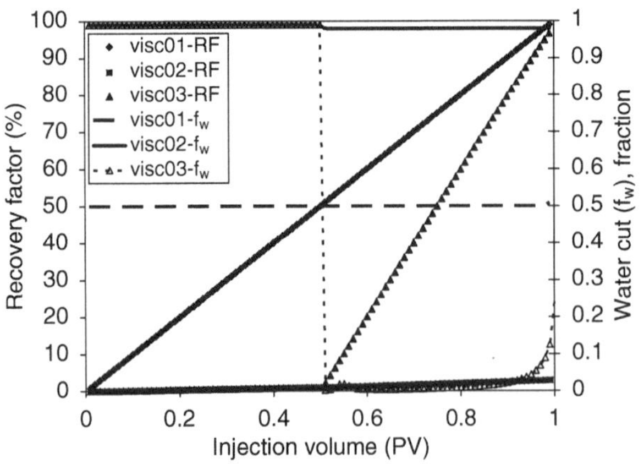

FIGURE 4.5 Recovery factors and water cuts for visc01, visc02, and visc03.

moves the oil forward, and the oil replaces the pore space evacuated by the mobile water ahead of it. After 0.5 PV, basically half of the pore space near the production end is fully occupied by oil. Therefore, the water cut after 0.5 PV is almost 0. At one PV injection, almost all the oil is displaced out of the pore.

The cases of visc04 and visc05 are the duplicates of visc02 and visc03, respectively, except that the oil viscosity is reduced from 100 mPa·s to 10 mPa·s. The results shown in Table 4.2 repeat the same observation as from visc02 and visc03. The recovery factor in visc04 is 28.2%, much lower than that of 98.34% in visc05. Next, we run Case visc08, which is the same as Case visc04 except that the initial water saturation is changed from 0.5 to 0 and the injection water viscosity is equal to the oil viscosity (10 mPa·s) so that the (total) mobility ratio is 1. The recovery factor at one PV injection is 99.79%, compared with the recovery factor of 28.2% for Case visc04. For the two cases, the total mobility ratios are the same (equal to 1). The only difference is the initial mobile water saturation. The comparison of these two cases shows that the mobility requirements for the flow systems with a single phase fluid and multiphase fluids ahead of the displacing front are different.

The water saturation distributions for the previous cases can further explain what would happen at different mobility ratios. The water saturation profile for Case visc01 at 0.5 PV injection is shown in Figure 4.6. Because the mobility ratio between the displacing fluid and displaced fluid is 1, the displacing front is stable. The finger is not further developed, and the displacing front is sharp.

In Case visc02, the oil viscosity is increased to 100 mPa·s. The oil mobility is then relatively small so that the water mobility is almost equal to the total

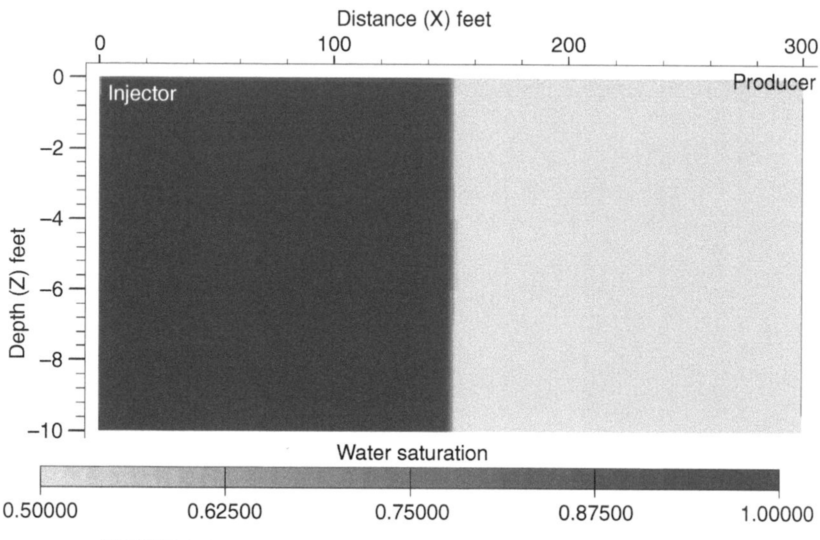

FIGURE 4.6 Water saturation profile at 0.5 PV injection (visc01 data).

mobility (oil and water) in the displaced zone. The polymer concentration is adjusted so that the polymer solution mobility is equal to the total mobility of the displaced water and oil. Interestingly, the water saturation in most of the displaced area shown in Figure 4.7 is about 0.495, less than the initial water saturation 0.5. The reason is that most of the oil is not produced, and some oil near the injector is displaced and spreads over the rest of the area. The initially existing water is produced immediately after the producer is opened, and the injected thickened water breaks through the producer after the initial water is produced. The observed phenomenon can also be verified in the recovery factor and water-cut curves in Figure 4.5.

Comparing the water saturation profiles of visc04 (Figure 4.8) and visc05 (Figure 4.9) is more convincing. In visc04, the injected polymer mobility is equal to the total mobility of water and oil. Because the oil viscosity is 10 times higher than the water, their total mobility is almost the same as the water mobility only. Therefore, the injected polymer mobility is actually almost the same as the displaced water mobility. The injected polymer and the initial water bypass the oil, leaving the oil saturation in the large middle area almost intact (0.562, a little bit higher than the initial oil saturation 0.5). While in visc05, the injected polymer mobility is the same as the oil mobility corrected by the initial oil saturation. It moves the oil forward, and the oil is banked ahead. The oil saturation in this oil bank is 0.999.

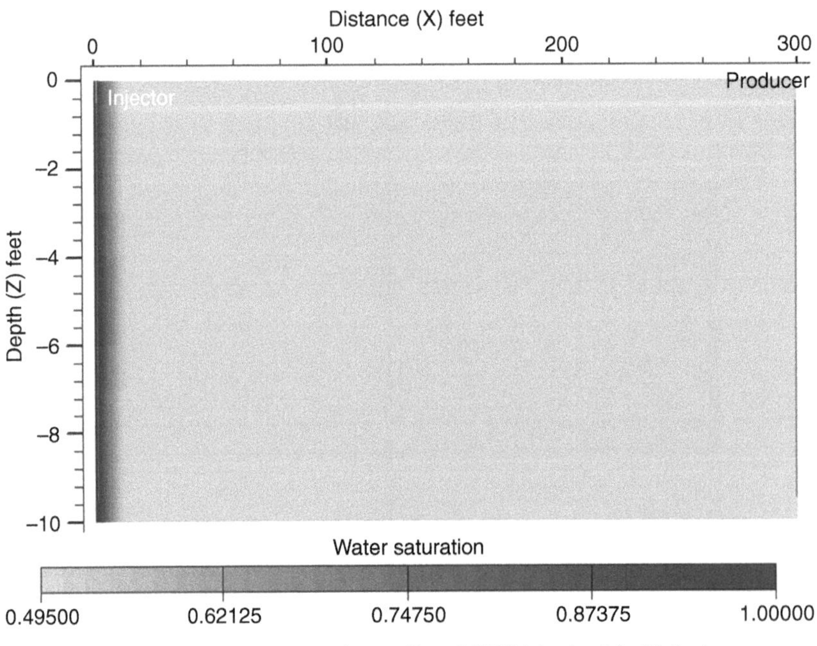

FIGURE 4.7 Water saturation profile at 0.5 PV injection (visc02 data).

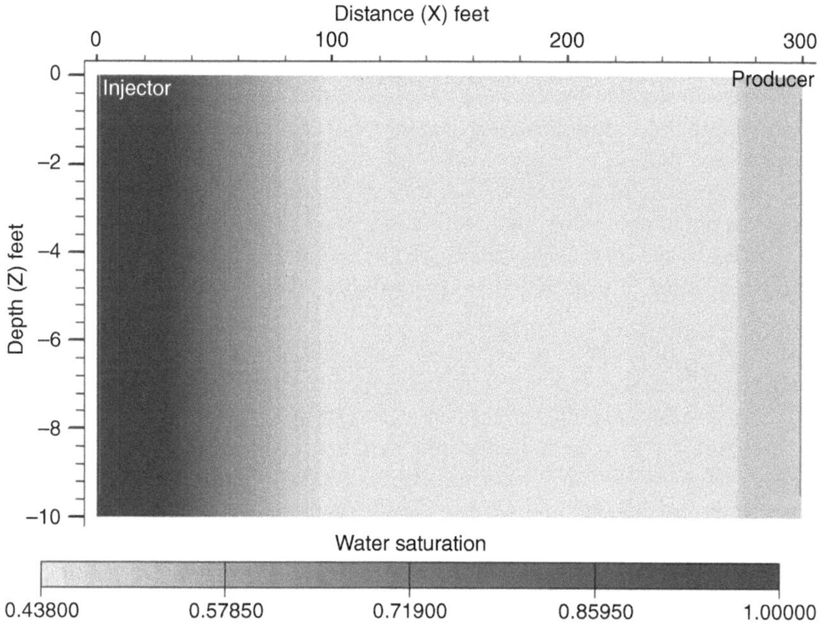

FIGURE 4.8 Water saturation profile at 0.5 PV injection (visc04 data). Oil viscosity is 10 mPa·s and $\lambda_{inj} = \lambda_t$.

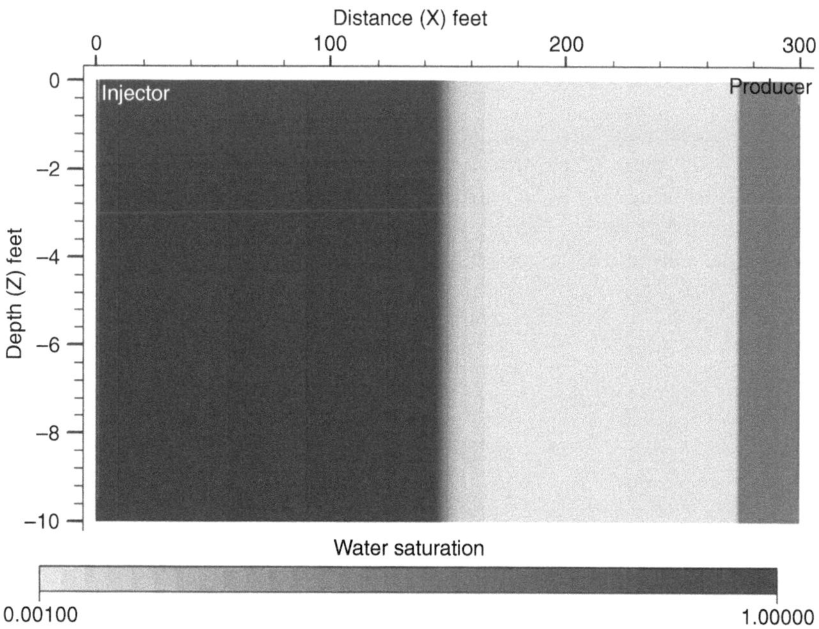

FIGURE 4.9 Water saturation profile at 0.5 PV injection (visc05 data). Oil viscosity is 10 mPa·s and $\lambda_{inj} = \lambda_o$.

4.5 THEORETICAL INVESTIGATION

The conventional mobility ratio in multiphase flow is defined as the displacing fluid mobility divided by the total mobility of displaced water and oil phases. From the previous section, we can see that the unit mobility ratio based on the conventional definition is not a valid criterion to distinguish "favorable" and "unfavorable" mobility control conditions. We have found that a better criterion should be the unit mobility ratio, which is defined as the displacing fluid mobility divided by the oil mobility multiplied by the oil saturation (Eq. 4.9). In this section, we attempt to justify the proposed idea from the stability of displacement front.

Let us assume that the displaced water and oil two-phase flow can be described by two separate flow channels: oil and water. The flow model therefore can be schematically represented as shown in Figure 4.10.

In Figure 4.10, p_i and p_o are the inlet pressure (injection pressure) and outlet pressure (flowing pressure), respectively; x_{of} and x_{wf} are assumed displacement fronts at the oil channel and water channel, respectively; q_1, q_2, q_o, and q_w are the injection rate in the oil channel, injection rate in the water channel, oil rate, and water rate, respectively, with $q_1 = q_o$ and $q_2 = q_w$; the cross-section areas of the oil and water channels are equal to their respective saturations in the displaced zones, S_o and S_w. The distance from the inlet to the outlet is L.

Now we consider the flow in the oil channel. We assume the displacement is piston-like, and no oil is left behind the displacement front. Accordingly, the displacing rate q_1 in the upstream swept zone is

$$q_1 = \frac{kk_{wr}A\bar{S}_o(p_i - p_{of})}{\mu_u x_{of}}, \tag{4.10}$$

where k is the absolute permeability, k_{wr} is the endpoint water relative permeability at the residual oil saturation S_{or}, A is the cross-section area, p_{of} is the pressure at the front x_{of}, and $\bar{S}_o$ is the normalized movable oil cross-section area (initial normalized oil saturation):

$$\bar{S}_o = \frac{S_o - S_{wc}}{1 - S_{or} - S_{wc}}. \tag{4.11}$$

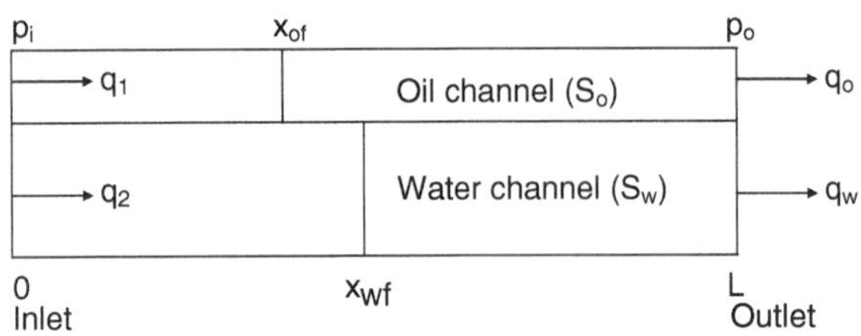

FIGURE 4.10 Schematic of flow channels.

Note that the water relative permeability here should be the upstream phase relative permeability (e.g., polymer solution relative permeability). To simplify the discussion, we just use water relative permeability. In the downstream unswept zone, the oil flow rate, q_o, is

$$q_o = \frac{kk_{ro}(S_w)A(p_{of} - p_o)}{\mu_o(L - x_{of})}. \tag{4.12}$$

For the oil channel,

$$\begin{aligned} p_i - p_o &= (p_i - p_{of}) + (p_{of} - p_o) = \frac{\mu_u x_{of} q_1}{kk_{wr}A\bar{S}_o} + \frac{\mu_o(L - x_{of})q_o}{kk_{ro}(S_w)A} \\ &= \frac{q_o\mu_u}{kk_{wr}A\bar{S}_o}[x_{of} + M_{roc}(L - x_{of})], \end{aligned} \tag{4.13}$$

where

$$M_{roc} = \frac{k_{wr}/\mu_u\bar{S}_o}{k_{ro}(S_w)/\mu_o} = \frac{\lambda_u}{\lambda_o}\bar{S}_o. \tag{4.14}$$

From Eq. 4.13 and the material balance of the injected water within dt,

$$\begin{aligned} \frac{dx_{of}}{dt} &= \frac{q_o}{A\bar{S}_o(1 - S_{wc} - S_{or})\phi} = \frac{kk_{wr}(p_i - p_o)}{\mu_u\phi(1 - S_{wc} - S_{or})[M_{roc}L + x_{of}(1 - M_{roc})]} \\ &= \frac{C}{[M_{roc}L + x_{of}(1 - M_{roc})]}, \end{aligned} \tag{4.15}$$

where C is a constant defined by

$$C = \frac{kk_{wr}(p_i - p_o)}{\mu_u\phi(1 - S_{wc} - S_{or})}. \tag{4.16}$$

Let us assume a small perturbation ε in x_{of},

$$\begin{aligned} \frac{d\varepsilon}{dt} &= \frac{d(x_{of} + \varepsilon)}{dt} - \frac{dx_{of}}{dt} \\ &= C\left[\frac{1}{M_{roc}L + (x_{of} + \varepsilon)(1 - M_{roc})} - \frac{1}{M_{roc}L + x_{of}(1 - M_{roc})}\right] \\ &\approx \frac{C\varepsilon(M_{roc} - 1)}{[M_{roc}L + x_{of}(1 - M_{roc})]^2} = C'(M_{roc} - 1)\varepsilon, \end{aligned} \tag{4.17}$$

where

$$C' = \frac{C}{[M_{roc}L + x_{of}(1 - M_{roc})]^2}. \tag{4.18}$$

From Eq. 4.17, we have

$$\varepsilon = \varepsilon_i \exp[(M_{roc} - 1)C'(t - t_i)]. \tag{4.19}$$

Equation 4.19 shows that ε grows exponentially with time when $M_{roc} > 1$, is unchanged when $M_{roc} = 1$, and decays exponentially when $M_{roc} < 1$. From the stability of displacement front, M_{roc} should be equal to or less than 1. In other words, the criterion for the mobility control requirement in EOR processes should be

$$M_{roc} \equiv \frac{k_{wr}/\mu_u \bar{S}_o}{k_{ro}(S_w)/\mu_o} \le 1. \tag{4.20}$$

The physical meaning of M_{roc} defined by Eq. 4.14 is the mobility ratio of the displacing fluid to the displaced oil phase in the assumed oil channel. This mobility ratio in the assumed oil channel is the mobility ratio of the displacing fluid to the displaced oil phase multiplied by the normalized movable oil saturation, $\bar{S}_o$.

Prats (1982) stated that a parameter commonly used to measure the stability of a displacement front is the ratio of the pressure gradient $\partial p/\partial n$ normal to and on the downstream side of the displacement front to that on the upstream side. In this case, the upstream pressure gradient in the oil channel based on Eq. 4.10 is

$$\left(\frac{\partial p}{\partial n}\right)_u = \frac{(p_i - p_{of})}{x_{of}} = \frac{q_1 \mu_u}{k k_{wr} A \bar{S}_o}. \tag{4.21}$$

The downstream pressure gradient in the oil channel based on Eq. 4.12 is

$$\left(\frac{\partial p}{\partial n}\right)_d = \frac{(p_{of} - p_o)}{(L - x_{of})} = \frac{q_o \mu_o}{k k_{ro}(S_w) A}. \tag{4.22}$$

Then

$$\frac{\left(\frac{\partial p}{\partial n}\right)_d}{\left(\frac{\partial p}{\partial n}\right)_u} = \frac{k_{wr}/\mu_u \bar{S}_o}{k_{ro}(S_w)/\mu_o}. \tag{4.23}$$

Comparing Eq. 4.23 with Eq. 4.14, we can see that

$$M_{roc} = \frac{\left(\frac{\partial p}{\partial n}\right)_d}{\left(\frac{\partial p}{\partial n}\right)_u}. \tag{4.24}$$

In other words, according to Eq. 4.24, the physical meaning of M_{roc} is the ratio of the downstream pressure gradient to the upstream pressure gradient in the assumed oil channel. This ratio should be equal to or less than 1. From a practical design point of view, because pressure gradients are not available, we have to use the definition equation (Eq. 4.14) for M_{roc}. In this chapter, M_{roc} is simply called the mobility ratio. Keep in mind that the mobility ratio used in this chapter is in the assumed oil channel, or the conventional mobility ratio multiplied (or corrected) by the normalized oil saturation.

In the assumed water channel, generally the displacing fluid is a polymer solution, or the other aqueous phase, and its viscosity is higher than that of the existing water except in a thermal recovery process. In a thermal recovery process such as steam flooding, however, the displacement is actually stable (Harmsen, 1971; Miller, 1975; Hagoort et al., 1976). The reason is that small steam fingers, if formed, tend to lose heat at relatively high rates, ultimately resulting in condensation and disappearance of the steam fingers (Prats, 1982).

4.6 NUMERICAL INVESTIGATION

This section investigates mobility effect on oil recovery factor in different formations: homogeneous, two-layered heterogeneous, and heterogeneous with a random permeability distribution.

4.6.1 Effect of Mobility Ratio in a Homogeneous Formation

After a discussion of the mobility control requirement using the simplified flow model, this section moves to a model with realistic water and oil relative permeability curves. Now the interstitial (connate) water saturation and residual oil saturation are 0.2. The endpoint relative permeabilities of oil and water are 0.85 and 0.3, respectively. The Corey exponents of relative permeabilities for oil and water are 2. Others are the same as those in the simplified model discussed earlier; particularly, the initial water saturation is 0.5. Again, capillary and gravity are not included.

Figure 4.11 shows the simulation results of the recovery factors after one PV injection versus mobility ratio, which is defined as the injection fluid mobility divided by the oil phase mobility multiplied by the normalized oil saturation (Eq. 4.14). This figure clearly shows that with the mobility ratio less than 1, the recovery factors are insensitive to the mobility ratio; with the mobility ratio greater than 1, the recovery factors decrease steeply with the mobility ratio. The unit mobility ratio is a kind of turning point. Figure 4.12 is similar to Figure 4.11, except that the mobility ratio in the horizontal axis is defined as the injection fluid mobility divided by the total mobility (Eq. 4.5). The turning point in this figure is around 0.2, not 1.

When the initial water saturation is 0.7, which is more representative in a tertiary recovery process, the recovery factor versus the two different mobility ratios are shown in Figures 4.13 and 4.14. These two figures more clearly show that if we define the mobility ratio using the oil mobility (Eq. 4.14), the unit mobility is a better turning point. In other words, when the mobility ratio is

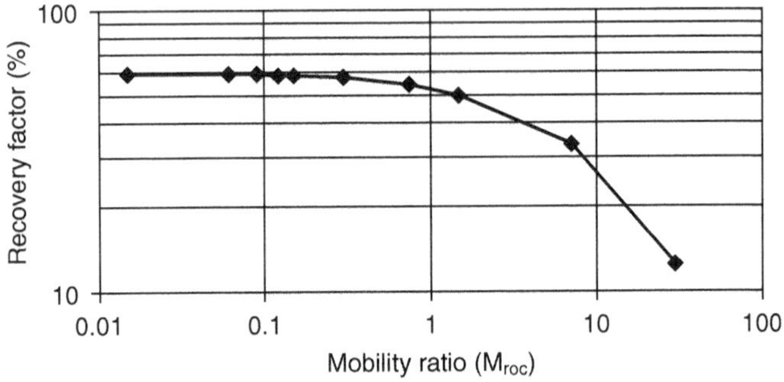

FIGURE 4.11 Recovery factors versus the mobility ratio defined in Eq. 4.14 for a homogeneous model ($S_{wi} = 0.5$).

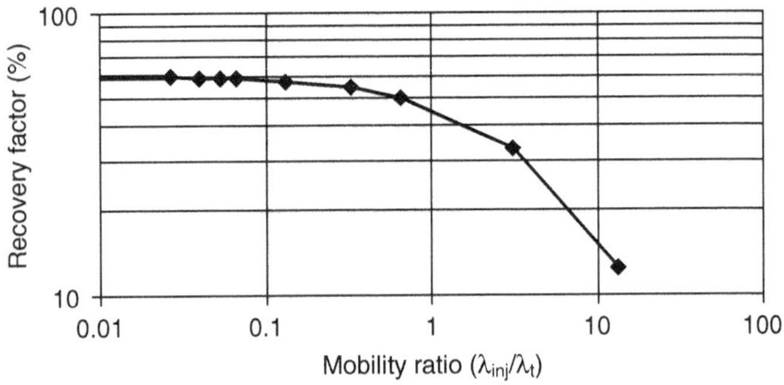

FIGURE 4.12 Recovery factors versus the mobility ratio defined in Eq. 4.5 for a homogeneous model ($S_{wi} = 0.5$).

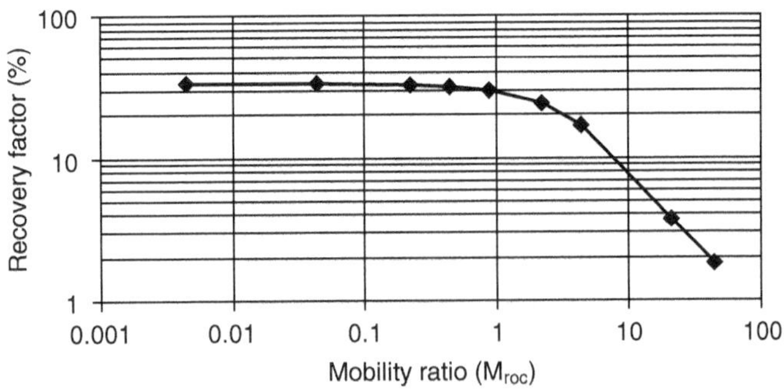

FIGURE 4.13 Recovery factors versus the mobility ratio defined in Eq. 4.14 for a homogeneous model ($S_{wi} = 0.7$).

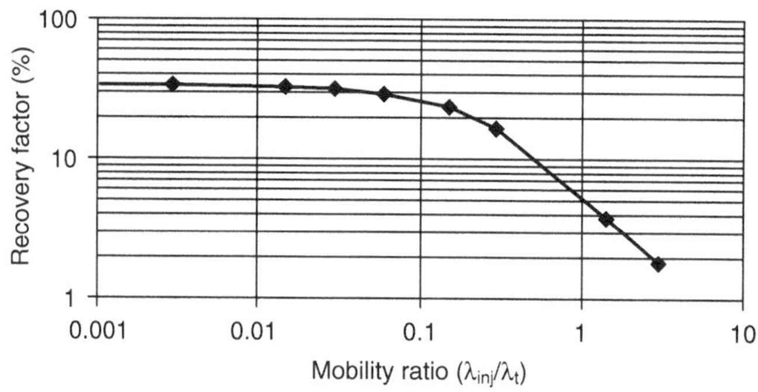

FIGURE 4.14 Recovery factors versus the mobility ratio defined in Eq. 4.5 for a homogeneous model ($S_{wi} = 0.7$).

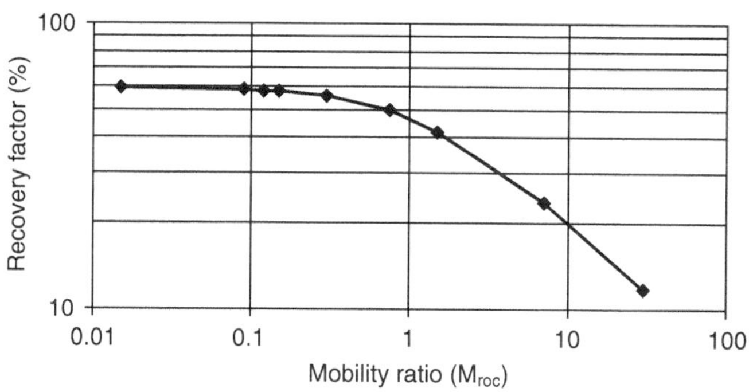

FIGURE 4.15 Recovery factors versus the mobility ratio defined in Eq. 4.14 for a two-layered model ($S_{wi} = 0.5$).

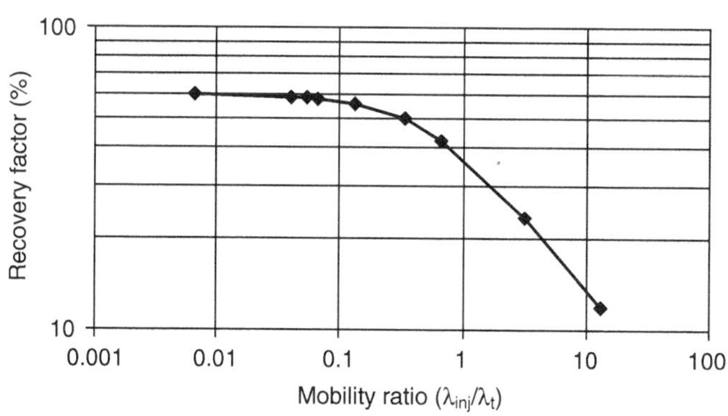

FIGURE 4.16 Recovery factors versus the mobility ratio defined in Eq. 4.5 for a two-layered model ($S_{wi} = 0.5$).

less than 1, the recovery factor will not be sensitive to the mobility ratio; when the mobility ratio is greater than 1, the recovery factor is very sensitive to the mobility ratio. Therefore, these results support our proposed idea that the mobility control requirement is as follows: the displacing fluid mobility should be equal to or less than the less-mobile phase mobility (generally, oil mobility) in the downstream multiplied by the normalized phase saturation in a homogeneous formation.

4.6.2 Effect of Mobility Ratio in a Layered Formation

The layered model discussed here is a two-layer model: top layer permeability is 5 md, and bottom layer permeability is 50 md. The ratio of vertical permeability to horizontal permeability is 0.1, and the total injection volume is 2 PV. The rest of the input data are the same as the homogeneous model. Figures 4.15 and 4.16 show the recovery factors versus M_{roc} and λ_{inj}/λ_t for the initial water saturation of 0.5.

Figures 4.17 and 4.18 show the recovery factors versus M_{roc} and λ_{inj}/λ_t for the initial water saturation of 0.7. From these figures we learn that if we define the mobility ratio as M_{roc} in Eq. 4.14, the unit mobility ratio is a much better criterion than the conventional one using the total mobility in Eq. 4.5.

4.6.3 Effect of Mobility Ratio in a Heterogeneous Formation

In the heterogeneous model, the random permeability distribution is generated using the geo-statistical software developed by Yang (1990). The input average permeability is 10 md; the coefficient of permeability variation (or simply permeability variation; Dykstra–Parsons, 1950) is 0.86; and the dimensionless correlation length is 0.67. The ratio of vertical permeability to horizontal permeability is 0.1, and the total injection volume is 2 PV. The rest of the input data are the same as the homogeneous model. Figures 4.19 and 4.20 show the recovery factors versus M_{roc} and λ_{inj}/λ_t for the initial water saturation of 0.5.

Figures 4.21 and 4.22 show the recovery factors versus M_{roc} and λ_{inj}/λ_t for the initial water saturation of 0.7. These figures show that the observations in the homogeneous model are still valid in the heterogeneous model. In other words, if we define the mobility ratio as the ratio of injection (displacing) fluid mobility to oil mobility multiplied by the normalized oil saturation, the unit mobility ratio is a much better criterion than the conventional one using the total mobility.

4.7 EXPERIMENTAL JUSTIFICATION

Wang et al. (2001c) performed polymer and ASP flooding tests after the cores were completely watered out. Increasing displacing fluid viscosity leads to a

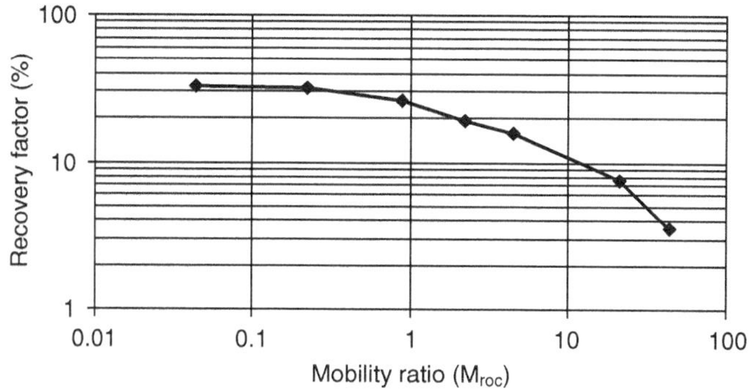

FIGURE 4.17 Recovery factors versus the mobility ratio defined in Eq. 4.14 for a two-layered model ($S_{wi} = 0.7$).

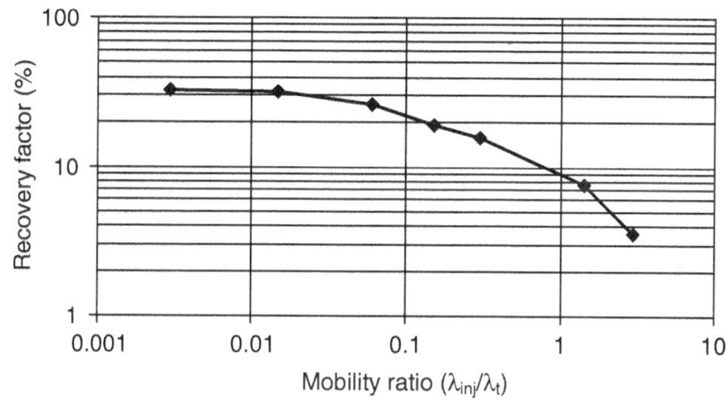

FIGURE 4.18 Recovery factors versus the mobility ratio defined in Eq. 4.5 for a two-layered model ($S_{wi} = 0.7$).

higher oil recovery factor. But Wang et al. wanted to know at what range of the viscosity ratio, μ_u/μ_o, the incremental oil recovery factor would be the most for the unit increase in the viscosity ratio. They found that the most effective range of viscosity ratio, μ_u/μ_o, was 2 to 4 for the watered-out cores. In these watered-out cores, the average oil saturation was about 0.45. The detailed data, especially relative permeability data, were not presented in their paper. We want to find the values of the mobility ratio M_{roc}, which correspond to their viscosity ratio of 2 to 4. We made the following estimates.

The water cut at the watered-out was 0.98. The oil viscosity was 9 mPa·s in the experiments by Wang et al. The water viscosity of 1 mPa·s is assumed. According to the fractional flow equation that follows,

$$f_w = \frac{1}{1+\frac{k_{ro}}{k_{rw}}\frac{\mu_w}{\mu_o}} = \frac{1}{1+\frac{k_{ro}}{k_{rw}}\frac{(1)}{(9)}} = 0.98,$$

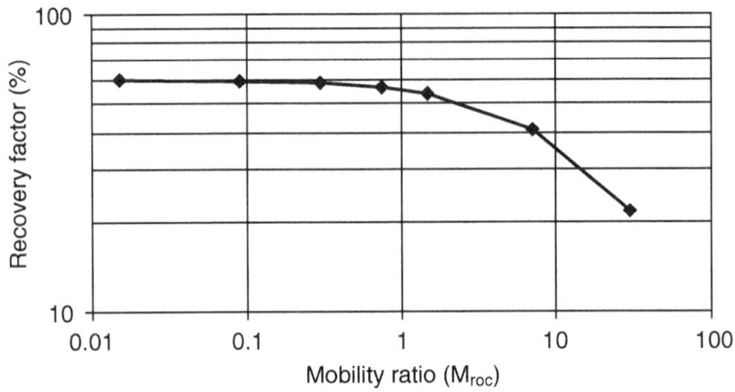

FIGURE 4.19 Recovery factors versus the mobility ratio defined in Eq. 4.14 for a random permeability model ($S_{wi} = 0.5$).

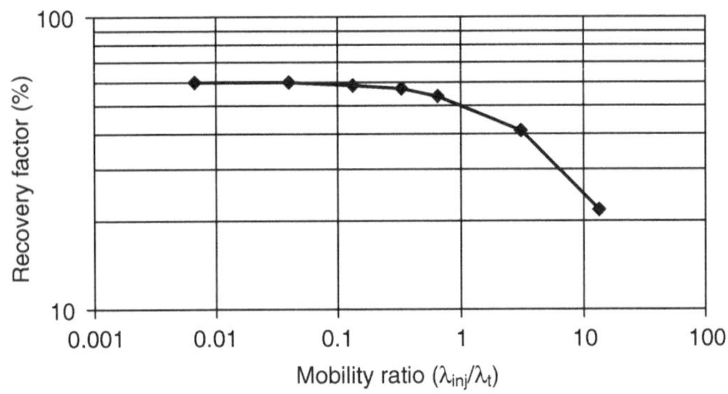

FIGURE 4.20 Recovery factors versus the mobility ratio defined in Eq. 4.5 for a random permeability model ($S_{wi} = 0.5$).

the estimated k_{rw}/k_{ro} is about 5. If we assume the endpoint k_{wr} is about twice the k_{rw} at the water saturation of 0.55, then $k_{wr}/k_{ro} = 10$.

If we further assume that $S_{wc} = 0.2$ and $S_{or} = 0.3$, it follows that $\bar{S}_o = (0.45 - 0.3)/(1 - 0.2 - 0.3) = 0.3$. Wang et al. also observed that the effective viscosity ratio was 2 to 4. When we use Eq. 4.14, the estimated mobility ratio is

$$M_{roc} = \frac{k_{wr}/\mu_u \bar{S}_o}{k_{ro}(S_w)/\mu_o} = \left(\frac{k_{wr}}{k_{ro}}\right)\left(\frac{\mu_o}{\mu_p}\right)\bar{S}_o = 10(1/4 - 1/2)(0.3) = 0.75 - 1.5.$$

Now we have found that M_{roc} is 0.75 to 1.5. The proposed $M_{roc} = 1$ is in the middle of their range.

In the simulated case with S_{wi} equal to 0.7, μ_u/μ_o is 2.1, which is consistent with the experimental data of Wang et al. (2001c). However, in the case with

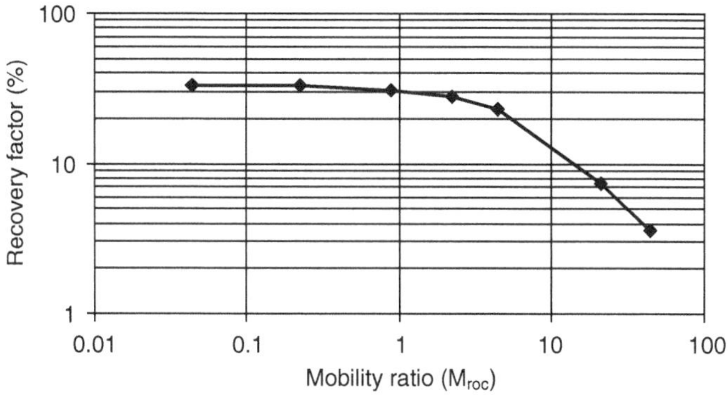

FIGURE 4.21 Recovery factors versus the mobility ratio defined in Eq. 4.14 for a random permeability model ($S_{wi} = 0.7$).

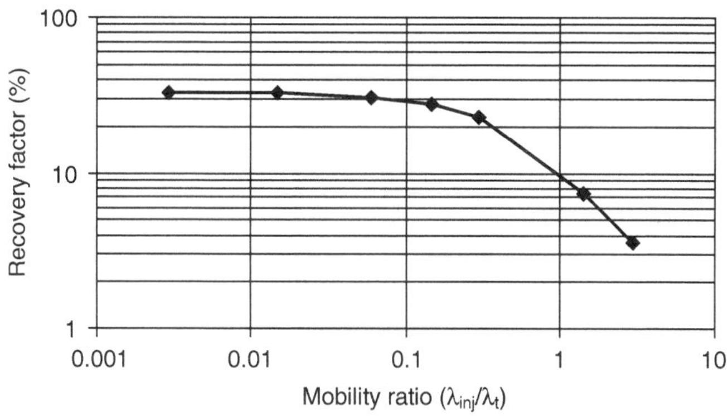

FIGURE 4.22 Recovery factors versus the mobility ratio defined in Eq. 4.5 for a random permeability model ($S_{wi} = 0.7$).

S_{wi} equal to 0.5, μ_u/μ_o is 0.71, which is outside their range. In this case, the initial water saturation is lower than theirs.

Wang et al. tried to define a viscosity ratio as a criterion for the mobility control requirement. However, we have to point out that the viscosity ratio μ_u/μ_o required for the mobility control should depend on relative permeabilities and fluid saturations.

4.8 FURTHER DISCUSSION

When we derived the mobility ratio (Eq. 4.14) for the mobility control requirement, we made several assumptions, as presented in Section 4.5. The subsequent numerical simulation results also show that in some cases the unit mobility ratio is not a perfect turning point in the plot of recovery factor versus mobility ratio M_{roc}. We have tried to improve the mobility ratio definition (Eq. 4.14) as a criterion for the mobility control requirement. For example, the initial oil cut at the outlet was used to replace the normalized oil saturation; also, a smaller relative permeability was used to replace the endpoint water relative permeability k_{wr} considering that the some movable oil is left behind the front. However, so far we have found that Equation 4.14 is the best formula.

The oil saturation in the downstream appears in Eq. 4.14. In an EOR process, such as surfactant flooding, an oil bank is built before the displacement front. Then the oil saturation in the oil bank may be used in Eq. 4.14. In the design of the mobility control requirement for an EOR process, the final criterion should be the economic parameters of the project such as net present value (NPV). Equation 4.14 can serve as a starting point for the economic evaluation. Based on the work presented in this chapter, we may conclude that the existing concept that the displacing fluid mobility should be equal to or less than the total mobility of the displaced multiphase fluids is invalid. Instead, the displacing fluid mobility should be equal to or less than the displaced oil mobility corrected by oil saturation. Such criterion should be used to design the concentration of the mobility control agent.

Chapter 5

Polymer Flooding

5.1 INTRODUCTION

As discussed in Chapter 4, the mobility control requirement is closely related to the ratio of displacing fluid mobility to displaced fluid mobility. Because changing displaced oil mobility (relative permeability and/or viscosity) often is not feasible without the injection of heat, most often we inject chemicals to change displacing fluid mobility. Primarily, the injected chemicals are polymers whose obvious function is to increase the displacing polymer solution viscosity, although other mechanisms are involved, as discussed in Chapter 6.

This chapter first introduces different types of polymers and polymer-related profile control systems used in enhanced oil recovery (EOR), although the list is in no way comprehensive. Then the chapter discusses several polymers developed in China, especially those used in field tests. Then it focuses on the polymer solution properties and polymer flow behavior in porous media. Numerous special subjects regarding polymer flooding (PF) are discussed, and field pilot tests and application cases are presented. Finally, the chapter summarizes the field experience and learning of polymer flooding.

5.2 TYPES OF POLYMERS AND POLYMER-RELATED SYSTEMS

The two main types of polymers are synthetic polymers such as hydrolyzed polyacrylamide (HPAM) and biopolymers such as xanthan gum. Less commonly used are natural polymers and their derivatives, such as guar gum, sodium carboxymethyl cellulose, and hydroxyl ethyl cellulose (HEC). Table 5.1 summarizes the characteristics of different polymer structures.

From Table 5.1, we learn that a good polymer should have the following properties:

- No –O– in the backbone (carbon chain) for thermal stability
- Negative ionic hydrophilic group to reduce adsorption on rock surfaces
- Good viscosifying powder
- Nonionic hydrophilic group for chemical stability

Based on these criteria, HPAM is a good polymer.

Modern Chemical Enhanced Oil Recovery. DOI: 10.1016/B978-1-85617-745-0.00005-X

TABLE 5.1 Polymer Structures and Their Characteristics

Structure	Characteristics	Sample Polymers
–O– in the backbone	Low thermal stability, thermal degradation at high T, only suitable at <80°C	Polyoxyethylene, sodium alginate, sodium carboxymethyl cellulose, HEC, xanthan gum
Carbon chain in the backbone	Good thermal stability, degradation not severe at <110°C	Polyvinyl, sodium polyacrylate, polyacrylamide, HPAM
$-COO^-$ in hydrophilic group	Good viscosifier, less adsorption on sandstones due to the repulsion between chain links, but precipitation with Ca^{2+} and Mg^{2+}, less chemical stability	Sodium alginate, sodium carboxymethyl cellulose, HPAM, xanthan gum
–OH or $-CONH_2$ in hydrophilic group	No precipitation with Ca^{2+} and Mg^{2+}, good chemical stability, but no repulsion between chain links, thus less viscosifying powder, high adsorption due to hydrogen bond formed on sandstone rocks	Polyvinyl, HEC, polyacrylamide, HPAM

Source: *Zhao (1991).*

5.2.1 Hydrolyzed Polyacrylamide

The most widely used polymer in EOR applications is HPAM (Manrique et al., 2007). For either a given polymer concentration or viscosity level, HPAM solutions have provided significantly greater oil recovery under Daqing conditions. The reason is that HPAM solutions exhibit significantly greater viscoelasticity than xanthan solutions (Wang et al., 2006a). Polyacrylamide adsorbs strongly on mineral surfaces. Thus, the polymer is partially hydrolyzed to reduce adsorption by reacting polyacrylamide with a base, such as sodium or potassium hydroxide or sodium carbonate. Hydrolysis converts some of the amide groups ($CONH_2$) to carboxyl groups (COO^-), as shown in the following structure:

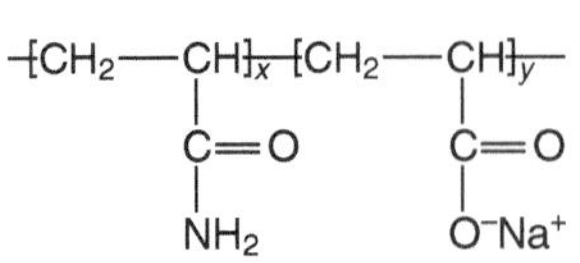

The degree of hydrolysis is the mole fraction of amide groups that are converted by hydrolysis. It ranges from 15 to 35% in commercial products. Hydrolysis of polyacrylamide introduces negative charges on the backbones of polymer chains that have a large effect on the rheological properties of the polymer solution. At low salinities, the negative charges on the polymer backbones repel each other and cause the polymer chains to stretch. When an electrolyte, such as NaCl, is added to a polymer solution, the repulsive forces are shielded by a double layer of electrolytes; thus, the stretch is reduced.

"Unhydrolyzed" polyacrylamide (PAM) also is used in some applications. Even unhydrolyzed PAM will have small percent (2–4%) of hydrolyzed groups unless exceptional precautions are taken in the manufacturing process. Polyacrylamide is mainly anionic, but could be nonionic or cationic (Green and Willhite, 1998). The reported molecular weights of HPAM used in EOR processes are up to higher than 20 million Daltons. A polymer with higher than 35 million Daltons was reportedly used in China's Daqing laboratory (Wang et al., 2006a).

HPAM is actually a copolymer of acrylamide and acrylic acid. In fresh water, because of the charge repulsion of the carboxylic group, the HPAM flexible chain structure is stretched so that the viscosity is high. In contrast, in saline water, the charge is neutralized or shielded. HPAM flexible chains are compressed, resulting in low viscosity. When the hydrolysis is higher (more carboxylic parts), COO- is increased so that the adsorption is reduced and the viscosity is increased, but chemical stability is reduced owing to less $CONH_2$. Low hydrolysis and more $CONH_2$ increase chemical stability, but adsorption is increased.

When hydrolysis is above 40%, the flexible chains are seriously compressed and distorted, and the viscosity is reduced. In a hard water (with high contents of Ca^{2+} and Mg^{2+}), when hydrolysis is above 40%, flocculation may occur. Because an EOR process is long, polymer stability is important. Generally, hydrolysis is required to be less than 40% after three months. However, hydrolysis of polyacrylamide is very fast under acidic and basic conditions. When the temperature is high, the hydrolysis is fast even under neutral conditions. In other words, HPAM is not tolerant to high temperature or high salinity (Wang et al., 2003a).

The monomer of acrylic acid must be kept between 15 and 20°C, and the monomer of acrylamide between 13 and 20°C. The polymer quality will be better if it is manufactured in dry places. The polymer can be supplied in emulsion or in powder; the emulsion is a water-in-oil type. The continuous phase is the oil with much lower viscosity than the polymer viscosity so that it can easily be transported. The emulsion must be transported from the manufacturer to the reservoir within six months to avoid significant degradation. The emulsion type of polymer is more expensive than the powder type both in manufacturing and in transportation. The price of one kilogram of emulsion is approximately equal to the price of one kilogram of powder, but the active material content in the emulsion (~50%) is approximately half of that in the powder (~90%) (Morel et al., 2008).

5.2.2 Xanthan Gum

Another widely used polymer, a biopolymer, is xanthan gum (corn sugar gum), or xanthan for short. The structure of a xanthan biopolymer is shown in the following figure. The polymer acts like a semirigid rod and is quite resistant to mechanical degradation. Average reported molecular weights of xanthan biopolymer used in EOR processes range from 1 million to 15 million. Xanthan biopolymers are supplied as a dry powder or as a concentrated broth (Green and Willhite, 1998). Generally, polyacrylamide copolymers are much more viscous than polysaccharide biopolymer at equivalent concentrations in fresh water, but these copolymers are much more sensitive to saline water than the biopolymers. The viscosity of copolymers is lower than that of biopolymers in the saline water (10,000 ppm TDS). Some permanent shear loss of viscosity could occur for polyacrylamide, but not for polysaccharide at the wellbore. However, the residual permeability reduction factor of polysaccharide polymers is low (Luo et al., 2006). In EOR processes, HPAM is much more widely used. Other potential EOR biopolymers are scleroglucan, simusan, AGBP, and so on (Luo et al., 2006).

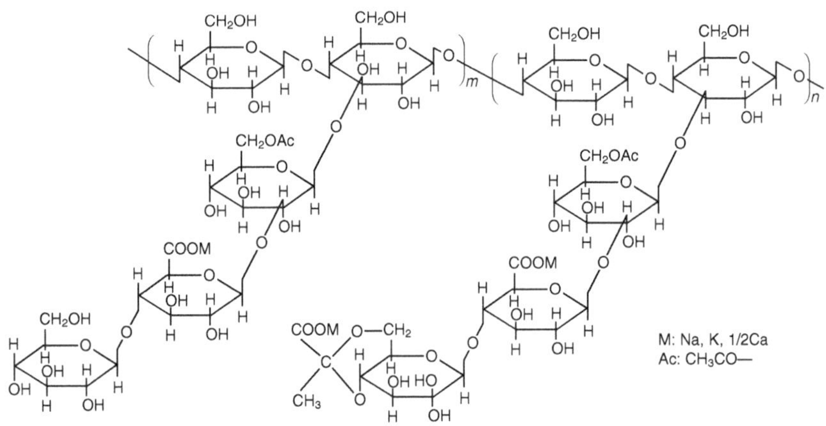

5.2.3 Salinity-Tolerant Polyacrylamide—KYPAM

KYPAM is the commercial name of a new Chinese product; its meaning in English is salinity-tolerant polyacrylamide, and its English translation is comb-shape polyacrylamide. There are several sample products of this type in the laboratory. RSP1 is used mainly in treating drilling fluids; RSP2 is used mainly in EOR; and RSP3 is used mainly in water shut-off or profile control. The commercial product RSP2, which is known as KYPAM in EOR, is produced by Beijing Hengju (Luo et al., 2002). This new copolymer incorporates a small fraction of functional monomers with acrylamide to form comb-like copolymers. The structure of a functional monomer, aromatic hydrocarbon with ethylene (AHPE), is

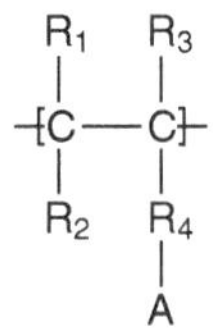

and the structure of KYPAM is

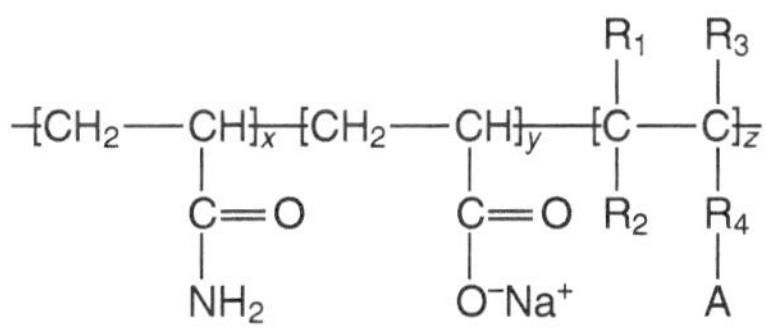

where R_1, R_2, and R_3 could be either H or C_1– C_{12} alkyl. A in the structure represents an ionic functional group that is tolerant to Ca^{2+} or Mg^{2+}. R_1, R_2, and R_3 mainly affect the elasticity of the polymer. As the carbon numbers increase, the elasticity increases. R_4 affects the polymer salinity tolerance. As the carbon number increases, the salinity tolerance is enhanced (Luo and Cheng, 1993). A microscope picture of KYPAM is shown in Figure 5.1.

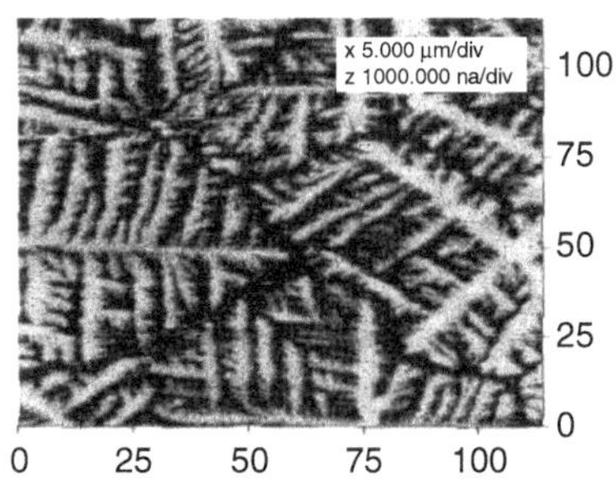

FIGURE 5.1 KYPAM microscope picture. *Source: Wang et al. (2006b).*

As discussed earlier, an HPAM flexible chain is compressed in saline water, resulting in low viscosity. When the hydrolysis is higher, the effect of ionic strength becomes stronger. In KYPAM, as the new functional monomer AHPE is introduced, the side chains of HPAM have both hydrophilic and hydrophobic groups. Because of the repulsion between the hydrophilic and hydrophobic groups, and the repulsion among the hydrophilic groups, the side chains are arranged in a comb shape. Thus, the flexible chains are stretched, and the KYPAM viscosity is relatively higher than the HPAM viscosity in more saline waters.

Table 5.2 compares KAPAM viscosity with the viscosities of the two HPAM polymers, HPAM 2B838 and MO-4000, in different saline waters (their salinities are shown in Table 5.3; Luo et al., 2002). MO-4000 is a Mitsubishi product. Some of the physiochemical properties of the polymers are shown in Table 5.4, according to Daqing Industry Specification Q/DQ0977-1996. We can see that KAPAM viscosities were 22 to 81% higher than those of the other two polymers at the same concentration of 1000 mg/L. In addition, the viscosities in Daqing waters showed that the higher the salinity, the higher the incremental percent of KYPAM viscosity over the others. Liu (2003) reported similar laboratory measurements and had similar results.

Laboratory measurements show that KYPAM is more temperature tolerant, and it has good shear and thermal stability. Core flood results were reported by

TABLE 5.2 Polymer Viscosities in Different Waters

Polymer Viscosity, mPa·s	HPAM 2B838	MO-4000	KYPAM	Higher than 2B838/ MO-4000 (%)
Daqing fresh water, 45°C	48.2	62.6	76.6	58/22
Daqing produced water, 45°C	26.2	27.8	47.5	81/70
Daqing ASP solution, 45°C	15.7	16.2	26.6	69/64
Daqing Gangdong produced water, 58°C	–	23.8	34.2	–/43
Daqing Guan produced water, 58°C	–	8.7	13.1	–/50
Gudong produced water, 70°C	–	14.5	24.4	–/68
Shengtuo produced water, 80°C	–	6.5	10.3	–/58

Source: *Luo et al. (2006).*

TABLE 5.3 Water Salinities Used in Measuring the Viscosities in Table 5.2

Water	Salinity, mg/L	$Ca^{2+} + Mg^{2+}$, mg/L
Daqing fresh water	1000	15
Daqing produced water	4000	60
Daqing ASP solution	Add 1.2% NaOH in Daqing fresh water	
Daqing Gangdong produced water	5700	105
Daqing Guan produced water	19334	514
Gudong produced water	5024	25
Shengtuo produced water	21636	476

TABLE 5.4 Physiochemical Properties of the Polymers

Parameter	KYPAM	HPAM 2B838	MO-4000
Appearance	White powder	White powder	White powder
Solid content (%)	90.0	90.2	92.4
Molecular weight, million	25.14	17	20.88
Hydrolysis (molar %)	26.4	26.8	31
Dissolution time (hours)	≤2	≤2	≤2
Insoluble (mass %)	0.115	0.19	0.14
Residual monomers (%)	0.0096	0.021	0.043
Filtration index	1.12	1.22	1.13
Screen coefficient	102.6	41.3	52.5

Source: *Luo et al. (2006).*

Luo et al (2002). The tested core permeability was 0.7 to 1.8 μm^2, and the porosity was 0.2. In addition, the displaced oil was 9.5 mPa·s at 70°C. Table 5.5 compares the performance of KYPAM with HPAM 1285 at a concentration of 1000 mg/L. A 0.4 PV injection volume was used. We can see that the flow behavior of KYPAM was better than HPAM 1285. KYPAM has been widely used in polymer flooding, ASP, and profile control projects in Daqing, Shengli, Huabei, and Xingjiang fields. Several field test cases are presented next.

TABLE 5.5 Comparison of Polymer Performance in Core Flood

Polymer	Water Used*	RF by Polymer (%)	k Reduction Factor F_{kr}	Residual k Reduction Factor F_{krr}
KYPAM	Produced	15.12	2.0	1.57
1285	Produced	11.96	1.5	1.65
1285	Fresh	13.16	1.75	1.42

*Water salinity not reported.
Source: *Luo et al. (2002).*

Lamadian Field, Daqing

A field test using KYPAM was conducted in the Northwestern block of Lamadian field in Daqing (Wang et al., 2003b). The test area was 3.45 km^2. There were 39 injectors and 44 producers with 25 wells in the middle area. A KYPAM solution was prepared using the produced water, and the injection was started in May 2001. Figure 5.2 compares the average water cut from the middle 25 wells in December 2002 with that in a neighbor block where fresh water was used for the HPAM solution. The average water cut was lower by 15.5%, but the average oil recovery factor was higher by 2%.

Xinger Block, Daqing

In the middle of the Xinger Block of 2.03 km^2, there were 45 wells (17 injectors, 27 producers, and 1 observation well), (Luo et al., 2006). Injection of ASP

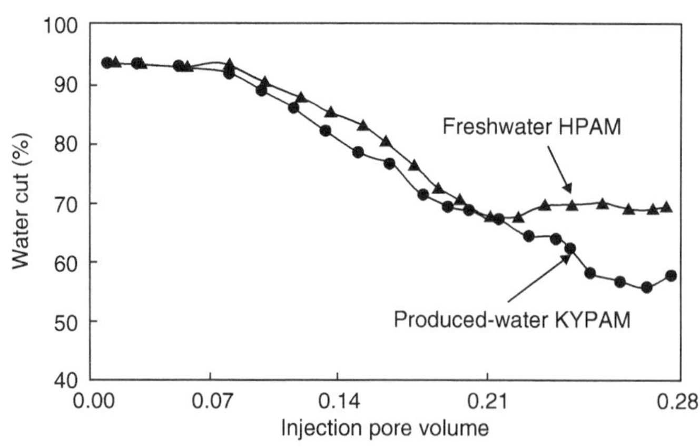

FIGURE 5.2 Comparison of average water cut under KAPAM injection with that under HPAM. *Source: Wang et al. (2003b).*

with KYPAM as a polymer was started in May 2001. By November 2001, 0.049 PV was injected, and the average water cut was reduced to 80.8% from 96% in September 2000. The water cut was reduced by 15.2%.

Shengtuo Field, Shengli

In the Shengtuo commercial polymer flooding test, KAPAM and MO-4000 were injected in two separate areas (Li, 2004a). The test areas are described in Table 5.6, and the water composition is shown in Table 5.7. The formation temperature was 80°C. The injection start dates were as follows:

- March 2002: MO-4000 injection using the river water.
- July 2002: KYPAM injection using the produced water.

The MO-4000 concentration was 1848 mg/L, and the KYPAM concentration was 1836 mg/L. Table 5.8 compares the performance of KYPAM and MO-4000 by August 2003. The water cut at the beginning of the KYPAM

TABLE 5.6 Description of Test Areas

	MO-4000	KYPAM
Area, km^2	2.39	1.63
Pore volume, m^3	1.219×10^7	5.97×10^6
OOIP, ton	7.513×10^6	3.6×10^6
Average net thickness, m	17	12
Injectors	22	10
Well spacing, m	220	300
Starting water cut, %	94.3	97.2
Water used	River	Produced

TABLE 5.7 Water Composition

Water	TDS, mg/L	$K^+ + Na^+$	Ca^{2+}	Mg^{2+}	Type
Formation	21000	7751	230	81	$CaCl_2$
Produced	12401	4600	106	35	$NaHCO_3$
River*	680	107	59	30	

**Including 118 mg/L Cl^- and 196 mg/L $HCO3^-$*

TABLE 5.8 Comparison of Polymer Performance (by August 2003)

Polymer	Injected Polymer (t)	Inj.	Prod.	Beneficial Prod. (%)	f_w Drop (%)	q_o Up (t/d/well)	RF Up (%)
MO-4000	3002.5	22	45	37.8	3.3	1.73	0.38
KYPAM	1175.1	10	16	43.9	2.4	2.12	0.41

injection was 97.2%, which was higher than that (94.3%) at the beginning of the MO-4000 injection. Plus, the well spacing in the KYPAM test area was larger than that in the MO-4000 test area. Note that well spacing in China is defined as the distance between an injection well and the adjacent production well. Table 5.8 shows that the oil rate increased and the incremental recovery factors for KYPAM were higher than those for MO-4000YPAM. KYPAM outperformed MO-4000.

5.2.4 Hydrophobically Associating Polymer

The polymer is hydrophobically associating water soluble, meaning it contains one or more water-soluble monomers (acrylamides) and a small fraction (0.5 to 4%) of water-insoluble (hydrophobic) monomers. A typical hydrophobically associating polymer (HAP) structure is

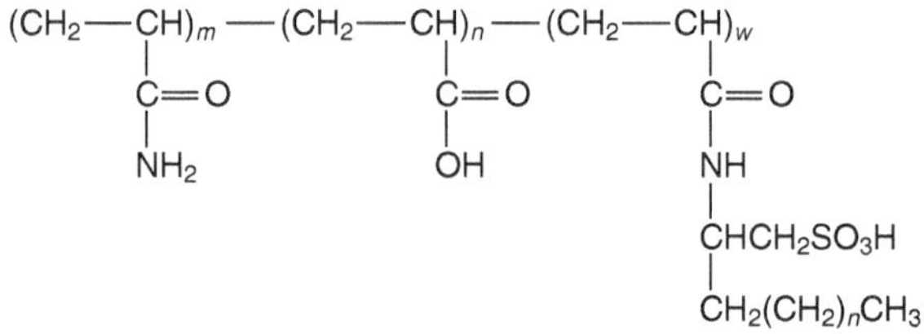

Figure 5.3 shows a picture of HAP taken with a microscope. Because of the hydrophobicity, there are two kinds of associations: one is intermolecular, and

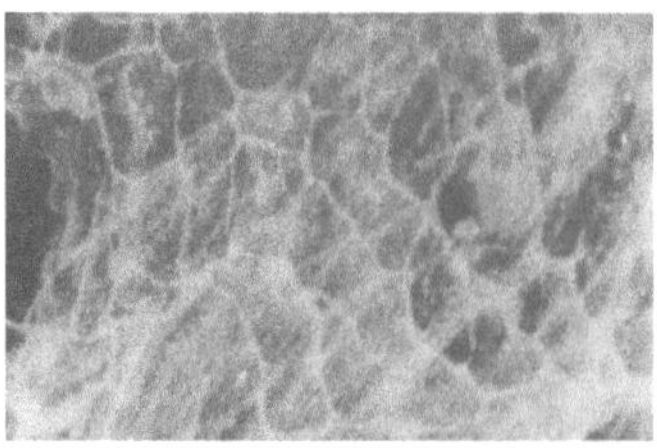

FIGURE 5.3 Microscopic picture of HAP. *Source: Wang et al. (2006b).*

the other is intramolecular. The intermolecular association increases the solution viscosity, whereas the intramolecular association decreases the viscosity. The viscosity change depends on the factors that affect the two associations. Several products have been developed: Ts-45, Ts-65, AP-P1, AP-P3, and AP-P4. The evaluation of this type of polymer is presented next.

Effect of Hydrophobic Molar Fraction on Polymer Viscosity

Figure 5.4 shows the effect of hydrophobic group octylacrylate (OA) on apparent viscosity of the copolymer of PAM and OA. We can see that Curve 3, PAMOA75 with 0.75 mol% OA, had the highest apparent viscosity; then Curve 2, PAMOA50 with 0.5 mol% OA; followed by Curve 4, PAMOA100 with 1 mol% OA. Curve 1, PAMOA25 with 0.25 mol% OA, almost overlaps Curve 0, PAMOA0 with 0 mol% OA. The apparent viscosities of PAMOA75 and PAMOA50 increased abruptly with the concentration higher than 0.15 g/dL. That means as the concentration is increased, the intermolecular association increases. However, if the OA molar fraction is increased further, the intramolecular association causes compression of the molecule; thus, the viscosity is decreased. Jiang et al. (2003a) showed that in dilute solutions, the solution with the higher hydrophobic fraction had a lower viscosity because the intramolecular association was dominated in a dilute solution. The intramolecular association compresses the molecular chain and thus reduces viscosity.

Salinity Effect on Viscosity

Figure 5.5 compares the viscosity versus polymer concentration of AP-P3 and MO-4000 at different salinities, at 80°C and 7.34 s^{-1}. The different salinities in ppm are shown in the legends as 10,000, 30,000, and 100,000. This figure

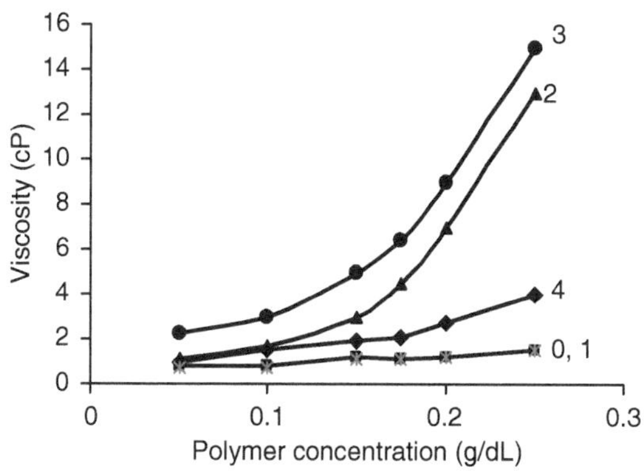

FIGURE 5.4 Effect of OA fraction on the copolymer AM/OA viscosity at different polymer concentrations (2000 mg/L NaCl, 25°C, 76.8 s^{-1}). OA mol% in the figure: 0, 0 mol% (PAM OA); 1, 0.25 mol% (PAM OA 25); 2, 0.5 mol% (PAM OA 50); 3, 0.75 mol% (PAM OA 75); and 4, 1 mol% (PAM OA 100). *Source: Zhou and Huang (1997).*

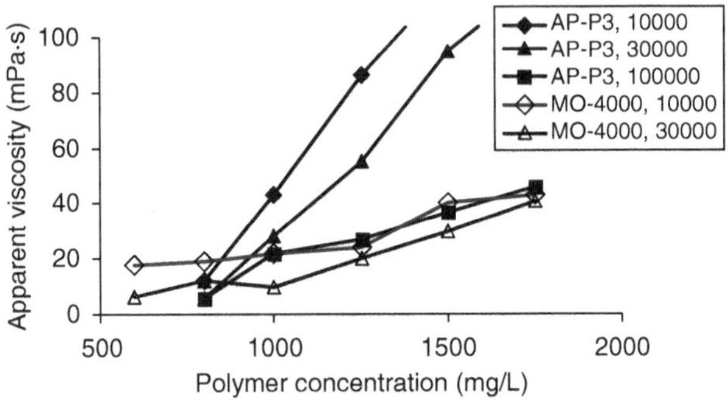

FIGURE 5.5 Viscosity versus concentration for AP-P3 and MO-4000. *Source: Data from Luo et al. (2006).*

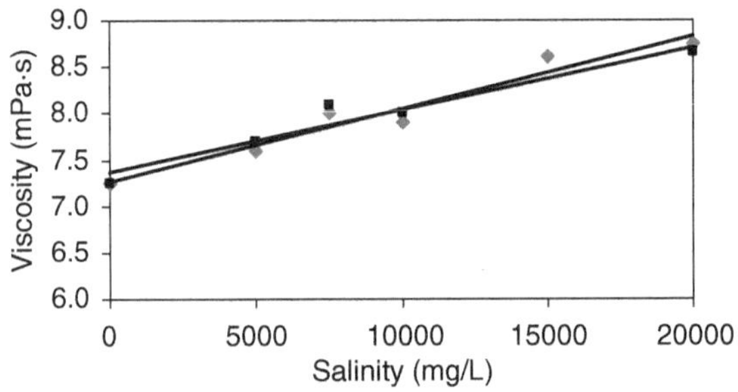

FIGURE 5.6 Salinity effect on the viscosity of the copolymer PAMOA50 at 25°C and 76.8 s^{-1}. *Source: Zhou and Huang (1997).*

shows that the viscosity of AP-P3 was higher than that of MO-4000 at the same high polymer concentrations, especially at high salinities. Wang et al. (1999c) made the same observation when they compared the viscosity of a water-soluble associating polymer with that of HPAM 333S (10 million molecular weight). Their polymer had 10.9 million molecular weight, 25 mol% anionic, and 0.25 mol% cationic. Figure 5.5 shows that the associating polymer viscosity decreased as the salinity was increased.

In contrast, Zhou and Huang (1997) presented the data shown in Figure 5.6, which shows that the copolymer viscosity was higher at a higher salinity. This phenomenon is called antipolyelectrolyte effect. In this case, the copolymer (PAMOA50) had 0.5 mol% octylacrylate (OA), and the polymer concentration was 2000 mg/L. In Figure 5.6, the square points represent $CaCl_2$ data, and diamond points represent NaCl data. This phenomenon was also reported by Bock et al. (1988) and McCormick (1988). The increase in salinity enhances the intermolecular association so that the viscosity increases. At a very high salinity solution, however, the intramolecular association also increases. Then the viscosity decreases.

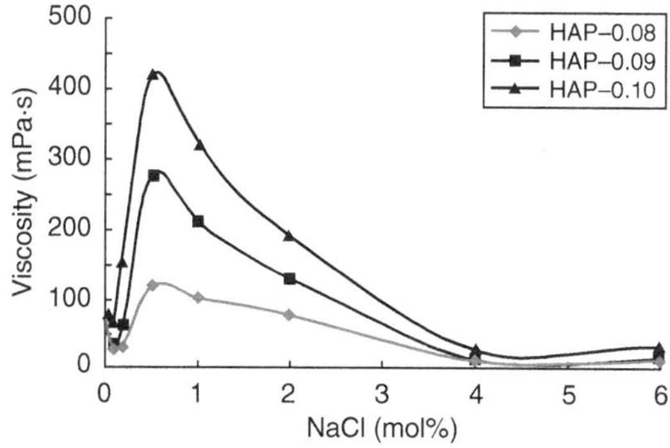

FIGURE 5.7 HAP viscosity changes with NaCl. *Source: Jiang et al. (2003a).*

Figure 5.7 is even more interesting. It shows the HAP viscosity change with salinity (NaCl wt.%). The HAP concentration was 1000 mg/L. In the figure, HAP-0.08 represents the HAP with 0.08 mol% hydrophobic composition, and similarly for HAP-0.09 with 0.09 mol% and HAP-0.10 with 0.10%. From the figure, we can see that the HAP apparent viscosity decreased, increased, decreased again, and finally stabilized as the salinity increased. Addition of salt has two effects. One is to shield ionic repulsion, which compresses the polymer molecular chains and lowers the viscosity. The other effect is to increase the solvent polarity, which enhances the hydrophobicity and prevents the chain from being compressed. Then the viscosity increases. These two effects together determine the increase or decrease of the viscosity.

Effects of Shear Rate and Polymer Concentration

Table 5.9 compares the effects of shear rate and polymer concentration on the viscosities of an associating polymer (AP) and a HAPM polymer (Luo et al., 2006). Several interesting observations can be made. Because of conformation of flexible coils, ordinary HAPM viscosity cannot be high in saline water even if its molecular weight is high. For hydrophobically associating water-soluble polymer, because of the hydrophobic groups in the HAPM chains, the strong association of molecules because of electrostatic force, hydrogen bond, and van der Waals force can build a large three-dimensional network structure. Such a network structure results in high structure viscosity, which is a result of the association.

In diluted AP polymer solution, the polymer exists in the form of single molecules, and the association is weak. Thus, the AP viscosity is not much different from HPAM viscosity. When AP concentration reaches a certain level, the molecules have more chance of contacting each other, thus the association becomes strong, and viscosity sharply increases. This concentration is called critical associating concentration. Table 5.9 shows that when the polymer concentration was below 400 mg/L, AP viscosities were lower than HPAM. Above

TABLE 5.9 Effects of Shear Rate and Concentration on Polymer Viscosities (mPa·s) at TDS 4000 mg/L, 45°C, 7.34 S^{-1}

		Polymer Concentration, mg/L				
Polymer	Rotation, r/min	*200*	*400*	*600*	*800*	*1000*
AP	0	3.9	7.6	30.0	68.2	109
	3000	3.8	6.6	18.1	53.1	100
		97%	87%	60%	78%	92%
	4000	3.2	5.7	14.9	46.3	85.6
		82%	75%	50%	68%	79%
HPAM	0	5.3	8.7	13.1	19.1	26.3
	3000	3.0	6.2	10.5	13.1	18.9
		57%	71%	80%	74%	72%
	4000	2.6	4.1	4.4	6.2	12.8
		49%	47%	47%	32%	49%

Source: *Luo et al. (2006).*

400 mg/L, AP viscosities were higher than HPAM. In other words, the critical associating concentration was 400 mg/L. The fact that below the critical associating concentration, its viscosity was lower than that of a HAPM could be a problem in a tertiary recovery process, because a high water saturation in the formation could dilute the injected polymer concentration below the critical associating concentration.

The viscosities at the two high rotations (3000 and 4000 r/min) are shown in Table 5.9. The viscosities at the high rotations are also represented by the percentages of their respective viscosities at the zero rotation. At the same rotation speed, the percent for AP was higher than that for HPAM at the same concentration. That means AP is more shear stable than HPAM.

Luo et al. (2001) reported that a hydrophobically associating polymer had a higher viscosity than an HPAM in an ASP solution (1.2% NaOH and 0.3% ORS41). They also studied shear stability. In an ASP solution (1.2% NaOH, 0.3% ORS41, and 769 mg/L associating polymer), at 1000 r/min using an OWC4060 mixer, the viscosity at 45°C was reduced from 40 to 16.5 mPa·s (41.2%). Fifteen minutes later, the viscosity was back to 35.8 mPa·s (89.5%). Twenty-four hours later, the viscosity was 36.1 mPa·s (90.25%). That means at a high shear rate, the associating polymer network structure was damaged and the viscosity was significantly reduced. When the shear rate was reduced, the network structure was restored, and the viscosity was back up again.

This shear reversibility is very beneficial in enhanced oil recovery field applications. It improves the well injectivity because of the shear thinning effect at the perforation and near the wellbore. Far away from the wellbore, flow velocity is reduced and viscosity is restored. For the hydrophobically associating polymers, both shear thickening and shear thinning were observed by Bock et al. (1988). It has also been reported that the viscosity after shearing was stopped was higher than that before shearing (McCormick et al., 1988).

Thermal Instability

Figure 5.8 shows that the apparent viscosity changed with temperature for PAMOA75 with 0.75 mol% octylacrylate (OA) and a polymer concentration of 2800 mg/L. The shear rate was 19.8 s^{-1}. Below 35°C, as the temperature increased, the viscosity increased slightly. Between 35°C and 45°C, the viscosity was almost unchanged. Above 50°C, the viscosity abruptly decreased. At 70°C, the viscosity was 15.8% of that at 20°C. Because of the dominated fraction of polyacrylamide, the copolymer cannot be tolerant to high temperature.

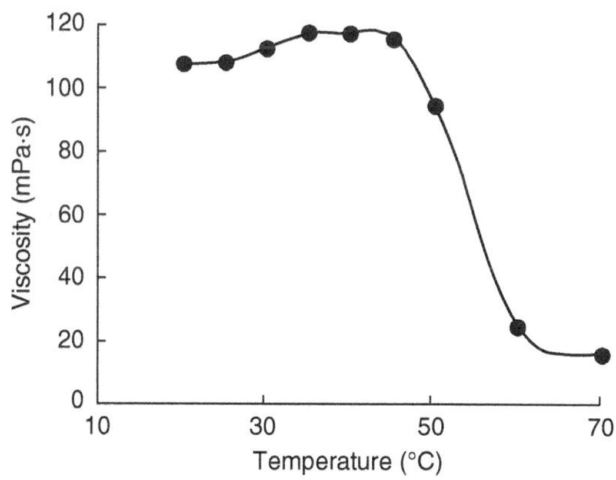

FIGURE 5.8 PAMOA75 viscosity versus temperature. *Source: Zhou and Huang (1997).*

Field Test Results

Core flood results show that the oil recovery factor using hydrophobically associating polymers was a few percent higher than that using the HPAM type of polymers (Wang et al., 1999c; Ou-Yang et al., 2004; Luo et al., 2006). The resistance factor and residual permeability reduction factor were also reported to be higher ((Luo et al., 2006; Ou-Yang et al., 2004). Apparently, hydrophobically associating polymer has been less tested in fields. The results of two field tests are presented next.

Crosslinkers were added in a hydrophobically associating polymer, AP-P4, for profile control in the Wengmingzhai and the Mazhai fields in Zhongyuan (Chen et al., 2004; Li et al., 2006a, 2006b). AP-P4 has 9 million MW (molecular water), 0.2% hydrophobic component, 20% hydrolysis, and 90% solid content. This polymer was made by the Sichuan Guang-Ya Science and Technology Company (Li et al., 2006a). The crosslinker components were MZ-YL, MZ-BE, and MZ-XS, and the acidity was adjusted by CS-2. The gel solutions were prepared using produced water from the Mazhai field (TDS 116,000 mg/L). The chemical formulation was 2500 to 3500 mg/L AP-P4 polymer, 400 to 500 mg/L MZ-YL, 600 to 700 MZ-BE, 1200 mg/L MZ-XS, and 12,000 mg/L CS-2. The prepared gel solutions had a gelation time of 10 to 12 hours and a viscosity of 35,000 to 61,000 mPa·s at 90°C.

In this test, the viscosity remained higher than 40,000 mPa·s after being aged at 90°C for 100 hours. That means the gel solutions were thermally stable by 90°C. Being sheared at 3000 r/min for 15 minutes, the gel solutions lost 87 to 89% of their viscosity. After shearing was stopped, the gel viscosities were restored to 70 to 85% of the unsheared viscosity. Using the reservoir cores of 973 md, the flood tests showed that the plugging rate of 88 to 96% and the residual resistance coefficient of 16.2 to 28.6 were obtained after 10 PV of gel injection. In a three-layered artificial core of 1000 md permeability and 0.72 permeability variation coefficient, the incremental recovery factor of gel treatment was 0.4 to 0.93% OOIP.

Gel treatments were made in the Wenmingzhai (80°C) and Mazhai (90°C) fields—24 gel treatments (well × number of treatment). The average injection volume was 942 m^3 per treatment, and the treatment radius was 8.9 m. The average effective time was 217 days, and the wells' average water intake was increased by 6.8 m in the injected 4 layers. The average gel viscosities before injection pump, after injection pump, and at the injection wellhead were 12,283, 11,070, and 9,900 mPa·s, respectively. The gelation time was 32 hours, and the gel viscosity was higher than 100,000 mPa·s on average.

For a typical well, Well 95-18, a gel solution was injected from July 26 to August 6, 2003, with the total injection volume of 1800 m^3. The well treatment radius was 10 m. After the treatment, the water intake thickness increased from 9.6 to 20.4 m, and the water intake layers increased from 6 layers to 8 layers. The effective time was 219 days. The cumulative incremental oil was 986 tons (Chen et al., 2004). A similar laboratory evaluation report of the polymer was done by Li et al. (2006b), and a more detailed report of the field trials was made by Li et al. (2006a).

5.2.5 2-Acrylamide-2-Methyl Propane-Sulfonate Copolymer

The structure of the AM and Na-AMPS copolymer is

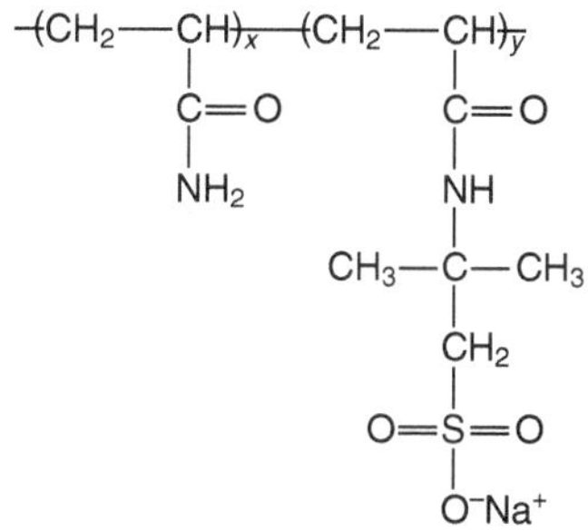

AMPS, or 2-Acrylamide-2-Methyl Propane-Sulfonate, has water-soluble anionic sulfonate, shielding acrylamide, and unsaturated double bond. Sulfonate makes it have good ionic exchange capability, electric conductivity, and good resistance to divalence and salinity in general. Acrylamide gives it good thermal stability and good resistance to hydrolysis, acid, and alkaline. Plus, the double bond leads to easy synthesis and polymerization. The rigid side chains, large chains, or chains of ring structure also give it good thermal stability. AMPS are combined with other monomers to produce copolymers that are used in many industries (Lu and Chen, 1996). In the oil industry, the main application is in drilling (Hou et al., 2003).

After aging at 90°C for 16 days, AMPS/AM copolymer viscosity was almost unchanged, and the hydrolysis was up to only 30%. Aging at 110°C, the fraction of AMPS was almost unchanged. Aging at 120°C for 10 days, the viscosity was reduced 50%, and the hydrolysis became 40%. AMPS/AM had better thermal stability than HPAM. For an Na-AMPS copolymer with a concentration of 3000 mg/L in a solution of 160,000 mg/L (3000 mg/L of Ca^{2+} and Mg^{2+}), the viscosity at 90°C was 14.5 mPa·s, which is 40.73% of its viscosity (35.6 mPa·s) at 25°C. At the same conditions, the HPAM viscosity (with molecular weight of 11.2 million) at 90°C was 18.8% of its viscosity at 25°C. Its decomposition and vitrification temperatures were 250°C and 195°C compared with 210°C and 188°C for HPAM, respectively.

The AMPS/AM copolymer can be used under 90°C (Luo et al., 2006). It also had better salinity tolerance. Zhao et al. (2006) reported that the AM/AMPS solution viscosity was about twice the HPAM viscosity at the same conditions, when the salinity was increased. However, AMPS products are more expensive than HPAM. Zhao et al. (2005b) reported that AMPS copolymer solubility was lower than HPAM (10 hours of dissolution time in fresh water compared with 2 to 4 hours for HPAM).

Figure 5.9 shows the rheological behavior of a 1.5 g/L AMPS solution at different salinities. The AMPS was developed by Zhongyuan Drilling Institute. It can be seen that at any salinity, when the shear rate was very low, the

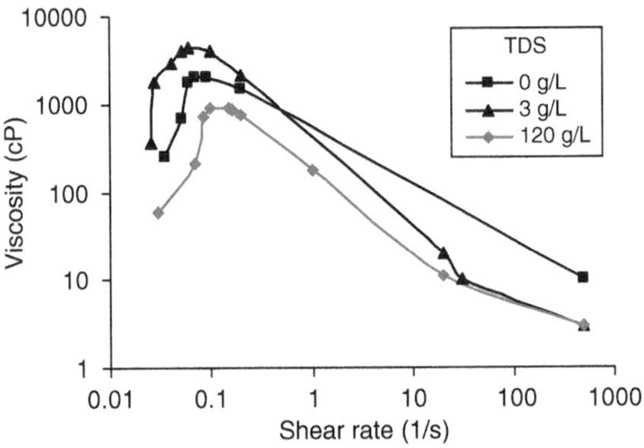

FIGURE 5.9 AMPS solution (1.5 g/L) rheological behavior at different salinities. *Source: Zhao et al. (2005b).*

solution viscosity increased with shear rate. When the shear rate became higher, the viscosity decreased. Comparing the viscosities at different salinities, we can see that when the shear rate was lower than 0.15 s^{-1}, the viscosity at 3 g/L TDS was the highest, even higher than that at 0 g/L TDS. That means the addition of salt can enhance molecular association at some range of polymer concentrations. However, at higher shear rates, such association deteriorated fast, and the viscosity sharply decreased below that at 0 g/L TDS. At very high shear rates (>10 s^{-1}), the shear effect became weaker, and the viscosity decreased at a lower rate. In the figure, discontinuity points are visible at the shear rates between 10 s^{-1} and 100 s^{-1}. Other copolymers developed for EOR include AM/DMDAAC/Na-AMPS (Fu, 1997), AM/DMAM/AMPS (Liang et al., 1997), and AM/C_{12}AM/AMPS (PL-1) (Luo and Lin, 2003).

Field Test: Mazhai Field, Zhongyuan

Cui (2004) reported on a case about the application of crosslinked AMPS. The formulation proposed by a laboratory study was 1000 to 2000 mg/L AMPS, 150 to 250 mg/L X-2 crosslinker, and 100 to 150 mg/L additive. The formulation could be used in 65 to 90°C, pH 5.5 to 10, and salinity 8000 to 160,000 mg/L. The actual formula used in the field test at the Mazhai field, Zhongyuan, was not reported, but some of the system data were: AMPS viscosity, 15 mPa·s at 1500 mg/L, and 26 mPa·s at 2,000 mg/L gelation time ≥ 5 days; and gel viscosity 60 to 2000 mPa·s. The viscosity was restored to 50% after shearing was stopped.

The test well pattern was Wei 95-159 in the Mazhai field. In this case, the test area was 0.55 km^2, 1 injector, and 5 producers. The OOIP was 6.07×10^5 tons, and the remaining recoverable reserve was 9.75×10^4 tons. The reservoir temperature was 80°C. The formation salinity was 160,000 mg/L, and the produced water salinity was 100,000 mg/L. Before testing, the water cut was 91.7%, and the sweep efficiency was 0.61.

The injected chemicals were 15,300 m^3 AMPS solution, 3.74 tons crosslinkers, and 12.68 tons additive. The pump pressure at the start and end were 14 MPa and 18 MPa, respectively; the maximum pump pressure was 19 MPa. After the profile control, the water injection pressure increased from 14.6 MPa to 19.6 MPa. Water entered 6 more layers of 7.9 m, equivalent to 38.3% of the total intake profile. The intake was reduced by 45.1% in high-permeability layers, whereas the intake was increased by 13.8% in low-permeability layers. The benefit was observed in all 5 producers. In addition, the water cut was reduced from 91.7 to 88.2%, and the daily rate increased from 10.3 to 19.8 tons.

5.2.6 Movable Gels

Research on injection of movable gels to improve conformance was carried out by Fletcher et al. (1992), Whittington and Naae (1992), Mack and Smith (1994), and others. Use of movable soft microgels was proposed as an effective method for deep profile control and/or relative permeability modification for water production control (Chauveteau et al., 2000, 2001, 2003, 2004; Feng et al., 2003; Rousseau et al., 2005; Zaitoun et al., 2007). Occidental Oil Company (Pyziak and Smith, 2007) and Kinder–Morgan also used similar products to control CO_2 breakthrough for their CO_2 flooding areas, and promising results were reported.

Crosslinked polymer-like bulk gel used in water shut-off has very poor flowability; the viscosity is very high (>10,000 mPa·s). Uncrosslinked polymer is used to increase water viscosity. A movable gel is used in between; it has the intermediate viscosity, and more importantly, it can flow under some pressure gradient. Colloidal dispersion gel (CDG) is a typical gel used in these situations. The mechanisms of a movable gel are (1) it has high viscosity to improve mobility ratio like an uncrosslinked polymer solution; (2) it has a high resistance factor and high residual permeability reduction factor; and (3) it has viscoelasticity so that the remaining oil in the rocks can be further reduced.

The mechanism of preferentially blocking high-permeability channels is controversial (Seright, 2006; Chang et al., 2006). The research work done so far has been focused more on formulation selection, improvement of properties, and determination of factors that affect gel performance, but less on flow behavior (Ma et al., 2005). The viscoelasticity of a low concentration HPAM/AlCit crosslinked system was mathematically described by Sun et al. (2005).

Colloidal dispersion gel (CDG) is made of low concentrations of polymer and crosslinkers. Crosslinkers are the metals, such as aluminum citrate and chromium, referred to as aggregates. Polymer concentrations range from 100 to 1200 mg/L, normally 400 to 800 mg/L. The ratio of polymer to crosslinkers is 30 to 60. Sometimes, this type of gel is called a low-concentration crosslinked polymer. In such concentration range, there is not enough polymer to form a continuous network, so a conventional bulk-type gel cannot form. Instead, a solution of separate gel bundles forms, in which a mixture of predominantly intramolecular and minimal intermolecular crosslinks connect relatively small numbers of molecules. By contrast, in a bulk gel, the crosslinks form a continuous network of polymer molecules, through predominantly intermolecular crosslinks. Colloidal dispersion gels get their name from the nature of the gel solutions, which are suspensions of individual bundles of crosslinked polymer molecules, or colloids. Because of the relatively low polymer and crosslinker concentrations, the gel formation rate is slow—on the order of days and weeks for gel solutions stored in the laboratory.

One important parameter of CDG is transition pressure. Below the transition pressure, the gel cannot flow through screen packs. Above the transition pressure, the gel flows like uncrosslinked polymer. That is why it sometimes is called movable gel. Other names, such as weak gel, microgel, weak viscoelastic fluid, crosslinked polymer, linked polymer solution, and deep diverting agent, are used as well. The transition pressure was 0.072 to 0.1 MPa when 7.5 to 60 mg/L Al^{3+} was used with 600 mg/L 3530S polymer (Niu et al., 2006). Other important parameters include gelation time and stability. The gelation time could be from hours to weeks; it must be long enough so that near an injection well, the gel behaves like an uncrosslinked polymer solution in that it can flow into deep formation. Partly, gel flowability could be achieved by the high-pressure gradient near the wellbore, which is higher than the transition pressure. The flow behavior of gels is generally evaluated by resistance factor and residual permeability reduction factor. The capability of conformance control is evaluated using heterogeneous layered models.

A number of field tests have been conducted in the Rocky Mountains region in the United States (Mack and Smith, 1994) and in China (Nie et al., 2005; Kong et al., 2005; Li et al., 2005a; Wu et al., 2005; Chang et al., 2006; Niu et al., 2006). These field applications demonstrated good or even better performance than uncrosslinked polymer injection, although the mechanisms are controversial. Some special cases are presented next.

CDG Tests in Daqing Fields

In a Daqing field test, three chemical slugs (0.18 PV CDG, 0.15 PV polymer, and 0.2 PV CDG) were injected from May 1999 through May 2003. The polymer concentration and polymer/aluminum ratio in the CDG slugs were 600 mg/L and 30 : 1, respectively. The polymer concentration in the polymer slug was also 600 mg/L. Compared with a typical polymer flooding, the CDG performed better than PF, with higher incremental oil recovery factor of approximately 14% over WF; and the CDG process used less polymer (Chang et al., 2006). Here low-concentration weak gels were injected before and after polymer slugs. Niu et al. (2006) observed that the performance of gel injection after polymer injection was not as good as that before polymer injection.

Weak Gel Application in a Low-Permeability Reservoir

Generally, weak gel is applied in high-permeability or fractured reservoirs. This target application was used in the Hong-Shan-Zui field of Karamay Oilfield, which had a permeability <50 md. The water cut was higher than 70%. The temperature was 42°C, and oil viscosity was 4.7 mPa·s. The injection water TDS was 14408 mg/L with 500 mg/L divalents. The well spacing was 250 to 350 m between the injection well and adjacent production wells. In this application, the formula selected was 600 to 1000 mg/L HPAM and 40 mg/L CXJ-II crosslinker. The weak gel viscosity in situ was 4.08 to 79.7 mPa·s. Nie et al. (2005) reported good performance.

5.2.7 pH-Sensitive Polymers

To control water production, or in general to improve reservoir sweep efficiency, polymer gels have been extensively used as conformance control agents. The most common gel placement method is the injection of polymer and crosslinking agent to the desired zone, letting them react to form a gel in situ. Movable gels belong to in situ or partially in situ formed gels. Because it is difficult to control the transport and reaction of chemicals in heterogeneous reservoirs, the success of field applications of in situ gelation has been mixed (Seright and Liang, 1994). Surface-preformed gels do not have the prior reaction control problem, but they are difficult to place deep in the reservoir because they cause a very high-pressure drop near the injection well and also tend to show mechanical trapping and filtration. Many of the surface-preformed gel applications have, therefore, been near wellbore treatments for vertical conformance control.

pH-sensitive crosslinked polymer (Al-Anazi and Sharma, 2002; Huh et al., 2005) has been proposed for a mobility control agent. Water-soluble ionic polymers generally form molecular networks by crosslinking. Crosslinked poly(acrylic acid) polymers, the Carbopol series from Noveon Inc., were used to prepare microgel dispersion in brine. The polymers, which are available in powder form, are crosslinked with polyalkenyl ethers or divinyl glycol and have carboxyl functional groups that control the rheological properties of the dispersion (Choi et al., 2006). Their key characteristic is the sensitive dependence of their molecular configuration on the ionic conditions in solution. Depending on pH, salinity, and temperature, they can swell, retaining a high volume of water, up to ~1000 times its own volume, or they may not swell at all in solution. Significant swelling and deswelling also bring about drastic changes in solution rheology, such as viscosity.

Choi et al. proposed to use pH change as a trigger for a poly(acrylic acid) solution because it exhibits a low viscosity at low pH, but at high pH above a critical value, the viscosity increases drastically and maintains more or less a plateau value with further increases in pH, as shown in Figure 5.10. Through injection of a polyelectrolyte at low pH, a low viscosity and good injectivity can be achieved. Once deep in the reservoir, the pH of water can increase because of reaction of injected acid with the carbonate and other minerals in the reservoir rock (and to a certain extent, because of mixing between injected

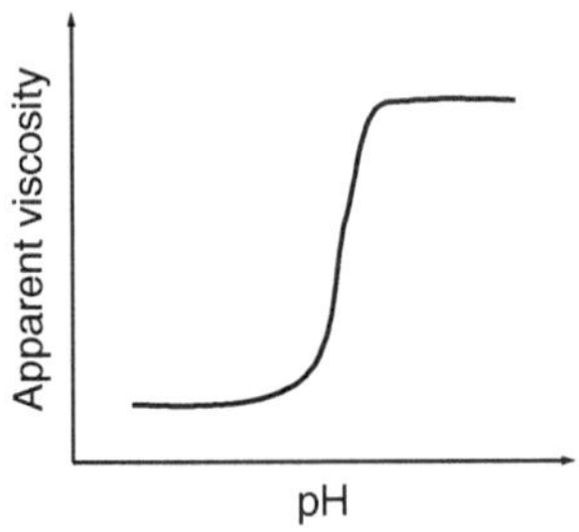

FIGURE 5.10 Schematic of pH-dependent polymer viscosity.

water and the bypassed resident water). For the other critical problem of chemical loss, an earlier preliminary study (Al-Anazi and Sharma, 2002) indicated that adsorption of poly(acrylic acid) on rock was small compared with the conventional polymers used in EOR. As with other anionic polymers, sensitivity to divalent ions remains an issue.

5.2.8 BrightWater

Other in situ formed gels that have been proposed include swelling mm-sized polymer grains (e.g., Diamond Seal) and swelling micron-sized polymers (e.g., BrightWater—see Pritchett et al., 2003; Frampton et al., 2004). The in situ formed microgels swell when exposed to a high temperature, instead of the exposure to a pH change as discussed previously. Unswelled microgel globules of submicron size ("kernels") are injected into a reservoir and, upon contact with the high-temperature resident fluids, swell with water ("pop") to provide in-depth waterflood conformance control (Pritchett et al., 2003; Frampton et al., 2004). The material is a highly crosslinked sulfonate-containing polyacrylamide microparticle in which the conformation is constrained by both labile and stable internal crosslinkers. When subject to elevated temperature, the rate of de-crosslinking of the labile crosslinker accelerates. This reduces the crosslink density of the particles and allows the particles to expand by absorbing the surrounding water. The particles are applied in a constrained state—called kernel particles for convenience. After heating, they are able to swell to adopt a much-expanded configuration called popcorn. The reduction in reversible crosslink density is time dependent and can be affected by the pH of the fluid.

Different labile crosslinkers have different rates of bond cleavage at different temperatures. The temperature required and the mechanism of bond dissociation depend on the chemical structure of the crosslinker. For example, when the labile crosslinker is a diacrylate ester, hydrolysis of the ester linkage becomes the most likely mechanism of de-crosslinking. Proper selection of crosslinker can give particles with different activation temperature. The product called BrightWater was developed by an industry consortium formed by BP, Chevron, Mobil, and Nalco. There are several commercial products for various downhole conditions. The product lines are applicable in wide ranges of reservoir temperature (35–140°C) and salinity (up to 120,000 ppm TDS). If the reservoir temperature is above 90°C, special attention is needed. The main advantage of this product is that it can be deployed by adding directly to the injection line. No rig or coiled tubing is needed.

The kernel microparticles are prepared using an inverse emulsion polymerization process to assure a preselected particle size range. Depending on the synthetic method, the particle diameters can range from about 0.1 to about 3 microns. Alternatively, the polymeric microparticles cross linked with labile crosslinks are prepared by internally crosslinking polymer particles that contain polymers with pendant carboxylic acid and hydroxyl groups. Filtration theory suggests that formation damage can happen even if the particles are smaller than the pore throat size of the medium being tested. Particles will typically pass through filters or porous media if the minimum pore throat diameter is at least about seven times larger than the particle diameter (Veizen and Leeriooljer, 1992). Figure 5.11 compares the size distributions of the kernel particle diameters with typical pore throat size distributions of several reservoir rocks that used in the tests. The kernel particle sizes are more than one magnitude of order smaller than the pore throat sizes, allowing the particles to flow through the matrix.

The microparticles prepared in inverse emulsion form can be dispersed into aqueous media with agitation and the aid of inverting surfactants. Because of their highly crosslinked nature, the size of the kernel microparticles changes very little in solutions of different salinity. Consequently, the rheological properties of the carrier fluid are not affected by the salinity change encountered in

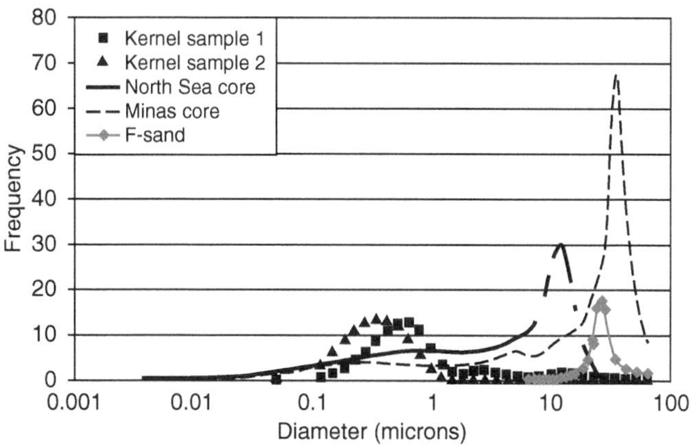

FIGURE 5.11 BrightWater particle size distributions compared with typical pore size distributions of several reservoir rocks. *Source: Chang et al. (2007).*

a subterranean formation. Accordingly, no special carrier fluid is needed for treatment.

The bottle test results in Figure 5.12 show that the onset and rate of particle expansion represented by the solution viscosity changes depended on temperature and time. Figure 5.13 shows the change in resistance factor (RF) with the test time for a 2250 ppm BrightWater solution in a 3.5 darcy silica sand at 85°C (185°F).

The following are several guidelines for field applications (Pitchett et al., 2003; Frampton et al., 2004):

- The selected reservoir should have earlier water breakthrough, high water cut, and low current recovery, and it must have great permeability contrast in layers.
- The selected BrightWater product is able to travel halfway to the producer before popping at the reservoir temperature.

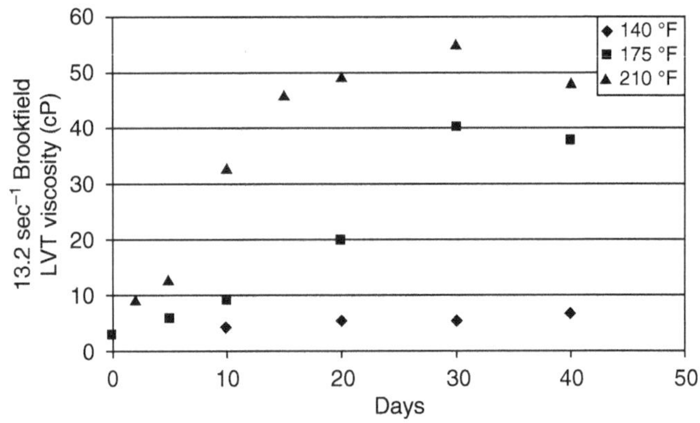

FIGURE 5.12 Solution viscosity changes with temperature and time. *Source: Chang et al. (2007).*

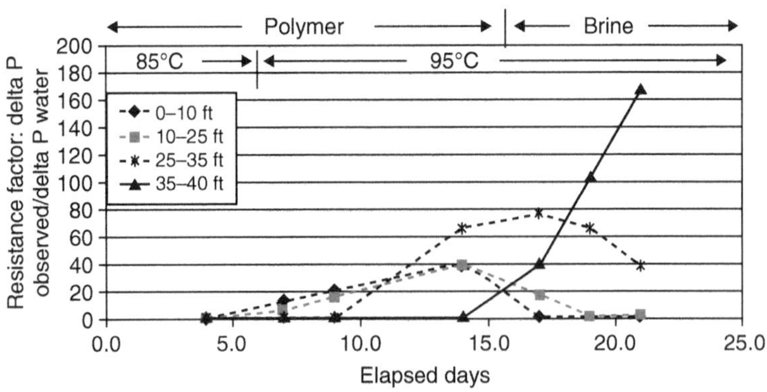

FIGURE 5.13 Resistance factor (RF) changes with the test time for a 2250 ppm BrightWater solution in a 3.5 Darcy silica sand at 85°C. *Source: Chang et al. (2007).*

- The residual resistance factor is expected to be 30 to 100 in the high permeability streaks.
- The permeability in the thief zone > 100 md and the porosity > 17%.
- Temperature is from 50°C (122°F) to 150°C (302°F).
- Expected injector–producer transient time > 30 days.
- Water salinity < 120,000 ppm.

The product has been successfully implemented in land-based, offshore, and subsea applications in the United States, Asia, Europe, and South America (Chang et al., 2007). Only the application in the Minas field is summarized next.

The first field test for BrightWater was conducted in the Chevron-operated Minas field in Indonesia in 2001 (Pritchett et al., 2003). The Minas field had OOIP of 8.7 billion barrels. The recovery was nearly 50%, and the water cut was greater than 97%. The formation permeability was 400 to 600 md. The profile control treatment of 42,000 barrels of water containing 4500 ppm of active materials was pumped into the injector 7E-12, targeting the A1 sand. The popping time was designed for 15 days. The injection started on November 11, 2001, and lasted for 9 days. In this test, 1500 ppm of surfactant was injected, and 50% caustic soda was added to keep the injection water pH at 9.5 in the first half of the injection period and the pH at 10 in the second half of the injection period. The injection wellhead was "on vacuum" throughout pumping, except during the last 14 hours, 10 psig wellhead pressure was observed.

After polymer injection, the polymer slug was overflushed with 68,800 barrels of field water in the following two weeks, and the injector was then shut in. In addition, at the end of the two-week overflush period, offset producers were shut in for up to three weeks to allow the polymer slug to "cook" and assure it would gel. A shut-in schedule was also developed to prevent polymer breakthrough and to minimize oil production loss during the shut-in period (i.e., higher water-cut wells were shut in for a longer period).

Analysis of the injection well falloff test showed that the permeability was decreased at a distance 125 ft from the injection well. The volume within the 125 ft radius was about half of the volume of 68,800 barrels, which was the overflushed volume after the treatment. However, the volume of the incremental oil attributable to this BrightWater treatment was uncertain because many factors could contribute to the outcome.

5.2.9 Microball

A deep profile control agent should be able to enter deep formation, block high permeability water channels, and be movable. To be able to enter deep formation, its size must be smaller than formation pore throats. To be able to block high permeability water channels, it must be able to expand and crosslink with polymer to form a highly viscous solution. To be movable, it must have some elastic properties so that it can move under some pressure gradient. The microball is designed based on these requirements, a concept similar to BrightWater.

The schematic structure of a microball is shown in Figure 5.14. The outermost layer of a microball is a hydration layer that makes the microball stable in water so that it will not precipitate. The middle, crosslinked polymer layer gives the microball some elasticity and deformability. The inner layer is a core that gives the microball some strength when it blocks a pore throat (Sun et al., 2006).

Initially, the microball is in oil external emulsion. When it hydrates, the microball gradually expands or swells. Figure 5.15 shows the microscopic pictures of microballs before and after hydration at room temperature. Before hydration (Figure 5.14) occurred, the sizes of microballs were tens of nanometers. After hydration (Figure 5.15), the microballs swelled, and the sizes were in the order of micrometers. The swelling was also confirmed at 100°C and in the water of 13,000 mg/L TDS (Sun et al., 2006).

A field test was conducted in the well 1-14 pattern in the eastern block of the Gudao field (Wang et al., 2005). The test formation was Ng3-4, and the formation thickness was 13 m. The clay content was 11.8%. In this test, the air permeability was 250 to 3165 md, with an average 1782 md, and the porosity was 30 to 32%. The initial reservoir temperature was 71°C. Before the test, the temperature was 64°C. The oil viscosity at reservoir conditions was 50 to 150 mPa·s, and the formation TDS was 3850 mg/L. In the test well pattern, there were 1 injector, D1-14, and 11 producers around the injector. In April 2004, infill drilling and conformance control were conducted. The water cut decreased only 2%. By January 2005, the water cut reached 96%.

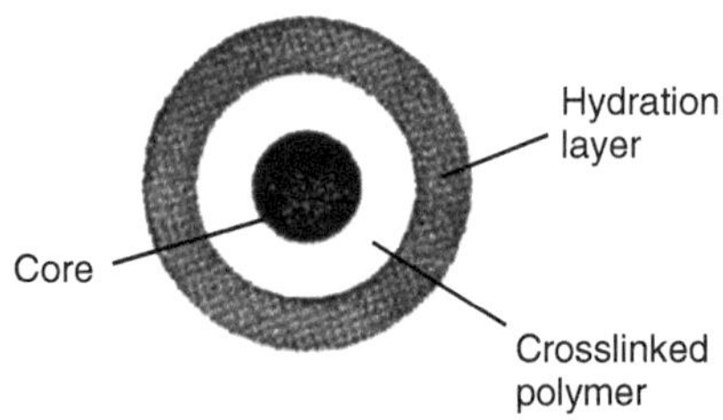

FIGURE 5.14 Schematic of microball structure. *Source: Sun et al. (2006).*

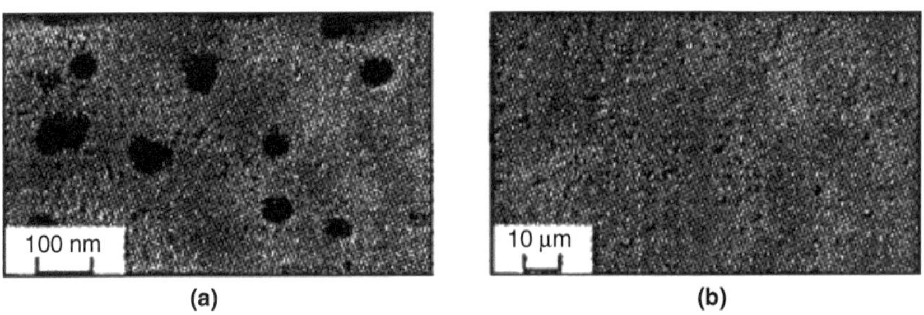

FIGURE 5.15 Microscopic pictures of microballs (a) before hydration and (b) after hydration for 30 days. *Source: Sun et al. (2006).*

In this field test, which started on December 2, 2004, four slugs were injected: (1) 300 mg/L polymer injection for 10 days, (2) 600 mg/L polymer injection for 50 days, (3) 1000 mg/L polymer injection for 30 days, and (4) 600 mg/L polymer injection for 33 days. The injection was stopped on April 5, 2005, followed by water injection. The median size of the injected microballs was 600 nm; microballs swelled to 4 μm at 60°C. Eight out of 11 wells responded to the injection. The water injection profile was improved, showing decreased water intake in high permeability layers. The oil rate started to increase by April 2005 and increased significantly during May. In May, the water cut decreased, and by October 2005, incremental oil production reached 1560.1 tons.

5.2.10 Inverse Polymer Emulsion

Another system similar to the microball is called inverse polymer emulsion. In this case, the polymer used is polyacrylamide (PAM). The inverse PAM emulsion is a W/O type of emulsion. The dispersed phase contains 6.4 to 10.5 million Daltons PAM and 1000 mg/L crosslinkers for a sample product. The external continuous phase is white oil. There is a surfactant interfacial film between the disperse phase and continuous phase, as shown in Figure 5.16. The emulsion is stable at the surface. When it is injected into a target formation, it is inverted into an O/W type of emulsion under certain temperature and salinity, with the help of a phase inversion agent. Thus, the name inverse emulsion is used.

The inverse emulsion has the following advantages:

1. Polymer concentration in the dispersed phase could be as high as > 25%, but it can still be easily carried to the deep formation.

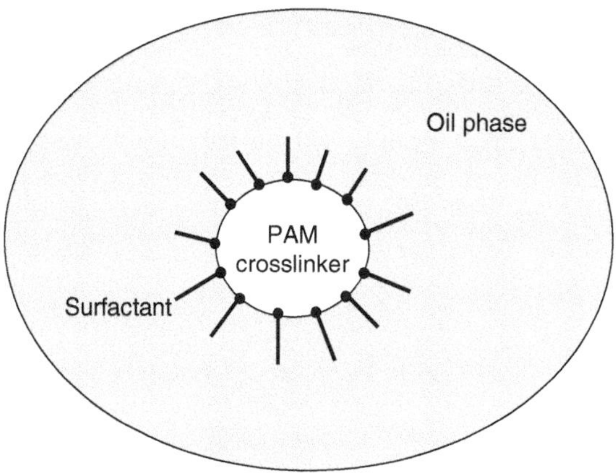

FIGURE 5.16 Schematic of the inverse emulsion.

2. The dissolution is fast after phase inversion (< 20 minutes).
3. The polymer and crosslinker will not adsorb on the rock surface during transport because it is protected by the oil phase. Also, the concentration ratio of polymer to crosslinker will not change, so the gelation time and strength can be controlled.
4. Inversion can occur in the deep formation, so the deep profile can be controlled.
5. Under a certain pressure gradient, the emulsion can flow and displace residual oil.

The inverse emulsion was tested in the southern Zhong-2 block in the Gudao field. The target formation was the Ng3-4 zone with a net thickness of 11.23 m. It was unconsolidated sand with porosity of 31.8% and air permeability of 2950 md. The test area was 0.782 km^2 with an OOIP of 1.6821 million tons. The reservoir temperature was 69°C, and the oil viscosity at the reservoir temperature was 20 to 100 mPa·s. In this case, the injected produced water salinity was 7246 mg/L.

Before inverse emulsion was injected, the field went through primary depletion, waterflooding, polymer flooding, and post-polymer waterflooding. By July 2004, the water cut in the test area was 90.64%, with a recovery factor of 50.1%. With 1 injector, Well 21-4, and 5 producers, the injection of inverse emulsion was started in December 2004 at one injection well pattern. The injection program was 10 m^3 polymer solution of 8000 mg/L concentration, 15 m^3 inversion emulsion with 6000 mg/L polymer, and 1167 mg/L phase inversion agent, followed by chase water drive. Four producers out of 5 wells responded to the injection in this test. The injection pressure increased from 7.5 to 9.5 MPa, the water cut reduced from 92.5 to 91.4%, the oil rate increased from 31.9 to 44 t/d, and the liquid rate increased from 423.1 to 513.2 t/d for the well pattern (Lei et al., 2006).

5.2.11 Preformed Particle Gel

One preformed gel is preformed particle gel (PPG)—see Coste et al. (2000) and Bai et al. (2007). A preformed gel on the surface can overcome some distinct drawbacks inherent in in situ gelation systems, such as lack of gelation time control, uncertainty of gelling due to shear degradation, chromatographic fractionation, change of gelant compositions, and dilution by formation water. Other products that have been developed in this area include preformed bulk gels (Seright, 2004) and partially preformed gels (Sydansk et al., 2004, 2005).

PPG is an improved super absorbent polymer (SAP). SAPs are a unique group of materials that can absorb over a hundred times their weight in liquids, and they do not easily release the absorbed fluids under pressure. Super absorbent polymers are used primarily as an absorbent for water and aqueous solutions for diapers, adult incontinence products, feminine hygiene products, and the agriculture industry. However, the traditional SAPs in the market do not meet the requirements for conformance control due to their fast swelling time, low strength, and instability at high temperature. A series of new SAPs called preformed particle gels were developed for the purpose of conformance control.

Preformed particle gels have been applied in about 2000 wells in China to reduce fluid channels in waterfloods and polymer floods (Liu et al., 2006c). PPG treatment has been widely accepted and is seeing more and more use by operators because of its unique advantages over traditional in situ gel including

- Particle gels are synthesized prior to formation contact, thus overcoming distinct drawbacks inherent in in situ gelation systems, such as uncontrolled gelation times, variations in gelation due to shear degradation, and gelant changes from contact with reservoir minerals and fluids.
- PPGs are strength- and size-controlled (μm–cm), environment-friendly, and stable up to 110°C in the presence of almost all reservoir minerals and formation water salinities.
- PPGs can preferentially enter into fractures or high-permeability channels while minimizing gel penetration into low-permeability hydrocarbon zones/matrices. Gel particles with the appropriate size and properties should transport through fractures or high permeability channels, but they do not penetrate into matrix sands.
- These gels usually have only one component during injection. Thus, using them is a simpler process and does not require many of the injection facilities and instruments that often are needed to dissolve and mix polymer and crosslinkers for conventional in situ gels.

  PPGs can be prepared with produced water without influencing gel stability. In contrast, traditional in situ gels are often very sensitive to salinity, multivalent cations, and H_2S in the produced water. Using PPG can not only save fresh water but also protect the environment.

5.3 PROPERTIES OF POLYMER SOLUTIONS

The properties of some specific polymer solutions were presented in the previous section, which introduced different types of polymers and polymer-related systems. This section discusses the general properties of polymer solutions.

5.3.1 Polymer Viscosity

Viscosity is the most important parameter for polymer solution. As mentioned earlier, hydrolyzed PAM, or HPAM, is the most used polymer in enhanced oil recovery. Some of factors which affect polymer viscosity are discussed next.

Salinity and Concentration Effects

The intrinsic viscosity of a homogeneous PAM solution increases when NaCl is added to the solution. When $CaCl_2$ is added, the viscosity increase is even more obvious. However, HPAM viscosity decreases when a monovalent salt (e.g., NaCl) is added. The reason is that the added salt neutralizes the charge in HPAM side chains. When HPAM is dissolved in water, Na^+ dissipates in the water. $-COO^-$ in the high molecular chains repel each other, which makes them stretch, hydrodynamic volume increase, and viscosity increase. When the salt is added, $-COO^-$ is surrounded by some Na^+, which shields the charge. Then $-COO^-$ repulsion is reduced, the hydrodynamic volume becomes smaller, and the viscosity decreases. When divalent salts—$CaCl_2$, $MgCl_2$, and/or $BaCl_2$—are added in an HPAM solution, their effect is complex. At low hydrolysis, the solution viscosity increases after it reaches the minimum. At high hydrolysis, the solution viscosity decreases sharply until precipitation occurs.

The dependence of polymer solution viscosity at zero shear rate (μ_p^0) on the polymer concentration and on salinity may be described by the Flory–Huggins equation (Flory, 1953),

$$\mu_p^0 = \mu_w\left(1 + \left(A_{p1}C_p + A_{p2}C_p^2 + A_{p3}C_p^3\right)C_{sep}^{S_p}\right), \tag{5.1}$$

where μ_w is the water viscosity with its unit being the same as μ_p^0; C_p is the polymer concentration in water; A_{p1}, A_{p2}, A_{p3}, and S_p are fitting constants; and C_{sep} is the effective salinity for polymer. Be careful about the units in Eq. 5.1. The items in the parentheses must be dimensionless.

The simple way to avoid any mistakes when fitting the laboratory data to Eq. 5.1 is to use the same units as those in the prediction model (e.g., a simulator) you are going to use. Then those fitting constants obtained by matching experimental data can be directly used in the prediction model. The factor $C_{sep}^{S_p}$ allows for dependence of polymer viscosity on salinity and hardness. The effective salinity for polymer, C_{sep}, is given in UTCHEM-9.0 (2000) by

$$C_{sep} = \frac{C_{51} + (\beta_p - 1)C_{61}}{C_{11}}, \tag{5.2}$$

where C_{51}, C_{61}, and C_{11} are the anion, divalent, and water concentrations in the aqueous phase; and β_p, whose typical value is about 10, is measured in the laboratory.

The unit for C_{51} and C_{61} is meq/mL, and the unit for C_{11} is water volume fraction in the aqueous phase. The commonly used laboratory units for salinity are wt.% and ppm (mg/L). These units should be converted to meq/mL using Eq. 5.2. In principle, any units could be used, as long as they are used consistently in a study. It is suggested that one unit be used throughout a study. The unit meq/mL is a good scientific unit of salinity because it considers the effects of different ions with different electrolyte strength. In most cases, the unit is chosen based on convenience, not science. For example, the salinity is reported as total dissolved solids expressed in total weight percent (wt.%) of ions, or in ppm, mg/L, in a solution or in water. Sometimes, the total amount of anions in a solution or water, expressed in ppm (mg/L), is reported.

Most often, the total amount of chloride is used because NaCl is the most common salt. The justification of using it is that the current technology really cannot describe the effect of every single ion on chemical EOR. For example, when HPAM reacts with multivalent metal ions, such as Al^{3+}, Cr^{3+}, and Ti^{3+}, in a solution, a weak gel is formed. In this case, we cannot simply use Eq. 5.2 to calculate effective salinity. Equation 5.2 shows that divalents have a larger effect on the effective salinity than monovalents at the same concentration. In general, the order of effect is $Mg^{2+} > Ca^{2+} > Na^+ > K^+$. The activity of these ions is 10 to 20 kJ/mol, which is much less than the value for chemical reactions (about 200 kJ/mol). Therefore, the salt effect on polymer solution is a reversible electrostatic effect (Niu et al., 2006).

Note that electrolyte concentrations in the laboratory are commonly expressed in terms of the aqueous phase volume, which includes the volume of surfactant and cosolvent in addition to water. C_{11} in Eq. 5.2 is used to correct the aqueous volume. In Eq. 5.1, S_p is the slope of

$$\frac{\mu_p^0 - \mu_w}{\mu_w}$$

versus C_{sep} on a log–log plot, as shown in Figure 5.17. S_p in this case is –0.2398.

Figure 5.18 shows an example of the viscosity data measured in the laboratory and then calculated using Eq. 5.1. In this case, A_{p1}, A_{p2}, and A_{p3} are 9.45, 0, and 1298, respectively; C_{sep} is 0.68 meq/mL; the water viscosity is 1 mPa·s; and S_p is –0.2398.

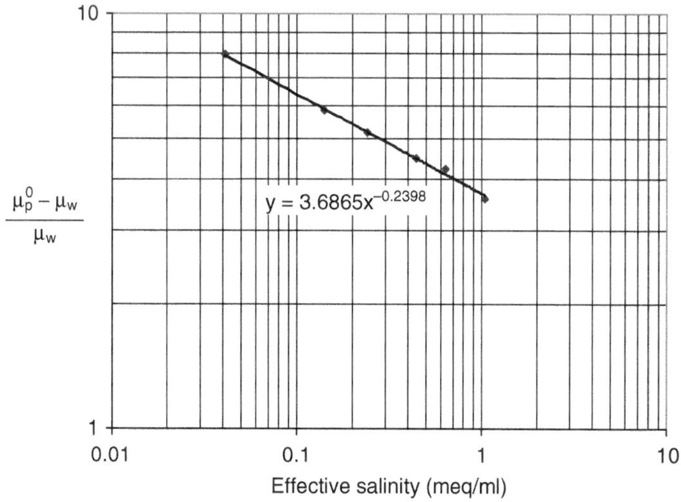

FIGURE 5.17 A log–log plot based on Eq. 5.1.

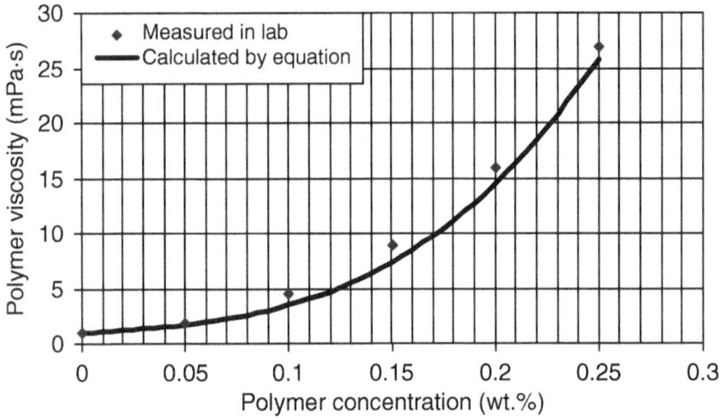

FIGURE 5.18 Polymer viscosity versus polymer concentration.

Shear Effect

In general, a polymer solution behaves like a pseudoplastic fluid. The reduction in polymer solution viscosity as a function of shear rate ($\dot{\gamma}$) is described by the power-law model (Bird et al., 1960), which is given by

$$\mu_p = K\dot{\gamma}^{(n-1)}, \tag{5.3}$$

where K is the flow consistency index, and n is the flow behavior index. In the pseudoplastic regime $n \leq 1$ (typically $n = 0.4$ to 0.7). At different concentrations, n hardly changes, but K changes. For a Newtonian fluid, $n = 1$ and K is simply the constant viscosity, μ. Although the preceding equation is quite satisfactory to describe the pseudoplastic regime, it is unsuitable at high and low shear rates (Sorbie, 1991). When the shear rate approaches zero, shear stress does not approach zero. Some yield stress exists. In other words, the HPAM solution behaves like a Bingham fluid (Luo et al., 2006).

A more satisfactory model for these shear regimes is Meter's equation (Meter and Bird, 1964),

$$\mu_p = \mu_w + \frac{\mu_p^0 - \mu_w}{1 + \left(\dfrac{\dot{\gamma}}{\dot{\gamma}_{1/2}}\right)^{(p_\alpha - 1)}}, \tag{5.4}$$

where p_α is an empirical parameter, or is obtained by matching laboratory-measured viscosity data; μ_p^0 is the limiting viscosity at the low (approaching zero) shear limit; and $\dot{\gamma}_{1/2}$ is the shear rate at which viscosity is the average of μ_p^0 and μ_w. This equation is used in UTCHEM. In Eq. 5.4, it is assumed that the polymer viscosity at infinite shear rate ($\dot{\gamma} \rightarrow \infty$) is equal to the water viscosity.

A more general model is the Carreau equation (Carreau, 1972; Bird et al., 1987),

$$\mu_p - \mu_\infty = (\mu_p^0 - \mu_\infty)\left[1 + (\lambda\dot{\gamma})^\alpha\right]^{(n-1)/\alpha}, \quad (5.5)$$

where μ_∞ is the limiting viscosity at the high (approaching infinite) shear limit and is generally taken as water viscosity μ_w; λ and n are polymer-specific empirical constants; and α is generally taken to be 2. μ_p^0 and $\dot{\gamma}$ are as defined earlier. For intermediate shear rates, the Carreau equation represents a power-law relation (Sorbie, 1991). Practically, μ_p and μ_p^0 are much higher than μ_∞, and $(\lambda\dot{\gamma})^\alpha$ is much larger than 1. Then Eq. 5.5 becomes the power-law equation of the form $\mu_p = \mu_p^0(\lambda\dot{\gamma})^{n-1}$, which describes the viscosity at the intermediate and high shear rate regimes. At the low shear rate regime, $\mu_p = \mu_p^0$, as shown in Figure 5.19. At the intersection of these two regimes, the viscosity and shear rate are the same. Then we must have $\dot{\gamma} = \lambda^{-1}$. This intersection is the first critical shear rate at which the fluid deviates from the Newtonian behavior.

Figure 5.20 is an example of the polymer viscosity at different shear rates. The laboratory data were used for fitting in Eq. 5.4. In this case, the water viscosity at zero shear rate was 6 mPa·s, and the two fitting parameters, p_α and $\dot{\gamma}_{1/2}$, were 1.8 and 450 s^{-1}, respectively. Note that according to Eq. 5.4, the calculated μ_p at any shear rate can never be higher than μ_p^0 at zero shear rate.

pH Effect

It is known that pH affects hydrolysis. Therefore, HPAM viscosity is pH-dependent. pH increases initially when alkali is added. However, adding alkali eventually will result in the decrease of HPAM viscosity owing to the salt effect. Mungan (1969) reported the effect of pH on HPAM viscosity. HCl was titrated against the original stock polymer solution with pH about 9.8 (pH of oilfield brines is usually in the range 7.5–9.5). The polymer concentration was

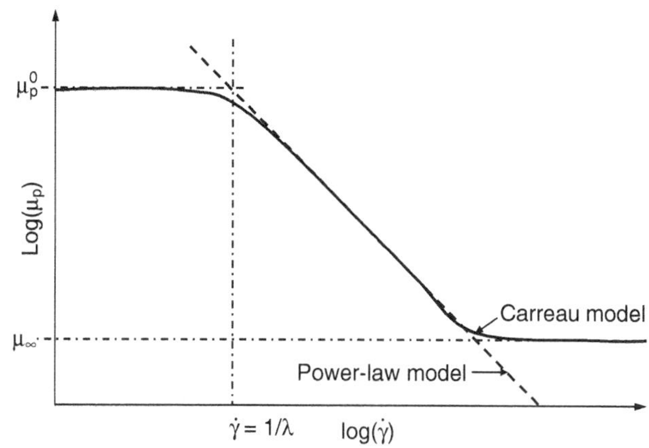

FIGURE 5.19 Comparison of the Carreau and power models.

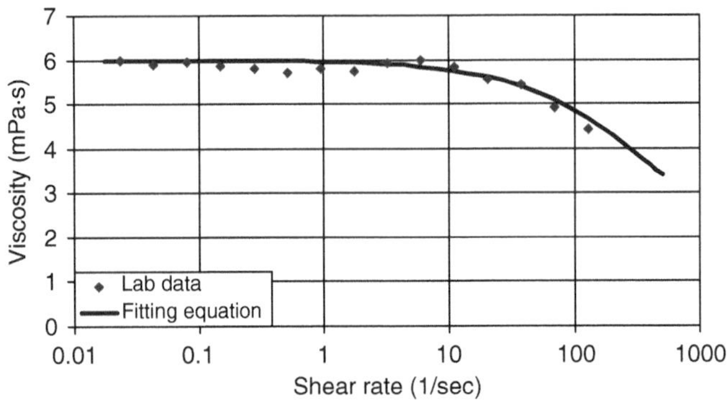

FIGURE 5.20 Polymer viscosity at different shear rates.

2500 mg/L. Interestingly, the HAPM viscosity at 50 s^{-1} shear rate significantly decreased when lowering pH.

Szabo (1979) reported the increases in the viscosity of AM/AMPS copolymer solution when NaOH was added. All these observations are probably related to early time hydrolysis effect. Adding alkali also increases electrolytes, which should decrease polymer solution viscosity. Even without alkali, hydrolysis will occur. Therefore, for the long term, the effect used to increase hydrolysis will become less important than the salt effect, and the polymer viscosity will decrease. These statements are consistent with those reported by Flournoy et al. (1977) in that the apparent viscosity was very dependent on pH, with the maximum apparent viscosity occurring at a pH of about 6 to 10 for polyacrylamide and at a pH of about 4 to 9 for polysaccharide. Considering the aging effect, the relationship between polymer viscosity and pH or alkali becomes more complex; this issue is discussed in more detail in Section 11.2.

Temperature Effect

At a low shear rate, the polymer solution apparent viscosity decreases with temperature according to the Arrhenius equation,

$$\mu_p = A_p \exp\left[\frac{E_a}{RT}\right], \tag{5.6}$$

where A_p is the frequency factor, E_a is the activity energy of the polymer solution, R is the universal gas constant, and T is the absolute temperature. Eq. 5.6 shows that the viscosity decreases rapidly as the temperature increases. As the temperature increases, the activity of polymer chains and molecules is enhanced, and the friction between the molecules is reduced; thus, the flow resistance is reduced and the viscosity decreases. Different polymers have different E_a. With a higher E_a, the viscosity is more sensitive to temperature. HPAM has two E_as. When the temperature is less than 35°C, E_a is low, and the viscosity does not change too much as the temperature increases. When the temperature is higher than 35°C, E_a is high, and the viscosity is more sensitive to the variations in temperature (Luo et al., 2006).

Equation 5.6 can be rewritten as

$$\mu_p = \mu_{p,ref} \exp\left[E_a\left(\frac{1}{T} - \frac{1}{T_{ref}}\right)\right], \tag{5.7}$$

where $\mu_{p,ref}$ is the viscosity at the reference temperature, T_{ref}. When measurements are made at different temperatures, the preceding equation may be used to fit the measurement data by adjusting E_a if E_a does not change at different temperatures.

5.3.2 Polymer Stability

Polymer degradation refers to any process that breaks down the molecular structure of macromolecules. The main degradation pathways of concern in oil recovery applications are chemical, mechanical, and biological. The research work on polymer stability from the mid-1970s to late-1980s is summarized in Sorbie (1991).

Chemical Stability

Chemical degradation refers to the breakdown of polymer molecules, either through short-term attack by contaminants, such as oxygen and iron, or through longer-term attack to the molecular backbone by processes such as hydrolysis. The latter is caused by the intrinsic instability of molecules even in the absence of oxygen or other attacking species. In other words, polymer chemical stability is mainly controlled by oxidation-reduction reactions and hydrolysis.

Oxidation Reduction

The presence of oxygen virtually always leads to oxidative degradation of the polyacrylamide polymer. However, at a low temperature, the effect of dissolved oxygen on HPAM solution viscosity is not significant, and the polymer solution could be stable for a long time. As the temperature increases, even if a small amount of oxygen exists, HPAM solution viscosity quickly decreases with time. For example, the half-lives for a polymer at 50°C, 70°C, and 90°C are 117, 20, and 2.6 hours, respectively. As the oxygen concentration increases, the viscosity decreases faster (Luo et al., 2006).

Yang and Treiber (1985) studied the chemical stability of polyacrylamide solution under simulated field conditions. They identified the main variables encountered by a polymer solution in the field as oxygen, temperature, oxygen scavengers, metal/metal ions, hydrogen sulfide, pH, salinity/hardness, chemical additives, and biocide. Their main finding was that the rate and extent of polymer degradation were governed mainly by the oxygen content of the

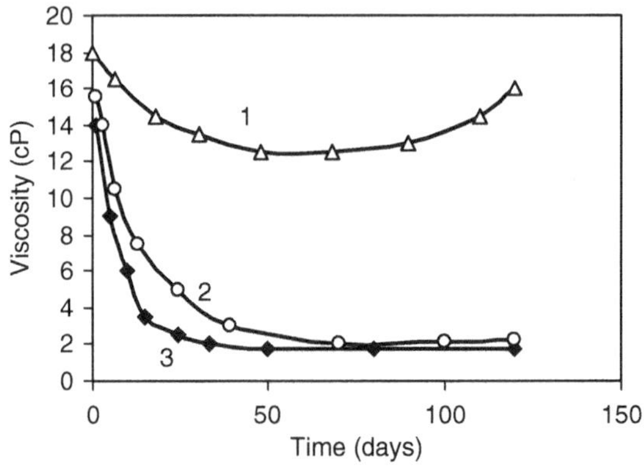

FIGURE 5.21 Oxygen's effect on HPAM stability at 90°C: 1, low level of oxygen; 2, air; 3, oxygen. *Source: Luo et al. (2006).*

solution and temperature, although they remarked that limited levels of oxygen produced only limited polymer degradation. At low oxygen levels (1 part per billion, ppb), they found that their polyacrylamides were stable over 500 days up to 93.3°C and indeed showed an increase in viscosity over this time. This increase had been reported previously by Ryles (1983), later by Luo et al. (2006) and by Han et al. (2006a). (See Figure 5.21.) This behavior is thought to be the result of the increasing degree of hydrolysis that occurs at elevated temperatures. When the oxygen was completely consumed, the degradation reaction stopped; this behavior is contrary to the general suspicion that after the reaction is initiated by oxygen, it will proceed without further oxygen supply (Luo et al., 2006).

Figure 5.21 shows that the degradation is severe when air or oxygen exists. Therefore, the amount of oxygen in the solution should be minimized by using oxygen scavengers, possibly along with some methanol or thiourea to protect the polymer from any further oxygen ingress into the solution (Yang and Treiber, 1985). Wellington (1980) found that the most effective formulation contained thiourea as the radical transfer agent, isopropyl alcohol as the sacrificial oxidizable alcohol, sodium sulfite as the oxygen scavenger, either tri- or pentachlorophenol, and a sufficient brine concentration. Luo et al. (2006) reported that the combination of thiourea and cobalt salt could prevent oxidation reduction more effectively. Using such a combination retained a polymer viscosity at 69% after 360 hours, while at 20% if using thiourea only, and at 27% if using cobalt salt only. Sorbie (1991) listed the effects of some additives and their combinations on the stability of HPAM.

Effects of Ironic Ions Figure 5.22 shows the ferric ion (Fe^{3+}) effect on an HPAM viscosity at room temperature. The initial viscosity was 72.9 mPa·s. We can see that when the Fe^{3+} concentration was low, the viscosity loss was not

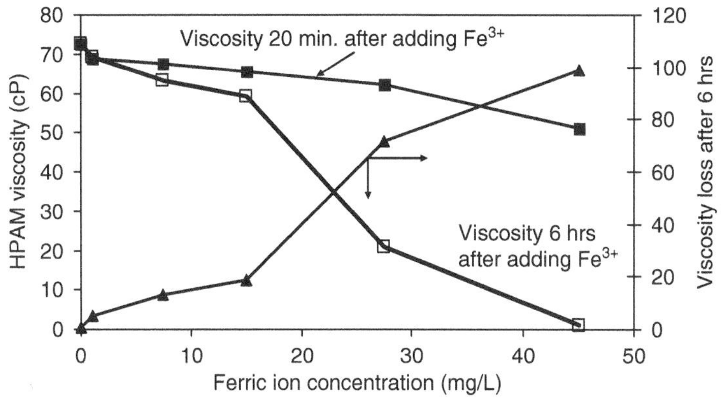

FIGURE 5.22 Effect of Fe^{3+} concentration on HPAM viscosity. *Source: Data from Luo et al. (2006).*

significant in a short time. The viscosity loss was caused by the salinity effect. When the Fe^{3+} concentration was high (>15 mg/L), a brown precipitate, $Fe(OH)_3$, was observed. Because the amount of the precipitate was small, the viscosity loss was not significant. When the Fe^{3+} concentration was high enough, Fe^{3+} crosslinked with HPAM to form insoluble gel; as a result, the viscosity loss was significant.

Figure 5.23 shows the 1000 mg/L HPAM viscosity in a closed system without oxygen at 30°C and 3 hours after adding Fe^{2+}. When Fe^{2+} concentration was lower than 10 mg/L, the viscosity loss was less than 10% owing to the salinity effect. However, when the HPAM solution was put in an open system, the viscosity was significantly lost, as shown in Figure 5.23. In the open system, Fe^{2+} was oxidized to Fe^{3+}. For comparison, the viscosity loss 6 hours after

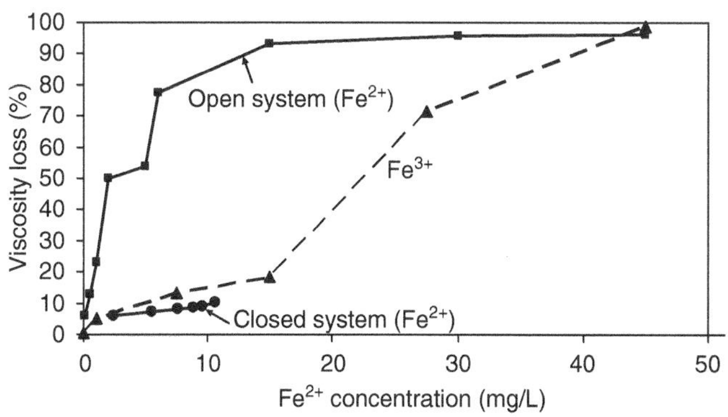

FIGURE 5.23 Effect of Fe^{2+} concentration on HPAM viscosity. *Source: Data from Luo et al. (2006).*

adding Fe^{3+} in Figure 5.22 is also shown in Figure 5.23. We can see that the viscosity loss caused by oxidized Fe^{3+} (3 hours after adding Fe^{2+}) was much higher than that caused by pure Fe^{3+}. The reason is that when Fe^{2+} is oxidized to Fe^{3+}, the free radical O_2^- is produced. O_2^- reacts with HPAM to produce peroxide and break the backbones of HPAM. The free radical produced from the reaction further reacts with Fe^{3+} to generate Fe^{2+}, which is further oxidized to produce Fe^{3+} and O_2^-. A chain reaction occurs, and the polymer viscosity is significantly reduced. In this chain reaction, Fe^{2+} virtually works as a catalyst. Fe^{2+} is the only element discovered so far that can reduce the polymer viscosity to almost water viscosity within seconds. Fe^{2+} concentration should be controlled below 0.5 mg/L (Luo et al., 2006). Levitt et al. (2010) reported that sodium carbonate and bicarbonate are demonstrated to play a key role in stabilizing polymer against multiple reported sources of degradation, and it seems likely that this is due to their effect on iron stability.

Hydrolysis

This section reviews the effects of temperature and divalent on hydrolysis.

Effect of Temperature In the absence of oxidative degradation, the backbone chain of vinyl polymers, such as polyacrylamide, is quite thermally stable to temperatures as high as 120°C (Ryles, 1983). Indeed, Ryles (1988) found that polyacrylamide was stable at 90°C for at least 20 months under controlled conditions. At elevated temperatures, however, the pendant amide groups tend to hydrolyze, therefore increasing the total carboxylate content of the polymer. This increase results in significant changes in solution properties, rheology, and phase behavior because the primary mechanism of polyacrylamide degradation is found to be amide group hydrolysis.

Thermal stability tests performed by Ryles (1988) showed that the dissolved salts had just a minor effect on the hydrolysis rate and that the temperature was the main determining factor. From his data, we can see the following:

- The higher the temperature, the faster the rate of hydrolysis.
- The higher the temperature, the higher the degree of hydrolysis.
- Hydrolysis was significantly affected by temperature.
- The divalent concentration strongly affected viscosity reduction.
- The highest viscosity retention occurred at 40 to 50% hydrolysis. This observation is consistent with the observation by Kong (1996).

The preceding observations are consistent with those by Moradi-Araghi and Doe (1984). In alkaline conditions, initially hydrolysis is fast. As the hydrolysis reaches a certain level, the electrostatic repulsion between carboxyl group and OH^- limits further hydrolysis at pH > 13. Finally, hydrolysis is stopped. Therefore, pH has been found to have a minimum effect.

At a high temperature, acrylamide is progressively hydrolyzed into acrylic acid; thus, hydrolysis is increased, as shown in Figure 5.24. In the beginning, hydrolysis increased almost linearly with aging time. After hydrolysis of 44%,

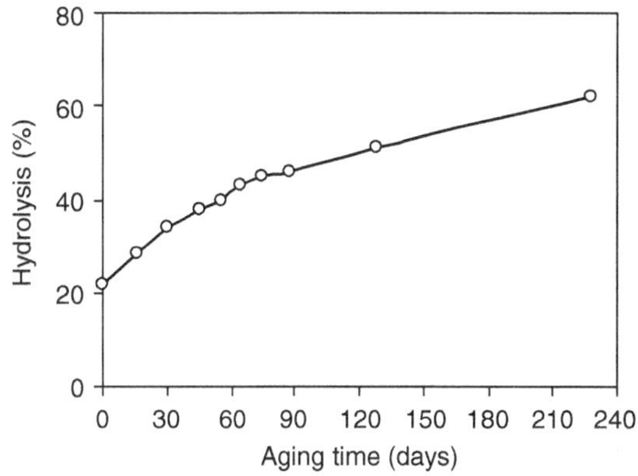

FIGURE 5.24 HPAM-A525 hydrolysis at different aging times at 75°C. *Source: Kong (1996).*

the rate of increase slowed down. One of the HPAM characteristics is that hydrolysis quickly increases at high temperatures. Consequently, hydrolysis directly affects HPAM stability.

The preceding observations can be supported by data from Han et al. (2006a). They investigated the effect of initial hydrolysis and found that the rate of hydrolysis was higher at a higher initial hydrolysis. Therefore, a higher initial hydrolysis is needed in a low-temperature (e.g., 55°C) reservoir so that high viscosity can be quickly reached. In a high-temperature reservoir, the HPAM viscosity near the wellbore will be lower if a polymer with a lower initial hydrolysis is used. This technique will improve polymer injectivity. As the polymer moves deep into the reservoir, hydrolysis increases and viscosity also increases.

Tan (1998) investigated the effect of temperature gradient near wellbore on HPAM polymer thermal stability. For the reservoir he studied, there was a temperature gradient from 40°C near the injection wellbore to 75°C deep in the reservoir. He observed that when the polymer was under thermal degradation gradually from 40°C to 75°C, the polymer had higher viscosity retention than when the polymer was under 75°C thermal degradation right from the beginning (see Figure 5.25). During the early stages of thermal degradation, oxygen is consumed, and no oxygen is available during the later stages.

Tan's experiments showed that the polymer would be more stable if it is under thermal degradation when the temperature is gradually increased so that the oxygen is consumed at low temperatures. However, the initial polymer viscosities were different in his experiments (57.8 mPa·s at 75°C constant temperature compared with 77.5 mPa·s under the temperature gradient). The water TDS was 362.6 mg/L, and sand was used in the tests. Tan tried to imitate the actual thermal degradation conditions. He also observed that oil did not affect the polymer thermal stability.

Tan (1998) also investigated the effect of oil sand on HPAM thermal stability. He observed that oil sand improved the polymer thermal stability.

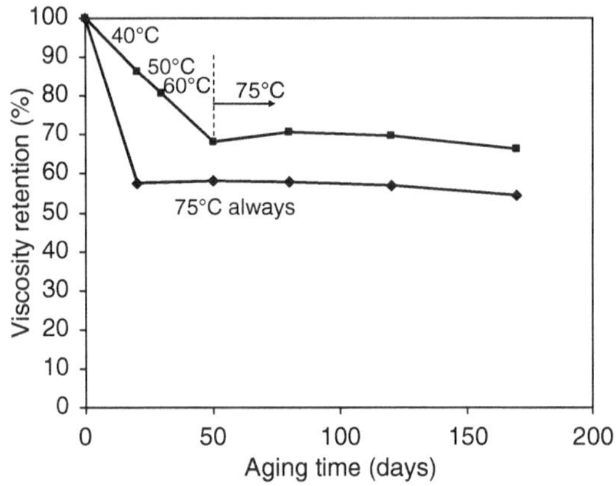

FIGURE 5.25 The effect of temperature gradient on polymer thermal stability. *Source: Data from Tan (1998).*

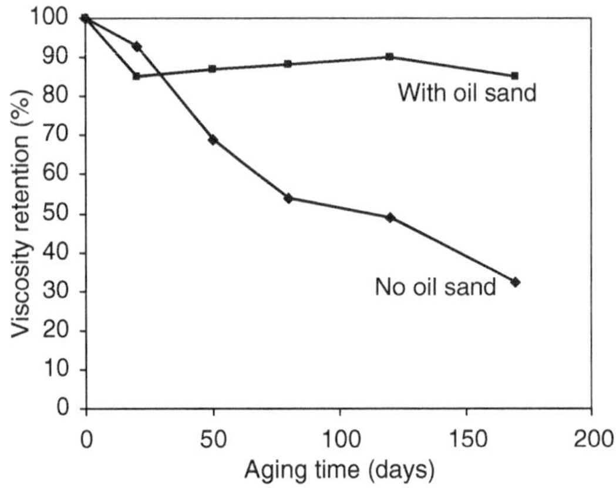

FIGURE 5.26 A plot of polymer thermal stability tests (750 mg/L HPAM S525, 0.5 mg/L oxygen). *Source: Data from Tan (1998).*

Figure 5.26 shows the retained viscosity in the percent of its initial value at 75°C, after the polymer went through the preshearing at a velocity equivalent to the flow velocity through perforation. The HPAM S525 polymer had 15 million MW and 25% hydrolysis. The TDS was 4002 mg/L. The figure shows that the polymer viscosity retained 85% of the initial viscosity after 170 days with oil sand in the solution, whereas the solution retained only 25% without oil sand.

Ryles (1988) investigated the stability of xanthan and found that its stability followed a pattern similar to that of polyacrylamide in terms of temperature; however, the mechanisms are quite different. Xanthan's stability was independent of divalent metal ion concentration, but apparently it was related to the conformational transition temperature. The degradation of xanthan was a function of temperature. Xanthan was totally degraded in the alkaline media at >50°C.

Divalent Effect In the brine of low to medium salinities (monovalent content), the viscosity of polyacrylamide solution increases as hydrolysis proceeds (increases). However, in the presence of divalents, the viscosity behavior will be determined largely by the divalent metal ion concentrations. As hydrolysis increases, more acrylic acid exists in the solution. Hydrolyzed polyacrylamides (negative carboxyl groups) interact strongly with divalent metal cations such as Ca^{2+} and Mg^{2+}. This phenomenon is commonly associated with reduction in solution viscosity, formation of gels or precipitates.

Ryles (1988) observed that the HPAM solution viscosity remained stable at >100% retention until the polyacrylamide was hydrolyzed to about 60 mol%, when the Ca^{2+} concentration was below 200 ppm. Between 60 and 80 mol%, polymer solutions lost almost one half of their original viscosity. Thus, when hydrolysis is limited to less than about 60 mol%, excellent long-term stability can be achieved. This observation is supported by data from Han et al. (2006a). Increasing the Ca^{2+} concentration to 500 ppm had a more pronounced effect on viscosity retention, even though the polyacrylamide remained soluble. Mg^{2+} had similar but less effect than Ca^{2+}. At 50°C, the rate of hydrolysis was so slow that viscosity was retained essentially intact after 21 months of aging. Note that the tests were under anaerobic condition.

Davison and Mentzer (1980) found that the precipitation time for the HPAM in seawater at 90°C depended on the initial degree of the polymer's hydrolysis; the higher the initial hydrolysis, the shorter the time before precipitation was observed. Zaitoun and Potie (1983) also noted that precipitation was affected quite strongly by the degree of hydrolysis of the polymer. Thus, the central role of divalent ions (Ca^{2+} and Mg^{2+} mainly) and the initial degree of hydrolysis of the polyacrylamide are known to affect and limit the stability of HPAM in solution at elevated temperatures, even under anaerobic conditions. Addition of calcium chloride causes polymer to precipitate. However, when sufficiently large quantities of calcium chloride are added—for example, above 55 g/L calcium chloride for a sodium chloride concentration of 20 g/L—then the precipitated polymer redissolves. Zaitoun and Potie (1983) discussed various theoretical interpretations of this precipitation–redissolution phenomenon (Michaeli, 1978; Kaczmar, 1980). Zaitoun and Potie noted that the precipitation reaction is reversible with temperature.

Moradi-Araghi and Doe (1984) also showed that the cloudy solutions resulting from the polyacrylamide precipitation led to severe plugging of porous media, and this therefore indicates that only clear solutions are useful for polymer flooding. They presented an extensive amount of data on the cloud point/hardness/temperature behavior of a range of polyacrylamides with molecular weights up to about 34 million. Their data indicated a limit of about 75°C for brines containing 2000 ppm hardness and higher. This temperature limit increased to around 88°C for 500 ppm hardness, 96°C at 270 ppm, and about 204°C at 20 ppm hardness and lower. The temperature limit is called cloud point (T_c), at which phase separation occurs. At a given Ca^{2+} concentration, T_c decreases with hydrolysis. For example, for a 1% $CaCl_2$ solution, when hydrolysis is at 15% and 48%, T_c is at 45.6°C and the room temperature, respectively. However, T_c will not increase linearly with hydrolysis.

Figures 5.27 and 5.28 show an HPAM solution viscosity versus an NaCl concentration and an $CaCl_2$ concentration, respectively. Figure 5.27 shows that the HAPM viscosity increased with hydrolysis. However, viscosity decreased with hydrolysis when hydrolysis was above 40%. Of course, HPAM viscosity decreased with NaCl concentration because of the salt effect. Figure 5.28 shows that in a $CaCl_2$ solution, as the concentration of $CaCl_2$ was increased, the HPAM

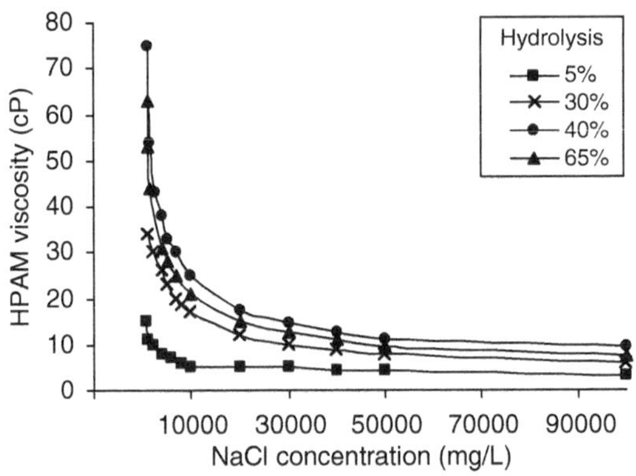

FIGURE 5.27 Effect of NaCl concentration on HPAM viscosity with different levels of hydrolysis. *Source: Luo et al. (2006).*

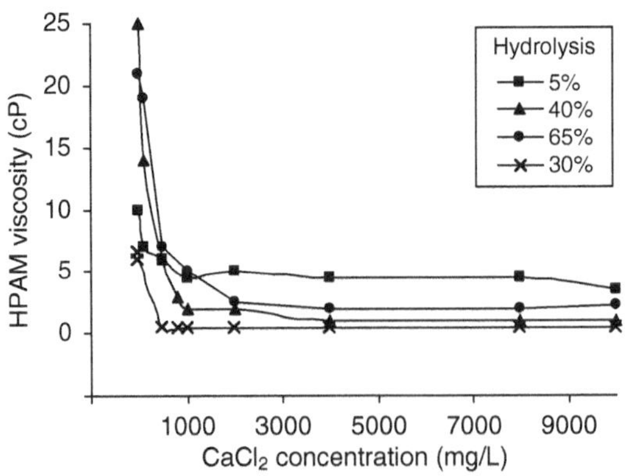

FIGURE 5.28 Effect of $CaCl_2$ concentration on HPAM viscosity with different levels of hydrolysis. *Source: Luo et al. (2006).*

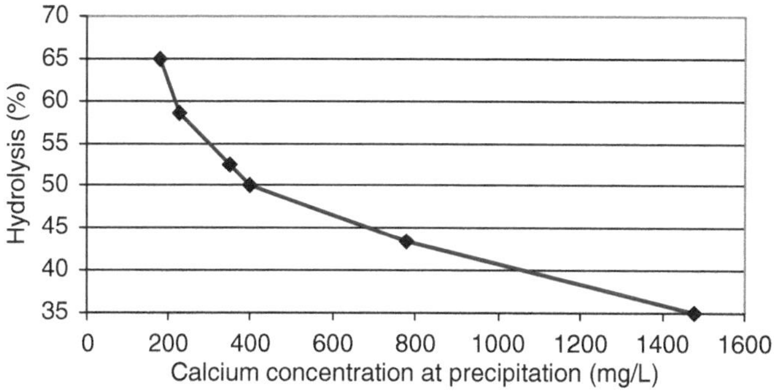

FIGURE 5.29 Degree of hydrolysis versus calcium ion concentration at which precipitation occurs. *Source: Luo et al. (2006).*

viscosity decreased rapidly at different levels of hydrolysis. At high $CaCl_2$ concentrations, the viscosity at a lower hydrolysis was higher, which is different from the solution with NaCl. The reason is that the divalent Ca^{2+} crosslinks with the acrylic group in HPAM, resulting in coagulation of the molecules, in addition to the compression function of monovalent Na^+.

The higher the hydrolysis, the lower the $CaCl_2$ concentration at which the coagulation occurs, as shown in Figure 5.29. Therefore, in a reservoir of low temperature and low hardness, HAPM viscosity may not change within some period of time, as hydrolysis increases gradually. Sometimes, the viscosity may increase initially. In a reservoir of low temperature and high hardness, HAPM viscosity decreases slowly, as hydrolysis increases gradually. Finally, precipitation may occur. In a reservoir of high temperature and low hardness, HAPM viscosity decreases sharply as hydrolysis increases rapidly due to the strong temperature effect, but precipitation may not occur. In a reservoir of high temperature and high salinity, HAPM viscosity decreases sharply as hydrolysis increases rapidly, and precipitation may occur.

Mechanical Degradation

Mechanical degradation describes the breakdown of molecules in the high flow rate region close to a well as a result of high mechanical stresses on the macromolecules. This short-term effect is important only in the reservoir near the wellbore (and also in some of the polymer handling equipment, in chokes, and so on).

Figure 5.30 compares HPAM viscosity versus shear rate in a core when it was unsheared and presheared before core flood, whereas Figure 5.31 shows the viscosity for a xanthan solution. Presheared solutions were used to investigate mechanical degradation. When the shearing effect on the two polymers is compared, the striking difference is that xanthan appeared to be extremely shear stable because of the rigid rod structures, whereas

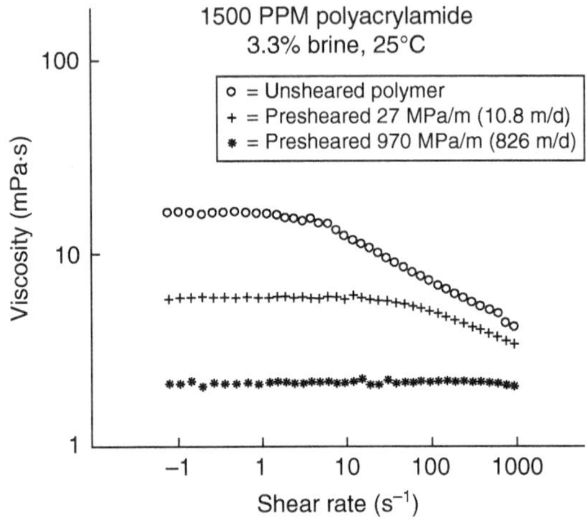

FIGURE 5.30 Effect of severe shearing and resulting mechanical degradation in a Berea core on the viscosity of an HPAM sample. *Source: Seright et al. (1983).*

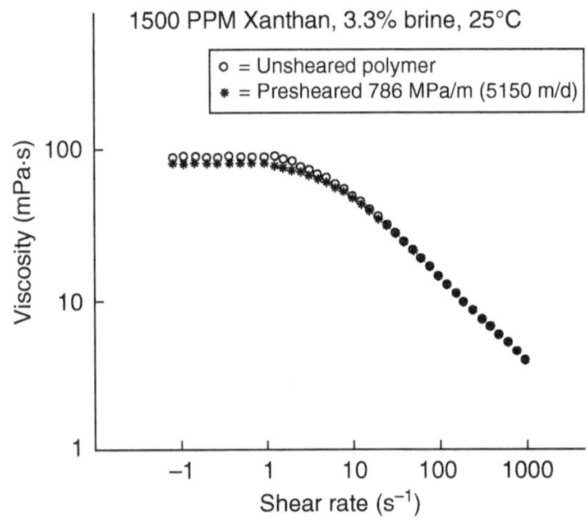

FIGURE 5.31 Effect of severe shearing in a Berea core on the viscosity of a xanthan solution; very little mechanical degradation was evident. *Source: Seright et al. (1983).*

polyacrylamide was very sensitive to shear degradation because of the flexible coil molecules.

In Figure 5.30, the viscosity versus shear rate curves of a given polyacrylamide solution are shown before and after different levels of shearing through a consolidated sandstone core. Even after fairly modest levels of shearing for the polyacrylamide solution (10.8 m/d), the viscosity was considerably reduced; after extreme shearing at a very high flow rate through the sandstone core, the viscosity was only slightly above that of the brine. The corresponding results for a 1500 ppm xanthan solution in the same brine are shown in Figure 5.31. These results demonstrate that the xanthan was extremely stable to mechanical degradation, even at very high flow rates through the porous medium.

Mechanical degradation of polymer is much more severe at higher flow rates, longer flow distances, and lower brine permeabilities of porous media. In a lower-permeability porous medium, the average pore throat diameter is smaller, and the stress acting on the polymer is larger. Thus, it is more probable for the polymer chains to be broken and the viscosity to be more heavily reduced. Similarly, we can understand the effects of flow rate and flow distance.

The rate of polymer chain rupture in high shear flow depends on the molecular weight. Larger molecules offer more resistance to flow, consequently experience larger shearing or elongational stresses, and are therefore more likely to break (Sorbie, 1991; Luo et al., 2006). The higher molecular weight species in the molecular weight distribution (MWD) are broken down into some combination of the lower molecular weight fragments, leading to a redistributed MWD after shearing. Work by Seright et al. (1981) confirmed that, when HPAM is sheared at high flow rates through Berea cores, the MWD is altered as shown for an HPAM sample in Figure 5.32. The initial MWD of the polymer is altered to a final MWD showing a higher peak at lower molecular weights. For a given fluid shear stress, there is a "critical" molecular weight, M_c, below which no mechanical degradation will occur (Sorbie, 1991). Akstinat (1980) found that the average molecular mass of an HPAM is not the decisive factor for the shear stability; instead, the MWD is more important.

Maerker (1975, 1976) found that the mechanical degradation of the polymer was more severe in higher salinity brines and that the presence of calcium ions (Ca^{2+}) had a particularly damaging effect over and above that expected

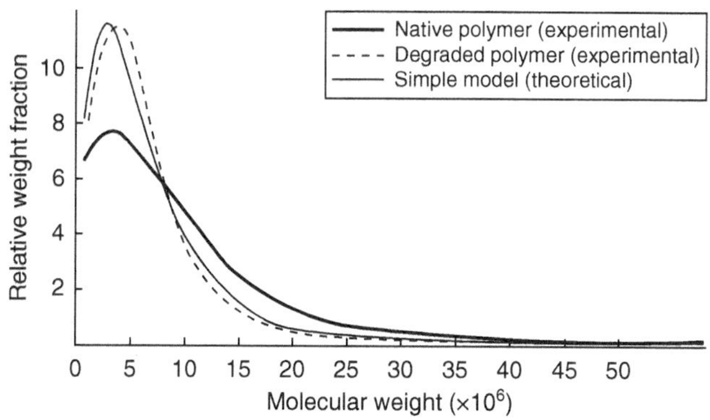

FIGURE 5.32 The measured changes in the MWD of an HPAM sample after mechanical degradation in a sandstone core. *Source: Seright et al. (1981).*

from the simple increase in the solution's ionic strength. Maerker suggested that softening the injection water may significantly reduce the mechanical degradation.

A simple and useful device for characterizing polymer solution mechanical degradation, known as the screen viscometer, was introduced by Jennings et al. (1971). This device consists of three or five 100-mesh screens, as shown in Figure 5.33. Such a viscometer has several modifications (for example, by Lim et al., 1986). The screen factor (SF) is defined as the ratio of the flow time for the polymer solution through the screen viscometer to the flow time for the same volume of solvent. The screen factor is a measure of the viscoelastic response of the polymer that is able to sustain sudden elongational deformation and the resulting normal stresses. Elongational deformation occurs as the solution accelerates through the screen openings of the screen viscometer.

Similar deformation will occur in porous media during the flow of fluids through the constrictions of the convergent-divergent channels. The induced normal stresses can be large enough to break the polymer bonds and mechanically degrade the polymer. Thus, the screen factor is a better measure of mechanical degradation than its viscosity, which responds to shear stresses. This measurement, however, is polymer specific and cannot be used to compare different polymers (Castor et al., 1981a). The flow through the screen is a complex shear/elongational flow and is not amenable to simple analysis. The flow rates in the screen viscometer experiments are generally very high compared with typical interstitial flow rates in porous media. Thus, it is suggested that the SF measurement is a simple, straightforward, and useful qualitative characterization for polymer solutions.

Another parameter used to describe polymer solution (polysaccharides) filterability is millipore filter ratio. Tests for the filterability of polysaccharides through millipore filters are similar in procedure to the screen factor tests for polyacrylamides but yield different interpretations. The millipore filter ratio is defined as the ratio of the time required for the last 250 mL to that for the first 250 mL of 1000 mL of 500 ppm polymer solution to flow through a presaturated 1.2 micron millipore filter under a constant gas pressure of, for example,

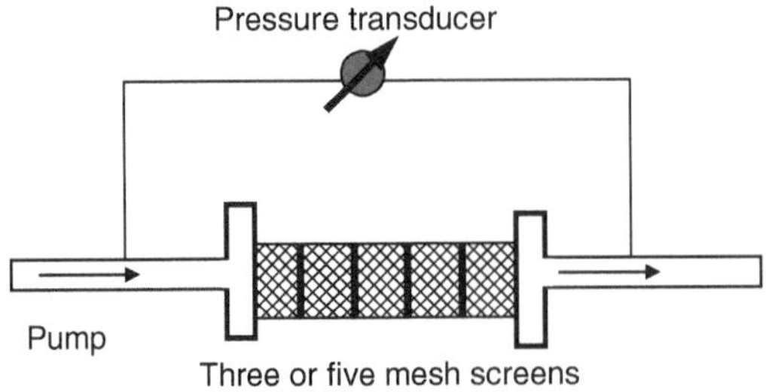

FIGURE 5.33 Schematic of screen viscometer apparatus for measuring screen factor.

40 psi. The millipore filter ratio is a measure of the propensity of polysaccharide molecules to plug the filter and, by inference, the oil formation. Solutions with millipore filter ratios greater than 1.5 are considered unacceptable (Castor et al., 1981a).

To characterize polymer stability, the University of Texas at Austin uses the filtration ratio (FR), which is defined as

$$FR = \frac{t_{200mL} - t_{180mL}}{t_{80mL} - t_{60mL}}, \tag{5.8}$$

where t_{200mL} is the flow time when 200 mL of polymer solution passes through the filter, and similar definitions for t_{180mL}, t_{80mL}, and t_{60mL}. FR should be less than 1.2 for the polymer to pass this screen test.

Biological Degradation

Biological degradation refers to the microbial breakdown of macromolecules of polymers by bacteria during storage or in the reservoir. Although the problem is more prevalent for biopolymers, biological attack may also occur for synthetic polymers. It has been found that HPAM can provide nutrition to sulfate-reducing bacteria (SRB). As the number of SRB increases, HPAM viscosity decreases. For example, when the number of SRB reaches 36000/mL, the viscosity loss of HPAM of 1000 mg/L is 19.6% (Luo et al., 2006).

There are four bacteria in reservoirs, and their concentrations are in the order of TGB-O > TGB-A > HOB > SRB. Their effect on polymer viscosity loss is shown in Figure 5.34. Their concentrations were 2% with 10^5 bacteria in 1 mL liquid (these concentrations are much higher than typical values in reservoirs, though). The polymer concentration was 1000 mg/L.

Biological degradation is important only at low temperatures or in the absence of effective biocides. The use of a biocide is the almost universal

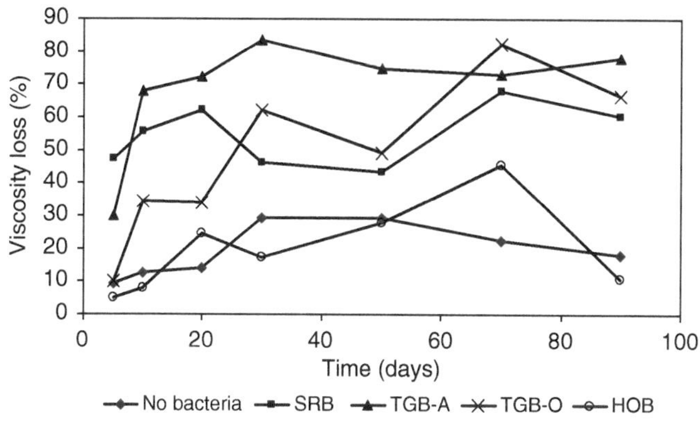

FIGURE 5.34 The effect of bacteria on polymer viscosity. *Source: Data from Niu et al. (2006).*

answer to biological degradation. Probably, the most common biocide used in oilfield applications in the past was formaldehyde (HCHO) diluted in aqueous solution (O'Leary et al., 1985; Luo et al., 2006). Because formaldehyde is toxic, it has limited applications these days. Also, if such a biocide is used, it may affect other chemicals in the package that are used to protect the polymer; for example, it may interact with the oxygen scavengers. Bacterial attack has been observed in at least two field tests (van Horn, 1981; Bragg et al., 1982).

Viscosity Loss

The viscosity loss from the mixing tank to the static mixer was observed to be about 6% (Liu et al., 1998). Pang et al. (1998a) found that the HPAM polymer viscosity losses at the static mixer, injection well, observation well, and production well in Daqing were cumulatively 10%, 18%, 55%, and 73% of the viscosity at the storage tank, respectively. The initial polymer concentration was 1000 mg/L. The molecular weights decreased from 10 million at the storage to 7, 6, 5, and 3.5 million at the static mixer, injection well, observation well, and production well, respectively. The viscosity losses due to *mechanic* shearing at the high-pressure metering pump, transportation pump, and filter were about 5, 2, and 1%, respectively. The viscosity loss at perforation due to *mechanic* shearing was about 9%. A 42.95% loss was reported after polymer solution was returned from an injection well (i.e., after the polymer solution flowed through perforation twice; Zhang and Yang, 1998). Zhu and Zheng (1998) measured the viscosity losses at several parts of an injection system. The operation requirement in Chinese oil companies is that the shear loss in the surface system should be less than 20% (Liu et al., 1998).

The viscosity loss from the static mixer to the injection wellhead was mainly caused by chemical degradation due to F^{2+}. It was found that F^{2+} concentrations at the static mixer, injection well, and in the solution returned from the injection well were 0.3, 0.6, and 10 mg/L. Experimental data showed that the viscosity loss reached 77% when the solution had 2 mg/L Fe^{2+}. If 100, 400, or 800 mg/L formaldehyde was added, the viscosity loss was 67%, 56%, or 36%, respectively (Pang et al., 1998a).

5.4 POLYMER FLOW BEHAVIOR IN POROUS MEDIA

This section discusses polymer rheology, polymer retention in porous media, and rock permeability reduction.

5.4.1 Polymer Rheology in Porous Media

In a discussion of rheology, one important parameter is viscosity. First, we should be aware of the different terminologies related to viscosity. Bulk viscosity is the viscosity measured in a viscometer, which was discussed previously. In situ viscosity in porous media, which is one of the subjects in this section, is not directly measured. Instead, it is calculated according to the Darcy equation using core flood experimental data. This calculated viscosity is called apparent viscosity. Sorbie (1991) used the terms apparent viscosity (the symbol η_{app}) to describe polymer solution viscosity in porous media and effective viscosity (the symbol η_{eff}) to describe polymer viscosity in a single capillary tube. For bulk viscosity, he used the symbol μ to describe Newtonian viscosity, η to describe non-Newtonian viscosity, and $\bar{\eta}$ to describe elongational viscosity. This book simply uses the symbol μ with some appropriate subscripts to describe viscosities. To calculate in situ polymer viscosity, we must first calculate equivalent shear rate in porous media.

Equivalent Shear Rate in Porous Media

Polymer viscosity is strongly shear dependent. If we use the bulk viscosity measured at different shear rates to describe the flow behavior in porous media, our first task is to calculate the shear rate which is equivalent to that in the bulk viscometer. To do that, we start with the capillary flow of a non-Newtonian fluid.

Figure 5.35 provides a schematic of capillary flow. Here, we take a small element that is of a length ΔL in **x** direction and from the axis to r in **r** direction. According to the force balance for this small element, we have

$$2\pi r(\Delta L)\tau_r = \pi r^2(\Delta p), \tag{5.9}$$

where τ_r is the shear stress at r, and Δp is pressure drop within ΔL. If the non-Newtonian rheology can be described by a power-law, then

$$\tau_r = K(\dot{\gamma})^n = -K\left(\frac{dv}{dr}\right)^n, \tag{5.10}$$

where v is the flow velocity, and K and n are the constants fitting the power-law equation to describe the bulk viscosity. We now substitute Eq. 5.10 for τ_r in Eq. 5.9 and integrate from r to R with the corresponding u(r) from u to 0 because the velocity at the capillary wall r = R is 0:

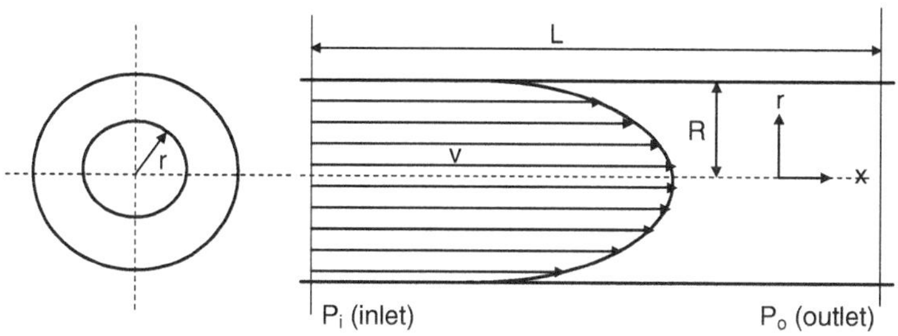

FIGURE 5.35 Schematic of capillary flow.

$$\int_{u}^{0} dv = \int_{r}^{R} \left(\frac{\Delta p}{2K(\Delta L)} \right)^{1/n} r^{1/n} dr. \tag{5.11}$$

We now have

$$v(r) = \left(\frac{\Delta p}{2K(\Delta L)} \right)^{1/n} \frac{n}{n+1} R^{(n+1)/n} \left[1 - \left(\frac{r}{R} \right)^{(n+1)/n} \right]. \tag{5.12}$$

From the preceding equation, it can be easily shown that the average velocity is

$$\overline{v} = \frac{n}{3n+1} R^{(n+1)/n} \left(\frac{\Delta p}{2K(\Delta L)} \right)^{1/n}. \tag{5.13}$$

For a Newtonian fluid, if n = 1 and K = μ, then the previous equation becomes

$$\overline{v} = \frac{R^2 \Delta p}{8\mu(\Delta L)}. \tag{5.14}$$

This is the well-known Hagen–Poiseuille equation.

From Eq. 5.12, the shear rate is

$$\dot{\gamma}(r) \equiv \frac{dv(r)}{dr} = -\left(\frac{\Delta p}{2K(\Delta L)} r \right)^{1/n}. \tag{5.15}$$

The shear rate at the capillary *wall* (r = R) is

$$\dot{\gamma}_w = -\left(\frac{\Delta p}{2K(\Delta L)} R \right)^{1/n}. \tag{5.16}$$

According to Eq. 5.13, the preceding equation becomes

$$|\dot{\gamma}_w| = \left(\frac{3n+1}{n} \right) \frac{\overline{v}}{R}, \tag{5.17}$$

According to Eq. 5.15, it can be shown that the average shear rate is

$$\overline{\dot{\gamma}} = -\frac{2n}{(2n+1)} \left(\frac{\Delta p}{2K(\Delta L)} R \right)^{1/n} = \frac{2n}{(2n+1)} \dot{\gamma}_w. \tag{5.18}$$

In the capillary flow, the flow capacity is defined by capillary radius R. In porous media, the flow capacity is defined by permeability k and porosity ϕ. Now we have to correlate these parameters.

In porous media, according to the Darcy equation,

$$\overline{v} = \frac{kA(\Delta p)}{\phi\mu(\Delta L)}, \tag{5.19}$$

where ϕ is the porosity, and $\bar{v}$ is the average interstitial pore velocity. Comparing Eq. 5.19 with Eq. 5.14 for a Newtonian fluid, we have

$$R = \sqrt{\frac{8k}{\phi}}. \tag{5.20}$$

The preceding relationship is derived based on a Newtonian fluid.

If we calculate the equivalent shear rate in porous media using the formula at the capillary wall, according to Eq. 5.17, the formula is

$$\dot{\gamma}_{eq} = \left(\frac{3n+1}{4n}\right)\alpha\frac{4u}{\sqrt{8k\phi}}, \tag{5.21}$$

where u (= $\bar{v}\phi$) is the Darcy velocity in porous media. In the preceding equation, a parameter α is added to fit experimental data. The unit should be consistent. For example, if the shear rate is in s^{-1}, u is in m/s, and k is in m^2.

The procedures to calculate the equivalent sear rate in porous media can be summarized as follows:

1. Measure the bulk viscosity of a polymer solution at different shear rates. Then we have μ versus $\dot{\gamma}$. Obtain K and n by fitting the data into the power-law equation:

$$\mu = K(\dot{\gamma})^{(n-1)}. \tag{5.22}$$

2. Conduct core flood tests with the polymer solution at different injection rates. Measure the pressure drop, Δp, corresponding to each injection rate (velocity u). The core permeability and porosity are measured before the core flood tests.
3. Calculate the apparent viscosity, μ_{app}, using the Darcy equation at each injection rate (u), and shear rates, $\dot{\gamma}_{eq}$, according to Eq. 5.21, by setting α to an empirical value.
4. Adjust α to calculate $\dot{\gamma}_{eq}$ so that the core flood data, μ_{app} versus $\dot{\gamma}_{eq}$, match the viscometric bulk viscosity data, μ versus $\dot{\gamma}$.

Different researchers have proposed formulae similar to Eq. 5.21 to calculate $\dot{\gamma}_{eq}$ in porous media—for example, Christopher and Middleman (1965), Hirasaki and Pope (1974), Teeuw and Hesselink (1980), Willhite and Uhl (1986), and Cannella et al. (1988). Cannella et al. used the following equations to estimate the equivalent shear rate (s^{-1}), $\dot{\gamma}_{eq}$, and apparent viscosity (mPa·s), μ_{app}, in porous media:

$$\dot{\gamma}_{eq} = C\left[\frac{3n+1}{4n}\right]^{n/(n-1)}\left[\frac{u_p}{\sqrt{kk_{rw}S_w\phi}}\right], \tag{5.23}$$

$$\mu_{app} = \mu_\infty + K\dot{\gamma}_{eq}^{(n-1)}. \tag{5.24}$$

Here, the constants, K ($mPa{\cdot}s^n$) and n (dimensionless), are the consistency index and the exponent, respectively; u_p is the Darcy velocity (m/s) of the polymer-containing water phase; $\bar{k}$ is the average permeability in m^2; k_{rw} is the water phase relative permeability; S_w is water saturation (fraction); ϕ is porosity (fraction); μ_∞ is the viscosity at infinite shear rate; and C is an empirical constant. Note that Eq. 5.23 is made more general by including the nonunit water saturation, S_w, and the water relative permeability, k_{rw}, as was done previously by Hirasaki and Pope (1974). To consider the polymer permeability reduction factor F_{kr} explicitly (to be discussed later), we should divide the permeability $\bar{k}$ by F_{kr}, and u_w is substituted for u_p. Then Eq. 5.23 becomes

$$\dot{\gamma}_{eq} = C\left[\frac{3n+1}{4n}\right]^{n/(n-1)}\left[\frac{u_w}{\sqrt{\bar{k}k_{rw}S_w\phi/F_{kr}}}\right]. \tag{5.25}$$

Cannella et al. (1988) reported that C = 6 in Eq. 5.23 fits a wide variety of core flood data well. In UTCHEM, the coefficient of Eq. 5.23 is lumped into one coefficient $\dot{\gamma}_c$,

$$\dot{\gamma}_c = C\left[\frac{3n+1}{4n}\right]^{n/(n-1)}, \tag{5.26}$$

where $\dot{\gamma}_c$ is an empirical parameter (dimensionless) used to consider nonideal effects such as slip at the pore walls. $\bar{k}$ is given by

$$\bar{k} = \left[\frac{1}{k_x}\left(\frac{u_{xj}}{u_j}\right)^2 + \frac{1}{k_y}\left(\frac{u_{yj}}{u_j}\right)^2 + \frac{1}{k_z}\left(\frac{u_{zj}}{u_j}\right)^2\right]^{-1}, \tag{5.27}$$

for phase j in UTCHEM. Wreath et al. (1990) made a comprehensive review of different expressions for the equivalent shear rate, including Eq. 5.27.

There is no consensus regarding which formula gives the best prediction. Ideally, we need to have two sets of data: viscometric and core flood. Generally, we have viscometric data. If we lack core flood data, we have to rely on matching other available data by adjusting α in Eq. 5.21 or C in Eq. 5.23 or Eq. 5.25. According to the review by Sorbie (1991), there appears to be less change in n between the bulk and the porous medium, compared with the change in K.

When comparing the viscometric and core flood data, the reader should be reminded that several factors could lead to incorrectly estimated values of μ_{app} in the core flood tests. The polymer may be adsorbed and retained in the porous media, or there is microgel, which would lead to reduced permeability. Thus, if the permeability reduction is not considered, the estimated μ_{app} using the Darcy equation could be higher than the actual viscosity values because the shear rate is underestimated (see Eq. 5.25). There is also the slip effect (Sorbie, 1991), which occurs in a low-shear regime and in a low-concentration polymer solution. The slip effect leads to a layer at the pore wall that is depleted in polymer. In the depleted layer, the solution viscosity is lower (close to solvent viscosity) than the bulk solution viscosity; therefore, it works as a lubricant layer. If the μ_{app} is estimated using the Darcy equation, it would be lower than the bulk viscosity. Actually, inaccessible pore volume (IPV) and the slip effect represent the same phenomenon. The former focuses on macroscope, the latter on microscope in low-concentration polymers. Other terms, such as surface exclusion mechanism (Chauveteau, 1982), velocity enhancement, and excluded pore volume (Sorbie, 1991), are also used.

A typical value of n in the power-law model is 0.4 to 0.7 (Sorbie, 1991). If n = 0.5 and C = 6, $\dot{\gamma}_c$ is 4.8 without a unit conversion coefficient. With these empirical constants and the velocity of 1 ft/day, the shear rates will be 13.6 to 136 s^{-1} for the permeability from 5000 md to 50 md and the porosity of 0.3. The C values from different researchers were summarized by Cannella et al. (1988); their C = 6 is the largest. The other values are 0.98 from Christopher and Middleman (1965), 1.414 from Teeuw and Hesselink (1980), and 2.041 from Bird et al. (1960). Chen et al. (1998) and Tang et al. (1998) used C equal to 1.8, but their velocity was defined as $v_w/[\phi \times (S_w - S_{wc})]$. According to Sorbie (1991), $\dot{\gamma}_c$ should be 4.5 to 6. $\dot{\gamma}_c$ equals 0 means without considering shear rate effect (polymer viscosity equal to that at zero shear rate).

5.4.2 Polymer Retention

Polymer retention includes adsorption, mechanical trapping, and hydrodynamic retention. These different mechanisms were discussed by Willhite and Dominguez (1977). Mechanical entrapment and hydrodynamic retention are related and occur only in flow-through porous media. They play no part in free powder/bulk solution experiments. Retention by mechanical entrapment is viewed as occurring when larger polymer molecules become lodged in narrow flow channels (Willhite and Dominguez, 1977).

The significance of the mechanical entrapment depends on the pore size distribution. It is a more likely mechanism for polymer retention in low-permeability formation (Szabo, 1975; Dominguez and Willhite, 1977). If the entrapment process acts on polymer molecules down to about the average size in the distribution, it will inevitably lead to a buildup of material close to the injection well, which gives an approximately exponential penetration profile into the formation, as shown in Figure 5.36. This will ultimately lead to pore blocking and well plugging, which is, of course, totally unsatisfactory. This is one reason that the polymer flood should be used in a high permeability formation.

In the experiment using sand pack (see Figure 5.36), when the injected polymer concentration was 600 ppm, the dynamic polymer retention at the inlet was 24.5 μg/g and retention at the exit it was 6 μg/g. If the curve's trend was extended to the infinite distance, then the retention was 3.31 μg/g; the

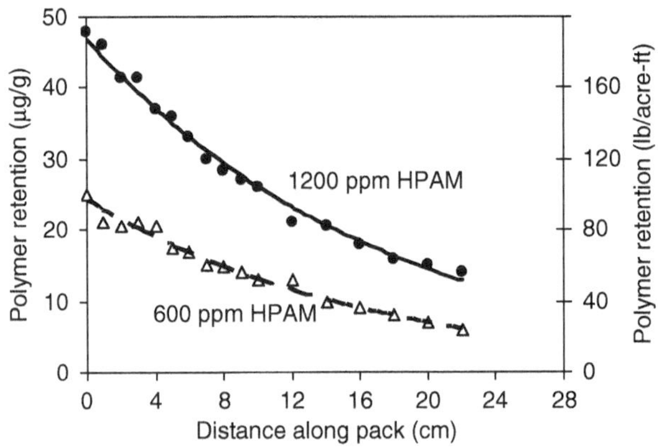

FIGURE 5.36 Distribution of retained HPAM along a sand pack after a polymer flood. *Source: Szabo (1975).*

corresponding static adsorption was 2.5 μg/g. The static adsorption was independent of the polymer concentration, while dynamic retention depends on polymer concentration. This concentration dependency arises when multiple particles or molecules arrive simultaneously at a pore throat large enough to admit one particle, but not several particles. Experimental data show that the dynamic retention was much higher than the static retention. This indicates that the mechanical trapping of polymer molecules plays an important role in polymer retention.

After a steady state is reached in a polymer retention experiment in a core, the total level of retention increases when the fluid flow rate is increased (Chauveteau and Kohler, 1974). This type of rate-dependent retention, called hydrodynamic retention, is not understood as well. Fortunately, it is generally thought to give a small contribution to the total retained material (Sorbie, 1991).

Adsorption refers to the interaction between polymer molecules and the solid surface. This interaction causes polymer molecules to be bound to the surface of the solid, mainly by physical adsorption, van der Waals forces, and hydrogen bonding. Essentially, the polymer occupies surface adsorption sites. Adsorption depends on the surface area exposed to the polymer solution, and it is the only mechanism that removes polymer from the bulk solution if a free solid powder, such as silica sand or latex beads, is introduced into the bulk solution and stirred until equilibrium is reached.

For the preceding three mechanisms of polymer retention, mechanical entrapment can be avoided by prefiltering or preshearing the polymer or by applying the polymer in a high permeability formation. Hydrodynamic retention is probably not a large contributor in the total retention and can be neglected in field applications. Adsorption is a fundamental property of the polymer-rock surface-solvent system and is the most important mechanism. Compared with alkaline and surfactant, because of large molecules, polymer mechanical trapping and hydrodynamic retention are more significant. However, because it is difficult to differentiate these three mechanisms in dynamic flood tests, we may simply use the term retention to describe the polymer loss, sometimes just using the term adsorption. Apparently, adsorption is discussed more often in literature on the topic.

Units

The laboratory unit used to define polymer retention, $\hat{C}_p$, is in mass of polymer per unit mass of solid, usually in micrograms per gram of rock (μg/g). Sometimes (e.g., in UTCHEM), the unit is in grams per 100 milliliter (cm^3) of pore volume (PV), g/100 mL PV, which is equivalent to weight percent (wt.%) if the solvent (water) density is 1 g/mL and the pore volume is filled up by the solvent (water) only. In bulk static adsorption, a more fundamental measure of adsorption is the mass of polymer per unit surface area of solid, which is referred to as the surface excess, $\hat{C}_{ps}$, usually in milligrams or micrograms per square meter (mg/m^2 or μg/m^2). Sometimes, in field applications, the retention unit is in mass of polymer per unit volume of rock, usually in lb/acre-foot.

These units can be converted to each other according to the relationships.

$$\begin{aligned}
\hat{C}_p\left[\frac{\text{g polymer}}{\text{100g water}}\right] &\equiv \hat{C}_p[\text{wt\%}] \approx \hat{C}_p\left[\frac{\text{g polymer}}{100\text{cm}^3\ \text{PV}}\right] \\
&= \hat{C}_p\left[\frac{\mu\text{g polymer}}{\text{g rock}}\right]\left[\frac{10^{-6}\text{g polymer}}{1\mu\text{g polymer}}\right]\left[\frac{\rho_r\text{g rock}}{1\text{cm}^3\ \text{rock}}\right] \\
&\quad \left[\frac{(1-\phi)\text{cm}^3\ \text{rock}}{1\text{cm}^3\ \text{bulk rock}}\right]\times\left[\frac{1\text{cm}^3\ \text{bulk rock}}{\phi\text{cm}^3\ (\text{PV})}\right]\left[\frac{100\text{cm}^3\ \text{water}}{100\rho_w\ \text{gram water}}\right] \\
&= \hat{C}_p\left[\frac{\mu\text{g polymer}}{\text{g rock}}\right]\left[\frac{10^{-4}(1-\phi)\rho_r}{\phi\rho_w}\right],
\end{aligned} \tag{5.28}$$

where ρ_r is the rock density in g/cm^3, and ϕ is the porosity in fraction. The conversion factor is 9.4×10^{-4} if $\phi = 0.22$, $\rho_r = 2.65$ g/cm^3, and $\rho_r =$ 1 g/cm^3.

$$\begin{aligned}
\hat{C}_{ps}\left[\frac{\mu\text{g polymer}}{\text{m}^2}\right] &= \hat{C}_p\left[\frac{\mu\text{g polymer}}{\text{g rock}}\right]\left[\frac{1\text{g rock}}{S_r\ \text{m}^2}\right] \\
&= \hat{C}_p\left[\frac{\mu\text{g polymer}}{\text{g rock}}\right]\left[\frac{1}{S_r}\right],
\end{aligned} \tag{5.29}$$

In the preceding equation, S_r is the pore surface area in m^2 per gram of rock.

$$\hat{C}_p\left[\frac{\text{lb polymer}}{\text{acre-ft}}\right]=\hat{C}_p\left[\frac{\mu\text{g polymer}}{\text{g rock}}\right]\left[\frac{10^{-6}\text{g polymer}}{1\mu\text{g polymer}}\right]\left[\frac{0.0022\text{lb polymer}}{1\text{g polymer}}\right]$$
$$\times\left[\frac{\rho_r\text{g rock}}{1\text{cm}^3\text{ rock}}\right]\left[\frac{(1-\phi)\text{cm}^3\text{ rock}}{1\text{cm}^3\text{ bulk rock}}\right]$$
$$\left[\frac{1\,233\,481\,855.3\text{cm}^3\text{ bulk rock}}{1\text{acre-ft}}\right]$$
$$=\hat{C}_p\left[\frac{\mu\text{g polymer}}{\text{g rock}}\right][2.714(1-\phi)\rho_r]. \tag{5.30}$$

The conversion factor is 5.6 if $\phi = 0.22$ and $\rho_r = 2.65$ g/cm^3.

Equation to Define Polymer Adsorption

The Langmuir-type isotherm can be used (Lakatos et al., 1979), as it is in UTCHEM, to describe polymer adsorption. The Langmuir-type isotherm is given by

$$\hat{C}_p=\min\left(C_p,\frac{a_p\left(C_p-\hat{C}_p\right)}{1+b_p\left(C_p-\hat{C}_p\right)}\right), \tag{5.31}$$

where C_p is the injected polymer concentration, or in general, the polymer concentration before adsorption. $C_p - \hat{C}_p$ is actually the equilibrium concentration in the rock-polymer solution system. a_p and b_p are empirical constants. The unit of b_p must be the reciprocal of the unit of C_p. a_p is dimensionless. Note that C_p and $\hat{C}_p$ must be in the same unit. Because the C_p unit is usually in wt.%, using the unit for $\hat{C}_p$ in wt.% has some advantages. a_p is defined as

$$a_p=(a_{p1}+a_{p2}C_{sep})\left(\frac{k_{ref}}{k}\right)^{0.5}, \tag{5.32}$$

where a_{p1} and a_{p2} are input or fitting parameters, C_{sep} is the effective salinity, k is the permeability, and k_{ref} is the reference permeability of the rock used in the laboratory measurement. The effective salinity for polymer (C_{sep}) is defined in Eq. 5.2.

The reference permeability (k_{ref}) is the permeability at which the input adsorption parameters are specified. Eqs. 5.31 and 5.32 take into account the salinity, polymer concentration, and permeability.

Note that the Langmuir model is an equilibrium relationship, and its application assumes adsorption is instantaneous and reversible in terms of polymer concentration. When polymer adsorption is considered to be irreversible, the Langmuir model cannot be used *directly* when the polymer concentration is declining. An additional parameter, $\hat{C}_{p,max}$, must be used to track the adsorption history so that Eq. 5.33 applies.

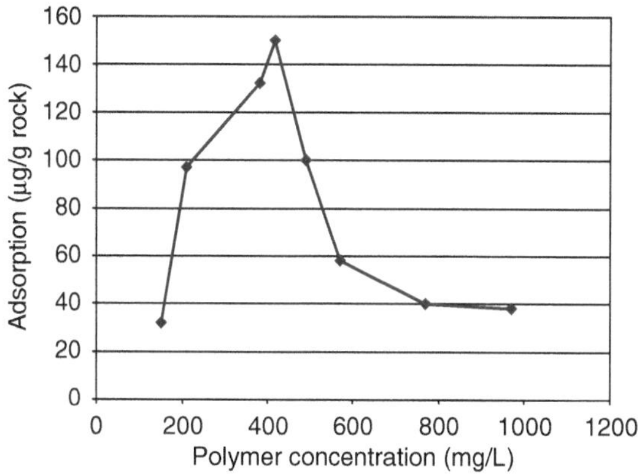

FIGURE 5.37 AP-2 polymer adsorption on Daqing sand. *Source: Data from Li (2007).*

$$\hat{C}_{p,max} = \max\left\{\left(\hat{C}_p\right)^1, \left(\hat{C}_p\right)^2, \ldots \left(\hat{C}_p\right)^n\right\}, \tag{5.33}$$

where 1, 2, …, n indicate time steps, with the current time step being n. $\hat{C}_{p,max}$ is not greater than the adsorption capacity. At very low concentrations, the Longmuir isotherm may be simplified to

$$\hat{C}_p = \min\left(C_p, a_p\left(C_p - \hat{C}_p\right)\right). \tag{5.34}$$

Li (2007) observed that the adsorption of a hydrophobically associating water-soluble polymer, AP-2, did not follow the Langmuir-type isotherm. Figure 5.37 shows that the adsorption increased to a maximum and then decreased as the polymer concentration was increased. The reason is probably that the hydrophobic polymer has an adsorption layer of multiple molecules on rock surfaces. When the polymer concentration is increased, the adsorption layer becomes thicker because of more adsorption. When the polymer concentration is further increased, the molecular interaction in the liquid is stronger than that between the adsorbed molecules and rock surfaces. Then the adsorbed molecules may leave the rock surfaces and redissolve into the liquid. Thus, the adsorption decreases.

Some Observations on Polymer Retention

Polymer adsorption/retention depends on the polymer type, solvent (salinity), and rock surface. Figures 5.38 and 5.39 present the cumulative distributions of some published adsorption data for HPAM and biopolymers, respectively. From Figure 5.38, the presented data show the median adsorption (at 50% cumulative distribution) for the HPAM type of polymer is 24 μg/g rock. The figure also shows that 70% of the adsorption data are below 30 μg/g rock.

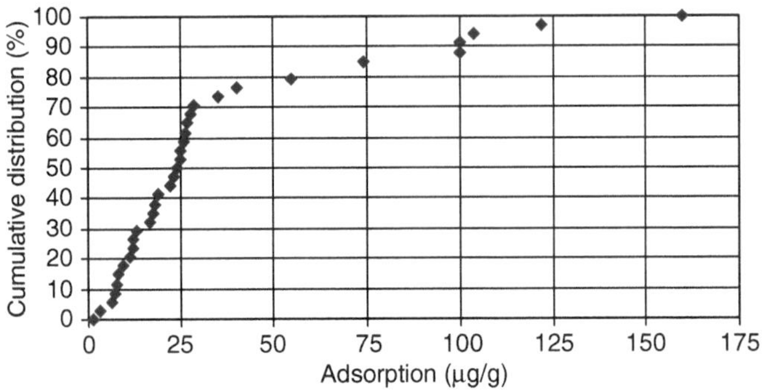

FIGURE 5.38 Cumulative distribution of synthetic polymer adsorption.

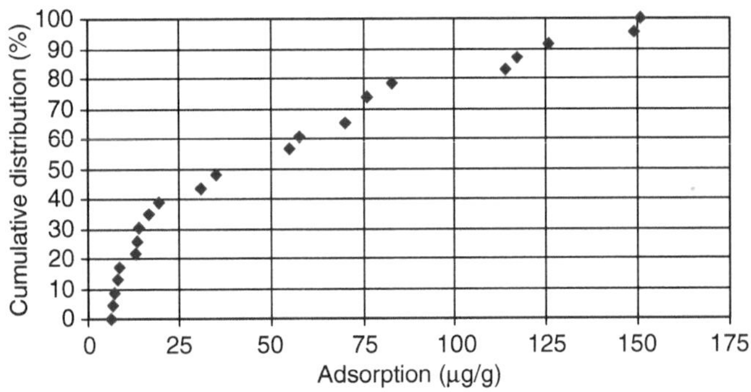

FIGURE 5.39 Cumulative distribution of biopolymer adsorption.

Figure 5.39 shows the median adsorption for biopolymers is 35 μg/g rock. Note that synthetic polymer adsorption is lower than biopolymer adsorption for the data analyzed. These median data may be used as a reference in cases without experimental data for a particular project. Some of the observations from the literature on the polymer adsorption/retention in flow-through porous media are discussed next.

Static Bulk Adsorption versus Dynamic Core Flood

There are large differences between the level of static adsorption of HPAM and dynamically retained level in a core or pack (Lakatos et al., 1979). These differences are the result of changes in the specific surface area of consolidated and unconsolidated packs and also the accessibility of certain portions of the pore space. These differences also depend on the extent of mechanical retention that is present in the dynamic core flood experiment. Polymer retention in consolidated porous media cannot be determined with static bulk adsorption (batch adsorption techniques) because the process of disaggregation to obtain

TABLE 5.10 Retention of Synthetic Polymers in Flow Tests

		Adsorption			
Reference	**Rock**	**μg/g**	**μg/m²**	**lb/acre-foot**	**Additional Data**
Smith, 1970	Silica		50		10% TDS
	Carbonate		300		10% TDS
	Carbonate		450		10% TDS+0.04% Ca^{2+}
Szabo, 1979	Berea			35–72	AMPS
	Berea			88–196	
	Sand pack	1.2–1.5			AMPS
	Silica flour	22.5			AMPS
	Silica flour	55	28		0.03% C_p
	Sand pack	3.3	27		0.03% C_p
	Carbonate	100	380		0.06% C_p

representative granular material generates a significant amount of new surface area, and polymer adsorption is usually excessive (Green and Willhite, 1998).

Data from Szabo (1979) in Table 5.10 show that the adsorption of silica flour (measured in static bulk adsorption) is 55 μg/g, which is higher than that (3.3 μg/g) of sand pack (dynamic flow test) because the surface area in the silica flour is higher than that in the sand pack. When the adsorption is defined in $\mu g/m^2$, which takes into account the surface area, the adsorption data from the two types of measurements are almost the same (28 $\mu g/m^2$ versus 27 $\mu g/m^2$).

Reversible or Irreversible Process

In most cases, polymer adsorption is considered irreversible; that is, it does not decrease as polymer concentration decreases (Szabo, 1979; Lakatos et al., 1979; Gramain and Myard, 1981). The irreversible effect is caused by polymer adsorption on rock. However, this is not exactly true because small amounts of polymer can be removed from porous rock using prolonged exposure to water or brine injection. Usually, however, the rate of release is so small that it is not possible to measure the concentrations accurately. It is thus more accurate to state that the rate of polymer retention is much greater than the rate of polymer removal. Retention also may occur when flow rates are suddenly increased. This process is called hydrodynamic retention, which is reversible (Green and Willhite, 1998).

Rock Surface Effect

The adsorption level of HPAM on calcium carbonate is much higher than that on the silica surface. The higher adsorption may be attributed to the strong interactions between the surface Ca^{2+} and the carboxylate groups on the HPAM (Smith, 1970; Szabo, 1979; Lakotos et al., 1979), as shown by the data reported by Smith (1970) and Szabo (1979) in Table 5.10.

Salinity Effect

The effect of increasing salinity (NaCl) concentration is to increase the level of polymer adsorption, as shown in Figure 5.40, where the adsorption at 2% total dissolved solids (TDS) is higher than that at 0.1% TDS for each pair of data (Martin et al., 1983). This observation is consistent with the prediction made by Eq. 5.31. Adding a low concentration of divalent calcium ion, Ca^{2+}, promotes HPAM adsorption on silica, as shown by data from Smith (1970) in Table 5.10, because the divalent ions compress the size of the flexible HPAM molecules and reduce the static repulsion between the polymer carboxyl group and silica surface.

Polymer Effect

AMPS adsorption is found to be lower than HPAM, shown by data from Szabo (1979) in Table 5.10, where the polymer used is HPAM if not marked with AMPS. Broadly, xanthan adsorption in porous media is rather less than that of HPAM and also tends to show less sensitivity to the salinity/hardness conditions of the solvent (Sorbie, 1991; Green and Willhite, 1998). However, this conclusion is not supported by the data shown in Figures 5.38 and 5.39, which show that the median adsorption for synthetic polymers (24 μg/g) is lower than that for biopolymers (35 μg/g).

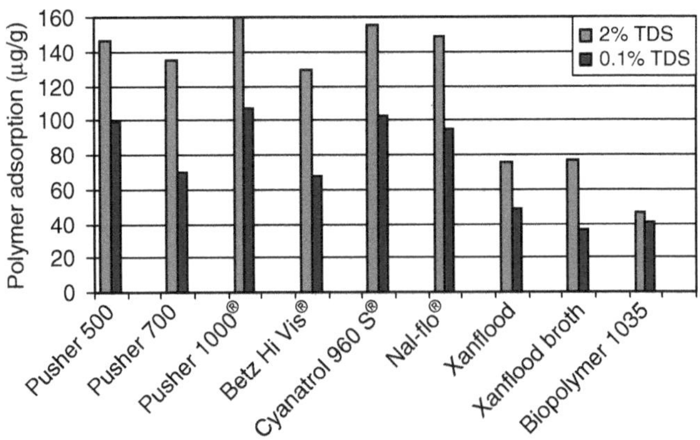

FIGURE 5.40 Salinity effect on polymer adsorption.

Molecular Weight Effect

Sometimes higher levels of adsorption are seen for higher molecular weight polymers (Lipatov and Sergeeva, 1974; Gramain and Myard, 1981), but this adsorption levels off after a value of molecular weight (Gramain and Myard, 1981). Figures 5.41 and 5.42 show the polymer (HPAM with 30% hydrolysis) adsorption on calcium-montmorillonite and sodium-montmorillonite, respectively. Figure 5.41 shows that the polymer adsorption seems to decrease with molecular weight. The adsorption reaches its plateau after 500 mg/L equilibrium concentration.

From Figure 5.42 for the adsorption on sodium-montmorillonite, the molecular weight effect is not obvious. The adsorption reaches its plateau after about 600 mg/L equilibrium concentration. Comparing the two figures, we can see that the adsorption on sodium–montmorillonite is much higher than calcium–montmorillonite. Probably, it is caused by the clay swelling in

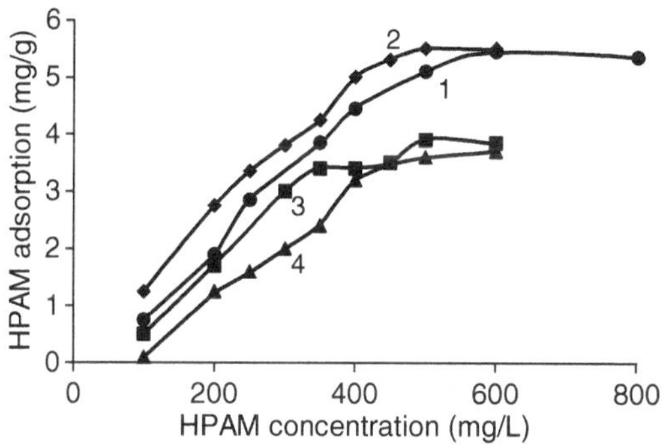

FIGURE 5.41 HPAM adsorption on calcium-montmorillonite (25°C). Molecular weight: curve 1, 6 million; curve 2, 9 million; curve 3, 10 million; and curve 4, 15 million. *Source: Yang et al. (2002a).*

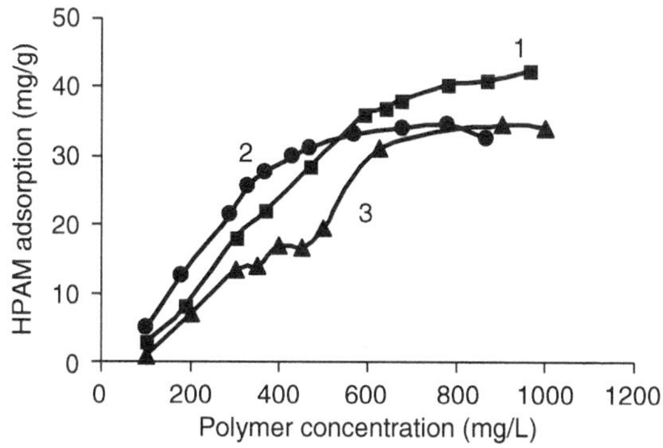

FIGURE 5.42 HPAM adsorption on sodium–montmorillonite (25°C). Molecular weight: curve 1, 6 million; curve 2, 9 million; and curve 3, 15 million. *Source: Yang et al. (2002a).*

sodium–montmorillonite. The clay swelling results in more adsorption sites. Lakatos et al. (1979, 1980), however, noted that dynamic adsorption in a silica sand decreased with increasing molecular weight, although the effect was not large. Chen and Chen (2002) made the same observation.

Polymer Concentration Effect

Equation 5.31 appears to show that polymer adsorption is a strong function of polymer concentration. Actually, polymer adsorption has weak concentration dependence (Vela et al., 1976; Shah, 1978). Figure 5.43 shows a sample curve defined by Eq. 5.31. This figure shows that as the polymer concentration increases, the adsorption curve quickly levels off to the value of a_p/b_p. Data from Shah's experiment (not shown here) showed the adsorption curve approached the plateau even more quickly than that in Figure 5.43. However, data from Lötsch (1988), shown in Figure 5.44, show that a higher biopolymer concentration led to a higher adsorption almost linearly.

Hydrolysis Effect

The level of HPAM retention in the sand pack decreases as the degree of hydrolysis increases (Lakatos et al., 1979; Chen and Chen, 2002). Figure 5.45 presents the HPAM adsorption data on unconsolidated Miocene sand. It shows that the adsorption decreased with hydrolysis, but there was a degree of hydrolysis at which the adsorption was at a minimum. The minimum adsorption is related to the charge interaction between the negatively charged silica surface

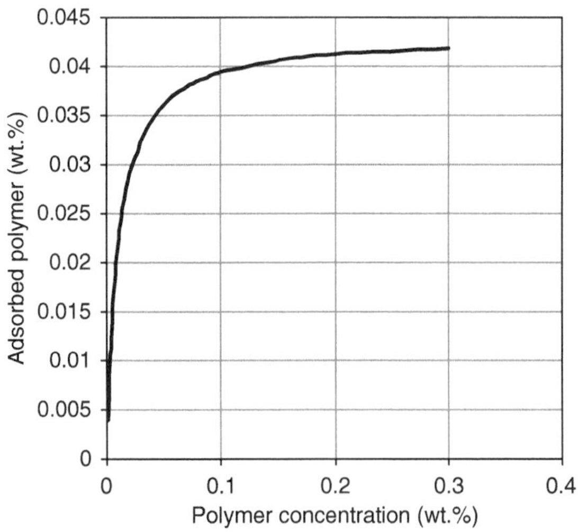

FIGURE 5.43 Polymer adsorption at different polymer concentration.

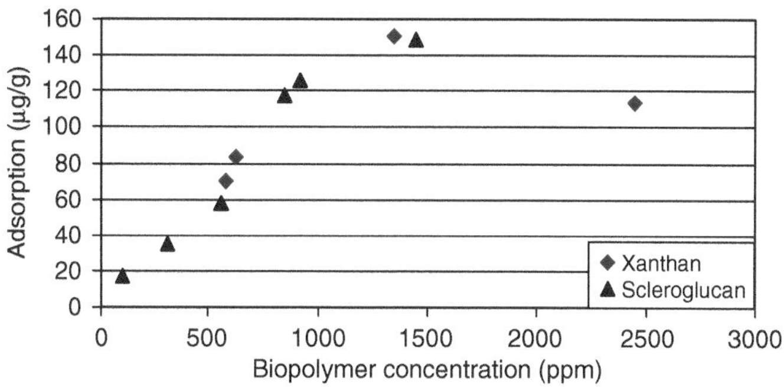

FIGURE 5.44 Biopolymer concentration effect on adsorption. *Source: Data from Lötsch (1988).*

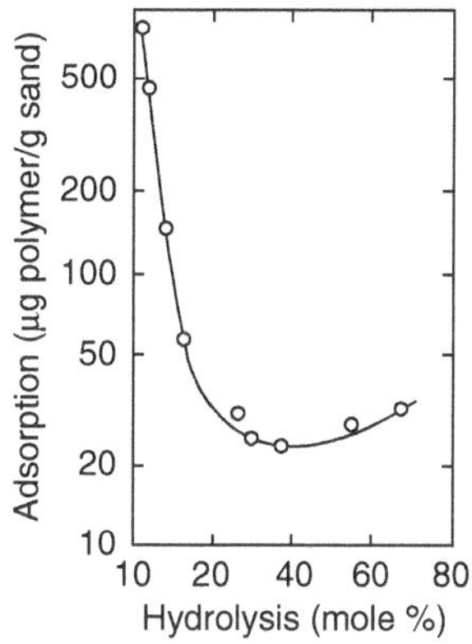

FIGURE 5.45 Adsorption of hydrolyzed polyacrylamide in 2.2% NaCl solutions on Miocene sand. *Source: MacWilliams (1973).*

on the sand and the negatively charged carboxyl group on the polymer. Because of the electrostatic repulsion, the adsorption decreases. For this example, the minimum adsorption was at 38% hydrolysis. That is one reason we need to have a maximum hydrolysis of 35 to 40%. This figure also indicates that the polymer must be partially hydrolyzed in order to reduce adsorption. For the same reason, adding alkali in polymer solution will reduce polymer adsorption.

Permeability Effect

As shown in Eqs. 5.31 and 5.32, polymer retention decreases with permeability. Figure 5.46 shows an example for HPAM adsorption in Berea. It is obvious that mechanical trapping in a low-permeability rock is higher than that in a high-permeability rock. Another possible explanation is high clay content in low-permeability rocks.

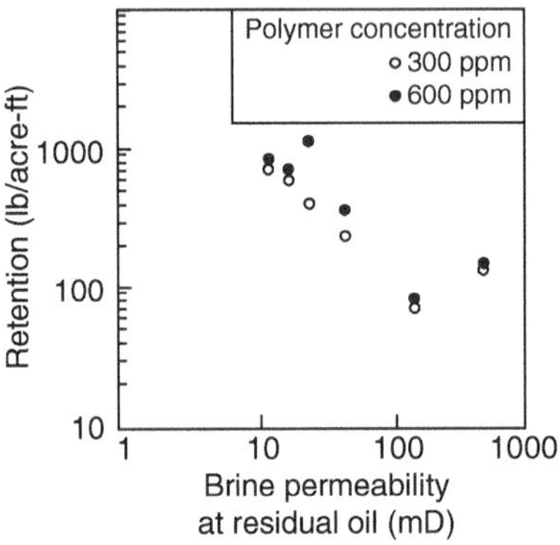

FIGURE 5.46 Variation of HPAM retention with initial brine permeability of a Berea core. *Source: Vela et al. (1976).*

Effect of Temperature

Adsorption for nonionic and anionic polymers decreases with temperature because of combined electrostatic repulsion and molecular forces that include van der Waals, hydrogen bond, hydrophobicity, and so on. For a nonionic polymer such as PAM, adsorption is more related to hydrogen bond. For an ionic polymer such as HPAM, adsorption is more related to electrostatic repulsion. When the temperature is increased, it is easier for the hydrogen bond to break up, causing PAM adsorption to decrease. When the temperature is increased, the negative charge on the rock surface is increased, resulting in higher electrostatic repulsion. Thus, the ionic polymer HPAM adsorption is reduced (Chen and Chen, 2002).

5.4.3 Inaccessible Pore Volume

When polymer molecular sizes are larger than some pores in a porous medium, the polymer molecules cannot flow through those pores. The volume of those pores that cannot be accessed by polymer molecules is called inaccessible pore volume (IPV). In an aqueous polymer solution with tracer, the polymer molecules will run faster than the tracer because they flow only through the pores that are larger than their sizes. This results in earlier polymer breakthrough in the effluent end. On the other hand, because of polymer retention, the polymer breakthrough is delayed. In other words, if only polymer retention is considered, the polymer will arrive in the effluent later than the tracer.

These two factors can be best explained by Figure 5.47. In the displacement shown in the figure, a constant polymer concentration is injected into the core, which has not been contacted previously by polymer. Retention causes the effluent concentration to lag behind the tracer, as shown in the first flood in the

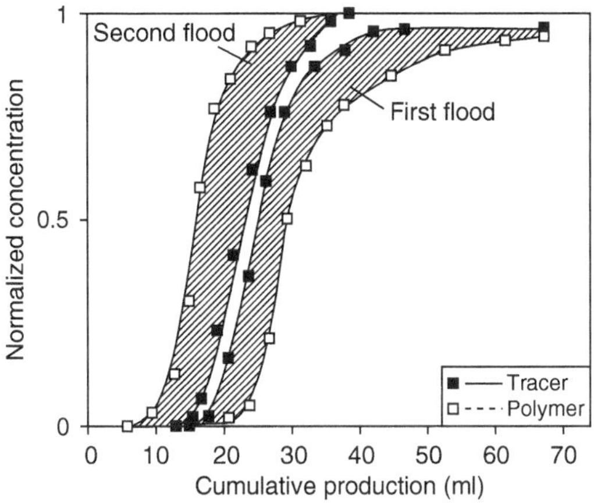

FIGURE 5.47 Comparison of tracer and polymer concentration profiles in the effluent when the polymer retention mechanism is dominated (first flood) and when IPV is significant (second flood). *Source: Hughes et al. (1990).*

figure. In the second flood, the adsorption sites may be partly or fully occupied by the previously injected polymer in the first flood, and inaccessible PV can offset the lag so that the polymer breakthrough is earlier than the tracer.

Another fact is that both polymer molecules and pores have a wide range of size distribution. Some small polymer molecules can flow through small pores, which tends to help the polymer flow with the tracer. However, IPV has been observed in all types of porous media for both synthetic polymers and biopolymers and is considered to be a general characteristic of polymer flow in porous media. Several models have been offered to explain why IPV occurs (DiMarzio and Guttman, 1970; Chauvetean, 1982; Chauveteau and Kohler, 1984; Kolodziej, 1987), but none has gained universal acceptance (Green and Willhite, 1998). The effect of IPV is modeled in UTCHEM by multiplying the porosity in the conservation equation for polymer by the input parameter defined as the ratio of effective porosity to the initial porosity. Inaccessible PV could be 1 to 30% PV. Laboratory data indicate that inaccessible pore volume is usually greater than adsorption loss for polymers following a micellar solution. The inaccessible pore volume in laboratory cores typically is 20% (Trushenski et al., 1974).

5.4.4 Permeability Reduction

Permeability reduction, or pore blocking, is caused by polymer adsorption. Therefore, rock permeability is reduced when a polymer solution is flowing through it, compared with the permeability when water is flowing. This permeability reduction is defined by the permeability reduction factor (F_{kr}):

$$F_{kr} = \frac{\text{Rock perm. when water flows}}{\text{Rock perm. when aqueous polymer solution flows}} = \frac{k_w}{k_p}. \tag{5.35}$$

The permeability reduction factor in UTCHEM is modeled as

$$F_{kr} = 1 + (F_{kr,max} - 1)\frac{b_{kr}C_p}{1 + b_{kr}C_p}, \tag{5.36}$$

where

$$F_{kr,max} = \max\left\{\left[1 - \frac{c_{kr}\left(A_{p1}C_{sep}^{S_p}\right)^{1/3}}{\sqrt{\frac{k}{\phi}}}\right]^{-4}, 10\right\}, \tag{5.37}$$

where b_{kr} and c_{kr} are input parameters derived from fitting core flood data, A_{p1} is the constant in Eq. 5.1, C_{sep} is calculated using Eq. 5.2, and S_p is from Figure 5.17. Note the units in the preceding two equations. The item $b_{kr}C_p$ in Eq. 5.36 must be dimensionless. The units of the items in Eq. 5.37 must be consistent so that $F_{kr,max}$ is dimensionless. A simple way to avoid any mistake is to use the same units as those in the prediction model when fitting the laboratory data to the preceding two equations. For example, if the unit of k in a UTCHEM model is Darcy, Darcy should be used when fitting experimental data.

In Eq. 5.37, it is assumed that the maximum permeability reduction, $F_{kr,max}$, is 10, which is an empirical value. However, a permeability reduction factor higher than 10 was observed in low-permeability formations.

Bondor et al. (1972) assumed that the permeability reduction is caused by polymer adsorption, and the adsorption process is irreversible. They further assumed the maximum permeability reduction corresponds to the polymer adsorptive capacity on the rock, AdC. The permeability reduction factor is linearly interpolated based on the ratio of the amount of polymer adsorbed to the adsorptive capacity:

$$F_{kr} = 1 + (F_{kr,max} - 1)\frac{\hat{C}_p}{AdC}. \tag{5.38}$$

If we assume that the permeability reduction is caused by polymer adsorption/retention, let us check whether the prediction from the previous equations is consistent with some observations on polymer adsorption/retention discussed earlier in Section 5.4.2.

Figure 5.48 shows the permeability effect on the maximum permeability reduction factor, $F_{kr,max}$, predicted from Eq. 5.37. This figure shows that $F_{kr,max}$ decreases with permeability, which is consistent with the observation that the polymer retention decreases with permeability, as shown in Figure 5.46. The laboratory-measured permeability reduction data at the polymer concentration

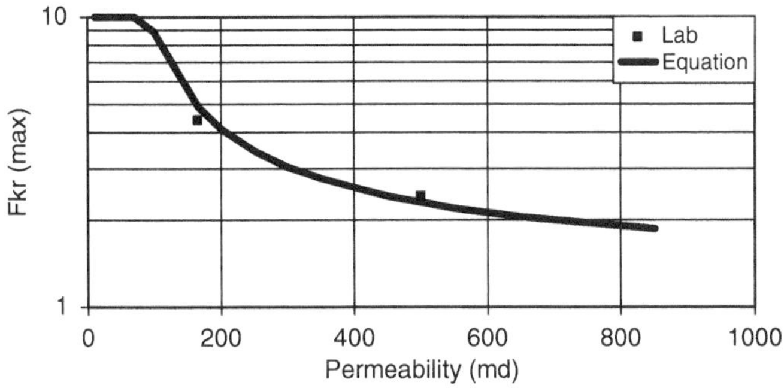

FIGURE 5.48 Permeability effect on $F_{kr,max}$ predicted from Eq. 5.37.

TABLE 5.11 Data Used in Generating Figure 5.48

C_{sep} (meq/mL)	0.68	From Figure 5.18
A_{P1}	9.45	From Figure 5.18
S_p	−0.2398	From Figure 5.17
b_{kr}	4.11	Derived from fitting lab data
c_{kr}	100	Derived from fitting lab data
porosity	0.22	Lab data
$F_{kr,max}$ at 166 md	4.4	Lab data
$F_{kr,max}$ at 500 md	2.4	Lab data
C_p (wt.%)	0.1	Lab data

of 0.1% are also shown in Figure 5.48. The data used to generate Figure 5.48 are shown in Table 5.11. Pang et al. (1998b) also showed that the higher the permeability, the lower the permeability reduction factor.

Figure 5.49 shows the polymer concentration effect on the permeability reduction factor, F_{kr}, predicted from Eq. 5.36. This figure shows that F_{kr} is a weak function of polymer concentration, and it increases slightly within a low concentration range. Concentration quickly reaches a plateau. This effect is consistent with the polymer adsorption shown in Figure 5.43.

Figure 5.50 shows the salinity effect on the permeability reduction factor, F_{kr}, predicted from Eq. 5.36. This figure shows that F_{kr} decreases with salinity. However, Figure 5.40 shows that higher salinity leads to higher polymer adsorption. Therefore, the salinity effect on permeability reduction factor is different from that on polymer adsorption. In other words, because of the salinity effect, the permeability reduction based on Eqs. 5.36 and 5.37 does not

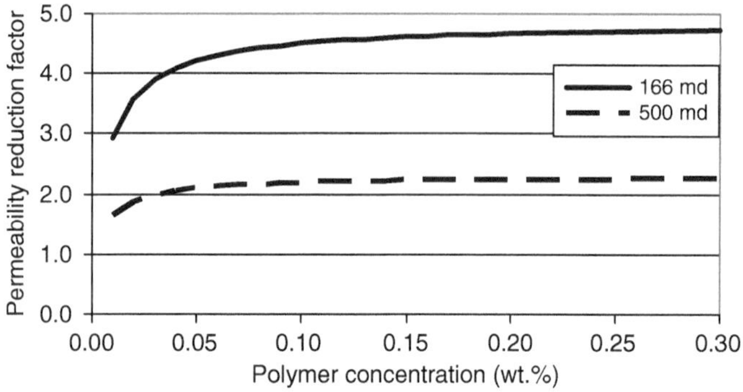

FIGURE 5.49 Polymer concentration effect on F_{kr} predicted from Eq. 5.36.

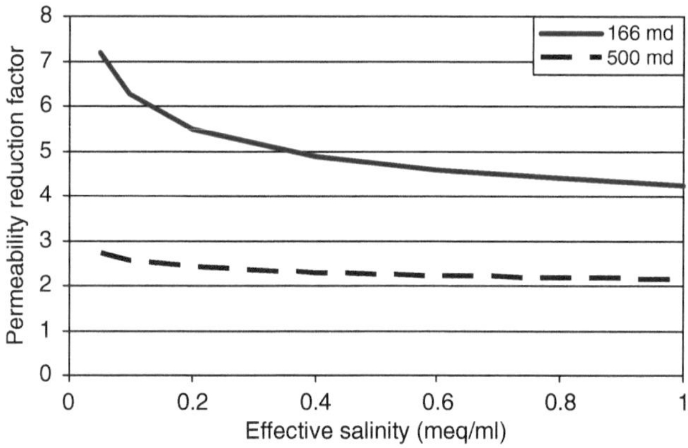

FIGURE 5.50 Salinity effect on F_{kr} predicted from Eq. 5.36.

correlate with the increase of polymer adsorption. However, the prediction from Eqs. 5.36 and 5.37 is in line with the data shown in Figure 5.51, where most of the residual permeability reduction factors at 2% TDS are lower than those at 0.1% TDS.

Eqs. 5.36 to 5.38 can be used to describe the irreversible and reversible processes of polymer permeability reduction. If they are used to describe an irreversible process, an additional parameter called residual permeability reduction factor, F_{krr}, must be used to track the history of F_{kr} so that

$$F_{krr} = \max\left\{(F_{kr})^1, (F_{kr})^2, \ldots (F_{kr})^n\right\}, \tag{5.39}$$

where 1, 2, …, n indicate time steps with the current time step being n. Also, F_{krr} is not greater than $F_{kr,max}$.

Because the polymer permeability reduction process is considered to be an irreversible process, even when the polymer solution is fully displaced by

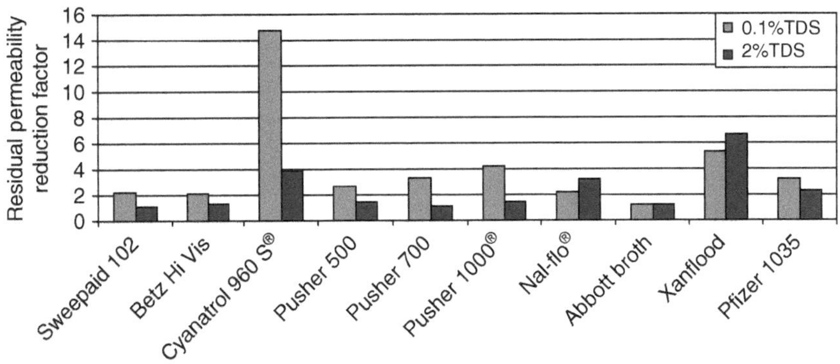

FIGURE 5.51 Salinity effect on F_{krr}. *Source: Data from Martin et al. (1983).*

water, the reduced polymer permeability still exists. The residual permeability reduction factor is defined as

$$F_{krr} = \frac{\text{rock perm. to water before polymer flow}}{\text{rock perm. to water after polymer flow}}. \tag{5.40}$$

Note Eq. 5.40 defines the term *residual permeability reduction factor*. In the literature (Jennings et al., 1971; Bondor et al., 1972; Sorbie, 1991; Green and Willhite, 1998; UTCHEM-9.0, 2000), the term *residual resistance factor* (F_{rr}) is used to represent the residual permeability reduction factor (F_{krr}). Their residual resistance factor is defined as:

$$F_{rr} = \frac{\text{water mobility before polymer flow}}{\text{water mobility after polymer flow}}. \tag{5.41}$$

Resistance is related to mobility, which includes the effects of both permeability reduction and viscosity increase. Obviously, the viscosity effect is not included in the residual resistance factor defined in Eq. 5.41 because water viscosity is used before and after polymer flow. Such a name convention is confusing. Therefore, we suggest the terms "permeability reduction factor" and "residual permeability reduction factor" be used. If the process were considered reversible, there would be no need for the term of residual permeability reduction factor. To include both permeability reduction and viscosity increase, we define another parameter, resistance factor (F_r):

$$F_r = \frac{\text{polymer mobility during polymer flow}}{\text{water mobility during polymer flow}}. \tag{5.42}$$

The permeability to the aqueous phase is reduced by polymer injection, but it is hardly reduced to the other components or other phases (White et al., 1973; Schneider and Owens, 1982). Therefore, we do not change permeability in a numerical simulator. Instead, we modify the polymer solution viscosity by F_{kr}, or the water viscosity after polymer flow by F_{krr} (Bondor et al., 1972; UTCHEM-9.0, 2000) to include the permeability reduction. Because permeability reduction is considered to be irreversible—that is, it does not decrease as polymer concentration decreases, as described by Eq. 5.39. The irreversible effect is caused by polymer adsorption on rock.

As discussed earlier, however, polymer adsorption is not a fully irreversible process. Prolonged water injection will reduce the polymer adsorption. Then the rock permeability to the water after polymer flood will not be the same as that to the polymer solution. It will gradually come back to the initial water permeability. In general, $F_{krr} \leq F_{kr}$, but the process may take many pore volumes of water flush (Gogarty, 1967).

Baijal (1981) studied HPAM transport in porous media and noted that permeability reduction in high-permeability sand packs showed a maximum at a hydrolysis between 20 and 30%. That is the reason a certain optimum degree of chain flexibility is required to give a satisfactory permeability reduction. He indicated that the mobility in the porous medium depended on the optimum degree of interaction between the polymer and the porous matrix. This interaction may be weaker than electrostatic, but it is certainly stronger than van der Waals forces. It was suggested that this may be a dipole-dipole interaction between the polymer and adsorbent surface. However, Huang et al. (1998a) found that F_{kr} decreased with hydrolysis. This result is probably related to lower adsorption as hydrolysis is increased.

Pang et al. (1998b) found the higher the polymer molecular weight (MW), the higher F_{krr} was. When the MW was the same, F_{krr} was higher when the polymer had a wide MW distribution. Because F_{krr} is related to polymer retention, the previous effects on F_{krr} also apply to polymer retention and polymer adsorption thickness. Huang et al. (1998a) observed that the permeability reduction factor (F_{kr}) increased with higher injection velocity and lower temperature.

Although higher molecular weight results in higher F_{krr} and even higher oil recovery factor, the molecular weight used must be limited by formation permeability. Figure 5.52 shows the highest molecular weight at different permeabilities. The data connected with solid lines are from Zhang and Yang (1998), the data with the empty triangle points are from Wang et al. (2006c), and the data marked with solid points unconnected are from Niu et al. (2006). These data show that lower molecular weight polymer is needed for a low-permeability formation. For example, Wang et al. (2006c) showed that polymer with molecular weight of 2.4 million can satisfy the need of reservoirs with permeability of 20 mD, 5.5 million can meet the demand of reservoirs with permeability of 50 mD, and 10 million is suitable for reservoirs with permeability of 200 mD. The data from Zhang and Yang (1998) are more conservative. Zhang and Yang considered the rates flowing through perforation. The rates 50, 100, 200, and 400 m/ d correspond to the rates 2.25, 4.5, 9.0, and 18 $m^3/d \cdot m$, respectively, through 0.008 m diameter holes with 10 holes per meter.

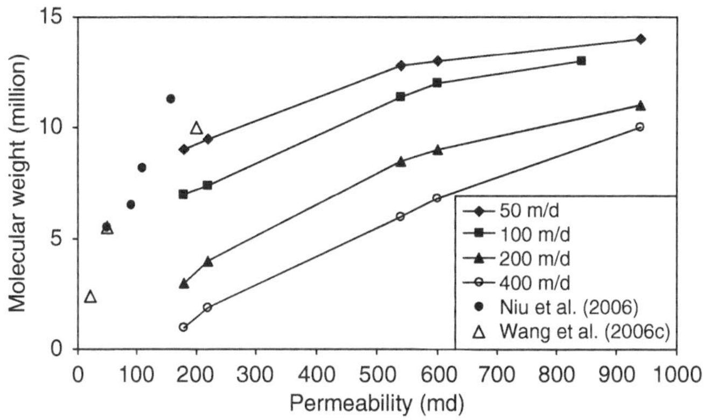

FIGURE 5.52 Polymer molecular weight limits.

In an earlier paper, Wang et al. (1998b) stated that laboratory test results showed when five times the gyration radius of the polymer molecule was smaller than the median size (radius) of the pore space of a reservoir, the polymer molecule would not plug the formation pore space. A variation of the relationship (Wang et al., 2009) between permeability (k) and molecular weight (MW) based on Daqing data is k = 9 MW – 5, where k is in md and MW is in million Daltons. This equation works better in the low-MW range. In the high-MW range, k is underestimated.

As pore sizes are widely distributed, it could be expected that a polymer with a broad molecular weight distribution should be beneficial because polymer particles of different gyration radii can flow through different pore throats. This type of polymer may enter and propagate more effectively through pores and reduce the inaccessible volume. This expectation has been verified by core flood results (He and Chen, 1998).

5.4.5 Relative Permeabilities in Polymer Flooding

The conventional belief is that polymer flooding does not reduce residual oil saturation in a micro scale. The polymer effect is to increase displacing fluid viscosity and thus to increase sweep efficiency. It is also acceptable that fluid viscosities do not affect relative permeability curves. Therefore, it is logical to believe that the relative permeabilities in polymer flooding and in waterflooding after polymer flooding are the same as those measured in waterflooding before polymer flooding if we take into account the resistance factor for the k_{rw} in polymer flooding and the residual permeability reduction factor for the k_{rw} after polymer flooding. This belief has been supported by some experiments.

Schneider and Owens (1982) conducted experiments to determine the effect of polymer on relative permeability in a displacement sequence in which polymer solution was injected into a reservoir that was at waterflood residual oil saturation. Steady-state relative permeability data were obtained for Berea sandstone and reservoir cores having a range of permeabilities and wettabilities. All the tests were conducted with polyacrylamides. Two-phase flow of oil and polymer solution was studied in water-wet cores on the secondary drainage path. That is, the polymer solution was displaced by oil from its maximum saturation $1-S_{or}$ to its minimum saturation S_{wi}. The relative permeability to oil was essentially unaffected by the polymer flow. The relative permeability curve for polymer solution, however, was significantly lower than the corresponding relative permeability curve for water before polymer contact of the core.

Relative permeability curves were also determined for the displacement of oil by water following the polymer/oil tests. Figure 5.53 compares the relative permeability data for the oil and water phases before (with the subscript l) and after (with subscript p) polymer contact. RRF in the figure denotes F_{krr} in the text. In the water-wet rocks, there was little difference between the residual oil saturation obtained before and after polymer contact, as would be expected. Oil

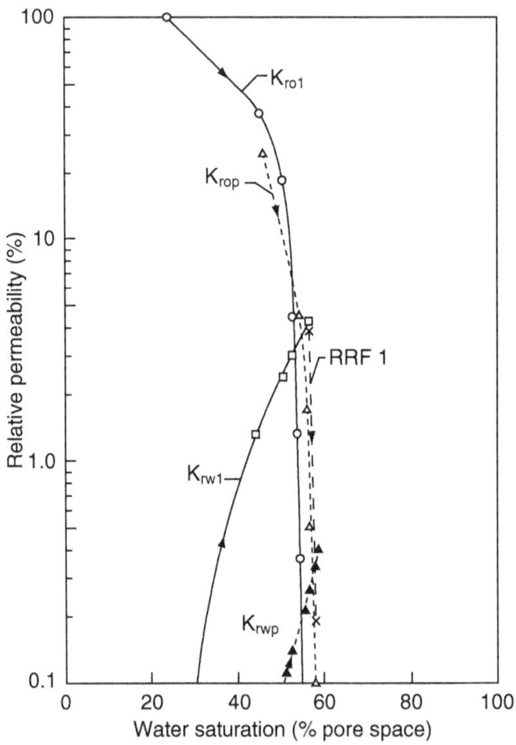

FIGURE 5.53 Water/oil relative permeabilities before and after contact with Dow Pusher 1000 (Sample Berea-3). *Source: Schneider and Owens (1982).*

relative permeabilities were relatively unaffected. The relative permeability to water after polymer contact, k_{rwp}, was reduced significantly compared with the relative permeability to water before polymer contact, k_{rw1}, as seen in Figure 5.53. The parallelism of k_{rw1} and k_{rwp} indicates that the reduction in k_{rwp} after polymer contact was caused by the permeability reduction by polymer adsorption. The parallelism seen in these curves and the lack of any large effect on residual oil saturation provide excellent confirmation of the flow channel concept (Standard Oil, 1951); that is, immiscible phases flowing simultaneously flow largely in separate pore networks. Polymer adsorption occurs only in the pore networks transporting the aqueous phase.

Figure 5.54 shows an example of relative permeability curves in an oil-wet rock. The water relative permeability curve after polymer contact, k_{rwp}, was parallel but significantly lower than the water relative permeability curve before polymer flood, k_{rw1}. k_{rw1} with S_w increasing and k_{rw2} with S_w decreasing were different owing to hysteresis. The residual oil saturation decreased in the polymer/oil test as the k_{ro3} shifted toward higher water saturation, as shown

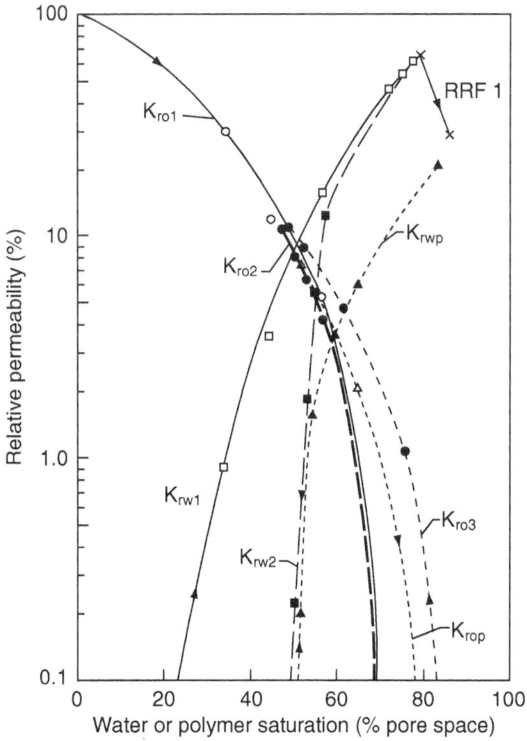

FIGURE 5.54 Water/oil/Dow Pusher 500 relative permeabilities (Sample Tensleep-1). *Source: Schneider and Owens (1982).*

in Figure 5.54, because the oil saturation in an oil-wet rock exists as thin films and in small pores. Injection of a viscous fluid will decrease the oil saturation in oil-wet rocks. Note that the k_{ro3} curve in the figure was obtained during simultaneous injection of oil and polymer solution, and the polymer relative permeability curve is not shown in the figure.

Chen and Chen (2002) showed similar experimental data and made similar conclusions, except that they also observed increased immobile water saturation in water-wet cores and decreased residual oil saturation.

In UTCHEM, the viscosity of the aqueous phase that contains the polymer is multiplied by the value of the polymer permeability reduction factor, F_{kr}, to account for the mobility reduction. In other words, water relative permeability, k_{rw}, is reduced, whereas oil relative permeability, k_{ro}, is sometimes considered almost unchanged. The reason is that polymer is not soluble in oil, so it will not reduce effective oil permeability. The mechanism of disproportionate permeability reduction is widely used in gel treatment for water shut-off. Many polymers and gels can reduce permeability to water more than to oil or gas.

For adsorbing polymers and weak gels, permeability reduction factors and residual permeability reduction factors increase with decreased permeability (Seright, 2006).

Liang et al. (1995) suggested from their experimental data that the segregation of oil and water pathways through a porous medium (on a microscopic scale) may play a dominant role in the disproportionate permeability reduction. Because of the separation of water and oil paths, polymer solution preferentially flows through water paths, particularly in high water saturation zones, whereas oil flows through oil paths that could remain connected after polymer injection. The experimental results of Liang et al. indicate the disproportionate permeability reduction is not caused by gravity or lubrication effects. Although wettability may play a role in the disproportionate permeability reduction, it does not appear to be the main cause for water permeability being reduced more than oil permeability. Taber and Martin (1983) also reported that polymer can alter the flow path by reducing water effective permeability permanently, whereas oil effective permeability remains relatively unchanged.

Tang et al. (1998) measured polymer solution/oil (P/O) relative permeability curves using the steady-state flow method. To calculate the polymer solution viscosity, they used

$$\mu_p = K\left[\frac{3n+1}{4n}\right]^n \left[\frac{Cv_w}{\sqrt{k_w\phi/F_{krr}}}\right]^{n-1}, \tag{5.43}$$

where C = 1.8, and

$$v_w = \frac{Q_w}{\phi A(S_w - S_{wc})}. \tag{5.44}$$

TABLE 5.12 P/O k_r Curve Carey-Type Parameters Compared with those of W/O

C_p,		k_o at	k_{rw} at S_{or}		S_{or}		n_w		n_o	
mg/L	S_{wi}	S_{wi}, md	*W/O*	*P/O*	*W/O*	*P/O*	*W/O*	*P/O*	*W/O*	*P/O*
800	0.344	185	0.088	0.0486	0.271	0.211	1.641	1.346	2.182	4.881
1000	0.36	242	0.145	0.0402	0.262	0.205	1.449	1.084	3.724	3.250
1200	0.34	254	0.058	0.0218	0.252	0.202	1.086	0.524	2.908	2.446

Source: *Tang et al. (1998).*

Note that they used interstitial velocity in Eq. 5.43. Table 5.12 shows the measured Corey-type parameters. From these data, we can see that the residual oil saturations in P/O were reduced by 0.05 to 0.06 compared with water/oil (W/O) k_r curves, and the water relative permeabilities at S_{or}, k_{wr}, were reduced by 0.036 to 0.105. The oil relative permeability curves were not much changed.

According to the previous discussion, water relative permeability, k_{rw}, in polymer flooding is reduced, whereas oil relative permeability, k_{ro}, is little changed. There are several reasons as summarized here:

- Polymer is soluble in water phase but not in oil phase. When polymer solution flows through pore throats, high molecular weight polymer is retained at the throats. Then the polymer blocks water flowing through, and k_{rw} is reduced.
- Polymer molecules can form a hydrogen bond with water molecules; this capability enhances the affinity between the adsorption layer and water molecules. Rock surfaces become more water-wet. Thus, k_{rw} is reduced (Huang and Yu, 2002).
- Polymer and oil have segregated flow paths. Therefore, polymer reduces k_{rw} but not k_{ro} (Liang et al., 1995).
- Other factors related to disproportionate permeability reduction could also be used to explain reduced k_{rw}.

For oil and water relative permeability curves after polymer injection, Huang and Yu (2002), and Chen and Cheng (2002) reported their observations, which were similar to residual permeability reduction after polymer flooding. Compared with the relative permeability curves before polymer flooding, the relative permeability curves had the following three characteristics: (1) k_{rw} was reduced at the same water saturation, and corresponding to the same k_{rw}, water saturation was larger; (2) immobile water saturation was increased; and (3) residual oil saturation was reduced. It is believed this result was caused by polymer adsorption, which made a rock surface more water-wet.

5.5 DISPLACEMENT MECHANISMS IN POLYMER FLOODING

One obvious mechanism in polymer flooding is the reduced mobility ratio of displacing fluid to the displaced fluid so that viscous fingering is reduced. When viscous fingering is reduced, the sweep efficiency is improved, as shown in Figure 1.2. This mechanism is discussed extensively in the waterflooding literature; it is also discussed in Chapter 4. When polymer is injected in vertical heterogeneous layers, crossflow between layers improves polymer allocation in the vertical layers so that vertical sweep efficiency is improved. This mechanism is detailed in Sorbie (1991).

One economic impact of polymer flooding that has been less discussed is the reduced amount of water injected and produced compared with waterflooding. Because polymer improves mobility ratio and sweep efficiency, less water is injected and less water is produced. In some situations such as offshore environments and desert areas, water and the treatment of water could be costly.

Polymer is also used to shut off water channeling through high-permeability layers and water coning from bottom aquifers. In these types of applications, if the injected polymer volume is not large, or practically, a large volume may not be injected because of high injection pressure constraints or short gelation time, blocking water channeling or water coning is temporary. Eventually, water will bypass the injected polymer zone and crossflow to high permeability zones or bypass the polymer zone to the producing wellbores. To avoid this kind of problem, a weak gel that has high resistance to flow but is still able to flow can be injected deep into reservoir. Thus, a large volume or large area of polymer zone is formed to block water thief zones or channels. In polymer and gel treatment, another mechanism is called disproportionate permeability reduction (DPR). Through the use of this mechanism, polymer and gel can reduce water permeability much more than oil permeability.

In a very heterogeneous reservoir, an injected viscous polymer solution may still break through producers early. An idea similar to weak gel was proposed to attack this problem (Yang and Ni, 1998). Instead of injecting crosslinkers through injections, cationic polymer is injected through producers. The injected cationic polymer has high adsorption on the rock. When the anionic polymer injected through an injection well meets with the cationic polymer, they crosslink to form a water-insoluble gel to block water channeling or fingering.

Another mechanism is related to polymer viscoelastic behavior. The interfacial viscosity between polymer and oil is higher than that between oil and water. The shear stress is proportional to the interfacial viscosity. Because of polymer's viscoelastic properties, there is normal stress between oil and the polymer solution, in addition to shear stress. Thus, polymer exerts a larger pull force on oil droplets or oil films. Oil therefore can be "pushed and pulled" out of dead-end pores. Thus, residual oil saturation is decreased. This mechanism is detailed in Chapter 6.

5.6 AMOUNT OF POLYMER INJECTED

In general, a larger amount of polymer was preferred in Daqing. However, when the product of polymer concentration (mg/L) and injected pore volume (fraction) is larger than 400 (mg/L PV), incremental oil recovery becomes less sensitive to the amount of polymer injected (Niu et al., 2006). Figure 5.55 shows the history of amount of polymer injected in the polymer flooding projects in China. From the 1970s to 1980s, the amount of polymer injected was 100 to 200 mg/L·PV. During the early 1990s, 500 to 600 mg/L·PV were tried in a few projects. In the early years of 2000, 400 to 500 mg/L·PV have been used consistently.

Table 5.13 shows more results based on the economic analysis of simulation data. This table shows that more polymer injection corresponds to higher incremental oil recovery but lower economic parameters (tons of incremental oil per ton of polymer injection; Qi and Feng, 1998).

TABLE 5.13 Amount of Polymer Injection vs. Economic Return

Injected Polymer, mg/L·PV	Incremental RF, %	Tons of Oil/Ton of Polymer Injected
380	10	150
570	12	120
1500	19.7	80

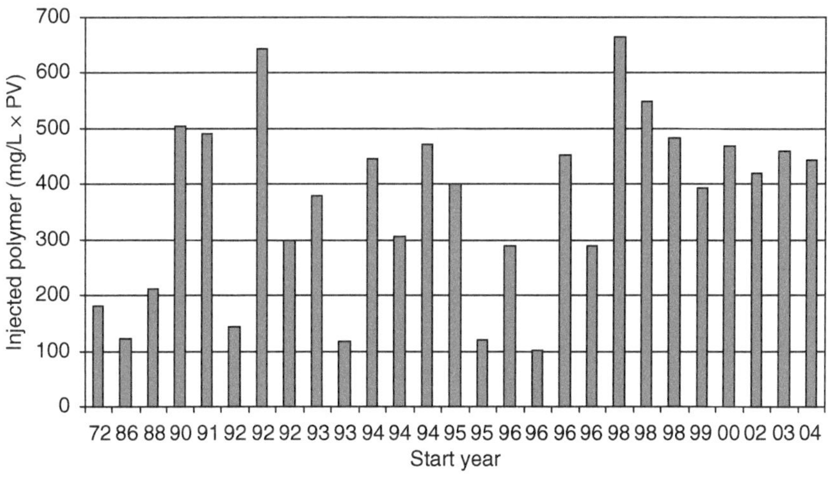

FIGURE 5.55 History of amount of polymer injected in the polymer flooding projects in China.

For the same amount of polymer fixed, the question of optimization remains. Which option is better: a higher concentration with a smaller injection pore volume or a lower concentration with a larger injection pore volume? Generally, the ultimate incremental oil recovery mainly depends on the total amount of polymer injected. A higher concentration could result in more initial water-cut reduction due to polymer injection. However, a high concentration may be limited by the allowable injection pressure. From a mobility control point of view, a higher concentration should be injected at the front to counteract dilution. A commonly used concentration in China is around 1200 mg/L.

5.7 PERFORMANCE ANALYSIS BY HALL PLOT

The Darcy equation for single-phase water in waterflooding is

$$q(t)=\frac{kk_{rw}h[p_{wf}(t)-p_e]}{141.2B_w\mu_w[\ln(r_e/r_w)+s]}, \tag{5.45}$$

where these field units are used: q–STB/day, p–psi, μ_w–mPa·s, k–md, h–ft. In Eq. 5.45, q is the injection rate, p_e is the formation pressure at the interface between the original reservoir fluid and injected fluid, and p_{wf} is the wellbore injection pressure:

$$p_{wf}=p_{tf}-\Delta p_f+\rho gD. \tag{5.46}$$

In the preceeding equation, p_{tf} is the wellhead pressure, Δp_f is the friction loss, and D is the reservoir depth.

Integrating both sides of Eq. 5.45 with respect to time by assuming only pressure difference and injection rate q are time-dependent, we have

$$\int[p_{wf}(t)-p_e]dt=\frac{141.2B_w\mu_w[\ln(r_e/r_w)+s]}{kk_{rw}h}W_i, \tag{5.47}$$

where W_i is the cumulative injection equal to $\int q(t)dt$. Equation 5.47 uses the bottom hole flowing pressure p_{wf}. In practice, p_{wf} is not readily measured if a downhole gauge is not installed while the wellhead pressure, p_{tf}, is available. Therefore, practically, we use p_{tf} in Eq. 5.47:

$$\int p_{tf}(t)dt=\frac{141.2B_w\mu_w[\ln(r_e/r_w)+s]}{kk_{rw}h}W_i+\int[p_e+\Delta p_f-\rho gD]dt. \tag{5.48}$$

If the second term is considered unchanged with time, it may be dropped when plotting the integral of wellhead pressures with respect to time versus the cumulative injection. This plot, known as the "Hall plot," was originally developed by Hall (1963) to analyze waterflood performance. According to Eq. 5.48, the slope of the Hall plot is

$$m_H = \frac{141.2B_w\mu_w[\ln(r_e/r_w)+s]}{kk_{rw}h}. \tag{5.49}$$

If an injection well is stimulated (s becomes smaller), the slope decreases; if a well is damaged (s becomes larger), the slope increases. When the Hall plot is applied to a polymer injection well, if we assume s does not change, the slope increases because of higher polymer solution viscosity. Buell et al. (1990) pointed out that injection data must be plotted in the form of Eq. 5.47 to make valid quantitative calculations, and Eq. 5.49 is not appropriate when multiple fluid banks with significantly different properties exist in the reservoir. However, in practice, we may still use Eq. 5.48 in polymer injection for approximation.

If the reservoir is initially oil saturated, there are several zones: post-water, polymer, polymer denuded, water and oil two-phase, and oil only, as schematically shown in Figure 5.56, which is similar to Figure 2.19. The total resistance may be described by series flow. Then the slope is

$$\begin{aligned} m_H = {} & \frac{141.2B_w\mu_w[\ln(r_{w2}/r_w)+s]}{kk_{rw}^e h/F_{krr}} + \frac{141.2B_w\mu_p\ln(r_p/r_{w2})}{kk_{rp}(\bar{S}_{wp})h} \\ & + \frac{141.2B_o\mu_o\ln(r_p/r_{w2})}{kk_{ro}(\bar{S}_{wp})h} + \frac{141.2B_o\mu_o\ln(r_{w1}/r_p)}{kk_{ro}(S_{w1})h} \\ & + \frac{141.2B_w\mu_w\ln(r_{w1}/r_p)}{kk_{rw}(S_{w1})h} + \frac{141.2B_o\mu_o\ln(r_{wf}/r_{w1})}{kk_{ro}(\bar{S}_{w1})h} \\ & + \frac{141.2B_w\mu_w\ln(r_{wf}/r_{w1})}{kk_{rw}(\bar{S}_{w1})h} + \frac{141.2B_o\mu_o\ln(r_e/r_{wf})}{kk_{ro}^e h}. \end{aligned} \tag{5.50}$$

Equations similar to Eq. 5.50 can be written with the saturation distributions similar to Figures 2.20 and 2.21.

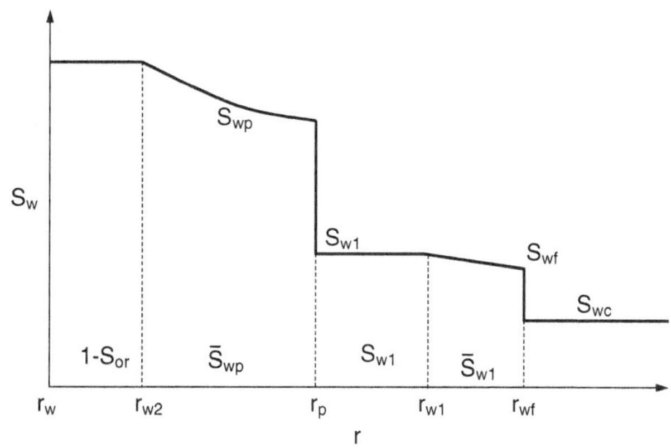

FIGURE 5.56 Saturation profile for polymer flood started at interstitial water saturation when $S_{w1} > S_{wf}$ with post-water drive added (not scaled).

Equation 5.50 shows that m_H is a function of fluid viscosities, fluid saturations, relative permeabilities, the length of each zone, and so on. Several flowing zones are assumed in Eq. 5.50. If the real fluid saturation is considered, it will be a more complex function.

To discuss the usefulness of the Hall plot, let us first look at a simple polymer injection case. There are only two zones: polymer and oil. $B_o = B_w = 1$, and $s = 0$. The slope is

$$m_H = \frac{141.2\mu_p \ln(r_p/r_w)}{kk_{rp}h} + \frac{141.2\mu_o \ln(r_e/r_p)}{kk_{ro}h}, \tag{5.51}$$

Assume the polymer flooding is designed according to the mobility control requirement defined by Eq. 4.14 and the average oil saturation is 0.5. In this case, we have

$$\frac{\mu_p}{k_{rp}} = \frac{\mu_o}{2k_{ro}}. \tag{5.52}$$

Using the relationship (5.52), the ratio of the first term (m_{H1}) to the second term (m_{H2}) in the right side of Eq. 5.51 is

$$\frac{m_{H1}}{m_{H2}} = \frac{\ln(r_p/r_w)}{2\ln(r_e/r_p)}. \tag{5.53}$$

If $r_e = 200$ m, $r_p = 100$ m, and $r_w = 0.15$ m, $m_{H1}/m_{H2} = 9.4$. m_{H1} is almost 10 times as large as m_{H2}. This example shows that the contribution to the slope (m_H) from the near wellbore term is much larger than that from the term far away from the wellbore. This observation leads to the following application.

After a long time of polymer injection (r_p is substantially large), m_H is mainly contributed from the polymer slug—that is,

$$m_{Hp} = \frac{141.2B_w\mu_p[\ln(r_p/r_w)+s]}{kk_{rp}h} = \frac{141.2B_w\mu_w F_r[\ln(r_p/r_w)+s]}{kk_{rw}h}, \tag{5.54}$$

where F_r is the polymer resistance factor, and r_p is the radius of injected polymer solution. r_p may be estimated from the Buckley–Leverett equation in radial coordinates (Collins, 1961):

$$r_p^2 = \frac{5.615W_i}{\pi\phi h}\left(\frac{\partial f_w}{\partial S_w}\right)_f + r_w^2. \tag{5.55}$$

The quantity $(\partial f_w/\partial S_w)_f$ is the derivative of the fractional flow curve at the polymer front. The water saturation and the derivative at the front are determined by Welge's method (1952). Here, r_p, r_w, and h are in ft, and W_i is the cumulative injection during the period in bbl. If all the other parameters in Eq. 5.54 are known, F_r can be estimated.

After polymer injection, water will be injected to drive the polymer. Similar to Eq. 5.54, we have

$$m_{w2} = \frac{141.2 B_w \mu_w F_{krr}[\ln(r_{w2}/r_w) + s]}{kkr_w h}. \tag{5.56}$$

Now we may use Eq. 5.55 to estimate r_p for r_{w2}. Using Eq. 5.56, we can estimate F_{krr}.

In the preceding approach, we assume the fluid near the wellbore dominates the flow resistance. We also assume the velocity in the near wellbore zone is constant. In real radial flow, the flow velocity changes as the displacement front moves. Fortunately, Buell et al. (1990) found that those assumptions were valid in their real cases. Another justification for neglecting the velocity gradient is that when the displacement front is substantially away from the wellbore, the velocity changes are not significant.

To include the velocity change in polymer flooding, we have to consider the velocity-dependent viscosity in the Darcy equation 5.45. For the power-law viscosity model, the polymer viscosity is defined by Eq. 5.3, and the shear rate is defined by Eq. 5.23. Then Eq. 5.45 becomes

$$q(t) = \frac{h(kk_{rw})^{(n+1)/2}\phi^{(n-1)/2}[p_{wf}(t) - p_e]}{141.2 B_w K C^{n-1}\left[\frac{3n+1}{4n}\right]^n \left(\frac{q}{A}\right)^{n-1} F_{kr}[\ln(r_p/r_w) + s]}. \tag{5.57}$$

According to Eq. 5.57, we can plot $\int[p_{wf} - p_e]dt$ versus $\int q^n dt$ and m_H becomes

$$m_H = \frac{141.2 B_w K C^{n-1}\left[\frac{3n+1}{4n}\right]^n A^{1-n} F_{kr}[\ln(r_p/r_w) + s]}{h(kk_{rw})^{(n+1)/2}\phi^{(n-1)/2}}. \tag{5.58}$$

5.8 POLYMER MIXING AND WELL OPERATIONS RELATED TO POLYMER INJECTION

This section briefly discusses polymer mixing and well operations related to polymer injection.

5.8.1 Mixing

Polymer can be delivered in liquid emulsion, water solution, or solid powders. When polymer is in liquid emulsion or water solution, it can be added to injection water using a pump. When it is in solid powder, several processes are needed to prepare the polymer solution: proration, dispersion, maturation, transportation, filtration, and storage (see Figure 5.57).

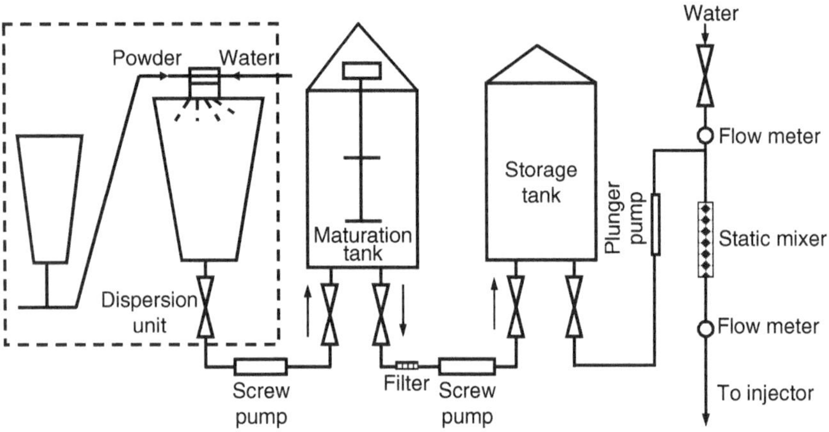

FIGURE 5.57 Schematic of a typical facility to prepare polymer solution.

Proration is metering solid polymer and water to be dispersed. Polymer is delivered to the feeder, which can filter impurities. Dispersion is a process to dissolve high molecular weight polymer into water. Two kinds of dispersing units can be used. In one type, air is blasted to carry polymer powder to mix with the water spray. The other one is a Venturi type in which polymer powder is sucked in the mixing unit because of the negative pressure caused by flowing water and then is mixed with water. The Venturi type is a better unit to control oxygen content (Huang et al., 1998c). The dispersed polymer (concentrated solution) is transported to a maturation tank where the mixer is rotating. The maturation takes 0.5 to 24 hours (Huang et al., 1998b; Liu et al., 2006a). The concentrated solution is transported to the storage tank through two filters to remove impurities and "fish eyes." A screw pump is used for transporting polymer solution to reduce mechanic shearing; a plunger displacement pump is used to inject polymer solution. Another unit is called a static mixer; this special unit is installed in pipes to change fluid flow direction so that the fluids can be fully mixed. Unlike a dynamic (rotary) mixer, the static mixer does not move, as the name implies.

For the polymer mixing and injection system in Daqing, polymer solution was transferred in the mode of one transfer pump–one pipeline–one injection station in the early days. These days it is in the mode of one transfer pump–one pipeline–two injection stations. Thus, the cost to deliver polymer solution is reduced. Plus, the system of one injection pump (station)–one well is replaced by the system of one injection pump (station)–several injection wells (Li et al., 2005d).

5.8.2 Completion

Completion techniques in polymer injection wells are similar to those used in water injection wells. The main objective in polymer completion is to reduce shear loss. Polymer injection wells are commonly completed through perforation. Therefore, the perforation done should be high-density, deep-penetration, and large-diameter holes. In some cases, hydraulic fracturing may be used to reduce mechanic shearing near the wellbore.

5.8.3 Injection Velocity

The injection velocity should mainly depend on reservoir injectivity and allowable injection pressure. Some simulation results, however, show that higher injection velocity increased the ultimate recovery factor a little bit (maximum 2%). A recommended injection velocity range in Daqing is 0.1 to 0.16 PV per year (Niu et al., 2006). However, injection rates several times higher have been seen in practical cases in the Chinese literature.

5.8.4 Separate Layer Injection

For most polymer injection wells, injection does not have to be allocated among different layers because polymer can adjust the injection profile itself. However, if there is large injectivity difference in different layers, separate layer injection has to be implemented. The techniques can be classified into single string and multiple strings.

5.8.5 Removing Plugging

When formation is plugged for some reason, such as fish eyes or precipitation, workover must be carried out to remove the plugging. Acidizing can remove inorganic plugging. For organic plugging, special chemical treatments and hydraulic fracturing have to be carried out. In chemical treatments, oxidants such as chloride dioxide and hydrogen peroxide are generally used. For hydraulic fracturing, silicon sand has been found to be an ineffective proppant because it can be carried to the deep formation by high viscous polymer solution. Thus, the sand is not packed near wellbore, and the fractures are closed up after treatment. Resin-coated sand may be used because resin can become soft due to high formation temperature and make solid particles connected (Liu et al., 2006a).

5.9 SPECIAL CASES, PILOT TESTS, AND FIELD APPLICATIONS OF POLYMER FLOODING

This section presents a number of special cases, pilot tests, and field applications. Effort has been made to select these cases so that each case covers special issues or topics.

5.9.1 Profile Control by Injection of Polymers with Different Molecular Weights

When a layered reservoir has high permeability contrast in vertically different layers, polymer can be injected through separate layers to control the injection profile, as mentioned previously. Another method is alternate injection of polymers with different molecular weights (MW). As discussed in Section 5.4.4, high MW polymer can be used in a high-permeability reservoir, and low MW polymer must be used in a low-permeability reservoir. For the alternate injection, the layers are grouped into different permeability layers: high, intermediate, and low.

First, the polymer with higher MW that is suitable for the high-permeability layer is injected, and no polymer is injected into low- and intermediate-permeability layers. When the water cut rises to the level prior to the polymer injection, polymer injection is changed into intermediate and low permeability layers using lower MW polymer. The procedures can be alternately repeated if needed. Of course, high permeability layers are grouped into a high-permeability layer, and low permeability layers are grouped into a low-permeability layer. The intermediate permeability layers may be grouped into either the high- or low-permeability layer. The orders of injection are not unchangeable. This method is similar to separate layer injection. In addition, different MW polymers are used in different permeability layers. An example is presented next (Yan et al., 2005).

The target layers are in Sa and Pu II from 830 to 1040 m subsea. A total of 421 layers were defined with a total gross thickness of 326.7 m and net thickness of 114.2 m. The net thickness of the layers < 100 md was 30.74%; 100 to 250 md, 39.4%; 250 to 500 md, 14.36%; and > 500 md, 15.5%. The defined injection interval and injection parameters are summarized in Table 5.14. The parameters for an earlier lump injection (injection into all the layers) and a separate layer injection are also included in the table for comparison.

TABLE 5.14 Injection Parameters

Injection Mode	Injection Period	$\bar{k}$, md	MW (million)	Injection PV
Lump	Dec. 2000–Mar. 2001	145	7–8	0.087
Separate	Apr. 2001–Sept. 2002	145	8–10	0.607
Alternate:				
High to intermediate k	Oct. 2002–Aug. 2003	162	14	0.341
Intermediate to low k	Sept. 2003–Jun. 2004	70	10	0.24
High k	Jul.–Oct. 2004	337	17	0.069

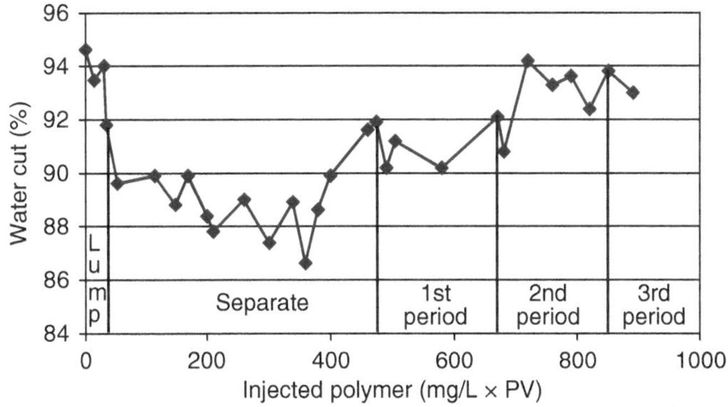

FIGURE 5.58 Water-cut curve of several center wells in the tested zone. *Source: Yan et al. (2005).*

Figure 5.58 shows the water-cut curve of several central wells in the tested zone. This figure shows that water cut dropped after each alternate injection, but that the magnitude of the drop decreased, indicating lower potential for improvement.

Similar to (but not the same as) the concept to inject different MW polymers, different polymer concentrations can be injected for profile control. Yang et al. (2006) presented laboratory and pilot test results showing that the recovery increased by injecting high-concentration polymer solution in the early slugs. In this case, the high concentration used was 1500 to 2500 ppm, and the incremental oil recovery over waterflooding was about 20%.

5.9.2 Polymer Injection in Viscous Oil Reservoirs

This section presents two cases of polymer injection in viscous oil reservoirs: Xia-er-men field and Godong Block 8.

Xia-er-men Field, Henan

In the Xia-er-men field operated by Henan Oilfield, Sinopec, the produced water was used to make a polymer solution. Because of the high viscous oil and very heterogeneous reservoir, a normal polymer solution was not good enough to reach desired sweep efficiency. Profile control was tried instead. Because of small injection volume, however, water soon bypassed the injected gel. Therefore, a large volume of weak gel (deep profile control) was tested in a pilot.

The pilot test was conducted in the H_2 layer. Some reservoir and fluid properties in the H_2 II layer are shown in Table 5.15 (Xie et al., 2001; Zhang et al., 2003; Fan et al., 2004). The other tested layers, H_2 III and H_2 IV, were similar. The distance between injector and producer was 190 to 340 m. Tests showed that the tracer broke through in 4 to 6 days. The fastest velocity could be up to 80 m/d with the average velocity of about 40 m/d. By 1995, the water cut of

TABLE 5.15 Xia-er-men H2 II Reservoir and Fluid Data

OOIP, million tons	2.406
Formation depth, m	928–1050
Area, km^2	1.32
Thickness, m	14.2
Porosity, %	23.7
Average permeability, md	2330
Permeability variation coefficient	0.91
Temperature,°C	46
Oil viscosity, mPa·s	76
Formation water TDS, mg/L	2127

about 60% of the wells was about 90%, and the recovery factor was 24%. The sweep efficiency was 0.5 to 0.6. In July 1996, weak gel was injected in 7 wells. The injection volume was 1500 to 2500 m^3. The injection velocity was 4.5 to 5 m^3/h. After injection, the water intake index was reduced by 53.2%, and the injection pressure was increased by 1.9 MPa. The water injection profile was improved by 37.4 to 89.5% with an average 61.3%. The effectiveness lasted up to one year.

When the injection volume was up to 0.1 PV, the water-cut reduction and oil rate increase slowed down, and polymer stopped breaking through high-permeability channels (Xie et al., 2001). The weak gel was made of 400 mg/L HPAM and 60 mg/L Cr^{3+}. Its viscosity was 130 mPa·s after aging 180 days at 50°C. Compared with 1000 mg/L polymer solution that was used before gel injection, the weak gel cost was reduced by 21% (Fan et al., 2004). Produced water was used to make the polymer solution. It was observed that if the produced water from producers was used immediately, polymer solution viscosity loss was up to 60%. However, if the produced water was used some time after it was produced, the viscosity loss was significantly reduced (Xie et al., 2001).

Polymer was injected in three layers: H_2 II, H_2 III, and H_2 IV. The performance presented previously was mainly from H_2 II, which was better than the other two layers. The better performance was caused by the following factors (Zhang et al., 2003).

- Polymer injection was started in the H_2 II layer when 60% of wells had water cut of about 90%, compared with 27% for H_2 III and 13% for H_2 IV. The rest of the wells had water cut higher than 90%. In other words, polymer injection was started in the H_2 II layer earlier when the water cut was lower than in the other two layers.

- H_2 II temperature was lower than the other two layers—46°C compared with 52.8°C in H_2 III and 58.1°C in H_2 IV.
- The producers in H_2 II had better reservoir connection with injection wells.
- The injection volume of weak gel for profile control in H_2 II was larger than in the other two layers.
- Better polymer and better imported equipment were used in H_2 II.

Gudong Block 8, Shengli

In Godong Block 8, which belonged to the Godong field, Sinopec, polymer was injected into two separate layer groups: Ng 3–4 and Ng 5–6. There were 44 injection wells and 90 production wells in a line-drive pattern. The distance between wells was 106 m, and the distance between injection line and producer line was 212 m. Some reservoir and fluid data are shown in Table 5.16 (Jiang et al., 2001).

The polymer injection was started on August 8, 1997. A unique polymer injection scheme was designed well by well. The injected polymer concentration in each well depended on the injection pressure, as shown in Table 5.17 (Yi et al., 1999). The overall design for the graded injection scheme was 1700 mg/L and 0.06 PV followed by 1300 mg/L and 0.27 PV. The total amount of polymer injection was then 453 mg/L·PV. In this case, the following three observations were made: (1) polymer injection was effective, resulting in increased injection pressure, decreased liquid offtake rate, decreased water cut, and increased oil rate; (2) the average well sand production increased from 0.69 m^3 to 1.22 m^3; and (3) the vertical injection profile was improved.

TABLE 5.16 Block 8 Reservoir and Fluid Data

OOIP, million tons	11.77
Formation depth, m	1220–1460
Porosity, %	34
Average permeability, md	1783, 3072
Temperature,°C	63.5
Oil viscosity, mPa·s	37.4–91.6
Formation water TDS, mg/L	7962
Ca^{2+} and Mg^{2+}, mg/L	290

TABLE 5.17 Polymer Injection Schemes

P_{inj} (tubing), MPa	Polymer Concentration, mg/L	Crosslinker Concentration, mg/L	No. of Wells
< 7.5	2000–3000	1000	9
7.5–10	1500–25000	800–1000	9
10–12.5	1000–1500	800	21
> 12.5	800–1200	0	5

5.9.3 Profile Control in a Strong Bottom and Edge Water Drive Reservoir

Polymer flooding would be more effective in a reservoir that does not have strong natural aquifer drive energy. In the case discussed here, the Gao-Qian-Bei block in the Ji-Dong field, Shengli, there is an active bottom and edge aquifer. The combination of high viscous oil and heterogeneous formation resulted in a low oil recovery. In this case, profile control was carried out by injecting crosslinked polymer to increase oil recovery (Li et al., 2005c).

Reservoir Description

For the target group of layers, NgIV, the depth was 1800 to 1900 m, the porosity was 30%, permeability was 602 to 1622 md, and the oil viscosity was 90.34 mPa·s at the reservoir temperature of 65°C. The permeability variation coefficient was 0.8. In this case, strong bottom and edge water flowed through high-permeability channels. By 1999, the water cut was 80.4%, and the recovery factor was 8.16%. The ultimate oil recovery factor was estimated to be 15%.

Laboratory and Numerical Simulation Studies

Both 2D and 3D physical models were built to study the effectiveness of the profile control. In the 2D model, the incremental oil recovery factor was 8.19% over aquifer drive. In the 3D model, the incremental oil recovery factor was 6.2% (Li et al., 2005c). In the 3D model, 0.08 PV of 3000 mg/L polymer was injected. When crosslinked polymer was injected, high permeability channels were immediately blocked, the injection pressure rose, and the water cut fell. However, because of strong edge water, water bypassed the blocked zone, the injection pressure fell, and the water cut quickly rose again. A numerical simulation was carried out to study the feasibility of polymer injection and optimize the program (Yao et al., 2005). The optimum concentrations from the laboratory results were 0.3 to 0.5% polymer, 0.2% crosslinker concentration, pH 5, and 0.004% coagulation accelerator. The gelation time for the accelerator was 16 hours to 9 days.

Field Implementation

Eight producers were selected for the pilot test. The polymer injection was started in May 2000 and ended in November 2000, with a total injection of 19,012 m^3. The average injection for one well was 2376.5 m^3. A total of 12 wells had a positive response to the injection, and the effectiveness lasted for about a year. The water cut decreased from 86.3% to 82.7%, and the total incremental oil was 8384 tons. From April to August 2002, 5 more producers were treated. From July 2003 to May 2004, an additional 11 wells were treated.

Positive results similar to the pilot test were observed from these treatments. The total incremental oil recovery factor was 5.8%. In this case, the following measures were taken in the implementation:

- Upstream wells were selected for crosslinked polymer injection in the directions of edge water invasion.
- Sequential treatments were carried from the edge to the inner areas.
- Multiple slugs were injected. A slug injected at a later time had a longer gelation time so that it could bypass earlier slugs to increase sweep efficiency.
- The injection velocity was limited to 15 m^3/h. Injection was stopped when a significant increase in injection pressure was observed. Then water was injected. When the injection pressure decreased, polymer was injected again. After polymer injection was over, water was injected to displace the polymer in the tubing and annulus into the reservoir. Thus the injected polymer was farther displaced about 4 m deep into the reservoir.

Azri et al. (2010) and Brooks et al. (2010) also studied the feasibility of polymer injection in a heavy oil reservoir (250–500 cP) under strong bottom water drive.

5.9.4 Offshore Polymer Flooding

A pilot test in the SZ36-1 field in Bohai Bay, China, was summarized based on papers from Xiang et al. (2005), Zhou et al. (2006), and Han et al. (2006b). Some of the reservoir and fluid properties are listed in Table 5.18. The formation was unconsolidated and poorly cemented. Gravel packing was completed in the wells. The permeability was from 10s up to 8541 md, with an average of 3798.7 md. The oil viscosity in situ varied from 13 to 380 mPa·s with an average of 70 mPa·s. The well locations in the test area are shown in Figure 5.59. This pilot test had only 1 injector (J3) and 5 producers (J16, A2, A7, A12, and A13). The average well spacing in the test area was about 370 m.

Seawater with a TDS of 32,423 mg/L was injected for 8 years. The history of water injection made the salinity of produced water higher than the formation water salinity, especially the divalent. The water used in the polymer solution was initially Guantao formation water followed by a mixture of Guantao

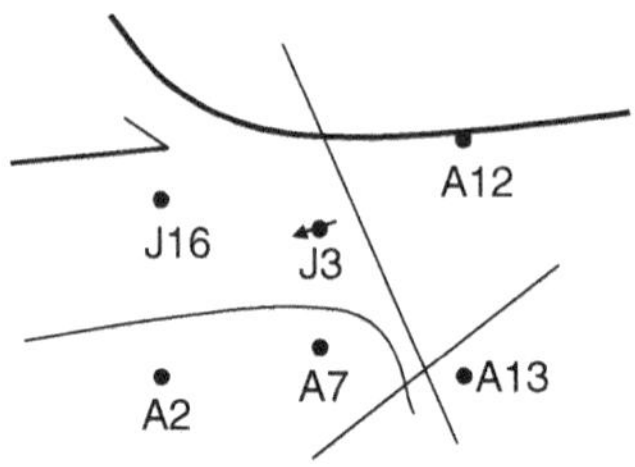

FIGURE 5.59 Schematic of J3 pilot pattern.

TABLE 5.18 SZ36-1 Pilot Reservoir and Fluid Data

ASP area, km^2	0.396
Formation depth (subsea), m	1300–1500
OOIP, tons	737,000
Porosity, %	19.6
Permeability, air, md	3798.7
Permeability variation coefficient	0.76
Formation temperature,°C	65
Average thickness, m	61.5
Oil viscosity at reservoir temperature, mPa·s	70

formation water and produced water. The TDS of Guantao formation water was 9048 mg/L, and Ca^{2+} and Mg^{2+} concentration was 800 mg/L. This test used the hydrophobically associating polymer AP-P4 made in China. The pilot test was started in September 2003 and ended in May 2005. The injection rate was 500 m^3/d. The slug concentrations were 3000 mg/L and 1750 mg/L, respectively. Only 0.037 PV was injected. The viscosity of the sheared polymer solution was about 25 mPa·s. This sheared viscosity corresponded to 1750 mg/L injection concentration with 78 mPa·s before shearing. After polymer injection, the injection pressure was stabilized at 6 to 7 MPa. The producers responded after 10 months of polymer injection. It was reported that water cut decreased and oil rate increased, but the increase in oil rate was not significant except in Well J16.

The limitations in implementing offshore polymer flooding are large well spacing and limited spacing in the platform. Due to the large well spacing, high injection pressure and late response were observed. The limited spacing problem in the platform was solved by using a portable automatic skid injection unit. A detailed description of the skid was given by Chen (2005). In a current separate deep offshore polymer injection ongoing pilot test, the Dalia Angola case, it was proposed to store polymer on board a barge or a dry powder carrier located at the field, or on a skid on the floating production storage and offloading (FPSO). The barge or carrier would be equipped for processing the polymer solution and transferring the solution to the FPSO (Morel et al., 2008).

5.9.5 Uses of Produced Water

In a large-scale polymer injection field application, a large amount of water is needed to mix polymer solution, and a large amount of produced water needs to be dumped somewhere. If we can make use of the produced water to mix a polymer solution, then we can accomplish two things at once.

The problem is that if produced water with higher salinity is used, the viscosity of polymer solution will be lower. For example, if produced water is used instead of fresh water, the polymer concentration has to be increased by 55% to reach the same viscosity. To solve this problem, we need a polymer that can tolerate high salinity. A polymer with 30 million MW was used for this purpose in a Daqing pilot test (Niu et al., 2006).

The pilot test was in the Northwestern block of the Lamadian field. The target formation was in $PI_{1\text{-}2}$. Some of the reservoir, fluid, and well data are shown in Table 5.19. The produced water came from the La400 produced water treatment station; it was transported into a pressure container and put in contact with air for 5 minutes before it was transported into a reaction container for 2 hours. Then it was pumped into a water pipeline to mix with a mother polymer

TABLE 5.19 Reservoir, Fluid, and Well Data in the Northwestern Lamadian Block

Test area, km^2	3.45
OOIP, 10^6 tons	7.3
Pore volume, 10^6 m^3	13.43
Number of injectors	39
Number of producers	44
Injector-producer distance, m	237
Formation temperature,°C	45
Formation water TDS, mg/L	7150
Produced water salinity, mg/L	3500–5000
Oil viscosity in place, mPa·s	10.28

solution and injected into the target formation. The mother polymer solution was mixed using fresh water.

The polymer injection was started in May and ended in December 2002. A 0.248 PV polymer solution with 1248 mg/L concentration (total 310 mg/L·PV) was injected; the injection solution viscosity was 54.8 mPa·s. In November 2002, the injection pressure was 9.5 MPa compared with 5.4 MPa before injection. The increase was 0.6 MPa lower than that in the compared area where a polymer solution mixed with fresh water was injected. The water cut decreased by 38.6%, 9.4% more than that in the compared area. During the polymer injection period, the incremental oil recovery factor was 6%. The produced water was injected in other areas of the Lamadian field. Thus no produced water was disposed.

The produced water had some polymer. The produced water may be injected before or after the main polymer slug as preflush or postflush slugs. A calculation showed that 2.5% incremental oil recovery could be obtained if the produced water was injected as preflush and 0.9% as postflush. A laboratory test showed that if the produced water had 400 mg/L polymer (with solution viscosity of 2.5 mPa·s), the incremental oil recovery factor was about 3% (Zhang, 1998).

5.9.6 Early Pilot Tests in Daqing

The first pilot test in Daqing was started on August 30, 1972, and ended on September 24 in the same year, a total of 26 days. This test was conducted in the $SaII_{7+8}$ layer. One inverted four-spot pattern was used with the injector, Well 501, in the center. Thus, it was called the Well 501 pattern. The distance between the injector and a producer was 75 m. The formation thickness was 5.2 m, and the permeability was 631 md. The reservoir temperature was 45°C. A 0.163 PV of polymer solution was injected having a concentration from 1000 to 1800 mg/L. The three producers started to respond after 12 days of polymer injection. The water cut at one producer (Well 503) was reduced from 99 to 60.4%, and the well pattern incremental oil recovery was about 5%. The well injection pressure increased, and the liquid production rate was reduced significantly. In this first pilot test, low molecular weight polymer (3 to 5 million Daltons) was used (Liu, 1995; Yang et al., 1996).

On February 10, 1988, another pilot test was started in Bei-3-Qu-Xi PI1-3 in the Saertu field, Daqing, and ended on September 4, 1990. There were 4 injectors and 9 producers, with the distance between injector and producer being 200 m. The incremental oil recovery was 3% (Yang et al., 1996).

5.9.7 PO and PT Pilot Tests in Daqing

The two pilot tests, PO (Polymer One) and PT (Polymer Two), were extensively studied and frequently reported (Chauveteau et al., 1988; Corlay et al., 1992;

Delamaide et al., 1994; Liu, 1995; Yang et al., 1996; Niu et al., 2006). The project for these two tests was started in 1984 for selecting test areas, and polymer injection was ended in February 1992. The project lasted 7 years. The two tests were conducted in the west central area (Zhong-Qu-Xi-Bu in Chinese) of the Saertu field, Daqing. In the PO pilot test, polymer was injected into the single layer zone PI_{1-4}, whereas in the PT test, polymer was injected into the two layer zones PI_{1-4} and SII_{1-3}. The two test areas were separated by 150 m. Each test had four inverted five-spot patterns with 4 injectors and 9 producers, as shown in Figure 5.60. The distance between injector and producer was 106 m. Some of the reservoir and fluid data are shown in Table 5.20. The polymer used was FLOPAAM3330S with a viscosity of 31 to 38 mPa·s at 1000 mg/L and at 45°C.

For the PO test, polymer injection was started on August 5, 1990, and ended on February 20, 1992. The total injected polymer was 504 mg/L·PV. For the PT test, polymer injection was started on November 7, 1990, and ended on February 24, 1992. The total injected polymer was 491 mg/L·PV. The concentrations were decreased in the subsequent slugs, as shown in Table 5.21.

After polymer injection, the water intake indices decreased by 23.6 to 36.9% for the wells in PO and PT, and the liquid production indices decreased by 69.6% for PO and 59.9% for PT. The water cuts decreased from 95.2 to 79.4% for PO, and from 94.7 to 84.4% for PT. The incremental oil recovery factor reached 14% for PO by July 1992 and 11.6% for PT by November 1992.

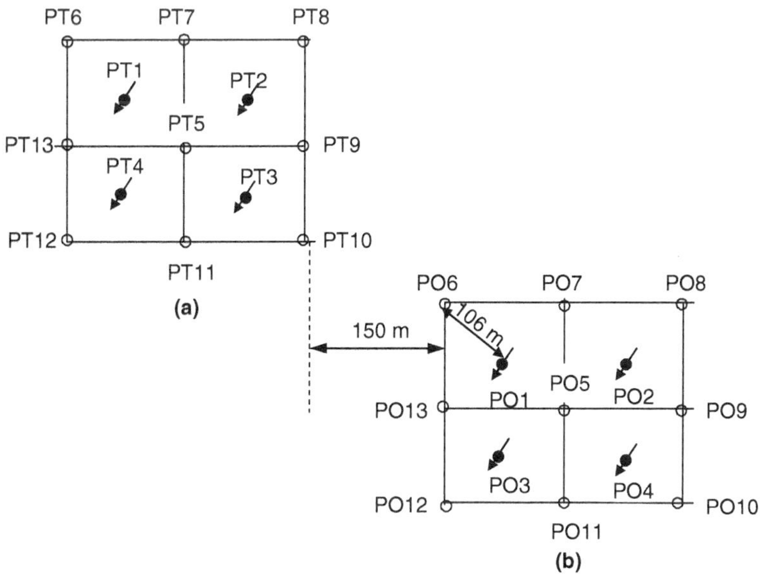

FIGURE 5.60 Schematic of well patterns for the (a) PO and (b) PT pilot tests.

TABLE 5.20 PO and PT Reservoir and Fluid Data

	PO	PT	
	$PI_{1\text{-}4}$	$PI_{1\text{-}4}$	$SII_{1\text{-}3}$
PV, m^3	319568	412435	169586
Porosity, %	31.11	31.05	31.39
Average permeability, md	1150	1100	937
Effective thickness, m	11.6	15	6.1
Permeability variation coefficient	0.6–0.8	0.6–0.8	0.5–0.7
Formation temperature,°C	45		
Formation water TDS, mg/L	7000		
Injection water salinity, mg/L	800		
Produced water salinity, mg/L	3000		
Oil viscosity in place, mPa·s	9.5		

TABLE 5.21 PO and PT Injection Schemes

	Slug 1		Slug 2		Slug 3		Total
	PV	C_p, mg/L	PV	C_p, mg/L	PV	C_p, mg/L	mg/L·PV
PO	0.527	837	0.0381	683	0.0103	400	504
PT	0.504	908	0.0715	461			491

One ton of polymer injected increased 241 tons of oil recovered for PO and 209 tons for PT. The polymer solution viscosity loss was 12% at injection pump, 30% at injection wellhead, 57% 30 m away from the injector, and 70% cumulatively 106 m away from the injector (Yang et al., 1996). These data show that the viscosity loss occurred mainly from the injection pump to the reservoir near the injection well. The viscosity loss in the reservoir was also caused by higher salinity. Therefore, we can see that the viscosity shear loss in the reservoir was less significant.

5.9.8 Large-Scale Field Applications

There are a number of ongoing large-scale polymer flood field applications; several of them are described in the next subsections.

Bei-1-Qu-Duan-Xi (B1-FBX), Daqing

Bei-1-Qu-Duan-Xi was the first large-scale polymer flooding field application in the northern Saertu field, Daqing. There were 25 injectors and 37 producers in the test area in five-spot patterns. The target layers were PI_{1-4}. The well spacing from injector to producer was 250 to 300 m. Some of the reservoir and fluid data are shown in Table 5.22 (Chang et al., 2006; Yan et al., 2006).

Before polymer flooding, 0.66 PV water had been injected with a recovery factor of 28.5%. The water cut was 88%. Polymer injection was started in January 1993 and ended in April 1997 with a total 592 mg/L·PV. Approximately 40% of the polymer used in the first slug was high MW polymer (17 to 19 million); the MW in the main slug was 11 to 12 million. The polymer concentration was 800 to 1000 mg/L. The post-PF water drive was completed in October 1998. Some observations regarding this test are summarized here:

1. Initial production response to polymer injection was observed after about 0.1 PV injection, showing increased oil rate and reduced water cut. The oil peaked at 0.64 PV injection. Low water cut lasted about one year.
2. Tracer broke-through production wells from 32 days to 200 days. Polymer broke-through from 102 to 124 days. The produced polymer concentration stayed at 400 mg/L and peaked at 600 to 800 mg/L. Seventy percent of injected polymer was retained in the reservoir.
3. The residual permeability reduction factor in the test area was about 2.
4. High MW polymer injection resulted in 1 MPa increase in injection pressure.
5. The producers connected with injectors in more directions performed better than those with fewer connections.

TABLE 5.22 B1-FBX Reservoir and Fluid Data

Test area, km^2	3.13
Formation depth (subsea), m	1036–1117
Average permeability, md	347–1182
Effective thickness, m	49
Formation temperature,°C	45
Formation water TDS, mg/L	7000
Injection water salinity, mg/L	1275
Injection water divalents, mg/L	<50
Oil viscosity in place, mPa·s	9–10

6. The layers with lower waterflooding sweep efficiency performed better in polymer flooding.

Gudao Zhong-1-Ng3, Shengli

The application took place in the Zhong-1 block of the Gudao field, Sinopec. There were 40 injectors and 85 producers in five-spot patterns. The target layers were Ng3, and the well spacing from injector to producer was 200 to 300 m. Some of the reservoir and fluid data are shown in Table 5.23 (Li, 2004b; Chang et al., 2006; Yan et al., 2006).

Before PF, the water cut was 93.7% with a recovery factor of 38.1%, and the predicted final WF recovery was 43.8% at 98% water cut. Polymer injection was started in December 1994 and ended in late 1996 with a total of 432 mg/L·PV of polymer injected. In the first slug, 0.027 PV 1739 mg/L polymer was used before the second main slug, which was 0.249 PV of 1544 mg/L solution. Some observations are summarized here:

1. Initial production response to polymer injection was observed after about 0.12 PV injection, showing the water cut decreased from 93.7% to 89%.
2. The produced water was used to dilute the mother polymer solution. The polymer solution viscosity (12.8 mPa·s at 1500 mg/L) was about half of the viscosity (23.6 mPa·s) if fresh water was used.
3. The residual permeability reduction factors using the Hall plot in the test area were 1.5 to 2.5.

Other large-scale polymer flooding applications in China include

- Daqing Bei-2-qu-xi-bu-dong-kuai with 52 injectors and 58 producers (Wang et al., 1997a),

TABLE 5.23 Zhong-1-Ng3 Reservoir and Fluid Data

Test area, km^2	4.5
Average permeability, md	1500–2500
Porosity, fraction	0.33
Effective thickness, m	12.1
Formation temperature,°C	69.5
Formation water TDS, mg/L	3898
Injection water salinity, mg/L	7093
Oil viscosity in place, mPa·s	46

- Shengli Shengtuo Block 1 with 49 injectors and 89 producers (Liu et al., 2002; Li, 2004c; Yan et al., 2006),
- The Shengli Feiyantan field with 35 injectors and 86 producers (Lin et al., 2005),
- Shengli Gudong Block 6 with 27 injectors and 43 producers (Wang, 2006),
- Shengli Gudao Zhong-2-Nan with 28 injectors and 62 producers (Guan et al., 2002; Jiang et al., 2003c),
- Shengli Gudong Block 8 with 44 injectors and 90 producers (Yi et al., 1999),
- The Western Block 7 in Gudong, Shengli, with 42 injectors and 54 producers (Dou et al., 1999),
- The Shuanghe field of Henan field with 30 injectors and 48 producers (Li et al., 2004).

5.9.9 Polymer Flooding Pilot in the Sabei Transition Zone

Most of polymer floods in Daqing were conducted in oil zones. There are a significant amount of reserves in transition zones. In 1995, the Sabei transition zone was selected for a polymer flooding pilot test. The target formation was $PI_{1\text{-}4}$. The basic reservoir, fluid, and well data are shown in Table 5.24 (Niu et al., 2006). The well pattern was irregular four-spot.

TABLE 5.24 Reservoir, Fluid, and Well Data in the Sabei Transition Zone

Test area, km^2	0.84
OOIP, 10^6 tons	1.657
Pore volume, 10^6 m^3	2.8691
Average initial water saturation	0.424
Effective thickness, m	13.2
Average permeability, md	587
Permeability variation coefficient	0.647
Number of injectors	10
Number of producers	9
Injector-producer distance, m	175–200
Formation temperature,°C	45
Oil viscosity in place, mPa·s	1.5–20

Before PF, 0.1153 PV water had been injected from October 1996 to August 1997. A total polymer injection of 597.64 mg/L·PV was started on August 28, 1997, and ended in October 2002. The average injected polymer concentration was 892 mg/L, and the injection PV was 0.67. The viscosity at the injection wellhead was 20 mPa·s. In this case, the incremental oil recovery factor was 8.55%.

Basically, the polymer flooding performance in the transition zone was similar to that in other oil zones. Some of the observations in performance compared with the oil-zone polymer floods in Daqing (Niu et al., 2006) follow.

- The injection pressure increased by 51.2% from 7.87 MPa to 11.9 MPa, compared with a 20% increase in a thick layer pilot, 49.3% increase in Bei-1-Qu-Duan-Xi, and 31.3% increase in Bei-3-XI-XI-Kuai oil zones.
- The liquid rate and liquid production index were lower than those in the oil zones. The liquid production index decreased by 82% from 2.55 to 0.46 m^3/(d·m·Mpa).
- The response to polymer injection at a producer was observed after 46 mg/L·PV of polymer was injected, earlier than in the other oil zones (58, 53.6, and 90 mg/L·PV for Bei-2-XI-XI-Kuai, Bei-3-XI-XI-Kuai, and Bei-3-XI-Dong-Kuai, respectively).
- The incremental oil recovery factor was 8.55%, which is lower than 10% in the oil zones. This result might be due to the higher oil viscosity of 15 to 20 mPa·s compared with 9 to 10 mPa·s in the oil zones.
- The water-cut reduction in the middle wells in the test area was higher. This response was also observed at the external wells outside the injection patterns.

5.9.10 Dagang Gangxi Block 4 PF Pilot with Profile Control

This section presents a pilot test of polymer injection after profile control (Yan et al., 2005). The pilot area, Gangxi Block 4, belonged to the Dagang Petroleum Administration Bureau (Dagang Oilfield). The basic reservoir, fluid, and well data are shown in Table 5.25. From this table, we can see the permeability variation coefficient is high, showing this block is very heterogeneous. Before polymer injection, a profile control was needed.

The profile control agent was aluminum citrate. The penetration radius was 16.3 to 36.7 m, which is about 1/6 to 1/4 of the injector-producer distance. The AT-430 polymer was used at a concentration of 1000 mg/L. Injection of the profile control agent was started on March 28, 1986. After the control agent was injected, polymer was injected starting from December 4, 1986. A total of 124 mg/L PV was injected. The incremental oil recovery factor was 11.5%. For one ton of polymer injected, 514 tons of oil were recovered. However, it was realized that the amount of polymer injected in this project was not large enough. The produced polymer showed that the polymer MW was reduced

TABLE 5.25 Reservoir, Fluid, and Well Data in the Gangxi Block 4

Test area, km^2	0.59
OOIP, 10^6 tons	0.753
Porosity, %	31
Average permeability, md	234
Permeability variation coefficient	0.77–0.81
Number of injectors	3
Number of producers	10
Formation water salinity, mg/L	5731
Formation temperature,°C	51
Oil viscosity in place, mPa·s	20

TABLE 5.26 Reservoir and Fluid Data in the Block 3-2 the Ke-Shang Layer

Test area, km^2	3.64
Formation subsea depth, m	475
Porosity, %	21.2
Average permeability, md	329
Permeability variation coefficient	0.7
Thickness, m	4.7
Oil viscosity in place, mPa·s	30

from 10 million to 3 million Daltons. Higher MW polymer should have been used.

5.9.11 Karamay Crosslinked Polymer Solution Pilot

This section presents a pilot test of crosslinked low polymer solution (LPS) injection (Liu et al., 2005b). The pilot area, Block 3-2 Ke-Shang layer, belonged to the Karamay Petroleum Administration Bureau (Karamay Oilfield). The basic reservoir, fluid, and well data are shown in Table 5.26.

The reservoir was very heterogeneous, and waterflooding was not efficient. By September 2003, the water cut was 66.4%, and the oil recovery factor was 22.39%. In 2001, colloidal dispersion gel (CDG) injection and other profile controls were carried out in several well patterns. The injected agents had high molecular weight and high concentration polymers. However, they had poor injectivity and could not penetrate deep into the reservoir. Because crosslinked polymer solution has low viscosity, good injectivity, and deep penetration, it was chosen for the test. Aluminum citrate was the crosslinker used. Through laboratory and simulation studies, the designed LPS system was 225 to 250 ppm HPAM and 18 ppm Al^{3+}. Fresh water was used in preparing the mother polymer solution, whereas produced water was used in dilution. There were two injection slugs. In the first 0.034 PV slug, 225 ppm polymer was used. In the second slug, 250 ppm polymer and 250 ppm oil-displacing agent (surfactant) were injected.

The LPS injection was started in October 2003. By June 2006, injection of 0.031 PV was completed for the first slug. However, the water cut in some high water-cut wells was not reduced. Therefore, an additional 500 ppm polymer solution was injected for 15 days. Positive responses were observed. By July 2005, 15,000 tons of incremental oil had been recovered.

5.9.12 Polymer Flooding in a High-Temperature and High-Salinity Reservoir

Another polymer flooding pilot test was undertaken in the Wangchang field by the Jianghan Petroleum Administration Bureau (Jianghan Oilfield; Yang and Feng, 2001). The target formation was the Chang 3I zone in the northern fault block. In this case, the temperature was 70.4°C, and the total dissolved salinity was 326,000 mg/L. More data are shown in Table 5.27.

TABLE 5.27 Reservoir, Fluid, and Well Data in the Wangchang Field Pilot

Test area, km^2	1.545
OOIP, 10^6 tons	2.9636
Reservoir depth, m	1490.5
Porosity, %	23.3
Thickness, m	16.5
Average permeability, md	457.1
Formation water salinity, mg/L	326000
Injection water salinity, mg/L	129000
Formation temperature,°C	70.4

Because of the high salinity, a xanthan gum type of polymer, Flocon 4800, was selected. The viscosity versus polymer concentrations at 30°C and 75°C are shown in Figure 5.61. Figure 5.62 shows the viscosity versus salinity at 800 mg/L and 30°C. This figure shows that the viscosity was not sensitive to salinity; viscosity was even higher at a higher salinity. It was not very sensitive to pH either. Therefore, the produced water with 12.9 wt.% salinity was used in mixing polymer. The produced water was controlled under 0.4 to 0.6 mg/L oxygen by a closed nitrogen system. Iron ion concentration was less than 0.1 mg/L, and 0.2% formaldehyde was added as a biocide. Before polymer injection, the water cut was 94%, and the recovery factor was 42.7%. The estimated incremental oil recovery factor was 10 to 11%. The economic calculation showed that the cash flow became positive starting from the third year of polymer injection.

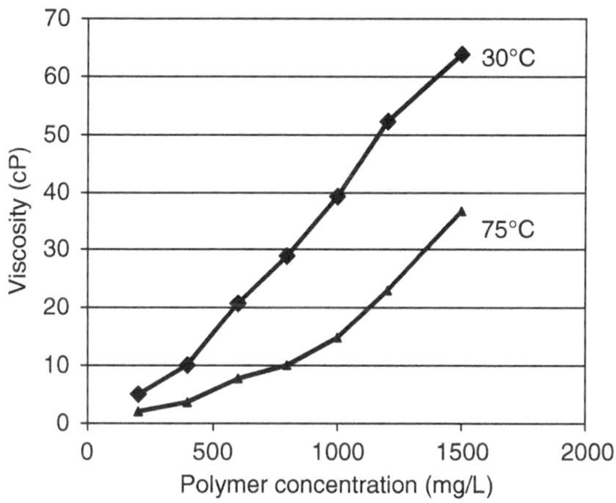

FIGURE 5.61 Flocon 4800 viscosity versus xanthan concentration at 30°C and 75°C.

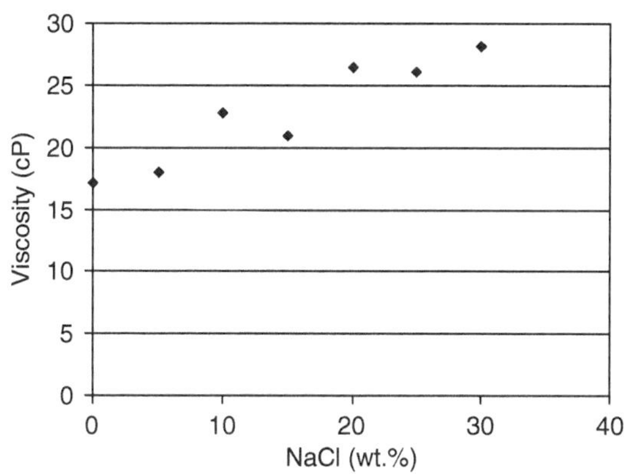

FIGURE 5.62 Flocon 4800 viscosity versus salinity at 800 mg/L and 30°C.

5.10 POLYMER FLOODING EXPERIENCE AND LEARNING IN CHINA

This section summarizes the polymer flooding experience and what was learned in China. Before that, the performance characteristics during different periods of polymer flooding are summarized.

5.10.1 Performance Characteristics during Different Periods

According to its dynamic performance, polymer flooding can be divided into five periods (Shao et al., 2005).

1. *Initial period.* During this period, the injected polymer volume is 0 to 0.04 PV. The injected polymer mainly flows through high-permeability channels. Injection profile and mobility are controlled. Injection pressure increases, but water cut continues increasing.
2. *Decreasing water cut.* During 0.05 to 0.15 PV injection, injection pressure continues increasing, water cut sharply decreases, and lower permeability formations start to intake injected water. About 17% of the cumulative oil is produced in this period.
3. *Low water cut.* During 0.15 to 0.35 PV injection, the water cut is low, and the minimum water cut and higher oil rate appear in this period. The liquid production rate declines, polymer concentration starts to decrease, and polymer starts to be produced. The injection pressure increases slowly. The water intake in low permeability layers starts to decrease, and the injection profile starts to return to its initial profile. About 39% of the cumulative oil is produced in this period.
4. *Water cut starts to increase.* After 0.35 PV is injected, the water cut increases, oil rate decreases, and produced polymer concentration and injection pressure are high. About 33% of the cumulative oil is produced in this period.
5. *Post-water drive.* The water cut continues increasing and injection pressure decreases. Water breaks through high permeability channels. Polymer concentration decreases and liquid offtake rate increases. About 11% of the cumulative oil is produced in this period.

Note that we refer to the injection PV values and cumulative oil production in each period for illustration purposes. These values should not be applied to other fields. Even for the Daqing fields, different authors reported quite different values—for example, Liu et al. (2005a).

5.10.2 Experience and Learning

This section summarizes the experience and learning on several subjects gained during more than 20 years of pilot testing and large-scale commercial applications in polymer flooding in China.

High Molecular Weight Polymer

High molecular weight (MW) polymer has a higher viscosifying power and higher permeability reduction factor than low molecular weight polymer. For the same amount of polymer injected, the polymer with higher molecular weight would result in a higher recovery. For the same recovery factor, a higher polymer solution requires less polymer. When polymer of 25 million MW is dissolved in produced water, the viscosity could reach 40 to 50 mPa·s at 800 to 900 mg/L concentration (Niu et al., 2006). However, the molecular weight is limited by formation permeability (see Figure 5.52). High molecular weight polymer cannot be injected in low-permeability formations.

The injection sequence of different molecular weights also affects oil recovery. Table 5.28 shows the incremental oil recovery factors when high molecular weight polymer was placed in the front end of a polymer slug (Niu et al., 2006). In the experiments, the total amount of polymer injected was 570 mg/L·PV, and the polymer concentration was 1000 mg/L. The rock permeability variation was 0.72. The molecular weight in the main slug was 12 million, and the high molecular weight placed in the front of slug was 17 million. From this table, we can see that the maximum recovery factor using the high MW polymer was 3% higher than that using the low MW polymer; when the high MW slug was about 33% of the total polymer injection, 2.9% incremental oil recovery factor was obtained; this amount is close to that obtained when 100% high MW was used. Zhang (1998) reported much higher incremental oil recovery factors using high MW polymers than using low MW polymers.

In the northern Lamadian field, high molecular weight polymer was used in the northern block before the main slug, but not in the eastern block. The injection pressure in the northern block increased by 7.5 MPa compared with 6.1 MPa in the eastern block at 0.44 PV injection; the injectivity index decreased

TABLE 5.28 Effect of High MW Slug Placed in the Front End on Oil Recovery

% of High MW Polymer	Incremental RF over WF	Incremental RF over Low MW
0	20.7	0
10	21.6	0.9
20	22.7	2.0
33	23.6	2.9
50	23.6	2.9
100	23.7	3.0

by 48% in the northern block compared with 25.2% in the eastern block; the water-cut reduction in the northern block was also higher (Shao et al., 2005). Similar to the effect of high MW polymer, higher concentration polymer solution injected at the very stage of polymer flooding should also be beneficial because it can reduce mobility ratio at the displacing front and decrease polymer dilution.

High Sweep Efficiency

High sweep efficiency by polymer fluid is the key to the success of polymer flooding. This effect can be achieved by lowering well spacing, adjusting well patterns, separate layer injection, profile modification/control, and so on. A detailed discussion regarding sweep efficiency was provided by Wang et al. (2006a).

The data from the Daqing polymer flooding suggest that the permeability variation coefficient significantly affects polymer sweep efficiency. When the coefficient is less than 0.72, the polymer performance (in terms of incremental oil recovery) becomes better as it gets higher. When the coefficient is greater than 0.72, the polymer performance significantly becomes worse as it gets higher. At 0.72, the polymer performance is at its optimum.

As described in Chapter 4, the higher the displacing fluid viscosity, the higher the sweep efficiency (recovery factor). Chapter 4 proposed that the displacing fluid mobility should be equal to the displaced oil mobility corrected by the oil saturation. In Daqing, however, the polymer solution viscosity of three to five times the oil viscosity was used so that a high oil recovery factor can be obtained in heterogeneous reservoirs.

For very heterogeneous reservoirs, the profile control deep in the reservoir was tested using different technologies, such as crosslinked polymer. The deep profile control agents could be injected before, during, or after polymer injection. Colloidal dispersion gels were claimed to be successful (Mack and Smith, 1994; Fielding et al., 1994; Smith et al., 2000; Chang et al., 2006). However, whether colloidal dispersion gels could be a viable option is controversial (Wang et al., 2006a; Seright, 2006).

Amount of Injected Polymer

Studies of more than 200 polymer floods reported that the total amount of polymer injected on average was 19 to 150 lb/acre-ft, which is equivalent to 23.3 to 184.2 mg/L·PV if a porosity of 0.3 is used. The polymer concentrations were 50 to 3700 mg/L in the early days (Manrique et al., 2007). The 1976 U.S. National Petroleum Council (NPC) study reported 125 mg/L·PV. In contrast, the amount of polymer reported in the 1984 NPC study was increased to 240 mg/L·PV, but it is still much lower than that used in China (Chang et al., 2006).

Daqing polymer flooding performance showed that incremental oil recovery factor peaked when the amount of polymer injected was 180 to 210 mg/L·PV.

The best comprehensive economics occurred when the injected polymer was about 380 mg/L·PV. However, there seems to be a trend to inject larger amounts of polymer in Daqing. As discussed in Section 5.6, 400 to 500 mg/L·PV were used according to the author's analysis. More references (e.g., Shao et al., 2005) reported that more than 600 mg/L·PV were used in Daqing's large-scale applications; the highest amount was 771 mg/L·PV. Because surface facility investment is almost independent of the amount of polymer injected, a larger amount of polymer injection would result in a higher incremental oil recovery and thus better economics. A detailed analysis is needed for an individual project.

Conventionally, a graded or tapered scheme that is an empirical model (Claridge, 1978) is used to reduce the amount of polymer injected. The simplest model assumes that the polymer concentration declines exponentially. Claridge (1978) and Stoneberger and Claridge (1988) developed a method based on the method by Koval (1963) to design graded viscosity banks. However, as the amount of polymer injected becomes larger, such a graded scheme becomes less important. The reason is that the chase water would have less opportunity to break through the polymer slug ahead of it if the polymer slug is large. Thus, polymer injection time is reduced, and the operation cost will be reduced as well.

Time to Shift Waterflooding to Polymer Flooding

According to polymer drive and water drive fractional flow curves, at the same oil recovery (saturation), the water-cut increase rate in waterflooding is much higher than in polymer flooding at low water cuts. As the water cut increases, the difference in the water cuts becomes smaller. When the water cut is above 92%, the water-cut increase rates are almost the same. Therefore, polymer injection should be stopped when the water cut is about 92 to 94% in Daqing (Shao et al., 2005). However, PF was still carried out in some large-scale field applications—for example, Gudao Zhong-1-Ng3, discussed in Section 5.9.8, and Shengtuo Block 1 (Li, 2004c; Liu et al., 2002). Chang et al. (2006) concluded that polymer flooding can be applied effectively to reservoirs with water cuts > 95%.

Hydraulic Fracturing

For low-permeability injection and production wells, hydraulic fracturing improves the polymer injectivity and the productivity of liquid. Daqing practices show that the improvement was more effective when fracturing was conducted during the early low water-cut periods (Niu et al., 2006).

Reinjection of Produced Water and Polymer

Less saline water should be used in mixing polymer solution (Wang et al., 2006a). If produced water is used to make the polymer solution, the cost of the polymer will be increased by 55% (Niu et al., 2006). However, desalination is expensive. Ayirala et al. (2010) presented results using low-salinity water to mix polymer solution so that a low polymer concentration is needed to achieve the target viscosity compared with using seawater. Their data indicate about 5 to 10 times lower consumption of polymer using low-salinity water when compared with using seawater when their "designer water" desalination scheme is used. The incremental cost of water desalination and hardness removal can be paid out within a four-year project time frame because of the large savings associated with chemical and polymer facility costs in low-salinity waterflood polymer flooding in an offshore environment.

Daqing polymer flooding performance shows that oil rate increased before produced polymer concentration increased. Produced polymer concentrations peaked at 400 to 900 mg/L, approximately half of the injected concentration. As mentioned earlier, the produced water with polymer may be re-injected to save water cost and polymer cost.

Chapter 6

Polymer Viscoelastic Behavior and Its Effect on Field Facilities and Operations

6.1 INTRODUCTION

The conventional belief is that polymer flooding can improve only sweep efficiency, but it cannot increase displacement efficiency. In other words, waterflooding residual oil saturation cannot be reduced by polymer flooding, and polymer flooding is expected to increase the oil recovery factor over waterflooding only by about 5%. The recent polymer flooding practices in Daqing, however, have increased the recovery factor by up to 12% (Wang, 2001). Chinese researchers attribute such high performance to polymer viscoelastic behavior.

This chapter discusses this new concept. It first reviews some fluid viscoelastic properties. Then it presents the evidence of polymer viscoelastic behavior in the laboratory and in the field. In addition, this chapter discusses the displacement mechanisms of polymer solution and the effect of viscoelastic polymer solution on field facilities and operations.

6.2 VISCOELASTICITY

Viscoelasticity is the property of materials that exhibit both viscous and elastic characteristics when undergoing deformation. To understand fluid viscoelasticity, we need to start with the fluid viscous and solid elastic behaviors that are well known.

For a simple viscous fluid, the viscous behavior of it is described by the equation

$$\tau = \mu\dot{\gamma}, \tag{6.1}$$

where τ is the stress, μ is the viscosity, and $\dot{\gamma}$ is the shear rate. Note that when μ is not a constant for a non-Newtonian fluid, some authors (e.g., Sorbie, 1991) use η to represent μ in the preceding equation. η is called a viscosity function $\eta(\dot{\gamma})$ that depends on the shear rate. This book does not differentiate μ and η.

Modern Chemical Enhanced Oil Recovery. DOI: 10.1016/B978-1-85617-745-0.00006-1

When elastic materials (e.g., sponge or spring) are deformed through a small displacement, they tend to return to their original configuration. If a shear stress is applied to an ideal solid, then for a small displacement, Hooke's law is valid:

$$\tau = G'\gamma, \tag{6.2}$$

Here, G' is the elastic modules, and γ is the displacement (strain) that is the ratio of the change caused by the stress to the original state of the object. The unit of γ is dimensionless. So G' (also G'' to be defined later) and τ have the same unit.

Note that the simple Hooke's law behavior of the stress in a solid is analogous to Newton's law for the stress of a fluid. For a simple Newtonian fluid, the shear stress is proportional to the rate of strain, $\dot{\gamma}$ (shear rate), whereas in a Hookian solid, it is proportional to the strain, γ, itself. For a fluid that shares both viscous and elastic behavior, the equation for the shear stress must incorporate both of these laws—Newton's and Hooke's. A possible constitutive relationship between the stress in a fluid and the strain is described by the Maxwell model (Eq. 6.3), which assumes that a purely viscous damper described by Eq. 6.1 and a pure spring described by Eq. 6.2 are connected in series (i.e., the two $\dot{\gamma}$ from Eqs. 6.1 and 6.2 are additive).

$$\frac{\tau}{\mu} + \frac{1}{G'}\left(\frac{\partial \tau}{\partial t}\right) = \dot{\gamma}. \tag{6.3}$$

This equation has the correct limiting behavior: it reduces to an equation for a simple Newtonian fluid when $\partial\tau/\partial t$ approaches to 0 for steady shear flow. When the stress changes rapidly with time, and τ is negligible compared with $\partial\tau/\partial t$, it reduces to the constitutive equation of a Hookian solid.

Traditionally, shear viscosity measurements are used to rheologically characterize fluids. Figure 6.1 shows the principle for shear viscosity measurement; this figure shows a steady shear flow field between two parallel plates, one of which is moving with a velocity v. The measured quantities are the velocity of the top plate, the separation gap d, and the force in the direction of shear experienced by the stationary plate. Equation 6.1 is used to calculate the shear viscosity of the fluid, and the shear rate is calculated $\dot{\gamma} = v/d$ (velocity/distance between the two parallel plates). Shear rate is also called velocity gradient. We can see that this shear rate or velocity gradient is constant. In this case, the displacement (strain) is

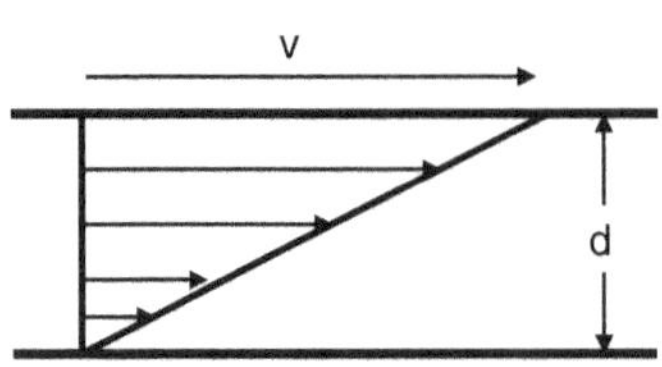

FIGURE 6.1 Schematic of the principle to measure shear viscosity.

$$\gamma = \text{a constant} \times \text{t (time)}. \tag{6.4}$$

Steady shear flow measurements, however, can measure only viscosity and the first normal stress difference, and it is difficult to derive information about fluid structure from such measurements. Instead, dynamic oscillatory rheological measurements are used to characterize both enhanced oil recovery polymer solutions and polymer crosslinker gel systems (Prud'Homme et al., 1983; Knoll and Prud'Homme, 1987). Dynamic oscillatory measurements differ from steady shear viscosity measurements in that a sinusoidal movement is imposed on the fluid system rather than a continuous, unidirectional movement. In other words, the following displacement is imposed:

$$\gamma = \text{a constant} \times \sin(\omega t). \tag{6.5}$$

Remember that a viscoelastic fluid has two components related to $\dot{\gamma}$ by Eq. 6.1 and γ by Eq. 6.2. From Eq. 6.5, it is clear that for such dynamic oscillatory displacement, the measured stress response has two components: an in-phase component ($\sin\omega t$) and an out-of-phase component ($\cos\omega t$). Viscoelastic materials produce this two-component stress response when they undergo mechanical deformation because some of the energy is stored elastically and some is dissipated or lost. The stress response, which is in-phase with the mechanical displacement, defines a storage or elastic modulus, G', and the out-of-phase stress response defines a loss or viscous modulus, G''. The storage modulus (G') provides information about the fluid's elasticity and network structure.

Through use of classical network theories of macromolecules, G' has been shown to be proportional to crosslink density by $G' = nKT + G_{en}$, where n is the number density of crosslinkers, K is the Boltzmann's constant, T is the absolute temperature, and G_{en} is the contribution to the modulus because of polymer chain entanglement (Knoll and Prud'Homme, 1987). The loss modulus (G'') gives information about the viscous properties of the fluid. The stress response for a viscous Newtonian fluid would be 90 degrees out-of-phase with the displacement but in-phase with the shear rate. So, for an elastic material, all the information is in the storage modulus, G', and for a viscous material, all the information is in the loss modulus, G''. Refer to Figure 6.2, the dynamic viscosities μ' and μ'' are defined as

$$\mu' = G'/\omega = G^* \cdot \sin\theta/\omega, \tag{6.6}$$

$$\mu'' = G''/\omega = G^* \cdot \cos\theta/\omega, \tag{6.7}$$

where G^* is the complex dynamic modulus, and θ is the phase shift. When the stress components G' and G'' are combined, the complex viscosity μ^* may be calculated using (Knoll and Prud'Homme, 1987)

$$\mu^* = \frac{\sqrt{(G')^2 + (G'')^2}}{\omega}, \tag{6.8}$$

where ω is frequency in radians per second, and the moduli have the same unit as stress (e.g., dyne/cm^2).

Polymer solutions generally exhibit viscous behavior when flowing in capillary tubes with constant diameters. However, in porous media where capillary diameters change rapidly, polymer chains are pulled or contracted to exhibit elastic behavior. The elastic behavior leads to a higher apparent viscosity, as described by Eq. 6.8.

Another viscoelastic parameter is the relaxation time, t_r, which is defined as the time for the viscoelastic polymer fluid to respond to the changing flow field in the porous medium. Because G′ and G″ represent the elastic and viscous components of viscoelasticity, it has been suggested that the inverse of the frequency at which G′ and G″ intersects is the characteristic relaxation time of the polymer solution (Volpert et al., 1998; Castelletto et al., 2004), as shown in Figure 6.3. This intersection has also been described as an indication of the onset of a phenomenon called entanglement coupling. In this phenomenon there is a strong coupling of neighboring molecules to molecular motion along the

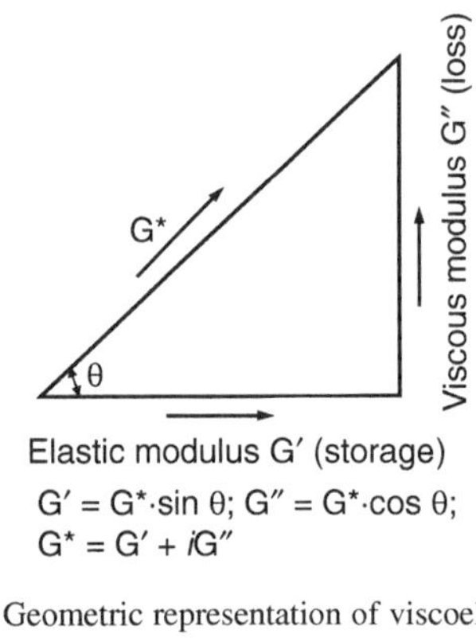

FIGURE 6.2 Geometric representation of viscoelastic parameters.

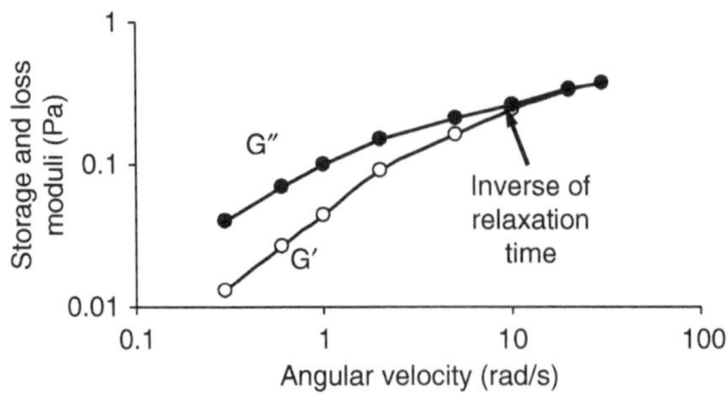

FIGURE 6.3 Principle using a dynamic oscillatory test to determine relaxation time.

chain. Delshad et al. (2008) listed some models for polymer molecule relaxation time.

Figure 6.4 shows an example of steady shear flow and dynamic oscillatory flow measurements of an HPAM solution using a HAAKE RS150 rheometer. From the steady shear flow measurements (Figure 6.4a), we can see that the HPAM is a shear-thinning solution, and the first normal stress difference increases with shear rate. As the molecular weight increases, the viscosity and first normal stress difference increase, indicating higher viscoelastic characteristics. From the dynamic oscillatory measurements (Figure 6.4b), we can see that both the storage modulus and loss modulus increase as the molecular weight increases, indicating that the higher viscosity corresponds to the higher elasticity. According to Knoll and Prud'Homme (1987), the material is more solid-like if G' is higher than G''. For the fluid 2 in Figure 6.4b as an example, at lower frequencies, HPAM behaves more like a fluid, whereas at higher frequencies, it behaves more like a solid.

The normal stress difference measures the difference in the normal stresses in the direction of elongation and that normal to it. The magnitudes of the stresses of a "particle" or "point" are not the same in the various directions of

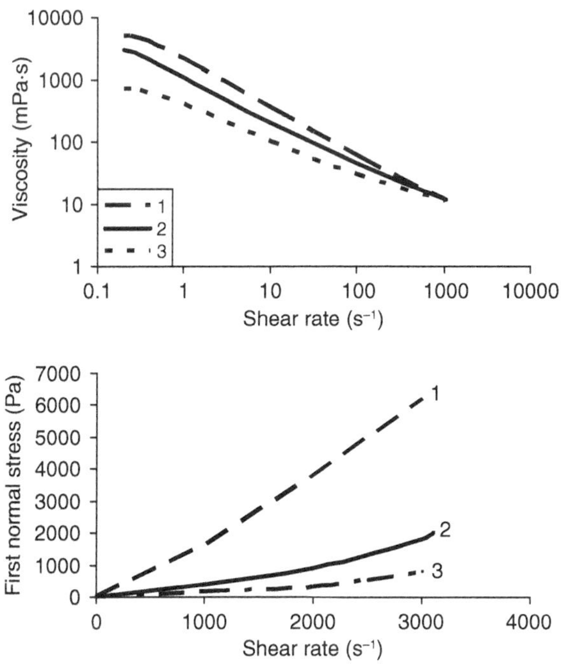

FIGURE 6.4a Steady shear flow measurements. Molecular weight (millions): 1, 21; 2, 12; and 3, 7.5. *Source: Xia et al. (2001).*

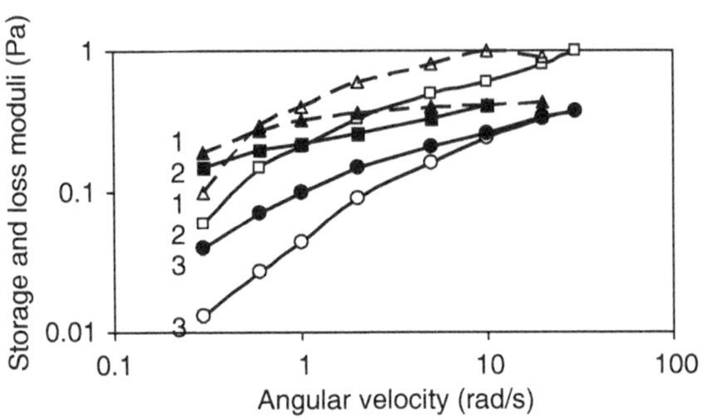

FIGURE 6.4b Dynamic oscillatory flow measurements. *Source: Xia et al. (2001).*

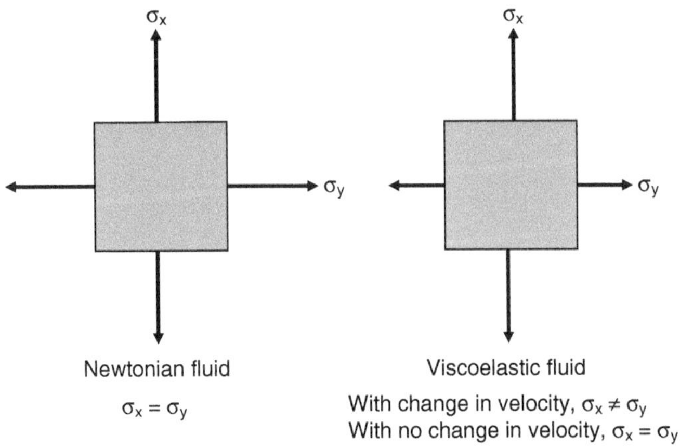

FIGURE 6.5 Comparison of normal stress difference on a Newtonian fluid and viscoelastic fluid.

a viscoelastic fluid, whereas for a Newtonian fluid they are always the same in all directions, as illustrated in Figure 6.5. The flow direction of these "points" in a fluid is determined by their stresses (or ratio of normal stresses). The stresses on these "points" are different for fluids with or without elastic properties. Therefore, the streamline of fluids with or without elastic properties should also be different.

6.3 POLYMER VISCOELASTIC BEHAVIOR

This section discusses viscoelastic fluid viscosity, pressure drop during viscoelastic flow, and factors affecting viscoelastic behavior.

6.3.1 Shear-Thickening Viscosity

In simple shear flow, the vast majority of polymer solutions are pseudoplastic in nature, which means that the viscosity is decreased as the shear rate is increased. The viscosity related to this type of flow is shear-thinning viscosity. Generally, dilute polymer solutions with low molecular weights belong to this category. Another type of flow is elongational, or extensional, flow. In this type of flow, the fluid is stretched—for example, as the fluid flows through a series of pore bodies and pore throats in a porous medium. In such flow, the apparent viscosity is increased as the shear rate is increased. The viscosity related to this type of flow is shear-thickening viscosity.

In other words, the fluid has dilatant behavior. The dilatant behavior can be explained by the coil-stretch transition of macromolecules in elongational flow that results from varying flow geometry and high flow velocity. The stretching gives a detectable increase in viscous friction and thus the onset of dilatancy. At higher stretch rates, the highly elongated state of macromolecules induces very high viscous friction, causing the strong dilatancy observed.

When describing dilatant behavior, the maximum stretch rate, $\dot{\varepsilon}$, in the converging flow at the contraction is a better parameter, but more difficult to be calculated. Instead of the term stretch rate, other authors also used deformation rate (e.g., Chauveteau, 1981) or elongational rate (e.g., Sorbie, 1991). The shear-thickening viscosity is also called elongational viscosity (often referred to as the Trouton viscosity; Sorbie, 1991) or extensional viscosity in the literature. James and McLaren (1975) reported that for a solution of polyethylene oxide (a flexible coil, water-soluble polymer physically similar to HPAM), the onset of elastic behavior at maximum stretch rates was of the order of 100 s^{-1} and shear rates of the order of 1000 s^{-1}. In this instance, the stretch rate is about 10 times lower than the shear rate. However, some authors use shear rate instead of stretch rate in defining the Deborah number—for example, Delshad et al. (2008).

Shear thickening is caused by the viscoelastic nature of polyacrylamide, which has a flexible coil conformation in solution. When the flexible polyacrylamide molecule flows from pore to pore, it deforms (i.e., stretches) to adjust to the flow field. If the average flow time from one constriction to the next is large relative to the time required for the polymer molecule to relax and assume the random coil configuration, the polymer remains shear thinning. The characteristic time required for the polymer molecule to relax is called the relaxation time and can be measured with a specially designed rheometer. At high flow rates, however, the transient time between pore throats (i.e., successive deformations) is of the same order of magnitude as the relaxation time of the polymer, and the polymer chains remain elongated during flow, increasing the apparent viscosity of the flow fluid. Shear thickening of polyacrylamide is a characteristic of flow in porous materials and is not observed in rheological measurements of polyacrylamide at comparable shear rates (Green and

Willhite, 1998). The polymer viscoelastic properties must be measured in an oscillation flow meter and included in the Maxwell equation.

To describe the elongational viscosity, μ_{el}, Hirasaki and Pope (1974) proposed a model of the form

$$\mu_{el} = \frac{\mu_{sh}}{[1 - N_{De}]}, \tag{6.9}$$

where μ_{sh} is the shear-thinning viscosity, and N_{De} is the Deborah number, which is defined at the end of this section.

Masuda et al. (1992) proposed

$$\mu_{el} = \mu_{sh} C_c (N_{De})^{m_c}, \tag{6.10}$$

where C_c and m_c are empirical constants. Delshad et al. (2008) proposed another model that is not scaled by μ_{sh}:

$$\mu_{el} = \mu_{max}\left[1 - \exp\left(-(\lambda_2 N_{De})^{n_2 - 1}\right)\right], \tag{6.11}$$

In this model, μ_{max}, λ_2 and n_2 are empirical constants. One notable distinction between the earlier models (Eqs. 6.9 and 6.10) and this model (Eq. 6.11) is that this model provides the plateau value of μ_{max}, whereas the maximum μ_{el} values from the earlier models could increase indefinitely as N_{De} increases.

The Deborah number is a dimensionless number used in rheology to characterize how "fluid" a material is. Even some apparent solids "flow" if they are observed long enough. The origin of the name, coined by Markus Reiner, is the line "The mountains flow before the Lord" in a song by the prophetess Deborah recorded in the Bible (Judges 5:5).

Formally, the Deborah number is defined as the ratio of a relaxation time (t_r), characterizing the intrinsic fluidity of a material, to the characteristic time scale of an experiment (or a computer simulation) probing the response of the material. It is calculated by

$$N_{De} = \frac{t_r}{t_c}, \tag{6.12}$$

where t_r refers to the relaxation time scale, and t_c refers to the time scale of observation (characteristic time or process time). The relaxation time represents the time required for the stress decays to the 1/e times of its initial value under the constant strain condition. It is calculated by

$$t_r = \frac{\mu}{G'}. \tag{6.13}$$

From Eq. 6.13, we can see that the relaxation time is determined by the viscous and elastic properties. It represents the total of viscous and elastic behaviors. The larger the fluid elasticity, the longer the relaxation will be. The characteristic time is often considered to be equal to the inverse of the stretch rate (elongation rate), $\dot{\varepsilon}$:

$$t_c = \frac{1}{\dot{\varepsilon}}. \tag{6.14}$$

Different researchers use different formulae to calculate characteristic time. A general formula is (Savins, 1969)

$$t_c = \frac{C_{el}\phi d_p}{u}. \tag{6.15}$$

Here, C_{el} is a constant. Different researchers used different values for the constant C_{el}; for example, Marshall and Mentzner (1964) considered C_{el} to be 1 or 0.5; Sadowski and Bird (1965) considered both C_{el} and ϕ equal to 1. Also in this formula, u is the Darcy velocity in m/s, ϕ is the porosity in fraction, and d_p is the grain particle diameter.

In polymer flooding, some authors—for example, Masuda et al. (1992) and Delshad et al. (2008)—defined the Deborah number as the ratio of a polymer molecule's relaxation time to its average residence time (t_{re}) between pore body and pore throat, and the residence time is defined as the inverse of the equivalent shear rate ($\dot{\gamma}_{eq}$):

$$N_{De} = \frac{t_r}{t_{re}} = t_r\dot{\gamma}_{eq}. \tag{6.16}$$

Hirasaki and Pope (1974) defined the Deborah number as

$$N_{De} = \frac{\mu u}{G'\phi d_p}. \tag{6.17}$$

Basically, Eq. 6.17 results from combining Eqs. 6.12, 6.13, and 6.15 with $C_{el} = 1$. According to Marshall and Mentzner (1964), the onset of viscoelastic behavior occurs at a Deborah number around 0.1. From the work of Durst et al. (1982), the Deborah number is 0.5. The smaller the Deborah number, the more the material appears like a fluid.

6.3.2 Apparent Viscosity Model for a Full Velocity Range

Theoretically, a polymer solution viscosity at different shear rates could have three regimes, as shown schematically in Figure 6.6. At very low shear rates that are below the first critical shear rate, the polymer solution behaves like a Newtonian fluid. The viscosity is independent of shear rate. At intermediate shear rates that are above the first critical shear rate and below the second critical shear rate, the polymer solution behaves like a pseudoplastic fluid. Here, the viscosity decreases with shear rate. At high shear rates that are above the

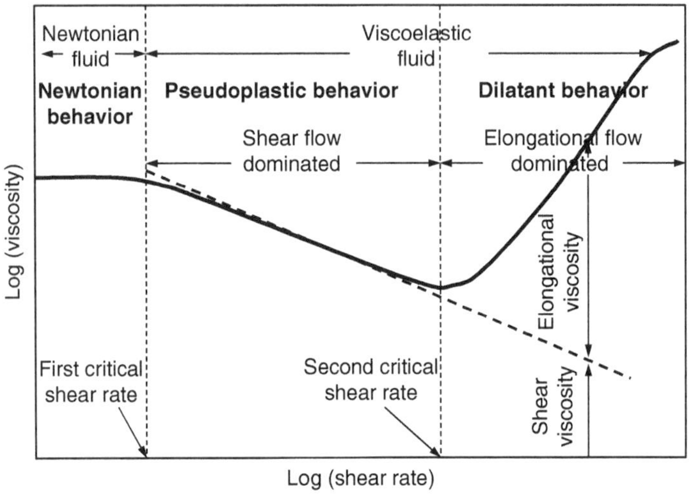

FIGURE 6.6 Schematic illustration of viscoelastic fluid flow behavior.

second critical shear rate, the polymer solution behaves like a dilatant fluid. In this case, the viscosity increases with shear rate.

The second critical shear rate is much higher (i.e., 100 times; Chauveteau, 1981) than the first one. The first critical shear rate is equal to the inverse of the longest rotational relaxation time t_r in the solution. Dilatancy starts as soon as the product of Rouse relaxation time and the maximum stretch rate, $\dot{\varepsilon}$, is greater than 4 (Chauveteau, 1981). The Rouse relaxation time demarcates the onset of entanglement effects (Roland et al., 2004). Chauveteau reported that the ratio of shear rate $\dot{\gamma}$ to the maximum stretch rate $\dot{\varepsilon}$ at the contraction was about 2.5 by laser anemometry for similar polymer solutions and flow geometries. Therefore, the second stretch rate (elongation rate) corresponds to the product of shear rate and Rouse relaxation time equal to 10.

Jennings et al. (1971) reported that in the usual case of medium permeability and medium polymer molecular weight, significant increases in viscosity due to viscoelasticity were seen only at rates in excess of 1.5 to 3.0 m/d. The velocity range of 1.74 to 3.30 m/d reported by Han et al. (1995) is in line with that of Jennings et al. Han et al. reported that the range increases with increasing permeability of cores in their experiments.

It is difficult to use a single equation to describe the viscosity in the entire shear rate range. In developing a comprehensive model for apparent viscosity, μ_{app}, Delshad et al. (2008) assumed that its dependence on Darcy velocity (or equivalent shear rate) consists of two parts: the shear–viscosity–dominant part, μ_{sh}, and elongational–viscosity–dominant part, μ_{el}:

$$\mu_{app} = \mu_{sh} + \mu_{el}. \tag{6.18}$$

Then the apparent viscosity expression that covers the entire range of Darcy velocity is

$$\mu_{app} = \mu_\infty + (\mu_p^0 - \mu_\infty)\left[1 + (\lambda\dot{\gamma}_{eq})^\alpha\right]^{(n-1)/\alpha} + \mu_{max}\left[1 - \exp\left(-(\lambda_2 t_r \dot{\gamma}_{eq})^{n_2-1}\right)\right], \quad (6.19)$$

where μ_p^0 and μ_∞ are the limiting Newtonian viscosities at the low and high shear limits, respectively; λ and n are polymer-specific empirical constants; and α is generally taken to be 2. $\dot{\gamma}_{eq}$ is the equivalent shear rate calculated by Eq. 5.23.

Eq. 6.19 basically results from adding the Carreau model (Eq. 5.5) for shear thinning viscosity and the elongational viscosity (the third component in Eq. 6.19 that is formulated by Eq. 6.11). As shown in Figure 6.6, the shear viscosity (bulk viscosity) decreases with increasing shear rate (or flow rate), whereas elongational viscosity (i.e., apparent viscosity when flowing through the core) increases with shear rate. Hirasaki and Pope (1974) and Kang (2001) derived alternative expressions similar to Eq. 6.19. All these equations need experimental data to tune some empirical constants.

Chen et al. (1998) used the power–law equation to describe the apparent viscosity in shear-thinning and shear-thickening regimes:

$$\mu_{app} = K\dot{\gamma}_{eq}^n \quad \begin{matrix} n \le 1 & \text{shear-thinning} \\ n > 1 & \text{shear-thickening} \end{matrix}. \quad (6.20)$$

In the shear-thickening regime, K increases with polymer concentration and molecular weight, but it is not sensitive to permeability. In this equation, n could be 1.311 to 1.437.

6.3.3 Total Pressure Drop of Viscoelastic Fluids

Viscoelastic flow may be described by the two separate types of flow: shear-thinning and elongational. Then the total pressure drop is the addition of the pressure drops caused by these two types of flow,

$$\Delta p_t = \Delta p_{sh} + \Delta p_{el}, \quad (6.21)$$

where Δp_t is the total pressure drop, Δp_{sh} is the pressure drop caused by viscous behavior, and Δp_{el} is the pressure drop caused by elastic behavior.

Δp_{sh} can be estimated according to viscous flow equation. Δp_{el} has the following relationship with the first normal stress difference (τ_{11}–τ_{22}):

$$\Delta p_{el} = C'(\tau_{11} - \tau_{22}). \quad (6.22)$$

For steady laminar flow, the relaxation time (t_r) has the following relationship with shear rate ($\dot{\gamma}$), shear stress τ_{12}, and the first normal stress difference:

$$t_r = \frac{1}{2\dot{\gamma}}\left(\frac{\tau_{11}-\tau_{22}}{\tau_{12}}\right). \tag{6.23}$$

Combining Eqs. 6.22 and 6.23, we have

$$\Delta p_{el} = 2C' t_r \dot{\gamma}\tau_{12}. \tag{6.24}$$

Then the total pressure drop is (Kang, 2001)

$$\Delta p_t = \Delta p_{sh} + 2C' t_r \dot{\gamma}\tau_{12}. \tag{6.25}$$

6.3.4 Factors Affecting Polymer Viscoelastic Behavior

This section presents the factors that affect polymer viscoelastic behavior. These factors include polymer concentration, salinity, surfactant, and temperature. The viscoelastic behavior in a typical Daqing solution is also presented.

Effect of Polymer Concentration

Figure 6.7 shows that the elastic modulus and relaxation time of the polymer solution increased with polymer concentration. ω was 1.351 radians per second. The HPAM 1275A and HPAM 1255 polymers were used. As polymer concentration increases, the distance between polymer molecules decreases. The entanglement of the long flexible chains will be more severe, and the van der Waals force will become larger so that it is more difficult for polymer molecules to deform. When the external force is removed, polymer molecules quickly return to their curling state.

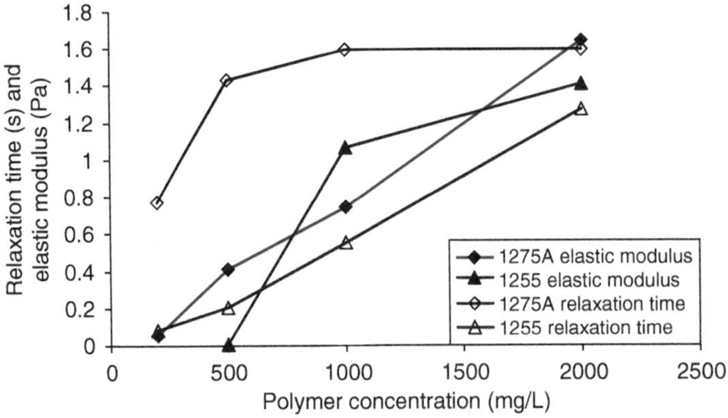

FIGURE 6.7 Effect of polymer concentration and type on elastic modulus and relaxation time. *Source: Data from Kang (2001).*

Salinity Effect

Figure 6.8 shows that the elastic modulus and relaxation time decreased with NaCl concentration. The polymer concentration was 2000 mg/L, and ω was 1.351 radians per second. This result was caused by the ionic shield effect. As the ionic strength is increased, the ionic shield effect increases. Then polymer molecules cannot crimp freely. Figure 6.9 shows that as the divalent (Ca^{2+}) molar fraction was increased with the same ionic strength I = 0.03, the elastic modulus and relaxation time decreased.

Surfactant Effect

Figure 6.10 shows that the elastic modulus and relaxation time slightly decreased with surfactant concentration. The polymer concentration was 2000 mg/L, and ω was 1.351 radians per second. The ORS-41 surfactant was used, and the temperature was at 45°C.

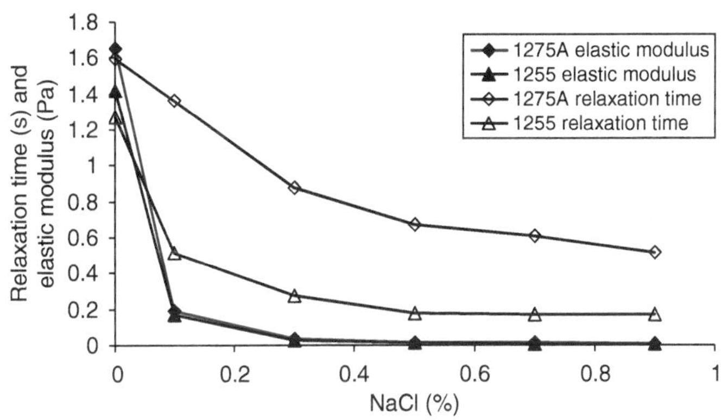

FIGURE 6.8 Effect of NaCl concentration on elastic modulus and relaxation time. *Source: Data from Kang (2001).*

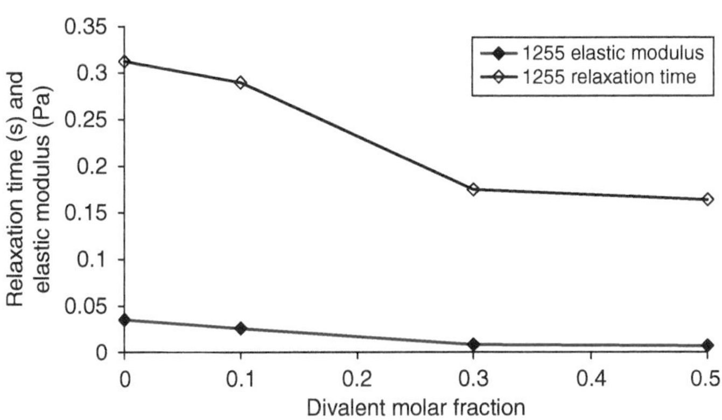

FIGURE 6.9 Effect of divalent molar fraction on elastic modulus and relaxation time. *Source: Data from Kang (2001).*

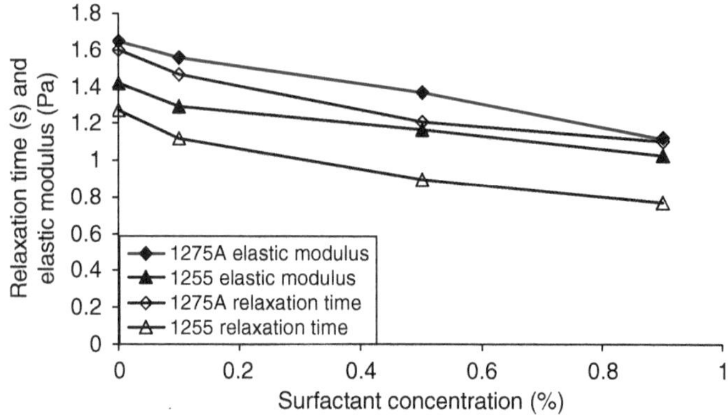

FIGURE 6.10 Effect of surfactant concentration on elastic modulus and relaxation time. *Source: Data from Kang (2001).*

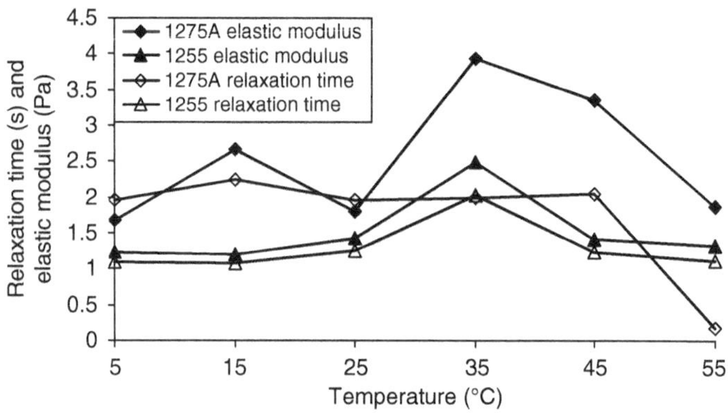

FIGURE 6.11 Effect of temperature on elastic modulus and relaxation time. *Source: Data from Kang (2001).*

Temperature Effect

Figure 6.11 shows that the temperature effect on elastic modulus and relaxation time of the polymer solution was not significant. The polymer concentration was 2000 mg/L, and ω was 1.351 radians per second. The reason is that the increase in temperature increases the molecular thermal motion, but cannot change the polymer curling state within 5 to 55°C. Interestingly, the elastic modulus for both polymers and the relaxation time of HPAM 1255 peaked at 35°C.

Viscoelastic Behavior in a Typical Daqing ASP Solution

Figure 6.12 shows the elastic modulus and relaxation time of a typical Daqing ASP solution that had 1200 mg/L polymer, 0.3% surfactant, and 1.2% alkali

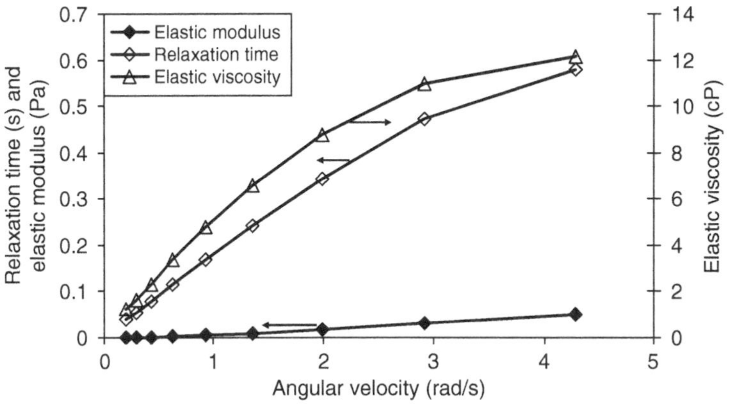

FIGURE 6.12 Relaxation time, elastic modulus, and elastic viscosity versus angular velocity for a typical Daqing ASP solution. *Source: Data from Kang (2001).*

(NaOH). For such solution without alkali shown in Figure 6.10, the average elastic modulus for the two polymers was 1.35 Pa, and the average relaxation time was 1.17 s. Because of the presence of NaOH and lower polymer concentration, the elastic module was 0.008866 Pa which was reduced by more than two orders of magnitude, and the relaxation time was 0.243 s which was reduced by five times. The alkali significantly reduced elastic modules and relaxation time because polymer concentration cannot affect their values so much according to Figure 6.7. In addition, Figure 6.12 shows the elastic viscosity.

6.4 OBSERVATIONS OF VISCOELASTIC EFFECT

This section discusses viscoelastic effect observed from core floods, its effect on relative permeabilities, and its effect on polymer flooding.

6.4.1 Core Flood Observations

In the Daqing laboratory, the cores were flooded with water, glycerin, and an HPAM polymer solution. The flood by each fluid was not stopped until no oil appeared at the outlet. The oil recovery from the two injection sequences was compared: (1) water, glycerin, and HPAM; (2) water, HPAM, and glycerin. The viscosity of glycerin and HPAM fluid were the same (30 cP). The results for sequence 1 are shown in Figure 6.13. We can see that both glycerin and polymer flooding increased the recovery after waterflooding because higher viscosity could increase the volumetric sweep efficiency. Glycerin increased the recovery by 6 to 8% OOIP. After glycerin flooding, polymer flooding further increased the recovery by 6 to 7% OOIP. When the flooding sequence was water, polymer, and glycerin (sequence 2), however, glycerin flooding did not further increase the recovery, as shown in Figure 6.14.

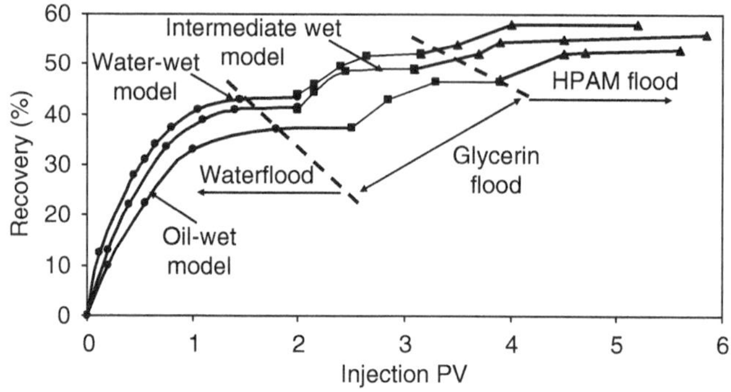

FIGURE 6.13 Oil recovery curves in the sequence of waterflood → glycerin flood → HPAM polymer flood. *Source: Niu et al. (2006).*

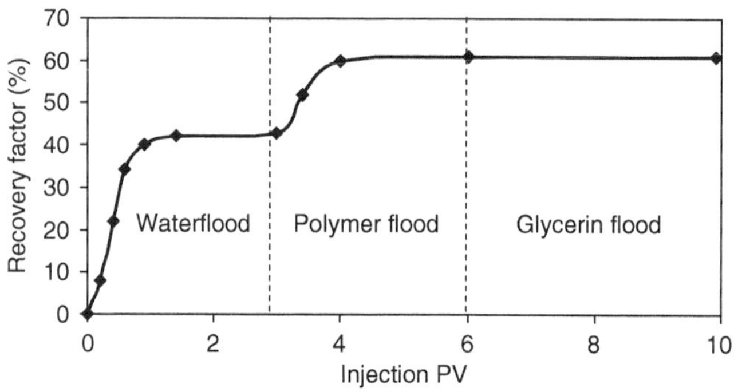

FIGURE 6.14 Oil recovery curve in the sequence of waterflood → HPAM polymer flood → glycerin flood. *Source: Niu et al. (2006).*

Figure 6.15 shows the difference in residual oil saturations after glycerin flood and after polymer flood. Because their viscosity and interfacial tension to oil were about the same, the further significant reduction in residual oil saturation by polymer flooding was probably caused by the polymer elasticity. Wang et al. (2000b) showed that the initially oil-wet surfaces became more water-wet after polymer flood, indicating that polymer flood can "strip" off more oil films from rock surfaces.

In the "dead ends" (inaccessible pore ends) with the normal line of its oil–water interface perpendicular to the flow direction, the residual oil is immovable because it is constrained by the rock configuration. In the experiments shown in Figure 6.16, the cores were flooded with water, glycerin, and HPAM. The pore diameter along the flow streamline was 250 μm. The viscosity of the glycerin or polymer was 30 mPa·s. We can see that the portions (depth) of the dead pore flushed by water and glycerin were about the same, although the

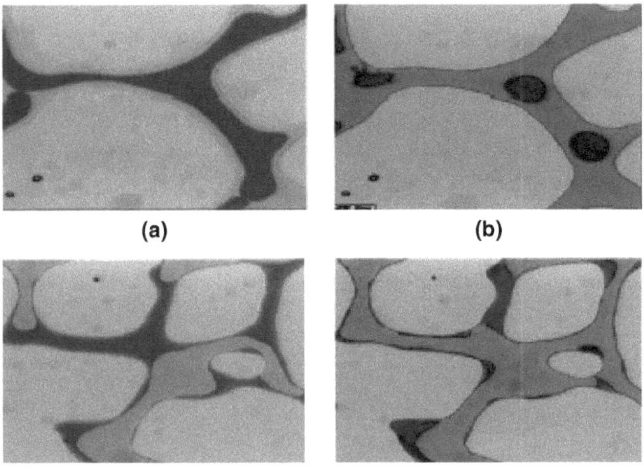

(a) (b) (c) (d)

FIGURE 6.15 Residual oil (darker color) distributions after glycerin flood and polymer flood: (a) after glycerin flood, (b) after polymer flood, (c) after glycerin flood, and (d) after polymer flood. *Source: Niu et al. (2006).*

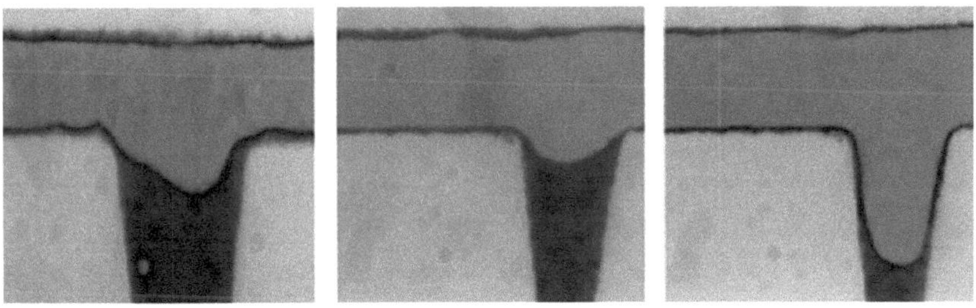

(a) (b) (c)

FIGURE 6.16 Residual oil (darker color) in "dead ends" after (a) water, (b) glycerin, and (c) HPAM floods. *Source: Wang (2001).*

glycerin viscosity was 30 times the water viscosity. The portion of the dead pore flushed by HPAM polymer, however, was much deeper. Quantitatively, the depths "penetrated" by water and glycerin were about 80 to 100 μm, whereas the depth penetrated by polymer was about 320 μm.

From taped videos (not shown here), it was seen that oil was "pulled" out of the dead ends by polymer solution. Apparently, the viscoelastic polymer solution not only pushes the fluids ahead, but also pulls the fluids beside and behind. This "pulling" phenomenon may be explained by polymer elastic properties; the effect may be caused by the long molecular chains of polymer, which can entangle and pull molecular chains behind and beside it. Water or glycerin does not have elastic characteristics; therefore, no residual oil was "pulled" out.

Figure 6.17 shows that the polymer flooding recovery was higher than that from xanthan. Kang (2001) analyzed the core flood data based on the relationship between the friction factor and Reynolds number, and found that the increase in pressure drop at high rates was not caused by inertial flow. It must be caused by polymer viscoelasticity. Kang also found that the pressure drop was higher when HPAM 1225 flowed through the core than that when the biopolymer xanthan gum flowed through. Xanthan gum has rigid structure and has pseudoplastic characteristics but not elastic characteristics.

Figure 6.18 shows the residual oil saturation (lower six curves) and displacement efficiency (upper six curves) versus capillary number in a slightly

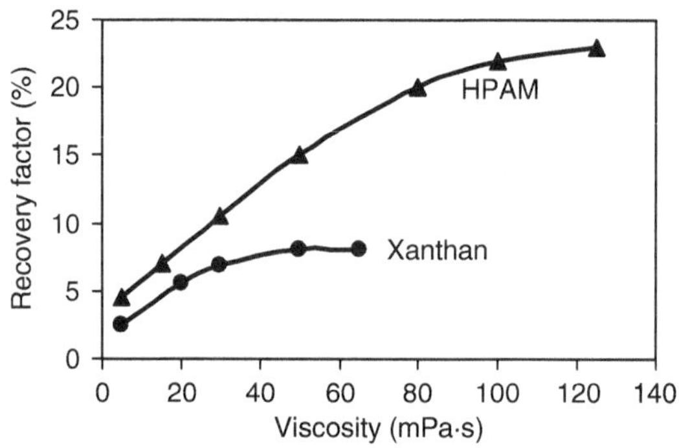

FIGURE 6.17 Displacement efficiency of homogeneous cores flooded by xanthan and HPAM. *Source: Wang et al. (2007).*

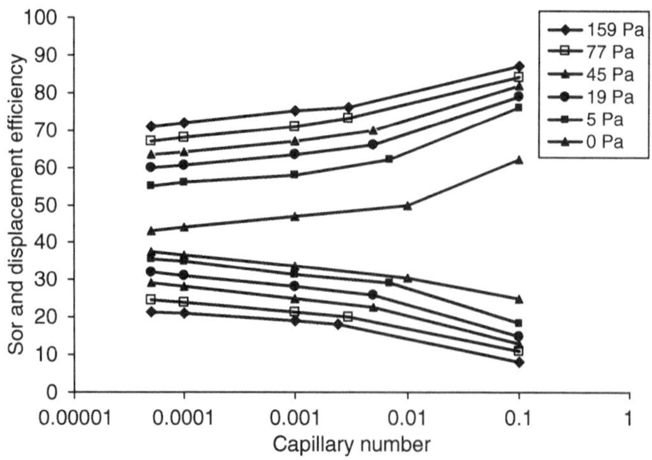

FIGURE 6.18 Residual oil saturation and displacement efficiency versus capillary number. *Source: Wang et al. (2007).*

oil-wet core flooded by polymers with different first normal stress differences (marked in legends). With a higher first normal stress difference, the viscoelasticity was higher. From Figure 6.18, we can see that for the same capillary number, the displacement efficiency was higher for a driving fluid with a higher elasticity indicated by the higher stress difference.

6.4.2 Relative Permeability Curves

Figure 6.19 shows a set of relative permeability curves for waterflooding and polymer flooding. The following observations can be made:

- The residual oil saturation of polymer flooding was lower than that of the waterflooding. On the average of the numerous curves not shown here, it was 6 to 7% lower.
- At the same water saturation, the permeability to polymer was significantly lower than that to water. However, the oil relative permeabilities in polymer flooding and waterflooding are not very different.
- At the same saturation, the water cut of polymer flooding was significantly lower than that of waterflooding. At the same water cut, the oil saturation was significantly lower.

Similar relative permeability measurements were reported earlier in the literature. However, the results were not consistent regarding whether polymer flood residual oil saturation is lower than waterflood.

Zaitoun and Kohler (1987, 1988) observed lower residual oil saturation during polyacrylamide polymer flood but not during xanthan flood, while the opposite observation was made by Pusch et al. (1987). Bakhitov et al. (1980)

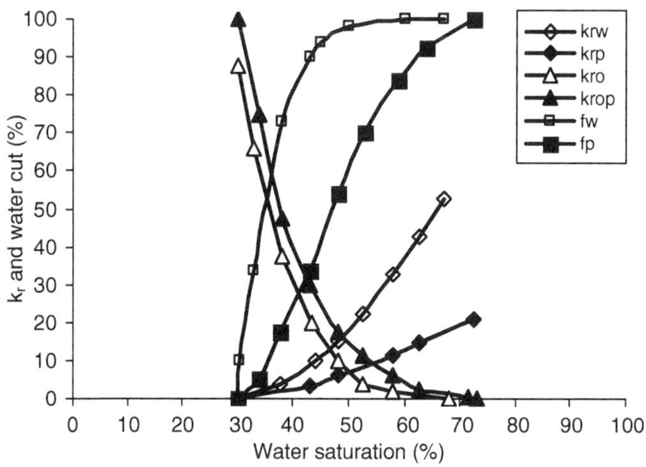

FIGURE 6.19 Relative permeability curves of HPAM/oil and water/oil systems. *Source: Niu et al. (2006).*

observed lower residual oil saturation in polyacrylamide polymer flood in water-wet media. Schneider and Owens (1982) observed lower residual oil saturation in water-wet media, but did not make this observation in oil-wet media.

Sherborne et al. (1967) observed a 15% reduction in residual oil saturation in a HPAM flood. Wreath (1989) did not observe reduction in residual oil saturation in tertiary polymer flooding in Berea and Antolini sandstones. Wreath (1989) did not observe reduction in residual oil saturation in one Berea core even in secondary polymer flood mode, but did observe reduction in one Antolini core that was a heterogeneous and eolian sandstone. For more discussion about k_r curves in polymer flooding, see Section 5.4.5.

6.4.3 Secondary and Tertiary Polymer Flooding

Huh and Pope (2008) reported that a tertiary polymer flood did not mobilize the waterflood residual oil saturation, whereas a secondary polymer flood displaced oil below the waterflood residual oil saturation, based on the Antolini core flood experiment data that show the secondary polymer flood residual oil saturations were below the corresponding waterflood residual oil saturation by 0.02 to 0.22 (Wang, 1995). The main argument proposed by Huh and Pope was that the secondary polymer flood results could not be matched if the same residual oil saturation was used as that for waterflood based on the simulation work by Lu (1994). In this author's opinion, in testing whether polymer flooding could reduce residual oil saturation over waterflooding, homogenous cores would be a better option. Through such use, the effect of difference in volumetric sweep efficiency is removed.

With a simplified pore-level modeling study, Huh and Pope (2008) also showed that when a viscoelastic polymer solution surrounded a mobile, funicular oil column in a chain of pores and slowly drained it, the breakage of oil column into oil ganglia was delayed by polymer elasticity resisting the deformation of the oil/water interface. Wang et al. (2001b) also observed the phenomenon for the polymer solution to "pull" the residual oil in the dead ends after waterflooding. As reported by Xia et al. (2001), micro-model videos showed that at the beginning, the oil droplet in the dead end only deformed; it could not move out of the dead end. When there was movable oil in the dead end, however, an oil droplet flowed down and merged with the residual oil in the dead end, and a larger movable oil droplet was formed. When the movable oil was flooded with a viscoelastic fluid, the residual oil saturation was lower than that with waterflooding. That means the residual oil in dead ends first has to be changed into movable oil before it can be mobilized by a viscoelastic fluid. In a tertiary waterflood, less residual oil in dead ends can be mobilized by the other movable oil droplets, which themselves are less likely available.

6.5 DISPLACEMENT MECHANISMS OF VISCOELASTIC POLYMERS

The relationship between capillary number and residual oil saturation is well established, as reviewed by Stegemeier (1977) and Lake (1989). It is known that to obtain a substantial decrease in residual oil saturation at a micro scale in cores, the capillary number needs to be increased to two or three orders of magnitude above typical waterflood values, but the increase in polymer flood is usually less than 100. Therefore, it was believed that polymer flooding did not reduce residual oil saturation in a micro scale. However, the recovery factors from natural and artificial consolidated cores in the laboratory and in fields were generally higher when polymer flooded than waterflooded, as reviewed by Huh and Pope (2008).

Figure 6.20 schematically shows four types of residual oil saturation distribution after waterflooding:

(a) Oil lodges at the rock crevices and the "dead ends" of flow channels. This type of distribution is what is generally observed in oil-wet and mixed-wet rocks.
(b) Oil film coats rock surfaces, commonly observed in oil-wet rocks.
(c) Oil droplets (oil globules) are trapped at pore throats by capillary forces, commonly observed in strongly water-wet rocks.
(d) Oil droplets or oil clusters are trapped in microscopic pores when the rock has very small-scale heterogeneity.

Because of the viscoelastic behavior of the polymer solution, the percolation characteristics of polymer solution are different from those of water. The following subsections discuss several mechanisms of the viscoelastic polymer solutions.

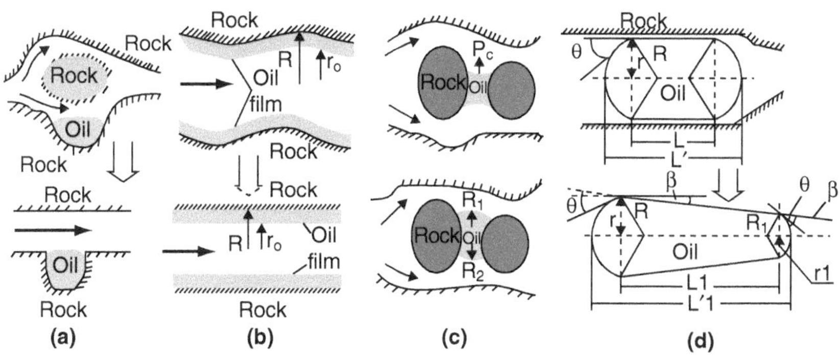

FIGURE 6.20 Simplified models of residual oil distribution after waterflooding. *Source: Wang et al. (2001b).*

6.5.1 Pulling Mechanism

The oil lodged at rock crevices and dead ends cannot be displaced by waterflooding because the oil is constrained by the configuration. The normal line of the oil–water interface is perpendicular to the streamlines. Figure 6.21 shows pictures from experiments using microscopic oil displacement with polymer solutions in a glass-etched model. In the model, the dead end was a wedge-shaped hole. The entrance diameter of the hole was 0.2568 mm, and the depth was 0.2944 mm. The velocity of polymer solution flowing streamline was 0.0385 mm/s. The residual oil after waterflooding remained in the dead oil, as shown in Figure 6.21a. Figure 6.21b shows the distribution of remaining oil after waterflood followed by polymer flood.

The effect of polymer solution "pulling" the remaining oil out of the dead end after waterflood was not significant. Figure 6.21c and 6.21d show the distribution of residual oil after the *mobile* oil was displaced by polymer solutions. We can see that with the increase in viscoelasticity of the polymer solution (represented by G′/G″), the residual oil in the dead end decreased. In the figure,

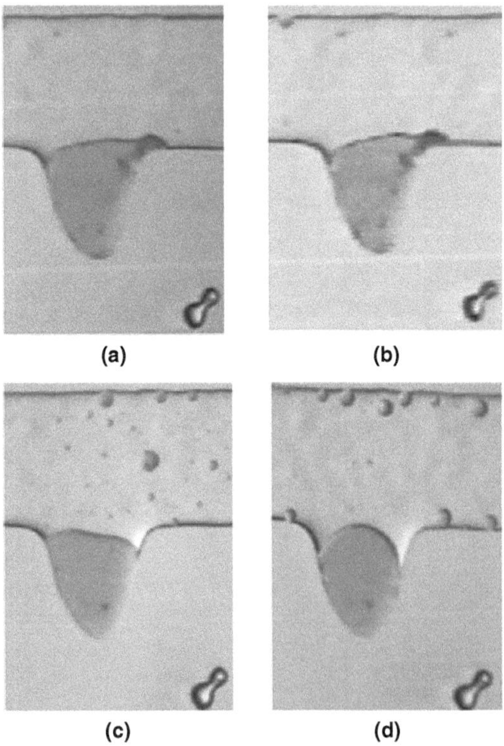

FIGURE 6.21 Distribution of residual oil by water and polymer solutions with different viscoelastic properties: (a) G′/G″ = 0, E_D = 0.0; (b) G′/G″ = 0.9167, E_D = 0.0; (c) G′/G″ = 1.746, E_D = 0.13; and (d) G′/G″ = 2.7245, E_D = 0.18. *Source: Luo et al. (2006).*

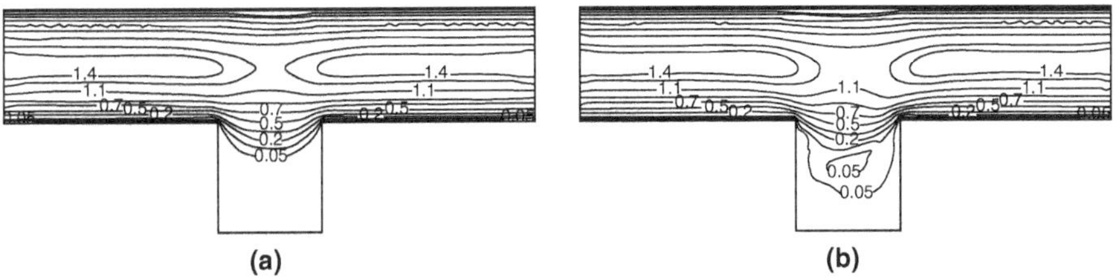

FIGURE 6.22 Velocity contours for (a) a Newtonian fluid (We = 0) and (b) a viscoelastic fluid (We = 0.35). *Source: Xia et al. (2008).*

displacement efficiency, E_D, was defined as the displaced residual oil volume divided by the dead end volume. The higher the G′/G″, the higher the displacement efficiency.

Yin et al. (2006) qualitatively showed this mechanism by solving relevant flow equations numerically. Xia et al. (2008) also developed a simplified pore scale model to describe polymer flow. The numerical solutions from Xia et al. have verified the proposed mechanism. Figure 6.22 shows the velocity contours of a Newtonian fluid with Weissenberg number (We) = 0 and a viscoelastic fluid with We = 0.35 in a flow channel with a dead end when the Reynolds number (Re) = 0.001. We can see that the velocity (m/day) of the viscoelastic fluid is higher than that of the Newtonian fluid at the same position of the dead end. This pulling mechanism also works in the case shown in Figure 6.20c, where the residual oil is trapped at the pore throats by capillary force.

6.5.2 Stripping Mechanism

Within oil-wet cores, residual oil sticks to the rock surface in the form of a continuous oil film, as shown in Figure 6.20b. Wang (2001) reported the measured velocity profiles of water and polyacrylamide solution in a capillary tube, as shown in Figure 6.23. We can seen that the "peak" velocity of HPAM was lower than that of water. The polymer velocity profile was "flatter," and the velocity gradient near the pipe wall was higher. During the experiment using the polymer solution flowing through a capillary tube, it was observed that small particles near the wall moved, but not for the experiment using water. That result indicates the force to "strip" the oil film off the tube wall was stronger for the polymer solution. The numerical solution proposed by Xia et al. (2008) shows that the polymer displacement efficiency is higher than glycerin.

For the type of oil trapped by heterogeneous microscopic pores and capillary forces, as shown in Figure 6.20d, Wang et al. (2001b) showed that it could barely be pushed out by normal polymer solutions. A fraction of the trapped oil may be displaced by polymer solution by combination of pulling and stripping mechanisms.

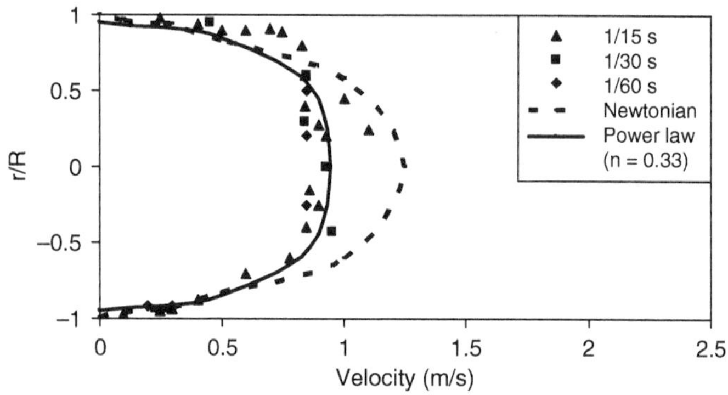

FIGURE 6.23 Velocity profiles of water and an HPAM solution in a capillary tube. *Source: Wang (2001).*

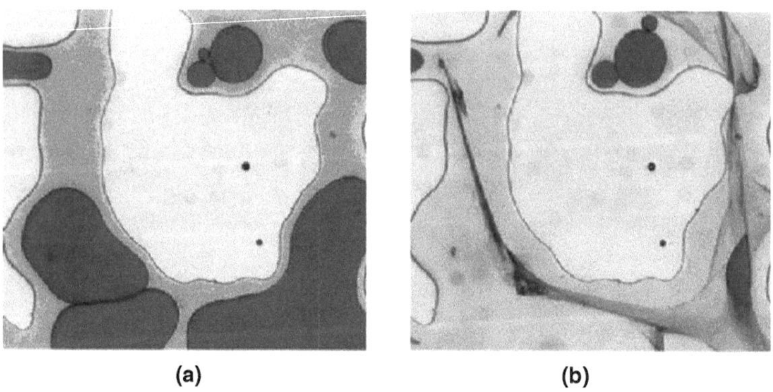

FIGURE 6.24 Residual oil droplets (dark color) pulled by polymer into threads: (a) after waterflood and (b) after HPAM flood. *Source: Luo et al. (2006).*

6.5.3 Mechanism of Oil Thread Flow

From the previous discussions, the residual oil was pulled and stripped from the rock surfaces. As shown in a 2D glass-etched model (see Figure 6.24), the residual oil after waterflood became isolated oil droplets. The polymer solution pulled the oil into oil columns. These oil columns became thinner and longer to form "oil threads" as they met the residual oil downstream. The oil upstream flowed along these oil threads to meet the residual oil downstream so that an oil bank was built. In the process of residual oil flowing along the oil threads, because of the cohesive force of the oil/water interfaces, it was also possible to form new oil droplets, which flowed downstream and coalesced with other oils. Now we are ready to discuss the role viscoelasticity plays.

Assume that an oil thread has the shape of a slim cylinder, as represented by the two dashed parallel lines in Figure 6.25. Under different kinds of forces, concave or convex oil/polymer solution interfaces might form along the oil thread. In this case, consider the oil thread as the axis of the cylindrical coordinate. Polymer solution flows coaxially in the horizontal direction.

The flow velocity is in the order of 10^{-5} m/s. The radius of an oil thread is about 10^{-6} m. The relaxation time of polymer solution used in the oil displacement process is about 10^{-1} to 10^{-3} s. Under these conditions, the range of Deborah number, N_{De}, is between 0.1 and 10. Figure 6.26 shows the normal stress of the viscoelastic fluid with different Deborah numbers. The stress acts on the undulated oil/water interface. When the representation in Figure 6.26 was constructed, the fluid velocity of 3.47×10^{-5} m/s and the relaxation time of 0.247 s were used. In the figure, negative stress represents that the stress direction is opposite to the external normal line of the acting surface. We can

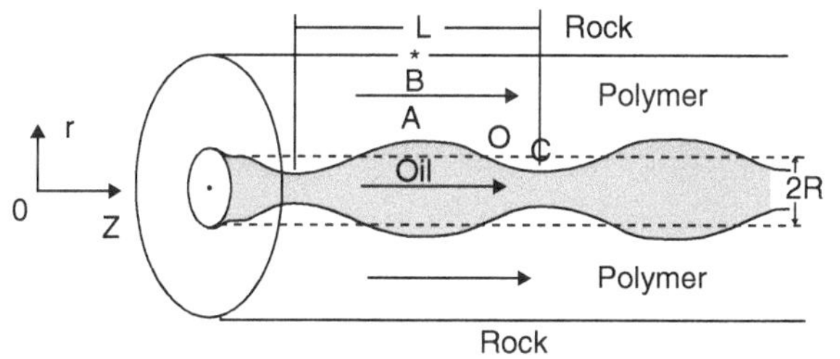

FIGURE 6.25 Schematic to analyze "oil thread" stability. *Source: Luo et al. (2006).*

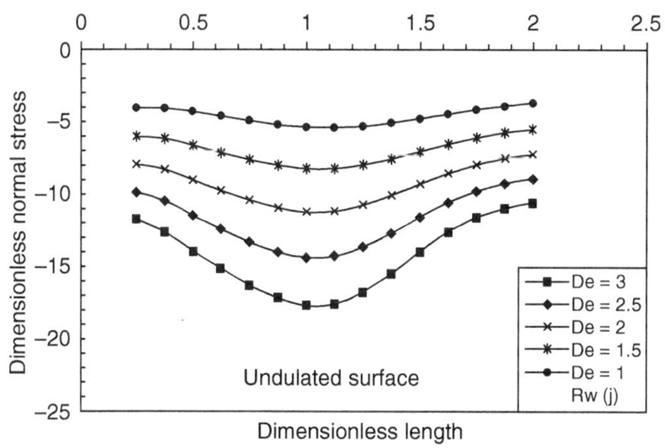

FIGURE 6.26 Distribution of normal force along the undulated surface. *Source: Luo et al. (2006).*

see that the normal force acting on the undulated convex surface is larger, and the normal force acting on the undulated concave surface is smaller. In other words, a larger pressure is imposed on the convex surface, whereas a smaller pressure is imposed on the concave surface. The larger the Deborah number, the larger the corresponding normal stress and the larger the difference between the normal stress on the convex interface and that on the concave interface. Thus, we can see that the essential function of the normal stress is to prevent the oil thread from changing its shape, which makes the streamlines of polymer solution stable. In other words, the normal stress of the viscoelastic polymer solution can stabilize the oil threads.

6.5.4 Mechanism of Shear-Thickening Effect

Another possible explanation for the improvement in oil recovery is that the polymer shear-thickening behavior helps to displace the still-mobile but hard-to-displace oil near the residual oil condition, or displace the bypassed oil in small-scale heterogeneous areas more effectively (Delshad et al., 2008). A simple analysis elaborating this point was made by Jones (1980). He showed that except for the near wellbore, the presence of a shear-thickening fluid for the bulk of reservoir volume is more effective in displacing the bypassed oil from the low-permeability zone. The apparent viscosity of polymer solution in the high-permeability zones could then become high, fulfilling the objective of improving the sweep efficiency.

Ideal characteristics for polymer solution, therefore, are that its viscosity is low near the injection wellbore, but otherwise it is shear-thickening for a high viscosity. The shear-thickening viscosity can be caused by polymer viscoelasticity. As the polymer molecules flow through series of pore bodies and pore throats in reservoir rock, flow field elongation and contraction occur. Accordingly, the polymer molecules repeatedly stretch and recoil to adjust to the flow field. If the flow velocity is too high, the polymer molecules do not have sufficient relaxation time to stretch and recoil to adjust to the flow. The resultant elastic strain leads to high apparent viscosity, represented as shear-thickening behavior.

6.6 EFFECT OF POLYMER SOLUTION VISCOELASTICITY ON INJECTION AND PRODUCTION FACILITIES

This section briefly summarizes the problems and solutions with injection and production facilities that are related to polymer solution viscoelastic properties. For more details, see Wang (2001) and Wang et al. (2004a, 2004c).

6.6.1 Vibration Problem with Flow Lines

Referring to Figure 6.27, HPAM polymer solution, which is a viscoelastic fluid, has an extension viscosity and a normal-stress difference. When the polymer

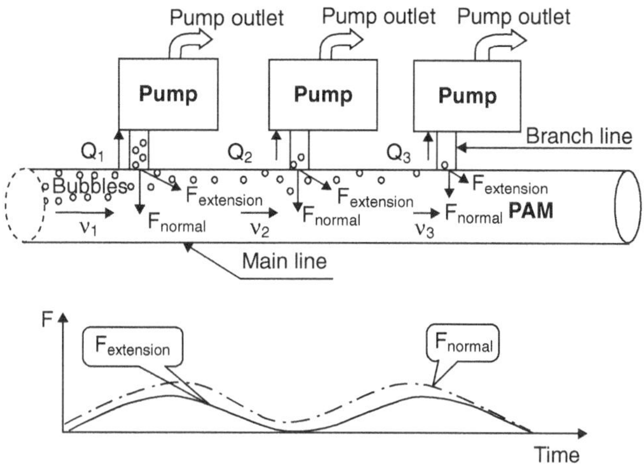

FIGURE 6.27 Schematic to show the vibration problem caused by the oscillation of normal and extension forces. *Source: Wang (2001).*

solution flows into a branch line (at a tee section), a "pulling force" tries to pull the solution back into the main supply line. This pulling force increases with the increase in velocities of the branch and main supply lines. The velocity in the branch line oscillates, when the triplex pump pumps. The oscillation of the velocity changes normal stress and extension viscosity, thus causing the pump vibration. The fluid velocity in the upstream of the supply line is higher than the velocity in the downstream. Therefore, pressure fluctuations and vibrations were more serious in the upstream. It was common that there was no tee section at the outlet side of a pump (the fluid from a pump flowed directly to the wellhead). In addition, it was observed that the vibration at the outlet of the pump was less than at the inlet.

Because of the high viscosity of polymer solution, it was difficult for the gas bubbles, formed when preparing the polymer solution, to separate from the solution. The gas bubbles rose to the top of the main supply line and entered the branch line. Because more gas bubbles entered the upstream branch lines, the pumps situated in the upstream vibrated more seriously.

Major modifications were made to the fluid-supply system to solve the vibration problem. The most important one was to increase the diameter of the main flow line. Consequently, the fluid velocity in the main line was practically zero, and the extension viscosity and normal-stress difference at the inlet of the fluid-supply line were also practically zero. There was sufficient time for the gas bubbles to rise to the top of the main line and be discharged before entering the pump inlet. The vibration was decreased after the modification. With lower vibration, the service life of the pumps was greatly increased. Before the modifications, two or three pumps per injection well were required to maintain continuous injection of polymer solution. Only one pump (no standby pump) was required after the modification. With the decrease in the oscillating velocity, shear degradation associated with the triplex pumps was also reduced.

6.6.2 Problems with Pump Valves

The change in the streamlines is greatest at the valve seats. The normal force at the valve seats will be the largest. This large normal force could prevent the valve from closing (see Figure 6.28). This normal force increases the requirement for closing pressure. The valve seat area was decreased to increase the closing pressure, and the angle of the valve seat was adjusted. After it was adjusted to its optimum value, the pump volumetric efficiency to polymer fluid was increased from less than 85% to more than 92%.

6.6.3 Problems with Maturation Tanks

Figure 6.29 compares the flow streamlines in a mixing tank of a Newtonian fluid and a viscoelastic fluid. The flow streamlines of polymer solution are different from those of water. The blind area at the bottom of the tank is much larger for polymer solution. There is a difference in speed between the blade tip and the surrounding polymer solution, but approximately the same for water. These problems increase the energy requirement to rotate the blade and the time to dissolve polymer powder. After the blade was redesigned, the energy

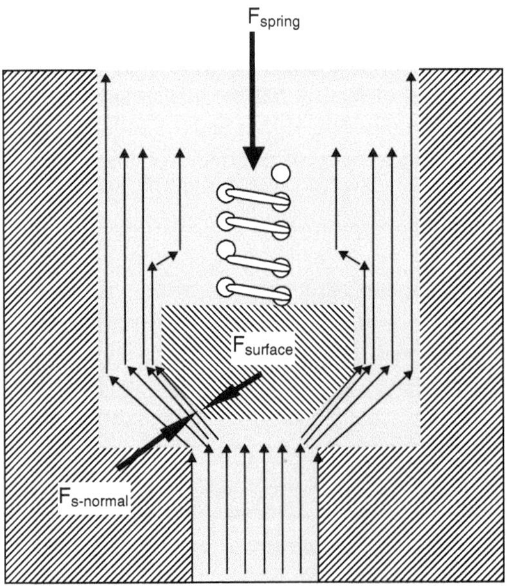

FIGURE 6.28 Schematic of force acting on valve seat at suction. *Source: Wang (2001).*

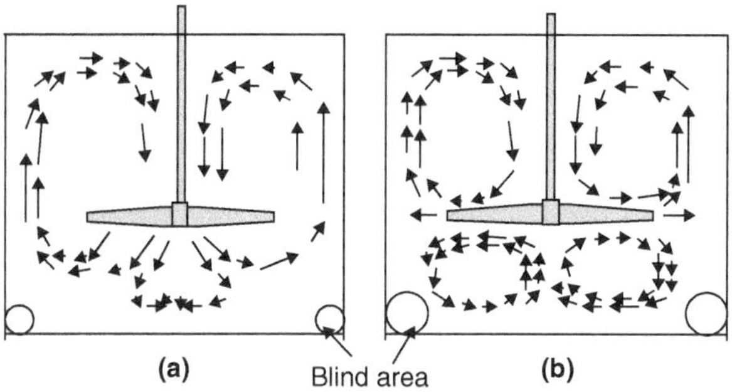

FIGURE 6.29 Comparison of flow streamlines in a mixing tank with (a) a Newtonian fluid and (b) a viscoelastic fluid. *Source: Wang (2001).*

requirement to rotate the blade was decreased by 80%, and the maturation time was shortened from 2 to 1.5 hours.

6.6.4 Problems with Beam Pumps

Wang et al. (2004a) reported that excessive eccentric sucker-rod wear with beam pumps occurred when the produced fluid contained more than 100 mg/L polymer. The service life of rods was approximately only half a year, a 75% reduction compared to the rod life in waterflooding (waterflood producers with beam pumps had a service life of approximately 2 years). Wang et al. (2004c) also reported that based on the statistical data in the year 2000, 880 wells had been eccentrically worn in 2,365 rod-pumped wells, producing fluids containing polymer. Compared to the wells with no polymer before the year 1997, the number of wells with eccentric wear had increased by 840. Thirty-six percent of wells suffered the eccentric wear. The pump service work wells had also increased from 3.2% in 1997 to 27.1% in 2000 because of eccentric wear in rod-pumped wells. The main cause of eccentric wear on sucker rods was the normal stress of polymer solution, as explained next.

To calculate stress, Wang et al. (2004c) used the equation

$$\tau_{rr} = t_r \mu (\dot{\gamma})^2, \tag{6.26}$$

where τ_{rr} is the normal stress, t_r is the relaxation time, μ is the viscosity, and $\dot{\gamma}$ is the fluid velocity gradient. From Eq. 6.26, the normal force is in direct proportion to the squared velocity gradient. When the sucker rod is placed in the middle of the tubing, the value of the resultant force is zero because the flow streamlines and the normal forces are symmetrical around the rod. However, the sucker rod is not always aligned in the middle of the tubing; the velocity gradient exists in the annulus between the sucker rod and the tubing. The closer the sucker rod is to the tubing, the higher the velocity gradient is, and the higher the normal force will be. The normal force on the side of the high-velocity gradient is higher than that at the other side of low-velocity gradient. The difference in the normal forces points to the side of smaller annulus spacing between the sucker rod and tubing (Liu et al., 2006a). Therefore, the resulting force pushes the rod toward the tubing until the sucker rod ultimately touches the tubing wall.

The normal force versus eccentricity and sucker rod velocity is shown in Figures 6.30 and 6.31, respectively. Figure 6.30 shows that as the eccentricity is increased, the normal force is increased significantly. In the case in which the initial eccentricity is small, it will be increased because of the effect of the normal force. If the eccentricity is increased, the normal force will become larger, and the eccentricity will be increased further. Finally, the sucker rod will touch the tubing to cause eccentric wear. The positive values in the abscissa in Figure 6.31 represent the downstroke of the rod, and the negative values

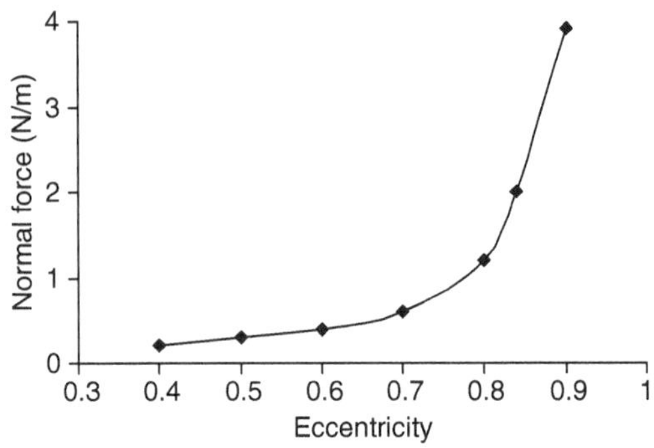

FIGURE 6.30 Relationship between normal force and eccentricity. In the figure, tubing diameter = $2^7/_8$ in., rod diameter = 1 in., relaxation time = 0.019 s, fluid sheared viscosity = 20 cP, and eccentricity = 0.8. *Source: Liu et al. (2006a).*

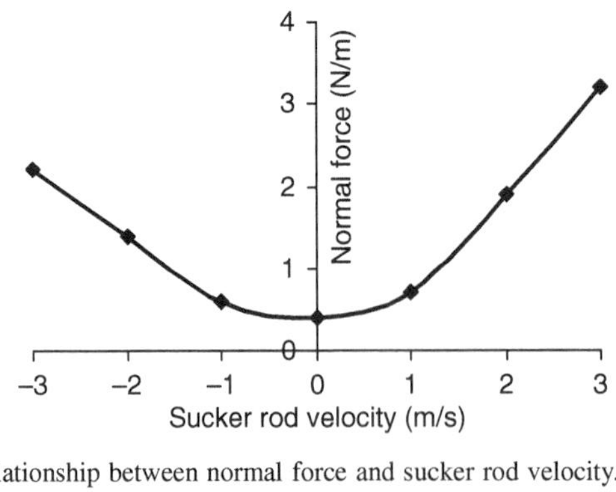

FIGURE 6.31 Relationship between normal force and sucker rod velocity. In the figure, tubing diameter = $2^7/_8$ in., rod diameter = 1 in., relaxation time = 0.019 s, fluid sheared viscosity = 20 cP, and eccentricity = 0.8. *Source: Liu et al. (2006a).*

represent the upstroke. The normal force exists in both the downstroke and upstroke. The velocity of the sucker rod is zero when the rod is at the upper and lower dead points. At these points, the difference of the normal forces at the sides of the sucker rod is at a minimum. The rod returns to the initial states from the eccentric wear state. The normal force reaches its maximum during downstroke or upstroke.

Thus, the normal force has two highs and two lows during one stroke, which means that the rod will bend twice in one stroke. For a beam pump pumping at 6 strokes per minute, the sucker rod will bend more than 6 million times a year, greatly increasing the chance it will break (snap) at its joint. The centralizer can be used to counteract this normal force. Calculation shows that one centralizer is need for each 8 m sucker rod. More than 1,700 pumping wells in Daqing had centralizers attached to each sucker rod. After the centralizers were installed, the rod failure was reduced by more than 90% (Wang et al., 2004a).

Furthermore, Wang et al. (2004c) presented an equation to calculate the minimum uniform load that causes eccentric wear. Their calculation shows that the minimum load is much lower than the actual load. In addition to the method to centralize sucker rods, other methods include using large clearance pumps with large channels, using uniform-diameter sucker rods in the whole wellbore, and regularly rotating the sucker rod string. According to Wang et al. (2004c), these measures have been applied to all the pumping wells in Daqing fields that lift fluid containing polymer using beam pumps.

6.6.5 Problems with Centrifugal Pumps

Daqing polymer flooding practices show that the energy consumption of submergible pumps is the highest, compared with beam pumps and screw pumps (Wang et al., 2004a). Centrifugal pumps rotate at a high speed to produce centrifugal force to pump the fluids out of the well. When polymer fluids are

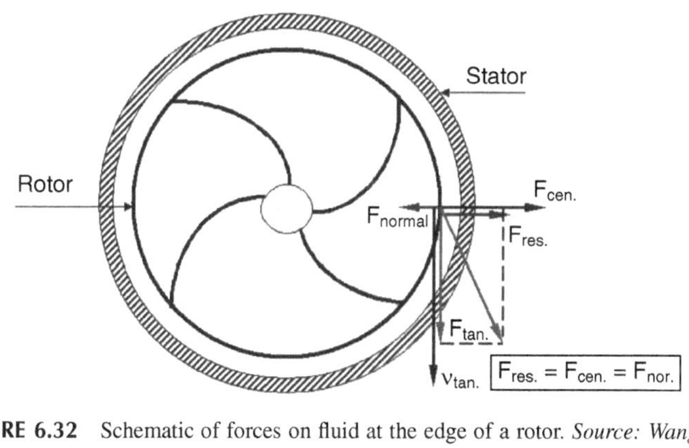

FIGURE 6.32 Schematic of forces on fluid at the edge of a rotor. *Source: Wang (2001).*

pumped, there is also a normal force pointing in the opposite direction of the centrifugal force, resulting in a lower force (see Figure 6.32). Therefore, more energy is needed for submergible pumps. To increase the energy efficiency of centrifugal pumps, the best method is to use frequency modulators, but the normal force will also be increased. Then the energy requirement will increase. Because of that increase, only plunger pumps and screw pumps are used in surface systems that transport polymer solutions.

Chapter 7

Surfactant Flooding

7.1 INTRODUCTION

This chapter covers the fundamentals of surfactant flooding, which include microemulsion properties, phase behavior, interfacial tension, capillary desaturation, surfactant adsorption and retention, and relative permeabilities in surfactant flooding. It provides the basic theories for surfactant flooding and presents new concepts and views about capillary number (trapping number), relative permeabilities, two-phase approximation of the microemulsion phase behavior, and interfacial tension. This chapter also presents an experimental study of surfactant flooding in a low-permeability reservoir.

7.2 SURFACTANTS

This section presents types of surfactants and the methods to characterize surfactants.

7.2.1 Types of Surfactants

The term surfactant is a blend of surface acting agents. Surfactants are usually organic compounds that are amphiphilic, meaning they are composed of a hydrocarbon chain (hydrophobic group, the "tail") and a polar hydrophilic group (the "head"). Therefore, they are soluble in both organic solvents and water. They adsorb on or concentrate at a surface or fluid/fluid interface to alter the surface properties significantly; in particular, they reduce surface tension or interfacial tension (IFT).

Surfactants may be classified according to the ionic nature of the head group as anionic, cationic, nonionic, and zwitterionic (Ottewill, 1984). Anionic surfactants are most widely used in chemical EOR processes because they exhibit relatively low adsorption on sandstone rocks whose surface charge is negative. Nonionic surfactants primarily serve as cosurfactants to improve system phase behavior. Although they are more tolerant of high salinity, their function to reduce IFT is not as good as anionic surfactants. Quite often, a mixture of anionic and nonionic is used to increase the tolerance to salinity. Cationic surfactants can strongly adsorb in sandstone rocks; therefore, they are generally

Modern Chemical Enhanced Oil Recovery. DOI: 10.1016/B978-1-85617-745-0.00007-3

not used in sandstone reservoirs, but they can be used in carbonate rocks to change wettability from oil-wet to water-wet. Zwitterionic surfactants contain two active groups. The types of zwitterionic surfactants can be nonionic-anionic, nonionic-cationic, or anionic-cationic. Such surfactants are temperature- and salinity-tolerant, but they are expensive. A term amphoteric is also used elsewhere for such surfactants (Lake, 1989). Sometimes surfactants are grouped into low-molecular and high-molecular according to their weight.

Within any class, there is a huge variety of possible surfactants. For more surfactants used in oil recovery, see Akstinat (1981). For more details on the effect of structure on surfactant properties, see Graciaa et al. (1982) and Barakat et al. (1983).

7.2.3 Methods to Characterize Surfactants

The most common surfactants used in surfactant flooding are sulfonated hydrocarbons. The term crude oil sulfonates refers to the product when a crude oil is sulfonated after it has been topped. Petroleum sulfonates are sulfonates produced when an intermediate-molecular-weight refinery stream is sulfonated, and synthetic sulfonates are the product when a relatively pure organic compound is sulfonated (Green and Willhite, 1998). Surfactants stable above 200°C are, almost exclusively, sulfonate groups, while sulfate moieties decompose rapidly at temperatures above 100°C (Isaacs et al., 1994). Thermal stability of sulfonates increases in the following order (Ziegler, 1988):

petroleum sulfonates < alpha olefin sulfonates < alkylarylsulfonates

In general, sulfate surfactants have greater availability and tolerance to divalent ions, but they have an ester linkage and are subject to hydrolysis at high temperatures and low pH. Above pH 8, modest amounts of calcium can cause severe degradation. The propylene oxide (PO) group is more lipophilic, and the ethylene oxide (EO) group is more hydrophilic. Sulfonates are stable at high temperatures but sensitive to divalent ions. Internal olefin sulfonates (IOS) have a double-bond structure and, after sulfonation, become branched, which makes them less viscous than those with a linear structure. Alpha olefin sulfonates (AOS), which have a linear structure, are especially sensitive to oxygen, which affects the unsaturated species (Labrid, 1991). Several methods to characterize surfactants are introduced next.

Hydrophile–Lipophile Balance

The hydrophile–lipophile balance (HLB) has been used to characterize surfactants. This number indicates relatively the tendency to solubilize in oil or water and thus the tendency to form water-in-oil or oil-in-water emulsions. Low HLB numbers are assigned to surfactants that tend to be more soluble in oil and to form water-in-oil emulsions. When the formation salinity is low, a low HLB surfactant should be selected. Such a surfactant can make middle-phase microemulsion at low salinity. When the formation salinity is high, a high HLB surfactant should be selected. Such a surfactant is more hydrophilic and can make middle-phase microemulsion at high salinity.

HLB is determined by calculating values for the different regions of the molecule, as described by Griffin (1949, 1954). Other methods have been suggested, notably by Davies (1957). Griffin's equation to calculate HLB for nonionic surfactants is

$$HLB = 20MW_h/MW, \tag{7.1}$$

where MW_h is the molecular mass of the hydrophilic portion of the molecule, and MW is the molecular mass of the whole molecule, giving a result on an arbitrary scale of 0 to 20.

An HLB value of 0 corresponds to a completely hydrophobic molecule, and a value of 20 corresponds to a molecule made up completely of hydrophilic components. The HLB value can be used to predict the following surfactant properties:

- A value from 0 to 3 indicates an antifoaming agent.
- A value from 4 to 6 indicates a W/O emulsifier.
- A value from 7 to 9 indicates a wetting agent.
- A value from 8 to 18 indicates an O/W emulsifier.
- A value from 13 to 15 is typical of detergents.
- A value of 10 to 18 indicates a solubilizer or hydrotrope.

In 1957, Davies suggested a method for calculating a value based on the chemical groups of the molecule. The advantage of this method is that it takes into account the effect of strongly and less strongly hydrophilic groups. The equation is

$$HLB = 7 + mH_h - nH_l, \tag{7.2}$$

where m is the number of hydrophilic groups in the molecule, H_h is the value of the hydrophilic groups, n is the number of lipophilic groups in the molecule, and H_l is the value of the lipophilic groups. For ethoxylated amphiphiles, the HLB is one-fifth the weight of the ethylene oxide portion of the molecule (Bourrel and Schechter, 1988).

Critical Micelle Concentration and Kraff Point

Another important characteristic of a surfactant is critical micelle concentration (CMC). CMC is defined as the concentration of surfactants above which micelles are spontaneously formed. Upon introduction of surfactants (or any surface active materials) into the system, they will initially partition into the interface, reducing the system free energy by (a) lowering the energy of the interface (calculated as area times surface tension) and (b) removing the hydrophobic parts of the surfactant from contact with water. Subsequently, when the surface coverage by the surfactants increases and the surface free energy

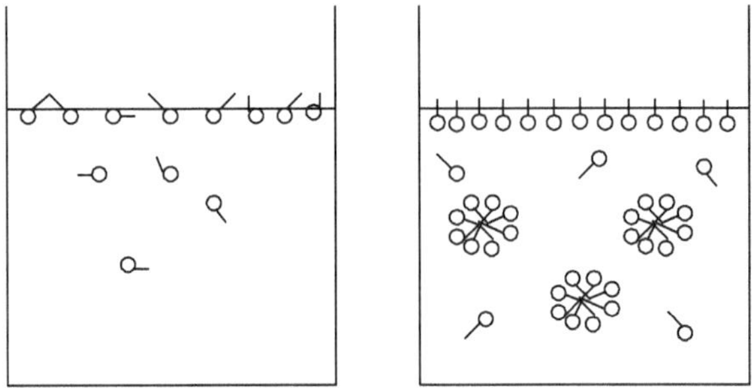

FIGURE 7.1 Distribution of surfactant molecules in solution at concentrations (a) below and (b) above CMC.

(surface tension) decreases, the surfactants start aggregating into micelles, thus again decreasing the system free energy by decreasing the contact area of hydrophobic parts of the surfactant with water.

Upon reaching CMC, any further addition of surfactants will just increase the number of micelles (in the ideal case), as shown in Figure 7.1. In other words, before reaching the CMC, the surface tension decreases sharply with the concentration of the surfactant. After reaching the CMC, the surface tension stays more or less constant. For a given system, micellization occurs over a narrow concentration range. This concentration is small—about 10^{-5} to 10^{-4} mol/L for surfactants used in EOR (Green and Willhite, 1998). Therefore, CMC is in the range of a few ppm to tens of ppm.

One parameter related to CMC is Krafft temperature, or critical micelle temperature. This is the minimum temperature at which surfactants form micelles. Below the Krafft temperature, there is no value for the critical micelle concentration; that is, micelles cannot form.

A parameter related to a nonionic surfactant is cloud point; that is, the temperature at which phase separation occurs, thus becoming cloudy. This behavior is characteristic of nonionic surfactants containing polyoxyethylene chains that exhibit reverse solubility versus temperature behavior in water and therefore "cloud out" at some point as the temperature is raised. For a nonionic surfactant, the hydrophilic group is a function group with oxygen. Its solubility is caused by the hydrogen–oxygen bond. As the temperature is raised, the bond is broken due to high surfactant molecular activity. Thus, surfactant molecules are separated and the solution becomes cloudy, or even precipitation occurs.

Solubilization Ratio

There are several theories to guide new surfactant design and explain surfactant phase behavior. These theories are solubilization ratio (SR), R-ratio, and packing factor. Solubilization ratio is used in this book. Solubilization ratio for oil (water) is defined as the ratio of the solubilized oil (water) volume to the surfactant volume in the microemulsion phase. Solubilization ratio is closely related to IFT, as formulated by Huh (1979). When the solubilization ratio for oil is equal to that for water, the IFT reaches its minimum.

R-ratio

R-ratio was discussed by Bourrel and Schechter (1988). Let us consider the interfacial zone (the C layer in their term) of a finite thickness. There are hydrophilic heads (H) and lipophilic tails (L) of surfactant molecules, in addition to oil and water molecules. If the interaction between oil molecules and surfactant molecules is strongly attractive, the surfactant has affinity to (is miscible with) the oil phase. This interaction, denoted by A_{CO}, should include the interaction of oil molecules with both lipophilic tails (A_{LCO}) and hydrophilic heads (A_{HCO}). In mathematical formula, it is

$$A_{CO} = A_{LCO} + A_{HCO}. \tag{7.3}$$

Similarly, if the interaction between water molecules and surfactant molecules is strongly attractive, the surfactant has affinity to (is miscible with) the water phase. Therefore, the interaction is denoted by A_{CW}:

$$A_{CW} = A_{HCW} + A_{LCW}. \tag{7.4}$$

Because the lipophilic tails are oriented in the oil phase, A_{HCO} in Eq. 7.3 may be neglected in many cases. Similarly, because the hydrophilic heads are oriented to the water phase, A_{LCW} in Eq. 7.4 may be neglected. Then the surfactant affinity to oil or water phase may be described, as proposed initially by Winsor (1948), by the R-ratio:

$$R = \frac{A_{CO}}{A_{CW}}. \tag{7.5a}$$

Equation 7.5a does not take into account the repulsive interactions between oil molecules A_{oo}, between water molecules A_{ww}, between lipophilic tails A_{ll}, or between hydrophilic heads A_{hh}. Bourrel and Schechter (1988) extended Eq. 7.5a to include these interactions:

$$R = \frac{A_{CO} - A_{oo} - A_{ll}}{A_{CW} - A_{ww} - A_{hh}}. \tag{7.5b}$$

Consequently, when $R < 1$, the relative miscibility with water has increased and/or that with oil has decreased. When $R > 1$, the relative miscibility with oil has increased and/or that with water has decreased. When $R \ll 1$, the characteristic system is type I, and when $R \gg 1$, the characteristic system is type II. $R = 1$ corresponds to the optimum system in Winsor type III.

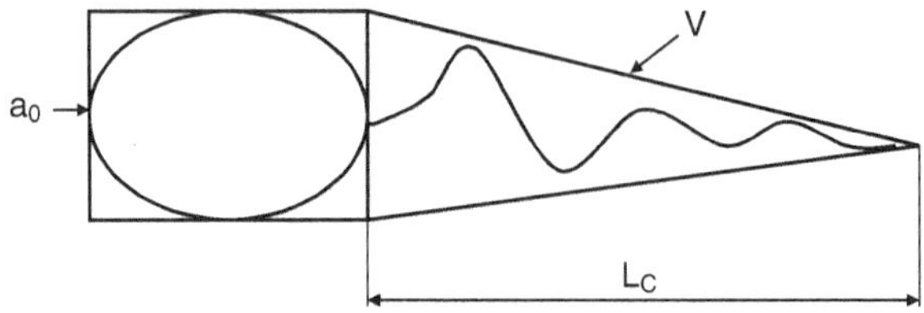

FIGURE 7.2 The parameters in the packing factor.

TABLE 7.1 Packing Factors for Aggregate Structures

< 0.33	Spherical, ellipsoidal micelles
0.33–0.5	Rod-like micelles
0.5–1.0	Vesicles, bilayers
1.0	Planar bilayers
> 1.0	Reverse micelles (small head and large tail)

Packing Factor

Packing factor (Φ) is defined as (Wang et al., 2006b)

$$\Phi = \frac{V}{a_0 L_c}, \tag{7.6}$$

where V is the volume occupied by the hydrophobic group in the micellar core, a_0 is the cross-sectional area occupied by the hydrophilic group at the micelle surface, and L_c is the length of the hydrophobic group (see Figure 7.2).

Packing factors for several aggregate structures are listed in Table 7.1. The minimum IFT is at the packing factor equal to 1.

7.3 TYPES OF MICROEMULSIONS

Surfactant solution phase behavior is strongly affected by the salinity of the brine. In general, increasing the salinity of the brine decreases the solubility of the anionic surfactant in the brine. The surfactant is driven out of the brine as the electrolyte concentration increases. Figure 7.3 shows that as the salinity is increased, the surfactant moves from the aqueous phase to the oleic phase. At a low salinity, the typical surfactant exhibits good aqueous-phase solubility. The oil phase, then, is essentially free of surfactant. Some oil is solubilized in the cores of micelles.

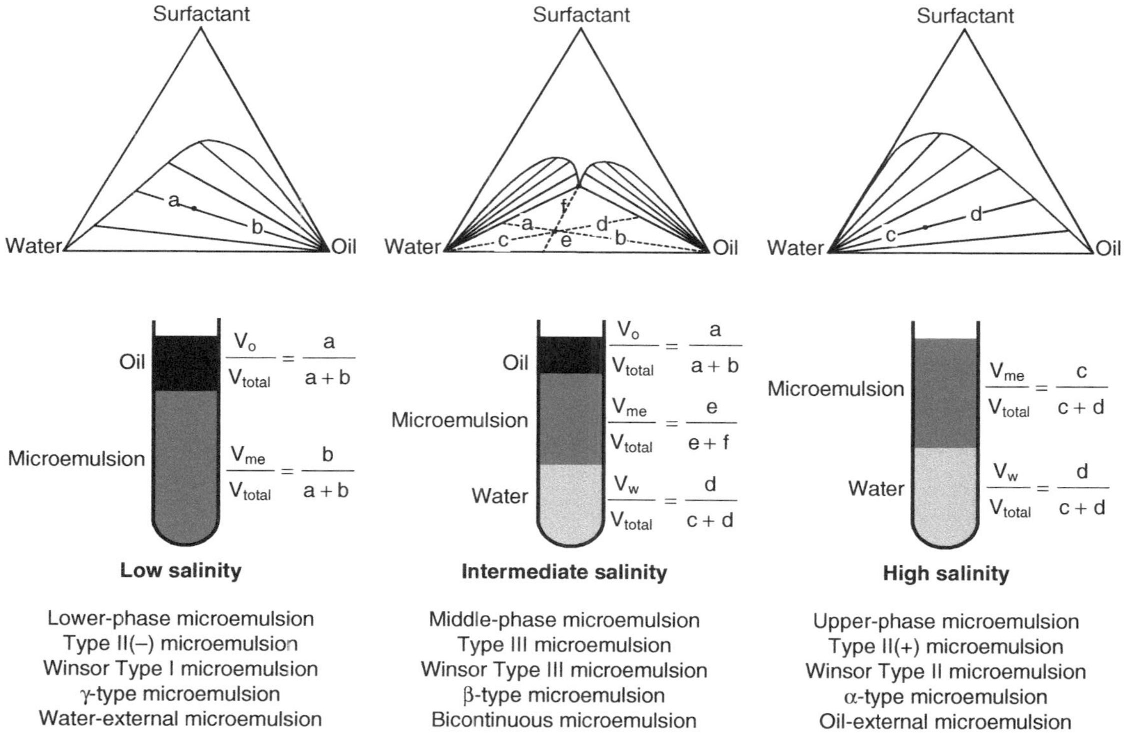

FIGURE 7.3 Three types of microemulsions and the effect of salinity on phase behavior.

The system has two phases: an excess oil phase and a water-external microemulsion phase. Because microemulsion is the aqueous phase and is denser than the oil phase, it resides below the oil phase and is called a lower-phase microemulsion. At a high salinity, the system separates into an oil-external microemulsion and an excess water phase. In this case, the microemulsion is called an upper-phase microemulsion. At some intermediate range of salinities, the system could have three phases: excess oil, microemulsion, and excess water. In this case, the microemulsion phase resides in the middle and is called a middle-phase microemulsion (Healy et al., 1976). Such terminology is consistent with their relative positions in a test tube (pipette) with the water being the dense liquid. In the environmental sciences and engineering, however, a dense nonaqueous phase liquid (DNAPL) could be denser than water (UTCHEM-9.0, 2000). Fleming et al. (1978) used γ, β, and α to name the lower-phase, middle-phase, and upper-phase microemulsions, respectively.

Surfactant–brine–oil phase behavior is conventionally illustrated on a ternary diagram, as shown in Figure 7.3. *If* the top apex of the ternary diagram represents the surfactant pseudocomponent, the lower left represents water, and the lower right represents oil, then the tie lines within the lower-microemulsion environment have negative slopes. Therefore, the phase environment is called type II(–) because there are two phases in the system and the slopes of tie lines are negative. Similarly, type II(+) and type III are used to describe the upper- and middle-phase environments, respectively (Nelson and Pope, 1978). Their names refer to the phase environment types. As Nelson (1982) emphasized, phase environment type refers to a type of phase diagram, not to a type of microemulsion. If the apex representations are changed—for example, if the water and oil positions are exchanged—the original representations of type II(–) and type II(+) will be changed.

The terminologies originally given by Winsor (1954)—Winsor type I, II, and III microemulsions—are also presented in Figure 7.3. The single-phase region above the multiphase boundaries with relatively high surfactant concentration is termed the type IV region (Meyers and Salter, 1981). These terms are not consistent with the *maximum number of possible phases* existing in the system, and it is difficult to link the terms with any characteristics of microemulsions, but these terms are used outside the petroleum literature.

In the author's opinion, we should use O/W, bicontinuous, and W/O microemulsions to describe water-external, bicontinuous, and oil-external microemulsions to be consistent with the terms used in emulsion. In this case, the left lobe (node) and right lobe (node) in a type III phase environment are termed O/W-lobe and W/O-lobe. This book mainly uses two naming systems—(1) type II(–), type III, and type II(+); (2) Winsor I, Winsor III, and Winsor II—even though other names are sometimes used. The book does not differentiate the name of a microemulsion type from that of the corresponding type of phase environment.

Microemulsion and macroemulsion need to be distinguished. Macroemulsion is a mixture of two or more immiscible (unblendable) liquids. One liquid (the dispersed phase) is dispersed in the other (the continuous phase). A macroemulsion tends to have a cloudy appearance because phases scatter the light that passes through it. The term ordinary emulsion or simply emulsion generally means macroemulsion. Emulsions are part of a more general class of two-phase systems of matter called colloids. Although the terms colloid and emulsion are sometimes used interchangeably, emulsion tends to imply that both the dispersed and continuous phases are liquid.

Emulsions are thermodynamically unstable and thus do not form spontaneously. Energy input through shaking, stirring, homogenizing, or spray processes are needed to form an emulsion. Over time, emulsions tend to revert to the stable state of the phases comprising the emulsion. There are three types of emulsion instability: flocculation, where the particles form clumps; creaming, where the particles concentrate toward the surface (or bottom, depending on the relative density of the two phases) of the mixture while staying separated; and breaking and coalescence, where the particles coalesce and form a layer of liquid.

Microemulsions are clear (transparent and translucent are also used in the literature), thermodynamically stable, isotropic liquid mixtures of oil, water, and surfactant, frequently in combination with a cosurfactant. The aqueous phase may contain salt(s) and/or other ingredients, and the "oil" may actually be a complex mixture of different hydrocarbons and olefins. In contrast to ordinary emulsions, microemulsions form upon simple mixing of the components and do not require high shear conditions generally used in the formation of ordinary emulsions. Microemulsions tend to appear clear due to the small size of the disperse phase. However, clear appearance (transparency) may not be a fundamental property. Sometimes microemulsion may not look clear to the naked eye in the case where dark viscous oil exists. The solution may not be purely transparent because it contains aggregates of micelles. Quite often, we still use these terms, even in this book. Probably we should simply use the term homogeneous solution.

One primary difference between microemulsion and macroemulsion may be drop size. The size of macroemulsion drops is generally orders of magnitude larger than the size of microemulsion drops. The difference in size explains their difference in properties and appearance; however, their fundamental difference is thermodynamic stability (Bourrel and Schechter, 1988).

7.4 PHASE BEHAVIOR TESTS

Phase behavior tests are conducted in small tubes that are called pipettes. Therefore, the phase behavior tests sometimes are called pipette tests. For pipette tests, the tips of—for example, 5 mL—glass pipettes from Fisher or similar pipettes are sealed by acetylene and oxygen flame with a Victor torch. Phase behavior tests include the aqueous stability test, salinity scan, and oil scan. The main objective of phase behavior tests is to find the chemical formula

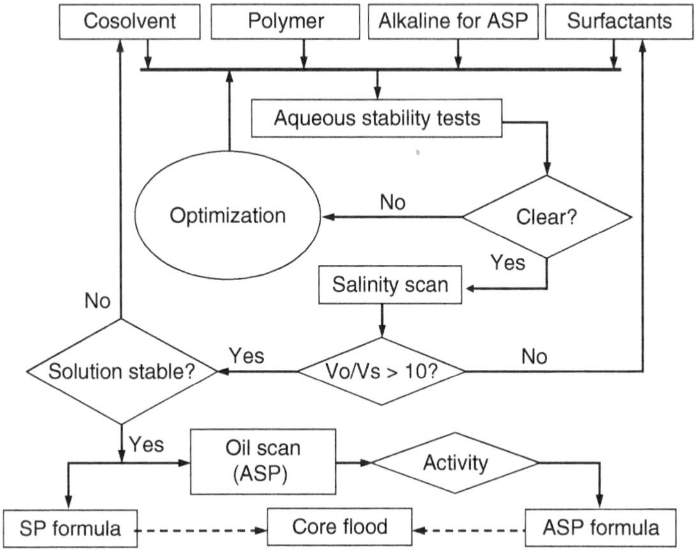

FIGURE 7.4 Flow chart of phase behavior tests.

for a specific application. A typical flow chart of phase behavior tests is shown in Figure 7.4.

Injection of a single-phase solution is important because formation of precipitate, liquid crystal, or a second liquid phase can lead to nonuniform distribution of injected materials and nonuniform transport owing to phase trapping or different mobilities of coexisting phases. When polymer is added to increase slug viscosity, it is essential to prevent separation into polymer-rich and surfactant-rich phases. The separation also yields highly viscous phases unsuitable for either injection or propagation through the formation. Therefore, we first need to check whether the aqueous solution is transparent without adding oil. The solution should be transparent (clear) up to or higher than the salinity at which we intend to inject the solution. Most likely, this is the optimum salinity at which both water/microemulsion IFT (σ_{wm}) and oil/microemulsion IFT (σ_{om}) are at minimum.

If the solution is clear up to this salinity, any problem mentioned previously would not appear because the solution will be more stable after mixing with the oil in situ. If the solution is hazy or any precipitation appears, chemicals must be reselected. Such a test is an aqueous stability test. Generally, the salinity limit in an aqueous stability test is close to the optimum salinity of microemulsion. When an alkali is added for screening ASP formula, sometimes precipitation is seen. This result is probably due to an alkali reaction with the glass tube so that silicate forms (Sheng et al., 1994). Figure 7.5 shows an example of an aqueous stability test.

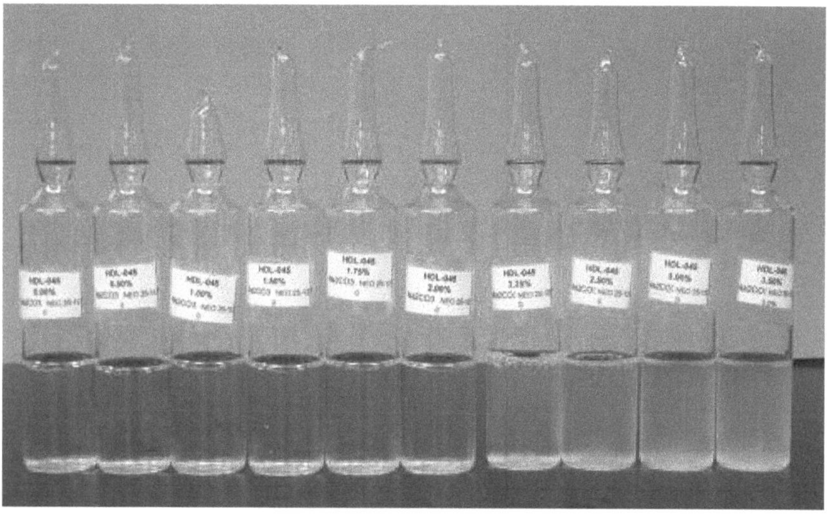

FIGURE 7.5 Aqueous stability tests.

If the solution is clear, we add oil in the solution and change the salinity. In the pipette tests, the temperature and concentrations of surfactant(s) and cosolvent are fixed, whereas the concentration for the electrolyte is varied between various test tubes. Pressure is assumed to be a minor effect, and it is generally atmospheric. As discussed earlier, the microemulsion changes from type II(–) to type III to type II(+) as the salinity is increased. These tests are referred to as salinity scans. Generally, the water/oil ratio (WOR) in a salinity scan is 1 or a fixed value.

Table 7.2 shows an example of salinity scan test data. Figure 7.6 may help facilitate understanding of how the data in the different columns of Table 7.2 are calculated. A photo of a real scan test facility is shown in Figure 7.7. Evidently, increasing salinity causes the microemulsion phase to undergo the transitions from type II(–) to type III to type II(+). The transition of microemulsion can be represented by a volume fraction diagram, which provides an understanding of the sensitivity of the surfactant solution behavior to additional electrolytes. In addition, such a diagram provides information on the solubilization of oil in the microemulsion and the optimum salinity.

A schematic of change in the type of microemulsion with the salinity is shown in Figure 7.8, and a volume fraction diagram of the data presented in Table 7.2 is shown in Figure 7.9. The volume fraction information can also be represented by a solubility plot, as shown in Figure 7.10 (see page 254). We will see later that the solubilization ratio is a very important parameter in interfacial tension calculation.

To reach a low IFT required to increase the capillary number, the oil solubilization ratio (Vo/Vs) may need to be higher than 10. If it is too low, we must

TABLE 7.2 Salinity Scan Pipette Test Data

Salinity (meq/mL)	Aqueous Level	Oil Level	Top of ME	Bottom of ME	Type+	Sol. Oil (mL)	Sol. Water (mL)	Oil Sol. Ratio (mL/mL)	Water Sol. Ratio (mL/mL)	Vol. Fraction of Oil	Vol. Fraction of ME	Vol. Fraction of Water
	At Start of Test		*In Equilibrium*									
A	B	C	D	E	F	G = B – D	H = E – B	I = G/[(5 – B)*1%]*	J = H/ [(5 – B) *1%]*	K = (D – C) /(5 – C)*	L = (5 – D)/ (5 – C), or (E – D)/(5 – C)*	M = 1 – K – L
0.091	2.88	0.80	2.85		I	0.03		1.4		0.488	0.512	
0.141	2.86	0.88	2.80		I	0.06		2.8		0.466	0.534	
0.190	2.82	0.75	2.80		I	0.02		0.9		0.482	0.518	
0.240	2.85	0.85	2.73		I	0.12		5.6		0.453	0.547	
0.290	2.87	0.82	2.70		I	0.17		8.0		0.450	0.550	
0.315	2.84	0.79	2.60	3.50	III	0.24	0.66	11.1	30.6	0.430	0.214	0.356

0.340	2.86	0.84	2.55	3.33	III	0.31	0.47	14.5	22.0	0.411	0.188	0.401
0.365	2.86	0.80	2.50	3.22	III	0.36	0.36	16.8	16.8	0.405	0.171	0.424
0.390	2.84	0.80		3.20	II		0.36		16.7		0.571	0.429
0.415	2.87	0.80		3.17	II		0.30		14.1		0.564	0.436
0.439	2.82	0.80		3.03	II		0.21		9.6		0.531	0.469
0.464	2.87	0.80		3.00	II		0.13		6.1		0.524	0.476
0.489	2.83	0.78		2.95	II		0.12		5.5		0.514	0.486
0.539	2.85	0.80		2.95	II		0.10		4.7		0.512	0.488
0.589	2.86	0.80		2.95	II		0.09		4.2		0.512	0.488

**Surfactant concentration: 1 wt.%; Pipette size: 5 mL.*
+Based on visual observation of the liquid levels.

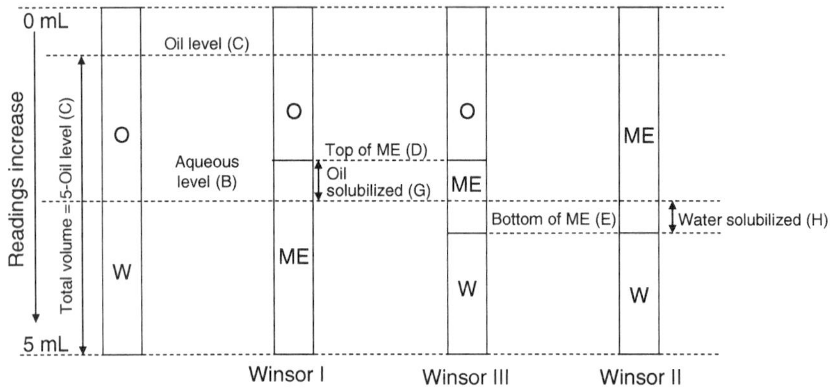

FIGURE 7.6 Schematic to show how salinity scan test data are measured and calculated.

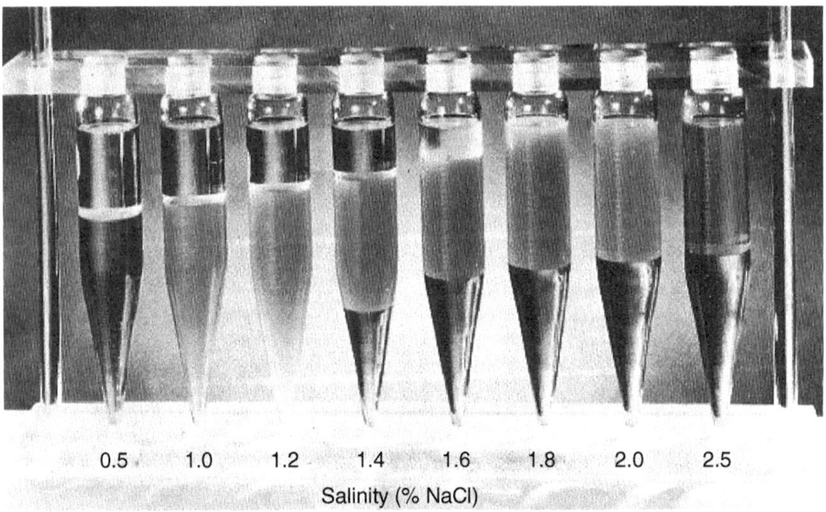

FIGURE 7.7 Photograph of salinity scan for phase behavior with WOR = 1, 4% 63/37 MEA-C12OXS/TAA, 48% 90/10 I/H, 48% x% NaCl. *Source: Healy et al. (1976).*

select other surfactants and repeat the preceding phase behavior tests. In practice, even if the oil solubilization ratio is high enough, but the optimum salinity identified is too far away from the salinity of the injection water we intend to use, we have to find other surfactants because changing the injection water salinity could be costly. It is suggested that the optimum salinity identified from screening tests be close to the salinity of an 80/20 mixture of the injection water and formation water because injected chemical solutions are expected to contact some portion of formation brine, even after a very efficient freshwater or alkaline preflush.

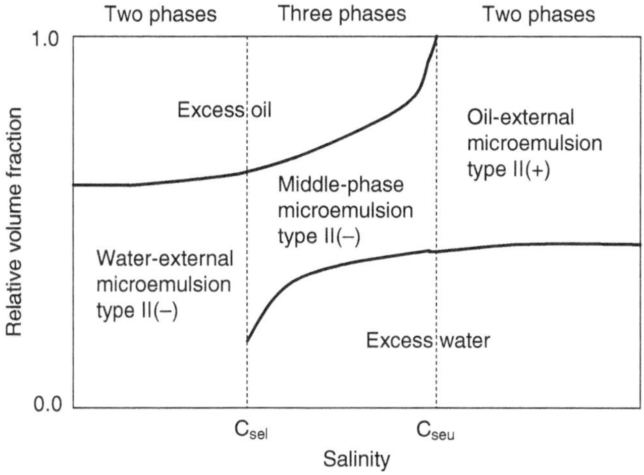

FIGURE 7.8 Schematic of volume fraction diagram.

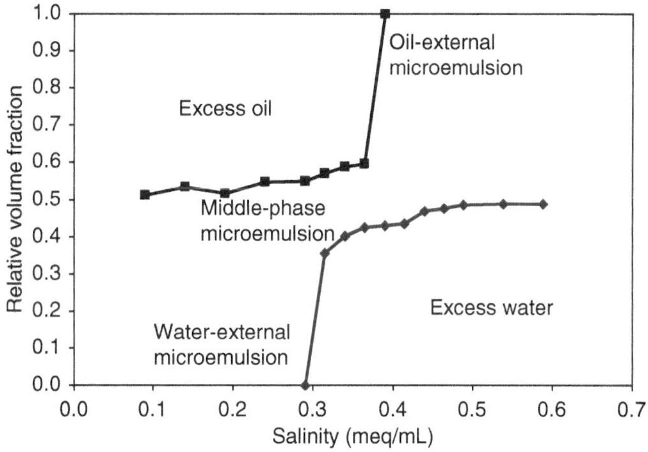

FIGURE 7.9 Volume fraction diagram of a salinity scan test. *(Data from Table 7.2.)*

In salinity scanning, there are two methods to change salinity. One method is to fix a salt concentration and change the alkaline concentration because the alkali can also work to adjust salinity. Another method is to fix an alkaline concentration and change the salt concentration. The negative effect of high alkaline concentration is to reduce polymer viscosity and possibly other negative alkaline effects such as scaling and emulsion problems. The negative side to increasing salinity may be to increase surfactant adsorption. There seems to be no criterion to determine the optimum portions of alkali and salt to be used.

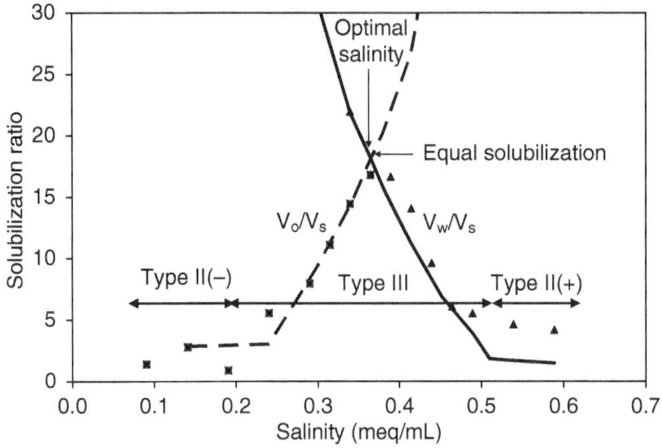

FIGURE 7.10 Solubilization plot (V_o/V_s and V_w/V_s) as a function of salinity. *(Data from Table 7.2, the two curves are calculated data from a model, and the points are test data.)*

7.5 SURFACTANT PHASE BEHAVIOR OF MICROEMULSIONS AND IFT

The phase behavior of microemulsions is complex and depends on a number of parameters, including the types and concentrations of surfactants, cosolvents, hydrocarbons, brine salinity, temperature, and to a much lesser degree, pressure. There is no universal equation of state even for a simple microemulsion. Therefore, phase behavior for a particular microemulsion system has to be measured experimentally. The phase behavior of microemulsions is typically presented using a ternary diagram and empirical correlations such as Hand's rule.

7.5.1 Ternary Diagrams

Before presenting a ternary diagram, we must discuss several concepts, including pseudocomponents. A microemulsion usually is composed of many components: surfactants, cosurfactants, cosolvents or alcohols, hydrocarbon, water, and electrolytes. To describe phase behavior rigorously, we need to include the effect of each component. The complexity of the system and the current technology, and sometimes the time and economic constraints, do not allow us to do so. In some practical situations, including the detailed effect of each component may not be necessary. Consequently, the number of components is reduced by combining similar components into pseudocomponents. A pseudocomponent is a true pseudocomponent if its compositions partition equally (in the exact ratio) in every phase. Typically, ternary and quaternary diagrams are used—although much more often, ternary diagrams. In a typical ternary

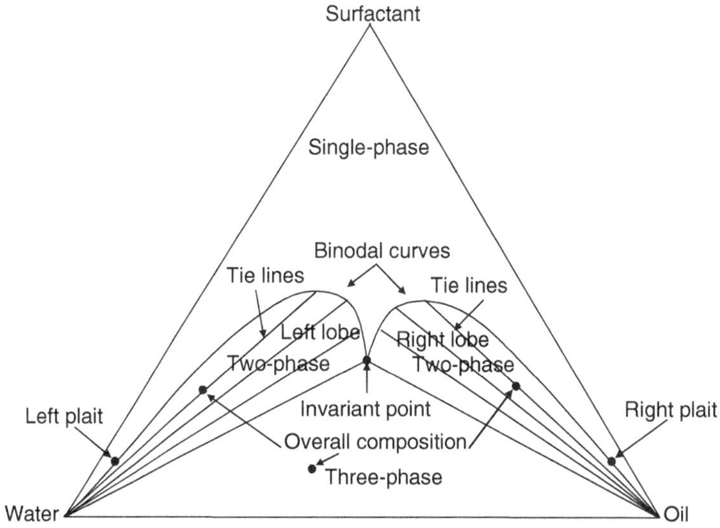

FIGURE 7.11 Schematic of a ternary diagram (not scaled).

diagram, there are three pseudocomponents: water, oil, and surfactant. In the ternary diagram, the system temperature and pressure are fixed. A ternary diagram is an extremely useful tool in EOR because it can simultaneously represent phase and overall compositions as well as relative amounts.

Figure 7.11 shows the schematic of a ternary diagram. Its apex locations on the equilateral triangle represent 100% water, oil, and surfactant components of a solution. The concentrations may be expressed in mole, mass, or volume fractions. The single-phase region is in the high surfactant concentration zone. The three-phase region exists in the middle zone. The two-phase lobes (nodes) exist in the upper right and upper left of the three-phase triangle. There is a third two-phase region located at very low surfactant concentrations below the three-phase region. This region typically is quite small and therefore is not included on the diagram. Because all substances are in principle at least slightly soluble, any region that has one or more corners of the triangle as part of its boundary is single-phase. Practically, the single-phase region may disappear when the mutual solubilities are very low—for example, at the water corner or the oil corner.

A line on a diagram separates two regions that differ by unity in the number of phases. At any point that is common to three regions, three phases coexist. Any region in which three phases coexist is necessarily bounded by a straight-sided triangle, as shown in Figure 7.11 where the three-phase region is bounded by the triangle formed by water–invariant–oil points. The sides of the triangle are tie lines, also called connodals, connecting all mixtures of the phases at the ends. A tie line connects the compositions of the two equilibrium phases at its two ends. The tie lines must be straight, as explained by Lake (1989). Type III phase behavior ternaries consist of an area close to the brine/oil axis bounded by a triangle. Compositions within this area will result in three phases, the composition of each phase being equal to the composition of the apexes of the bounding triangle.

The relationship among the number of components N_C, number of phases N_P, and number (degree) of freedom N_F of the system is given by the Gibbs' phase rule:

$$N_F = N_C - N_P + 2. \tag{7.7}$$

At the invariant point, $N_C = 3$, $N_P = 3$, the N_F is 2. Because the system temperature and pressure are fixed, the final degree of freedom is 0. In other words, the composition at the invariant point does not change as a function of total composition in the three-phase region of a true ternary diagram for a given salinity. That is why the name invariant point is used. The invariant point moves from the left water corner toward the right oil corner as the salinity is increased. The lower (C_{sel}) and upper (C_{seu}) limits of effective salinity are the effective salinity in which three phases form or disappear. Up to the lower effective salinity, the invariant point is still at the water corner of the ternary diagram. At the higher effective salinity, the invariant point is at the oil corner of the ternary diagram.

The binodal curves (phase boundary) separate the one- and two-phase regions. Below the binodal curves are two-phase regions, and above the curves is a single-phase region. At the plait point of the binodal curve, all phase compositions are equal. The right plait point is usually located very close to the oil apex, and the left plait point is usually located very close to the water apex. In a two-phase region, the compositions of phases in equilibrium are connected with tie lines, along which may be found all possible proportions of the two phases, as explained in Example 7.1. Before that, we first need to review the lever rule.

The lever rule is used to determine quantitatively the relative composition of a mixture in a two-phase region in a phase diagram. The distances from the mixture point along the horizontal tie line to both phase boundaries give the composition

$$n_\alpha l_\alpha = n_\beta l_\beta, \tag{7.8}$$

where n_α represents the amount of phase α and n_β represents the amount of phase β. Based on this lever rule, we have

$$\frac{n_\alpha}{n_\alpha + n_\beta} = \frac{l_\beta}{l_\alpha + l_\beta}. \tag{7.9}$$

Applying Eq. 7.9 to different types of microemulsions, we can determine the relative volume of each phase, as shown in Figure 7.12. Note that for the

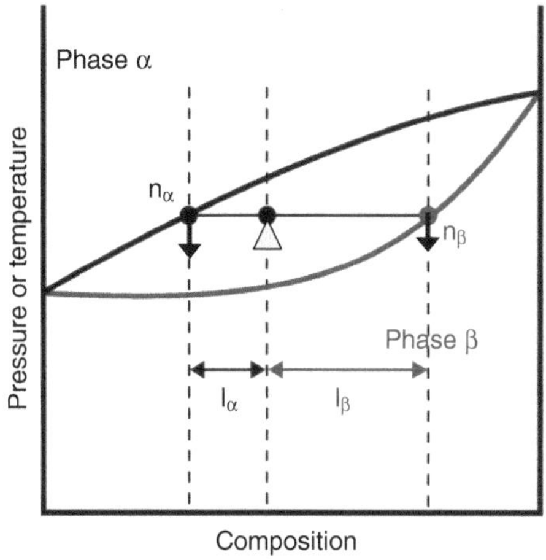

FIGURE 7.12 Schematic of lever rule.

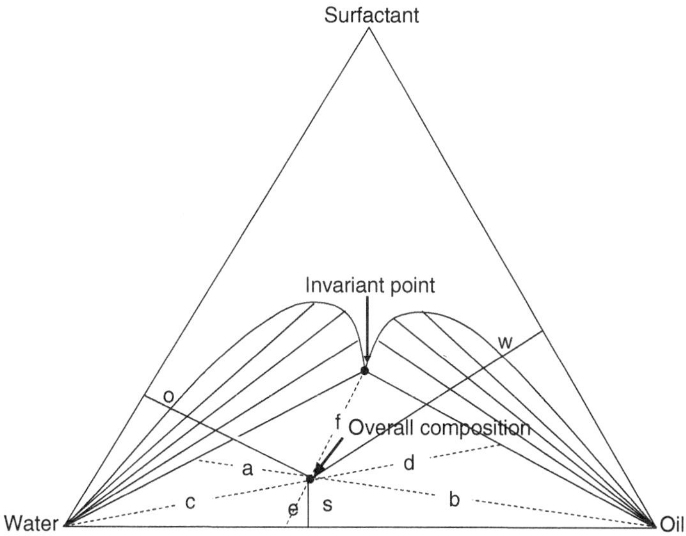

FIGURE 7.13 Ternary diagram for Example 7.1.

type III system, the following equation can be validated based on the fact that the total area of the large triangle is the sum of the three small triangles in Figure 7.13:

$$\frac{a}{a+b}+\frac{d}{c+d}+\frac{e}{e+f}=1. \tag{7.10}$$

Example 7.1 Determine Phase Compositions and Total Compositions from a Ternary Diagram

Refer to Figure 7.13. The total volume of the fluids in the system V_t is 10 cm^3. The positions of the overall composition and invariant point are marked in the figure. Assuming the equilibrium is reached, what are the volumes of the equilibrium phases? What is the volume of each component in the microemulsion phase?

Solution

Refer again to Figure 7.13. First, measure the perpendiculars from the overall composition point to the three sides of the triangle—s, o, and w. Then the total surfactant volume (V_{st}), total water volume (V_{wt}), and total oil volume (V_{wt}) in the system are

$$V_{st} = s/(s+o+w) \times V_t = 0.8 \text{ cm}^3,$$

$$V_{wt} = w/(s+o+w) \times V_t = 5.6 \text{ cm}^3, \text{ and}$$

$$V_{ot} = o/(s+o+w) \times V_t = 3.6 \text{ cm}^3.$$

Here, the subscripts i and j in V_{ij} represent component i and phase j, respectively. The subscript t represents the total system.

Next, find the equilibrium phase volumes of the excess water (V_{ww}), excess oil (V_{oo}), and microemulsion (V_{mm}) using the level rule:

$$V_{ww} = d/(c+d) \times V_t = 4.5 \text{ cm}^3,$$

$$V_{oo} = a/(a+b) \times V_t = 2.5 \text{ cm}^3, \text{ and}$$

$$V_{mm} = e/(e+f) \times V_t = 3 \text{ cm}^3.$$

Finally, find the volume of each component in the microemulsion phase at the invariant point. The volumes of water component (V_{wm}), oil component (V_{om}), and surfactant component (V_{sm}) are

$$V_{wm} = V_{wt} - V_{ww} = 5.6 - 4.5 = 1.1 \text{ cm}^3,$$

$$V_{om} = V_{ot} - V_{oo} = 3.6 - 2.5 = 1.1 \text{ cm}^3, \text{ and}$$

$$V_{sm} = V_{mm} - V_{wm} - V_{om} = 3 - 1.1 - 1.1 = 0.8 \text{ cm}^3.$$

Note that the surfactant volume in the microemulsion phase (V_{sm}) is the same as that in the total system (V_{st}). In other words, all the surfactant is in the microemulsion phase.

This example shows that the ternary diagram can represent the composition of the phases as well as the overall composition on the same diagram.

A further note should be made regarding the ternary diagram, especially the type III phase environment with the two lobes. As shown in Figure 7.3, as the salinity is increased, the negative slope of the tie lines in type II(−) is changed to the positive slope in type II(+). We may naturally think there is a zero slope between. The physical meaning of the zero slope is that the solubilities of the surfactant in the water- and oil-rich phases are exactly equal. However, Nelson and Pope (1978) reported that they had not seen such phase behavior for microemulsions used in EOR processes. The transition from type II(−) to type II(+) always occurred through a type III environment. More than one microemulsion type can be found within a single-phase environment type. The type of microemulsion observed depends on the overall surfactant/brine/oil composition. Furthermore, a lower-phase microemulsion can be from either a type II(−) or type III phase environment, and an upper-phase microemulsion can be from either a type II(+) or type III system. A middle-phase microemulsion is always from a type III phase environment.

Knickerbocker et al. (1982) laid out the general features of the progression of the three-phase system as salinity is increased. Figure 7.14 shows a series of pseudoternary diagrams with oil, surfactant, and water as the vertices. Each pseudoternary diagram represents a constant salinity. At low salinities, a two-phase system is present in which an oil phase is in equilibrium with an aqueous phase with surfactant. A plait point is shown in the figure. As salinity is

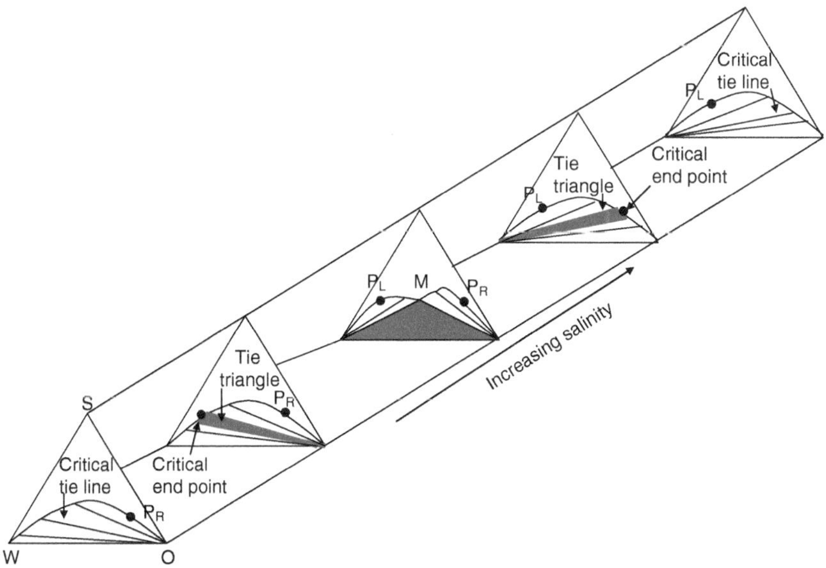

FIGURE 7.14 Schematic showing a sequence of ternary system transitions from type I to type III to type II as the salinity increases.

increased, the lower aqueous phase becomes saturated with surfactant at the salinity, and a third phase erupts at this critical tie line. In other words, the critical tie line, under slightly altered conditions, broadens into a tie triangle for which two micellar solutions, both water-rich, are in equilibrium with the oil phase. A type III system exhibiting a tie triangle having one short leg also is shown in the figure. The two vertices are so closely spaced that these two phases virtually have the same composition. These two phases are therefore near their critical composition, which exists at the point where the tie line initially opens into a triangle (Bourrel and Schechter, 1988). The critical end point (CEP) of this tie line is indicated in the figure. The shaded three-phase region is surrounded by the three two-phase regions.

One corner (marked with M in the figure) of the new three-phase region is the middle-phase composition, a phase that has surfactant, water, and oil at a specific salinity. As salinity is increased, the middle-phase composition moves toward the oil vertex as the capability of the middle phase to solubilize water is reduced. At a specific salinity, the three-phase region merges into a two-phase region where the surfactant-rich oleic phase is in equilibrium with a surfactant-poor aqueous phase. The three-phase region disappears at the CEP indicated in the upper part of the figure, where the surfactant-rich oleic hydrocarbon phase becomes indistinguishable from the middle phase.

In a type III system, a left lobe or right lobe microemulsion cannot coexist with the middle-phase microemulsion. The total composition determines the existence of a lobe or the middle-phase microemulsion. Gary A. Pope (Personal communication on February 17, 2009) pointed out that, as a practical matter, we rarely measure a sufficient number of points in the ternary system to clearly define two-phase and three-phase regions. When cosolvent and/or Ca^{++} is used, or when soap forms, a ternary diagram does not accurately represent the phase behavior. When typical salinity scans at WOR = 1 and a low surfactant concentration are performed, almost all the cases in a type III environment will be three phases. So there is little, if any, practical issue involved in a typical phase behavior experiment.

However, on rare occasions we might mislabel a sample as type II or type I when it is really type III, so our estimate of the type III salinity window might be a little off. Typically, however, other practical problems when using crude oils would be of much greater concern; for example, interfaces are often hard to read, emulsions often form at interfaces and take a long time to coalesce, we do not usually account for the partitioning and volume of the cosolvent, and so on. In most cases, the goal is to understand the overall trends and select the best formulation for a core flood, and we can usually do that without taking into account all these complexities.

Changing either the surfactant concentration or the water/oil ratio (WOR) in the system also may modify the phase behavior to a great extent. For example, for the left lobe, the micellar structure is such that the microemulsion is more related to a middle-phase microemulsion than to the ordinary type II(+)

microemulsion. In fact, it will often become a type III microemulsion simply through addition of oil; for this reason, it may be called an oil-deficient type III microemulsion or a "pseudo type III" microemulsion. Similarly, the right lobe may be called a water-deficient type III microemulsion (Reed and Carpenter, 1982). Concentrations could change phase behavior, especially when a cosolvent or cosurfactant exists in the system. This limits the possibilities of using pseudoternary representation (Baviere et al., 1981).

7.5.2 Hand's Rule

This section describes how to use Hand's rule to represent binodal curves and tie lines. The surfactant–oil–water phase behavior can be represented as a function of effective salinity after the binodal curves and tie lines are described. Binodal curves and tie lines can be described by Hand's rule (Hand, 1939), which is based on the empirical observation that equilibrium phase concentration ratios are straight lines on a log-log scale. Figures 7.15a and 7.15b show the ternary diagram for a type II(−) environment with equilibrium phases numbered 2 and 3 and the corresponding Hand plot, respectively. The line segments AP and PB represent the binodal curve portions for phase 2 and phase 3, respectively, and the curve CP represents the tie line (distribution curve) of the indicated components between the two phases. C_{ij} is the concentration (volume fraction) of component i in phase j (i or j = 1, 2, or 3), and 1, 2, and 3 represent water, oil, and microemulsion, respectively. As the salinity is increased, the type of microemulsion is changed from type II(−) to type III to type II(+). C_i represents the total amount of composition i.

The binodal curves for all three types of phase behavior are represented by the Hand equation:

$$\frac{C_{3j}}{C_{2j}} = A_H \left(\frac{C_{3j}}{C_{1j}} \right)^{B_H} \quad j = 1, 2, \text{or } 3. \tag{7.11}$$

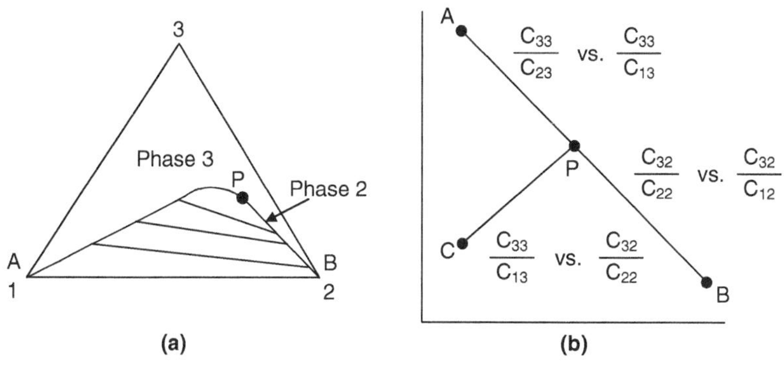

FIGURE 7.15 Correspondence between (a) a ternary diagram and (b) a log scale Hand plot.

Here, A_H and B_H are empirical parameters. For a symmetric binodal curve, $B_H = -1$, which is the current formulation used in UTCHEM. Then A_H is estimated from

$$A_H = \frac{(C_{3j})^2}{C_{1j}C_{2j}} \quad j = 1, 2, \text{or } 3. \tag{7.12}$$

In the preceding equation, j = 2, 3 for type II(–), and j = 1, 3 for type II(+).

Theoretically, Eq. 7.12 applies at any point, including the plait point in the binodal curve, which covers phases 2 and 3 for the type II(–) environment. Therefore, the value of A_H for phase 1 is the same as that for phase 3. Similarly, the value of A_H for phase 1 is the same as that for phase 3 for the type II(+) environment. However, we assume that the surfactant is in phase 3 (the microemulsion phase). When we use Eq. 7.12 to calculate A_H, j = 3, in general, we use experimental data to calculate A_H at different effective salinities (C_{se}), as shown in Figure 7.16 (dot points), assuming that the surfactant is in phase 3 (microemulsion phase).

We fit the data with two lines: one with $C_{se} < C_{seop}$ (optimum salinity) and the other one with $C_{se} > C_{seop}$. A_H can be simply interpolated at any C_{se}:

$$\begin{aligned} A_H &= A_{H0} + (A_{H1} - A_{H0})C_{seD}, \quad C_{seD} \le 1 \\ A_H &= A_{H1} + (A_{H2} - A_{H1})(C_{seD} - 1), \quad C_{seD} > 1. \end{aligned} \tag{7.13}$$

Here, C_{seD} is C_{se}/C_{seop}. From Figure 7.16, we can see that at any salinity, A_H can be defined. Thus, we should be able to perform phase behavior calculation with A_H. However, UTCHEM requires different input parameters: C_{33max0}, C_{33max1}, and C_{33max2}. These parameters are calculated in Eq. 7.14 from A_H.

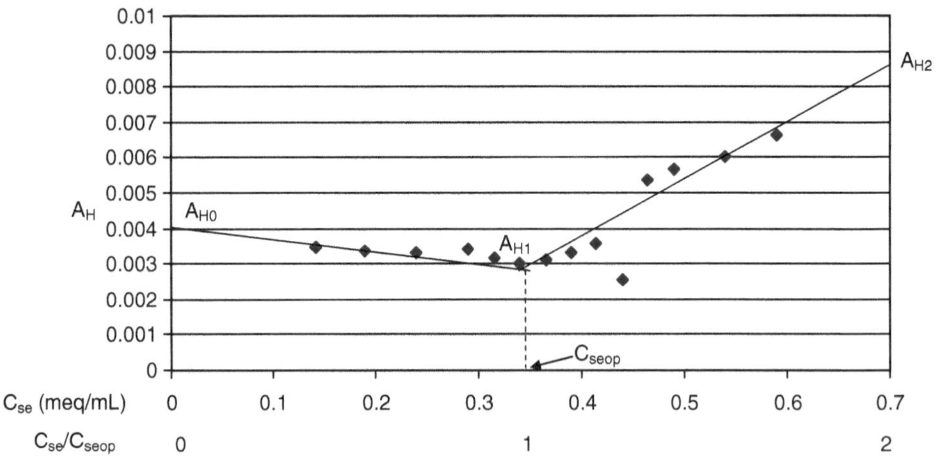

FIGURE 7.16 Hand parameter A_H versus salinity.

$$C_{33\max} = \frac{\sqrt{A_H}}{2+\sqrt{A_H}}. \tag{7.14}$$

At the optimum salinity (normalized salinity $C_{seD} = 1$), $C_{13} = C_{23} = (1 - C_{33})/2$ for j = 3 in Eq. 7.12, which means the binodal curve is symmetric. Thus, Eq. 7.14 is readily derived at the optimum salinity, and $C_{33\max 1}$ can be calculated from Eq. 7.14 using A_{H1}. The physical meaning of $C_{33\max 1}$ is the maximum height of the binodal curve at the optimum salinity. At the zero optimum salinity (practically very low salinity, $C_{seD} = 0$) and twice optimum salinity ($C_{seD} = 2$), we also *assume* the binodal curves are symmetric and use Eq. 7.14 to calculate $C_{33\max 0}$ and $C_{33\max 2}$ corresponding to A_{H0} and A_{H2}, respectively.

At these salinities, A_{H0} and A_{H2} are obtained from the extrapolated lines in Figure 7.16 or from Eq. 7.13. However, although $C_{33\max 1}$ is the maximum height of the binodal curve at the optimum salinity, $C_{33\max 0}$ and $C_{33\max 2}$ do not represent the maximum heights of binodal curves because the condition that $C_{13} = C_{23} = (1 - C_{33})/2$ is not physically satisfied any more at these salinities. Therefore, $C_{33\max 0}$ and $C_{33\max 2}$ are the *hypothetical* maximum heights of binodal curves. In this book, $C_{33\max 0}$ and $C_{33\max 2}$ are also termed the UTCHEM input parameters calculated using Eq. 7.14 at C_{seD} equal to 0 and 2, respectively.

To use Hand's rule for phase behavior calculation, we need the values of the Hand parameter A_H. Although A_H is defined in Figure 7.16, UCHEM does not use A_H. Instead, $C_{33\max 0}$, $C_{33\max 1}$, and $C_{33\max 2}$ are the required input parameters, and they may be used to back-calculate A_{H0}, A_{H1}, and A_{H2}, respectively, in UTCHEM. It would be less confusing had A_{Hi}, instead of $C_{33\max i}$, been used directly in UTCHEM.

Another confusing convention in the literature is to have mixed $C_{33\max 0}$ and $C_{33\max 2}$ with C_{33} at $C_{seD} = 0$ and 2. $C_{33\max 0}$ and $C_{33\max 2}$ are the calculated parameters using Eq. 7.14, while C_{33} should be calculated as follows.

Based on material balance, at the optimum salinity ($C_{seD} = 1$), we have

$$C_{33} = C_{33\max 1} = C_{3M} = \frac{C_3}{S_3}, \tag{7.15}$$

where S_3 is the volume fraction (saturation) of the microemulsion phase, and C_3 is the total surfactant concentration in the system. Here, we assume the excess phases are free of surfactant. In this case, C_{33} at $C_{seD} = 1$ equals $C_{33\max 1}$. At zero salinity (practically very low salinity), we have

$$C_{33}\big|_{C_{seD0}} = \frac{C_3}{S_3}\bigg|_{C_{seD0}} \neq C_{33\max 0}. \tag{7.16}$$

At twice salinity (high salinity), we have

$$C_{33}\big|_{C_{seD2}} = \frac{C_3}{S_3}\bigg|_{C_{seD2}} \neq C_{33\max 2}. \tag{7.17}$$

We know that at a very low salinity or at a high salinity, the microemulsion is usually Winsor type I or Winsor type II. The microemulsion volumes (S_3) at these salinities are generally higher than that at the optimum salinity. Then according to Eqs. 7.15 through 7.17, $C_{33}|_{C_{seD0}}$ and $C_{33}|_{C_{seD2}}$ will be lower than C_{33max1}. However, C_{33max0} and C_{33max2} are higher than C_{33max1} because A_{H0} and A_{H2} are higher than A_{H1}, as shown in Figure 7.16, and C_{33max} is higher as A_H is higher according to Eq. 7.14. Therefore, C_{33max0} and C_{33max2} estimated from Eq. 7.14 do not represent $C_{33}|_{C_{seD0}}$ and $C_{33}|_{C_{seD2}}$, respectively. $C_{33}|_{C_{seD0}}$ and $C_{33}|_{C_{seD2}}$ are the physical heights of binodal curves at their respective salinities, while C_{33max0} and C_{33max2} are not. In general, the physical height of a binodal curve is estimated from

$$C_{33} = \frac{C_3}{S_3}. \tag{7.18}$$

When A_H is prepared using Eq. 7.13 for a UTCHEM model, C_{se0} must be sufficiently small, although theoretically it should be 0 but cannot be 0 in practice. Also, C_{se2} must be two times C_{seop}. A more general form of Eq. 7.13 is

$$\begin{aligned} A_H &= A_{H1} + (A_{H0} - A_{H1})\frac{C_{se} - C_{seop}}{C_{se0} - C_{seop}}, \quad C_{seD} \le 1 \\ A_H &= A_{H1} + (A_{H2} - A_{H1})\frac{C_{se} - C_{seop}}{C_{se2} - C_{seop}}, \quad C_{seD} > 1, \end{aligned} \tag{7.19}$$

where C_{se0} and C_{se2} are arbitrary salinities at the left and right side of the optimum salinity (C_{seop}), respectively.

The tie lines of type II(−) and II(+) phase behavior are also represented by the Hand equation

$$\frac{C_{3j}}{C_{2j}} = E_H\left(\frac{C_{33}}{C_{13}}\right)^{F_H} \quad j = 1, \text{or } 2, \tag{7.20}$$

where E_H and F_H are empirical parameters. For type II(−), j = 2, and for type II(+), j = 1.

For a simple case, $F = -B^{-1} = 1$. Eq. 7.20 applies at the plait point and we have

$$E_H = \frac{C_{1P}}{C_{2P}} = \frac{1 - C_{2P} - C_{3P}}{C_{2P}}, \tag{7.21}$$

where the subscript P could be the left plait point (PL) or the right plait point (PR). Applying the binodal curve equation (Eq. 7.12) to the plait point, we have

$$C_{3P} = \frac{1}{2}\left[-A_H C_{2P} + \sqrt{(A_H C_{2P})^2 + 4A_H C_{2P}(1 - C_{2P})}\right], \tag{7.22}$$

where C_{2P} is the oil phase composition at the plait point (left or right) and is an input parameter. C_{2P} in Eq. 7.22 is substituted in Eq. 7.21 to calculate E_H, which is salinity-dependent.

For a type III phase environment, there possibly exist the left lobe [type II(+)] and the right lobe [type II(−)]. The plait point must vary between 0 and C^*_{2PL}, the left plait point for type II(+), or between 1 and C^*_{2PR}, the right plait point for II(−). The idea to calculate the phase compositions in the lobes is to follow the approach for type II(−) and type II(+) phase environments with the transformed concentrations. Before that, however, we need to define how the plait points and invariant point move.

Refer to Figure 7.17 and Table 7.3, before the salinity increases up to C_{sel}, it is a type II(−) environment with the plait point C^*_{2PR} (shown as PR* in Figure 7.17). The superscript * refers to a low or high salinity limiting case. At C_{sel}, type II(−) starts to become type III and the right lobe. From C_{sel} to C_{seu}, it is a type III environment. The left and right lobes have developed. At C_{seu}, type III and the left lobe start to become type II(+). When the salinity is greater than C_{seu}, it is a type II(+) with the plait point C^*_{2PL} (shown as PL* in Figure 7.17). These changes with salinity are summarized in Table 7.3.

For the II(+) left lobe, the plait point is calculated by interpolation on effective salinity:

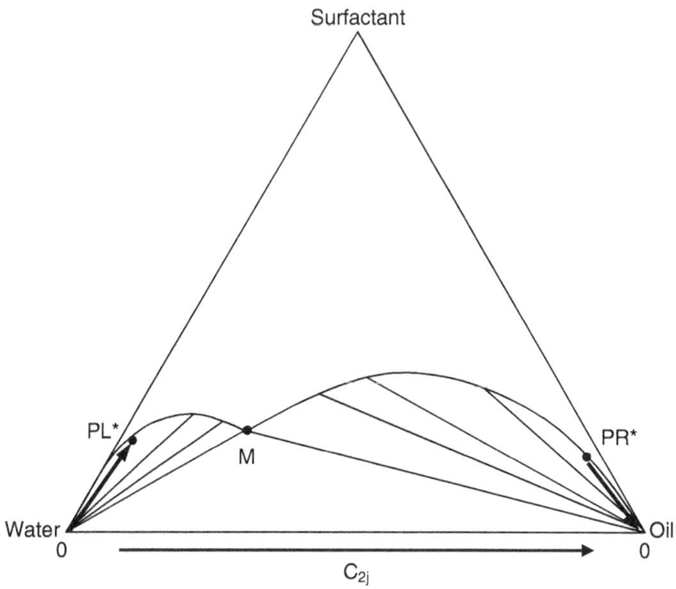

FIGURE 7.17 A schematic illustrating plait point and invariant point migration as salinity is increased.

TABLE 7.3 Summary of Phase Type Change, Invariant and Plait Point Migration

Salinity C_{se}	Phase Type	Plait Point	Invariant C_{2M}
$< C_{sel}$	II(−)	C^*_{2PR}	N/A
$= C_{sel}$	II(−) → III, II(−) → right lobe	C^*_{2PR}	0
$C_{sel} < C_{se} < C_{seu}$	III, left + right lobes	C_{2PR}: $C^*_{2PR} \to 1$, C_{2PL}: $0 \to C^*_{2PL}$	0 → 1
$= C_{seu}$	III → II(+), left lobe → II(+)	C^*_{2PL}	1
$> C_{seu}$	II(+)	C^*_{2PL}	N/A

$$C_{2PL} = \frac{C_{se} - C_{sel}}{C_{seu} - C_{sel}} C^*_{2PL}. \tag{7.23}$$

As shown in Figure 7.18, the transformed concentrations (denoted by superscript prime) are (UTCHEM-9.0, 2000)

$$C'_{2j} = C_{2j} \sec\theta, \tag{7.24}$$

$$C'_{3j} = C_{3j} - C_{2j} \tan\theta, \tag{7.25}$$

$$C'_{1j} = 1 - C'_{2j} - C'_{3j}, \quad j = 1 \text{ or } 3. \tag{7.26}$$

The angle θ is

$$\tan\theta = \frac{C_{3M}}{C_{2M}}. \tag{7.27}$$

Alternatively,

$$\sec\theta = \frac{\sqrt{C_{2M}^2 + C_{3M}^2}}{C_{2M}}. \tag{7.28}$$

The parameter E_H of the Hand equation is now calculated in terms of untransformed coordinates of the plait point as

$$E_H = \frac{C'_{1P}}{C'_{2P}} = \frac{1 - (\sec\theta - \tan\theta)C_{2PL} - C_{3PL}}{C_{2PL}\sec\theta}, \tag{7.29}$$

where C_{3PL} is given by C_{3P} in Eq. 7.22 and C_{2PL} is given by Eq. 7.23.

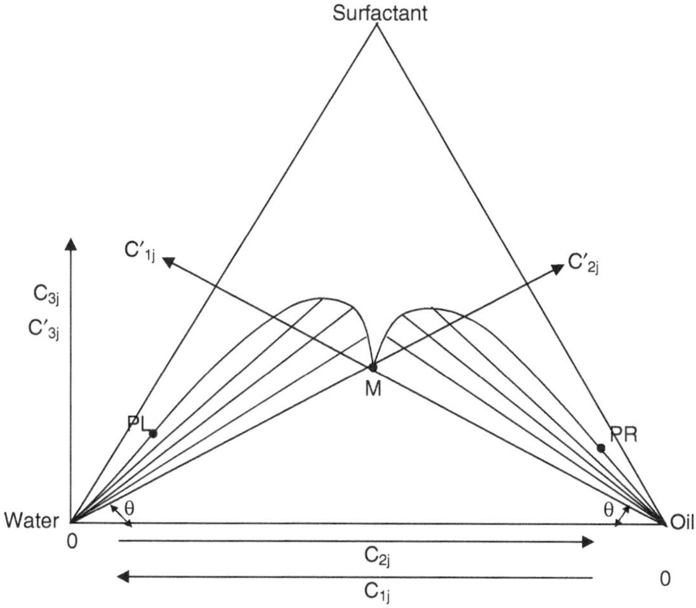

FIGURE 7.18 Coordinate transformation for the two-phase calculation in type III lobes.

For the II(–) right lobe, the plait point is calculated by interpolation on effective salinity:

$$C_{2PR} = C^*_{2PR} + \frac{C_{se} - C_{sel}}{C_{seu} - C_{sel}}(1 - C^*_{2PR}). \tag{7.30}$$

As shown in Figure 7.18, the transformed concentrations (denoted by superscript prime) are (UTCHEM-9.0, 2000)

$$C'_{1j} = C_{1j}\sec\theta, \tag{7.31}$$

$$C'_{3j} = C_{3j} - C_{1j}\tan\theta, \tag{7.32}$$

$$C'_{2j} = 1 - C'_{1j} - C'_{3j}, \quad j = 2 \text{ or } 3. \tag{7.33}$$

The angle θ is

$$\tan\theta = \frac{C_{3M}}{C_{1M}}. \tag{7.34}$$

Alternatively,

$$\sec\theta = \frac{\sqrt{C^2_{2M} + C^2_{3M}}}{C_{1M}}. \tag{7.35}$$

The parameter E_H of the Hand equation is now calculated in terms of untransformed coordinates of the plait point as

$$E_H = \frac{C'_{1P}}{C'_{2P}} = \frac{C_{1PR}\sec\theta}{1-(\sec\theta-\tan\theta)C_{1PR}-C_{3PR}}. \tag{7.36}$$

where C_{3PR} is given by C_{3P} in Eq. 7.22 and $C_{1PR} = 1 - C_{2PR} - C_{3PR}$.

In the preceding description, a basic assumption is made that the binodal curve is the same function of salinity for all three types of phase behavior. So, for a type III diagram, the left and right lobes are described by a continuous function of the same form as for type II(+) and type II(–). Therefore, A_H for the two lobes follows Eq. 7.13 or Eq. 7.19.

In summary, this section has introduced Hand's rule to describe phase compositions and discussed how to estimate Hand parameters. From the preceding discussion, we know that seven parameters are needed to describe phase behavior: C_{selD}, C_{seuD}, C_{33max0}, C_{33max1}, C_{33max2}, C_{2PL}, and C_{2PR}.

7.5.3 Quantitative Representation of Phase Behavior

To follow the preceding section's discussion of the ternary diagram and Hand's rule, this section discusses how to calculate phase compositions C_{ij}. The general approach is to represent the binodal and distribution curves (tie lines) as a function of the total concentration of water, oil, and surfactant (i.e., C_1, C_2, C_3—of which only two are independent) with the electrolyte concentration as a parameter, based on an idea outlined by Lake (1989). A similar approach was proposed by Pope and Nelson (1978) and Camilleri (1983).

We start with a type II(–) system. In this two-phase system, there are six unknowns, the phase concentrations C_{ij} ($i = 1, 2, 3$, $j = 2, 3$). However, there are five equations—two from Eq. 7.11 ($j = 2, 3$), one from Eq. 7.20 ($j = 2$), and two consistency constraints:

$$\frac{C_{32}}{C_{22}} = A_H\left(\frac{C_{32}}{C_{12}}\right)^{B_H}, \tag{7.37}$$

$$\frac{C_{33}}{C_{23}} = A_H\left(\frac{C_{33}}{C_{13}}\right)^{B_H}, \tag{7.38}$$

$$\frac{C_{32}}{C_{22}} = E_H\left(\frac{C_{33}}{C_{13}}\right)^{F_H}, \tag{7.39}$$

$$\sum_{i=1}^{3} C_{i2} = 1, \tag{7.40}$$

$$\sum_{i=1}^{3} C_{i3} = 1. \tag{7.41}$$

Note that A_H and B_H for phases 2 and 3 are the same, as discussed earlier. We need more conditions to fix the problem. Because the total compositions are known and $S_2 + S_3 = 1$, we have additional independent equations:

$$C_1 = C_{12}S_2 + C_{13}S_3, \tag{7.42}$$

$$C_2 = C_{22}S_2 + C_{23}S_3, \tag{7.43}$$

$$S_2 + S_3 = 1. \tag{7.44}$$

Now we have eight equations (7.37 through 7.44) to solve eight unknowns, C_{ij} and S_j ($i = 1, 2, 3, j = 2, 3$). In principle, the solution is determined. However, these equations are not all linear; the solution procedures are not straightforward. In the following, we provide the detailed iterative procedures for a simple case in which B_H is equal to -1 and $F_H = 1$ for the symmetric binodal curves.

1. Pick a phase concentration, say, C_{33}. Parameter C_{33} was chosen because it is very sensitive and its initial value can be easily estimated. From Eq. 7.38, we have

$$C_{23} = \frac{1}{2}\left[(1-C_{33}) - \sqrt{(1-C_{33})^2 - 4C_{33}^2/A_H}\,\right], \tag{7.45}$$

$$C_{13} = 1 - C_{23} - C_{33}. \tag{7.46}$$

From Eq. 7.39,

$$C_{32} = E_H\left(\frac{C_{33}}{C_{13}}\right)^{F_H} C_{22}. \tag{7.47}$$

From Eq. 7.37,

$$(C_{32})^2 = A_H C_{12} C_{22}. \tag{7.48}$$

2. From Eqs. 7.47 and 7.48, we have

$$C_{32} = \frac{A_H E_H\left(\dfrac{C_{33}}{C_{13}}\right)^{F_H}}{A_H + A_H E_H\left(\dfrac{C_{33}}{C_{13}}\right)^{F_H} + E_H^2\left(\dfrac{C_{33}}{C_{13}}\right)^{2F_H}}, \tag{7.49}$$

$$C_{22} = \frac{C_{32}}{E_H\left(\dfrac{C_{33}}{C_{13}}\right)^{F_H}}, \tag{7.50}$$

$$C_{12} = 1 - C_{22} - C_{32}. \tag{7.51}$$

Now we have obtained all C_{ij}. From Eqs. 7.42 through 7.44, we have

$$\frac{C_1 - C_{13}}{C_{12} - C_{13}} = \frac{C_2 - C_{23}}{C_{22} - C_{23}}. \tag{7.52}$$

Check whether Eq. 7.52 is satisfied within a limited error. If it is, the picked C_{33} is correct, and the other phase concentrations can be calculated as described previously. If not, use the just-calculated C_{33} as the initial picked value, and repeat steps 1 and 2.

After all C_{ij} have been calculated, the saturations can be readily estimated using the following equations:

$$S_2 = \frac{C_1 - C_{13}}{C_{12} - C_{13}}, \tag{7.53}$$

or

$$S_2 = \frac{C_2 - C_{23}}{C_{22} - C_{23}}, \tag{7.54}$$

$$S_3 = 1 - S_2. \tag{7.55}$$

The material balance already has been made on the concentrations. The calculated saturations (S_2) using Eqs. 7.53 and 7.54 should be close to each other.

For the type II(+) phase environment, the equations and calculation procedures for the type II(–) system can be repeated, except that phase 2 is changed to phase 1.

The preceding procedures are applied at one salinity point. To quantify a phase behavior, we need to repeat these procedures at selected salinity points to cover the whole range of salinity for a particular case.

Here, we described the general procedure to quantify phase behavior for type II(–) and type II(+) systems. For the type III phase environment with middle-phase microemulsion, the phase compositions are fixed at the invariant point (C_{1M}, C_{2M}, C_{3M}). The subscript M denotes the invariant point, and C_{iM} ($i = 1, 2, 3$) is the composition at the invariant point. What are the invariant compositions then?

As we can see in Figure 7.11 and Table 7.3, the invariant point moves from C_{2M} equal to 0 to C_{2M} equal to 1 as the salinity is increased from C_{sel} to C_{seu}. C_{sel} to C_{seu} are the lower and upper effective salinity limits for type III microemulsion. Based on this general observation, it has been proposed that C_{2M} is interpolated linearly as a function of salinity from C_{sel} to C_{seu} (L.W. Lake, personal communication on June 25, 2009):

$$C_{2M} = \frac{C_{se} - C_{sel}}{C_{seu} - C_{sel}}. \tag{7.56}$$

Note that Eq. 7.56 is an approximation only which shows that C_{2M} is $\frac{1}{2}$ at the optimum salinity. However, as we know, C_{2M} is actually equal to $\frac{1}{2}(1 - C_{3M})$

at the optimum salinity. Because the surfactant concentration C_{3M} is low, the error is not large.

Now we can calculate C_{1M} and C_{3M} from Eq. 7.37 by replacing C_{i2} with C_{2M} and by using the condition:

$$C_{1M} + C_{2M} + C_{3M} = 1. \tag{7.57}$$

For a simple case with $B_H = -1$, C_{3M} and C_{1M} are

$$C_{3M} = \frac{1}{2}\left[\sqrt{(A_H C_{2M})^2 + 4A_H C_{2M}(1 - C_{2M})} - A_H C_{2M}\right], \tag{7.58}$$

$$C_{1M} = 1 - C_{2M} - C_{3M}. \tag{7.59}$$

The compositions at the other two phases for the three-phase region of type III are (1, 0, 0) at the aqueous phase and (0, 1, 0) at the oleic phase if we assume the excess oleic and aqueous phases are pure.

For the left and right lobes in a type III phase environment, calculation of the phase compositions follows the approach for type II(+) and type II(−) phase environments with the transformed concentrations that are described earlier in the previous section.

So far, we have described quantification of microemulsion phase behavior. The procedures described here can be coded in a small program or even in an Excel spreadsheet. More practically, we can use a sample UTCHEM simulation file called batch.txt to simulate phase behavior pipette tests. The idea of the batch.txt file is to treat a pipette as a core plug with porosity 1.0 and a very high permeability (e.g., 1,000,000 darcies). When we inject many pore volumes of water, oil, and surfactant whose compositions are the same as those in the pipette test, the flow becomes a steady-state. In such a steady-state flow, the component concentrations from the simulation should be the same as their respective volumetric fractions in the pipette test.

The general procedures to run batch simulation for matching experimental data follow:

1. Set up a base model like batch.txt for surfactant flooding with injection compositions (water/oil ratio, surfactant concentration, and so on) being the same as the phase behavior tests. The initial lower and upper salinities, C_{sel} and C_{seu}, may be the same as those in the pipette tests.
2. In principle, for each pipette test (each salinity), we need to input seven parameters: C_{33max0}, C_{33max1}, C_{33max2}, C_{2PL}, C_{2PR}, C_{sel}, and C_{seu}.
3. Run a batch simulation for many injection pore volumes. Check whether the solubilization ratios match the experimental data. The solubilization ratios can be calculated from the concentrations in the microemulsion phase (.COMP_ME). If matched, the input parameters are correct, and these input parameters can be used in other simulation studies. Otherwise, repeat steps 2 and 3 with new values of those seven parameters. Sometimes, C_{sel}, and

C_{seu} must be fine-tuned because these values may not be exactly the same as read from the test tubes.

4. Repeat steps 2 and 3 for other salinities. The matched seven parameters obtained by matching experimental data at different salinities may be different; their final values have to be compromised. For more details, see Example 7.2.

Example 7.2 Run a Batch Simulation to Obtain Surfactant Phase Behavior Parameters

The experimental data for this example are those shown earlier in Table 7.2, and the solubilization data ($V_o/V_s = C_{23}/C_{33}$ and $V_w/V_s = C_{13}/C_{33}$) are shown in Figure 7.10. In the phase behavior tests, the surfactant concentration is 1 wt.% that is treated approximately as vol.%. The water/oil ratio is 1. Find the surfactant phase behavior parameters required in simulation: C_{33max0}, C_{33max1}, C_{33max2}, C_{2PL}, C_{2PR}, C_{sel}, and C_{seu}.

Solution

We first have to set up a simulation model; we can modify the UTCHEM sample file batch.txt. Table 7.4 lists key parameters in a phase behavior simulation model and provides some comments to help set up the model. These comments should be helpful even if other simulators are used or a model is built from scratch.

We start by matching the solubilization ratios at the optimum salinity. From Figure 7.10, the optimum salinity is 0.365 meq/mL. From the test data in Table 7.2, C_{sel} and C_{seu} (the salinities for the microemulsion phase to appear and disappear) are around 0.31 meq/mL and 0.42 meq/mL, respectively. Note that in UTCHEM, the optimum salinity C_{seop} is equal to $(C_{sel} + C_{seu})/2$. Although it may not be generally true, we have to adjust these salinities to satisfy this condition. At the optimum salinity, $C_{23op} = C_{2M} = \frac{1}{2}(1 - C_{33}) \cong \frac{1}{2}$. Then the C_{33} concentration at the optimum salinity, which is C_{33max1}, can be approximately estimated from Eq. 7.60 because C_{23op} is approximately equal to 1/2. In the equation, $(C_{23}/C_{33})_{op}$ is the solubilization ratio at the optimum salinity:

$$C_{33max1} \approx \frac{C_{33max1}}{2C_{23op}} = \frac{1}{2(C_{23op}/C_{33})_{op}}. \tag{7.60}$$

Note that the parameters required in UTCHEM, C_{33max0} and C_{33max2}, are at zero salinity and twice optimum salinity, respectively, not at C_{sel} and C_{seu}.

Using Eq. 7.60, we can estimate an initial C_{33max1} that is 1/[(2)(16.8)] = 0.03. Here, 16.8 is the solubilization ratio at the optimum salinity. Interestingly, if we use Eq. 7.15, we have $C_{33max1} = C_{31}(WOR)/(1+WOR)/S_3 = 0.01(1)/(1+1)/0.171 = 0.029$, very close to what is estimated using Eq. 7.60. In this case, $S_3 = 0.171$ at 0.365 meq/mL from Table 7.2, WOR =1, and $C_{31} = 1\% = 0.01$. Here, $C_3 = C_{31}(WOR)/(1+WOR)$. For now, we can assign initial $C_{33max0} = C_{33max2} = 0.06$, which should not affect matching the optimum salinity point too much. In addition, we assume $C_{2PL} = 0.0$ and $C_{2PR} = 1.0$. Because we are attempting to match the solubilization ratios at the optimum salinity, the input parameters—initial and injected salinities—in the simulation model are the same as the optimum salinity.

TABLE 7.4 Key Parameters in a Phase Behavior Simulation Model

Parameters	UTCHEM Parameter	Parameter Value	Comments
Grid blocks	NX, NY, NZ	5, 1, 1	No change needed
Grid block size	DX1, DY1, DZ1	1, 1, 1 (ft)	No change needed
Components	W, O, S, P, Cl, Ca, alcohol		Minimum 5 components
Flag to output the profile of KCth component	IPRFLG(KC)	1 for W, O, S, Cl	
Flag to output component concentrations	ICKL	1	
Flag to output effective salinity	ICSE	1	
Total injection period	TMAX	120 days	Adjustable until steady flow
Rock compressibility	COMPR	0	
Porosity	PORC1	1	Must be 1
Permeability	PERMXC, PERMYC, PERMZC	1000000 md	Large value
Initial water saturation	SWI	1	Not critical, better 1
Initial pressure	PRESS1	1 psia	Not critical
Initial salinities and harness	C50, C60	meq/mL	**Same as those of each test tube**
Oil concentration at left plait point	C2PLC	0	
Oil concentration at right plait point	C2PRC	1	
Flag to input binodal curve	IFGHBN	0	Must be 0
Slope of C_{33maxi} versus alcohol 1	HBNS70, HNBS71, HBNS72	0	0 unless have test data to match
C_{33max0}, C_{33max1}, C_{33max2} for alcohol 1	HBNC70, HNBC71, HBNC72	volume fraction	**Tuning parameters**
Slope of C_{33maxi} versus alcohol 2	HBNS80, HNBS81, HBNS82	0	0 unless have test data to match

Continued

Example 7.2 Run a Batch Simulation to Obtain Surfactant Phase Behavior Parameters—*Continued*

TABLE 7.4 Key Parameters in a Phase Behavior Simulation Model—*Continued*

Parameters	UTCHEM Parameter	Parameter Value	Comments
C_{33max0}, C_{33max1}, C_{33max2} for alcohol 2	HBNC80, HNBC81, HBNC82	volume fraction	0 unless have test data to match
C_{sel}, C_{seu}	CSEL7, CSEU7		**Tuning parameters**
Flag for capillary number dependency	ITRAP	0	Not considered
Residual saturations	S1RWC, S2RWC, S3RWC	0	Must be 0
Endpoint relative permeabilities	P1RWC, P2RWC, P3RWC	1	Must be 1
Relative permeability exponents	E1WC, E2WC, E3WC	1	Must be 1
Densities	DEN1, DEN2, ...	0.433 psi/ft	
Capillary pressure	CPC0, EPC0	0	Must be 0
Viscosities	VIS1, VIS2	1 cP	VIS1 = VIS2
Diffusion	D(KC,i), i =1–3	0	Not considered
Dispersivity	ALPHAL(i), i = 1–3	0	Not considered
Flag for adsorption	IADSO	0	Not considered
Flag for constant potential boundaries	IBOUND	0	No boundary specified
Flag for aquifer	IZONE	0	No aquifer
Number of wells	NWELL	2	One injector and one producer
Flag to specify rate or pressure constraint	ICHEK	0	No check
Production pressure	PWF(M)	1 psia	
Flag for rate or pressure constraint	IFLAG	1 for injection rate constraint; 2 for pressure constraint for producer	
Injection rate	QI(M,L)	0.05 ft^3/day or arbitrary; QI(M,1)/QI(M,2) must equal WOR	
Injection fluid concentrations	C(M,KC,L)	Units depend on each component; same values as for each test tube	

Example 7.2 *Continued*

In this way, we can maintain exactly the same salinity as in the pipette test (optimum salinity now), although the initial salinity could be arbitrary because it will be displaced by a large volume of injected solution, and eventually it will be replaced by the injected salinity. The injected salinity must be the same as that in the pipette test. We should always check the resulting effective salinity (in the .SALT file) to confirm that the effective salinity has not been changed after the simulation is completed. The other phase behavior parameters can be left the same as the default numbers in the batch.txt because they may not affect the results at the optimum salinity. Note that the input salinity in the injection solution, C(M,KC,L), is the salinity in the injected aqueous phase, C_{51}, which is *not effective* salinity. However, in .SALT, the output is the effective salinity, which is defined as

$$C_{se} = \frac{C_{51} + (\beta_6 - 1)C_{61}}{C_{11}}. \quad (7.61)$$

The input parameters—C50 (initial brine salinity), C60 (initial brine divalents), CSEL7 and CSEU7 (C_{sel} and C_{seu} when alcohol and divalents are 0) in UTCHEM input—are *effective* salinities in meq/mL water.

If we input C_{33max1} = 0.03, C_{33max0} = C_{33max2} = 0.06, and injected salinity C(M,KC,L) = $C_{11} \times C_{se}$ = 0.99 × 0.365 = 0.3614 meq/mL solution (not water), the effective salinity in .SALT is then exactly equal to 0.365 meq/mL water. Here, C_{11} = 1 − C_{31} = 1 − 0.01 = 0.99 because the surfactant concentration is 1%. In C(M,KC,L), M denotes the well number, which is 1 for the injector in this simulation model; KC denotes the component number, which is 5 for anion; and L denotes the phase number, which is 1 for the injected aqueous phase. The solubilization ratios C_{23}/C_{33} and C_{13}/C_{33} from the simulation are the same—16.2. This solubilization ratio is lower than the experimental data—16.8. To improve this ratio, we reduce C_{33max1} to 0.03 × 16.2/16.8 = 0.0289 and keep the other parameters unchanged. Then we have the solubilization ratios equal to 16.8. Thus, we have matched the point at the optimum salinity.

Now we try to match a point in the Winsor type I region. We pick the salinity 0.141 meq/mL at which the test solubilization ratio, C_{23}/C_{33}, is 2.8. By simply changing the salinity to 0.141 meq/mL, we get a solubilization ratio of 0.88, which is too small. Remember that C_{33max0} must be higher than C_{33max1}. Therefore, we may use the following criteria to progressively search for a suitable value of C_{33max0}:

$$C_{33max0}^{n+1} = \frac{C_{33max1} + C_{33max0}^{n}}{2}. \quad (7.62)$$

Here, the superscript n and n + 1 represent the previous and current trials, respectively. Based on this approach, we find C_{33max0} = C_{33max2} = 0.03, at which the solubilization ratio is 2.7. This value is close to the test value of 2.8, so we can leave it for the moment and move to a point in the Winsor type II region.

Example 7.2 Run a Batch Simulation to Obtain Surfactant Phase Behavior Parameters—*Continued*

We next pick the salinity 0.539 meq/mL at which the experimental solubilization ratio C_{13}/C_{33} is 4.7. We set $C_{33max2} = C_{33max0} = 0.03$. From the simulation, we get the solubilization ratio C_{13}/C_{33} of 2.7, which is too low. Then we use $C_{33max0} = C_{33max2} = C_{33max1} = 0.0289$. The solubilization ratio becomes 2.82, which is still too low. Because C_{33max2} or C_{33max0} must be greater than C_{33max1}, we cannot reduce them any more. What we can do next is change C_{sel} and C_{seu}. By changing the range of the type III region to 0.21 to 0.52, the solubilization ratio is still 2.82. Then we change C_{33max0} and C_{33max2} to 0.03 and 0.05, respectively. We have the solubilization ratio C_{13}/C_{33} of 1.4. It seems as though we cannot easily match this point.

We therefore ignore that point and choose another salinity, 0.489 meq/mL, at which the experimental solubilization ratio, C_{13}/C_{33}, is 5.5. We go back to use $C_{33max2} = C_{33max0} = 0.03$, $C_{sel} = 0.31$ meq/mL, and $C_{seu} = 0.42$ meq/mL. The simulated solubilization ratio becomes 2.75. Then we change C_{sel} and C_{seu} to 0.21 meq/mL and 0.52 meq/mL, respectively. Consequently, the simulation solubilization ratio becomes 5.3, which is close to the experimental value of 5.5. It seems as though C_{sel} and C_{seu} are very sensitive parameters in this example.

Table 7.5 summarizes the fitting parameters we have obtained. These parameters are used to calculate solubilization ratios using UTCHEM. The calculated ratios (in curves) are compared with the experimental ratios (in points) in Figure 7.10.

TABLE 7.5 Fitted Phase Behavior Parameters

Phase Behavior Parameter	Symbol in Text	Symbol in UTCHEM	Value
Lower salinity limit, meq/mL water	C_{sel}	CSEL	0.21
Upper salinity limit, meq/mL water	C_{seu}	CSEU	0.52
Input parameter defined in Eq. 7.14 at zero salinity	C_{33max0}	HBNC70	0.03
Maximum height of binodal curve at optimum salinity	C_{33max1}	HBNC71	0.0289
Input parameter defined in Eq. 7.14 at twice optimum salinity	C_{33max2}	HBNC72	0.03
Oil concentration at the left plait point, volume fraction	C_{2PR}	C2PLC	0.0
Oil concentration at the right plait point, volume fraction	C_{2PL}	C2PRC	1.0

7.5.4 Effect of Cosolvent (Alcohol) on Phase Behavior

In general, surfactant is more effective (higher solubilization) without an added cosolvent such as alcohol, because cosolvent or cosurfactant is such a chemical that its molecules exhibit a substantial presence within the interfacial layers (Bourrel and Schechter, 1988). However, cosolvents are almost always added (Gary A. Pope, personal communication on July 30, 2008) to surfactant formulations to minimize the occurrence of gels, liquid crystals, emulsions or polymer-rich phase separating from the surfactant solution, to lower the equilibration time, and/or to reduce microemulsion viscosity. Usually, the ratio of surfactant to cosolvent is about 2 to 3.

For micellar flooding of a high surfactant concentration, the polymer and surfactant are sometimes incompatible. Micellar solutions develop viscosity due to structuring of the micelles that often requires the addition of cosurfactants and/or alcohols (Wyatt et al., 2008). Alcohol has another function: it can stabilize a microemulsion. When a microemulsion is generated using a surfactant without an alcohol, the micelles have unlimited solubilization capability. Then it is possible for the microemulsion type to be reversed due to the expansion of the inner phase. With the presence of alcohol, the microemulsion can remain the desired type, and the inner phase cannot expand without control. A middle-phase microemulsion could exist only at proper concentrations. Sometimes, alcohol can assist surfactant to have a low IFT system by adjusting the surfactant HLB. However, generally, when alcohol is added, although system compatibility may be improved, the IFT becomes higher, as shown in Figure 7.19.

Hirasaki et al. (2008) demonstrated an alternative to the use of alcohol by blending two dissimilar surfactants: a branched alkoxylated sulfate and a double-tailed, internal olefin sulfonate. The presence of cosolvent affects the effective salinity and causes a shift in phase boundaries. Alcohol is an organic compound with a functional group of –OH. In aqueous solutions, the hydrogen can become detached, producing slightly acidic solutions. Alcohols with short

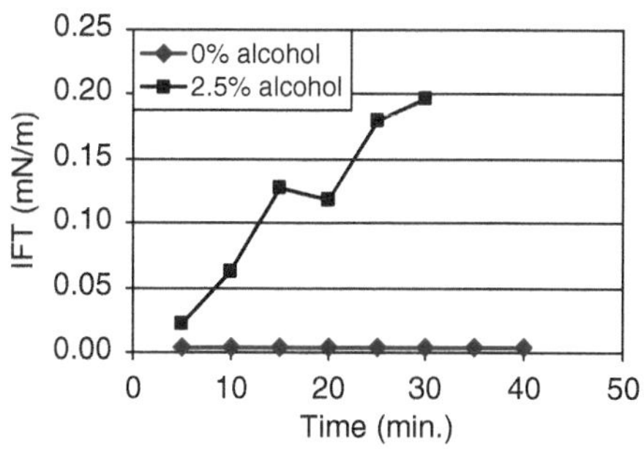

FIGURE 7.19 Alcohol effect on ASP/crude oil IFT. *Source: Kang (2001).*

chains such as propanol increase optimal salinity for sulfonate surfactants, whereas longer-chain alcohols such as pentanol and hexanol decrease optimal salinity. For petroleum sulfonates and synthetic alkyl/aryl sulfonates with light crude oils, it has been found that 2-butanol (SBA) acts as a cosolvent but has less effect on optimal salinity than other alcohols.

A portion of alcohol is also involved in the interfacial structure of sulfonate-rich micellar phases. For example, the addition of isopropanol increases the solubility of sulfonate in the aqueous phase relative to the oil phase (Baviere et al., 1981). However, alcohol itself with short tails like IPA, having only three carbons, cannot form micelles. The length of carbon tail should be about 8, 10, or more. Plus, the OH group in alcohol is not polar enough to act as a good hydrophilic group (Larry Britton, University of Texas at Austin, personal communication in 2008).

Hirasaki (1982a) used Gibbs' phase rule to show that for a mixture of four pure components—oil, surfactant, water, and NaC1—a unique value of optimum salinity exists. He then concluded that when alcohol and divalent ions are added to the four pure components, the optimum salinity must be a function of at least two additional intensive variables. He chose these two to describe the optimum salinity: f_7^s, the fraction of alcohol associated with the surfactant–plus–alcohol pseudocomponent, and f_6^s, the fraction of total divalent cations (calcium) bounded to surfactant micelles. He used the following empirical relation to define the optimum salinity:

$$C_{51,op} = C^*_{51,op}\left(1+\beta_7 f_7^s\right)\left(1-\beta_6 f_6^s\right). \tag{7.63}$$

Here, β_6 and β_7 are constants for a particular formulation, and $C^*_{51,op}$ is the optimum anion concentration in the absence of cosolvent (alcohol) or divalents. Camilleri et al. (1987) extended Eq. 7.63 to the entire salinity range:

$$C_{51} = C^*_{51}\left(1+\beta_7 f_7^s\right)\left(1-\beta_6 f_6^s\right). \tag{7.64}$$

The effective salinity in anion concentration in the presence of alcohol and divalents is defined as:

$$C_{se} = \frac{C_{51}}{\left(1+\beta_7 f_7^s\right)\left(1-\beta_6 f_6^s\right)}. \tag{7.65}$$

Be aware that the effect of alcohol and divalents on the optimum salinity and the effect on the effective salinity are opposite, as shown by Eqs. 7.63 and 7.65. When a divalent exists in the system, the optimum salinity in terms of monovalent concentration (C_{51}) of the system should be lower than that had the system not had the divalent. However, when a divalent does exist in the system, because of the divalent contribution to the salinity effect, the effective salinity will become higher than the salinity of C_{51}. The alcohol effect (contribution) is opposite to the divalent effect.

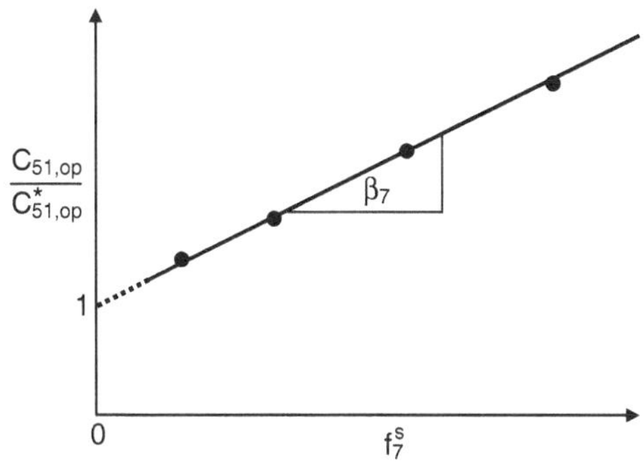

FIGURE 7.20 Schematic to determine β_7.

Experimental data suggest that the optimum salinity varies linearly with the cosolvent concentration. Therefore, β_7 can be estimated from the slope of the straight line of normalized optimum salinity ($C_{51,op}/C^*_{51,op}$) versus f^s_7 in the case without divalent cations, as schematically shown in Figure 7.20. To obtain the effect of cosolvent on the shift in optimum salinity, β_7, we need to measure the volume fraction diagram for at least two different cosolvent concentrations and must know $C^*_{51,op}$. According to the definition, f^s_7 is defined as $V_7/(V_7 + V_3)$.

When we calculate phase behavior using UTCHEM, we need to input C_{33maxm} at three salinities (m = 0, 1, 2). The following linear relationship between the C_{33maxm} and f^s_k is assumed for the case in which one cosolvent exists:

$$C_{33max,km} = m_{km} f^s_k + C_{33maxm} \quad \text{for} \quad m = 0, 1, 2; k = 7. \tag{7.66}$$

Here, m = 0 means at the zero salinity (practically very small salinity), 1 at the optimum salinity, and 2 at two times the optimum salinity. m_{km} is the slope for the $C_{33max,km}$ versus f^s_7 at m times the optimum salinity for the cosolvent k = 7; C_{33maxm} is the intercept at zero fraction of cosolvent at m times the optimum salinity, as shown in Figure 7.21. Here, we need to define six parameters: m_{km} and C_{33maxm} for m = 0, 1, 2. Theoretically, we need to conduct tests without cosolvent to obtain C_{33maxm} and conduct tests with cosolvent to obtain m_{km}. Practically, these six parameters are obtained by matching the volume fraction diagrams corresponding to at least three different total chemical (alcohol + surfactant) compositions.

For the first iteration, the slope parameters (m_{km}) are set to 0, and the intercept parameters (C_{33maxm}) are adjusted to obtain a reasonable match of the volume fraction diagrams; then the slope parameters are obtained. After obtaining the slope parameters, we repeat the matching procedure for further improvements (UTCHEM-9.0, 2000). For a two-cosolvent case, UTCHEM requires input of 12 parameters: m_{km} and C_{33maxm} for m = 0, 1, 2, and k = 7, 8. We

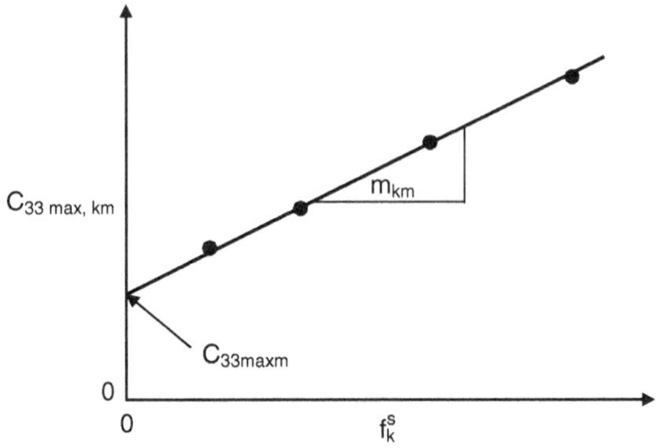

FIGURE 7.21 Schematic to determine C_{33maxm} and m_{km}.

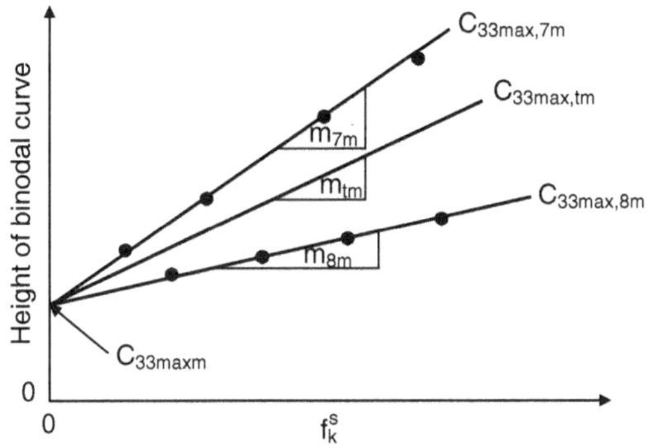

FIGURE 7.22 Concept to determine C_{33maxm} and m_{km} for a two-cosolvent case.

actually need to input only 9 parameters because C_{33maxm} for k = 7 must be the same as that for k = 8.

The following procedures are suggested to determine those 9 parameters. The concept is schematically shown in Figure 7.22.

1. Assume Eq. 7.67 holds for k = 7 and 8. For the first iteration, set the slope parameters (m_{tm}) to 0 and adjust the intercept parameters (C_{33maxm}) to obtain a reasonable match of the volume fraction diagrams.
2. Obtain the slope parameters (m_{km}) by matching the volume fraction diagrams based on Eq. 7.68, where $C_{33max,tm}$ is affected by the two cosolvents, 7 and 8.
3. Fine-tune the nine parameters obtained for the improved matching of volume fraction diagrams.

$$C_{33max,tm} = m_{tm}\left(f_7^s + f_8^s\right) + C_{33max\,m} \quad \text{for} \quad m = 0, 1, 2;\ k = 7 \text{ and } 8. \qquad (7.67)$$

$$C_{33\max,tm} = C_{33\max,8m} + (C_{33\max,7m} - C_{33\max,8m}) \frac{f_7^s}{f_7^s + f_8^s} \\ = \left(m_{7m}\left(f_7^s\right)^2 + m_{8m}\left(f_8^s\right)^2\right) \Big/ \left(f_7^s + f_8^s\right) + C_{33\max m}. \quad (7.68)$$

As we can see from the preceding discussion, including the cosolvent (alcohol) effect requires not only more experimental work, but also simulation work to find the parameters to describe the effect. In most practical cases, we just add the minimum amount of cosolvent, or alcohol (generally less than the amount of the surfactant used), when we select chemicals for a project. Alcohol adsorption is thought to be less than surfactant (Camilleri, 1983). Trushenski et al. (1974) found that the adsorption loss of the isopropyl alcohol cosurfactant is negligible. Alcohol can work as a tracer. Thus, the chromatographic separation between surfactant and alcohol makes it more complex to include the alcohol effect in phase behavior calculation.

The phase behavior, however, could be extremely sensitive to alcohol concentration and not to surfactant concentration (Salager et al., 1979b). At a specific concentration where the partition coefficient is 1, the alcohol partition does not depend on the water/oil ratio of the system. If we choose such a concentration, it is easy to optimize the chemical formula for an application (Eisenzimmer and Desmarquest, 1981). However, sometimes, such a concentration may be higher than necessary.

7.5.5 Two-Phase Approximation of Phase Behavior without Type III Environment

It has been observed that at low surfactant concentrations, the oil/brine/surfactant/alcohol systems form two phases, whereas often at high surfactant concentrations, middle-phase microemulsions form in equilibrium with excess oil and brine. For example, in a petroleum sulfonate/isobutanol/dodecane/brine system, in the low surfactant concentration range (0.1–0.2%), it is a two-phase system, but in the higher concentration range (4–10%), it becomes a three-phase system (Chan and Shah, 1981). Some researchers stated that a minimum of 1% surfactant is needed to have a Winsor type III microemulsion (Kang, 2001). Others have stated that a higher surfactant concentration (e.g., 4–10%) is needed. When the surfactant concentration is low, no evidence of an intermediate phase could be found, and the amount of surfactant in the excess phases would become a more significant fraction of the total (Nelson, 1981). A model described as an adsorbed surfactant monolayer separating the two equilibrium phases (oil and aqueous phase) has been used to account for all the interfacial tension minimum phenomena (Chan and Shah, 1979).

Sometimes (especially in the past), surfactant flooding using low concentrations is called dilute surfactant flooding or simply surfactant flooding, whereas the surfactant flooding using high concentrations is called micellar or microemulsion flooding. In low-concentration surfactant flooding, we use the two-phase model (oil and aqueous phases) for phase behavior; in high-concentration micellar flooding, we use the three-phase phase behavior model instead. These days, we generally use the same term, surfactant flooding, to refer to both low-concentration surfactant flooding and high-concentration micellar flooding. Apparently, in dilute surfactant flooding applications, the surfactant concentration is around 0.1% (Michels et al., 1996; Babadagli et al., 2002). Wyatt et al. (2008) classified the surfactant-related processes into several groups according to the surfactant concentration: 0.1 to 2% in surfactant-polymer (SP), 2 to 12% in micellar-polymer (MP), and 0.05 to 0.5% in alkaline-surfactant-polymer (ASP).

Many of the field projects conducted in the 1970s and 1980s were MP projects in spite of the complexity of designing these solutions compared with the relative ease of SP formulations. The reason is partly that adsorption by reservoir rock strips the surfactant from low-concentration SP formulations, rendering them ineffective as the solution surfactant concentration decreases while advancing through the reservoir. In ASP projects, because alkaline agents can reduce the surfactant adsorption, lower concentrations are used. Therefore, the trend is to use low-concentration surfactant flooding. Thus, a simple two-phase model is needed. From the discussion of the phase behavior model in the previous section, we can see that microemulsion phase behavior in a type III environment is complex and difficult to quantify. Therefore, researchers tried to simplify the quantification using a two-phase model without a type III environment (Adibhatla et al., 2005; Liu et al., 2008).

In Liu's two-phase model, a partition coefficient was used to describe the allocation of surfactant between the aqueous and oleic phases. The partition coefficient of the surfactant component 3 is defined as

$$K_s = \frac{C_{32}}{C_{31}}, \tag{7.69}$$

where C_{31} is the concentration of surfactant component 3 in the aqueous phase 1, and C_{32} is the concentration of surfactant component 3 in the oleic phase 2. Liu et al. (2008) assumed that K_s is unity at the optimum salinity. Above the optimum salinity, most of the surfactant is in the oleic phase, so the partition coefficient is much larger than unity. Below the optimum salinity, the partition coefficient is much smaller than unity. The partition coefficient between the aqueous and oleic phases for surfactant is calculated using the empirical equation

$$K_s = \begin{cases} 10^{2(C_{se}/C_{seop}-1)}, & C_{se} > C_{seop} \\ 10^{2(1-C_{seop}/C_{se})}, & C_{se} < C_{seop} \end{cases}, \tag{7.70}$$

where C_{se} and C_{seop} are effective salinity and optimum salinity, respectively.

Equation 7.69 needs further discussion, however. As we know, the allocation of surfactant in aqueous and oleic phases depends mainly on the salinity (type of microemulsion), not water/oil ratio. For example, in a type II(−) environment, almost all the surfactant is in the aqueous phase, regardless of WOR. Let us assume the surfactant concentration in the aqueous phase is C_{31}, and the surfactant concentration in the oleic phase (C_{32}) is K_sC_{31} according to Eq. 7.69. Then the amount of surfactant in the aqueous phase is $C_{31}S_1 = C_{31}(WOR/(1+WOR)$, and the amount of surfactant in the oleic phase is $K_sC_{31}S_2 = K_sC_{31}/(1+WOR)$. Here, S_1 and S_2 are the aqueous and oleic phase saturations, respectively. The total amount of surfactant in the system, C_3, is

$$C_3 = C_{31}S_1 + C_{32}S_2 = C_{31}\frac{WOR}{1+WOR} + K_sC_{31}\frac{1}{1+WOR}. \quad (7.71)$$

From the preceding equation, we have

$$C_{31} = \frac{C_3(1+WOR)}{WOR+K_s}. \quad (7.72)$$

Then the total amount of surfactant in the aqueous phase, V_{31}, is

$$V_{31} = C_{31}S_1 = \frac{C_3(WOR)}{WOR+K_s}. \quad (7.73)$$

Equation 7.73 shows that the surfactant allocation in the aqueous phase depends on WOR, which is not consistent with the fact or the assumption. We propose that K_s should be modified as

$$K_s = \frac{V_{32}}{V_{31}} = \frac{C_{32}S_2}{C_{31}S_1}. \quad (7.74)$$

By the preceding definition, K_s will be independent of WOR and depend only on salinity, as we would expect. According to Eq. 7.74, when S_1 (i.e., WOR) increases, C_{31}, the surfactant concentration in the water phase, will decrease so that the total amount of surfactant in the water phase, V_{31}, is independent of WOR but dependent on the salinity. At the optimum salinity, $V_{32} = V_{31}$, the surfactant has equal partition in the aqueous and oleic phases. However, Eq. 7.74 has not been tested using experimental data.

The argument to use a two-phase model to represent surfactant phase behavior without type III microemulsion is that experiments (Seethepalli et al., 2004; Zhang et al., 2006; Liu et al., 2008) indicate that the volume of type III microemulsion phase is small if the overall surfactant concentration is low (<0.1 wt.%). In the cases of low surfactant concentration, a type III microemulsion system was not observed by Salager et al. (1979b). The reason is that if we cannot make a sufficient number of salinity scans, and the volume of the type III microemulsion phase is small, the equilibrium phase behavior *appears* to go from a lower-phase microemulsion to an upper-phase microemulsion over a "masked" narrow salinity range. Actually, even for low-concentration surfactant solutions, a Winsor type III microemulsion could also exist when the salinity is in the type III range. Only the solubilized oil and water volumes in type III are so small that the middle-phase microemulsion cannot be clearly seen.

Liao (1998) used thin glass tubes to confirm the existence of a type III microemulsion at low surfactant concentrations. Whether a type III microemulsion can exist in low surfactant concentrations is a confusing topic in the literature. In the case of a very low surfactant concentration, the author believes that it is better to have a type I microemulsion because the small volume of a type III microemulsion with highly concentrated surfactant could be easily bypassed (trapped).

In the cases in which surfactant concentrations are low, the actual salinity range for type III most likely would be wider than we measure in the salinity scans. Thus, IFT measurements for salinity scans for low surfactant concentration are usually between the upper and lower phases observed in the sample tubes. The phases may be (1) lower-phase microemulsion and excess oil [type II(−)], (2) excess brine and upper-phase microemulsion [type II (+)], or (3) excess brine and excess oil (type III).

Pope and Nelson (1978) pointed out that representing the complex phase behavior of surfactant/brine/oil systems in a practical way is the biggest single problem involved in a completely compositional chemical flood simulator. The surfactant cannot be considered a component that partitions between phases and lowers the oil/water interfacial tension without changing the volume of the water-rich (aqueous) or oil-rich (oleic) phase. During a chemical flood, the distribution of chemicals between the phases is not even approximately constant and therefore is not represented conveniently by conventional partition coefficients. Note that K_s in Eq. 7.70 changes with salinity. Furthermore, as discussed by Nelson and Pope (1978), three-phase behavior often is a key determinant in the performance of a chemical flood and must be accounted for in any general compositional model. Whether the approximation of a two-phase model is sufficiently accurate in practical cases needs to be investigated. A two-phase model significantly simplifies the quantification of surfactant phase behavior, though.

In quantification of phase behavior, we generally assume that all the surfactant is in the microemulsion phase. In other words, the amount of surfactant in the excess oil or water phase is negligible. Actually, surfactants such as petroleum sulfonate are amphiphilic; they are partitioned in oil and water phases. For example, the petroleum sulfonate GTSP is partitioned in oil and water phases, as shown in Figure 7.23. This figure shows that as the surfactant concentration is increased, the partition coefficient becomes lower, indicating less surfactant is partitioned in the oil phase.

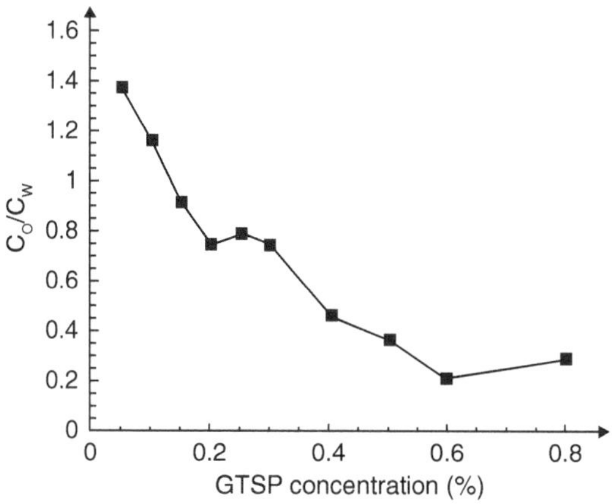

FIGURE 7.23 GTSP partition in a water/oil system. *Source: Yang et al. (2002a).*

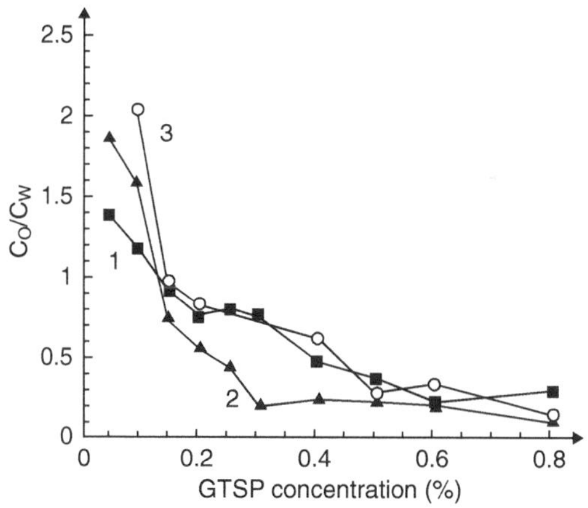

FIGURE 7.24 GTSP partition in an alkali/oil system: 1, GTSP; 2, GTSP + 0.75% MA (alkai); and 3, GTSP + 1.5% MA. *Source: Yang et al. (2002a).*

Figure 7.24 shows the GTSP partition when alkali is added in water. In low GTSP concentrations, as alkali concentration is increased, GTSP has more partition in the oil phase. In high GTSP concentrations, the alkali concentration effect is not significant.

Figure 7.25 shows the GTSP partition when polymer is added. We can see that the addition of polymer reduces the GTSP partition in the oil phase, but the polymer concentration effect is not significant. In an ASP/oil system (see Figure 7.26), the GTSP partition coefficient is between an alkali/oil and a polymer/oil system and is lower than in a water/oil system.

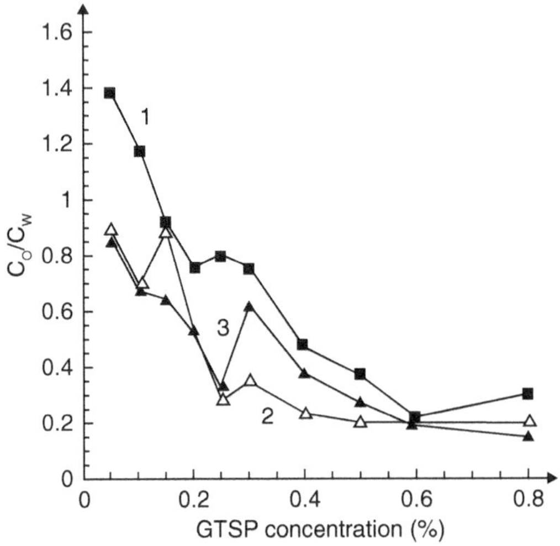

FIGURE 7.25 GTSP partition in a polymer/oil system: 1, GTSP; 2, GTSP + 0.06% 1275A (polymer); and 3, GTSP + 0.12% 1275A. *Source: Yang et al. (2002a).*

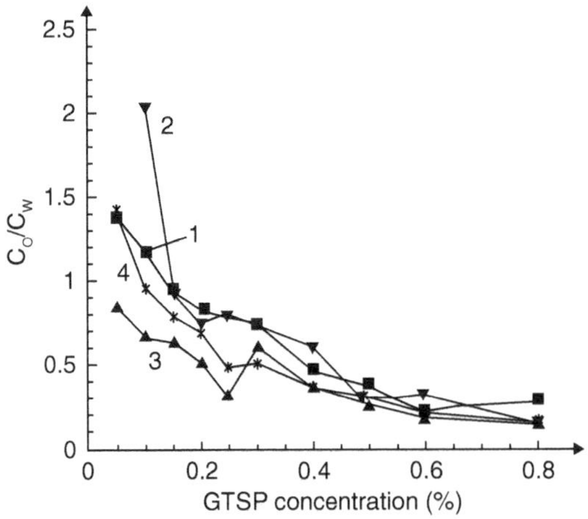

FIGURE 7.26 GTSP partition in an ASP/oil system: 1, GTSP; 2, GTSP + 1.5% MA; 3, GTSP + 0.12% 1275A; and 4, ASP system. *Source: Yang et al. (2002a).*

7.5.6 Quantitative Representation of Interfacial Tension

Healy et al. (1976) found that a large number of anionic surfactant systems exhibited good correlations between interfacial tension and solubilization parameter. Healy and Reed (1977a, 1977b) proposed the following correlation by fitting experimental data, as schematically shown in Figure 7.27:

$$\log \sigma_{p3} = H_{p2} + \frac{H_{p1}}{1 + H_{p3} R_{p3}}, \quad \text{for} \quad p = 1, 2. \tag{7.75}$$

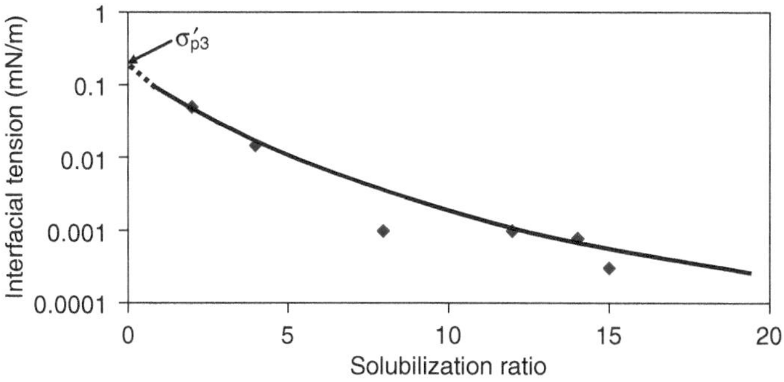

FIGURE 7.27 Schematic of a correlation fit between IFT and solubilization.

Here, H_{p1}, H_{p2}, and H_{p3} are the fitting parameters; R_{p3} is the solubilization ratio (C_{p3}/C_{33}) in the microemulsion phase numbered 3. p = 1 is for the microemulsion/water IFT (σ_{wm}); p = 2 is for the microemulsion/oil IFT (σ_{om}). H_{p2} is obtained from the intercept σ'_{p3}:

$$H_{p2} = \log\sigma'_{p3} - H_{p1}, \quad \text{for } p = 1, 2. \tag{7.76}$$

The parameters H_{p1}, H_{p2}, and H_{p3} for p = 1 should be similar to those for p = 2. Without experimental data to fit the correlation, it is difficult to choose the values of those parameters. From this point of view, the equations from Huh (1979) are preferred; they are introduced next.

Huh developed a theoretical relationship between the solubilization parameter and IFT for a middle-phase microemulsion (type III). His equations are

$$\sigma_{mw} = \frac{C_H \cos[(\pi/2)\varphi_w]}{(V_{wm}/V_{sm})^2} = \frac{C_{Hw}}{(V_{wm}/V_{sm})^2}, \tag{7.77}$$

$$\sigma_{mo} = \frac{C_H \cos[(\pi/2)\varphi_o]}{(V_{om}/V_{sm})^2} = \frac{C_{Ho}}{(V_{om}/V_{sm})^2}, \tag{7.78}$$

where V_{sm} is the surfactant volume in the microemulsion phase; C_H is an empirical constant usually determined experimentally, mN/m; $\varphi_w = V_{wm}/(V_{wm} + V_{om})$ and $\varphi_o = V_{om}/(V_{wm} + V_{om})$; $C_{Hw} = C_H \cos[(\pi/2)\varphi_w]$ and $C_{Ho} = C_H \cos[(\pi/2)\varphi_o]$. In practice, we may ignore the cosine term and treat C_{Hw} equal to C_{Ho}. C_H ranges from 0.1 to 0.35, and a typical value may be 0.3 (UTCHEM-9.0, 2000). Therefore, even without experimental data, we can pick a typical value within the range. Other reported C_H values determined experimentally are (0.48 ± 0.05) mN/m for alkylbenzene sulfonates and (0.34 ± 0.06) mN/m for ethoxylated alkylphenols (Graciaa et al., 1982). Barakat et al. (1983) reported a value of (0.4 ± 0.15) mN/m for alkane and alpha-olefin sulfonates.

The Huh equations are much simpler than the Healy and Reed correlation. Note that a large number of papers published since 1979 have shown the Huh equation accurately models the IFT between equilibrium microemulsions and oil or water for numerous combinations of surfactants and crude oils over a wide range of concentration, salinity, temperature, and other conditions typical of oil reservoirs. Although IFT in the range of interest can be measured using the spinning drop method, using the Huh equation to calculate it from phase behavior data affords several significant advantages. The most significant advantage is that we do not have to physically measure IFT for each test tube during the initial screening process. Measuring IFT could be difficult for some oils and can be time consuming. Thus, simply observing the phase volumes in pipettes rather than measuring IFT for each pipette is really an advantage.

7.5.7 Factors Affecting Phase Behavior and IFT

One of the main mechanisms of surfactant-related EOR is reducing IFT, which is closely linked to water and oil solubilization (phase behavior). Several mechanisms for ultralow IFT have been proposed. Rosen (1978) proposed that for ultralow IFT to occur, not only the oil/water interface adsorbs surfactant molecules, but also a third phase exists. Chan and Shah (1981) proposed that surfactant concentrations in water and oil phases must reach the critical micelle concentration, and the partition coefficient must equal 1. Chen et al. (1999a) proposed that the ultralow IFT in an AS or ASP system is attributed to the synergy between surfactant and in situ generated soap.

Variables identified as important in the achievement of the low IFT in a W/O/S/electrolyte system are the surfactant average MW and MW distribution, surfactant molecular structure, surfactant concentration, electrolyte concentration and type, oil phase average MW and structure, temperature, and the age of the system. Salager et al. (1979b) classified the variables that affect surfactant phase behavior in three groups: (1) formulation variables: those factors related to the components of the system–surfactant structure, oil carbon number, salinity, and alcohol type and concentration; (2) external variables: temperature and pressure; (3) two-position variables: surfactant concentration and water/oil ratio. Some of the factors affecting IFT-related parameters are briefly discussed in this section. Some other factors, such as cosolvent, salinity, and divalent, are discussed in Section 7.4 on phase behavior. Healy et al. (1976) presented experimental results on the effects of a number of parameters.

Effect of Oil

Both oil compositions and surfactant constitutes are complex. Selecting a surfactant for an oil requires a lot of screening work. To make the screening work easier, Cayias et al. (1977) proposed the concept of equivalent alkane carbon number (EACN). EACN is the sum of the mole–fraction–weighted alkane carbon number (ACN) of each pure species, calculated from

$$(\text{EACN})_{\text{oil}} = \sum_{\text{i}} (\text{EACN})_{\text{i}} \text{X}_{\text{i}}, \tag{7.79}$$

where X_i is the mole fraction of the component i of the oil (Morgan et al., 1977). Values of EACN are not necessarily integers. As a practical matter, oils having characteristics similar to the lower straight chain hydrocarbons tend to form three-phase regions over a narrow salt concentration range and at a relatively low salt concentration, but the relative uptake of oil and brine per unit volume of surfactant into type III phase increases (Glinsmann, 1978).

Based on this concept, a crude oil/surfactant/brine system should have phase behavior (e.g., optimum salinity, IFT minima) similar to that of the pure alkane/surfactant/brine system whose ACN is the same as the crude EACN. However, the concept of EACN is not practically applicable for several reasons. First, all the hydrocarbon compositions of a crude oil are not readily identified. Thus, the EACN of a crude oil cannot be calculated directly using Eq. 7.79. Second, measurement of the EACN of a crude oil requires a series of surfactant solutions to be tested to obtain individual minimum IFT. Then these surfactant solutions are tested against increasing alkane carbon numbers to find minimum IFTs. The ACN at which a surfactant solution also gives the lowest IFT for the crude oil is the EACN of the oil. Finding it is not an easy task. Third, several parameters affect the value of the EACN. Variations in EACN with alcohol cosolvent type, total WOR of the sample, and crude oil composition have been observed (Tham and Lorenz, 1981). In practice, we always select surfactants by scan tests using the actual crude oil for a specific application.

Effect of Surfactant Structure, Surfactant Concentration, and WOR

Extensive research on surfactants has established a clear relationship between surfactant structure and fluid properties related to EOR performance (Bourrel and Schechter, 1988). For example, the salinity range in which a surfactant is interfacially most active is primarily a function of surfactant molecular weight (MW). As the surfactant MW increases, the optimum salinity for production of a low IFT decreases (Morgan et al., 1977). With increasing hydrophobe length, the solubilization ratio increases, and optimum salinity decreases. Weakly hydrophobic functional groups such as propylene oxide (PO) have been characterized as having interface affinity and, as such, increase the breadth of the ultralow IFT region. The addition of these hydrophobic groups lowers the optimum salinity and adds calcium tolerance, so the degree of propoxylation can be used to tailor the surfactant to a given crude oil, temperature, and salinity. Similar statements can be made with respect to the addition of ethylene oxide (EO) or both EO and PO groups to the surfactant. Fortunately, both EO and PO are relatively inexpensive chemicals. Therefore, they are among the most practical ways to tailor a surfactant to the desired conditions as well as to improve its performance (Levitt et al., 2006).

If the surfactant concentration is above the critical micelle concentration (CMC), the IFT stays constant. Below the CMC, the IFT changes extensively with the concentration of the surfactant. If the pseudocomponents represent the

TABLE 7.6 WOR Effect on Optimum Salinity

	Optimum Salinity (g/L)		
WOR	*System A*	*System B*	*System C*
0.27	41.5	18.5	10.6
1.10	31.5	19.5	9.4
5.23	29.0	20.1	7.7

system well, and the system has only sodium (no divalent), optimum salinity should be independent of overall surfactant concentration and WOR (Hirasaki, 1982a). For a more complex system, the optimum salinity may be dependent on overall surfactant concentration (Nelson, 1982). It has been observed that IFT is a function of surfactant concentration. It decreases with surfactant concentration, but after a certain concentration, IFT increases slightly with the concentration or stays steady. However, this is not universal, as discussed in Section 8.2.

When an alcohol is added, because alcohol has different partition coefficients in oil and water phases, generally optimum salinity will be affected by WOR. For example, Table 7.6 shows the WOR effect on optimum salinity (Baviere et al., 1981). Three systems were used:

- System A – octane, 1 wt.% C12OXSO3Na, and 3 wt.% isopropanol
- System B – octane, 1 wt.% C12OXSO3Na, and 3 wt.% 2-butanol (SBA)
- System C – octane, 1 wt.% C12OXSO3Na, 1 wt.% isopropanol, and 3 wt.% 2-butanol (SBA)

The results are mixed. For Systems A and C, the optimum salinity decreases with WOR, whereas for System B, the optimum salinity increases with WOR.

Effect of Divalents

IFT increases when anionic surfactants form nonactive complexes with divalents (Ca^{2+} and Mg^{2+}). However, Kang et al. (1998) found that adding a small amount of phosphorous complexing agent increased system interfacial activity and tolerance to divalents. Li et al. (1999a) found that natural mixed carboxylates have high interfacial activity and high resistance to divalents. Synthetic carboxylates, on the other hand, have relatively poor resistance to divalents.

Effect of Temperature

An increase in temperature causes water and oil solubilization to decrease at optimum salinity, IFT to increase (Healy et al., 1976; Austad et al., 1997), and the optimum salinity to shift to a higher value (Healy et al., 1976). However, no temperature dependence of the EACN of any crude oil has been observed (Morgan et al., 1977).

Effect of Pressure

As with liquid systems, the general effect of pressure on phase behavior is negligible (Nelson, 1983). In practice, dead oils are used in phase behavior tests. Therefore, pressure effect is not investigated, although the reservoir temperature is maintained in phase behavior tests. However, different pressure causes a different amount of gas dissolved in the oil. In such cases, the pressure would have some effect. High-pressure PVT cells are needed for such phase behavior tests.

Dynamic Behavior of IFT

Liu et al. (2004) observed that the IFTs of the equilibrium oil and water phases were much higher than the dynamic IFTs between the fresh oil and water phases. They explained that the low dynamic IFT was the result of a reaction of the alkali with the acid in the oil and the mass transfer of the surfactant across the oil/water interface.

After the reaction and mass transfer across the oil/water interface ended, the IFT of the equilibrium system rose to a higher value.

7.6 VISCOSITY OF MICROEMULSION

One important characteristic of microemulsion viscosity is that it is a strong function of phase compositions. In UTCHEM, liquid phase viscosities are modeled as a function of pure component viscosities and the phase concentrations of the organic, water, and surfactant:

$$\begin{aligned}\mu_j = {} & C_{1j}\mu_w \exp[\alpha_1(C_{2j}+C_{3j})] + C_{2j}\mu_o \exp[\alpha_2(C_{1j}+C_{3j})] + \\ & C_{3j}\alpha_3 \exp[\alpha_4 C_{1j} + \alpha_5 C_{2j}].\end{aligned} \tag{7.80}$$

Here, j = 1 for the aqueous phase, 2 for the oleic phase, and 3 for the microemulsion phase. The α parameters are determined by matching laboratory microemulsion viscosities at several compositions. In the absence of surfactant and polymer, aqueous and oleic phase viscosities reduce to pure water and oil viscosities (μ_w and μ_o), respectively. When polymer is present, μ_w is replaced by polymer viscosity (μ_p). Figure 7.28 shows an example of microemulsion viscosity expressed in the preceding equation, where α = (2, 3, 0, 0.9, 0.7), μ_w = 1 cP, and μ_o = 5 cP.

Note that when C_{23} approaches 0, the microemulsion approaches water viscosity (μ_w); when C_{23} approaches 1, the microemulsion approaches oil viscosity (μ_o). The microemulsion viscosity in the middle range of C_{23} is higher than its pure component viscosities (μ_w and μ_o); it cannot be linearly interpolated.

Surfactants are highly prone to forming viscous microemulsions, gels, complexes, and liquid crystals under different conditions. Trushenski et al. (1974) showed that microemulsion viscosity increases linearly with increased divalent

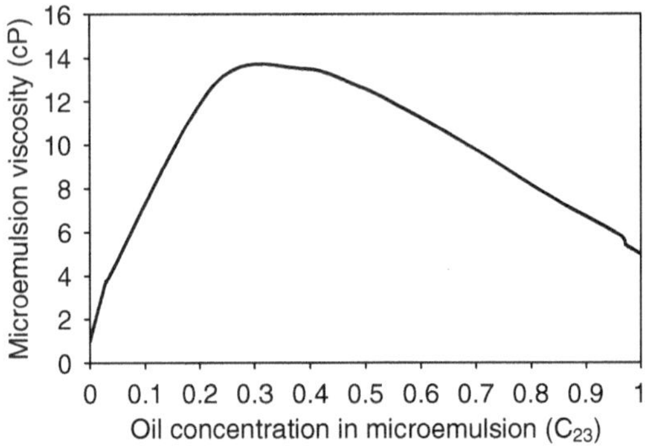

FIGURE 7.28 Microemulsion viscosity as a function of phase composition (C_{23}).

ion content. A very viscous microemulsion will be difficult to pump through perforation and may lead to an injectivity problem. A very high viscous microemulsion will also create local fluid heterogeneities that will be bypassed by subsequent chase fluid; thus, it correlates with high surfactant phase trapping. An oil bank is formed by the coalescence of displaced oil ganglia. The high interfacial viscosity would prevent the coalescence of the oil ganglia (drops). Therefore, when selecting a chemical formula, we must prevent a very high viscous microemulsion phase.

One approach is to add a cosolvent such as alcohol. An alternative to the use of alcohol is to blend two dissimilar surfactants (Hirasaki et al., 2008). During the surfactant screening process in the laboratory, we must make sure the formulation produces low viscosity microemulsions with fluid interfaces and little tendency to exhibit gels or macroemulsions. This can be observed qualitatively by tilting pipettes and observing the interface fluidity (the movement or lack of movement of the interface; Levitt et al., 2006). This is the only estimate of viscosity during the salinity scan period, and it does require some experience to make a judgment.

Salinas et al. (2009) presented a novel tool to measure microemulsion viscosity. It is a falling-sphere viscometer with multiple, ring-shaped, inductive proximity sensors. The device uses 0.78 mm, gold-coated, paramagnetic, 440 stainless steel spheres. With this size sphere, it is possible to accurately estimate the viscosity of fluids from 1 to 1,200 cp. The spheres are paramagnetic, so they can be lifted to the top of the tube with a magnet, and meet the constraints of a Reynolds number less than 800 for low viscosity fluids, and a velocity fast enough to be detected for a high viscosity fluid. A data acquisition system controls a set of four sensors for signal conditioning and time recording. When the same phase spans the space between a pair of sensors, its viscosity can be estimated directly from the transit time of the sphere. When two or three phases span the space between a pair of sensors, the viscosity of the undetermined phase can be estimated from the transit time across it after correcting for the transit times across the upper and/or lower phases. Thus, the viscosity of a middle-phase microemulsion can be estimated even if it spans only a small fraction of the distance between sensors.

A high viscous ASP system should help oil displacement. Then often this question is raised: why do we need a low interface viscosity (short equilibrium time)? A high viscous ASP system helps oil displacement because of more favorable mobility ratio. The oil displacement is "macroscopic." In a chemical flood process, the isolated oil drops need to coalescence to form an oil bank. A low interface viscosity between the oil drop and the displacing ASP fluid will ease such a coalescence process, and this process is "microscopic." Therefore, we need a low interfacial viscosity to build up an oil bank and a high viscous ASP fluid to displace the oil bank ahead.

For the emulsification mechanism to work, however, the interfacial films must be stable. Plus, higher interfacial viscosity increases the stability of emulsions and oil lamellae. Cooke et al. (1974) reported that their qualitative data indicated an increase in interfacial viscosity could increase oil recovery under certain conditions. Regardless, experiences tell us a low interfacial viscosity is needed for a higher recovery.

7.7 CAPILLARY NUMBER

This section discusses the definitions of capillary number and how to calculate capillary numbers.

7.7.1 Definitions: Which One to Use

The main objective of surfactant flooding is to reduce residual oil saturation, which is closely related to capillary number. Therefore, the concept of capillary number is discussed first. Analysis of the pore-doublet model yields the following dimensionless grouping of parameters (Moore and Slobod 1955), which is a ratio of the viscous-to-capillary force:

$$N_C = \frac{F_v}{F_c} = \frac{v\mu}{\sigma \cos\theta}. \tag{7.81}$$

Here, F_v and F_c are viscous and capillary forces, respectively; v is the pore flow velocity of the *displacing* fluid in their derivation; μ is the *displacing* fluid viscosity; and σ is the interfacial tension between the displacing and displaced phases. The dimensionless group is called a capillary number, N_C. A set of consistent units is used so that the dimensionless group is dimensionless. For example, v is in m/s, μ in mPa·s, and σ in mN/m or dyne/cm.

Numerous experiments have been conducted to correlate the capillary number with the residual saturation in porous media. Most of the data consist of measurements of residual saturations when a nonwetting phase (oil) is displaced by a wetting phase (water without surfactant). Fewer data exist for trapping of a wetting phase displaced by a nonwetting phase.

The capillary number defined by Eq. 7.81 is a semi-empirical parameter. Different forms have been used in the literature. Foster (1973) and Green and Willhite (1998) used interstitial velocity and omitted the cosθ term in Eq. 7.81:

$$N_C = \frac{u\mu}{\phi\sigma}. \tag{7.82}$$

In this equation, ϕ is the porosity in fraction, and u is the Darcy velocity of the displacing fluid. The velocity used by Abrams (1975) is $v/[\phi(S_{oi} - S_{or})]$. He also modified the capillary number by multiplying the viscosity ratio $(\mu_w/\mu_o)^{0.4}$:

$$N_C = \frac{u\mu_w}{\phi\sigma\cos\theta(S_{oi} - S_{or})}(\mu_w/\mu_o)^{0.4}. \tag{7.83}$$

Abrams actually assumed cosθ = 1 for his strongly water-wetting permeable media.

In Eq. 7.82, if the Darcy velocity (u) is used, then

$$N_C = \frac{u\mu}{\sigma}. \tag{7.84}$$

If the Darcy equation is used to describe the velocity and the cosθ term is omitted, Eq. 7.81 becomes (Brownell and Katz, 1947)

$$N_C = \frac{-k\Delta\Phi_p/\Delta L}{\sigma} = \frac{k|\vec{\nabla}\Phi_p|}{\sigma}, \tag{7.85}$$

where Φ_p is the potential of displacing fluid, and $\Delta\Phi_p/\Delta L$ is the potential gradient.

Dombrowski and Brownell (1954) added cosθ in the denominator of Eq. 7.85:

$$N_C = \frac{k|\vec{\nabla}\Phi|}{\sigma\cos\theta}. \tag{7.86}$$

Pennell et al. (1996) included relative permeability k_{rw} explicitly:

$$N_C = \frac{kk_{rw}|\vec{\nabla}\Phi|}{\sigma\cos\theta}. \tag{7.87}$$

The relative permeability behavior at reduced residual oil saturation was discussed by Morrow et al. (1983). Note that it could be difficult to include relative permeability in a simulator because this relative permeability is affected by the reduced residual saturation, which depends on capillary number.

As presented previously, different forms of capillary numbers have been introduced in the literature. In the following, we discuss which form would be less system specific.

When an oil blob is stopped at a pore neck, the pressure difference required to overcome the capillary pressure is

$$\Delta p = 2\sigma\cos\theta\left(\frac{1}{r_n} - \frac{1}{r_p}\right), \tag{7.88}$$

where r_n and r_p are the radii of pore neck and pore body, respectively; σ is the interfacial tension; and θ is the contact angle. Here, the wettability hysteresis is not included. The preceding equation can be rewritten as

$$\frac{\Delta p\bar{r}^2}{\sigma\cos\theta} = 2\bar{r}^2\left(\frac{1}{r_n} - \frac{1}{r_p}\right), \tag{7.89}$$

where $\bar{r}$ is the characteristic pore size radius. According to Eq. 5.20, we have $\bar{r}^2 = 8k/\phi$.

Then Eq. 7.89 can be rewritten as

$$\frac{k\Delta p/L}{\phi\sigma\cos\theta} = \frac{\bar{r}^2}{4L}\left(\frac{1}{r_n} - \frac{1}{r_p}\right), \tag{7.90}$$

where L is the system characteristic length. This equation shows that if the left side is defined as a capillary number, this capillary number is unique for geometrically similar systems. The definition also includes wettability. If the wettability is properly included, the hysteresis should be considered. For a simple definition, the following capillary number may be used:

$$N_C = \frac{k\Delta p/L}{\phi\sigma}. \tag{7.91}$$

Note that in this definition, the porosity term is included. If the velocity is used, the preceding equation becomes Eq. 7.82. It is expected that for a group of rocks with different porosities, if the porosities are included, the calculated capillary numbers should be closer to their average. However, the data that is shown in Table 7.7 do not consistently support this expectation. The ratio of the average to the standard deviation decreases for the data from Chatzis and Morrow (1984) if the porosity is included, but it increases for data from Taber et al. (1973). From these two data sets, it seems as though the capillary numbers that do and do not include porosity are equally good.

The difference between Eqs. 7.84 and 7.85 is that the Darcy velocity is used in Eq. 7.84, whereas the pressure gradient and permeability are used in Eq. 7.85. When core flood tests were run *at a constant velocity* (thus, a constant $v\mu/\sigma$), it appeared that the oil recovery increased for those rocks with lower permeabilities (Taber et al., 1973). Clearly, the higher recovery was not a

TABLE 7.7 Comparison of Calculated N_C with/without Including Porosity

	Chatzis and Morrow (1984)		Taber et al. (1973)	
	$k_a\Delta p/(L\sigma)$	$k_a\Delta p/(\phi L\sigma)$	$k_a\Delta p/(L\sigma)$	$k_a\Delta p/(\phi L\sigma)$
Average	3.78×10^{-5}	17.9×10^{-5}	3.40×10^{-5}	18.27×10^{-5}
Standard deviation	1.17×10^{-5}	4.84×10^{-5}	1.77×10^{-5}	10.03×10^{-5}
Std. deviation/avg.	0.31	0.27	0.52	0.55

"reverse permeability effect," but it came about because of higher pressure gradient in the tighter rocks. In a reservoir simulation model, a constant velocity is more or less conserved between neighboring blocks because liquid compressibility is small. To avoid this false "reverse permeability effect," we should use Eq. 7.85 in a reservoir simulator. If laboratory core flood tests are run in a constant pressure drop mode, both Eqs. 7.84 and Eq. 7.85 give lower capillary numbers in tight rocks, which lead to lower oil recovery factors. Thus, the "reverse permeability effect" would not be seen. Therefore, Eq. 7.85 is preferred in any situation.

The filled circles in Figure 7.29 represent the experimental data of $\Delta p/(L\sigma)$ at different permeabilities reported by Taber et al. (1973). These data, the critical values at which the residual oil saturation started to decrease, show a declining trend with permeability. However, when these data are converted to $k_a\Delta p/(L\sigma)$, which is the capillary number defined in Eq. 7.85, and plotted in the same figure (empty circles), we can see that all the data are near a horizontal line at 1500 (md·psi/ft)/(mN/m), which is equivalent to the dimensionless capillary number 3.4×10^{-5}, as reported in Table 7.7. This result further suggests that we should use Eq. 7.85 because the capillary number required to mobilize

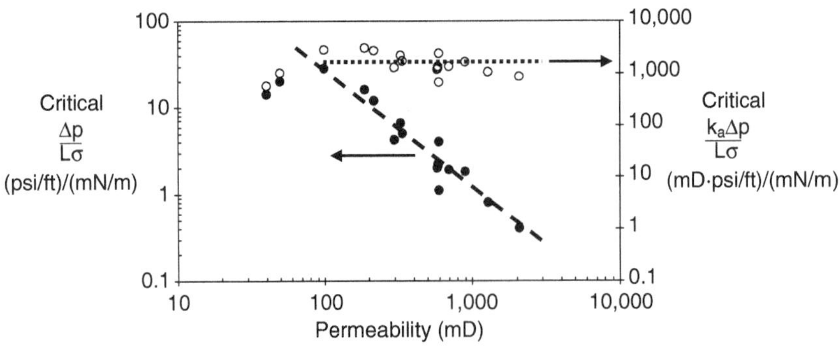

FIGURE 7.29 Graphic of the relationship between permeability and the critical values of $\Delta p/(L\sigma)$ and $k\Delta p/(L\sigma)$.

residual oil is independent of permeability so that the result can be applied more universally (without considering the difference in permeability). The data shown in Figure 7.29 also show that $\Delta p/L\sigma$ is not a good parameter to correlate residual saturation, but $k\Delta p/L\sigma$ is.

Note that according to Eq. 7.90, the capillary number required to mobilize oil is

$$N_C = \frac{\bar{r}^2}{4L}\left(\frac{1}{r_n} - \frac{1}{r_p}\right). \tag{7.92}$$

In the case of discontinuous oil, L may be equal to D_b, which is the diameter of a single oil blob. The capillary number required to mobilize the single oil blob is calculated using the preceding equation with $L = D_b$. In the case of continuous oil whose size could be several times of D_b, and L would be several times of D_b, then the capillary number required to mobilize the continuous oil would be several times lower than that required to mobilize a single oil blob. In other words, the critical capillary number required to mobilize discontinuous oil is higher than that to mobilize continuous oil. This is another justification that chemical flood should be conducted early in the secondary recovery mode instead of in the tertiary recovery mode.

7.7.2 How to Calculate Capillary Number

This section discusses how to select the parameters to calculate capillary number. Initially, capillary number was proposed to correlate the residual saturation of the fluid (oil) displaced by another fluid (water) in the two-phase system. In surfactant-related flooding, there is multiphase flow (water, oil, and microemulsion), especially at the displacing front. If we use $u\mu/\sigma$ to define the relationship between capillary number and residual oil saturation, which phase u and μ and which σ should be used then? To the best of the author's knowledge, this issue has not been discussed in the literature. The following is what we propose.

In general, when we discuss a parameter between two phases, we must define the conjugate phases. In any two-phase flow, it is obvious that both of the phases are conjugate phases. In a three-phase flow—for example, involving water, oil, and microemulsion three conjugates exist: water–oil, water–microemulsion, and oil–microemulsion. There are also three interfacial tensions: σ_{wo}, σ_{wm}, and σ_{om}. Each phase has two capillary numbers. For example, the oil phase has two capillary numbers: $(N_C)_{ow}$ for the water phase displacing the oil phase and $(N_C)_{om}$ for the microemulsion phase displacing the oil phase.

When the definition $N_C = u\mu/\sigma$ is used to calculate the capillary number, u_w and μ_w of the water phase and σ_{wo} should be used for $(N_C)_{ow}$; and u_m and μ_m of the microemulsion phase and σ_{om} should be used for $(N_C)_{om}$. When the definition $N_C = k(\Delta p/L)/\sigma$ is used, k_w, Δp_w, and σ_{wo} should be used for $(N_C)_{ow}$;

and k_m, Δp_m, and σ_{om} should be used for $(N_C)_{om}$. Two residual oil saturations are calculated: S_{orw} (residual oil saturation in the water–oil conjugate phases) calculated using $(N_C)_{ow}$, and S_{orm} (residual oil saturation in the oil–microemulsion conjugates) calculated using $(N_C)_{om}$. The final residual oil saturation should be saturation-weighted, as in

$$S_{or} = S_{orw}\frac{S_w}{S_w + S_m} + S_{orm}\frac{S_m}{S_w + S_m}, \tag{7.93}$$

where S_w and S_m are the water and the microemulsion phase saturations, respectively.

Similarly, the water phase also has two capillary numbers: $(N_C)_{wo}$ for the oil phase displacing the water phase and $(N_C)_{wm}$ for the microemulsion phase displacing the water phase. When the definition $N_C = u\mu/\sigma$ is used to calculate the capillary number, u_o and μ_o of the oil phase and σ_{wo} should be used for $(N_C)_{wo}$; and u_m and μ_m of the microemulsion phase and σ_{wm} should be used for $(N_C)_{wm}$. When the definition $N_C = k(\Delta p/L)/\sigma$ is used, k_o, Δp_o, and σ_{wo} should be used for $(N_C)_{wo}$; and k_m, Δp_m, and σ_{wm} should be used for $(N_C)_{wm}$. Two residual oil saturations are calculated: S_{wro} (residual water saturation in the water–oil conjugates) calculated using $(N_C)_{wo}$, and S_{wm} (residual water saturation in the water–microemulsion conjugates) calculated using $(N_C)_{wm}$. The final residual water saturation should be saturation-weighted, as in

$$S_{wr} = S_{wro}\frac{S_o}{S_o + S_m} + S_{wrm}\frac{S_m}{S_o + S_m}, \tag{7.94}$$

where S_o is the oil saturations.

For the microemulsion phase, there are two capillary numbers: $(N_C)_{mo}$ for the oil phase displacing the microemulsion phase and $(N_C)_{mw}$ for the water phase displacing the microemulsion phase. When the definition $N_C = u\mu/\sigma$ is used to calculate the capillary number, u_o and μ_o of the oil phase and σ_{mo} should be used for $(N_C)_{mo}$; and u_w and μ_w of the microemulsion phase and σ_{wm} should be used for $(N_C)_{mw}$. When the definition $N_C = k(\Delta p/L)/\sigma$ is used, k_o, Δp_o, and σ_{mo} should be used for $(N_C)_{mo}$; and k_w, Δp_w, and σ_{wm} should be used for $(N_C)_{mw}$. Two residual oil saturations are calculated: S_{mro} (residual microemulsion saturation in the microemulsion–oil conjugates) calculated using $(N_C)_{mo}$, and S_{mrw} (residual microemulsion saturation in the water–microemulsion conjugates) calculated using $(N_C)_{mw}$. The final residual microemulsion saturation should be saturation-weighted, as in

$$S_{mr} = S_{mro}\frac{S_o}{S_o + S_w} + S_{mrw}\frac{S_w}{S_o + S_w}. \tag{7.95}$$

The reader may ask: is $(N_C)_{ow}$ for the water phase displacing the oil phase the same as $(N_C)_{wo}$ for the oil phase displacing the water phase, for example? In principle, they are different from the preceding discussion. However, the formulas to calculate capillary number are empirical. In most practical cases,

$(N_C)_{ow}$ and $(N_C)_{ow}$ are not differentiated, simply calculating the single form of N_C. This is more obvious when the definition $N_C = k(\Delta p/L)/\sigma$ is used to calculate capillary number; then we use the same absolute permeability (k), the pressure drop (Δp) along the core with the length L, and the interfacial tension (σ_{wo}).

A similar discussion can be applied to $(N_C)_{om}$ and $(N_C)_{mo}$, and $(N_C)_{wm}$ and $(N_C)_{mw}$. In practice, a further approximation may be made. For instance, in measuring three-phase relative permeabilities, Delshad et al. (1987) used $N_C = k(\Delta p/L)/\sigma$, where σ was the average of the two IFTs: σ_{mo} (IFT between the microemulsion and oil phases) and σ_{mw} (IFT between the microemulsion and water phases). In this case, only a single form of capillary number was calculated for the three-phase flow.

According to current practice, we generally use average parameters to calculate the capillary number for the whole system, regardless of a two-phase or three-phase system. When the definition $N_C = u\mu/\sigma$ is used to calculate the capillary number, the velocity (u) and viscosity (μ) are those of the injection fluid. When the definition $N_C = k(\Delta p/L)/\sigma$ is used, the absolute permeability (k) and the total pressure drop (Δp) along the distance L are used. In either case, we use σ_{mo} for a type II(−) system, σ_{mw} for a type II(+) system, and an average IFT for a type III system—for example, the arithmetic average of σ_{mo} and σ_{mw}, as Delshad et al. (1987) did.

The previously discussed saturation-weighted approach is consistent with the current practice to select σ in calculating N_C. The reason is that in a type II(−) system, only σ_{mo} exists and the saturation of the microemulsion/oil system is 1. In a type II(+) system, only σ_{mw} exists and the saturation of the microemulsion/water system is 1. In a type III system, two IFTs, σ_{mo} and σ_{mw}, exist and the solubilized oil volume and water volume are the same at the optimal salinity; thus, the average $\sigma = \frac{1}{2}(\sigma_{mo} + \sigma_{mw})$. We propose that at any salinity the average IFT ($\bar{\sigma}$) is calculated by

$$\bar{\sigma} = \begin{cases} \sigma_{mo} & C_{se} \le C_{sel} \\ \sigma_{mo} \dfrac{V_{om}}{V_{om} + V_{wm}} + \sigma_{mw} \dfrac{V_{wm}}{V_{om} + V_{wm}} & C_{sel} < C_{se} < C_{seu} \\ \sigma_{mw} & C_{se} \ge C_{seu}, \end{cases} \tag{7.96}$$

where V_{om} and V_{wm} are the solubilized oil volume and water volume in the microemulsion phase, respectively. An example of actual IFT data and the average calculated using Eq. 7.96 is shown in Figure 7.30. The concept of average interfacial tension is consistent with the concept of controlling IFT proposed by Reed and Healy (1977). However, the former is quantified, whereas the latter is qualitative.

The value of salinity at which $\sigma_{mo} = \sigma_{mw}$ is called the optimum salinity for IFT (Healy et al., 1976). The salinity is usually very close to the optimum salinity for phase behavior that is defined as the salinity at which $V_{wm} = V_{om}$.

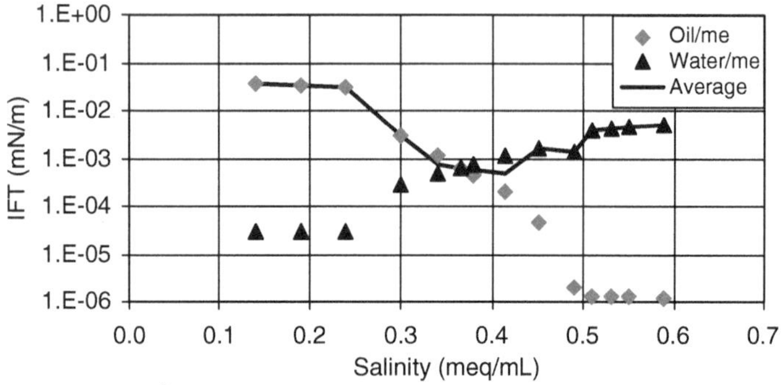

FIGURE 7.30 IFTs of oil/microemulsion (σ_{mo}) and water/microemulsion (σ_{mw}) and their average calculated using Eq. 7.96.

Plus, the good oil recovery usually corresponds to the optimum salinity (Glinsmann, 1978). According to the concept of capillary number, the minimum IFT corresponds to the maximum capillary number, which leads to the minimum residual saturation. Thus, good oil recovery usually corresponds to the minimum IFT. Now we want to see whether the minimum average IFT defined by Eq. 7.96 corresponds to the optimum salinity at which $V_{wm} = V_{om}$.

We start with the equations by Huh (1979) who developed a theoretical relationship between the solubilization parameter and IFT for a middle-phase microemulsion (type III). His equations are shown as Eqs. 7.77 and 7.78. According to Eq. 7.96,

$$\bar{\sigma} = \frac{C_{Hw}(V_{sm})^2}{V_{wm}(V_{om}+V_{wm})} + \frac{C_{Ho}(V_{sm})^2}{V_{om}(V_{om}+V_{wm})}. \tag{7.97}$$

We need to find at what condition $\bar{\sigma}$ is the minimum. The conditions are

$$\frac{\partial\bar{\sigma}}{\partial V_{om}} = 0, \tag{7.98}$$

$$\frac{\partial\bar{\sigma}}{\partial V_{wm}} = 0. \tag{7.99}$$

Because C_{Hw} and C_{Ho} are cosine functions, calculating the partial derivatives of $\bar{\sigma}$ based on Eqs. 7.98 and 7.99 is not straightforward. If we assume C_{Hw} and C_{Ho} are equal and constants, Eqs. 7.98 and 7.99 lead to

$$V_{wm} = V_{om}. \tag{7.100}$$

This condition proves that the minimum average IFT defined by Eq. 7.96 does correspond to the optimum salinity at which $V_{wm} = V_{om}$. Therefore, the concept of the average IFT defined by Eq. 7.96 is justified.

Note that the microemulsion/oil IFT (σ_{mo}) itself is not the minimum at the optimum salinity. Instead, it continuously decreases with the salinity until it disappears in the type II(+) environment. Because a lower IFT would lead to a higher capillary number, and a higher capillary number would result in a lower residual oil saturation, we would expect that the higher oil recovery should be obtained just before the microemulsion phase becomes type II(+) when σ_{mo} is the lowest. However, the reality is that the highest oil recovery is *usually* obtained at the optimum salinity. The discussion here shows that the highest oil recovery is not obtained when an individual IFT, including σ_{mo}, is the minimum; it is obtained when the average IFT of the system is the minimum. The average IFT of the system represents the total effects of every phase in the system.

7.8 TRAPPING NUMBER

A new dimensionless number called the trapping number has been defined; it includes both gravity and viscous forces (UTCHEM-9.0, 2000). The dependence of residual saturations on interfacial tension is modeled in UTCHEM as a function of the trapping number. This is a formulation necessary to model the combined effect of viscous and buoyancy forces in three dimensions. Buoyancy forces are much less important under enhanced oil recovery conditions than under typical surfactant-enhanced aquifer remediation (SEAR) conditions; therefore, it had not been carefully considered under three-dimensional surfactant flooding field conditions.

Logically, the gravity effect can be included if the term $g(\rho_p - \rho_{p'})\sin\alpha$ is added in the potential gradient term in Eq. 7.85, as in the waterflooding case (Leverett, 1941). In other words, the trapping may be defined as

$$N_T = \frac{-k\dfrac{\Delta\Phi_p}{\Delta L} + k\left[g(\rho_p - \rho_{p'})\sin\alpha\right]}{\sigma_{p'p}}. \tag{7.101}$$

Alternatively, in a general form,

$$N_T = \frac{\left|-\vec{\vec{k}}\cdot\vec{\nabla}\Phi_p + \vec{\vec{k}}\cdot\left[(\rho_p - \rho_{p'})g\vec{\nabla}h\right]\right|}{\sigma_{p'p}}, \tag{7.102}$$

where $\vec{\vec{k}}$ is permeability tensor, and $\vec{\nabla}$ is gradient symbol. p and p′ represent the displacing phase and displaced phase. α is the angle formed from the horizontal axis $\vec{x}$ to the local flow vector (counterclockwise).

For an arbitrary flow angle, the preceding trapping number can be rewritten as

$$N_T = |N_C + N_B \sin\alpha|, \tag{7.103}$$

where N_C is defined by Eq. 7.85, and N_B, the Bond number, is defined as

$$N_B = \frac{kg(\rho_p - \rho_{p'})}{\sigma_{p'p}} = \frac{kg(\Delta\rho)}{\sigma_{p'p}}. \tag{7.104}$$

For a *horizontal* flow ($\alpha = 0^o$), the expression for N_T reduces to

$$N_T = N_C. \tag{7.105}$$

For a vertical displacement ($\alpha = \pm 90^o$), Eq. 7.103 becomes

$$N_T = |N_C \pm N_B|. \tag{7.106}$$

The capillary and Bond numbers can be either additive or subtractive, depending on the direction of vertical displacement and sign of the density difference between the displacing and displaced fluids. Note that $\Delta\rho = \rho_p - \rho_{p'}$ (displacing fluid density ρ_p minus displaced fluid density $\rho_{p'}$). In an oil reservoir, generally, an aqueous fluid displaces oleic fluid. Thus, the density difference term is positive. For one-dimensional vertical upward displacement, the capillary number plus the Bond number equals the trapping number. For a one-dimensional vertical downward flow (where an aqueous phase displaces an oleic phase), the capillary number minus the Bond number equals the trapping number. Note that the capillary number is defined using the pressure gradient in Eq. 7.101. In a surfactant-enhanced aquifer remediation (SEAR) process, the dense nonaqueous phase liquid (DNAPL) is heavier than the water. Therefore, there are more cases of addition and subtraction of the capillary number and Bond number; Figure 7.31 summarizes all the possible cases. From this figure, we can see that if the lighter fluid is above the heavier fluid, the two numbers are additive; if the heavy fluid is above the lighter fluid, they are subtractive.

Pennell et al. (1996) managed to derive the trapping number defined earlier by applying a force balance on a *single* trapped nonaqueous phase liquid

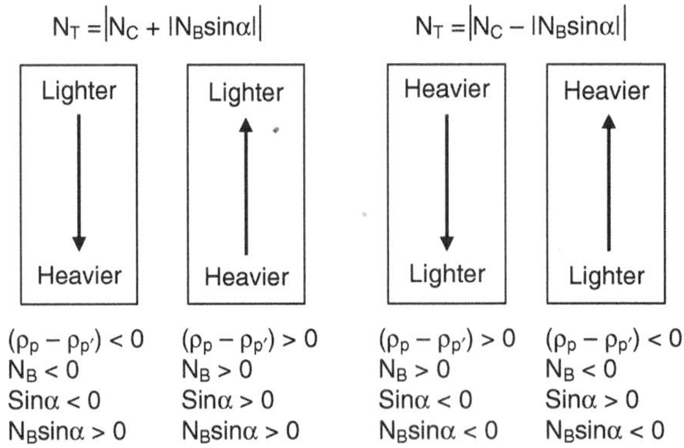

FIGURE 7.31 Summary of the additive and subtractive cases of N_C and N_B when calculating N_T.

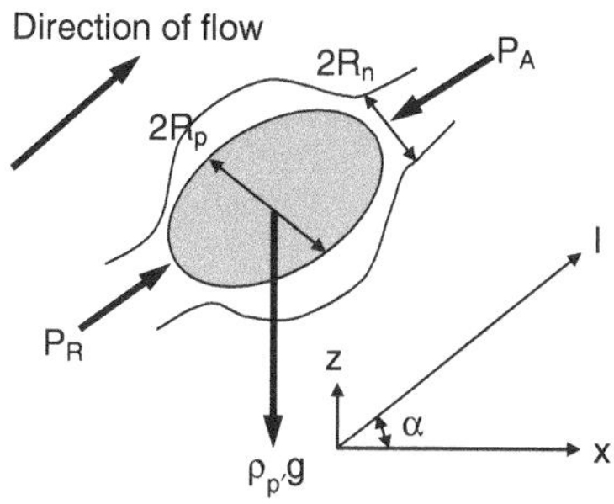

FIGURE 7.32 Schematic diagram of the single pore entrapment model.

(NAPL) globule entrapped in a single pore. As shown in Figure 7.32, the pore is oriented in a single line, **l**, which makes an arbitrary angle, α, with the horizontal axis. Within this pore, pressure and gravity forces, which act to mobilize the globule, are balanced by capillary forces acting to retain the globule.

A balance of forces in the direction of the pore permits a quantitative assessment of the conditions under which globule mobilization can occur within that pore. The conditions under which mobilization forces balance retention forces are termed the critical conditions for mobilization. Summation of the pressure and gravity forces acting on the globule along the **l** direction yields

$$(\pi R_p^2)(p_R - p_A) - (\pi R_p^2)\rho_{p'}g(\Delta l)(\sin\alpha), \tag{7.107}$$

where p_R is the pressure force on the receding side of the globule, p_A is the pressure force on the advancing foot, Δl is the average length of the globule, $\rho_{p'}$ is the density of the displaced organic liquid, g is the gravity acceleration constant, $\pi R_p{}^2$ is the globule area normal to the vector **l**, and the globule volume has been approximated as $\pi R_p{}^2\Delta l$. Under the critical conditions for mobilization, pressure and gravity forces are balanced by the maximum net capillary pressure force the globule can sustain within the pore.

This capillary pressure force can be approximated using the Laplace equation,

$$2\sigma_{p'p}\left(\frac{\cos\theta_A}{R_n} - \frac{\cos\theta_R}{R_p}\right)\pi R_p^2, \tag{7.108}$$

where $\sigma_{p'p}$ is the interfacial tension between the displaced nonaqueous phase (p′) and displacing aqueous phase (p); θ_A and θ_R are the advancing and receding contact angles, respectively; R_p is the radius of the pore body; and R_n is the radius of the pore neck. This expression assumes that the globule has a uniform internal pressure. Assuming that the contact angles of the advancing and receding ends of the globule are similar, Eq. 7.108 may be rewritten as

$$\frac{2\beta\sigma_{p'p}\cos\theta}{R_n}(\pi R_p^2) \quad \text{where} \quad \beta = 1 - \frac{R_n}{R_p}. \tag{7.109}$$

The critical condition for mobilization can be found by equating Eqs. 7.107 and 7.109 and dividing through by πR_p^2:

$$(p_R - p_A) - \rho_{p'} g(\Delta l)(\sin\alpha) = \frac{2\beta\sigma_{p'p}\cos\theta}{R_n}. \tag{7.110}$$

This equation can be rewritten as

$$\frac{(\Phi_R - \Phi_A)}{\Delta l} + \Delta\rho g(\sin\alpha) = \frac{2\beta\sigma_{p'p}\cos\theta}{R_n\Delta l}, \tag{7.111}$$

where $\Delta\rho = \rho_p - \rho_{p'}$, and

$$(\Phi_R - \Phi_A) = (p_R - p_A) - \rho_p g(\Delta l)\sin\alpha. \tag{7.112}$$

Equation 7.111 can be rewritten as

$$\frac{k(\Phi_R - \Phi_A)/\Delta l}{\sigma_{p'p}} + \frac{k\Delta\rho g(\sin\alpha)}{\sigma_{p'p}} = \frac{2\beta k\cos\theta}{R_n\Delta l}, \tag{7.113}$$

or

$$N_C + N_B\sin\alpha = \frac{2\beta k\cos\theta}{R_n\Delta l}, \tag{7.114}$$

where N_C and N_B are defined in Eqs. 7.85 and 7.104, respectively, and N_T is defined as

$$N_T = |N_C + N_B\sin\alpha|, \tag{7.103}$$

which was defined earlier.

The preceding model and corresponding equations describe the flow in a single direction (actually along a single line) and the two network throats. In natural porous media, however, pores are oriented in all directions, and there are many throats. Jin (1995) presented another equation to define the trapping number. The derivation of the trapping number is based on the following force balance:

$$|\vec{F}| = \sqrt{|\vec{F}_x|^2 + |\vec{F}_z|^2} = \frac{2\beta\sigma_{p'p}\cos\theta}{R_n}. \tag{7.115}$$

Here, the force $\vec{F}_x$ acting on the globule in the horizontal direction is

$$\vec{F}_x = -\frac{\partial p}{\partial x}\vec{i} = -\frac{\partial p}{\partial x}\vec{i}, \tag{7.116}$$

and the force $\vec{F}_z$ acting on the globule in the vertical direction is

$$\vec{F}_z = \left(-\frac{\partial p}{\partial z} - \rho_{p'} g\right)\vec{k} = \left(-\frac{\partial\Phi}{\partial z} + \Delta\rho g\right)\vec{k}. \tag{7.117}$$

The right side of Eq. 7.115 now represents average pore characteristics of the porous medium.

It can be shown that from the magnitude of the hydraulic and buoyancy forces, $|\vec{F}|$, the trapping number is

$$N_T = \sqrt{N_C^2 + 2N_C N_B \sin\alpha + N_B^2}. \tag{7.118}$$

Based on this equation, for the case of a *horizontal* flow ($\alpha = 0^o$), the expression for N_T reduces to

$$N_T = \sqrt{N_C^2 + N_B^2}, \tag{7.119}$$

and for a vertical flow ($\alpha = \pm 90^o$), Eq. 7.106 is obtained.

The paradox is that for a horizontal flow ($\alpha = 0^o$), both Eqs. 7.105 and 7.119 are obtained. Eq. 7.105 describes the flow of only a single pore with a single globule along a single line. A two-dimensional flow is not possible. However, if there are many globules in many pores, even the gross flow (injection direction) is in one horizontal direction; it therefore is possible that some globules move in the vertical direction owing to the buoyancy. In other words, buoyancy also plays an important rule in the determination of mobilization of the trapped residual phase even when the flow is in the horizontal direction.

Equation 7.119 seems to be consistent with the flow in real life. The explanation may be extended to the difference between Eq. 7.103 and Eq. 7.118 at an arbitrary flow angle. Equation 7.118 appears to model the two-dimensional flow in a homogeneous and isotropic porous media. The trapping numbers for two-dimensional and three-dimensional heterogeneous, anisotropic porous media were also derived in Jin (1995).

In the special horizontal flow ($\alpha = 0^o$) where the capillary number N_C is insignificant, according to Eq. 7.119, $N_T \approx N_B$. Also in the special vertical flow ($\alpha = \pm 90^o$) where the capillary number N_C is insignificant, according to Eq. 7.118, $N_T \approx N_B$. In other words, when the capillary number N_C is insignificant, N_T is always equal to N_B regardless of whether the flow is in a horizontal or vertical direction, which does not make sense.

The prediction from Eq. 7.118 is not consistent with the experimental data from Morrow and Songkran (1981) shown in Figure 7.33. The figure clearly shows that the trapped residual saturation depends on the dip angle, which means that the actual Bond numbers at different angles are different. In the experiments, the capillary number was very small (on the order of 10^{-6}), and its effect on trapping residual saturation was negligible.

The trapping number defined by Eq. 7.103 for an arbitrary dipping angle is consistent with the conventional Buckley–Leverett fractional flow theory. In the Buckley–Leverett fractional flow equation, the gravity term is multiplied by $\sin\alpha$ (Leverett, 1941). However, Figure 7.34 shows that the trapped residual saturation predicted by Eq. 7.103 is lower than the experimental data at the same trapping number. This figure compares the relationship between the

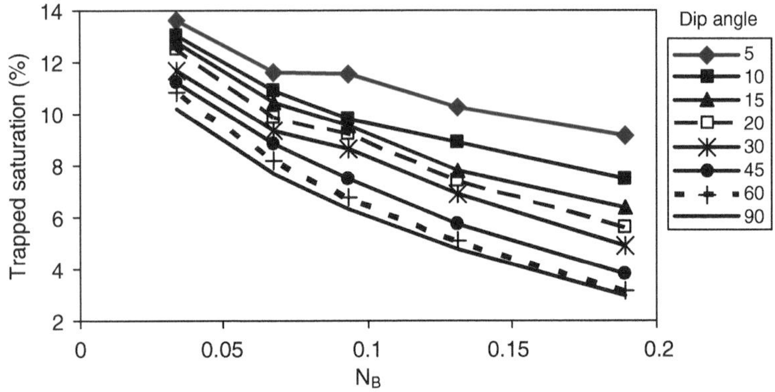

FIGURE 7.33 Effect of dip angles on trapping of residual saturation ($N_C = u\mu/\sigma = 2.82 \times 10^{-6}$). *Source: Data from Morrow and Songkran (1981).*

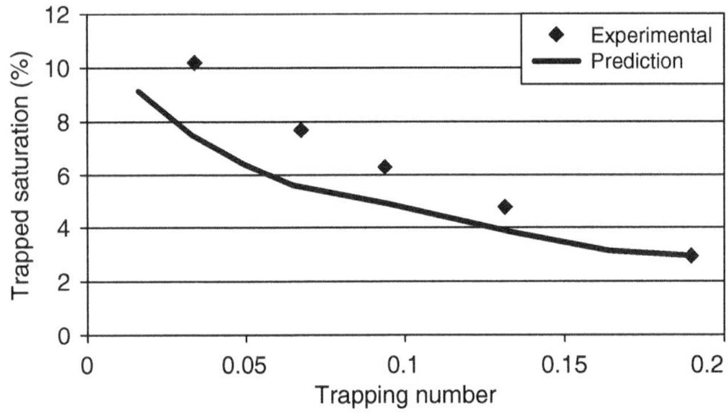

FIGURE 7.34 Comparison of the relationship of trapped residual saturation versus trapping number (with $N_C = u\mu/\sigma = 2.82 \times 10^{-6}$ not included in the trapping number).

trapped residual saturation and trapping number N_T (equal to N_B because the small capillary number of 2.82×10^{-6} was not actually included in the trapping number calculation). The experimental data of the residual saturation versus N_B are from the vertical flow tests ($\alpha = 90^o$) by Morrow and Songkran (1981), whereas the predicted trapping numbers are calculated using $N_B \cdot \sin\alpha$. The comparison in Figure 7.34 shows that Eq. 7.103 is not validated by the experimental data. Note that Morrow and Songkran (1981) calculated N_B for the glass bead packs using the equation

$$N_B = 0.001412(\Delta\rho g r_b^2/\sigma), \tag{7.120}$$

where $\Delta\rho$ is in g/cm^3, g is 980 cm/s^2, r_b (bead radius) is in cm, and σ is in mN/m (dyne/cm).

In summary, neither Eq. 7.103 nor Eq. 7.118, proposed to calculate the trapping number at an arbitrary dip angle, has been validated by the available experimental data. This is not a purely academic issue (L. W. Lake, personal communication on January 19, 2009) and needs to be investigated further.

7.9 CAPILLARY DESATURATION CURVE

Let us use the simple equation, Eq. 7.84, to calculate the capillary number in a typical waterflood case. Assume that injection velocity is 1 ft/day, which is 3.528×10^{-6} m/s, the water viscosity is 1 mPa·s, and the interfacial tension is 30 mN/m. The corresponding capillary number is then

$$N_C = \frac{u\mu}{\sigma} = \frac{(3.528 \times 10^{-6}\ \text{m/s})(1\ \text{mPa} \cdot \text{s})}{(30\ \text{mN/m})} \approx 10^{-7}.$$

To further reduce waterflood residual oil saturation, the capillary number must be higher than the preceding calculated value. In general, the capillary number must be higher than a critical capillary number, $(N_C)_c$, for a residual phase to start to mobilize. Practically, this $(N_C)_c$ is much higher than the capillary number at normal waterflooding conditions. Another parameter is maximum desaturation capillary number, $(N_C)_{max}$, above which the residual saturation would not be further reduced in practical conditions even if the capillary number is increased. Lake (1989) used the term total desaturation capillary number for $(N_C)_{max}$. In practical conditions, total desaturation (i.e., zero residual saturation) may not occur due to some films or blobs trapped in pores.

Morrow and coworkers (Morrow and Songkran, 1981; Morrow et al., 1988) used the terms capillary number for mobilization and capillary number for prevention of entrapment for $(N_C)_c$ and $(N_C)_{max}$, respectively. In UTCHEM, lower and higher critical capillary numbers are used for $(N_C)_c$ and $(N_C)_{max}$, respectively. Table 7.8 summarizes some of the published experimental data for these critical capillary numbers. In principle, the critical capillary numbers should be system specific. Experiments should always be conducted to determine the capillary desaturation curves (CDC) for the particular application whenever possible. The summarized data could be useful only when no experimental data are available. From Table 7.8, the following observations can be made regarding capillary number:

- The capillary number defined by Eq. 7.84 is more widely used, probably because it is the simplest form.
- Most of the data are about nonwetting phases. Only some data are about wetting phases. $(N_C)_c$ and $(N_C)_{max}$ for wetting phases are higher than those for nonwetting phases. That implies lower IFT (higher capillary number) is required for oil-wetting systems. However, these results cannot be interpolated for systems with intermediate wettability. It has been suggested that oil displacement would be most difficult in the intermediate wettability

TABLE 7.8 Summary of Experimental Work on Capillary Desaturation Curve

			Nonwetting Phase		Wetting Phase	
References	Medium	Definition of N_C	$(N_C)_c$	$(N_C)_{max}$	$(N_C)_c$	$(N_C)_{max}$
Dombrowski and Brownell (1954)	Synthetic	$k(\Delta\Phi/\Delta L)/(\sigma\cos\theta)$			10^{-2}	2
Moore and Slobod (1955)	Outcrop ss	$v\mu/(\sigma\cos\theta)$	10^{-7}	10^{-2}		
Taber (1969)*	Berea ss	$k(\Delta p/\Delta L)/\sigma$	$(2–6)10^{-5}$	$(1–5)10^{-2}$		
Foster (1973)	Berea ss	$v\mu/\sigma$	$10^{-5}–10^{-4}$	$10^{-2}–10^{-1}$		
Lefebvre du Prey (1973)	Synthetic	$u\mu/\sigma$	10^{-4}	$10^{-2}–10^{-1}$		0.5
Ehrlich et al. (1974)*	Outcrop ss	$u\mu/\sigma$	3×10^{-4}			
Abrams (1975)	Outcrop ss	$v\mu(\mu/\mu o)^{0.4}/(\sigma\cos\theta\Delta S)$	$10^{-5}–10^{-4}$	$10^{-2}–10^{-1}$		
Abrams (1975)	Limestone	$v\mu(\mu/\mu o)^{0.4}/(\sigma\cos\theta\Delta S)$	2.8×10^{-7}	10^{-2}		
Gupta and Trushenski (1979)	Berea ss	$u\mu/\sigma$	10^{-5}	10^{-2}	5×10^{-5}	2×10^{-2}
Gupta (1984)	Berea ss	$u\mu/\sigma$	$10^{-5}–10^{-4}$	$10^{-2}–10^{-1}$	10^{-4}	0.03
Morrow and Songkran (1981)	Bead pack	$u\mu/\sigma + 0.001412\Delta\rho g r_b^2/\sigma$	2×10^{-5}	10^{-3}		

Amaefule and Handy (1982)*	Berea ss	$u\mu/\sigma$	1.44×10^{-4}	10^{-2}	2×10^{-4}	
Mohanty and Salter (1983)	Berea ss	$k(\Delta p/\Delta L)/\sigma$	10^{-5}	10^{-3}		
Chatzis and Morrow (1984)	Berea ss	$k_w(\Delta p/\Delta L)/\sigma$	2×10^{-5}	1.5×10^{-3}		
Delshad et al. (1986)*	Berea ss	$k(\Delta p/\Delta L)/\sigma$	10^{-5}	10^{-3}	3×10^{-5}	10^{-2}
Morrow et al. (1986)	Berea ss	$k_w(\Delta p/\Delta L)/\sigma$	10^{-4}	$>10^{-2}$		
Morrow et al. (1988)	Bead pack	$u\mu/\sigma$	5×10^{-4}	10^{-2}		
Garnes et al. (1990)*	Reservoir cores	$k_w(\Delta p/\Delta L)/\sigma$	10^{-7}–10^{-5}	10^{-3}–10^{-2}		
Pennell et al. (1993)*	Ottawa	$u\mu/\sigma + \Delta\rho g k_w/\sigma$	4.6×10^{-5}			
Pennell et al. (1996)*	Ottawa	$u\mu/\sigma + \Delta\rho g k_w/\sigma$	$(2–5)10^{-5}$	10^{-3}		
Boom et al. (1995,1996)	Reservoir cores	$k(\Delta p/\Delta L)/\sigma$			10^{-5}–10^{-4}	10^{-2}
Dwarakanath (1997)*			10^{-5}–10^{-4}	10^{-3}		
Henderson et al. (1998)	Berea ss	$u\mu/\sigma$	10^{-6}			
Kamath et al. (2001)	Limestone	$u\mu/\sigma$	10^{-8}	$>10^{-6}$	10^{-7}	3×10^{-5}

**Surfactant, solvent (alcohol), or NaOH was used; ss denotes sandstone.*

systems because of contact angle hysteresis. Besides the reduced displacement efficiency in oil-wet systems, higher sulfonate loss also has been reported (Gupta and Trushenski, 1979).

- The critical capillary number for sandstones is in the order of 10^{-5} to 10^{-4}. The maximum desaturation capillary number is two to three orders of magnitude higher than the critical capillary number.
- The critical capillary numbers for limestones were found to be lower than those for sandstones. For some carbonate rocks, $(N_C)_c$ was not detected (Abrams, 1975; Kamath et al., 2001).
- The critical capillary number for gas/condensate cases (Henderson et al., 1998) was found in the order of 10^{-6}.
- Capillary number can be increased by increasing velocity or by lowering interfacial tension by surfactant or alkaline flooding. From Table 7.8, it seems that there is no distinct difference in the magnitude of critical capillary number based on these different approaches. However, we should note that for most of the tests in Table 7.8 the capillary number was increased by increasing flow velocity.
- Critical capillary numbers from weakly water-wet systems (on the order of 10^{-4}; Morrow et al., 1986) were higher than those for the corresponding strongly water-wet systems (on the order of 2×10^{-5}; Chatzis and Morrow, 1984).

Other observations regarding capillary number follow:

- The critical capillary number for bead packs (unconsolidated) is higher than that for sandstones (Morrow et al., 1988).
- The critical capillary number required to mobilize discontinuous oil is higher than that to mobilize continuous oil.
- Residual oil saturation and capillary number (thus IFT) have approximately a semi-log relationship. If we assume that an ultralow IFT of 10^{-3} order can increase oil recovery factor over waterflooding by 20%, then we may have a 5% increase in oil recovery factor when the water/oil IFT is on the order of 10 if the capillary number is greater than $(N_C)_c$.
- The displacement of the wetting phase requires a capillary number about 10 times higher than the one needed to displace the nonwetting phase to the same relative final saturation. The results cannot be interpolated for systems with intermediate wettability, as mentioned previously.

Now we have discussed the two important capillary numbers: critical and maximum. The general relationship between residual saturation of a nonaqueous or aqueous phase and a local capillary number is called capillary desaturation curve (CDC). The residual saturations start to decrease at the critical capillary number as the capillary number increases, and cannot be decreased further at the maximum capillary number. As discussed earlier, the range of capillary numbers for residual phases to be mobilized is, for example, 10^{-5} to

10^{-2}. The capillary number in a normal waterflood is on the order of 10^{-7}. To increase capillary number, according to Eq. 7.84, we may increase the displacing fluid viscosity or velocity.

However, it may be practically impossible to increase the viscosity or velocity by such a magnitude because doing so would require or result in a very high pressure difference between the injector and producer. Such high pressure difference would fracture the formation. Another way to increase capillary number is to reduce interfacial tension, which can be achieved through injection of surfactants. Recall that ultralow interfacial tension is one of the main mechanisms in surfactant-related processes.

The capillary desaturation curves presented in the literature and capillary numbers presented in Table 7.8 are restricted to mainly two-phase flow. Fluids are normally oil and water, sometimes with surfactant or alcohol additives to lower interfacial tension. Those surfactant systems are, or are treated as, type I systems. Most data are on residual oil saturation or nonaqueous petroleum liquid (nonwetting phase). Delshad et al. (1986) were the first to measure CDC in three-phase micellar solutions. They showed that the microemulsion phase was the most strongly trapped, implying that the microemulsion is the wetting phase. This conclusion was made elsewhere (Delshad et al., 1985; Delshad et al., 1987). Their conclusion is consistent with the observations in alkaline flooding. When alkaline flooding in a high-salinity environment, the wettability is changed from water-wet to oil-wet. This oil-wetness is consistent with the high-salinity environment that would likely result in an oil-external microemulsion phase. We hypothesize that in a type II environment, surfactant is in the oleic phase and adsorbs on the rock surface. The adsorbed surfactant, which is in the oil phase, has an affinity with oil so that the oil is in direct contact with the rock surface. Then the rock surface becomes more oil-wetting. Similar to a type I environment, water-external microemulsion would make the rock surface water-wet.

In a simulation model, we need to input a capillary desaturation curve model. Stegemeier (1977) presented a theoretical equation to calculate CDC based on the capillary number originally proposed by Brownell and Katz (1947). This equation requires several petrophysical quantities. Thus, it would probably be even more difficult to calculate a CDC using the Stegemeier equation than to obtain a CDC in the laboratory. In the laboratory, if several points of residual saturation versus capillary number are measured, we can use those measured points to fit a theoretical model. In UTCHEM, a form of Eq. 7.121 is used:

$$S_{pr} = S_{pr}^{(N_C)_{max}} + \left(S_{pr}^{(N_C)_c} - S_{pr}^{(N_C)_{max}}\right)\frac{1}{1+T_p N_C}. \quad (7.121)$$

In this equation, S_{pr} is the phase residual saturation; the subscript p means the phase that could be water, oil, or microemulsion; the superscript $(N_C)_c$ and

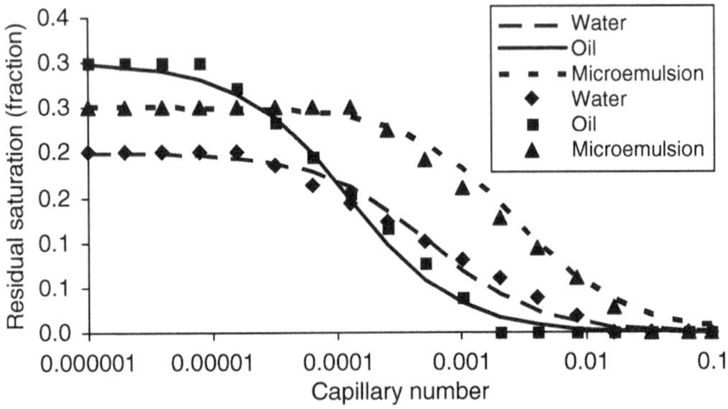

FIGURE 7.35 Example of capillary desaturation curve.

TABLE 7.9 Sample CDC Parameters

	T_p	$(S_{pr})^{(N_C)_C}$	$(S_{pr})^{(N_C)max}$	$(N_C)_c$	$(N_C)_{max}$
Water	1865	0.2	0	2×10^{-5}	1.5×10^{-2}
Oil	8000	0.3	0	1×10^{-5}	2×10^{-3}
Microemulsion	364	0.25	0	1.5×10^{-4}	3×10^{-2}

$(N_C)_{max}$ mean at critical capillary number and maximum desaturation capillary number; (N_C) is capillary number; and T_p is the parameter used to fit the laboratory measurements. The definition of capillary number used in the preceding equation must be the same as that used in the simulation model. One example of CDC using Eq. 7.121 is shown by the curves in Figure 7.35, and some of the CDC parameters are presented in Table 7.9. The data points in Figures 7.35 and 7.36 are calculated using Eq. 7.124, to be discussed later.

More generally, the normalized phase residual saturation is used:

$$\bar{S}_{pr} \equiv \frac{S_{pr} - S_{pr}^{(N_C)_{max}}}{S_{pr}^{(N_C)_c} - S_{pr}^{(N_C)_{max}}}. \tag{7.122}$$

Then Eq. 7.121 becomes

$$\bar{S}_{pr} = \frac{1}{1 + T_p N_C}, \tag{7.123}$$

and the curve in Figure 7.35 becomes the curve in Figure 7.36. $\bar{S}_{pr}$ not higher than 1 is warranted when Eq. 7.123 is used.

When CDC experimental measurements are not made in the beginning of a specific EOR project, we may use published data by analog for screening

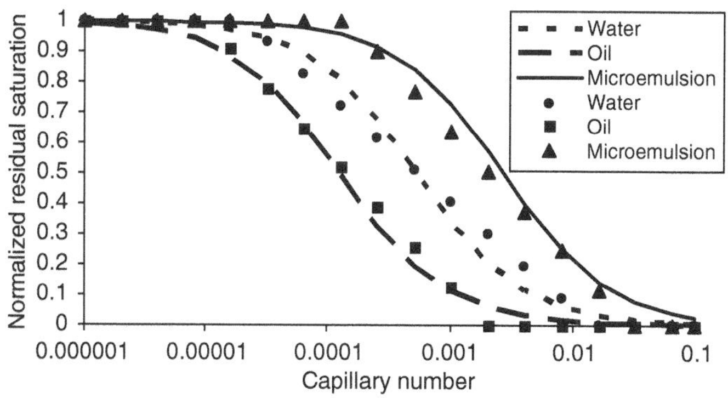

FIGURE 7.36 Example of normalized capillary desaturation curve.

studies. Some of the capillary number data are presented in Table 7.8, and the corresponding residual saturation data can be found from the references. Note that when published data are used, the relationship between normalized residual saturation and capillary number is preferred. However, to use those data, we cannot apply Eq. 7.121 or Eq. 7.123 directly. In other words, we have to normalize the data first.

From Figures 7.35 and 7.36, we can see that the CDC curves are approximately linear in the middle range of N_C where the residual saturation decreases steeply. Therefore, the CDC curves can be approximated using the following equation:

$$\overline{S}_{pr} \equiv \frac{S_{pr} - S_{pr}^{(N_C)_{max}}}{S_{pr}^{(N_C)_c} - S_{pr}^{(N_C)_{max}}} = \frac{\log(N_C)_{max} - \log(N_C)}{\log(N_C)_{max} - \log(N_C)_c}. \quad (7.124)$$

For the sample CDC curves in Figure 7.35 or Figure 7.36, the critical and maximum desaturation capillary numbers are identified from those figures and as presented in Table 7.9. The residual saturations at different capillary numbers are recalculated using those critical and maximum desaturation capillary numbers and their corresponding residual saturations in Table 7.9 according to Eq. 7.124. The recalculated data are shown in points in Figures 7.35 and 7.36. We can see that the data points match the original CDC curves very well.

As mentioned earlier in this section, the microemulsion is the most wetting phase, and the critical capillary number for a wetting phase is higher than that for a nonwetting phase. Therefore, as is shown in Figures 7.35 and 7.36, the microemulsion CDC lies on the right, the water CDC in the middle, and the oil CDC on the left. The effect of wettability on the CDC is also important. Because the rock surface tends to repel the nonwetting phase and attract the wetting phase, the nonwetting phase is easier to mobilize, and the reduction in its residual saturation will start to occur at a lower trapping number than for the wetting phase. Conversely, the rock surface has an affinity for the wetting phase and therefore would require a higher trapping number to mobilize its residual saturation.

In most cases, there is no clear-cut for the values of $(N_C)_c$ or $(N_C)_{max}$ from laboratory data. As shown previously in Figures 7.35 and 7.36, there are gradual change regimes near $(N_C)_c$ and $(N_C)_{max}$, and a sharp change regime in between. By checking the curves in Figure 7.35 or Figure 7.36 and the T_p parameters that are shown in Table 7.9, we can see that as T_p is smaller, the corresponding CDC curve moves to the right.

7.10 RELATIVE PERMEABILITIES IN SURFACTANT FLOODING

Relative permeability is probably one of the least-defined parameters in chemical flooding processes. The classical relative permeability curves represent a situation in which the fluid distribution in the system is controlled by capillary forces. When capillary forces become small compared to viscous forces, the whole concept of relative permeability becomes weak. This area has not been adequately researched, and theoretical understanding is rather inadequate (Brij Maini, University of Calgary in Canada, personal communication, 2007). This section discusses relative permeability models related to surfactant flooding and the IFT effect on relative permeabilities.

7.10.1 General Discussion of Relative Permeabilities

In surfactant-related processes, the interfacial tension is reduced. As IFT is reduced, the capillary number is increased, leading to reduced residual saturations. Obviously, residual saturation reduction directly changes relative permeabilities. A number of authors reported their research results, as reviewed by Amaefule and Handy (1982) and Cinar et al. (2007). The general observations were that the relative permeabilities tend to increase and have less curvature as the IFT decreases or the capillary number increases. However, Delshad et al. (1985) observed that even at IFT of 10^{-3} mN/m, k_r curves showed significant curvature.

Data from Fulcher et al. (1985) showed that above 5.5 mN/m, IFT seemed to have little effect on k_{ro} or k_{rw}. Chen and Chen (2002) observed that as water/oil IFT was reduced, both water and oil relative permeabilities were increased, their end points were raised, and residual saturations were decreased. These observations were obvious only when the IFT was below 0.1 mN/m. The imbibition curves were different from the drainage curves, even when the IFT was reduced below 0.02 mN/m.

When investigating low IFT relative permeabilities, most of the researchers treated the surfactant solution as the low IFT water phase (type I microemulsion). However, depending on the salinity, the surfactant solution could be a type I, type II, or type III microemulsion. When it is a type III microemulsion, the system becomes a three-phase system (aqueous, oleic, and microemulsion phases). A theoretical model of three-phase relative permeabilities depends on the wettability. Therefore, we first review the wettability of the microemulsion phase.

Hirasaki et al. (1983) assumed that if an excess water phase wets preferentially to a microemulsion phase and a microemulsion phase preferentially to an oleic phase, then (1) in the absence of an excess water phase, the microemulsion is the wetting phase; (2) in the absence of an excess oil phase, the microemulsion is the nonwetting phase; and (3) when all the three phases are present, the microemulsion is a spreading phase between the excess oil and excess water. Hirasaki et al. (2008) further pointed out that the current understanding of microemulsion phase behavior and wettability is that the system wettability is likely to be preferentially water-wet when the salinity is below the optimal salinity (Winsor I) and is likely to be preferentially oil-wet when the salinity is above the optimal salinity (Winsor II), even in the absence of alkali. Their view is supported by Nelson et al. (1984), Israelachvili and Drummond (1998), and Yang (2000).

Data from Reed and Healy (1979) bring in to question the preceding assumptions about the order of wetting. Their data showed that in low salinities, the microemulsion phase can approach being the wetting phase in preference to the excess water phase; at high salinities, the microemulsion phase can approach being the nonwetting phase in preference to the excess oil phase. Observations from Delshad et al. (1985, 1986, 1987) are neither in agreement with the assumption of Hirasaki et al. (1983, 2008) nor with the data from Reed and Healy (1979). Their data show that the microemulsion phase is the most strongly trapped, which implies that the microemulsion is the most wetting phase. One interesting observation from data by Delshad et al. (1987) is that the exponent of the type III microemulsion relative permeability was less than 1.0. That means the relative permeability was even higher than that in a miscible case.

Parlar and Yortsos (1987) showed that the exponent of the vapor phase k_r in steam/water relative permeability curves was less than 1. An exponent less than 1 was also observed in heavy oil k_r curves (Brij Maini, personal communication, 2007). Lake (1989) pointed out that such observation could be explained only as wall or interfacial slippage.

7.10.2 Relative Permeability Models

This section discusses prediction models of relative permeability. Based on their experimental data, which show that phase k_{rp} is a function of its own saturation S_p only, Delshad et al. (1987) proposed that the phase relative permeability k_{rp} for each phase p is

$$k_{rp} = k_{rp}^{e}\left(\overline{S}_p\right)^{n_p}, \tag{7.125}$$

where

$$\bar{S}_p = \frac{S_p - S_{pr}}{1 - \sum S_{pr}}, \tag{7.126}$$

where p could be w (aqueous phase), o (oleic phase), or m (microemulsion phase); k_{rp}^e is the end-point relative permeability of phase p at its maximum saturation (the superscript e means end point); n_p is the exponent of phase p; $\bar{S}_p$ is the normalized saturation; and S_{pr} is the irreducible or residual saturation of phase p. These parameters are capillary number dependent. Equations 7.125 and 7.126 are applied at a specific capillary number. At different capillary numbers, we need some kind of interpolation or extrapolation. For the ease of description and understanding, the following sections describe two-phase flow first and then three-phase flow.

Two-Phase Flow

Two-phase flow could be either a low capillary number waterflooding case (water and oil phases), type I system (excess oil and microemulsion phases), or type II system (excess water and microemulsion phases). For one two-phase flow, Eq. 7.126 becomes

$$\bar{S}_p = \frac{S_p - S_{pr}}{1 - S_{pr} - S_{p'r}}, \tag{7.127}$$

where $S_{p'r}$ is the residual saturation of the other phase (conjugate phase). In this equation, S_{pr} and $S_{p'r}$ are N_C-dependent and can be estimated according to Eq. 7.121.

It is assumed that the end-point relative permeabilities depend on the residual saturation of *the other conjugate* phase. If we assume that k_{rp}^e at any capillary number can be interpolated between those at the critical capillary number and at the maximum capillary number, we have

$$k_{rp}^e = \left(k_{rp}^e\right)^{(N_C)_c} + \frac{\left(S_{p'r}\right)^{(N_C)_c} - S_{p'r}}{\left(S_{p'r}\right)^{(N_C)_c} - \left(S_{p'r}\right)^{(N_C)_{max}}}\left[\left(k_{rp}^e\right)^{(N_C)_{max}} - \left(k_{rp}^e\right)^{(N_C)_c}\right]. \tag{7.128}$$

Similarly, for the exponent, we have

$$n_p = n_p^{(N_C)_c} + \frac{\left(S_{p'r}\right)^{(N_C)_c} - S_{p'r}}{\left(S_{p'r}\right)^{(N_C)_c} - \left(S_{p'r}\right)^{(N_C)_{max}}}\left[n_p^{(N_C)_{max}} - n_p^{(N_C)_c}\right]. \tag{7.129}$$

It is assumed that k_{rp}^e and n_p are correlated to the residual saturation of the conjugate phase through linear interpolation in Eqs. 7.128 and 7.129, although this assumption may not be exactly correct (Fulcher et al., 1985; Anderson, 1987; Masalmeh, 2002; Tang and Firoozabadi, 2002).

Amaefule and Handy (1982) proposed two empirical equations to calculate water and oil relative permeabilities at reduced IFT. Their model showed that the water relative permeability is a function of its own saturation and its own reduced residual saturation at low IFT, but the oil relative permeability is a function of the two phase saturations and the two reduced residual saturations. Henderson et al. (2000) proposed a correlation for the N_c-dependent k_r of each phase that is interpolated between the base relative permeability at a low capillary number and the miscible relative permeability.

Three-Phase Flow

Three-phase flow occurs in a type III system. For the three-phase flow, the relative permeability of phase p has the two residual saturations of the other two phases. For the excess water phase relative permeability, k_{rw} (i.e., here phase p is w), there are two residual saturations, S_{or} and S_{mr}. In this case, $\Sigma S_{\not{p}r}$ is the sum of all the residual saturations except the water phase in the normalized saturation, which is defined as

$$\sum S_{\not{w}r} = \min\{S_{or}, S_o\} + \min\{S_{mr}, S_m\}. \tag{7.130}$$

Here, S_o and S_m are the oil and microemulsion phase saturations, respectively, present in a specific location (a grid block in simulation). They are not necessarily equal or larger than their respective residual saturations. The normalized saturation is now

$$\bar{S}_w = \frac{S_w - S_{wr}}{1 - S_{wr} - \sum S_{\not{w}r}}, \tag{7.131}$$

and the end-point relative permeability k_{rw}^e is

$$k_{rw}^e = \left(k_{rw}^e\right)^{(N_C)_c} + \frac{\left(\sum S_{\not{w}r}\right)^{(N_C)_c} - \sum S_{\not{w}r}}{\left(\sum S_{\not{w}r}\right)^{(N_C)_c} - \left(\sum S_{\not{w}r}\right)^{(N_C)_{max}}}\left[\left(k_{rw}^e\right)^{(N_C)_{max}} - \left(k_{rw}^e\right)^{(N_C)_c}\right]. \tag{7.132}$$

The exponent is

$$n_w = n_w^{(N_C)_c} + \frac{\left(\sum S_{\not{w}r}\right)^{(N_C)_c} - \sum S_{\not{w}r}}{\left(\sum S_{\not{w}r}\right)^{(N_C)_c} - \left(\sum S_{\not{w}r}\right)^{(N_C)_{max}}}\left[n_w^{(N_C)_{max}} - n_w^{(N_C)_c}\right]. \tag{7.133}$$

For the excess oil phase, the sum of the other two residual saturations is

$$\sum S_{\not{o}r} = \min\{S_{wr}, S_w\} + \min\{S_{mr}, S_m\}. \tag{7.134}$$

The normalized oil saturation is

$$\bar{S}_o = \frac{S_o - S_{or}}{1 - S_{or} - \sum S_{\not{o}r}}, \tag{7.135}$$

and the end-point relative permeability k^e_{ro} is

$$k^e_{ro} = \left(k^e_{ro}\right)^{(N_C)_c} + \frac{\left(\sum S_{\not{or}}\right)^{(N_C)_c} - \sum S_{\not{or}}}{\left(\sum S_{\not{or}}\right)^{(N_C)_c} - \left(\sum S_{\not{or}}\right)^{(N_C)_{max}}}\left[\left(k^e_{ro}\right)^{(N_C)_{max}} - \left(k^e_{ro}\right)^{(N_C)_c}\right]. \quad (7.136)$$

The exponent of the excess oil phase is

$$n_o = n_o^{(N_C)_c} + \frac{\left(\sum S_{\not{or}}\right)^{(N_C)_c} - \sum S_{\not{or}}}{\left(\sum S_{\not{or}}\right)^{(N_C)_c} - \left(\sum S_{\not{or}}\right)^{(N_C)_{max}}}\left[n_o^{(N_C)_{max}} - n_o^{(N_C)_c}\right]. \quad (7.137)$$

For the microemulsion phase, the sum of the other two residual saturations is

$$\sum S_{\not{mr}} = \min\{S_{wr}, S_w\} + \min\{S_{or}, S_o\}. \quad (7.138)$$

The normalized microemulsion phase saturation is

$$\bar{S}_m = \frac{S_m - S_{mr}}{1 - S_{mr} - \sum S_{\not{mr}}}, \quad (7.139)$$

and the end-point relative permeability k^e_{rm} is

$$k^e_{rm} = \left(k^e_{rm}\right)^{(N_C)_c} + \frac{\left(\sum S_{\not{mr}}\right)^{(N_C)_c} - \sum S_{\not{mr}}}{\left(\sum S_{\not{mr}}\right)^{(N_C)_c} - \left(\sum S_{\not{mr}}\right)^{(N_C)_{max}}}\left[\left(k^e_{rm}\right)^{(N_C)_{max}} - \left(k^e_{rm}\right)^{(N_C)_c}\right]. \quad (7.140)$$

The exponent of the microemulsion phase is

$$n_m = n_m^{(N_C)_c} + \frac{\left(\sum S_{\not{mr}}\right)^{(N_C)_c} - \sum S_{\not{mr}}}{\left(\sum S_{\not{mr}}\right)^{(N_C)_c} - \left(\sum S_{\not{mr}}\right)^{(N_C)_{max}}}\left[n_m^{(N_C)_{max}} - n_m^{(N_C)_c}\right]. \quad (7.141)$$

According to the preceding formulation, the parameters S_{pr}, n_p, and k_{rp} for each phase at $(N_C)_c$ and $(N_C)_{max}$ are needed, as required in the current version of UTCHEM. In practice, depending on the assumptions made, some parameters may be estimated from others. For example, it is assumed that the end-point relative permeabilities at high capillary numbers are unity, and the residual saturations are 0 in the formulation by Hirasaki et al. (1983), although this is not supported by data from Delshad et al. (1985). In the formulation by Hirasaki et al., the end point and exponent of microemulsion relative permeability approach those of water or oil, depending on the trapping saturations (approaching water when the oil trapping saturation approaches 0, and vice versa), as in

$$k_{rm} = \left[\omega k^e_{rw} + (1-\omega)k^e_{ro}\right]\left(\bar{S}_m\right)^{n_m}, \quad (7.142)$$

where

$$n_m = \omega n_w + (1-\omega) n_o, \tag{7.143}$$

$$\omega = \frac{S_{oT}}{S_{wT} + S_{oT}}, \tag{7.144}$$

$$S_{wT} = \min\{S_{wr}, S_w\}, \tag{7.145}$$

$$S_{oT} = \min\{S_{or}, S_o\}. \tag{7.146}$$

In doing so, their formulation provides a prediction for microemulsion phase relative permeabilities from the water and oil phase data. If no oil exists in a simulation block (e.g., in an aquifer block), then S_{oT} is 0. However, it is not necessarily true that k_{rm} can be represented by k_{ro}. If Eqs. 7.142 through 7.146 are used to define microemulsion phase relative permeability, the microemulsion phase behavior should depend on composition, and we propose that ω is defined by the component fractions in the microemulsion phase such that

$$\omega = \frac{C_{wm}}{C_{wm} + C_{om}}, \tag{7.147}$$

where C_{wm} and C_{om} are the composition fractions of water and oil, respectively, in the microemulsion phase.

Equation 7.147 shows that when C_{om} is small, which corresponds to a type I microemulsion, ω is close to 1 and k_{rm} is close to k_{rw}; when C_{wm} is small, which corresponds to a type II microemulsion, ω is close to 0 and k_{rm} is close to k_{ro}. The prediction from Eq. 7.147 is consistent with the assumption of Hirasaki et al. (1983) that the microemulsion relative permeability approaches the water relative permeability at low salinities and oil relative permeability at high salinities.

Generally, it is assumed that S_{mr}, n_m, and k_{rm} for the microemulsion phase at the low capillary number, $(N_C)_c$, would be the same as those for the water at $(N_C)_c$. According to the preceding discussion, if the microemulsion is type II, k_{rm} is close to k_{ro}. Then S_{mr}, n_m, and k_{rm} for the microemulsion phase at $(N_C)_c$ should be close to those for the oil at $(N_C)_c$. In other words, S_{mr}, n_m, and k_{rm} for the microemulsion phase at $(N_C)_c$ should also be composition-dependent, as defined by Eq. 7.147, for example. The current version of UTCHEM does not take into account the composition-dependent k_{rm}. Instead, S_{mr}, n_m, and k_{rm} at $(N_C)_c$ and $(N_C)_t$ are required input, and the interpolation is performed in between based on the capillary number.

7.10.3 IFT Effect on the Relative Permeability Ratio k_{rw}/k_{ro}

As we know, relative permeabilities tend to increase as the IFT decreases or capillary number increases. With reduced IFT, the relative permeability curves become closer to straight lines (exponents close to 1), and the immobile saturations are closer to 0.

According to the theory by Buckley and Leverett (1942), the fraction of the displacing water (f_w) is

$$f_w = \frac{1}{1 + \frac{k_{ro}}{k_{rw}} \frac{\mu_w}{\mu_o}}. \quad (7.148)$$

This equation shows that the fraction is a function of the relative permeability ratio, k_{rw}/k_{ro}. It increases as the ratio is increased. We would be interested to know how the ratio changes as the IFT is decreased or the capillary number is increased.

This section uses the Corey-type (Brooks and Corey, 1966) equation (Eq. 7.125) to describe relative permeabilities. We first use this type of equation to see the effect of IFT on the k_r ratio in two-phase flow. Figure 7.37 compares the three k_r ratios: normal water/oil k_r, miscible k_r, and k_r between. Their parameters are presented in Table 7.10. The figure clearly shows that as the IFT is increased from the normal waterflooding high IFT to the intermediate IFT to the miscible flooding low IFT, the ratio of k_{rw} to k_{ro} increases in the low S_w

TABLE 7.10 Parameters to Generate k_r Ratios in Figure 7.37

	Water/Oil k_r	k_r Between	Miscible k_r
S_{or}	0.2	0.1	0
S_{wr}	0.3	0.1	0
n_w	2	1.5	1
n_o	2	1.5	1
k_{rw}	0.2	0.6	1
k_{ro}	0.85	0.925	1

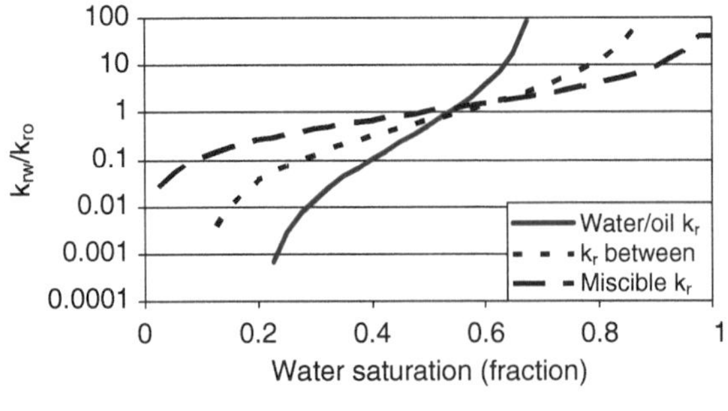

FIGURE 7.37 Effect of IFT on k_r ratio.

range ($S_w < 0.55$) but decreases in the high S_w range ($S_w > 0.55$). Chemical EOR processes are generally practiced in a high S_w range. Figure 7.37 shows the k_{rw}/k_{ro} ratio becomes lower in chemical flooding than in waterflooding. According to the fractional flow equation, the water cut becomes lower, thus improving the displacement performance.

Figure 7.38, with data from Delshad et al. (1985), compares the k_r ratio of a type I system (ME/O) with the k_r ratio of the corresponding water/oil system (W/O). In the type I system, the ratio is the k_r of the water-rich microemulsion phase to the k_r of the excess oil phase. In the figure, the dotted line is the extension of W/O data, which is above the k_r ratio of ME/O low IFT system. Figure 7.39 compares the k_r ratio of a type II system (ME/O) with the k_r ratio of the corresponding water/oil system (W/O). In the type II system, the ratio is the k_r of the excess water phase to the k_r of the oil-rich microemulsion phase. These data were generated using the parameters listed in Delshad et al. (1987). Those parameters were obtained by fitting their experimental data. The figure shows that the k_r ratio in the type II system (W/ME) is lower than that in the corresponding W/O system in the high saturation range.

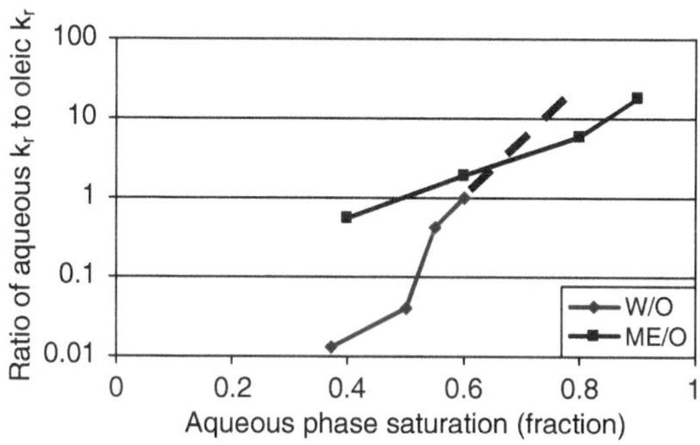

FIGURE 7.38 Comparison of k_r ratio in a type I system with k_r ratio in waterflooding.

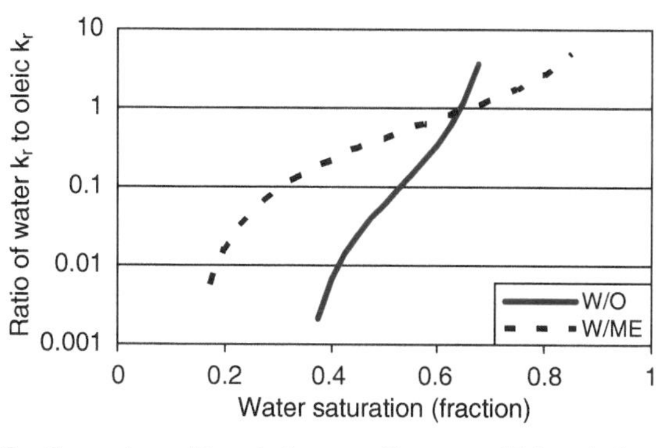

FIGURE 7.39 Comparison of k_r ratio in a type II system with k_r ratio in waterflooding.

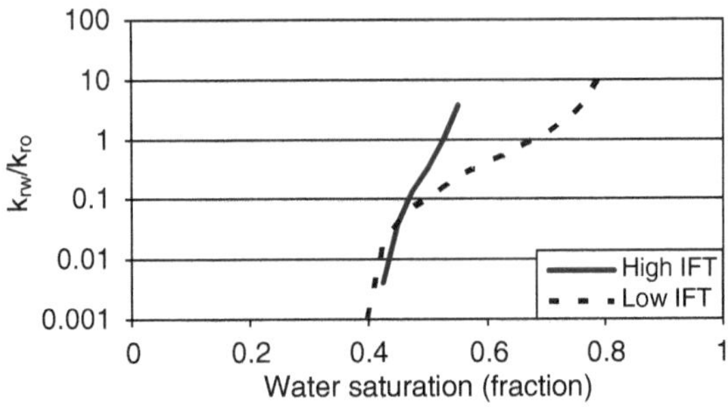

FIGURE 7.40 Effect of IFT on k_r ratio.

Similarly, we used k_r correlations developed by Fulcher et al. (1985) by fitting their experimental data to calculate the k_r ratio of water to oil. The k_r ratio of a high IFT system is compared with that of a lower IFT in Figure 7.40. The same observation can be made from this figure. We also checked other published data (not shown here to avoid tedious presentation), and they all show that the k_r ratio is decreased when IFT is lower; thus, the oil displacement efficiency is improved in the high aqueous phase saturation range as IFT is reduced.

7.11 SURFACTANT RETENTION

Control of surfactant retention in the reservoir is one of the most important factors in determining the success or failure of a surfactant flooding project. In a typical surfactant flood, chemical cost is usually half or more of the total project cost. Based on the mechanisms, surfactant retention can be broken down into precipitation, adsorption, and phase trapping. However, it is difficult to separate the surfactant losses from different mechanisms. Therefore, we usually report the total surfactant loss as surfactant retention without clearly specifying the losses from different mechanisms.

7.11.1 Precipitation

When we introduced phase behavior tests earlier, we mentioned aqueous stability tests. The main objective of aqueous stability tests is to eliminate the surfactant precipitation problem. As we already know, the solubility of surfactant decreases with salinity. During aqueous stability tests, the surfactant solution becomes opaque up to some salinity, showing the surfactant starts to aggregate or even precipitate. When divalent or multivalent ions exist in the solution, the salinity needed to start precipitation is much lower.

If the surfactant concentration is increased, the solution will also become opaque, as shown in Figure 7.41. In the figure, the reduction in light transmittance through the solution represents the degree of surfactant precipitation in

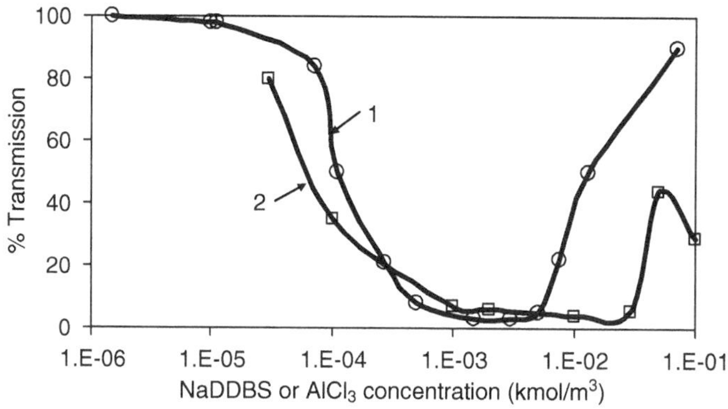

FIGURE 7.41 Light transmittance of Na-dodecylbenzenesulfonate solutions as a function of dodecylbenzenesulfonate (NaDDBS) and $AlCl_3$ concentrations. Deoiled NaDDBS (H Do 2), T = 24°C. Curve 1, NaDDBS 0.8 mol/m^3 $AlCl_3$, pH = 4.1 ± 0.1; and Curve 2, 0.77 mol/m^3 NaDDBS + $AlCl_3$. *Source: Somasundaran et al. (1984b).*

the solution. It also shows that at higher surfactant concentrations that are above about 5×10^{-3} kmol/m^3, precipitated surfactant redissolved. Some apparent redissolution of surfactant also is shown at a higher concentration of $AlCl_3$. The precipitate–dissolution phenomenon was also observed in the presence of divalent ions, but not with monovalent ions (Somasundaran et al., 1984a; Isaacs et al., 1994; Li, 2007; Wu et al., 2008).

Petroleum sulfonates are widely used in surfactant flooding. When there are divalents such as Ca^{2+} and Mg^{2+} in the solution, however, the divalent complex is formed. The complex has limited solubility in water and precipitation occurs. When the concentration of petroleum sulfonates is increased, the precipitates redissolve. When the concentration is further increased, the precipitation occurs again. In other words, as the surfactant concentration is increased, there is a phenomenon of precipitation–dissolution–reprecipitation. This means the following reaction is reversible:

$$2(R-SO_3)+M^{2+} \leftrightarrow M(R-SO_3)_2\downarrow. \qquad (7.149)$$

Here, M represents a divalent, and R-SO_3 represents a surfactant.

Figure 7.42 shows the light transmittance through the system of surfactant TRS10-80, 136.4 mg/L phosphate, and 0.0114 mol/L Ca^{2+} at 12°C. As is well known, phosphates are good chelating agents. Without phosphate (Curve 1), the solution became cloudy, represented by low light transmittance in the very low surfactant concentration range. KH_2PO_4 solution was similar to the solution without phosphate. Adding $Na_5P_3O_{10}$ and $(NaPO_3)_6$ greatly increased light transmittance, indicating that petroleum sulfonate TRS10-80 becomes more tolerant to Ca^{2+}. At the minimum light transmittance (corresponding to the maximum precipitation), when $Na_5P_3O_{10}$ was added, the maximum solubility

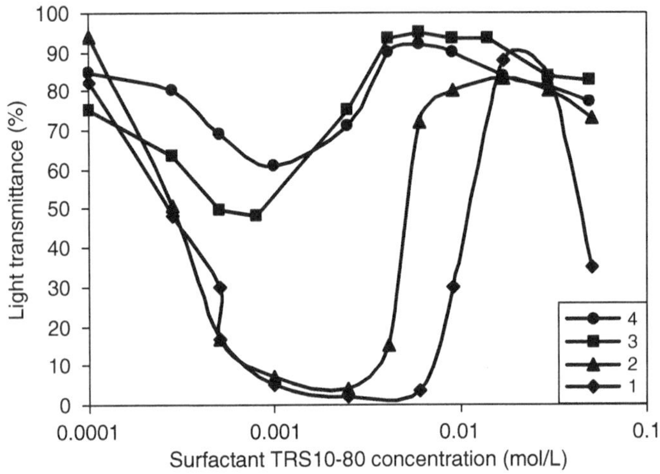

FIGURE 7.42 Light transmittance of the surfactant TRS10-80 solution with phosphates and Ca^{2+}. Curve 1, no phosphate; Curve 2, KH_2PO_4; Curve 3, $Na_5P_3O_{10}$; Curve 4, $(NaPO_3)_6$. *Source: Li (2007).*

of the surfactant was reduced to 9×10^{-4} mol/L; when $(NaPO_3)_6$ was added, the maximum solubility was reduced to 1×10^{-3} mol/L.

The results indicate that $(NaPO_3)_6$ was a better chelating agent than $Na_5P_3O_{10}$ at low surfactant concentrations. When the surfactant concentration was increased, the light transmittance became higher. The light transmittance reached the maximum at a concentration of 1×10^{-2} mol/L before it went down. From Curve 1 (without any chelating agent) shown in Figure 7.42, we can see that at the lowest light transmittance, the surfactant concentration was close to the Ca^{2+} equivalent concentration. At the highest light transmittance, the surfactant concentration was several times (1–3 times) that of Ca^{2+} concentration.

The mechanism of precipitation–dissolution–reprecipitation is not yet well understood. Li (2007) provided some explanations. When the petroleum sulfonate concentration is below its CMC, the single surfactant molecules in the solution increase as the surfactant concentration is increased. The concentrations of Na^+ and RSO_3^- also increase linearly with the surfactant concentration. Their concentrations reach their maximums at the CMC. Meanwhile, the concentrations of Ca^{2+} and Cl^- are constant. Ca^{2+} reacts with RSO_3^- to generate $Ca(RSO_3)_2$ precipitation; then Ca^{2+} concentration is reduced. $Ca(RSO_3)_2$ is suspended in the solution, resulting in lower light transmittance.

This precipitation continues until the surfactant concentration reaches its CMC and all the Ca^{2+} in the solution has been consumed. When the surfactant concentration is above the CMC, micelles are formed and single surfactant molecules cannot be increased any more, so no further precipitation can be generated. Owing to micelle solubilization, the existing $Ca(RSO_3)_2$ precipitates are solubilized, and the light transmittance is increased. As the surfactant concentration is further increased, more micelles are formed, and more $Ca(RSO_3)_2$ precipitates are solubilized, The light transmittance becomes higher until it reaches the maximum. Afterward, when the surfactant concentration is increased above a limit, the surfactant itself will precipitate because of its limited solubility. There is a possibility that, liquid crystals may form. So the light transmittance is decreased again.

In another possible mechanism, when the surfactant concentration is above the CMC, the micelles and single surfactant molecules are in equilibrium. The precipitates may form complexes with surfactant molecules. The complexes have electric charge:

$$Ca(RSO_3)_2 + RSO_3^- \leftrightarrow Ca(RSO_3)_2 \cdot RSO_3^-. \tag{7.150}$$

The charged precipitates can be redissolved as the surfactant concentration is increased.

As we discussed in Section 7.4 on phase behavior, alcohol can increase the solubility of a surfactant. Other factors could be temperature, chromatographic separation of surfactant species, and so on.

7.11.2 Adsorption

Adsorption of surfactant on reservoir rock can be determined by static tests (batch equilibrium tests on crushed core grains) and dynamic tests (core flood) in the laboratory. The units of surfactant adsorption in the laboratory can be mass of surfactant adsorbed per unit mass of rock (mg/g rock), mass per unit pore volume (mg/mL PV), moles per unit surface area (μeq/m^2), and moles per unit mass of rock (μeq/g rock). The units used in field applications could be volume of surfactant adsorbed per unit pore volume (mL/mL PV) or mass per unit pore volume (mg/mL PV). Some unit conversions follow:

$$\begin{aligned}\hat{C}_s\left[\frac{\text{mL}}{\text{mL PV}}\right] &= \hat{C}_s\left[\frac{\mu\text{eq}}{\text{m}^2}\right]\left[\frac{10^{-6}\,\text{eq}}{\mu\text{eq}}\right]\left[\frac{\text{MWg}}{\text{eq}}\right]\left[\frac{\text{mL}}{\rho_s\text{g}}\right]\left[\frac{S_r\text{m}^2}{\text{g rock grain}}\right]\\ &\quad\times\left[\frac{\rho_r\text{g rock grain}}{\text{mL rock grain}}\right]\left[\frac{(1-\phi)\text{mL rock grain}}{\text{mL rock bulk}}\right]\left[\frac{\text{mL rock bulk}}{\phi\text{mL PV}}\right]\\ &= \frac{10^{-6}(\text{MW})\rho_r S_r(1-\phi)}{\phi\rho_s}\hat{C}_s\left[\frac{\mu\text{eq}}{\text{m}^2}\right],\end{aligned} \tag{7.151}$$

$$\hat{C}_s\left[\frac{\text{mg}}{\text{mL PV}}\right] = \hat{C}_s\left[\frac{\text{mL}}{\text{mL PV}}\right]\left[\frac{\rho_s\text{g}}{\text{mL}}\right]\left[\frac{10^3\text{mg}}{\text{g}}\right] = 10^3\rho_s\hat{C}_s\left[\frac{\text{mL}}{\text{mL PV}}\right], \tag{7.152}$$

$$\hat{C}_s\left[\frac{\text{mg}}{\text{g rock}}\right] = \hat{C}_s\left[\frac{\text{mg}}{\text{mL PV}}\right]\left[\frac{\phi\text{mL PV}}{\text{mL bulk PV}}\right]\left[\frac{\text{mL bulk PV}}{(1-\phi)\text{mL rock PV}}\right]\left[\frac{\text{mL rock PV}}{\rho_r\text{g rock}}\right]$$
$$= \frac{\phi}{(1-\phi)\rho_r}\hat{C}_s\left[\frac{\text{mg}}{\text{mL PV}}\right]$$
$$= \frac{10^3\phi\rho_s}{(1-\phi)\rho_r}\hat{C}_s\left[\frac{\text{mL}}{\text{mL PV}}\right]$$
$$= 10^{-3}(\text{MW})S_r\hat{C}_s\left[\frac{\mu\text{eq}}{\text{m}^2}\right]. \tag{7.153}$$

Here, $\hat{C}_s$ is the surfactant adsorption, ρ_r is the density of rock grain in g/mL, ρ_s is the surfactant density in g/mL, MW is the surfactant molecular weight in g/(eq.), ϕ is the porosity in fraction, and S_r is the surface area of rock grain in m^2/g. If $\rho_r = 2.65$ g/mL, $\rho_s = 1.1$ g/mL, MW = 450 g/(eq.), $\phi = 0.3$, $S_r = 1$ m^2/g, then $\hat{C}_s$ in mL/mL PV = 2.53×10^{-3} $\hat{C}_s$ in μeq/m²; $\hat{C}_s$ in mg/mL PV = 2.78 $\hat{C}_s$ in μeq/m²; $\hat{C}_s$ in mg/g rock = 178 $\hat{C}_s$ in mL/mL PV; $\hat{C}_s$ in mg/g rock = 1/6.2 $\hat{C}_s$ in mg/mL PV.

Surface area of the porous media has a remarkable effect on surfactant adsorption. Liu (2007) measured surfactant adsorption in three rock samples of the same carbonate porous medium but with different surface areas. He used a TC blend surfactant—1 : 1 mixture by weight of dodecyl 3 ethoxylated sulfate and iso-tridecyl 4 propoxylated sulfate from Stepan. He found that the adsorptions of the TC blend on the three samples were close to each other if the adsorption was calculated by using surfactant adsorption amount per porous media surface area, as shown in Figure 7.43. However, if the adsorption was

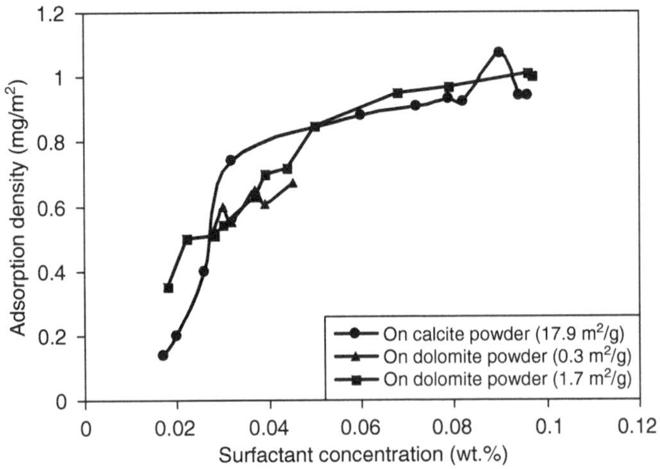

FIGURE 7.43 Graphic representation of adsorption of the TC blend on different samples expressed in mg/m². *Source: Liu (2007).*

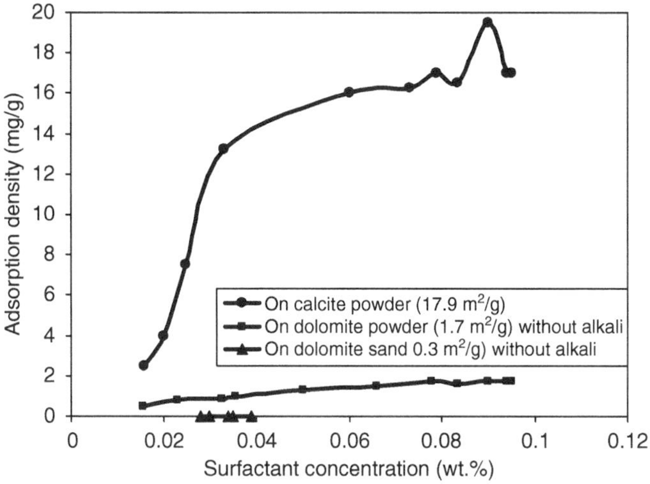

FIGURE 7.44 Adsorption of the TC blend expressed in mg/g on the same three samples as those in Figure 7.43. *Source: Liu (2007).*

represented by surfactant adsorption amount per weight of porous media, the adsorptions on the three samples were very different, as shown in Figure 7.44, even though the mineralogy of the three samples was similar. These results imply that it is the surface area, not the weight, of the porous media that should be used to compare the adsorption.

The surface area S_r is difficult to measure, however, and the correct values may not be readily available. So any unit with surface area S_r may not be a convenient unit. Probably a convenient unit for a field application is mL/(mL PV) because the reservoir pore volume is known, and the amount of injected surfactant is usually expressed in volume. The adsorption (retention) data in such a unit can provide a direct guide without unit conversion about the minimum surfactant to be injected.

Surfactant retention in reservoirs depends on surfactant type, surfactant equivalent weight, surfactant concentration, rock minerals, clay content, temperature, pH, redox condition, flow rate of the solution, and so on. As the equivalent weight of the surfactant increases, surfactant retention in general also increases (Glinsmann, 1978). Meyers and Salter (1981) summarized the surfactant retention data available in the literature published by the Society of Petroleum Engineers (SPE). Although surfactant retention is a function of many factors, a statistical average may be useful when laboratory measurements are not available. Figure 7.45 shows the statistical analysis of the data. The median value is 4.3 mg/mL PV. If we use a conversion factor of 1/6, this value of 4.3 mg/mL PV is equivalent to 0.7 mg/g rock.

This value seems to be on the higher side of the typical values of surfactant adsorption on Berea sandstone cores Green and Willhite (1998) summarized

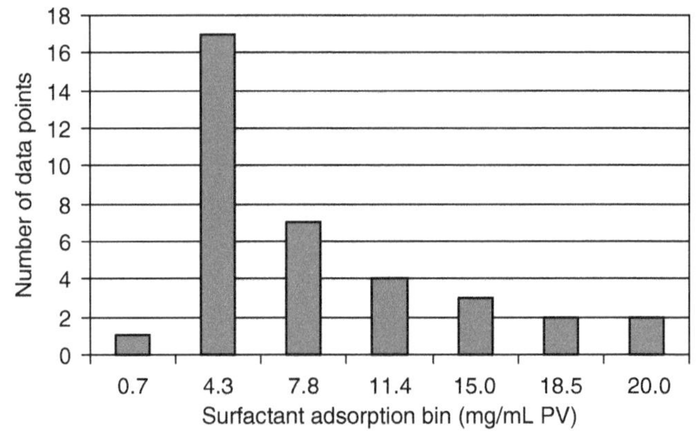

FIGURE 7.45 Statistical analysis of the surfactant retention data. *Source: Data summarized by Meyers and Salter (1981).*

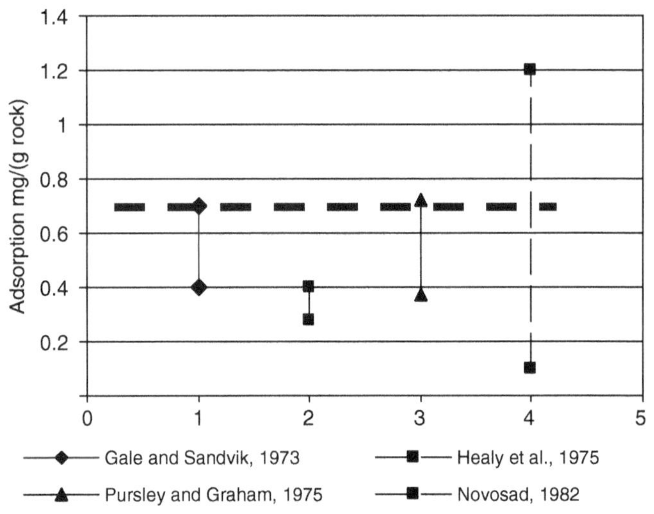

FIGURE 7.46 Typical values of surfactant adsorption from the SPE literature.

from the literature, as shown in Figure 7.46. It is also higher than the value of around 0.1 mg/g reported by Levitt et al. (2006) and Dwarakanath et al. (2008). In Figure 7.46, the horizontal dotted line represents the average from our statistical analysis.

Surfactant adsorption is strongly affected by the redox condition of the system. Laboratory cores typically have been exposed to oxygen and are in an aerobic state. Wang (1993) showed that the surfactant adsorption in anaerobic conditions was lower than that in aerobic conditions. Most of the data collected from the literature were obtained in aerobic conditions in the laboratory. It is possible that those data are higher than those in reservoir conditions. One important factor that could reduce surfactant adsorption is pH, discussed in Chapter 12 on alkaline-surfactant flooding.

Regarding the surfactant type and rock type, nonionic surfactants have much higher adsorption on a sandstone surface than anionic surfactants (Liu, 2007). However, Liu's initial experiments indicated that the adsorption of nonionic surfactant on calcite was much lower than that of anionic surfactant without the presence of Na_2CO_3 and was of the same order of magnitude as that of anionic surfactant with the presence of Na_2CO_3. Thus, nonionic surfactants might be candidates for use in carbonate formations from the adsorption point of view. The role of salinity is much less important, but the temperature effect is much more important for nonionics than for anionics (Salager et al., 1979a). More factors that affect adsorption were discussed by Somasundaran and Hanna (1977).

Surfactant adsorption isotherms are very complex in general. The amount adsorbed generally increases with surfactant concentration in the solution, and it reaches a plateau at some sufficiently large surfactant concentration. For pure surfactants, this concentration is in fact the CMC, which is often 100 times or more below the injected surfactant concentration. Thus, the complex detailed shape of the isotherm below the CMC has little practical impact on the transport and effectiveness of the surfactant. It has been found that a Langmuir-type isotherm can be used to capture the essential features of the adsorption isotherm, as in

$$\hat{C}_3 = \min\left(C_3, \frac{a_3\left(C_3 - \hat{C}_3\right)}{1 + b_3\left(C_3 - \hat{C}_3\right)} \right), \tag{7.154}$$

where C_3 is the injected surfactant concentration, or in general, the surfactant concentration before adsorption. $C_3 - \hat{C}_3$ is actually the equilibrium concentration in the solution system. Note that C_3 and $\hat{C}_3$ must be in the same unit. a_3 and b_3 are empirical constants. The unit of b_3 is the reciprocal of the unit of C_3, but a_3 is dimensionless. a_3 is defined as

$$a_3 = (a_{31} + a_{32}C_{se})\left(\frac{k_{ref}}{k}\right)^{0.5}, \tag{7.155}$$

where a_{31} and a_{32} are input or fitting parameters, C_{se} is the effective salinity, k is the permeability, and k_{ref} is the reference permeability of the rock used in the laboratory measurement.

According to the Langmuir isotherm equation, the degree of surfactant adsorption increases with concentration. Note that there are some special cases. Figure 7.47 shows the adsorption isotherm of petroleum sulfonate on kaolinite with or without Ca^{2+} present; there is a maximum adsorption in the plot of

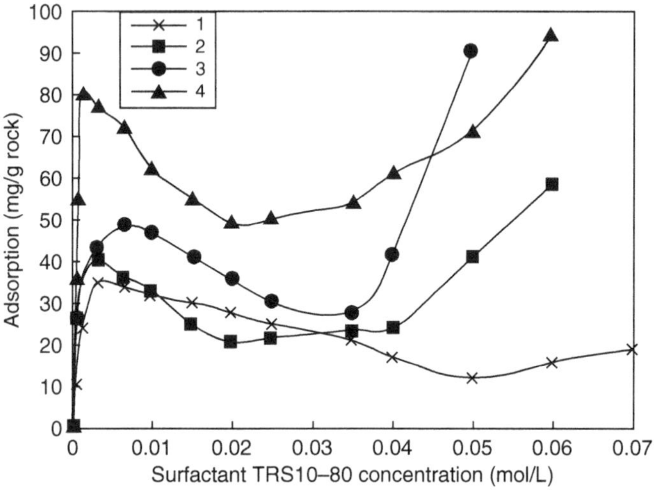

FIGURE 7.47 Effect of Ca^{2+} on the petroleum sulfonate adsorption on kaolinite. Curve 1, Na^+ only; Curve 2, 0.0114 mol/L Ca^{2+} and 310 mg/L phosphate; Curve 3, no Ca^{2+} or phosphate; and Curve 4, 0.0114 mol/L Ca^{2+} but no phosphate. The temperature was 30 °C. *Source: Li (2007).*

adsorption versus concentration when the divalents calcium and magnesium existed. This phenomenon is related to divalent association with the surfactant micelles. When the concentration is low, surfactant adsorption increases with its concentration according to the Langmuir isotherm. Because of divalent association with micelles, precipitation may occur. When the surfactant concentration is further increased, some of the adsorbed surfactant and precipitates may be solubilized because of strong micelle solubilization capability. Thus, adsorption decreases. When the surfactant concentration is increased further again, adsorption increases again. In other words, there is the precipitation–dissolution–reprecipitation phenomenon, which must have an effect on surfactant adsorption on rock surfaces.

Adsorption is considered to be irreversible with concentration. After adsorption, the concentration in the bulk is critical micelle concentration (CMC). The surfactant concentration on surfaces is above CMC. At this time, the chemical potential at the surface and in the bulk is the same. The surfactant concentration in the bulk can be reduced to lower than the original CMC by dilution. Then surfactant on the surface can be desorbed. However, a large volume of solvent is needed, so the desorption by dilution is not practical. Therefore, the adsorption is irreversible with surfactant concentration.

Surfactant adsorption is reversible with salinity, however, and it decreases with decreasing salinity (Somasundaran and Hanna, 1977). Hurd (1969) patented a method of desorbing and reusing surfactant by flooding a saline surfactant solution using less saline water. Figure 7.48 shows the surfactant history at the effluent end of a core flood (fraction of injected surfactant concentration

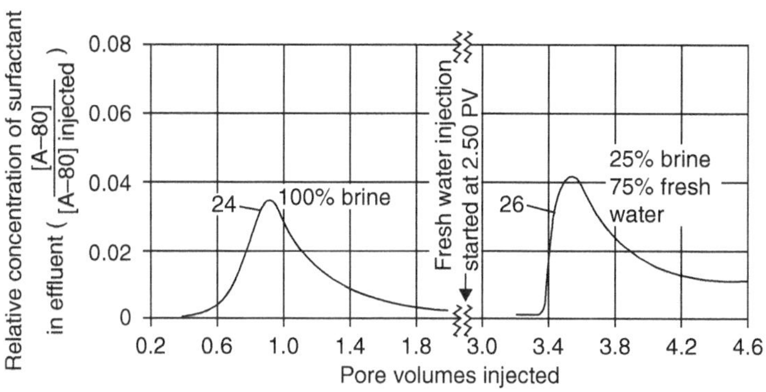

FIGURE 7.48 Produced surfactant concentration (fraction of injected) versus pore volume injected. *Source: Hurd (1969).*

versus injected pore volumes). In this case, a Lucite tube 1 inch in diameter by 12 inches long was packed with washed, disassociated core samples from the Loma Nova sand, Loma Novia field in Duval County, Texas. The injection sequences were as follows:

1. 0.1 PV Loma brine + 3 wt.% sodium carbonate
2. 0.03 PV Loma brine + 0.1 wt.% sodium carbonate + 0.1 wt.% sodium tripolyphosphate
3. 0.1 PV Loma brine + 0.05 wt.% sodium carbonate + 0.1 wt.% sodium tripolyphosphate + 1 wt.% Alconate 80 (A-80) surfactant
4. Up to 2.5 PV Loma brine + 0.05 wt.% sodium carbonate + 0.1 wt.% sodium tripolyphosphate
5. Up to 4.6 PV (25% Loma brine + 75% fresh water) + 0.05 wt.% sodium carbonate + 0.1 wt.% sodium tripolyphosphate

Sodium carbonate and sodium tripolyphosphate were added to the water to obtain the desired interfacial behavior. Figure 7.48 shows that the first peak (marked as 24) represents the maximum surfactant concentration at the effluent end from slug # 4 (saline water). The second peak (marked as 26) represents the maximum concentration obtained by less saline waterflooding. Note that peak 26 is higher than peak 24. The second bank of surfactant was formed from the desorption of surfactant left by the first bank of surfactant solution on the solid surfaces.

7.11.3 Phase Trapping

Phase trapping could be caused by mechanical trapping, phase partitioning, or hydrodynamical trapping. It is related to multiphase flow. The mechanisms are complex, and the magnitude of surfactant loss owing to phase trapping could be quite different depending on multiphase flow conditions. Surfactant phase trapping can be more significant than surfactant adsorption (Hirasaki et al., 2008). Although the phase trapping mechanism is complex, it is well known and accepted that phase trapping is related to types of microemulsion. In a Winsor II microemulsion system, the chase water behind the microemulsion is an aqueous phase, whereas the microemulsion is an oil-external phase whose viscosity could be higher than the chase water. Plus, the IFT in the rear of the microemulsion slug could be high so that the microemulsion and chase water are immiscible fluids. Thus, the chase water can easily bypass the microemulsion phase, resulting in phase trapping. It has been observed that surfactant phase trapping is much lower in a Winsor I environment where the microemulsion is the water-external phase, which can be displaced miscibly by the chase water.

7.12 DISPLACEMENT MECHANISMS

As mentioned earlier, traditionally, surfactant flooding can be grouped into dilute surfactant flooding and micellar flooding. Discussion of the displacement mechanisms may be made according to these two groups. The key mechanism for these surfactant floods is the low IFT effect. Surfactant flooding is principally an immiscible process in field applications, where slug size and surfactant concentration are limited by economics. Miscible displacement may occur in the early states of a flood, but the chemical slug quickly breaks down (multiple phases form) and the process becomes immiscible. Consequently, laboratory and process design must be based on immiscible flow (Green and Willhite, 1998). The miscible region is that portion of the phase diagram through which neither tie lines nor their extensions pass. In the development of miscibility by solubilization and swelling, what matters is not so much the partitioning of chemicals between aqueous and oleic phases, which are indexed by the location of the plait point, but whether or not the injected composition lies on a tie line or the extension of one (Larson et al., 1982).

7.12.1 Displacement Mechanisms in Dilute Surfactant Flooding

In dilute surfactant flooding a water-wet reservoir, when surfactant solution contacts residual oil droplets, the oil droplets are emulsified because of low IFT and entrained in surfactant solution. These entrained oil droplets are carried forward and are "pulled" to become long oil threads so that they can deform and pass through pore throats. When the salinity is low, oil-in-water (O/W) emulsions are formed. When the salinity is high, water-in-oil (W/O) emulsions are formed. These oil droplets are coalesced to form an oil bank ahead of the surfactant slug. As surfactant contacts rock surfaces, wettability may be changed.

In an oil-wet reservoir, oil sticks at pore throats or on pore walls. When surfactant solution flows through the pores, because of low IFT, oil droplets at pore throats are displaced. The oil droplets on pore walls are deformed and displaced along the walls by the dilute surfactant solution. These oil droplets are moved down to bridge with the oil droplets downstream to form continuous oil flow paths. They are pulled into long threads, and the oil threads flow downward. The oil threads could be broken during flow. Once broken, they become small droplets and are emulsified. These small droplets flow downward and lodge at the next throats to be coalesced with other oil droplets. Generally, these emulsions are W/O type. The emulsified oil and displaced oil coalesce to form an oil bank ahead. Wettability may be changed from oil-wet to water-wet owing to surfactant adsorption.

In dilute surfactant flooding, oil droplets must be able to deform to pass through pore throats. In other words, the deformation capability of oil droplets in dilute surfactant flooding is very important. This effect can be achieved by low IFT.

7.12.2 Displacement Mechanisms in Micellar Flooding

As discussed earlier, depending on salinity and compositions, there are three types of microemulsions: type I oil-in-water, type II water-in-oil, and type III bicontinuous middle-phase. The solubilization or swelling mechanism is related to the type of microemulsion. Solubilization corresponds to type I, similar to a vaporizing-gas drive (giving, from the original oil point of view); swelling corresponds to type II, similar to a condensing-gas drive (receiving; Giordano and Salter, 1984).

In type I microemulsion, the emulsified oil droplets are "carried" forward and are coalesced with the oil ahead to form an oil bank. In type II microemulsion, it is easy for the external oil to merge with residual oil to form an oil bank.

In middle-phase microemulsion, owing to the lowest IFT, oil and water can be solubilized in each other, and oil droplets can flow more easily through pore throats. The oil droplets move forward and merge with the oil downstream to form an oil bank. Because of the solubilization effect, water and oil volumes are expanded, leading to higher relative permeabilities and lower residual saturations. However, when k_{rw} increases faster than k_{ro} with decreasing IFT, the oil saturation in the oil bank and the oil recovery rate are deterioated, if no viscosity alteration is made.

In a water-wet reservoir, initially water film sticks to rock surfaces. Because middle-phase microemulsion can solubilize water, some water films will be replaced by the microemulsion. Thus, the rock surfaces will become less water-wet. Similarly, in an oil-wet reservoir, some oil films on rock surfaces will be replaced by microemulsion, and the rock will become less oil-wet. Therefore, microemulsion always behaves as the most wetting phase.

7.13 AMOUNT OF SURFACTANT NEEDED AND PROCESS OPTIMIZATION

Field test data show that a typical amount of surfactant injected in C_s (in %) × injection PV (in %) is about 10 to 12 (see Section 13.8). The minimum chemical mass injected should be just about the retention. The rule of thumb is about 20 to 50% more than the retention. According to the work of Trushenski et al. (1974), to prevent water breakthrough polymer and the polymer breakthrough micellar solution, the polymer volume requirement is about 0.5 PV.

Because a surfactant flood relies on phase behavior for its success, the answer to the injected surfactant concentration depends on the position of the slug composition relative to the boundary separating the miscible and semimiscible regions of the ternary diagram. The sensitivity to injected concentration is the greatest when the injected concentration lies near the boundary separating miscible and semimiscible regions (Pope and Nelson, 1978; Larson, 1979).

Todd et al. (1978) compared the two cases: (1) high concentration with small slug (soluble oil), and (2) large slug with low concentration. They found Case 1 to be preferable in the cases they investigated. They showed gravity was important. Based on published data, Gogarty (1976) also reported that a higher field recovery factor was obtained with a system of a high-concentration surfactant and low pore volume compared with a system of a low concentration and high pore volume. Murtada and Marx (1982) also observed that a low concentration slug did not bring the same recovery characteristics as a high concentration slug. For continuous surfactant injection, recovery was independent of surfactant concentration.

7.14 AN EXPERIMENTAL STUDY OF SURFACTANT FLOODING

Generally, chemical flooding is conducted in a high permeability reservoir. This section describes an experimental study of the application of surfactant flooding

TABLE 7.11 Effect of Surfactant Injection on Relative Permeabilities

	Before Surfactant Injection	After Surfactant Injection
S_{wi}	0.25	0.3
S_{or}	0.42	0.33
k_{rw} at S_{or}	0.12	0.2
k_{ro} at S_{wi}	1.0	1.0
Relative wettability index	0.23	0.42

in a low-permeability reservoir (He et al., 2006). The objective of using surfactant was to reduce injection pressure and thus enhance oil recovery in the low-permeability reservoir. The targeted field, Baolang, was operated by Henan Oilfield. Permeability was 19.8 md; porosity was 0.117; temperature was 93 to 103°C; and formation water TDS was 42,000 to 45,600 mg/L. Initially, water injection pressure was 22.0 MPa (3191 psi). Because the injected water was not enough, the injection pressure was increased to 32 MPa (4,641 psi). Such high pressure fractured the formation, and water quickly broke through production wells. To alleviate the problem, researchers investigated the option of surfactant flooding in the laboratory.

A 0.1% selected surfactant was then added to the injection water. The core flood experiments showed that injection pressure was reduced by 26.6%, and that the oil recovery was increased by 6.7%. This effect was a result of wettability alteration to more water-wet, reduced immobile water and oil saturations, and increased oil and water relative permeabilities. The data are shown in Table 7.11.

Chapter 8

Optimum Phase Type and Optimum Salinity Profile in Surfactant Flooding

8.1 INTRODUCTION

A microemulsion can exist in three types of systems—type II(−), type III , or type II(+)—depending on salinity. Below a certain salinity C_{sel}, the system is type II(−). Above a certain salinity C_{seu}, the system is type II(+). If the salinity is between C_{sel} and C_{seu}, the system is type III. In a type III system, the interfacial tension (IFT) of microemulsion/brine is lower than that in a type II(+) system, and the IFT of microemulsion/oil is lower than that in a type II(−) system. Thus, both IFTs are collectively low. At optimum salinity, which is defined as the middle of C_{sel} and C_{seu}, the two IFTs are equal. IFT is a very important parameter, with a lower value resulting in a higher capillary number (N_C). A higher capillary number will lead to lower residual oil saturation, thus higher oil recovery. Therefore, optimum salinity seems to be an obvious choice. Another area of contention is whether a type II(−) or type II(+) system is better for oil recovery (Larson, 1979).

The negative salinity gradient SG(−) means the salinities of preflush water, surfactant slug, and postflush (polymer solution and/or water drive) are in descending (decreasing) order. The negative salinity gradient was proposed based on the relationship that optimum salinity decreases as surfactant concentration is decreased (Nelson, 1982). Because of surfactant adsorption and retention, the surfactant concentration will be decreased as the surfactant solution moves forward. If the optimum salinity decreases with surfactant concentration, then the optimum salinity also decreases as the surfactant solution moves forward. Thus, the decreasing salinity will be consistent with the decreasing optimum salinity so that the optimum salinity is maintained as the surfactant solution moves forward. Therefore, the common belief is that the oil recovery factor in a type III system is higher than in a type II(−) or type II(+) system, and the recovery factor in the SG(−) system is the highest.

This chapter first reviews the literature's information on the subject. Then it presents extensive sensitivity results from UTCHEM simulations. Our

Modern Chemical Enhanced Oil Recovery. DOI: 10.1016/B978-1-85617-745-0.00008-5

argument is, if the common belief is correct, simulation should be able to demonstrate such results; and such simulation results should not be changed by varying the parameters that are not related to salinity. Our simulation results and the subsequent discussions show that the common belief cannot be held universally. Finally, this chapter proposes the concepts of optimum phase type and optimum salinity profile in surfactant-related flooding.

8.2 LITERATURE REVIEW

This section reviews information in the literature about optimum phase types, the relationship between optimum salinity and surfactant concentration, and optimum salinity gradients.

8.2.1 Optimum Phase Types

From the relationship between a capillary number and residual oil saturation, it is obvious that a low IFT will correspond to a high oil recovery. Even so, the first question is which IFT should be low—effluent IFT after core flood, equilibrium IFT before core flood, or even dynamic IFT? Equilibrium IFT is most commonly used. Gale and Sandvik (1973) used the IFT of the effluent fluids. Gerbacia and McMillen (1982) used the equilibrium IFT before flooding instead; they found that systems with low measured IFTs did not produce the highest recoveries.

Nelson and Pope (1978) discussed the possibility of reducing remaining oil to a low value by using the oil-swelling properties of a type II(+) phase environment. The recovered oil in their Experiment B of type II(+) was higher than that in their Experiment A of type II(−), but the surfactant in B was different from that in A. In a later paper, Nelson (1982) found that the surfactant concentration was the highest in a type II(+) phase environment. And the recovery in the environment was low, at least partly because of trapping or low mobility of surfactant-rich, oleic phases. A significant feature of type II(+) that should be considered to maximize oil recovery is that the composition path passes close to the plait point. Passing through the plait point implies a miscible displacement. Clearly, the displacement will be miscible for sufficiently high chemical concentration upstream. The chemical concentration will be high enough if a single-phase path exists from the injected composition to the plait point (Pope and Nelson, 1978). These authors also observed that little additional oil was produced after three-phase systems began to appear in the effluent. Therefore, they proposed that type III systems are the most active when displacing oil.

Larson (1979) showed that if the phase-volume effects of semimiscible flooding are to be relied on to recover oil (no chemical reduction in S_{or}) without requiring large quantities of chemical, then high-K_c (surfactant partition coefficient), type II(+) phase behavior is to be preferred over type II(−) phase

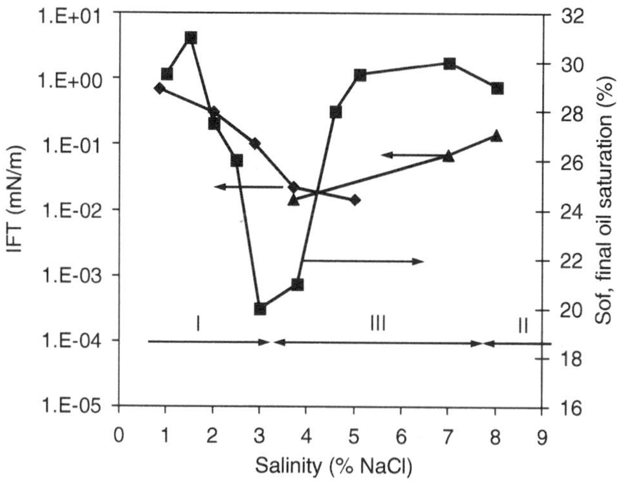

FIGURE 8.1 IFT and oil recovery versus salinity. *Source: Healy and Reed (1977a).*

behavior. However, high values of K_c delay chemical breakthrough and, therefore, delay oil recovery. If miscible, piston-like flooding is achieved, complete recovery at one PV injected is theoretically attainable for slug and continuous injection in the absence of dispersion and adsorption.

Healy and Reed (1977a) correlated the IFT with the oil recovery factor (final remaining oil saturation), as shown in Figure 8.1. From this figure, we can see that the final oil saturation followed the IFT trend. However, the minimum final oil saturation did not correspond to the minimum IFT. Healy and Reed did not find an obvious advantage attributable to either oil-external or water-external microemulsion from their 4-ft-long core floods.

8.2.2 Optimum Salinity versus Surfactant Concentration

The optimum salinity gradient depends on how the optimum salinity changes with surfactant concentration. Nelson (1982) proposed the negative salinity gradient concept based on the relationship that optimum salinity increased with surfactant concentration. For the relationship between optimum salinity and surfactant concentration, there are two groups. In one group, the optimum salinity increases with surfactant concentration, whereas in the other group, the optimum salinity decreases with surfactant concentration. Of course, there is another group in which the optimum salinity is independent of surfactant concentration. Hirasaki (1982a) pointed out that if the system actually contains three components plus sodium chloride, optimum salinity should be independent of overall surfactant concentration and water/oil ratio (WOR).

The change in optimum salinity is a consequence of divalent ions interacting with surfactant or of surfactant "pseudocomponents" partitioning in different proportions (Hirasaki et al., 1983). With NaCl brine, the electrolyte was partially excluded from the micelle. However, the opposite trend was observed with $CaCl_2$ brine because of the strong association of anionic surfactant with divalent cations. Therefore, decreasing surfactant concentration reduced interactions between the interfacial region and brine; then optimum salinity decreased (Pope and Baviere, 1991).

Glover et al. (1979) also discussed that the decreased optimum salinity with decreased surfactant concentration was caused by the exchange of divalent cations with monovalent cations and the existence of cosolvents in the surfactant solution. Puerto and Gale (1977) noted that increasing the alcohol's molecular weight decreased the optimum salinity. The same conclusions were reached by Hsieh and Shah (1977), who also noted that branched alcohols had higher optimum salinities than straight-chain alcohols of the same molecular weight.

Nelson (1981) reported that an increase in the surfactant concentration increased the optimum salinity (see Figure 8.2). The system was as follows: surfactant Petrostep 450, cosolvent NEODOL 25-3S, brine-to-oil ratio of 4, and the oil being 27% isooctane and 73% stock tank oil at 75.6°C. When multivalent ions were present, the surfactant concentration was more sensitive. For a system with multivalent cations, Nelson reported 64% decrease in optimum salinity when the surfactant concentration was lowered from 5% to 0.8%. He defined as favorable those conditions which cause reservoir clays to replace multivalent cations with monovalent cations in the region in which the surfactant is traveling. In other words, the conditions cause the brine in the surfactant bank to be "softened" partially by ion exchange with the reservoir clays. The actual optimum salinity at low surfactant concentrations will be a little higher than that read from a salinity requirement diagram if the reservoir clays can replace multivalent cations in the brine with monovalent cations. Nelson (1982)

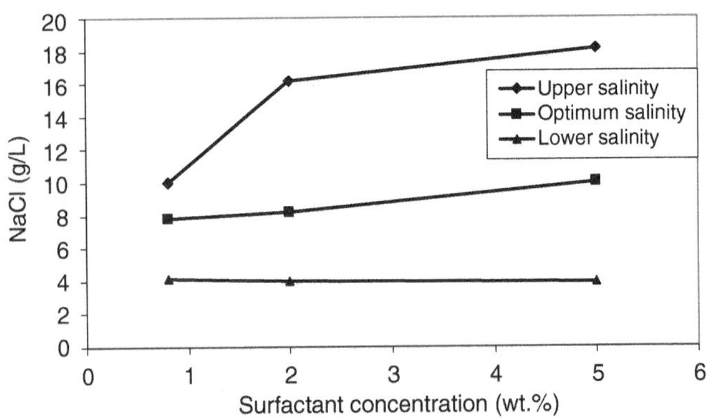

FIGURE 8.2 Salinity versus sulfonate concentration for the middle phase. *Source: Data from Nelson (1981).*

later stated that for most anionic surfactants, midpoint salinity decreases as surfactant concentration decreases, particularly in the presence of multivalent cations.

Another group of data shows that an increase in the sulfonate concentration decreases optimum salinity. For example, for the sulfonate solution [$C_{12}OXSO_3Na$, brine-to-octane ratio of 1.1, and 2-butanol (SBA) concentration of 3 wt.%], a 17 g/L NaCl brine gave a type I system at 0.1% sulfonate, but it gave a type II system at 5% sulfonate, as shown in Figure 8.3. This figure also shows that the salinity range of the middle phase (upper salinity minus lower salinity) decreased with the surfactant concentration. The middle phase volume increased with the surfactant concentration (data not shown). Baviere et al. (1981) further reported no shift in optimum salinity with petroleum sulfonate and a very small shift with pure sulfonate, but in an opposite direction to the one observed with sodium dodecyl orthoxylene sulfonate, as the surfactant concentration was increased.

Healy et al. (1976) and Reed and Healy (1977) reported that the dependence of optimum salinity on surfactant concentration was moderate, except for low concentration (<3%), where optimum salinity decreased as surfactant concentration increased. They also found that optimum salinity decreased with WOR.

Because alkalis provide an additional source of electrolytes, their presence in a surfactant solution will reduce the optimum salinity. When Martin and Oxley (1985) investigated the effects of alkalis on surfactant phase behavior, they found that for petroleum sulfonate, alkali anion had little or no effect on the phase behavior, whereas cations were effective for decreasing the optimum salinity in this order: potassium > sodium >> ammonium (see Figure 8.4). For the solutions with and without alkali, the optimum salinity decreased with the surfactant concentration. In the presence of alkaline chemicals, Martin and

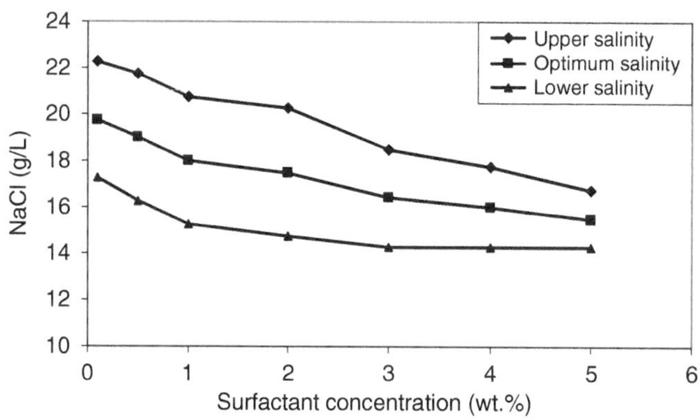

FIGURE 8.3 Salinity versus sulfonate concentration for the middle phase. *Source: Data from Baviere et al. (1981).*

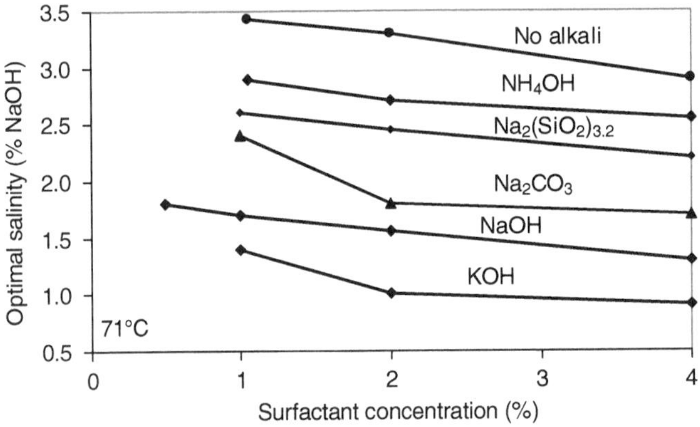

FIGURE 8.4 Effect of surfactant concentration on optimum salinity for 1 wt.% alkaline chemicals, brine, n-tetradecane, and surfactant Exxon 914-22. *Source: Martin and Oxley (1985).*

Oxley observed that the alkaline chemicals lowered the optimum salinity, which decreased with increasing petroleum sulfonate mixtures. In an $NaC_{12}OXS$/DN253S/IPA system, dilution led to a decrease in optimum salinity. Martin and Oxley attributed this to divalent cation sulfonate equilibria. In the second system, when only NaCl (without divalent) was in the brine, dependence of optimum salinity on surfactant concentration was much less, and the optimum salinity was higher. Martin and Oxley (1985) further discussed this interaction between divalent and sulfonate systems.

Glover et al. (1979) presented two types of phase behavior relationships that describe surfactant concentration versus optimum salinity. In one type, a decreasing surfactant concentration corresponded to an increasing optimum salinity. In the other type, a decreasing surfactant concentration corresponded to a decreasing salinity. In the $MEAC_{12}OXS$/TAA system, dilution (decreasing surfactant concentration) led to an increase in optimum salinity. Glover et al. proposed that the main factor to cause such change in direction was that TAA cosolvent concentration changed as the surfactant concentration changed. Such phase behavior was also reported by Bourdarot et al. (1984) and Rivenq et al. (1985).

Clearly, the relationship between the optimum salinity and surfactant concentration is complex (Salager et al., 1979b) and requires further investigation. The possibility of a shifting optimum salinity has to be taken into account to predict phase behavior during the oil recovery process.

8.2.3 Optimum Salinity Gradients

Gupta and Trushenski (1979) found that the most significant factor controlling oil recovery was the salinities of polymer and surfactant slugs (high oil recovery with a negative salinity gradient from the surfactant slug to the polymer slug and with a very low salinity in the polymer). Among their tests, a test with constant optimum salinity did not lead to the highest recovery. In that case, the surfactant loss was 100%. Whenever oil recovery was good, sulfonate loss was low, and oil/micellar IFT was low. However, low sulfonate loss did not ensure good oil recovery. In all cases, sulfonate loss was low when polymer salinity was low. Their injection scheme was waterflood, surfactant slug, and polymer. Note that their oil/micellar IFT showed a minimum but did not show the expected trend of decreasing oil/micellar IFT with increasing salinity.

Nelson (1982) performed extra experiments to support/propose the negative salinity gradient concept. He showed that all the salinities in brine, chemical slug, and drive water played a role. A negative salinity gradient should be used. He stated that it was because of the change (decrease) in salinity requirement as surfactant concentration dilutes during the flood. If the salinity requirement was increased, the opposite salinity gradient should work. However, there are no publications or results published on this subject.

To the best of our knowledge, the work by Gupta and Trushenski (1979) and experimental data from Nelson (1982) are the only data published so far to support the concept of a negative salinity gradient. Gupta and Trushenski, and Nelson used the same kind of surfactant with a special phase behavior. My explanation to their observation on salinity effect is that for the surfactant they used, the IFTs for both microemulsion/oil and microemulsion/water in the type III system were high. Therefore, when a lower salinity was in the drive water, low IFT was obtained because the lower salinity matched the lower optimum salinity of surfactant as the surfactant concentration was diluted.

Simulation results from Pope et al. (1979) showed that the best oil recovery for a given amount of injected surfactant occurred where a salinity higher than the optimum existed downstream of the slug and a salinity lower than the optimum existed upstream of the slug (in the polymer drive) and the slug itself traversed as much of the reservoir as possible in the low-tension type III environment. Generally, the chemical slug is small. Therefore, the initial and drive salinities matter most. Pope et al. observed that the low final salinity promoted low final retention of surfactant.

Experiments by Glover et al. (1979) for a type II(+) system showed that much of the surfactant retention could be caused by phase trapping. They also showed that much of this retained surfactant could be remobilized with a low-salinity drive. This view was supported by Hirasaki (1981). He pointed out that in a type II(+) environment, in the presence of dispersion, not only did the peak surfactant concentration decrease, but the location lagged behind with increased dispersion. These two factors resulted in a lower oil recovery and delay in oil production.

Hirasaki also explained why the negative salinity gradient works from the point of phase velocity. He stated that overoptimum salinity ahead of the surfactant bank is desirable because surfactant that mixes with high-salinity water will partition into the oleic phase, and because the phase velocity (phase cut/

phase saturation, f/S) of the oleic phase is less than unity, the surfactant will be retarded. Underoptimum salinity is not desirable ahead of the surfactant bank because the surfactant partitions into the aqueous phase, which has a phase velocity greater than unity. However, underoptimum salinity is desirable in the drive to propagate the surfactant. Thus, a system with overoptimum salinity ahead of the surfactant bank and underoptimum salinity in the drive will tend to accumulate the surfactant in the three-phase region where the lowest interfacial tensions generally occur.

Hirasaki et al. (1983) discussed the salinity effect on wetting phase, residual saturations, and relative permeabilities. The goal of their salinity gradient design was to keep as much surfactant as possible in the active region and minimize surfactant retention. Although they discussed the mechanisms to favor the gradient design, they did not show a case in which the recovery from a constant optimum salinity was lower than that from a gradient salinity.

Two characteristics of phase compositions in the transition zone for a type II(−) chemical flood should be emphasized. One is that the surfactant partitions into the brine-rich phase; the other is that for an aqueous-type chemical slug, the concentration of surfactant in the brine-rich phase never exceeds the concentration of surfactant in the original chemical slug. While in type II(+), the surfactant concentration in the microemulsion phase may be higher than that in the aqueous-type chemical slug, even though the slug is the only source of surfactant for the system. In type II(+), the "oil-rich" microemulsion is in equilibrium with essentially pure brine. Oil-rich is a relative term meaning a phase containing more oil and less brine than its conjugate phase; however, it may contain more brine than oil (Nelson and Pope, 1978). In type II(−), oil may be bypassed, whereas in type II(+), oil may be trapped (Gupta and Trushenski, 1979).

Simulation results in a heterogeneous field case by Wu (1996) showed that the recovery factor in a salinity gradient case was higher than that in a constant type III salinity only when the surfactant concentration was low. However, in another injection-scheme case (low concentration and large slug), the surfactant adsorption in the salinity gradient was even higher than that in an optimum salinity case, because in the salinity gradient case, surfactant contacted more reservoir volume. The oil recovery was higher in the salinity gradient case when polymer competitive adsorption was considered. Simulation results from Anderson et al. (2006) showed that the length of the slug changed the slope of the salinity gradient within the reservoir; however, this effect was outweighed by the changes in surfactant mass. They also found that increasing the polymer adsorption resulted in slightly higher oil recovery and better chemical efficiency. Where the salinity gradient could not be made, changing nonionic surfactant HLB or blending different nonionic surfactants with crude–oil–surfactants (anionic) optimized phase behavior (Kremesec et al., 1988).

The previous information taken from the literature may be summarized as follows:

- Because of low IFT in type III systems, the type III is the obvious choice. In surfactant flooding, low IFT is one important mechanism. Therefore, it is easy for people to accept this "obvious" choice. However, some data showed that the relationship between the IFT and oil recovery factor was not strong, or the oil recovery was not consistently higher with the low IFT. No further work has been done on the subject in recent years.
- The relationship between optimum salinity and surfactant concentration was system-dependent. In other words, the optimum salinity could decrease or increase with surfactant concentration, depending on surfactant, cosolvent, salinity, divalent contents, and so on.
- Negative salinity gradient was proposed based on very limited core flood data.

8.3 SENSITIVITY STUDY

A fine core-scale model is used to study the optimum phase type and optimum salinity profile in surfactant flooding.

8.3.1 Basic Model Parameters

The grid blocks used are 100 × 1 × 1, which is a 1D model, and the length is 0.75 ft. Some of the reservoir and fluid properties and some of the surfactant data are listed in Table 8.1. The viscosity of polymer solutions at different concentrations is presented in Figure 8.5. The polymer adsorption data are shown in Figure 8.6. The microemulsion viscosity is shown in Figure 8.7, and the capillary desaturation curves are shown in Figure 8.8.

8.3.2 Sensitivity Results

The following subsections discuss simulation results regarding the salinity and salinity gradient effect. When we investigate the effect of a factor, we generally compare the results of five cases (systems), as defined in Table 8.2. The base case injection scheme is 1.0 pore volume (PV) water, 0.1 PV 3 vol.% surfactant solution, 0.4 PV 0.07 wt.% polymer solution, followed by 1.0 PV water injection.

Effect of Relative Permeabilities (k_r Curves) in Continuous Injection of Surfactant

We start with simple cases of continuous injection of surfactant solution without polymer. The phase types for Cases C1 to C3 are III, II(+), and II(−), respectively. The salinities used in type III, II(+), and II(−) systems are 0.365, 0.415, and 0.335 meq/mL. The RFs by 2 PV injection for Case C2 of type II(+), Case C1 of type III, and Case C3 of type II(−) are in descending order, as shown in Figure 8.9. An interesting observation is that the RF in Case C2 of type II(+)

TABLE 8.1 Reservoir, Fluids, and Surfactant Data

Porosity	0.32
k_H, mD	180
k_V, mD	90
Initial water saturation	0.2
Water viscosity, cP	1
Oil viscosity, cP	5
Formation water salinity, meq/mL	0.4
Surfactant Data	
Optimum salinity, meq/mL	0.365
Lower salinity, meq/mL	0.345
Upper salinity, meq/mL	0.385
k_r curves at $(N_c)_c$:	
Residual Sat.: S_{1r}, S_{2r}, S_{3r}	0.2, 0.3, 0.2
End point k_r: K_{r1}^e, k_{r2}^e, k_{r3}^e	0.3, 0.8, 0.3
Exponents: n_1, n_2, n_3	2, 2, 2
k_r curves at $(N_c)_{max}$:	
Residual Sat.: S_{1r}, S_{2r}, S_{3r}	0, 0, 0
End point k_r: k_{r1}^e, k_{r2}^e, k_{r3}^e	1, 1, 1
Exponents: n_1, n_2, n_3	2, 2, 2
Solubilization Parameters	
C_{33max0}, C_{33max1}, C_{33max2}	0.065, 0.03, 0.065
C_{sel}, C_{seu}	0.345, 0.385
Surfactant Adsorption Parameters	
a_{31}, a_{32}, b_3	3, 0.25, 1000

is higher than in Case C1 of type III, demonstrating that oil is more effectively displaced in type II(+) systems. This observation was also made by Nelson and Pope (1978) from their experiments.

Earlier investigators generally attributed this kind of phenomenon to the effect of phase behavior. Figure 8.9 shows that in Case C2 [type II(+)], water break through later (longer low water-cut period) than in Case C1 (type III). In Case C3 [type II(−)], a high aqueous phase saturation in the two-phase flow

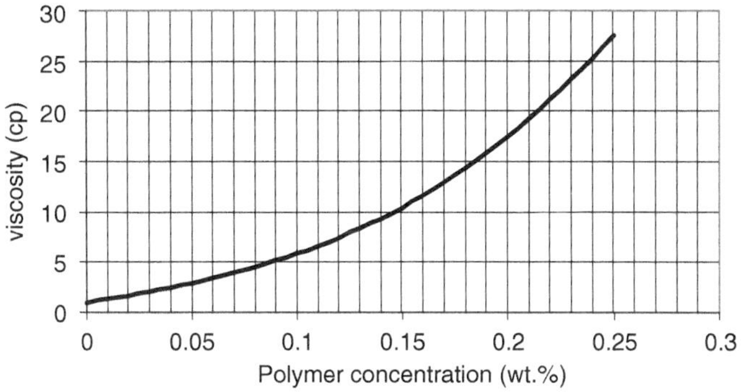

FIGURE 8.5 Viscosity of the used polymer solution.

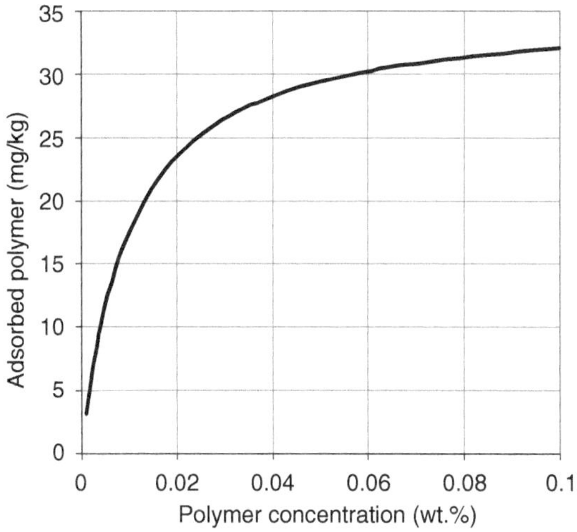

FIGURE 8.6 Polymer adsorption data.

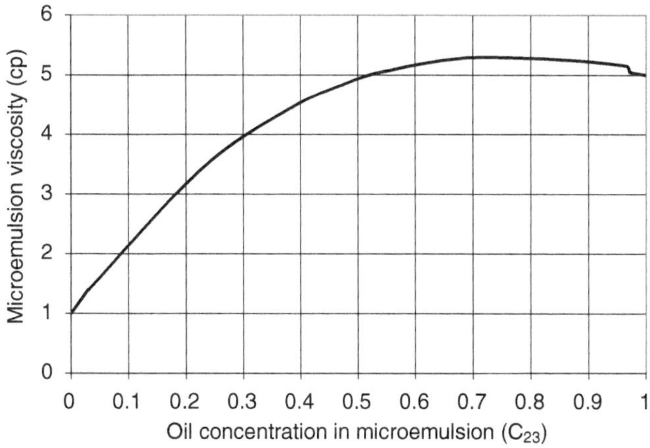

FIGURE 8.7 Microemulsion viscosity data.

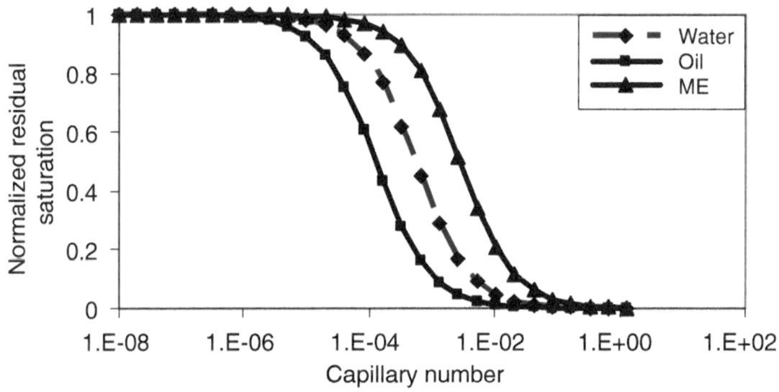

FIGURE 8.8 Capillary desaturation curves.

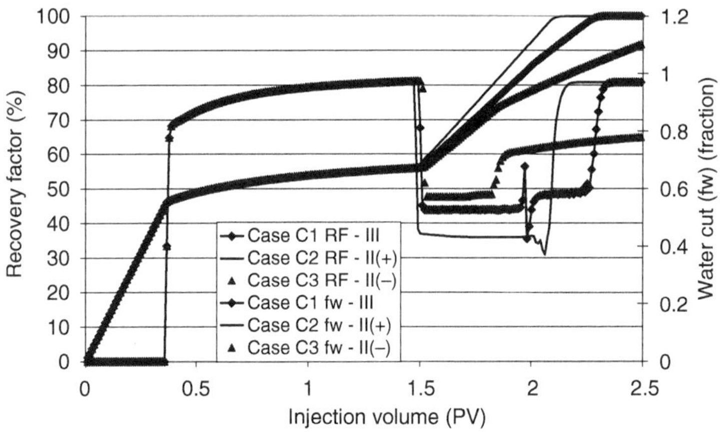

FIGURE 8.9 Recovery factors and water cuts for continuous injection cases of different microemulsion systems.

TABLE 8.2 Salinities in Different Phase Type Systems

System	Salinities
Type III	0.365 meq/mL in all injected fluids
Type II(+)	0.415 meq/mL in all injected fluids
Type II(–)	0.335 meq/mL in all injected fluids
Negative salinity gradient SG(–)	0.415 (W), 0.365 (S), 0.335 (P & W Drive)
Positive salinity gradient SG(+)	0.335 (W), 0.365 (S), 0.415 (P & W Drive)

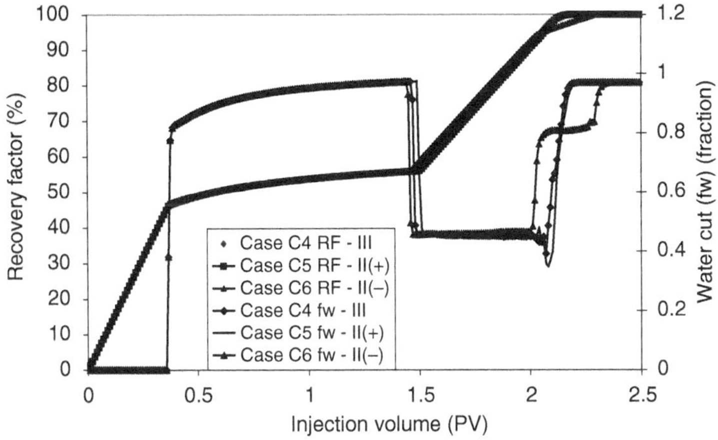

FIGURE 8.10 Recovery factors and water cuts for continuous injection cases of different microemulsion systems (k_r changed).

system results in the earliest water breakthrough and the lowest oil recovery at the same pore volume of injection. This figure shows that multiphase flow effect also plays an important role in determining the optimum phase type. In general, a three-phase flow provides less efficient displacement than a two-phase flow.

To verify the previous statement regarding the multiphase effect, we simply increase k_{r2} by reducing the exponent by half, and decrease k_{r3} by doubling the exponent and reducing the endpoint of k_{r3} by half at the high capillary number, $(N_C)_{max}$. The RF and water cut for these cases (Cases C4 to C6) are shown in Figure 8.10. This figure shows that the curves of RF versus PV of these cases are almost the same, which is different from Cases C1 to C3. Here, we can see that with the same phase behavior, by simply changing relative permeabilities, the performance has changed significantly. Needless to say, by changing other flow parameters, we could also change the performance, especially by changing capillary desaturation curves.

Effect of Relative Permeabilities (k_r Curves) in a Finite Surfactant Slug

Similar to the continuous injection cases C1 to C3, Cases kr1 to kr3 of a finite-size slug are in type III, II(+), and II(−), respectively. The salinities used in type III, II(+), and II(−) systems are 0.365, 0.415, and 0.335 meq/mL, the same as Cases C1 to C3. In these cases, 0.1 PV of surfactant slug is injected. The detailed injection scheme is 1 PV water, 0.1 PV 3 vol.% surfactant, 0.4 PV 0.07 wt.% polymer solution, and 1.0 PV water. A constant salinity is used in all the injection fluids for an individual type of system.

The recovery factors for these cases are shown in Table 8.3. Their order is the same as that of those in the continuous injection, that is II(+) > III > II(−),

TABLE 8.3 k_r Effect in Finite PV Surfactant Injection

	$e_2 = 2$, $e_3 = 2$, $k^e_{r3} = 1$		$e_2 = 1$, $e_3 = 4$, $k^e_{r3} = 0.5$		$e_2 = 2$, $e_3 = 4$, $k^e_{r3} = 0.5$	
Type	Case No.	RF, %	Case No.	RF, %	Case No.	RF, %
III	kr1	84.9	kr6	94.5	kr11	80.0
II(+)	kr2	97.0	kr7	76.6	kr12	76.8
II(−)	kr3	73.3	kr8	90.2	kr13	86.0
SG(−)	kr4	83.5	kr9	95.2	kr14	89.9
SG(+)	kr5	83.6	kr10	85.3	kr15	78.3

with the highest oil recovery being in the type II(+) case, not in type III. Note that the relative permeability parameters shown in Table 8.3 are those at the maximum desaturation capillary number, $(N_C)_{max}$. The relative permeability parameters at the low capillary number, $(N_C)c$, are the same as those in the base case shown in Table 8.1.

We further test the effect of the salinity gradient. In Case kr4, the salinities in the preflush water, surfactant solution, polymer, and water drive are 0.415, 0.365, 0.335, and 0.365 meq/mL, in descending order. Such a salinity gradient is called a negative salinity gradient, SG(−). In Case kr5, the salinities in the preflush water, surfactant solution, polymer, and water drive are 0.335, 0.365, 0.415, and 0.415 meq/mL, in increasing order. We call such a salinity gradient a positive salinity gradient, SG(+). The recovery factors for these two salinity gradient cases are also shown in Table 8.3. We can see that the RF from the negative salinity gradient case is lower than those from the type III and type II(+) cases.

Similar to the continuous injection Cases C4 to C6, we investigate the effect of relative permeability curves in Cases kr6 to kr8. For Cases kr6 to kr8, the data sets are the same as for Cases kr1 to kr3, except that e_2 is changed from 2 to 1 (k_{r2} increased), e_3 from 2 to 4, and the k^e_{r3} end point from 1 to 0.5 (k^e_{r3} reduced). The RFs for type III (Case kr6) and type II(−) (Case kr8) are increased, whereas the RF for type II(+) (Case kr7) is reduced. In Cases kr6 to kr8, the recovery factor from the type III (Case kr6) is the highest. In the similar cases kr1to kr3, the recovery factor from the type II(+) system (Case kr2) is the highest. We can see that by simply changing the relative permeability curves, we have a different observation regarding which type of microemulsion system can have the highest oil recovery.

In Cases kr9 and kr10, we use the negative salinity gradient and positive salinity gradient, respectively. Their recovery factors are shown Table 8.3. Now the RF from Case kr9 of SG(−) is higher than that from Case kr6 of type III, and this RF is the highest in the group. Again, we can see that by simply changing k_r curves, we have a different observation regarding which type of microemulsion system can have the highest oil recovery.

From the results of these cases, it seems that the k_r effect in Cases kr6 to kr8 is more pronounced than in Cases C4 to C6 of continuous surfactant injection because Cases kr6 to kr8 have quite different oil recovery factors, whereas the recovery factors of Cases C4 to C6 are very close.

In Cases kr11 to kr15, we repeat what is done in Cases kr6 to kr10, except that we change the relative permeability of the oil phase. In other words, we only reduce k_{r2} by changing e_2 from 1 to 2. Then for the constant salinity cases kr11 to kr13, Case kr13 of type II(−) has higher oil recovery than the other cases (see the results in Table 8.3). In this group, Case kr14 of SG(−) has the highest RF.

In the previous three groups of cases (Group 1: Cases kr1 to kr5; Group 2: Cases kr6 to kr10; Group 3: Cases kr11 to kr15), we change only the parameters related to relative permeabilities. We have seen that by changing only k_r curves, we could obtain different observations regarding which type of phase behavior [type II (−), type II(+), or type III] can have the highest oil recovery in constant salinity gradient systems, and whether a negative salinity gradient is preferred to a type III system. We have seen that we cannot simply make any general conclusion regarding which type is the best for oil recovery. The answer also depends on relative permeabilities. In the literature, the effect of phase behavior on oil recovery was more focused. Relative permeability was not much discussed. Interestingly, in the previous three groups, the RF in SG(−) is significantly higher than that in SG(+), regardless of which k_r curves are used. The exception is Group 1, in which the former is only slightly lower than the latter. However, the RF in the SG(+) case is not necessarily the lowest within each group.

For the tests described in the following subsections, we change some parameters of the reference Cases kr1 to kr5 to see whether we can obtain results consistent with the common belief or to see how sensitive these parameters are.

Effect of Microemulsion Residual Saturation, S_{3r}

We change microemulsion trapping saturation to see how sensitive this parameter can be. Cases kr16 to kr20 are based on Cases kr1 to kr5. The only change is to increase S_{3r} from 0.0 for the reference cases to 0.19 for the current cases at the high capillary number $(N_C)_{max}$. The results are shown in Table 8.4. The RF in Case kr19 of the negative salinity gradient is higher than that in Case kr16 of type III, but it is still lower than that in Case kr17 of type II(+). We further change the trapping saturation. In Cases kr21 to kr25, S_{3r} is changed from 0.0 for the reference cases to 0.29 for the current cases at $(N_C)_{max}$, and from 0.2 for the reference cases to 0.3 for the current cases at $(N_C)_c$. The RF in Case kr24 is the highest with the negative salinity gradient. In these cases, the microemulsion trapping saturation must be as high as 0.29 so that the

TABLE 8.4 S_{3r} Effect

Case No.	RF, %	Type	Case No.	RF, %
$S_{3r} = 0.19$ at $(N_C)_{max}$			$S_{3r} = 0.0$ at $(N_C)_{max}$	
kr16	78.8	III	Kr1	84.9
kr17	92.7	II(+)	Kr2	97.0
kr18	76.7	II(–)	Kr3	73.3
kr19	87.0	SG(–)	Kr4	83.5
kr20	79.0	SG(+)	Kr5	83.6
		$S_{3r} = 0.29$ at $(N_C)_{max}$		
kr21	77.9	III		
K22	76.1	II(+)		
kr23	80.8	II(–)		
kr24	88.1	SG(–)		
kr25	77.4	SG(+)		

negative salinity gradient becomes the most favorable case. Such S_{3r} at high N_C may not be realistic.

Effect of Salinity on Adsorption and Recovery Factor

UTCHEM uses the Langmuir-type isotherm equation to describe surfactant adsorption. The adsorption is directly proportional to the coefficient a_3 in the equation, which is defined as $a_3 = a_{31} + a_{32} \times C_{se}$, where C_{se} is the effective salinity. We can see that by changing a_{32}, we can change the level of salinity sensitivity. Higher salinity leads to higher surfactant adsorption.

The coefficient a_{32} is increased by 10 times in Cases Ad1 to Ad5 based on Cases kr1 to kr5. The results are shown in Table 8.5. Interestingly, the adsorption in Case Ad1 is the same as that in Case kr1 of type III, and the adsorption in Cases Ad2 to Ad4 is even lower than that in Cases kr2 to kr4. The recovery factors of Cases Ad1 to Ad5 are very close to those of the reference cases kr1 to kr5, respectively. The comparison of the results of these two groups shows that a_{32} is not a sensitive parameter because the adsorption plateau is a_3/b_3. In the data set, $b_3 = 1000$. When a_{32} is increased by 10 times, the incremental effect to adsorption is only 9/1000.

We suspect that the observed results are due to the small salinity contrast. Therefore, we further increase the upper salinity to 0.73 meq/mL (double of the optimum), reduce the lower salinity to 0.183 meq/mL (half of the optimum),

TABLE 8.5 Surfactant Adsorption Effect

Case No.	RF,%	Adsorption, mL/mL PV	Type	Case No.	RF,%	Adsorption, mL/mL PV
$a_{32} = 2.325$ ($C_{sel} = 0.335$, $C_{seu} = 0.415$)				$a_{32} = 23.25$ ($C_{sel} = 0.335$, $C_{seu} = 0.415$)		
kr1	84.9	1.9E-03	III	Ad1	84.9	1.9E-03
kr2	97.0	2.6E-03	II(+)	Ad2	95.0	2.5E-03
kr3	73.3	2.2E-03	II(−)	Ad3	73.5	2.1E-03
kr4	83.5	2.4E-03	SG(−)	Ad4	83.7	2.3E-03
kr5	83.6	1.9E-03	SG(+)	Ad5	83.4	1.8E-03
				($C_{sel} = 0.183$, $C_{seu} = 0.73$)		
			II(+)	Ad6	86.3	2.4E-03
			II(−)	Ad7	67.8	1.8E-03
			SG(−)	Ad8	81.7	2.1E-03
			SG(+)	Ad9	75.1	1.6E-03

and rerun Cases Ad1 to Ad5. The new case numbers are Ad6 to Ad9, respectively. The RF in Case Ad6 of the type II(+) is significantly reduced to 86.3% from 95% in the counterpart case Ad2, but it is still higher than that in Case Ad1 of the type III. Comparing the adsorption data and RF data of Cases Ad2 to Ad5 with those of Cases Ad6 to Ad9, respectively, we can see that the adsorption is lower and the RFs are lower when the salinity contrast is larger. Interestingly, it seems that the lower RF is correlated to lower adsorption. We also reduce the surfactant adsorption by changing b_3 from 10,000 to 100,000 and reduce the surfactant concentration to 0.5%. The observations remain the same.

The information from these simulation cases is that surfactant adsorption cannot be correlated to the oil recovery factor. Figures 8.11 and 8.12 show the published experimental data on surfactant retention. The data in Figure 8.11 are from Gupta and Trushenski (1979), and the data in Figure 8.12 are from Glover et al. (1979). These figures also show that the recovery factor could not be correlated with surfactant adsorption (retention).

Salinity Effect on Polymer Contribution

Experiments by Gupta and Trushenski (1979) were the first that supported the concept of the negative salinity gradient. Later, Nelson (1982) conducted

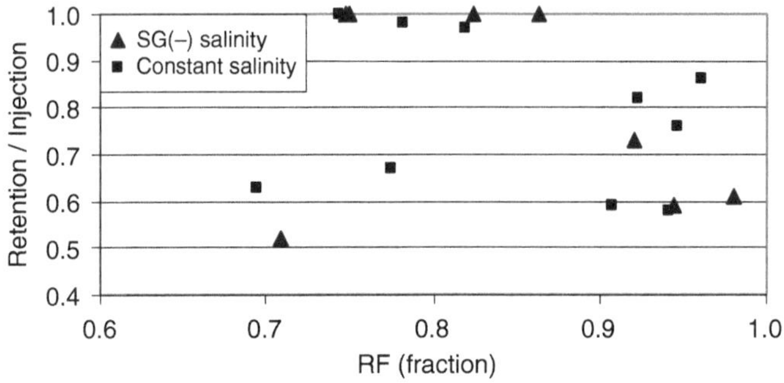

FIGURE 8.11 Surfactant retention versus oil recovery factor.

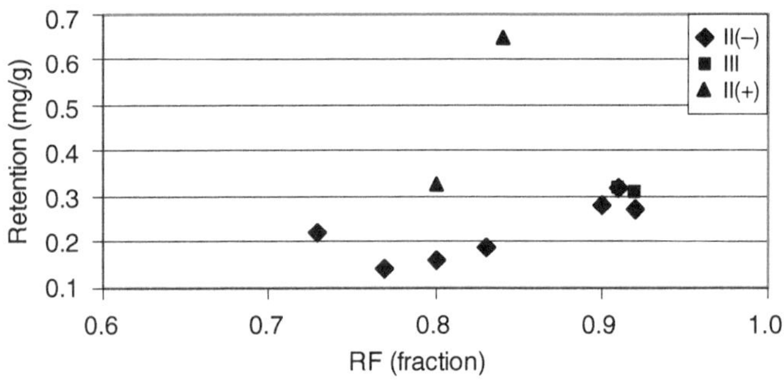

FIGURE 8.12 Surfactant retention versus oil recovery factor.

experiments using the same surfactants. In the Gupta and Trushenski experiments, the drive fluid was 2.0 PV polymer solution without chase water. We therefore rerun the reference cases kr1 to kr5 with an additional 2.0 PV 0.07 wt.% polymer solution but without water drive to be in line with the Gupta and Trushenski experiments; these new cases are identified as Cases P1 to P5. Our objective is to check whether the negative salinity gradient could greatly improve the high volume of polymer contribution. If it does when we use a high volume of polymer slug, we would expect higher oil recovery in the negative salinity case compared with the other cases.

The results of Cases P1 to P5 are presented in Table 8.6 and compared with those of the reference cases kr1 to kr5. The results from these cases have not changed the observations from the reference cases. In other words, the case of type II(+) gives the highest recovery. Note that the RF of SG(+) in Case P5 is higher than that of SG(–) in Case P4.

The results of Cases kr6 to kr10 are consistent with the belief that SG(–) is the most favorable gradient. We hypothesize that this favor is due to the effect of salinity gradient on polymer because lower salinity in the polymer drive is beneficial. Such favor will disappear if no polymer is used. Therefore, we

TABLE 8.6 Effect of a Large Polymer Drive Slug

Case No.	RF, %	Type	Case No.	RF, %
2 PV P drive			*1 PV W drive*	
P1	91.0	III	kr1	84.9
P2	99.9	II(+)	kr2	97.0
P3	73.5	II(−)	kr3	73.3
P4	83.5	SG(−)	kr4	83.5
P5	88.3	SG(+)	kr5	83.6

TABLE 8.7 Effect of Polymer

Case No.	RF, %	Type	Case No.	RF, %
2.9 PV water drive			*0.4 PV P and 1 PV W*	
P6	96.3	III	kr6	94.5
P7	78.7	II(+)	kr7	76.6
P8	90.2	II(−)	kr8	90.2
P9	95.2	SG(−)	kr9	95.2
P10	84.3	SG(+)	Kr10	85.3

replace a 0.4 PV polymer slug and a 1 PV water slug in Cases kr6 to kr10 by a 2.9 PV water drive slug to create Cases P6 to P10. Note that 2.9 PV water is used instead of 1.4 PV water because we have to use the larger PV of water to obtain the final recovery factor at >98% water cut. The RFs of these new cases with 2.9 PV water drive are shown in Table 8.7. We can see that without polymer, the RF in Case P9 of SG(−) become lower than that of type III in Case P6, although the difference is not significant. Therefore, we can conclude that the favorable result is caused by the salinity gradient effect on polymer.

Effect of Surfactant Concentration

In the cases discussed previously, the surfactant concentration is 3 vol.%. Most of the injected surfactant is retained (adsorbed and remaining). Next, we want to increase the surfactant concentration to 4 vol.% so that the total retained surfactant is less than the injected. By doing so, we can make sure that the previous observations are not caused by the insufficient surfactant injected. Cases Cs1 to Cs5 are the same as Cases kr6 to kr10, respectively, except the

TABLE 8.8 Effect of Surfactant Concentration

Case No.	RF, %	Type	Case No.	RF, %	Inc. RF
4% Surfactant			*3% Surfactant*		
Cs1	99.9	III	kr6	94.5	5.4
Cs2	90.9	II(+)	kr7	76.6	14.3
Cs3	96.8	II(−)	kr8	90.2	6.6
Cs4	99.1	SG(−)	kr9	95.2	3.9
Cs5	97.3	SG(+)	kr10	85.3	12.0
Cs6	99.8	II(+)	kr7	76.6	23.2

surfactant concentration is increased from 3% to 4%. We choose the group of Cases kr6 to kr10 as the reference cases because the salinity effect in this group is consistent with the belief that the RF in the type III is higher than that in type II(−) or II(+), and the RF in the SG(−) is the highest. We want to see whether any conclusions or observations we have made regarding the salinity effect can be changed by the amount of surfactant injected. The results from Cases Cs1 to Cs5 are presented in Table 8.8. We can see that the RF in Case Cs1 of type III is the highest among the three types, and it is higher than that in Case Cs4 of SG(−). However, the difference between these two cases is small. The table does show that the surfactant concentration can change the observation regarding which salinity works better.

In Case Cs2 of type II(+), the water cut is only 82.9%, and the recovery factor is 90.9% by the end of injection (total 2.5 PV injection). When we extend the injection to 3.5 PV with an additional 1 PV water drive in Case Cs6, the RF becomes 99.8%. Now the RF in Case Cs6 of type II(+) is higher than that in Case Cs3 of type II(−), whereas the RF in Case kr7 of type II(+)is lower than that in Case kr8 of type II(−). Again, the amount of surfactant injected may change the observation regarding the effect of salinity or salinity gradient, although some change may not be significant. Comparing the results of Case Cs1 to Case Cs6 with those of their reference Cases kr6 to kr10, we may say that with a higher surfactant concentration, the effect of different types diminishes.

The RFs of the reference cases kr6 to kr10 and the incremental RF due to 1 vol.% surfactant concentration increase are also shown in Table 8.8. We can see that the incremental RF from type II(+) is the highest, indicating that a higher surfactant concentration is more efficient in a type II(+) system.

Simulation results for a heterogeneous field case by Wu (1996) showed that the recovery factor in an SG(−) case was higher than that in a constant type III

TABLE 8.9 Effect of Low Surfactant Concentration

Case No.	RF, %	Adsorption, mL/mL PV	Type	Case No.	RF, %	Adsorption, mL/mL PV
1 vol.%, (C_{sel}, C_{seu}) = (0.335, 0.415)				*3 vol.%, (C_{sel}, C_{seu}) = (0.335, 0.415)*		
Cs7	70.1	6.4E-04	III	Ad1	84.9	1.9E-03
Cs8	71.8	8.9E-04	II(+)	Ad2	95.0	2.5E-03
Cs9	63.5	9.9E-04	II(−)	Ad3	73.5	2.1E-03
Cs10	67.3	9.9E-04	SG(−)	Ad4	83.7	2.3E-03
Cs11	69.2	6.2E-04	SG(+)	Ad5	83.4	1.8E-03
1 vol.%, (C_{sel}, C_{seu}) = (0.183, 0.73)				*3 vol.%, (C_{sel}, C_{seu}) = (0.183, 0.73)*		
Cs12	70.1	6.4E-04	III	Ad1	84.9	1.9E-03
Cs13	66.1	6.9E-04	II(+)	Ad6	86.3	2.4E-03
Cs14	62.9	1.1E-03	II(−)	Ad7	67.8	1.8E-03
Cs15	66.0	9.9E-04	SG(−)	Ad8	81.7	2.1E-03
Cs16	65.2	4.5E-04	SG(+)	Ad9	75.1	1.6E-03

salinity only when the surfactant concentration was low. In Cases Ad1 to Ad5, the RF in the SG(−) Case (Case Ad4) is not higher than that in the type III case (Case Ad1). We want to see whether we can reproduce Wu's results by reducing surfactant concentration. Cases Cs7 to Cs11 are based on Cases Ad1 to Ad5, except the surfactant concentration injected is 1 vol.% instead of 3 vol.%. The results are shown in Table 8.9. The RF in SG(−) (Case Cs10) is not higher than that in type III (Case Cs7). Interestingly, the RF in SG(+) (Case Cs11) is even higher than that in SG(−) (Case Cs10).

We run another group of Cases, Cs12 to Cs16, with a larger contrast in salinity; that is, C_{seu} is increased from 0.415 to 0.73, and C_{sel} is reduced from 0.335 to 0.183 meq/mL. These cases are based on Cases Ad1 and Ad6 to Ad9, respectively. As we can see in Table 8.9, the results from this group could not verify Wu's observation. Probably, Wu's data were due to the effect of heterogeneous formation.

Effect of Injection Scheme with Total Mass Unchanged

Based on Cases kr1 to kr4, we change the surfactant slug size from 0.1 PV to 0.2 PV, and the concentration from 3% to 1.5%. We also move 0.2 PV polymer into the surfactant slug. The resulting cases are I1 to I4. In these cases, we start with the residual oil saturation to study the tertiary recovery. The results are shown in Table 8.10. Note the recovery factor is the percentage based on 0.3 residual oil saturation. The results did not change the observation regarding which system gives a higher recovery factor.

Effect of Solubilization Data

Next, we try to change solubilization data to see whether the change would lead to a different observation regarding which system gives a higher recovery factor. In Cases SR1 to SR5, which are based on Cases kr1 to kr5, we change the maximum heights of binodal curves at zero salinity, at optimum salinity, and at twice optimum salinity (C_{3max0}, C_{3max1}, and C_{3max2}) from (0.065, 0.03, 0.065) to (0.035, 0.0268, 0.035). The results are shown in Table 8.11. The observation regarding which system gives a higher recovery factor is the same as the reference cases.

Based on Cases SR1 to SR5, we further change (C_{sel}, C_{seu}) in meq/mL from (0.345, 0.385) to (0.24, 0.49), in Cases SR6 to SR10. The RFs from Cases SR6 to SR10 become very close (see Table 8.12). After we change the C_{sel} and C_{seu},

TABLE 8.10 Effect of Injection Scheme

Case No.	RF, %	Type	Case No.	RF, %
I1	73.4	III	kr1	84.9
I2	93.9	II(+)	kr2	97.0
I3	58.9	II(−)	kr3	73.3
I4	75.9	SG(−)	kr4	83.5

TABLE 8.11 Effect of Heights of Binodal Curves

Case No.	RF, %	Type	Case No.	RF, %
(0.035, 0.0268, 0.035)			*(0.065, 0.03, 0.065)*	
SR1	86.2	III	kr1	84.9
SR2	99.0	II(+)	kr2	97.0
SR3	76.0	II(−)	kr3	73.3
SR4	84.6	SG(−)	kr4	83.5
SR5	87.9	SG(+)	kr5	83.6

TABLE 8.12 Effect of Lower and Upper Salinity Limits

Case No.	RF, %	Type	Type	Case No.	RF, %
(0.24, 0.49)				*(0.345, 0.385)*	
SR6	86.3	III	III	SR1	86.2
SR7	86.7	III	II(+)	SR2	99.0
SR8	86.0	III	II(−)	SR3	76.0
SR9	86.9	III	SG(−)	SR4	84.6
SR10	85.5	III	SG(+)	SR5	87.9

TABLE 8.13 Effect of Lower and Upper Salinity Limits, and Injected Salinities

Case No.	RF, %	Type	Case No.	RF, %
(0.18, 0.73) *(0.24, 0.49)*		*Injected C_{se} for II(−), II(+)* *Salinity Limits (C_{sel}, C_{seu})*	*(0.335, 0.415)* *(0.345, 0.385)*	
SR11	86.3	III	kr1	84.9
SR12	95.3	II(+)	kr2	97.0
SR13	74.7	II(−)	kr3	73.3
SR14	84.4	SG( )	kr4	83.5
SR15	84.1	SG(+)	kr5	83.6

all these cases are in a type III system. The results imply that if the system is of the same type, their RFs will be similar, regardless of their difference in salinities and/or salinity gradient. Again, the phase type is important.

In Cases SR6 to SR10, C_{sel} and C_{seu} in meq/mL are 0.24 and 0.49, respectively. Cases SR11 to SR15 are based on Cases SR6 to SR10 with the low and high salinities in the injection fluids changed to 0.183 and 0.73 meq/mL from 0.335 and 0.415 meq/mL, respectively, so that the systems in Cases SR11 to SR15 are of type III, type II(+), type II(−), SG(−), and SG(+). The RFs for these cases are shown in Table 8.13. The RFs for Cases kr1 to kr5 also are included in this table. Interestingly, although the injected low and high salinities are different, and C_{sel} and C_{seu} are also different in the two groups, the observation regarding which system gives a higher RF is the same, except that the RF of

TABLE 8.14 IFT Effect

Case No.	RF, %	Type	Case No.	RF, %
IFT1	83.7	III	kr1	84.9
IFT2	95.8	II(+)	kr2	97.0
IFT3	72.1	II(−)	kr3	73.3
IFT4	83.9	SG(−)	kr4	83.5

SG(+) in Case kr5 is a little bit higher than that of SG(−) in Case kr4. For each pair of the same type from these two groups, the RFs are close to each other.

The results of all the cases discussed in this section show that the type of phase behavior system is very important to the oil recovery factor, whereas the absolute salinity values are not important, at least for the data set used in these cases.

IFT Effect

We increase IFT by five times, based on the cases kr1 to kr4. The corresponding new cases are identified as Cases IFT1 to IFT4. The results are shown in Table 8.14. The results do not change our observation regarding which system gives the highest recovery factor.

We also tested the sensitivities of many other parameters. Our objective to run so many sensitivities was to see whether the recovery factor from SG(−) could be the highest, and the recovery factor from type III could be higher than that from type II(−) or type(+), by changing parameters. We had difficulty obtaining such results. Therefore, at least we can conclude that those conventional belief cannot be held universally. If we use the optimum salinity profile to be proposed later, however, all the tested cases show that the new concept can lead to the highest oil recovery.

8.4 FURTHER DISCUSSION

This section further discusses the effects of k_r curves, optimum phase type, and phase viscosity. The effect of negative salinity gradient is further discussed under conditions where different relationships between optimum salinity and surfactant concentrations occur.

8.4.1 Effect of k_r Curves and Optimum Phase Type

In most of the groups presented previously, the case of type II(+) has the highest recovery, and the case of type II(−) case has the lowest recovery. One important parameter is the relative permeability effect. For those k_r curves, k_{r1} and k_{r3} are increased more than k_{r2}. For example, the end point for k_{r1} is increased from 0.3 to 1, while the end point for k_{r2} is increased from 0.8 to 1. Then in a case of type II(−), the microemulsion phase is the aqueous phase with some oil solubilized. This phase k_r is increased more (from 0.3 to 1) than the excess oil phase k_r (from 0.8 to 1), resulting in earlier water breakthrough and higher water cut. Therefore, the oil recovery factor would be reduced compared with the case of type II(+). In the type II(+) case, the original oil phase becomes the microemulsion phase whose k_r is increased from 0.8 to 1. In addition to that, some water is solubilized in the oleic phase to increase its saturation; thus, k_{ro} is further increased. From the k_r point of view, a type II(+) case should work better than a type II(−) case, probably even better than a type III case. However, IFT must play an important role too. Other parameters may also contribute to the performance. Therefore, the optimum phase type is not simply type III based on the IFT value. The optimum phase type should be determined by corefloods using reservoir cores.

In Case kr2, the k_r parameters of microemulsion (e.g., k_{r3}) at $(N_C)_c$ in a type II(+) system are set to be the same as those of the aqueous phase (e.g., k_{r1}). However, we would intuitively think that k_{r3} should be the same as k_{r2} at $(N_C)_c$. Therefore, we set up such a case by modifying Case kr2 so that k_{r3} is the same as k_{r2} at $(N_C)_c$. The RF in this new case is 95.37% compared with 96.98% in Case kr2. A slightly lower recovery factor is observed in this case. Therefore, the possible wrong-assigned k_{r3} at $(N_C)_c$ is not the factor that causes the simulation results to be inconsistent with the common belief.

8.4.2 Effect of Phase Velocity

Hirasaki (1981) explained why the negative salinity gradient works from the point of phase velocity. He stated that an overoptimum salinity ahead of the surfactant bank is desirable because surfactant that mixes with the high-salinity water will partition into the oleic phase, and because the phase velocity (f_2/S) of the oleic phase is less than unity, the surfactant will be retarded. An underoptimum salinity is not desirable ahead of the surfactant bank because the surfactant partitions into the aqueous phase, which has a phase velocity greater than unity. However, an underoptimum salinity is desirable in the drive to propagate the surfactant. Thus, a system with overoptimum salinity ahead of the surfactant bank and underoptimum salinity in the drive will tend to accumulate the surfactant in the three-phase region where the lowest interfacial tensions generally occur.

According to Hirasaki's statement, if the oil velocity is reduced, then negative salinity gradient would work better. To verify his statement, we reduce oleic phase velocity by increasing oil viscosity. In Cases Vis1 to Vis5, which are based on Cases kr1 to kr5, we increase oil viscosity from 5 to 25 mPa·s. The oleic phase velocity is reduced about five times. The results are shown in

TABLE 8.15 Oil Viscosity Effect

Type	Case No.	RF, %	Case No.	RF, %	Case No.	RF, %
III	Vis1	71.9	Vis6	94.2	Vis11	99.8
II(+)	Vis2	88.5	Vis7	99.6	Vis12	99.9
II(−)	Vis3	65.4	Vis8	83.2	Vis13	88.4
SG(−)	Vis4	77.4	Vis9	92.4	Vis14	96.8
SG(+)	Vis5	72.5	Vis10	92.6	Vis15	99.4

Table 8.15. The RF in Case Vis4 [SG(−)] is higher than that in Case Vis5 [SG(+)], and also higher than that in Case Vis1 (type III). These results indicate that reducing oleic phase velocity does improve the performance of negative salinity gradient relatively. However, the recovery factor from type II(+) is still the highest.

We then reduce the oil viscosity to increase oil velocity in Cases Vis6 to Vis10 to 1 mPa·s. We would expect to see the opposite results regarding the oil recovery factor; that is, the RF from the SG(−) should be lower. The results, shown in Table 8.15, are not as expected. In this case, the oil viscosity probably is not reduced low enough. Therefore, we further reduce the oil viscosity in Cases Vis11 to Vis15 to 0.2 mPa·s. Then the recovery factor in Case Vis14 of SG(−) is lower than that in Case Vis11 of type III (see Table 8.15). And the RF of SG(+) in Case Vis15 is higher than that of SG(−) in Case Vis14. These results support Hirasaki's claim. However, the RF in Vis14 with SG(−) is not the lowest in the group, and the RF in Vis15 with SG(+) is not the highest. These results do not support Hirasaki's claim.

In these cases, the oil viscosity is reduced by 25 times to such a low value as 0.2 mPa·s. We suspect that the velocity effect could be the dominant effect. Even if the velocity effect is important, it certainly can be reduced or eliminated when polymer is added in the surfactant slug in surfactant-polymer flooding.

8.4.3 Negative Salinity Gradient

One justification to favor negative salinity gradient is the decrease in the salinity requirement as surfactant concentration is diluted, as shown in the salinity requirement diagram I (see Figure 8.13). In the figure, the two solid lines bracket the type III region. A dotted line with an arrow at its end represents the composition path for a specific system. In the figure, five composition paths [type II(+), III, II(−), SG(−), and SG(+)] are marked. Because of surfactant adsorption and phase trapping, the surfactant concentration decreases as it propagates.

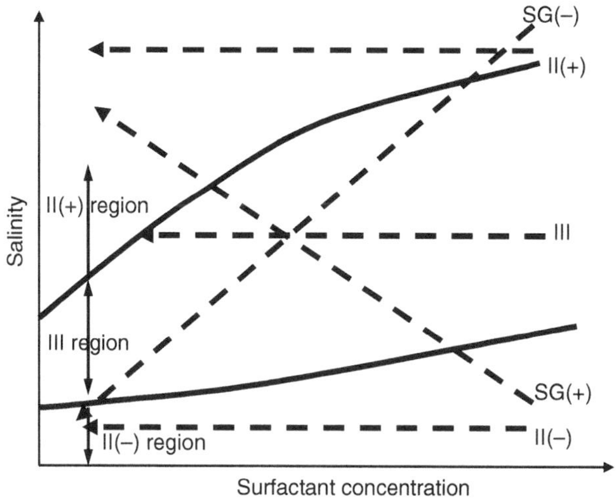

FIGURE 8.13 Salinity requirement diagram I.

According to the diagram, the salinity required to maintain the system in type III also decreases. As we can see from the diagram, the SG(−) path will cover most of the type III system. Plus, it is thought at first sight that this type of system would lead to the highest recovery, because of the low IFT in this type of system. However, along this SG(−) path, the IFT is not *always* at the lowest level. Actually, if the salinity contrast between type II(+) and type III at the front of the surfactant slug and the salinity contrast between type II(−) and type III at the back of surfactant slug are high, the lowest IFT at the optimum salinity can be obtained at only one surface from the displacing front to the upstream, because only at this surface is the salinity at the optimum, as shown in Figure 8.13. The IFT would be higher anywhere else.

Nelson (1982) discussed salinity requirement diagram I and proposed the concept of negative salinity gradient. As described in Section 8.2.2, however, there exists another kind of salinity requirement diagram, diagram II, as shown in Figure 8.14. In this diagram, the salinity required to maintain the system in type III increases as the surfactant concentration decreases. This relationship between salinity required and surfactant concentration is opposite to that in diagram I. As we can see in Figure 8.14, the SG(+) path will cover most of the type III system. Consequently, we would expect that an SG(+) system is favored to an SG(−) system. In this environment phase trapping could be a problem. Then the composite effect of salinity gradient and phase trapping becomes more complex.

From the previous discussion, we can see that which salinity system is favored depends on the salinity requirement diagram, and the diagram depends on the surfactant system. In diagram I, the SG(−) system may be the most

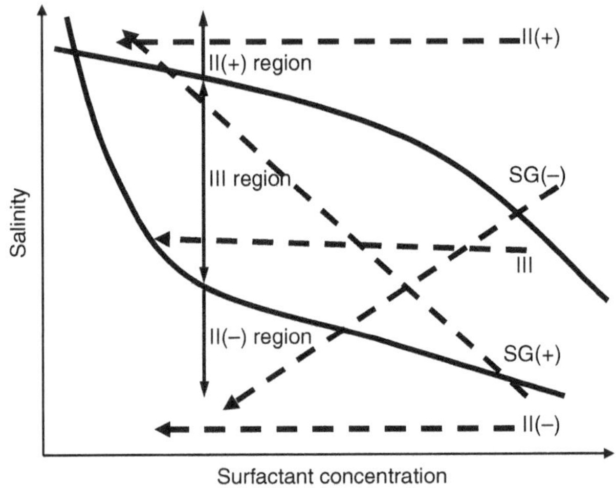

FIGURE 8.14 Salinity requirement diagram II.

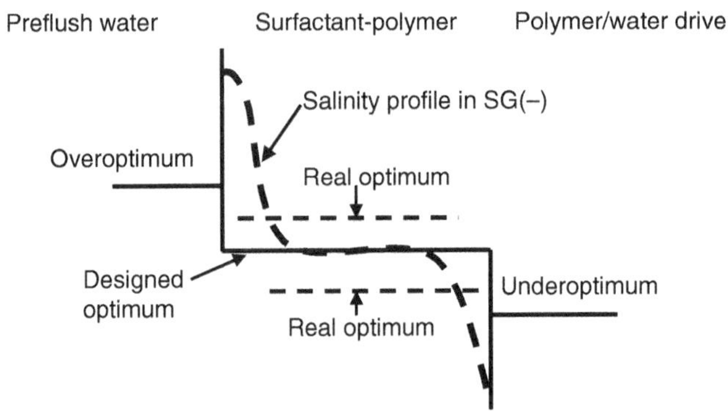

FIGURE 8.15 Salinity profiles in a negative salinity gradient.

favorable. In diagram II, the SG(+) system could be the most favorable. In some cases, another system (e.g., a type III system) could be the most favorable.

In a practical surfactant project, the designed optimum salinity from a laboratory study may not represent the real optimum salinity of the surfactant system. Another statement about the advantage of negative salinity gradient is that it can avoid the missing type III region because the salinity gradient covers the regions of all three types. Such a statement is questionable for two reasons. First, if the statement is valid, an opposite salinity gradient (positive salinity gradient) can also cover the three regions. Second, although the three regions are covered, it is possible that only a small portion of the surfactant slug is in the type III region, as can be seen in Figure 8.15. The dotted lines in the figure represent the salinity profile when negative salinity gradients are imposed. Only at the cross point of a dotted line and the real optimum salinity line (either above or below the designed optimum line in the figure) is the salinity at its optimum, which obviously is not desirable.

8.5 OPTIMUM PHASE TYPE AND OPTIMUM SALINITY PROFILE CONCEPTS

The following subsections describe the new concepts of optimum phase type and optimum salinity profile.

8.5.1 Optimum Phase Type

From the previous sensitivity results and discussions, we can see that phase type is very important in determining the final oil recovery. Table 8.16 lists some advantages and disadvantages of three types of microemulsion systems. The highest oil recovery could be from a type II(−), type III, or type II(+) system. Not only IFT, but many parameters, especially relative permeabilities, individually or in combination, may make any of type II(−), type III, and type II(+) microemulsion systems the optimum type. This is different from the conventional approach that focuses on interfacial tension as the determining parameter and consequently that the optimum phase type is, necessarily, type III.

The optimum phase type needs to be determined from core floods using reservoir cores. The phase type with the highest oil recovery factor is the optimum salinity type. It is not necessarily type III. Meanwhile, the optimum salinity is determined. It is not necessarily the middle salinity of type III or a salinity in type III. Core flood experiments take into account all parameters such as interfacial tension, relative permeability, phase trapping, and so on, because these experiments are essentially a replication of the flooding process that would occur during the EOR process in the field. Practically, we cannot afford to run many core floods to identify the optimum type, but we can run simulations to preselect the type.

TABLE 8.16 Advantages and Disadvantages of Three Types of Microemulsion Systems

Type	Advantages	Disadvantages
II(−)	Low phase trapping/adsorption	Bypassing excess oil due to its high velocity
III	Lowest IFT	Phase trapping due to three-phase k_r issues
II(+)	Favorable k_{ro}	Phase trapping due to its high viscosity

8.5.2 Optimum Salinity Profile

Because the highest oil recovery factor depends on the type of microemulsion, we must ensure the surfactant slug is in the phase type that leads to the highest oil recovery factor. In other words, the surfactant slug should be in the optimum phase behavior system. Therefore, we propose a concept of optimum salinity profile (OSP). The proposed optimum salinity profile is schematically shown in Figure 8.16. It can be described as follows:

- After the preflush slug (waterflood), a water/polymer slug as a salinity guard with the optimum salinity is preferred, but not necessary.
- An optimum salinity must be used in the surfactant slug.
- Immediately after the surfactant slug, a polymer or water drive slug with the same optimum salinity must be used as a salinity guard to make sure that salinity dispersion and diffusion cannot change the optimum salinity in the surfactant slug.
- The salinity in the post-water drive must be lower than C_{sel}.

Next, we further look at the sensitivity of salinity to the recovery factor. In the cases kr1 to kr5, kr4 is the SG(−) case. Case OSP1 is based on Case kr4. The salinity in 0.4 PV water preflush before the surfactant slug is changed from 0.415 to 0.365 meq/mL, and the salinity in 0.4 PV polymer drive after the surfactant slug is changed from 0.335 to 0.365 meq/mL. Then we have established the two salinity guard slugs. See Table 8.17 for the salinity profile for this case and other cases to be discussed in this section. The incremental RF in OSP1 over Case kr4 is 9.5%. This case demonstrates that using the two guard slugs in the OSP case outperforms the negative salinity SG(−) case.

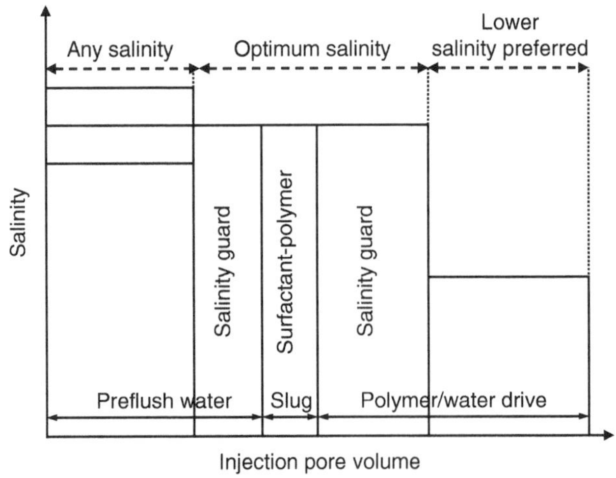

FIGURE 8.16 Schematic of optimum salinity profile.

TABLE 8.17 Simulation Cases to Test the OSP Concept

Case	RF, %	Salinity Profile
kr1	84.9	Type III, C_{se} = 0.365 in all slugs
kr2	97.0	Type II(+), C_{se} = 0.415 in all slugs
kr3	73.3	Type II(−), C_{se} = 0.335 in all slugs
kr4	83.5	SG(−), 1 PV 0.415 W, 0.1 PV 0.365 S, 0.4 PV 0.335 P, 1 PV 0.335 W
kr5	83.6	SG(+), 1 PV 0.335 W, 0.1 PV 0.365 S, 0.4 PV 0.415 P, 1 PV 0.415 W
OSP1	93.0	Based on kr4, 0.6 PV 0.415 W, 0.4 PV 0.365 W, 0.1 PV 0.365 S, 0.4 PV 0.365 P, 1 PV 0.335 W
OSP2	92.9	Based on OSP1, 1 PV 0.415 W, 0.1 PV 0.365 S, 0.4 PV 0.365 P, 1 PV 0.335 W
OSP3	74.0	Based on kr3, 0.6 PV 0.415 W, 0.4 PV 0.365 W, 0.1 PV 0.365 S, 0.4 PV 0.335 P, 1 PV 0.335 W
OSP4	91.4	Based on kr3, 1 PV 0.415 W, 0.1 PV 0.365 S, 0.4 PV 0.365 P, 1 PV 0.335 W
OSP5	93.0	Based on kr4, C_{se} = 0.365 in 0.4 PV P, C_{se} = 0.340 in 1.0 PV W
OSP6	84.9	Based on kr4, C_{se} = 0.365 in 0.4 PV P, C_{se} = 0.350 in 1.0 PV W
OSP7	98.5	Based on kr2, C_{se} in 1.0 PV W drive is changed from 0.415 to 0.335

Based on OSP1, we remove the 0.365 meq/mL guard slug immediately before the surfactant slug but keep the guard slug after in OSP2. The RF from OSP2 is 92.9%, which is almost the same as that from OSP1. This comparison shows that the effect of the guard before the surfactant slug is not significant. Probably, the salinity mixing is mainly caused by convection, or the diffusion is not significant compared with convection. In Case OSP3 based on Case kr3 [type II(−)], only the salinity in the guard slug before the surfactant slug is changed to 0.365 meq/mL. The RF of 74% is almost unchanged compared with the RF of 73.3% from Case kr3. In Case OSP4 based on Case kr3, however, only the salinity in the guard slug after the surfactant slug is changed to 0.365 meq/mL. The RF becomes 91.4%. These cases show that the effect of the salinity guard slug before the surfactant slug is not important, whereas the effect of the salinity immediately after the slug is very important.

In Case OSP1, the salinity in the chase water after the guard slug (polymer slog) is 0.335 meq/mL. Because the slug before the surfactant slug is not important, the difference in RF between OSP1 and kr1 (constant salinity of 0.365 meq/mL) is caused by only the salinity difference in the chase water. The RF from OSP1 is 8.1% higher than that from Case kr1. This result shows the salinity in the post-water drive should be lower than the salinity in the surfactant slug.

Cases OSP5 and OSP6 are really interesting. These cases are based on Case kr4. In Case OSP5, the salinity of 1.0 PV chase water is 0.340, just below C_{sel}, which is equal to 0.345. The RF is 92.94%. In this case, the chase water miscibly drive the type II(−) microemulsion. In Case OSP6, the salinity of chase water is 0.350, just above $C_{sel} = 0.345$, and the RF is 84.8%. These two cases show that the salinity in the chase water slug should be less than C_{sel}.

Based on the preceding cases, we understand that to find the optimum salinity profile, we need to find the optimum phase type and optimum salinity from constant salinity cases first. The optimum phase type and optimum salinity are from the highest RF case. For further explanation, the optimum phase type is not necessarily type III, and the optimum salinity is not necessarily the conventional middle point between C_{sel} and C_{seu}. So we use the optimum salinity in the surfactant slug, and add a salinity guard of the optimum salinity immediately after the surfactant slug, and use a salinity lower than C_{sel} (preferably much lower) in the slug after this guard. The guard slug of the optimum salinity immediately before the surfactant slug may not be necessary because the effect on the recovery is not significant, as is clear by comparing the RF of OSP2 with that of OSP1.

For the optimum salinity, in the cases kr1 to kr5, the optimum phase type is type II(+) in Case kr2, not type III in Case kr1, and the optimum salinity is 0.415 meq/mL, not the conventional optimum salinity of 0.365 meq/mL at the middle of C_{sel} and C_{seu}. In Case OSP7, we keep the optimum salinity of 0.415 meq/mL in the 0.4 PV polymer slug after the surfactant slug and in the preflush water, but change the salinity in the chase water from 0.415 to 0.335 meq/mL. Thus, the salinity profile follows the proposed optimum salinity profile. The RF from this case is higher than the RF from Case kr2, and actually higher than any RF from Cases kr1 to kr5. In other words, the RF from the OSP case is the highest.

We tested the OSP concept against many data sets. We found that the OSP concept was valid for every data set. In other words, the RF is always the highest in the case with the optimum salinity profile proposed here. Figure 8.17 compares some of the recovery factors from OSP and negative salinity gradient. Using this optimum salinity profile increases the oil recovery factor by an average 12.3% over the negative salinity gradient method.

One important point in the proposed OSP is that the salinity in the chase slug after the guard slug in OSP must be lower than C_{sel}. One of the main mechanisms to justify such salinity is surfactant desorption. Liu et al. (2004) found that, in an extended waterflood following an alkaline-surfactant slug

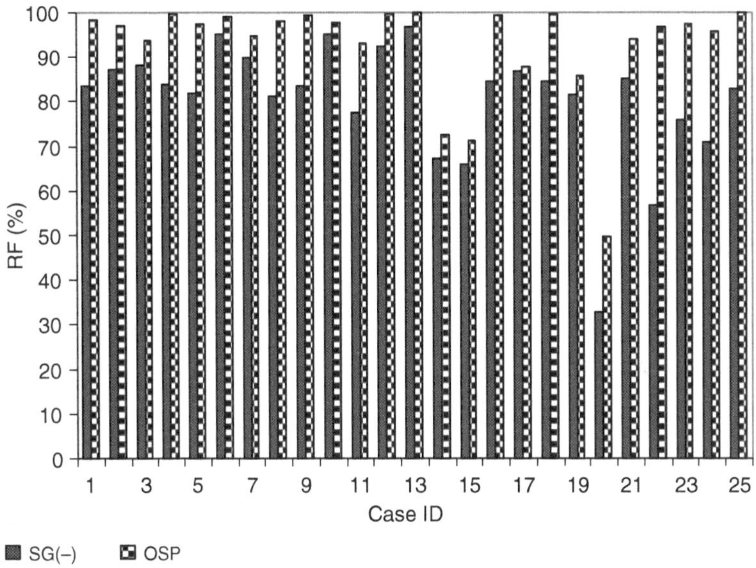

FIGURE 8.17 Comparison of recovery factors from SG(–) and OSP.

injection, surfactant desorbed into the water phase. This desorption of surfactant lasted for a long period of the waterflood. Although the concentration of the desorbed surfactant in the extended waterflood was very low, an ultralow oil/water IFT was obtained by using a suitable alkaline concentration. The added alkali probably provided necessary salt for phase behavior and required high pH to reduce surfactant adsorption. Their core flood results showed that an additional 13% of the initial oil in place (IOIP) was recovered after the alkaline-surfactant injection by the synergism of the desorbed surfactant and alkaline. This result indicates that the efficiency and economics of a chemical flood could be improved by utilizing the desorbed surfactant during extended waterflood processes.

8.6 SUMMARY

The preceding discussion shows that we cannot simply make any general conclusion regarding which type of microemulsion is the best type for oil recovery. The oil recovery depends on relative permeabilities and other parameters. The oil recovery from type III may not be higher than that from type II(–) or type II(+). The oil recovery factor in an SG(–) system may not always be the highest. However, the optimum salinity profile can always lead to the highest recovery factor. This concept has been tested in different data sets and found to be valid.

The main controlling parameters are relative permeability curves and types of microemulsion systems. Relative permeability curves control the multiphase flow, and the types of microemulsion systems dictate which relative permeability curves are sensitive.

In most of the simulated cases, the oil recovery factors in the cases of positive salinity gradient are lower than those in the corresponding cases of negative salinity gradient. Therefore, the negative salinity gradient is generally better than the positive salinity gradient. However, the recovery factor in a positive salinity gradient is not always the lowest in the five types of systems in the same group.

Chapter 9

Surfactant-Polymer Flooding

9.1 INTRODUCTION

Theories of surfactant flooding and polymer flooding are discussed in Chapters 5 to 7. This chapter focuses on surfactant-polymer (SP) interactions and compatibility. Optimization of surfactant-polymer injection schemes is also discussed. The methodology and even some conclusions in the presented optimization may be applied to other processes as well. Finally, this chapter presents a field example.

9.2 SURFACTANT-POLYMER COMPETITIVE ADSORPTION

The chromatographic separation of polymer and surfactant caused by the polymer's inaccessible pore volume cause the polymer to flow ahead of the surfactant; thus, polymer is sacrificed for adsorption. Because some adsorption sites are covered by polymer molecules, fewer of the sites are available for surfactant adsorption; this is called competitive adsorption. To consider competitive adsorption, we treat surfactant adsorption as a function of adsorbed polymer concentration.

In UTCHEM, a multiplier is applied to adjust the plateau value of surfactant adsorption (Wu, 1996). This multiplier is defined as

$$F_{SP} = 1 - \left(\frac{\hat{C}_p}{\hat{C}_{p,\max}} \right) F_{ADS}, \tag{9.1}$$

where $\hat{C}_p$ is the adsorbed polymer concentration; $\hat{C}_{p,\max} = a_p/b_p$, where a_p and b_p are the parameters in the Langmuir isotherm equation (Eq. 5.31); and F_{ADS} is a UTCHEM input parameter to adjust the surfactant adsorption due to polymer competitive adsorption. The new plateau of surfactant adsorption is calculated by

$$\left[\hat{C}_{3,\max} \right]_{new} = \hat{C}_{3,\max} F_{SP}. \tag{9.2}$$

Because the multiplier F_{SP} is always less than or equal to 1, the maximum surfactant adsorption is reduced or unchanged. Note that if a surfactant slug is injected ahead of a polymer slug, some adsorption sites are covered by

Modern Chemical Enhanced Oil Recovery. DOI: 10.1016/B978-1-85617-745-0.00009-7

surfactant molecules before the arrival of the polymer molecules. Therefore, polymer adsorption is reduced.

9.3 SURFACTANT-POLYMER INTERACTION AND COMPATIBILITY

When surfactant and polymer are injected in the same slug (SP flooding), their compatibility is an issue. Sometimes, polymer is injected before surfactant as a sacrificial agent for adsorption or for conformance improvement. Sometimes polymer is injected behind surfactant to avoid chase water fingering in the surfactant slug. Even though polymer is not injected with surfactant in the same slug, they will be mixed at their interface because of dispersion and diffusion. Polymer may also mix with surfactant owing to the inaccessible pore volume phenomenon when it is injected behind surfactant. Trushenski et al. (1974, 1977) and Szabo (1979) referred to these phenomena as surfactant-polymer interaction or incompatibility (SPI). The following subsections list some observations about SP and then discuss several factors affecting SPI.

9.3.1 Observations about Surfactant-Polymer Interaction

Some of literature information on surfactant-polymer interactions is summarized as follows.

- Surfactant can stay in the aqueous, oleic, or middle microemulsion phase. Essentially all the polymer in a surfactant-polymer solution, however, stays in the most aqueous phase, no matter where the surfactant is (Szabo, 1979; Nelson, 1981).
- Little difference is observed in the IFT values with and without polymer. The three-phase systems still exhibit ultralow IFT values. With the presence of polymer, the optimum salinity is decreased slightly (Healy et al., 1976; Pope et al., 1982).
- The IFT between the polymer-rich phase and surfactant-rich phase in an oil-free case could be very low, sometimes as low as 10^{-4} to 10^{-5} mN/m (Szabo, 1979). These low values of IFT indicate that the "trapping" of sulfonate, as discussed by Trushenski (1977), relates more to the difference in mobilities of the separated phases than to capillary force (IFT).
- The viscosity of the oil-free surfactant-rich phases (above the CEC, which is defined in the next section) is high; it is frequently higher than the polymer-rich phase, even though they apparently contain almost no polymer (Szabo, 1979; Pope et al., 1982). The surfactant-rich phase appears to expel polymer to the polymer-rich phase, and the sulfonate forms a complex with the polymer molecules within its phase (Szabo, 1979).
- The effect of polymer on systems with oil is to increase the viscosity of the water-rich phase only, with little effect on the microemulsion phase unless it is the water-rich phase (Pope et al., 1982).

- When a polymer is added in a surfactant system, there are two critical concentrations: CAC and CMC_2. CAC is the critical adsorption concentration at which surfactant starts to adsorb on the polymer chains; it is lower than the critical micelle concentration (CMC). CMC_2 is the surfactant concentration at which micelles are formed when polymer is present; it is higher than CMC (Li et al., 2002). Both CAC and CMC_2 are on the order of magnitude of CMC.
- Surfactant has two effects on polymer viscosity: (1) surfactant brings cations such as Na^+ to reduce polymer viscosity; (2) with surfactant added, aggregates can be formed so that polymer viscosity is increased. Practically, surfactant does not significantly change HPAM viscosity (see Figure 9.1). The two effects cancel each other. However, the viscosity of hydrophobic associating polymers is very sensitive to surfactant concentration. The reason is that the hydrophobic group in the polymer can be solubilized into micelles so that their molecular interaction becomes larger. The resulting viscosity is increased but varies with surfactant concentration, as shown in Figure 9.2. In this figure, the hydrophobic associating polymer was AP-P (600 mg/L), and HPAM was Alcoflood 1275A (800 mg/L). The water was Daqing injection water, and the temperature was 45°C.
- Gogarty (1983a, 1983b) reported that polymer preflush improved the vertical conformance of the surfactant solution so that recovery was increased. Murtada and Marx (1982) also observed that polymer preflush improved SP oil recovery when using a high-concentration surfactant solution. Experimental data from Chen and Pu (2006) showed that injection of polymer slug before surfactant slug led to a higher tertiary recovery factor than the mixing slug of surfactant and polymer. When polymer is injected before surfactant, the SPI problem appears to be alleviated.

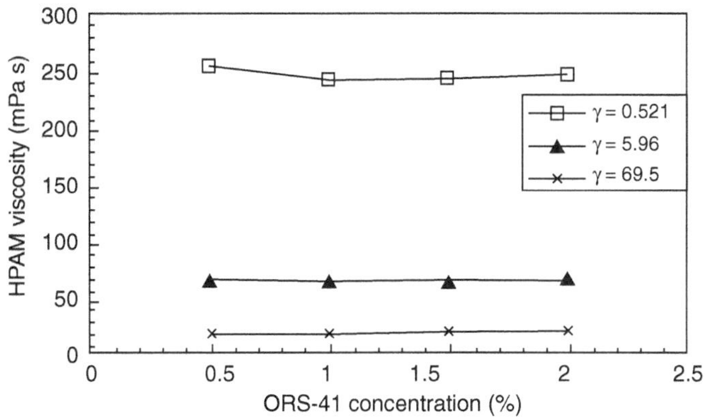

FIGURE 9.1 Surfactant effect on HPAM viscosity. *Source: Kang (2001).*

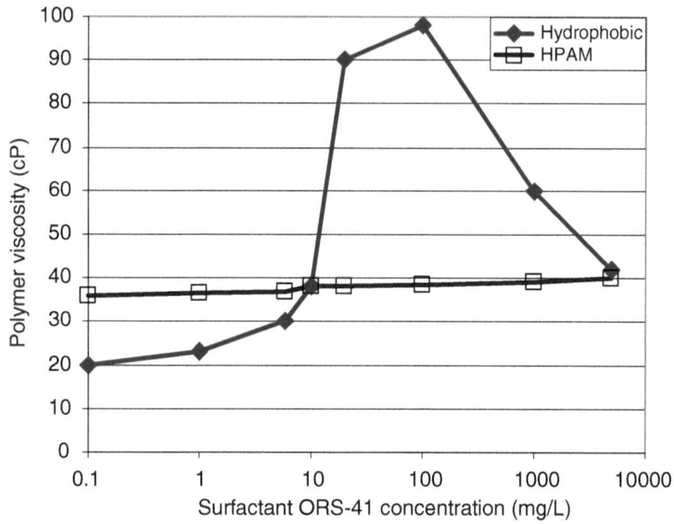

FIGURE 9.2 Surfactant effect on polymer viscosity. *Source: Li (2007).*

9.3.2 Factors Affecting Surfactant-Polymer Interaction

This section summarizes the factors affecting surfactant-polymer interaction, including electrolyte concentration, alcohol, oil, polymer concentration, competitive adsorption, and phase trapping.

Electrolyte Concentration

Pope et al. (1982) observed that when a polymer was mixed with a surfactant in an oil-free solution, there was a characteristic phase separation into an aqueous surfactant-rich phase and an aqueous polymer-rich phase at some sufficiently high salinity (NaCl concentration). They called this the critical electrolyte concentration (CEC). They reported that CEC was independent of the polymer type, polymer concentration, and surfactant concentration, but it was dependent on the used surfactant. This conclusion cannot be universally valid. Hou (1993) observed that CEC depended on polymer and surfactant concentrations in a HPAM-petroleum sulfonate solution. The CEC increased with increasing temperature for the anionic surfactants and decreased with increasing temperature for the nonionic surfactants. It also increased with alcohol concentration.

As the temperature was increased, the single-phase region was larger (Trushenski, 1977). The phase separation of polymer-sulfonate mixtures did not appear to be a polymer-induced flocculation of sulfonate molecules. In most systems, relatively large and stable volumes of the bottom sulfonate phase were obtained. To verify the stability of the bottom sulfonate phases, Szabo (1979) centrifuged many samples for several hours at 4000 rpm (66.67 rev/s), and no further volumetric change in the sulfonate-rich phase was observed.

The phase separation in a petroleum sulfonate-polymer solution can be explained by the DLVO theory introduced in Chapter 3. Both the sulfonate and polymer are negatively charged. When the electrolyte concentration is increased, the double layers of the negative-charged particles are compressed, and the zeta potential is reduced so that it will be easier for the particles to be aggregated and phase separation to occur. In addition, owing to the electrolyte's strong hydration effect, the water films between particles become thinner as the electrolyte concentration is increased, resulting in a less stable system.

Trushenski (1977) reported experimental data showing that lowering salinity polymer behind the surfactant slug was favorable to phase stability, and increasing polymer water salinity reduced oil displacement efficiency, a result also reported by Szabo (1979). The decreased oil recovery was partially caused by increased surfactant-polymer interaction. Trushenski showed that the two-phase region (surfactant-rich phase and polymer-rich phase) in the equal salinity surfactant-polymer system was much larger than that if the polymer salinity was lower. A larger two-phase region resulted in more significant phase trapping of surfactant. The addition of cosolvent to the polymer slug could eliminate phase separation. Trushenski also reported that the mobility in the SPI zone was higher than that of water flowing at residual oil; however, he did not give an explanation for this phenomenon.

Both the salinity and polymer concentration affect the volumetric ratios of separated phases and the fractionation of sulfonate and polymer in these phases. A proper selection of polymer concentration and salinity in a polymer-sulfonate mixture can result in equal viscosities of the separated phases (Szabo, 1979). This approach may not be practical, however, because a constant salinity or concentration cannot be maintained along the flow path because of adsorption and mixing, and so on. Therefore, we should try to select a formula that will not have an SPI problem.

Including polymer in a surfactant slug is essential for maintaining a favorable mobility ratio because the surfactant causes the water relative permeability to increase. This increase must be counterbalanced by decreasing the aqueous mobility with polymer (Hirasaki and Pope, 1974). Without polymer in the surfactant slug, the surfactant will finger into the oil bank, and the reservoir sweep will be very poor. Furthermore, the polymer in both the slug and drive helps mitigate the effects of permeability variation and improves the overall sweep efficiency in the reservoir. Floods in homogeneous cores show some but not all the benefits of adding polymer, so acceptable results in a core flood without polymer can be misleading with respect to performance in the field (Hirasaki et al., 2006).

Alcohol (Cosolvent)

In general, low-carbon alcohols can increase surfactant solubility, so SPI can be alleviated when an alcohol is added. However, the effect of alcohols on surfactant-polymer compatibility is complex. Not all alcohols can improve the compatibility. For example, addition of isopropanol and isobutanol can improve TRS10-80–HPAM compatibility (Hou, 1993), but n-pentanol cannot. The correct alcohol and proper concentration should be chosen.

Oil

When oil is added to the surfactant and polymer solution, the system follows the typical pattern of a system without polymer—that is, from type I to type III to type II as salinity increases. The three-phase region simply shifts a small distance to the left on the salinity scale, compared with the three-phase region without polymer. When polymer is present, it remains almost exclusively in the most aqueous phase whether the phase is lower-phase microemulsion or excess brine. Consequently, polymers affect relative mobility of the phases generated during a chemical flood, but they do not appear to affect phase equilibria significantly (Nelson, 1982). Some of the aqueous phases in the critical region of the shift (which is also just above oil-free CEC salinity) were found to be gel-like in nature.

Addition of oil yields an oil-in-water microemulsion with nearly spherical drops. Within limits, the higher the molecular weight of the oil added to produce an oil-in-water microemulsion, the less oil is needed to formulate single phases with polymer for mobility control (Hirasaki et al., 2008). During screening tests, a clear surfactant-polymer solution with oil added does not mean the corresponding aqueous solution without oil will be clear. Therefore, aqueous stability tests with polymer added in the surfactant solution are necessary and important.

HPAM Polymer Concentration

Hou (1993) reported that for a petroleum sulfonate–HPAM–mixed alcohol (isopropanol : isobutanol = 8 : 1) system, the addition of polymer did not change the three types of phase behavior, but the upper phase and lower phase volumes were increased very slightly and the middle phase volume was decreased accordingly. This volume changes were caused by the interaction of the alcohol with HPAM. HPAM brought some of alcohol from the middle phase into the aqueous phase, resulting in the decrease in the middle phase volume. As polymer concentration was increased, the aqueous phase viscosity was increased while the middle phase viscosity remained almost unchanged because very little polymer would go to the middle phase. Therefore, HPAM had little effect on the middle phase properties.

Competitive Adsorption and Polymer IPV

During a polymer flood, because of polymer adsorption, a polymer denuded zone forms at the front of polymer slug. If a surfactant slug is injected ahead of a polymer slug, however, adsorption sites are occupied by surfactant. In some cases, polymer loss is reduced to an insignificant level owing to the so-called competitive adsorption, discussed earlier. Thus, a polymer denuded zone may not develop. On the other hand, Kalpakci et al. (1990) presented a low-tension polymer flooding (LTPF) scheme and observed the reduction of surfactant retention due to the presence of the polymer in Type III to Type II(–) (salinity gradient) phase environments.

There is another phenomenon that is called polymer inaccessible pore volume (IPV). Laboratory data indicate that inaccessible pore volume is usually greater than the adsorption loss for polymers following a micellar solution (Trushenski et al., 1974). The competitive adsorption and IPV may make polymer penetrate the surfactant slug ahead of it. Therefore, surfactant-polymer interaction or incompatibility occurs not only in the surfactant-polymer process where the surfactant and the polymer are injected in the same slug, but also in the surfactant-polymer process where surfactant is injected before the polymer slug.

Phase Trapping

In a surfactant-polymer process, Trushenski (1977) reported that the presence of polymer in the surfactant slug caused an unexpected increase in surfactant loss. This increase was due to the bypass of surfactant by polymer (phase trapping). The trapping and remobilization of the micellar phase are shown in Figure 9.3. In this long core test, the water content of the micellar fluid was

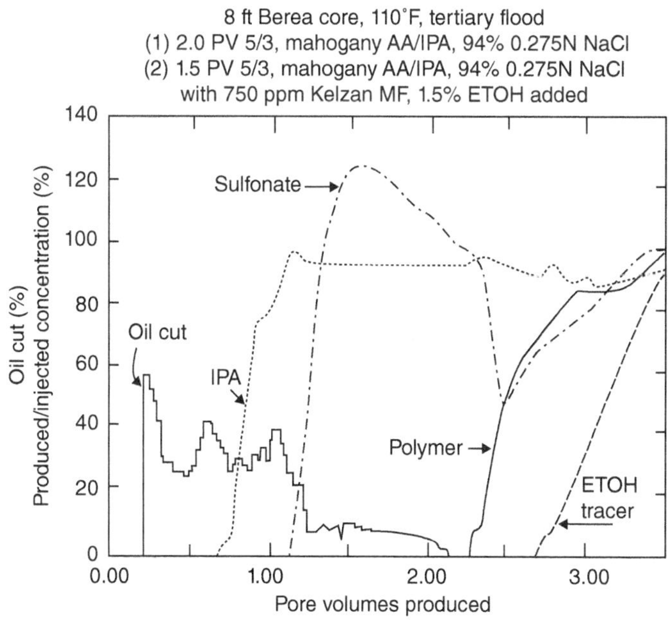

FIGURE 9.3 Trapped sulfonate was displaced at saturation limit. *Source: Trushenski (1977).*

94% (the Mahogany AA sulfonate : IPA ratio remained 5:3), and the salinity was 0.275 N NaCl. A mobility buffer bank was not injected. Instead, after 2 PV of micellar slug were injected, 750 mg/L Kelzan MF polymer plus ETOH tracer were added to the injected micellar fluid. Then any decrease in sulfonate concentration in the mixing region between the two micellar slugs was caused by polymer presence, not dilution.

Figure 9.3 shows the sulfonate concentration decreased when polymer was first produced. As the sulfonate concentration decreased in the zone of SPI, there was a marked change in effluent appearance. The fluid changed from turbid, caramel-colored to transparent amber. In this case, after the sulfonate concentration reached a minimum, it began increasing. As it increased, the second turbid, caramel-colored sulfonate phase was produced. The fractional volume of this phase increased as sulfonate concentration increased. This result indicates that SPI loss was caused by phase trapping.

The trapped sulfonate phase could be displaced by chase water behind the mobility buffer bank. In the long-core test shown in Figure 9.4, the salinity in the polymer slug was lower than that in the sulfonate slug ahead. Again, the polymer preceded the ethanol tracer, and SPI occurred. When the polymer concentration increased, the sulfonate concentration decreased. When the polymer concentration peaked, the sulfonate concentration decreased sharply. When the polymer concentration decreased, the sulfonate concentration increased, indicating that the sulfonate was remobilized. Although the trapped sulfonate could be displaced, it was not effective in displacing oil.

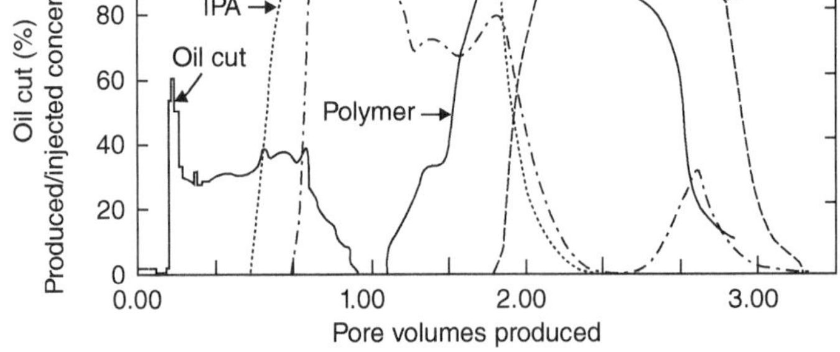

FIGURE 9.4 Trapped sulfonate was displaced when polymer concentration decreased. *Source: Trushenski (1977).*

9.4 OPTIMIZATION OF SURFACTANT-POLYMER INJECTION SCHEMES

In this section, simulation results are compared with the information from the literature for different polymer and surfactant-polymer injection schemes. We expect that UTCHEM simulation of a core-scale chemical process is the best simulation approach to study mechanisms. In this study, we use a 1D core flood model with 100 blocks to represent a 1-foot-long core. The permeability is 2000 md, and the water and oil viscosities are 1 and 2 mPa·s, respectively. To optimize injection schemes, we compare the incremental oil recovery factors over waterflooding and chemical costs. Chemical costs are evaluated using the amounts of chemicals injected per barrel of incremental oil (lb/bbl oil).

9.4.1 Placement of Polymer

Polymer can be placed in a mixed SP slug or in a polymer-only slug for mobility control. Table 9.1 compares the results from different schemes. In SIM 1, 0.25 PV 0.07 wt.% polymer is injected after the surfactant slug (0.1 PV 2% S). In SIM 2, 0.1 PV × 0.07% polymer is moved to the surfactant slug. In SIM 3, all the polymer in 0.25 PV, 0.07 % polymer slug (0.25 × 0.0007 = 0.000175 PV) is placed in the 0.1 PV surfactant slug. Then the polymer concentration in the 0.1 PV surfactant slug is 0.175%. The recovery factors and incremental recovery factors are almost the same in these three simulation cases. From these simulation cases, it seems that it does not matter where polymer is placed.

Based on experimental results, however, Yang and Me (2006) found that if polymer was injected separately from the alkaline and surfactant slug, the incremental oil recovery was higher than that with polymer, alkali, and surfactant placed in the same slug. They also reported that it was better to place polymer in the preflush slug than in the post-flush slug, and this conclusion is supported by the experiments reported by Li (2007). The preflush slug should be at least 0.12 PV, and the post-flush slug should be about 0.2 PV. Our

TABLE 9.1 Effect of Polymer Placement

SIM No.	S or S + P	P	Inc. RF, %	S, lb/bbl	P, lb/bbl
1	0.1PV 2% S	0.25PV 0.07 wt.% P	18.0	4.4	0.41
2	0.1PV 2% S + 0.07% P	0.15PV 0.07% P	18.2	4.3	0.40
3	0.1PV 2% S + 0.175% P	0.25PV W	18.1	4.4	0.40

simulation results are different from their reported laboratory observations, however, probably because our simulated cases are surfactant-polymer flooding, whereas their experiments were ASP cases. Another possibility is that their cores probably were heterogeneous, whereas our simulation results are from a 1D homogeneous model. Field results indicated that line-drive patterns were superior to five-spot patterns for SP (Gogarty, 1983a). For a given amount of polymer, we can have two injection schemes: (1) a small slug but a high concentration and (2) a large slug but a low concentration.

Yang and Me (2006) reported that Scheme 1 was better because a high-concentration polymer slug has a higher mobility ratio so that the sweep efficiency was better. When the polymer molecular weight was higher, the advantage of Scheme 1 was more obvious.

9.4.2 Effect of the Amounts of Polymer and Surfactant Injected

To investigate the effect of the amounts of polymer and surfactant injected, we use different concentrations and slug sizes and compare the incremental oil recovery factors and chemical costs (chemical lb/bbl incremental oil) for different amounts of polymer and surfactant injection. The amount of surfactant injected is commonly presented in concentration (%)·PV(%), and the amount of polymer is commonly presented in mg/L·PV(fraction). Note that the unit PV is in fraction of pore volume. This section presents both polymer and surfactant in concentration (%)·PV(%). The relationship between these two units is mg/L·PV(fraction) = 100 concentration (%)·PV(%). Figures 9.5 and 9.6 show the results. From these two figures, we can see that the more chemicals injected, the more incremental oil is recovered. However, in general, the chemical cost per barrel of incremental oil is also increased. Apparently, when a low load of chemicals is injected, the chemical cost per barrel of incremental oil is not sensitive to the amount of chemical injected. This observation is seen for both polymer and surfactant.

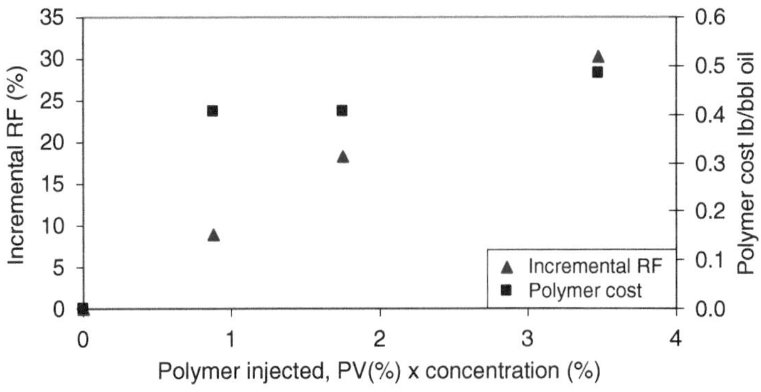

FIGURE 9.5 Effect of the amount of polymer injected.

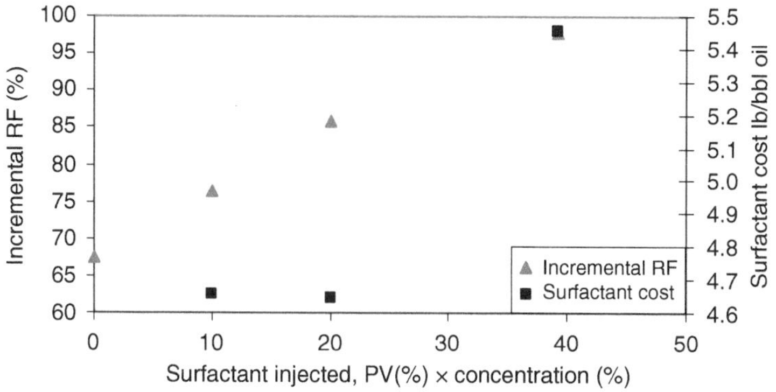

FIGURE 9.6 Effect of the amount of surfactant injected.

TABLE 9.2 Effect of the Time to Shift Waterflood to SP

SIM No.	WF before SP, PV	S_o before SP	Water Cut before SP, %	Total Injection PV	Incremental RF, %
1	0.15	0.67	0.0	1.5	18.0
4	0.35	0.47	0.0	1.7	18.0
5	0.55	0.30	92.5	2.1	18.0
6	0.85	0.28	95.5	2.3	17.9
7	1.50	0.26	97.7	3.0	17.8

9.4.3 Time to Shift Waterflood to SP

Almost all chemical flood projects are started after some waterflood history. We want to know whether early chemical injection could be a better option. To do that, we change the water injection PV before chemical injection so that average oil saturations (S_o) before SP are different. The results are shown in Table 9.2. We can see that different total injection PVs are required to achieve about the same incremental recovery factor. The incremental oil recovery factor (RF) is defined as the RF from an SP case minus the RF from the 1.5 PV waterflooding case. The later SP is started, the higher the total injection PV is required. Therefore, it is better to start surfactant-polymer flood earlier to accelerate production, and thus, less water will be injected. Such results have been confirmed by the ASP corefloods in Daqing (Li, 2007).

From fractional flow analysis (taking polymer flooding as an example in Figure 9.7), the displacement front velocity is

$$v_j = \frac{f_{wp} - f_{wj}}{S_{wp} - S_{wj}} \quad j = 1, 2. \tag{9.3}$$

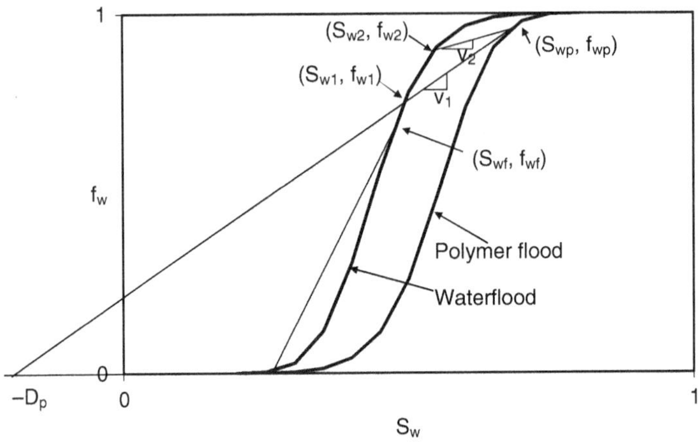

FIGURE 9.7 Schematic of frontal displacement velocities in polymer flooding at different initial oil saturations.

From Figure 9.7, we can see that as S_{wj} is increased from S_{w1} to S_{w2}, the velocity is decreased from v_1 to v_2. The oil bank breakthrough time is proportional to the reciprocal of the velocity. As the velocity is reduced, the oil bank breakthrough time is increased (Lake, 1989). Then recovering the remaining oil requires a longer time, as we predicted from the simulation runs. Therefore, it is better to start polymer flooding or chemical flooding in the earlier phase of field development.

Note that the simulation results and simple frontal flow analysis show that the final oil recovery factor is similar even though a chemical flood is started at different initial oil saturations. However, more water is needed to displace the residual oil because it will be easier for the remaining oil to be trapped or bypassed by displacing fluids to lose oil phase continuity if the initial oil saturation is lower. Therefore, in reality, when a chemical flood is started at a higher oil saturation, a higher oil recovery factor is expected because the production will be stopped at an economic water cut.

In spite of this fact, a chemical flood could never be started from the beginning of the field development for several reasons:

- A chemical flood requires a relatively long preparation time, including laboratory study and facility installment.
- More technical skills and competence are needed to run a chemical flood project. Designing the project takes longer.
- More time is needed to get the project approved.
- An early waterflood history is required for the reservoir characterization. This is the key justification for the late start of a chemical flood.

9.4.4 Optimization of the Chemical Flooding Process

To optimize the flooding processes, we first have to select which optimization criterion to use. Generally, we choose incremental oil recovery factor as a criterion. Alternatively, we may choose maximum NPV as a criterion with economic analysis. The latter choice is more proper because it takes into account discounted cash flow. However, performing economic analysis requires more economic data that are generally not available. The criterion to be used depends on the objective.

This section discusses both the incremental oil recovery factor and chemical cost per barrel of incremental oil recovered. We have seen that the two criteria sometimes give different answers regarding the optimum process.

Many parameters could affect the chemical flood performance, and it is impossible to find an absolutely optimum process. In the published literature, different authors have focused on different sets of parameters for optimization. Generally, only a few parameters were included in their optimization process. Anderson et al. (2006) investigated the effects of these parameters: slug size, polymer mass, adsorption, k_V/k_H, and permeability. Zerpa et al. (2005) mainly considered slug size and chemical concentration in their optimization of the ASP process. Delshad et al. (2004) considered chemical concentrations and slug sizes as optimization parameters using an experimental design approach. Obviously, if more parameters are included, the number of cases generated would be very large.

We notice that the main chemical cost is surfactant cost. The alkaline or polymer cost is relatively low. Therefore, we may reduce the amount of surfactant injected to reduce the cost but increase the amount of polymer injected to maximize oil recovery. Table 9.3 shows the results of some optimized cases.

Based on SIM 2 in Table 9.1, we reduce the amount of surfactant and increase the amount of polymer in SIM 8. When we use the surfactant price of $2.75/lb and polymer price of $1/lb, the chemical cost in SIM 8 for a surfactant-polymer process goes down from $13.1 to $8.0 per barrel of incremental oil. In SIM 9, no surfactant is injected, so the SP process is changed to a polymer

TABLE 9.3 Process Optimization by Reducing Surfactant Injection

SIM No.	0.1 PV SP	0.15 PV P	Inc. RF, %	S Cost, lb/bbl	P cost, lb/bbl	Chem. Cost, $/bbl
2	2% S + 0.07% P	0.07% P	18.2	4.3	0.4	13.1
8	1% S + 0.14% P	0.14% P	15.8	2.7	0.9	8.0
9	0% S + 0.07% P	0.07% P	6.4	0.0	1.1	1.0

flooding process. The chemical cost in SIM 9 goes down to \$1.0 per barrel of incremental oil. Apparently, polymer flooding is more efficient in terms of cost per barrel than surfactant-polymer flooding, but the recovery factor has to be sacrificed. Again, it is also demonstrated that the economic criteria and ultimate oil recovery criteria give different injection schemes. National oil companies would probably opt for ultimate oil recovery, but international oil companies would put more focus on economics.

9.5 A FIELD CASE OF SP FLOODING

This section presents an example using surfactant-polymer flooding in Layers $Ng5^4$ to $Ng6^1$ in the southwest part of the seventh zone of the Gudong field operated by the Shengli Administration Bureau (Shengli Oilfield), China (Zhang et al., 2004). Although ASP could bring higher incremental oil recovery, it also brings some problems such as scaling and stable emulsions that are difficult to break at surface facilities. Here, a pilot test of surfactant-polymer without alkaline flooding was presented.

Description of Pilot Area

The pilot area covered 0.94 km^2 with the original oil in place (OOIP) of 2.77 million tons. The reservoir depth was 1261 to 1294 m, and temperature was 68°C. The average permeability was 1320 md with the permeability variation coefficient 0.58. Oil viscosity was 45 mPa·s. Formation TDS was 8207 mg/L, and the sum of Ca^{2+} and Mg^{2+} was 231 mg/L.

The pilot test included 10 injectors, 16 producers, 2 observation wells, and 2 wells for coring. Before the test, the water cut was 97.4%, and the recovery factor was 34.3% with the expected waterflooding recovery factor of 36.3%. The residual oil saturation was 0.33, and the average sweep efficiency was 0.54. The injection pressure was about 11.9 MPa (1725.8 psi).

Prepilot Study

Surfactant screening tests were conducted in this prepilot study. Figure 9.8 shows the IFT contours at different surfactant and cosurfactant concentrations. It was found that the mixture of 0.3 wt.% SLPS (petroleum sulfonate made from Shengli oil) and 0.1 wt.% cosurfactant had IFT of 2.95×10^{-3} mN/m, whereas the petroleum sulfonate alone had IFT of 4×10^{-2} mN/m. In the lowest IFT zone, the surfactant and cosurfactant concentrations were 0.3 wt.% and 0.1 wt.%, respectively. Therefore, the mixed surfactants were selected.

Core flood tests were conducted to compare the performance from the surfactant-polymer option and polymer injection option at different injection schemes. The formula selected for use in the core flood was 0.3 wt.% SLPS + 0.1 wt.% cosurfactant + 1500 mg/L HPAM. For this formula, the chemical cost was about \$2.6/bbl incremental oil. The final selected injection scheme in the pilot is presented in Table 9.4. According to this scheme, the

TABLE 9.4 Pilot Injection Scheme

Slug No.	Slug Size, PV	Injection Liquid, 10^3 m^3	Surfactant		Cosurfactant		Polymer		Injection Time, days
			Conc., wt.%	*Mass, tons*	*Conc., wt.%*	*Mass, tons*	*Conc., mg/L*	*Mass, tons*	
1	0.05	218.4					2000	486	156
2	0.30	1310.1	0.45	5986	0.15	1966	1700	2475	936
3	0.05	218.4					1500	364	156
Total	0.40	175.0		5986		1966		3325	1248

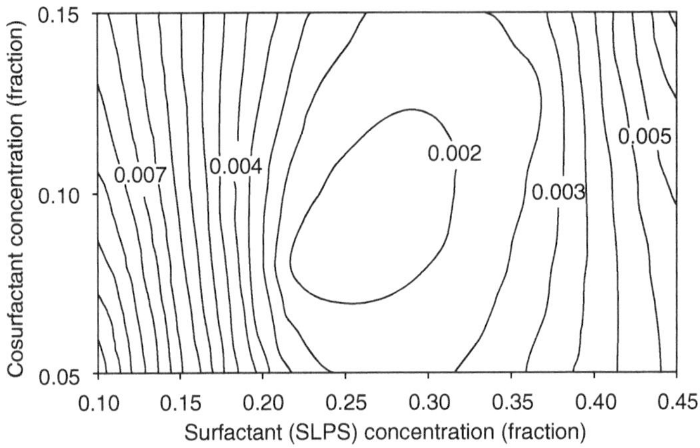

FIGURE 9.8 IFT contours at different surfactant and cosurfactant concentrations. *Source: Zhang et al. (2004).*

incremental oil recovery factor from a simulation study using the simulator called SLCHEM was 12%, equivalent to 0.33×10^6 m^3 oil. The optimum injection rate based on the simulation study was 0.11 PV/year. The total surfactant injected was concentration in wt.% × injection PV in PV% = 0.6 × 30 = 18 (wt.% × PV%). Similarly, the total polymer injected was 6.85 (wt.% × PV%).

Note that the final surfactant and cosurfactant concentrations were 0.15 wt.% and 0.05 wt.% higher than their optimum concentrations, respectively, to compensate for the loss due to adsorption. The polymer concentration was also increased by 200 mg/L in the pilot.

Pilot Test Results

The pilot test was started with a polymer preflush slug (Slug 1 in Table 9.4) for conformance modification on September 11, 2003. The main slug (Slug 2) was started on June 1, 2004. The polymer postflush was also used. After Slug

TABLE 9.5 Surfactant Concentration in Oil and Water Phases

Well No.	In Water Phase, mg/L	In Oil Phase, mg/L	Concentration Ratio (oil phase/water phase)
7-32-3135	130	34.9	0.27
7-33-12	100	22	0.22
7-36-195	900	48.2	0.05

1 polymer injection, injection pressure was increased 2.3 MPa from 8.2 MPa to 10.5 MPa because of higher resistance from polymer solution. The permeability reduction factor was 1.89. After the surfactant injection in Slug 2, injection pressure was reduced slightly. The improved oil production was observed after a 10-month injection (0.123 PV) in November 2004. The oil production in the pilot area was increased about 3 times from 34 to 95 tons/day. And the water cut was decreased 3.2% from 98.2% to 95%. The actual results matched the prediction from the simulation study.

Liquid samples were also collected from three producers to analyze the surfactant concentrations in oil and water phases. The results are presented in Table 9.5. Surfactants were seen in both water and oil phases (Sun, 2005).

Chapter 10

Alkaline Flooding

10.1 INTRODUCTION

The alkaline flooding method relies on a chemical reaction between chemicals such as sodium carbonate and sodium hydroxide (most common alkali agents) and organic acids (saponifiable components) in crude oil to produce in situ surfactants (soaps) that can lower interfacial tension. Another very important mechanism is emulsification, which is discussed in Section 13.5. The addition of the alkali increases pH and lowers the surfactant adsorption so that very low surfactant concentrations can be used to reduce cost; this issue is discussed in Section 12.7. This chapter focuses on alkaline reactions with crude oil and rock. Another main focus of this chapter is the simulation of alkaline flooding, which is probably the most complex task in modeling chemical processes. This chapter also discusses a surveillance and monitoring program and the application conditions of alkaline flooding.

10.2 COMPARISON OF ALKALIS USED IN ALKALINE FLOODING

This section compares different alkalis used in alkaline flooding and discussed their application advantages and disadvantages.

10.2.1 General Comparison and pH

Alkaline flooding is also called caustic flooding. Alkalis used for in situ formation of surfactants include sodium hydroxide, sodium carbonate, sodium orthosilicate, sodium tripolyphosphate, sodium metaborate, ammonium hydroxide, and ammonium carbonate. In the past, the first two were used most often. However, owing to the emulsion and scaling problems observed in Chinese field applications, the tendency now is not to use sodium hydroxide. The dissociation of an alkali results in high pH. For example, NaOH dissociates to yield OH^-:

$$NaOH \rightarrow Na^+ + OH^-. \quad (10.1)$$

Modern Chemical Enhanced Oil Recovery. DOI: 10.1016/B978-1-85617-745-0.00010-3

Sodium carbonate dissociates as

$$Na_2CO_3 \rightarrow 2Na + CO_3^{2-}, \quad (10.2)$$

followed by the hydrolysis reaction

$$CO_3^{2-} + H_2O \rightarrow HCO_3^- + OH^-. \quad (10.3)$$

The dissociation of sodium silicate is complex and cannot be described by a single reaction equation. The pH values of several commonly used alkaline agents are presented in Figure 10.1. Of course, the pH of the solutions varies with salt content. For instance, the pH of caustic solutions decreases from 13.2 to 12.5 when the salinity increases from 0 to 1% NaCl. By comparison, the pH of sodium carbonate solutions is less dependent on salinity (Labrid, 1991). In

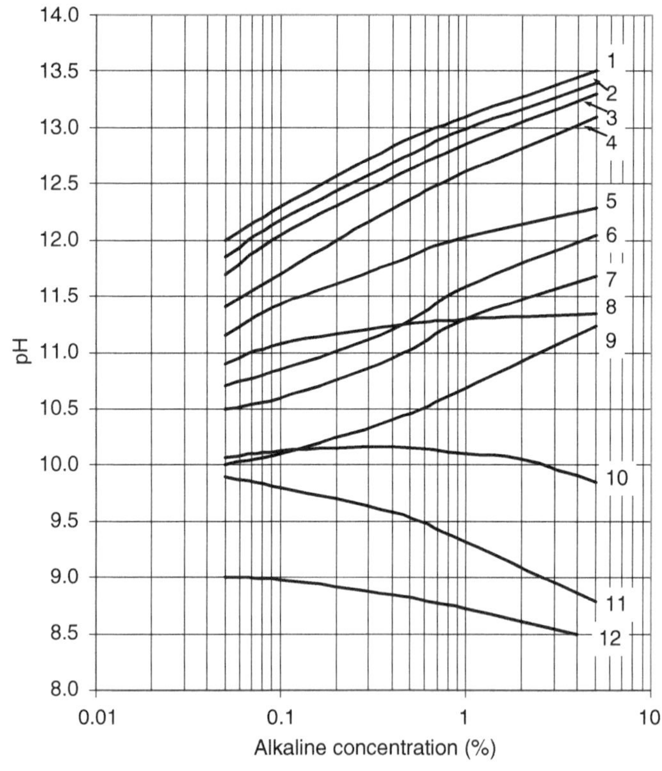

FIGURE 10.1 Graph of pH values of alkaline solutions at different concentrations at 25°C: 1, sodium hydroxide (NaOH); 2, sodium orthosilicate (Na_4SiO_4); 3, sodium metasilicate (water glass or liquid glass, Na_2SiO_3); 4, sodium silicate pentahydrate ($Na_2SiO_3{\cdot}5H_2O$); 5, sodium phosphate ($Na_3PO_4{\cdot}12H_2O$); 6, sodium silicate [$(Na_2O)(SiO_2)_n$, n = 2, where n is the weight ratio of SiO_2 to Na_2O.]; 7, sodium silicate [$(Na_2O)(SiO_2)_n$, n = 2.4]; 8, sodium carbonate (Na_2CO_3); 9, sodium silicate [$(Na_2O)(SiO_2)_n$, n = 3.22]; 10, sodium pyrophosphate ($Na_4P_2O_7$); 11, sodium tripolyphosphate ($Na_5P_3O_{10}$); and 12, sodium bicarbonate ($NaHCO_3$).

terms of effectiveness to reduce interfacial tension (IFT), it has been observed that there is little difference among the commonly used alkalis (Campbell, 1982; Burk, 1987). It has also been observed that the minimum IFT occurs over a narrow range of alkaline concentrations, typically 0.05 to 0.1 wt.% with a minimum IFT of 0.01 mN/m (Green and Willhite, 1998).

Table 10.1 shows a comparison of some of the properties of several common alkalis. Potassium-based alkalis, the price of which is higher than sodium-based alkalis, are not included. They are considered when clay swelling and injectivity problems are expected. Some alkalis are further discussed and compared in the following sections.

10.2.2 Polyphosphate

Chang (1976) showed that use of a polyphosphate, which is a buffer, improved recovery. Sodium tripolyphosphate (STPP) was used in laboratory tests for Cretaceous Upper Edwards reservoir (Central Texas). STPP was proposed to minimize divalent precipitation, for wettability alteration and emulsification (Olsen et al., 1990). Generally, it is not used as a primary alkali to generate soap for purposes of IFT reduction. Instead, it is used together with other alkalis such as sodium carbonate when divalents could be a problem (Harry Chang, Chemor Tech International, Plano, Texas, personal communication on June 16, 2009).

10.2.3 Silicate versus Carbonate

Campbell (1981) compared sodium orthosilicate and sodium hydroxide in recovering residual oil. The test results showed that the former was more effective than the latter under the conditions studied, both for continuous flooding and 0.5 PV slug. The mechanisms through which sodium orthosilicate produced higher recovery than sodium hydroxide in those tests were not concluded. Reduction in interfacial tension is similar for both chemicals. Other factors must play a more important role.

Radke and Somerton (1978) investigated the use of a sodium metasilicate (Na_2SiO_3) buffer in core floods. A metasilicate buffer at a pH of 11.2 showed breakthrough at 2.5 PV injection, whereas sodium hydroxide of the same pH did not appear until a 12 PV injection (Mayer et al., 1983). This result means that sodium metasilicate reaction with rock is much weaker than sodium hydroxide. Chang and Wasan (1980) indicated that there were differences in coalescence behavior and emulsion stability that favor sodium orthosilicate over sodium hydroxide.

Silicate precipitates, however, are generally hydrated, flocculent, and highly plugging even at low concentrations. Carbonate precipitates are relatively granular and less adhering on solid surfaces (Cheng, 1986). Thus, under equivalent experimental conditions of porosity and flow rate, sodium carbonate shows less degree of permeability damage in the presence of hard water (see Figure 10.2).

TABLE 10.1 Properties of Several Common Alkalis

Alkali Formula	Sodium Hydroxide NaOH	Sodium Carbonate Na_2CO_3	Sodium Orthosilicate Na_4SiO_4	Sodium Tripolyphosphate $Na_5P_3O_{10}$	Ammonium Hydroxide NH_4OH
Reduces IFT	Yes	Yes	Yes	No	No
Precipitates Ca^{2+}	Yes	Yes	Yes		
Precipitates Mg^{2+}	Easier than Ca^{2+}	No or much more difficult than Ca^{2+}	Yes		
Sequesters Ca^{2+}, Mg^{2+}				Yes	
Emulsifier	Good	Yes	Good	Good	
Wettability alteration	Yes	Yes	Yes	Yes	Good
Reduces S adsorption	Yes	Yes	Yes	Yes	Yes

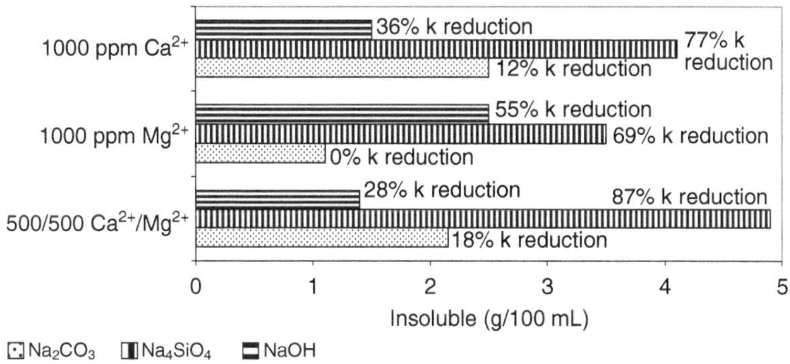

FIGURE 10.2 Effects of alkali precipitation with different hardness solutions. Extent of alkali precipitation is represented by the lengths of bars and permeability reduction is represented by % (shown beside bars) in flow experiments. *Source: Cheng (1986).*

Moreover, although calcium carbonate scales can be successfully removed at production wells by acidizing or by using inhibitors, no long-term treatment exists to control silicate-containing precipitation. This is probably one reason that sodium orthosilicate is not frequently used in chemical flooding. Because of the plugging function, sodium silicate is mixed with calcium chloride alternately to improve sweep efficiency.

Although solubilities of carbonate minerals could be lower than those of corresponding silicates or hydroxides, the continuous supply of fresh alkaline solution under dynamic field conditions may be expected to result in a continuous release of carbonate ions from rock minerals into the solution. This effect can be prevented by using carbonates such as alkali because carbonate ions brought by the solution oppose calcite and magnesite dissolution. Silicates do not profit from this kinetic effect, however. In another way, care must be taken when using silicates with rocks having high cation exchange capacity (CEC). Because of ion exchange processes, alkalinity loss is significant. This results in the formation of a region extending ahead of the pH front where the fluids are oversaturated with respect to silica. In contrast, for the carbonate injection, ion exchange results in the hydrolysis of CO_3^{2-} ions to the highly soluble form HCO_3^- (Labrid, 1991).

There are other reasons that sodium carbonate is often selected as the alkali used in chemical EOR:

- Sodium carbonate suppresses calcium ion concentration, but not magnesium's concentration.
- Sodium carbonate reduces the extent of ion exchange and mineral dissolution (in sandstones) as a weaker alkali compared with sodium hydroxide because mineral dissolution increases with pH. Owing to the buffer capacity of sodium carbonate, great changes in pH are not expected provided that the system is in chemical equilibrium. At some high concentrations, its pH

reaches a plateau. The preference of a weak alkali also comes from the concern of scale in production facilities.

- Sodium carbonate is an inexpensive alkali because it is mined as the sodium carbonate/bicarbonate mineral trona (Hirasaki and Zhang, 2004).
- In a carbonate reservoir, the carbonate/bicarbonate ion is a potential determining ion for carbonate minerals and thus is able to impart a negative zeta potential to the calcite/brine interface, even at neutral pH. A negative zeta potential is expected to promote water-wetness (Hirasaki and Zhang, 2004).
- Generally, ASP formulations use moderate pH chemicals such as sodium bicarbonate ($NaHCO_3$) or sodium carbonate (Na_2CO_3) rather than sodium hydroxide (NaOH) to reduce emulsion and scale problems. Chinese ASP projects have had difficulty in breaking emulsion when using a strong alkali such as NaOH.

10.2.4 Precipitation Problems

This section presents more problems than solutions regarding divalent precipitation because the existence of divalents is still a challenging problem in chemical flooding. In carbonate reservoirs where anhydrite $CaSO_4$ or gypsum $CaSO_4 \cdot 2H_2O$ exists, the $CaCO_3$ or $Ca(OH)_2$ precipitation occurs when Na_2CO_3 or NaOH is added. Carbonate reservoirs also contain brine with a higher concentration of divalents (Taber and Martin, 1983) and could cause precipitation. Liu (2007) discussed a couple of potential options. One option is to use $NaHCO_3$ and Na_2SO_4. $NaHCO_3$ has a much lower carbonate ion concentration, and additional sulfate ions can decrease calcium ion concentration in the solution. However, the concentration of CO_3^{2-} is about one hundredth of $NaHCO_3$ concentration in a $NaHCO_3$ solution, so a large amount of Na_2SO_4 is needed to avoid precipitation of $CaCO_3$. For example, in a 0.1M $NaHCO_3$ solution, the carbonate ion concentration is about 0.001M. Therefore, the amount of Na_2SO_4 needed to prevent the precipitation of $CaCO_3$ can be estimated as follows.

From the condition

$$[Ca^{2+}] = \frac{K_{sp}(CaCO_3)}{CO_3^{2-}} = \frac{K_{sp}(CaSO_4)}{SO_4^{2-}},$$

we have

$$[SO_4^{2-}] = \frac{K_{sp}(CaSO_4)}{K_{sp}(CaCO_3)}[CO_3^{2-}] = \frac{7.1 \cdot 10^{-5}}{8.7 \cdot 10^{-9}}(0.001) = 8.2.$$

So we need 8.2 M SO_4^{2-}. As a result, this option is not practical. Another option is to use NaOH and Na_2SO_4. Because the solubility product of $Ca(OH)_2$ is about 4.68×10^{-6}, the required $[SO_4^{2-}]$ is 15 times $[OH^-]^2$

$\left(= \frac{\text{Solubility product for CaSO}_4}{\text{Solubility product for (CaOH)}_2} = \frac{7.1 \cdot 10^{-5}}{4.68 \cdot 10^{-6}} \right)$. For example, for a 0.1 M NaOH solution, 1.5 M Na_2SO_4 is needed to suppress the calcium ion concentration so that no $Ca(OH)_2$ will precipitate. However, the adsorption of anionic surfactant on carbonates will not be decreased by NaOH solution because OH^- is not the determining ion.

To minimize the corrosion and scale problems associated with inorganic alkalis such as sodium hydroxide and sodium carbonate, Berger and Lee (2006) proposed an organic alkali. The organic alkali is derived from the sodium salts of certain weak polymer acids. They demonstrated the following benefits by using the organic alkali in the laboratory:

- Organic alkali performed equally well in softened and hard brines. It did not form precipitates with divalents such as calcium and magnesium.
- Organic alkali was as effective as inorganic alkali in obtaining low IFT.
- Organic alkali did not reduce the effect of polymer in increasing the injected fluid viscosity and improved polymer performance in hard waters.

Alkalis will react with rocks (ion exchange and dissolution) to result in precipitation. Some high-pH chelating agents such as Na_4EDTA and Na_3NTA (ethylenediaminetetraacetic acid and nitrilotriacetic acid salts) were added to replace alkalis to avoid this problem. Holm and Robertson (1981) reported that the addition of 7% Na_4EDTA in micellar solutions did not result in precipitation but improved oil recovery similar to the addition of Na_4SiO_4.

A benefit was observed from the addition of some chelating agents to the micellar solutions in protecting the slug from multivalent cations. The equilibrium constant for the formation of NTA/Ca/Na complex is about 10^6 times greater than that for the formation of the calcium/sulfonate complexes. This provides increased calcium ion tolerance for the surfactant solution. It was observed that the phase relationships between crude oil, brine, and sulfonate/solvent systems were also improved when these agents were added to the system. Less sulfonate/solvent was required to obtain clear microemulsions at high water concentrations when Na_3NTA or Na_4EDTA was present even though multivalent cations were not present. In addition, more oil was recovered by a micellar flood per unit of sulfonate/solvent injected (Holm and Robertson, 1981).

Metaborate was also proposed to sequester divalent cations such as Ca^{++} and to prevent precipitation (Flaaten et al., 2008). Apparently, it needs strict conditions to work. No field test was reported.

10.3 ALKALINE REACTION WITH CRUDE OIL

This section discusses the alkaline reaction with crude oil, which includes in situ soap generation, emulsification, and effect of ionic strength and pH on IFT.

10.3.1 In Situ Soap Generation

In alkaline flooding, the injected alkali reacts with the saponifiable components in the reservoir crude oil. These saponifiable components are described as petroleum acids (naphthenic acids). Naphthenic acid is the name for an unspecific mixture of several cyclopentyl and cyclohexyl carboxylic acids with molecular weight of 120 to well over 700. The main fractions are carboxylic acids (Shuler et al., 1989). Other fractions could be carboxyphenols (Seifert, 1975), porphyrins (Dunning et al., 1953), and asphaltene (Pasquarelli and Wasan, 1979). The naphtha fraction of the crude oil raffination is oxidized and yields naphthenic acid. The composition differs with the crude oil composition and the conditions during raffination and oxidation (Rudzinski et al., 2002).

This book does not discuss the details of alkali–oil chemistry related to saponification. It assumes a highly oil-soluble single pseudo-acid component (HA) in oil. The alkali–oil chemistry is described by partitioning of this pseudo-acid component between the oleic and aqueous phases and subsequent hydrolysis in the presence of alkali to produce a soluble anionic surfactant A^- (its component is conventionally denoted by $RCOO^-$), as shown in Figure 10.3.

The overall hydrolysis and extraction are given by

$$HA_o + NaOH \leftrightarrow NaA + H_2O, \tag{10.4}$$

and the extent of this reaction depends strongly on the aqueous solution pH. This reaction occurs at the water/oil interface. A fraction of organic acids in oil become ionized with the addition of an alkali, whereas others remained electronically neutral. The hydrogen-bonding interaction between the ionized and neutral acids can lead to the formation of a complex called acid soaps. Thus, the overall reaction, Eq. 10.4, is decomposed into a distribution of the molecular acid between the oleic and aqueous phases,

$$HA_o \leftrightarrow HA_w, \tag{10.5}$$

and an aqueous hydrolysis (deZabala et al., 1982),

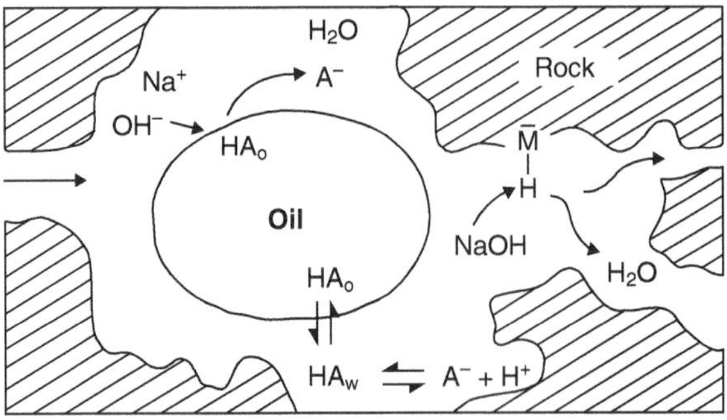

FIGURE 10.3 Schematic of alkaline recovery process. *Source: deZabala et al. (1982).*

$$HA_w \leftrightarrow H^- + A^-. \tag{10.6}$$

Here, HA denotes a single acid species, A denotes a long organic chain, and the subscripts o and w denote oleic and aqueous phases, respectively. The acid dissociation constant for Eq. 10.6 is

$$K_A = \frac{[H^+][A^-]}{[HA_w]}, \tag{10.7}$$

and the partition coefficient of the molecular acid is

$$K_D = \frac{[HA_w]}{[HA_o]}, \tag{10.8}$$

where brackets indicate molar concentrations. Additionally, the dissociation of water is

$$H_2O \leftrightarrow H^+ + OH^-, \tag{10.9}$$

and the dissociation constant of water is

$$K_w = [H^+][OH^-]. \tag{10.10}$$

Water concentration is essentially constant. An increase in $[OH^-]$ results in a decrease in $[H^+]$. pH is defined as $-\log[H^+]$. At high pH, the concentration of anionic surfactant (called soap in this book) in the aqueous phase is

$$[A^-] = \frac{K_A K_D [HA_o]}{[H^+]} = \frac{K_A K_D [HA_o][OH^-]}{K_w}. \tag{10.11}$$

Thus, for a fixed acid concentration in the oil phase and for a given pH, Eq. 10.11 estimates the amount of surface-active agent (A^-) present in the *aqueous* phase. This equation also reveals that K_A, K_D, $K_{w,}$ and pH regulate the amount of surface-active agent in the aqueous phase. K_D must be small enough so that the acid is not extracted into the aqueous phase by normal low-pH waterflooding. deZabala et al. (1982) used $K_w = 5 \times 10^{-14}$, $K_D = 10^{-4}$, and $K_A = 10^{-10}$.

When these numbers are used, for 1% NaOH, $[A^-]$ is only 5% of $[HA_o]$. Alternatively, a very high pH (close to 14, which is not practical) is required for the surface-active agent to be totally soluble in the aqueous phase. However, more $[A^-]$ is accumulated at the oil/water interface, which instantaneously reduces IFT. Sharma et al. (1989) took into account crude-oil/caustic interface (surface phase). They formulated the acid species dissociation and soap formation in the surface phase.

The acidic components in crude oil react with alkali to reduce IFT. Figure 10.4 shows the IFTs of a crude oil and its extracted oil with the same NaOH solution. It shows that the IFT of the extracted oil that loses active components is about three orders of magnitude higher than that of the crude

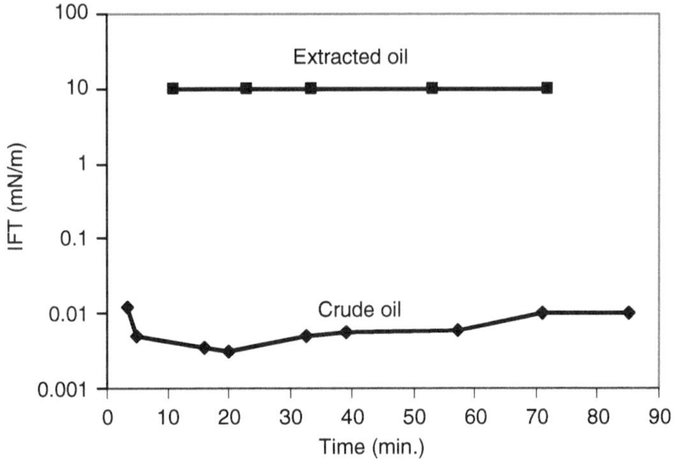

FIGURE 10.4 Dynamic IFT of a crude oil and its extracted oil with the same alkaline solution. *Source: Zhao et al. (2002).*

TABLE 10.2 Emulsibility of Different Oils with 0.2% Na_2CO_3

	Light Transmittance, %		
Time, min.	*Model Oil with 3% Acidic Components*	*Crude Oil*	*Extracted Oil*
0	0.6	0	91.7
30	2.8	0	96.8
90	6.1	0	97.2
1440	37.4	0.6	100

Source: *Zhang et al. (1998b)*

oil with the same alkaline solution. The active components are believed to cause the instantaneous low IFT, as discussed in Section 10.3.3.

10.3.2 Emulsification

Table 10.2 shows the light transmittances of different oils mixed with 0.2% Na_2CO_3 solution (the ratio of oil to the alkaline solution is 1/20). A higher light transmittance represents a lower emulsification. The table also shows that the extracted oil is not emulsified with the alkaline solution, whereas the model oil with 3% acidic components reacts strongly with the alkaline solution.

Table 10.3, however, shows that the extracted oil is well emulsified with the mixed solution of 0.2% petroleum sulfonate CY, 0.2% nonionic surfactant

TABLE 10.3 Emulsibility of Different Oils with Mixed the Solution

	Light Transmittance, %		
Time, min.	*Model Oil with 3% Acidic Components*	*Crude Oil*	*Extracted Oil*
0	0.2	0	0
30	0.4	0	0
90	0.7	0	0
1440	26.0	0	0

Source: *Zhang et al. (1998b)*

OP-10, and 1.5% Na_2CO_3. It seems that the extracted oil is better emulsified than the model oil with 3% acidic components.

Emulsification mainly depends on the water/oil IFT. The lower the IFT, the easier the emulsification occurs. The stability of an emulsion mainly depends on the film of the water/oil interface. The acidic components in the crude oil could reduce IFT to make emulsification occur easily, whereas the asphaltene surfactants adsorb on the interface to make the film stronger so that the stability of emulsion is enhanced. The extracted oil cannot be easily emulsified with alkaline solution because of the high IFT. However, the externally added surfactants can reduce the IFT between the extracted oil and mixed solution to a low value so that the emulsification can occur.

Huang and Yu (2002) observed that emulsification was not completely reversible. When the dynamic IFT reached ultralow, emulsification occurred. Even when dynamic IFT went up, emulsified oil droplets did not easily coalesce. In alkaline flooding, emulsification is instant, and emulsions are very stable. From this emulsification point of view, the dynamic minimum IFT plays an important role in enhanced oil recovery. From the low IFT point of view, we may think we should use equilibrium IFT because reservoir flow is a slow process. However, the coreflood results in the Daqing laboratory showed that when the minimum dynamic IFT reached 10^{-3} mN/m level and the equilibrium IFT was at 10^{-1} mN/m; the ASP incremental oil recovery factors were similar to those when the equilibrium IFT was 10^{-3} mN/m (Li, 2007). One explanation is that once the residual oil droplets become mobile owing to the instantaneous minimum IFT, they coalesce to form a continuous oil bank. This continuous oil bank can be move even when the IFT becomes high later. Then for this mechanism to work, the oil droplets must be able to coalesce before the IFT becomes high. It can be seen that it will be more difficult for such a mechanism to function in field conditions rather than in laboratory corefloods. This mechanism has not been universally accepted (G.J. Hirasaki, personal communication, October 2009).

10.3.3 Effects of Ionic Strength and pH on IFT

We have observed that the alkali concentration range in which the IFT between a crude oil and an alkaline solution is the minimum is very narrow. When the alkali concentration is out of this range, the IFT increases drastically. When we select chemicals for a project, either we perform a salinity scan by changing salinity while fixing an alkali concentration, or we perform an alkali scan by changing alkali concentration while fixing a salinity. Increasing alkali concentration (pH) also increases ionic strength (salinity). However, the effects of pH and salinity are different. Rudin and Wasan (1992a, 1992b) were among those who first recognized the difference of pH effect and salinity effect. Generally, salinity is higher than alkaline concentration used. In most practical applications, salinity is above several percent, whereas alkali concentration used is at most a few percent, most likely less than 1 to 2%. Therefore, their ionic strengths cannot be simply calculated (added) from salinity and alkali concentration to reflect the effect of ionic strength.

Figure 10.5 shows the dynamic IFT between a crude oil and NaOH solution at different concentrations and the fixed ionic strength of 1×10^{-2} mol/L. When NaOH concentration is very low (1×10^{-4} mol/L, Curve 1), the amount of soap generated at the oil/water interface is very small, and the IFT is above 10 mN/m.

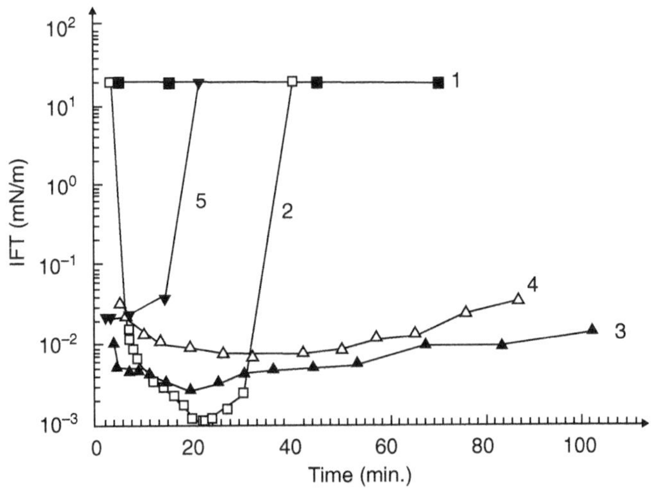

FIGURE 10.5 Dynamic IFT between a crude oil and NaOH solution at different concentrations with $[Na^+] = 0.01$ mol/L at 30°C. NaOH concentrations (10^{-3} mol/L): 1, 0.1; 2, 0.5; 3, 1; 4, 5; and 5, 10. *Source: Zhao et al. (2002).*

When NaOH concentration is not very low (5×10^{-4} mol/L, Curve 2), the IFT passes by a low value. As the soap leaves the interface and enters the aqueous phase, the IFT stays at a high value. At some optimum NaOH concentrations (1×10^{-3} mol/L, Curve 3, and 5×10^{-3} mol/L, Curve 4), the IFT will stay at a low value. At a very high NaOH concentration (1×10^{-2} mol/L, Curve 5), the soap quickly generates at the interface, and the IFT suddenly becomes low. However, as the soap leaves the interface, the IFT becomes high again.

Figure 10.6 shows the instantaneous (at 4 minutes), minimum, and equilibrium IFTs for the preceding systems. As NaOH concentration increases, the IFTs pass by a minimum value.

For a fixed alkali concentration, as the ionic strength increases from a very low value to a very high value, the dynamic IFT follows the trends of the curves in Figure 10.5, and instantaneous, minimum, and equilibrium IFTs follow the characteristics of the curves in Figure 10.6 (a "V" shape or concave curves).

In the preceding discussion, we used some experimental data to illustrate the dynamic IFT behavior at different ionic strength (salinity) and pH (alkali concentration). The reader should be aware that alkali concentration and salinity are much lower than what we inject in practical applications. The dynamic interfacial tension behavior at the salinity and alkali concentration of a particular application needs to be measured. Zhang et al. (1998a) presented similar observations about IFT changes with ionic strength and alkali concentrations for Gudong crude oil and extracted Gudong oil when external surfactants were added.

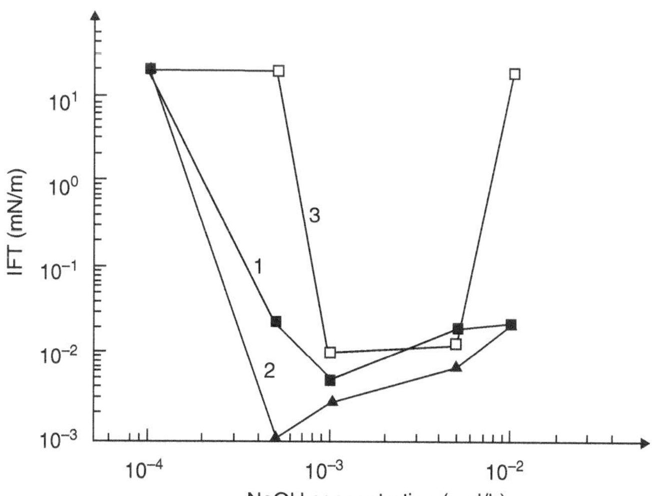

FIGURE 10.6 Dynamic IFT between a crude oil and NaOH solution at different concentrations with $[Na^+] = 0.01$ mol/L at 30°C: 1, instantaneous IFT; 2, minimum IFT; and 3, equilibrium IFT. *Source: Zhao et al. (2002).*

10.4 MEASUREMENT OF ACID NUMBER

A measure of the potential of a crude oil to form surfactants is given by the acid number (sometimes called total acid number, or TAN). This is the mass of potassium hydroxide (KOH) in milligrams that is required to neutralize one gram of crude oil. Usually, acid number determined by nonaqueous phase titration (Fan and Buckley, 2006) is used to estimate the soap amount. However, short chain acids, which also react with alkali, may not behave like surfactant because they are too hydrophilic. Also, phenolics and porphyrins in crude oil will consume alkali and will not change the interfacial properties as much as surfactant. Asphaltene and/or resin may have carboxylate functional groups but not be extracted into the aqueous phase. Total acid number determined by nonaqueous phase titration could not distinguish the acids that can generate natural soap and those that can consume alkali without producing surfactant. Another fact that could stimulate a question about nonaqueous phase titration is that acid number does not always correlate with oil recovery.

Figure 10.7 shows that even if the acid number of the oil was zero, oil/water IFT could be reduced by adding alkalis in the water. The organic acid was removed from the used Daqing oil (with zero acid number). The figure shows that at an equal alkaline concentration, different alkalis gave different IFTs. These different IFTs were not caused by different values of pH only, because at equal alkaline concentration, the pH value of sodium orthosilicate was higher than those of sodium carbonate and sodium bicarbonate. However, the IFT for sodium orthosilicate was higher. It is implied that some other factors could also

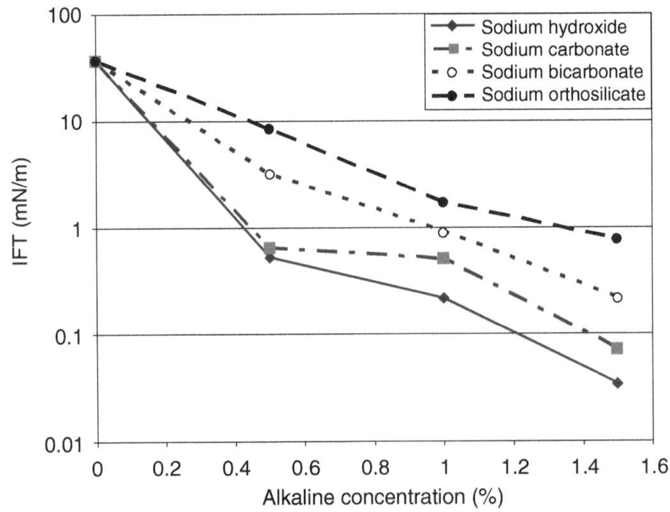

FIGURE 10.7 Oil–water IFT at different alkaline concentrations with zero acid number in the oil. *Source: Data from Li (2007).*

reduce IFT, or the acid number measured does not reflect all the factors that contribute to IFT reduction.

Figure 10.8 shows the dynamic IFT between 1.0% NaOH solution and gasoline engine oils with 0 to 2 acid numbers. We can see from this figure that the minimum dynamic and equilibrium IFTs were similar for the oils with 0 to 2 acid numbers, and the IFT reduction was not as significant as crude oils because the engine oils did not have active components such as asphaltene and resin.

Figure 10.9 compares the dynamic IFT of a light oil with that of a heavy oil. In this case, the oils had the same acid number and reacted with alkaline

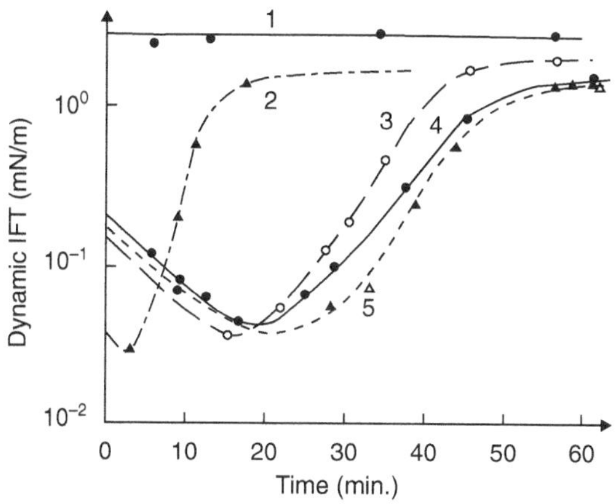

FIGURE 10.8 Effect of acid number on dynamic IFT (30°C). Acid number (mg KOH/g oil): 1, 0.0; 2, 0.1; 3, 0.5; 4, 1.0; and 5, 2.0. *Source: Yang et al. (1992).*

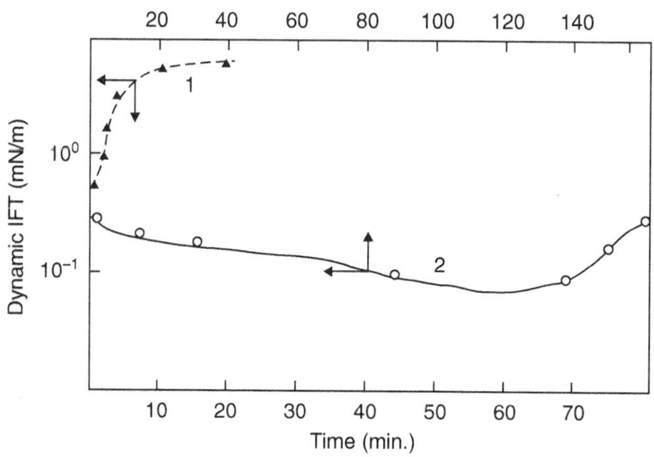

FIGURE 10.9 Effect of crude oil on dynamic IFT: 1, light oil; and 2, heavy oil. *Source: Huang et al. (1987).*

solution with the same concentration. The figure shows that the IFTs of these two oils had a difference of one order of magnitude. These two figures clearly demonstrate that IFT reduction or oil activity is related not only to acid number, but more importantly, to other active compounds. Some nonhydrocarbon compounds with sulfide, oxygen, and nitrogen also help emulsification (Cheng and Zheng, 1988).

Liu (2007) introduced another method called soap extraction to quantify acid number. Because the anionic surfactant can be accurately determined by potentiometric titration (see Appendix A in Liu, 2007) with benzethonium chloride (hyamine 1622), it is reasonable to use this method to find the natural soap amount. Because this potentiometric titration is for the aqueous phase, the soap should be extracted into the aqueous phase as the first step. As an anionic surfactant, the natural soap may stay in the oleic phase and form Winsor type II microemulsion when the electrolyte strength is high. To extract the soap into the aqueous phase, NaOH is used to keep the pH high with low electrolyte strength. Also, isopropyl alcohol is added to make the system hydrophilic so that soap will partition into the aqueous phase.

An oil's natural soap amount cannot be determined just by nonaqueous phase titration. Oils with high acid number by nonaqueous phase titration usually have high soap content; however, this is not always true (Liu, 2007). Because those acids that cannot generate soap will not be detected by the potentiometric titration, the acid numbers obtained by the soap extraction are less than the acid numbers determined by nonaqueous phase titration, as expected. There is no general ratio between those two acid numbers. Figure 10.10 compares the acid numbers measured from the two methods. The data in this figure show that the acid number from the soap extraction was about one half of the value from the nonaqueous phase titration.

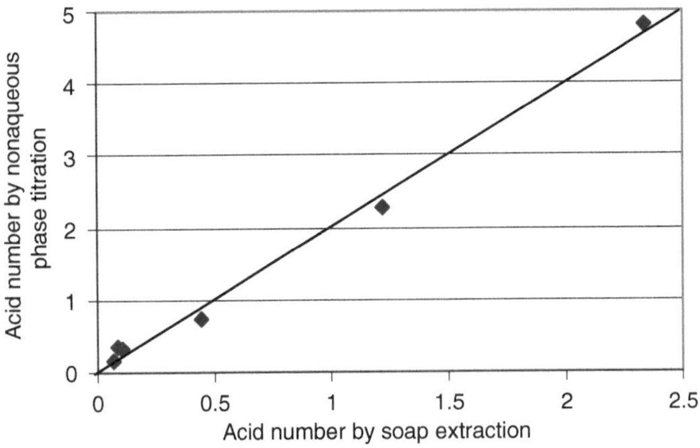

FIGURE 10.10 Comparison of acid numbers from the two methods. *Source: Data from Liu (2007).*

For acid numbers, greater than 1.0 is generally considered high, 0.3 to 1 is intermediate, and 0.1 to 0.25 is low. The acid numbers of Daqing oils are low, in the order of 0.1 mg/g. Most crude oils have an acid number lower than 5 mg KOH/g oil. Practically, the minimum acid number is 0.3 mg KOH/g for the generated soap to be effective in an ASP flooding (Chang et al., 2006).

When the acid number for a crude oil is known, we want to estimate how much soap can be generated, assuming (1) the required alkali is available, which is generally true; and (2) the total surface-active agents are converted into soap, which is generally not true, as discussed in Section 12.9.2. Based on the definition that acid number (AN) is the amount of potassium hydroxide in milligrams that is needed to neutralize the acids in one gram of oil, the soap concentration, C_{soap}, in meq/mL is

$$C_{soap}\left[\frac{\text{meq}}{\text{mL}}\right] = \frac{(\text{AN})\rho_o}{(\text{MW})_{\text{KOH}}(\text{WOR})}, \tag{10.12}$$

where $(\text{MW})_{\text{KOH}}$ is the molecular weight of KOH, which is 56 g/mole; ρ_o is the oil density in g/mL; WOR is the water/oil ratio in laboratory pipette tests, and it is the ratio of water saturation to oil saturation, S_w/S_o. Because surfactant concentration is generally expressed in volume percent in water, C_{soap} in vol.% is

$$\begin{aligned} C_{soap}[\text{vol.}\%] &= \frac{(\text{MW})_{soap}(\text{AN})\rho_o(10^{-3})}{(\text{MW})_{\text{KOH}}(\text{WOR})\rho_{soap}} \times 100\% \\ &= \frac{0.1(\text{AN})(\text{MW})_{soap}\,\rho_o}{(\text{MW})_{\text{KOH}}(\text{WOR})\rho_{soap}}\%, \end{aligned} \tag{10.13}$$

where $(\text{MW})_{soap}$ is the soap molecular weight. If $\rho_o/\rho_{soap} = 1$, and $(\text{MW})_{soap}/(\text{MW})_{\text{KOH}} - 10$, then the soap concentration in the volume percent in water is AN/WOR %. Further, if WOR = 1, the soap concentration is simply AN % without need of calculation. This calculation assumes that the surface-active agents are fully soluble in the aqueous phase, but they are not in reality.

10.5 ALKALI INTERACTIONS WITH ROCK

Alkali/rock reactions are probably the most difficult and least quantified aspect of alkaline flooding. Because of complex mineralogy in reservoirs, the number of possible reactions with alkalis is large. Because of the high surface area of clays, these materials play an important part in the alkaline solution displacement process. When clays originally in equilibrium with formation water are contacted with alkaline solution, the surface will attempt to equilibrate with its new environment, and ions will start exchanging between the solid surfaces and alkaline solution. Ions present on the clays originally include hydrogen. As the pH of the solution is increased, hydrogen ions on the surface react with hydroxide ions in the flood solution, lowering the pH of the alkaline solution. By this reaction, the base present in the alkaline solution is consumed as the alkaline solution moves through the reservoir. Calcium and magnesium ions are also present in clays. When calcium-free alkaline salt water contacts the clays, calcium ions on the rock surface will exchange for sodium ions in the alkaline solution.

Calcium ions are undesirable in alkaline flood solution, and their concentration must be kept to a very low value. This is ordinarily accomplished by using a sodium carbonate buffer, which removes calcium as it exchanges off the clay by precipitating it as insoluble calcium carbonate. In doing so, however, carbonate ions, which are the buffering agent in the system when sodium carbonate is used, are also removed. Thus, reaction with calcium on the clays also consumes the alkaline solution as it moves through the reservoir.

Interaction of alkali with rock minerals is complicated and can include ion exchange and hydrolysis, congruent and incongruent dissolution reactions, and insoluble salt formation by reaction with hardness ions in the pore fluids and exchanged from the rock surfaces. By congruent dissolution, we mean a mineral reaction that generates soluble aqueous species in the stoichiometric ratio of the mineral lattice. Conversely, incongruent dissolution refers to the dissolution of one mineral and the formation of a second, different mineral. The interactions may be classified into reversible or irreversible, and kinetic or instantaneous. This section briefly discusses these interactions. To facilitate the discussion, we first present the reaction equation for a single phase without including dispersion and diffusion:

$$\frac{\partial C}{\partial t}+\frac{u\partial C}{\phi\partial x}+R_E+R_D=0. \tag{10.14}$$

In Eq. 10.14, u is the superficial (Darcy) velocity; R_E is the net loss rate owing to ion exchange in miliequivalents per void volume (mL) per time to be consistent with the unit of $\partial C/\partial t$, with the subscript E denoting ion exchange; similarly, R_D is the net loss rate due to dissolution, with the subscript D denoting dissolution. In an acid-base system, it is not possible to distinguish by concentration measurements (e.g., by titration or by glass electrodes) between a reaction that consumes hydroxide ions and one that liberates hydrogen ions. Only a difference in concentration between these two ions has a well-defined zero value and is meaningful. Therefore, C in Eq. 10.14 indicates the difference between hydroxide-ion and hydrogen ion concentration; that is, $C=C_{OH^-}-C_{H^+}$ (Bunge and Radke, 1982).

10.5.1 Alkaline Ion Exchange with Rock

One rock-alkali interaction is described by the sodium/hydrogen-base exchange (hydroxide-exchange),

$$H\text{–}X+Na^{+}+OH^{-}\leftrightarrow Na\text{–}X+H_2O, \tag{10.15}$$

where X denotes mineral-base exchange sites (see Figure 10.3). In flowing through a reservoir rock, sodium ions must "fill" the available exchange sites before they can progress downstream. Equation 10.15 shows that both hydroxyl and sodium ions are consumed. Similarly, alkali also has cation exchange with the divalents in the rock. For example,

$$Ca\text{–}X_2 + 2Na^+ + 2OH^- \leftrightarrow 2(Na\text{–}X) + Ca(OH)_2, \tag{10.16}$$

$$Ca\text{–}X_2 + 2Na^+ + CO_3^{2-} \leftrightarrow 2(Na\text{–}X) + CaCO_3. \tag{10.17}$$

The hydroxide ion exchange based on the mass-action equilibrium of Eq. 10.15 leads to a Langmuir-type isotherm, as shown in Figure 10.11 (Somerton and Radke, 1983):

$$\frac{\hat{C}}{HEC} = \frac{K_eC}{1+K_eC}. \tag{10.18}$$

In Eq. 10.18, $\hat{C}$ is the difference between hydroxide-ion and hydrogen ion adsorption in the unit of miliequivalents per solid surface area (m^2); that is, $\hat{C} = \hat{C}_{OH^-} - \hat{C}_{H^+}$. HEC is the hydrogen exchange capacity (maximum uptake or maximum number of exchange sites) in the same unit as $\hat{C}$, and K_e is the ion-exchange equilibrium constant in the unit of inverse concentration. The

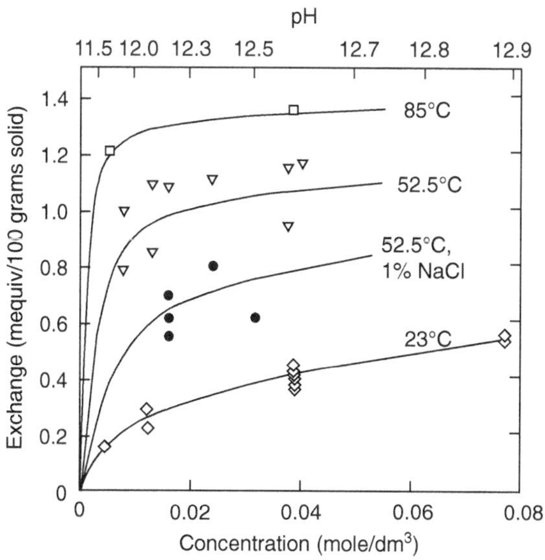

FIGURE 10.11 Hydroxide-exchange isotherms for Wilmington oil sand. *Source: Somerton and Radke (1983).*

hydroxide ion exchange is a fast-reversible process (the equilibrium time is less than 8 to 10 minutes). The ion exchange rate R_E is

$$R_E = \frac{S_r\rho_r(1-\phi)}{\phi}\frac{\partial\hat{C}}{\partial t} = \frac{S_r\rho_r(1-\phi)}{\phi}\left(\frac{\partial\hat{C}}{\partial C}\right)\frac{\partial C}{\partial t} = \frac{\alpha}{(1+K_eC)^2}\frac{\partial C}{\partial t}. \tag{10.19}$$

In Eq. 10.19, S_r is the specific solid surface area (m^2/g rock), ρ_r is the solid density (g/cm^3) when C is in meq/mL and R_E are in meq/mL/s, $\hat{C}$ and HEC are in meq/m^2, K_e is in mL/meq, and $\alpha = K_e(HEC)S_r\rho_r(1-\phi)/\phi$. S_r(HEC) is in the convenient CEC unit, meq/g rock.

Although the hydroxide exchange capacity is not large (about 10 meq/kg), it greatly retards the advance rate of alkali in a reservoir. Figure 10.12 shows the elution of hydroxide from tertiary oil floods in Wilmington sand for several injected pH values, all in 1 wt.% sodium chloride. As the injected pH was lowered, the alkali took progressively longer to elute from the core. With an injected pH of 11.2, hydroxide did not appear in the effluent even after 10 PV of flooding. The dashed lines in the figure are predicted a priori with standard equilibrium chromatography theory and the appropriate exchange isotherm (Somerton and Radke, 1983). Equation 10.19 shows that if $\partial\hat{C}/\partial C$ is small, the exchange rate will be small. Because the hydroxide ion exchange follows a Langmuir-type isotherm, the small values of $\partial\hat{C}/\partial C$ will occur at higher concentrations or higher pH values (see Figure 10.11). Then higher pH will yield earlier breakthrough, according to Figure 10.12.

For a fixed ion-exchange equilibrium constant, K_e, increasing α corresponds to increasing the hydrogen exchange capacity, (HEC)S_r. Figure 10.13 shows the fractional penetration lengths as a function of injection slug size for $\alpha = 10$ and 20 and for the two reaction orders (m = 0 and 1). The concentration is

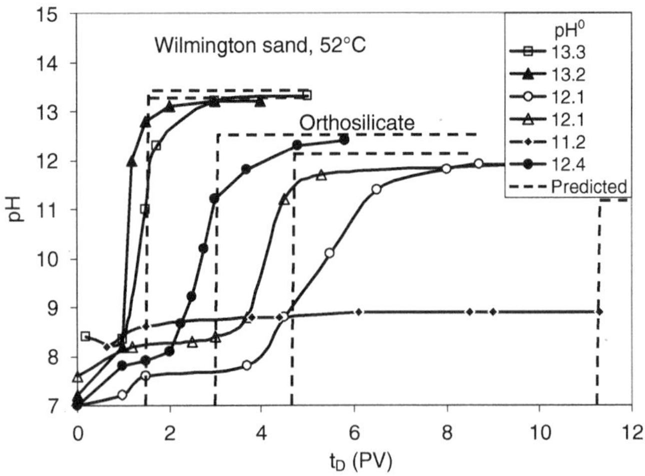

FIGURE 10.12 Hydroxide concentration histories for tertiary oil floods in Wilmington oil sands. *Source: Somerton and Radke (1983).*

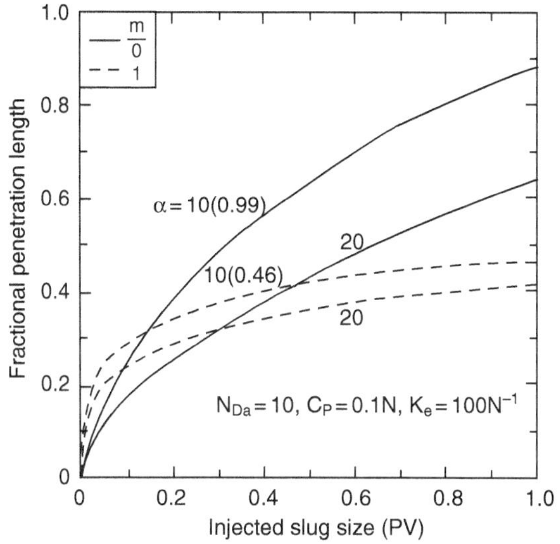

FIGURE 10.13 Fractional penetration length at two rock ion-exchange capacities (α). Numbers in parentheses are the fractional penetration lengths for continual chemical injection. *Source: Bunge and Radke (1982).*

0.1 N, and the Damköhler number, defined later, is 10. Figure 10.13 shows that increasing α decreases the penetration length.

Holm and Robertson (1981) estimated the amount of Na_4SiO_4 consumed by reaction with exchangeable divalent ions on Muddy sandstone to be 0.5 meq/kg rock (0.05 lb/bbl PV). Krumrine et al. (1982) found the NaOH consumption to be 40 to 160 meq/kg due to ion exchange using a mixture of 0.16% and 0.35% NaOH and NaCl, respectively.

10.5.2 Alkaline Reaction with Rock

In addition to ion exchange with rock surfaces, alkali can react directly with specific rock minerals. When divalents, Ca^{2+} and Mg^{2+}, exist, alkali will react with them and precipitation can occur. One example is the incongruent dissolution of anhydrite or gypsum in the rock to produce the less soluble calcium hydroxide ($CaSO_{4(s)} + NaOH \leftrightarrow Ca(OH)_{2(s)} + Na_2SO_4$). Another simple example is $Ca^{2+} + CO_3^{2-} \leftrightarrow CaCO_{3(s)}$. Alkali can also dissolve other minerals from a rock, for example, silica. These reactions could cause plugging.

Contrary to ion exchange, which is a fast-reversible process, the dissolution of rock minerals by alkalis is a long-term irreversible kinetic process. In alkaline solutions, soluble silica exists as several species. The exact speciation is not well established, but at lower concentrations it may be summarized by Eqs. 10.20 to 10.23. Table 10.4 summarizes the published rate constants of those equations collected by Bunge and Radke (1982).

TABLE 10.4 Silica Speciation Chemistry and Solubility ($\log_{10}K$)

Equation	0.5 N NaCl at 25°C	0.5 N $NaClO_4$ at 50°C	0.5 N $NaClO_4$ at 25°C	3.0 N $NaClO_4$ at 25°C	Unspecified Salt at 25°C
10.20	4.3	3.8	4.3	4.6	
10.21	1.0	1.1	1.2	1.3	
10.22					−1.0
10.23					−2.6
10.24					−3.7
10.25					−2.7
10.26	−13.7	13.0	−13.7	−14.0	

$$Si(OH)_4 + OH^- = Si(OH)_3O^- + H_2O, \tag{10.20}$$

$$Si(OH)_3O^- + OH^- = Si(OH)_2O_2^{2-} + H_2O, \tag{10.21}$$

$$Si(OH)_2O_2^{2-} + OH^- = Si(OH)O_3^{3-} + H_2O, \tag{10.22}$$

$$Si(OH)O_3^{3-} + OH^- = SiO_4^{4-} + H_2O. \tag{10.23}$$

The solubility reactions of both quartz and amorphous silica were given by Stumm and Morgan (1970):

$$SiO_2(\text{quartz}) + 2H_2O = Si(OH)_4, \tag{10.24}$$

$$SiO_2(\text{amorphous}) + 2H_2O = Si(OH)_4. \tag{10.25}$$

$$H_2O = H^+ + OH^-. \tag{10.26}$$

Figure 10.14 shows the calculated ionization states of soluble silica based on Eqs. 10.20 to 10.23 as a function of solution pH at 25°C. In neutral solutions, only silicic acid exists. For pH values from 10 to almost 13, the monovalent species dominates, whereas for pH values greater than 13, the divalent species prevails. Not until very high alkalinities do the tri- and tetravalent ions appear in solution.

Because the pH of typical alkaline floods falls within the range of the existence of $Si(OH)_3O^-$, Bunge and Radke (1982) presumed that solid silica dissolves according to the following rate-controlling reaction:

$$SiO_2(s) + H_2O + OH^- \rightarrow Si(OH)_3O^-. \tag{10.27}$$

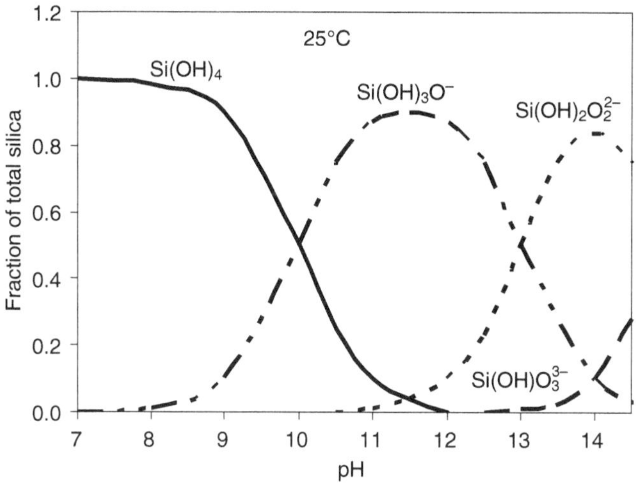

FIGURE 10.14 Soluble silica speciation calculated from Eqs. 10.20 to 10.23. *Source: Bunge and Radke (1982).*

For an m-order of reaction in general, R_D in Eq. 10.14 is

$$R_D = K_D C^m. \tag{10.28}$$

For mathematical simplicity, however, we may treat the kinetics to be first order (m = 1) or zero order (m = 0).

Using Eqs. 10.19 and 10.28, Eq. 10.14 becomes

$$\frac{\partial C}{\partial t} + \frac{u}{\phi}\frac{\partial C}{\partial x} + \frac{S_r \rho_r (1-\phi)}{\phi}\left(\frac{\partial \hat{C}}{\partial C}\right)\frac{\partial C}{\partial t} + K_D C^m = 0. \tag{10.29}$$

The dimensionless form of Eq. 10.29 is

$$(1+R_E)\frac{\partial C_D}{\partial t_D} + \frac{\partial C_D}{\partial x_D} + N_{Da} C_D^m = 0, \tag{10.30}$$

where R_E is the dimensionless retardation factor because of the reversible ion-exchange:

$$R_E = \frac{S_r \rho_r (1-\phi)}{\phi}\left(\frac{\partial \hat{C}}{\partial C}\right), \tag{10.31}$$

$$C_D = C/C_{inj}, \quad C_{inj} \text{ is the injection concentration,} \tag{10.32}$$

$$t_D = \frac{ut}{\phi L}, \tag{10.33}$$

$$x_D = x/L, \tag{10.34}$$

and the Damköhler number is

$$N_{Da} = \frac{K_D C_{inj}^{m-1} L \phi}{u}, \tag{10.35}$$

for the m-order of reaction. Note that the unit of $K_D C_{inj}^{m-1}$ is t^{-1} for example, s^{-1} in SI. For the first order of reaction, the Damköhler number is

$$N_{Da} = \frac{K_D L \phi}{u}. \tag{10.36}$$

For the silicate ions of reduced concentration $C_{siD} = C_{si}/C_{inj}$, the stoichiometry of the reaction 10.27 implies that (Somerton and Ranke, 1983)

$$C_{siD} = 1 - C_D. \tag{10.37}$$

Following is the solution to Eqs. 10.30 and 10.37 for continual injection of alkali (m = 1). The hydroxide effluent history is

$$C_D(t_D, x_D = 1) = \begin{cases} 0, & \text{for } t_D \le 1 + R_E \\ \exp(-N_{Da}), & \text{for } t_D > 1 + R_E \end{cases}. \tag{10.38}$$

The silicate concentration history is

$$C_{siD}(t_D, x_D = 1) = \begin{cases} 0, & \text{for } t_D \le 1 + R_E \\ 1 - \exp(-N_{Da}), & \text{for } t_D > 1 + R_E \end{cases}. \tag{10.39}$$

Bunge and Ranke (1982) reported the first-order rate constants of 1.3×10^{-6} s^{-1} at 74°C for Huntington Beach sand, 8.2×10^{-7} s^{-1} at 52°C for Wilmington sand, and 5.8×10^{-6} s^{-1} at 85°C for Berea sand.

The rate of most reactions depends highly on temperature. The rate constant changes with temperature may follow the Arrhenius equation,

$$K_D = A \cdot \exp\frac{-E_a}{RT}, \tag{10.40}$$

where A is the pre-exponential factor in the same unit as K_D, E_a is the activation energy in J/mol, R is the gas constant (8.314472 $J°K^{-1}mol^{-1}$), and T is the absolute temperature in °K.

In Eq. 10.30, the first term corresponds to accumulation in the fluid and the surfaces, the second term describes convective transport, and the third term indicates the loss by the kinetic dissolution reaction defined by Eq. 10.28. Equation 10.30 applies to any chemical transport process that includes fast and reversible ion-exchange, and slow and irreversible dissolution of the mth-order kinetics. In reservoir sands, both fine silica and clay minerals dissolve under attack by the alkali, yielding a complex distribution of soluble solution products

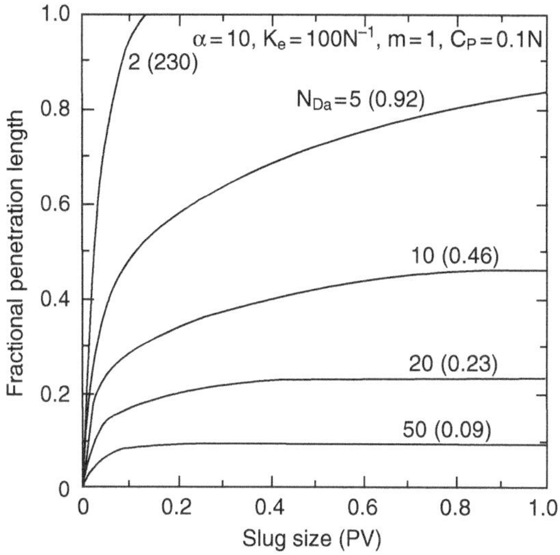

FIGURE 10.15 Fractional penetration length for alkaline slug injection. *Source: Bunge and Radke (1982).*

and new mineral species. In spite of these complications, slow hydroxide consumption is treated with a single, lumped-parameter reaction.

The Damköhler number provides a quick estimate of the degree of alkali penetration that can be achieved in continuous injection. We are interested in knowing how far the injected alkali can penetrate in 1D flow. Figure 10.15 shows the fractional penetration length (x_D) for alkaline slug injection assuming the first-order reaction with various Damköhler numbers. The numbers in parentheses give x_D for continuous alkaline injection. In the figure, C_p is the injection concentration, and the pH of the injection slug is 13.

Example 10.1 Find the Minimum Injected Alkali Concentration

In a laboratory test, the core length is 1 ft (0.3048 m), and the porosity is 0.3. At the end of alkaline flood, an oil bank will form ahead of the alkaline front and near the end of the core. If we assume the oil bank volume is 0.2 PV, then the minimum alkali penetration length (x_D) should be 0.8. If we further assume the dissolution reaction is of the first-order (m = 1), the injection concentration is 0.1 N, $\alpha = 10$, $K_e = 100\ N^{-1}$, and $K_D = 1.0 \times 10^{-6}\ s^{-1}$. Find the minimum injection alkali concentration.

Solution

First, check whether the diagram in Figure 10.15 can be used for this problem. In Figure 10.15, the concentration is 0.1 N, which is equivalent to 0.4 wt.% NaOH (0.1 N = 0.4% × 1000 mg/mL/40 mg/meq). From $\alpha = K_e(HEC)S_r\rho_r(1-\phi)/\phi$

Continued

Example 10.1 Find the Minimum Injected Alkali Concentration—*Continued*

= 10, we have S_r(HEC) = 10/(100 mL/meq × 2.65 g/mL × (1 − 0.3)/0.3) = 0.016 meq/g = 16 meq/kg, which is a reasonable cation exchange capacity. Therefore, the diagram in Figure 10.15 may be used.

Then calculate N_{Da}. For the first-order reaction, the Damköhler number is

$$N_{Da} = \frac{K_D L \phi}{u} = \frac{(1.0\times10^{-6}\ s^{-1})(1\ ft)(0.3)}{1\ ft/86400s} = 0.026$$

Such a small N_{Da} curve is not shown in Figure 10.15. From the figure, we can see that for a smaller N_{Da}, the penetration length is longer. For $N_{Da} = 0.026$ and $x_D = 0.8$, the volume of injection would be less than 0.02 PV injection if interpolation is used in the diagram. Generally, we use a larger slug volume but a lower concentration. If the total mass of chemical is maintained, and 0.2 PV injection volume is chosen, then the injected chemical concentration should be 0.01N (0.04 wt.%). This is the minimum alkali concentration for this problem.

Example 10.2 Scale the Laboratory Core Flood Test in Example 10.1 to a Field Test

Suppose we have run a core flood test and confirmed the parameters in Example 10.1. Now we want to plan a field pilot. Assume the well spacing is 100 m, and the field injection rate is the same as that used in the core flood test (1 ft/day). Find out the wt.% minimum injection concentration for the pilot.

Solution

To scale to field rates and length, we can use the Damköhler number. The Damköhler number for the field pilot is

$$N_{Da} = \frac{K_D L \phi}{u} = \frac{(1.0\times10^{-6}\ s^{-1})(328\ ft)(0.3)}{1\ ft/86400s} = 8.3$$

By interpolation from Figure 10.15, if we select the concentration of 0.1 N (0.4 wt.%), 1 PV injection would lead to a fractional penetration (x_D) of 0.65. If we inject 0.5 PV, to reach 0.65 fractional penetration, we need to use an injection concentration of 0.8 wt.%. For the first-order reaction, a higher concentration would result in a higher dissolution rate, which was not considered here. Therefore, to reach this fractional penetration, we probably should use an injection concentration of at least 1 wt.%.

Ehrlich and Wygal (1977) conducted static equilibration tests in which caustic solution contacted crushed samples of single-component minerals and reservoir rocks containing a number of mineral components. Table 10.5 shows their caustic consumption in alkalinity loss by pure minerals during one-week contact with 5% NaOH solution at room temperature. This table also presents

Table 10.5 Alkalinity Loss (meq/kg) by Minerals

Minerals	Ehrlich and Wygal (1977)	Mohnot and Bae (1989)[2]
Calcite	Insignificant	Insignificant
Chlorite		110, 140
Dolomite	Insignificant	610, 930
Gypsum (anhydrite)	11600[1]	
Gypsum (selenite)		1180, 1180
Illite	1360	720, 900
Kaolinite	130	1250, 1270
Labradorite (Ca-Na feldspar)		160, 210
Montmorillonite	2280	780, 1060
Quartz, fine		220, 450
Quartz, sand	Insignificant	Insignificant
Zeolite (Clinoptilolite)		670, 990

[1] *Calculated from stoichiometry assuming conversion to* $Ca(OH)_2$.
[2] *First value at pH 8.3, and second value at pH 10.*

the caustic consumption in hydroxide loss at the liquid-volume-to-solid-mass ratio of 1 mL/g, 5% NaOH, and 82°C for 11 days of contact (Mohnot and Bae, 1989). The table shows no measurable consumption for quartz and calcite, which is consistent with the report by Mohnot and Bae (1989).

Holm and Robertson (1981) also observed a very small amount of Na_4SiO_4 consumption by carbonate minerals. The reason is probably that there is little surface area associated with these minerals. Southwick (1985), however, reported that the loss of useful alkalinity for pure quartz sand via the slow dissolution of silica, for some typical alkaline flooding solutions, was about 10 to 20%. He also concluded that the dissolution of quartz can be eliminated by employing pre-equilibrated silicate solutions.

Appreciable exchange capacity has been measured for amorphous silica with high surface area (Culberson et al., 1975). The alkalinity losses for clays given by Ehrlich and Wygal (1977) were higher than those reported by Grim (1939) as the base-exchange capacity for those minerals, probably because the tests reported by Grim used lower-pH solutions for equilibration. These alkalinity loss data were also much higher (30–100 times!) than the NaOH consumptions, according to an unconfirmed source. Consumption has been shown to increase with increasing pH of the alkaline solution, increasing temperature, and increasing contact time (Cooke et al., 1974), and the ratio of solid to alkali solution in batch experiments (Mohnot et al., 1987, 1989).

The data from Ehrlich and Wygal also showed that the alkalinity loss by calcium sulfate minerals (gypsum, anhydrite) was one or two orders of magnitude higher than that by other minerals, which was not seen from Mohnot and Bae (1989) data. The alkalinity loss reported by kaolinite from Ehrlich and Wygal was one order of magnitude lower than that from Mohnot and Bae. This may have been caused by the test temperature difference (room temperature versus 82°C) because Johnson et al. (1988) showed the consumption of alkali by kaolinite and quartz increased considerably with increasing temperature.

In general, the order of consumption in which alkaline reacts with clays is as follows:

- Highest for kaolinite and gypsum
- Moderate for montmorillonite, illite, dolomite, and zeolite
- Moderately low for feldspar, chlorite, and fine quartz
- Lowest for quartz sand
- Insignificant for calcite

Shen and Chen (1996) put the alkaline consumption in this order: gypsum > montmorillonite > kaolinite > illite > anorthosite (plagioclasite) > microclinite > quartz > mica > dolomite > calcite. These orders are consistent with the general trend.

Alkalinity is a measure of the ability of a solution to neutralize acids to the equivalence point of carbonate or bicarbonate. It is closely related to the acid neutralizing capacity (ANC) of a solution, and ANC is often incorrectly used to refer to alkalinity. Alkalinity is equal to the stoichiometric sum of the concentrations of HCO_3^- and CO_3^{2-}—that is, $([HCO_3^-]+2[CO_3^{2-}])$ in mmol/L in most solutions. It is determined by titrating with acid down to a pH of about 4.5.

Alkalinity is sometimes incorrectly used interchangeably with basicity. For example, the pH of a solution can be lowered by the addition of CO_2. This will reduce the basicity; however, the alkalinity will remain unchanged. This is because the net reaction produces the same number of equivalents of positively contributing species (H^+) as negative contributing species (HCO_3^- and/or CO_3^{2-}):

At neutral pH,

$$CO_2 + H_2O \rightarrow HCO_3^- + H^+. \tag{10.41}$$

At high pH,

$$CO_2 + H_2O \rightarrow CO_3^{2-} + 2H^+. \tag{10.42}$$

The addition of CO_2 to a solution in contact with a solid can affect alkalinity, however, especially for carbonate minerals in contact with groundwater or seawater. The dissolution (or precipitation) of carbonate rock has a strong influence on the alkalinity. The reason is that carbonate rock is composed of $CaCO_3$, and its dissociation will add Ca^{2+} and CO_3^{2-} into solution. Ca^{2+} will not influence alkalinity, but CO_3^{2-} will increase alkalinity by 2 units.

Table 10.6 shows the caustic consumption by reservoir rocks during one-week contact with 5% NaOH solution. The measured values in parentheses were caustic consumption values attributable to clays only. Table 10.6 shows that a caustic consumption calculated based on the individual mineral consumption data in Table 10.5 reasonably agreed with the measured values, with two exceptions. The calculation overestimated the measured consumption in the Yates sand where clay content was high. This might result from the clay being present here as structural grains with not all its surface area accessible to the caustic solution. The calculation underestimated the measured consumption in the Queen sand, which contained trace quantities (a few tenths of a percent) of gypsum not detectable by the X-ray method. The consumption attributable only to the clays was in agreement with the calculated values (Ehrlich and Wygal, 1977). The consumption of caustic by calcium sulfate minerals resulted because calcium sulfate is more soluble in strongly alkaline solution than calcium hydroxide. In addition to Table 10.5, the data in Table 10.6 also show that the caustic consumption by quartz and dolomite was less significant than the measurement error.

Table 10.7 shows the caustic consumption by reservoir rocks during core flood tests with 5% NaOH solution (Ehrlich and Wygal, 1977). In continuous injection tests, caustic consumption was calculated from the frontal lag of produced alkalinity in the miscible displacement of connate water. In the slug injection tests, caustic consumption was calculated from the difference between the injected and the produced alkalinity. Table 10.7 shows that the consumptions from continuous injection were less than those given in Table 10.6 for Berea sand and Yates sand. The lower consumption probably resulted from the shorter contact time, which can be supported by comparing the consumption from continuous injection with that from the corresponding slug injection. For slug injection, the contact time is even less.

Experimental data from Cooke et al. (1974) showed that the base consumption was a strong function of time (increasing with time). Therefore, we may say that laboratory displacement tests are necessarily optimistic in their prediction of improved recovery because of the shorter contact time. The high consumption from Grayburg dolomite resulted from anhydrite content.

A core is flushed with alkaline solution to determine the minimum alkaline requirement. Then the alkali consumption includes ion exchange and dissolution. Holm and Robertson (1981) estimated the Na_4SiO_4 consumption for Muddy sandstone to be equal to 0.25 lb/bbl PV (2.5 meq/kg if the porosity and the rock density are taken to be 0.3 and 2.65 g/mL, respectively). It is also equivalent to the amount of Na_4SiO_4 in an 11% PV slug of a 0.7% active solution. Cheng (1986) found no significant consumption of Na_2CO_3 on

TABLE 10.6 Caustic Consumption by Reservoir Rocks

	X-Ray Diffraction Mineralogy		Caustic Consumption (meq NaOH/kg Rock)	
Formation	*Bulk*	*Clays*	*Calculated from Composition*	*Measured*
Berea sand	Quartz + 10–15% clay	75–80% kaolinite 15–20% illite	30–56	47
Grayburg dolomite	Dolomite + 5–10% anhydrite	None	740–1480	960
7,250 ft Miocene sand (Louisiana)	Quartz + 1–5% clay	50–55% kaolinite 40–45% illite	6–34	44
8,685 ft Miocene sand (Louisiana)	Quartz + 5–10% clay	55–60% kaolinite 20–25% illite 10–15% montmorillonite	29–76	67
2,630 ft Yates sand	Quartz, feldspar + 30–40% clay	55–60% montmorillonite 25–30% chlorite 5–10% illite	500–760	138
2,850 ft Yates sand	Quartz, feldspar, dolomite + 30–35% clay	55–60% montmorillonite 15–20% chlorite 15–20% illite	500–670	130
3,133 ft Queen sand	Quartz, feldspar, dolomite + 5–10% clay	35–40% montmorillonite 30–35% illite 20–25% kaolinite	61–142	255(55)
3,156 ft Queen sand	Quartz, feldspar, + 5–10% clay	50–60% montmorillonite 10–15% illite 20–25% chlorite	77–191	115(83)

Source: *Ehrlich and Wygal (1977)*

TABLE 10.7 Caustic Consumption by Flooding Reservoir Rocks

		Caustic Consumption (meq NaOH/kg Rock)	
Formation	Preflood Salts	*Continuous Injection*	*Slug Injection*
Berea sand	2% NaCl	5.1	0.8
Yates sand	9.5% TDS + 1.3% $CaCO_3$	34.5	8.6
Grayburg dolomite		956	

dolomite. Olsen et al. (1990) reported 5.8 and 6.8 meq of alkalinity consumed per kg of carbonate rock with an ASP system using Na_2CO_3 and sodium tripolyphosphate, respectively. NaOH consumption of 1.1 to 3.2 meq/kg rock was also reported for reservoir rocks.

10.5.3 Alkali–Water Reactions

The primary reaction of alkali with reservoir water is to reduce the activity of multivalent cations such as calcium and magnesium in oilfield brines. Upon contact of the alkali with these ions, precipitates of calcium and magnesium hydroxide, carbonate, or silicate may form, depending on pH, ion concentrations, temperature, and so on. If properly located, these precipitates can cause diversion of flow within the reservoir, leading to better contact of the injected fluid with the less-permeable and/or less-flooded flow channels. This then may contribute to improved recovery. Also, this reduction of reservoir brine cation activity will lead to more surfactant activity, resulting in lower IFT values (Mayer et al., 1983).

Novosad et al. (1981) carried out some comparisons between sodium hydroxide and sodium orthosilicate solutions. The results of the tests showed significantly lower brine hardness ion activity and IFT when sodium orthosilicate was used. These differences were attributed to the formation of calcium and magnesium silicates, which are much less soluble than calcium or magnesium hydroxides. Work by Campbell (1981) indicated that sodium orthosilicate might be more efficient in reducing IFT in hard-water systems. Because precipitation of hardness ions is largely a function of pH, the less basic alkalis, such as ammonium hydroxide and sodium carbonate, cannot be expected to be as effective in reducing hardness levels at equivalent weight percentages; however, specific anion interactions must be considered in the context of the overall chemistry to evaluate this (Mayer et al., 1983).

10.5.4 Total Alkali Consumption

From the previous discussion, total alkali consumption includes

$$C_i - C(t) = \Delta C_o + \Delta C_w + \Delta C_e + \Delta C_D, \qquad (10.43)$$

where C_i and C(t) are the initial and existing (current) concentrations, respectively; ΔC_o is the alkali consumption for the alkali to react with the crude oil to generate soap; ΔC_w is the alkali consumption caused by alkali reaction with the multivalent ions in the formation water; ΔC_e is the alkali consumption during the ion exchange between the alkali solution and the rock; and ΔC_D is the alkali consumption during dissolution reaction between the alkali and the rock.

The preceding four types of consumption must be determined experimentally in the laboratory and upscaled to field scales. The experimental conditions should be as close to the field conditions as possible. Field oil and water samples can be obtained, and experiments should be conducted at the field temperature. Ideally, reservoir rocks should be used. In practice, we may not be able to conduct all the necessary experiments because of the cost, available resources, and limited time. An approximation must be made to estimate the consumption for each type. For example, the consumption for alkali reaction with crude oil can be estimated from Eq. 10.12, assuming all the acidic components are consumed to react with the alkali. The alkali consumption ΔC_o in meq/mL is the same as the soap generated. ΔC_o is generally a small fraction of the total consumption. Because these consumptions involve complex chemical reactions, efforts have been made to collect some published experimental data and were presented earlier. A general rule is 0.05 to 2% alkali concentration and 0.1 to 0.23 PV injection volume. Note that alkali addition in an ASP system can reduce surfactant and polymer adsorption. However, addition of surfactant and/or polymer does not affect alkali consumption (Li, 2007). This is probably because the alkali molecules are smaller than the surfactant or polymer molecules, thus the existence of surfactant and polymer molecules will not affect the adsorption of alkali molecules, nor will their existence affect alkaline reactions.

10.6 RECOVERY MECHANISMS

This section summarizes alkaline flooding mechanisms and discusses some of the applications. The IFT function in alkaline flooding is further discussed.

10.6.1 A Brief Summary of Mechanisms

Johnson (1976) summarized several proposed mechanisms by which caustic waterflooding may improve oil recovery. In alkaline flooding, emulsification is an important mechanism. At least it is related to most of the other mechanisms. Several mechanisms are discussed further in the following subsections.

Emulsification and Entrainment

In emulsification and entrainment, the crude oil is emulsified in situ owing to IFT reduction, and it is entrained by the flowing aqueous alkaline solution (Subkow, 1942). The conditions for this mechanism to occur are high pH, low acid number, low salinity, and O/W emulsion size < pore throat diameter.

Emulsification and Entrapment

In emulsification and entrapment, the sweep efficiency is imposed by the action of emulsified oil droplets blocking the smaller pore throats (Jennings et al., 1974). The conditions for this mechanism to occur are high pH, moderate acid number, low salinity, and O/W emulsion size > pore throat diameter. This mechanism is especially important in waterflooding viscous oils where waterflood sweep efficiency is notoriously poor, but no significant reduction in residual oil is expected with this mechanism. Ehrlich and Wygal (1977) tested 19 crude oils and found only one viscous crude (44.2 cP at 25°C) with a high acid number (1.39 mg KOH per gram of oil) that showed evidence of emulsification as a recovery mechanism. They suggested that the minimum acid numbers ranging from 0.5 to 1.5 mg KOH per gram of oil are needed for the emulsification mechanism to be effective.

Wettability Reversal (Oil-Wet to Water-Wet)

When the wettability is changed from oil-wet to water-wet, oil production increases owing to favorable changes in permeabilities. Because residual oil in a water-wet porous medium is discontinuous and immobile, as compared with the continuous residual oil phase in an oil-wet porous medium, water-wet rocks could not respond to a wettability reversal mechanism (Wagner and Leach, 1959). Therefore, this mechanism is limited to oil-wet reservoirs where wettability could be reversed from oil-wet to water-wet. Mungan (1966a) demonstrated that alkaline floods lowered the water relative permeability, and later he (1966b) used Teflon cores (preferentially oil-wet material) in his experiments to demonstrate that higher oil recoveries could be achieved by the wettability reversal mechanism.

Mungan (1966a) also noted that the wettability reversal process for a given oil was dependent on temperature. Ehrlich and Wygal (1977) observed that regardless of initial wettability, the cores were indicated to be water-wet following NaOH waterflooding to a high water/oil ratio (WOR). Elsewhere several Russian researchers obtained similar results showing that the cores became more water-wet after contact by alkaline solutions. It was reported that oil and water relative permeability curves shifted to the right.

Wettability Reversal (Water-Wet to Oil-Wet)

In the water-wet to oil-wet type of wettability reversal, low residual oil saturation is attained through low IFT and viscous water-in-oil emulsions working together to result in a high capillary number. Obviously, the salinity in alkaline water should be high so that W/O emulsion can be generated with the help of low IFT caused by soap, and the rock surfaces are made to be oil-wet (Cooke et al., 1974).

The wettability reversal from water-wet to oil-wet needs to be discussed further because it is opposite to our perception that rock surfaces should be made more water-wet for improved oil recovery. To the best of our knowledge, only Cooke et al. provided a detailed discussion of this mechanism.

The mechanics of the process involve first the conversion of water-wet rock to oil-wet. Here, a discontinuous, nonwetting residual oil is converted to a continuous wetting phase, providing a flow path for what otherwise would be trapped oil. At the same time, low interfacial tension induces formation of an oil-external emulsion of water droplets in the continuous, wetting oil phase. These emulsion droplets tend to block flow and induce a high-pressure gradient in the region where they form. The high-pressure gradient, in turn, is said to overcome the capillary forces already decreased by low interfacial tension, thus reducing residual oil saturation further. Drainage of oil from the volume between emulsified alkaline water drops leaves behind a high water–content emulsion in which residual oil saturation may be as low as 5% PV. Figure 10.16 illustrates the distribution of oil and water in a pore near the displacement front.

Figure 10.17 illustrates the pressure and saturation changes that occur during an alkaline waterflood. Shown are typical pressure gradients and oil saturations during a flood of a sand-packed column previously waterflooded to residual oil saturation. Oil and water flow simultaneously ahead of the alkaline water front. Note the sharp gradient in oil saturation that occurs at the front (Figure 10.17c), and the rise and fall in pressure gradient behind the alkaline water front (Figure 10.17b). The schematic (Figure 10.17d) illustrates what is observed microscopically.

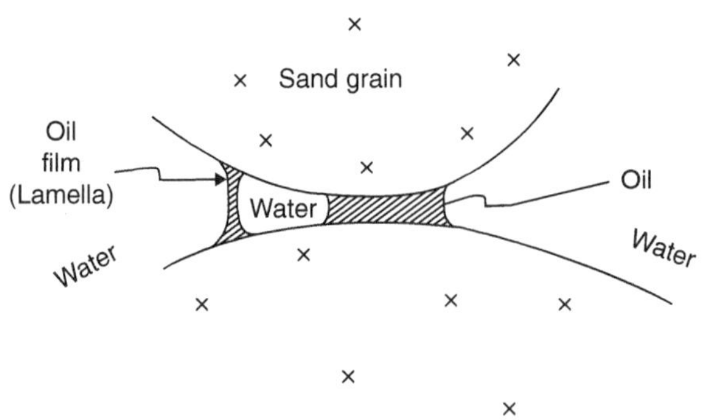

FIGURE 10.16 Distribution of oil in an oil-wet pore. *Source: Cooke et al. (1974).*

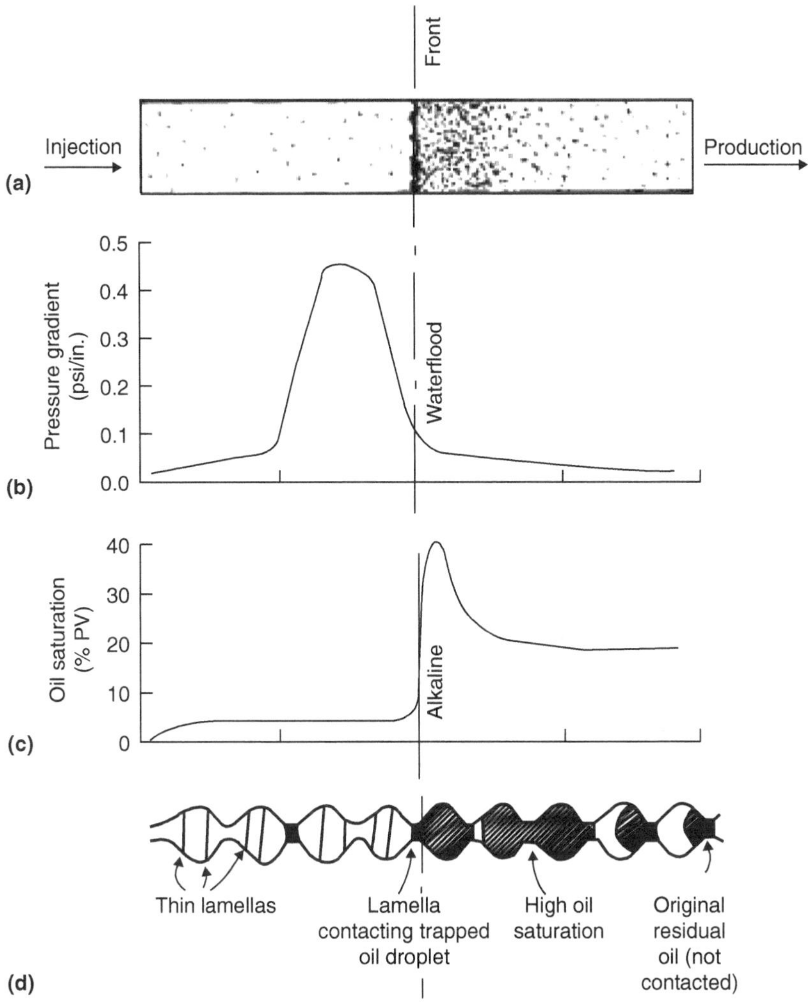

FIGURE 10.17 Pressure and fluid distribution in a sand column during an alkaline waterflood. (a) oil being displaced from a sand-packed column by alkaline water, (b) pressure distribution within the sand-packed column during the alkaline waterflood, (c) distribution of oil within the sand-packed column during the alkaline waterflood, and (d) schematic representation of the disposition of oil and water in the porous medium during the alkaline waterflood. *Source: Cooke et al., (1974).*

Emulsification and Coalescence

Emulsification and coalescence are related to spontaneously formed unstable W/O emulsion (Castor et al., 1981b) or mixed emulsion. Isolated oil droplets are emulsified after contacting with alkaline solution. The emulsified droplets coalesce with each other to become larger droplets while they move in the pores; this occurs because the films of W/O emulsion are not rigid and can be easily ruptured and coalesce to become larger. Some of the emulsified droplets are stopped at pore throats. Therefore, the mechanisms of oil recovery are to increase sweep efficiency and increase coalescence of oil drops into a continuous oil bank. The dynamic displacement experiments by Castor et al. show that alkaline flooding of acidic oils with hydroxides of certain divalent cations increased the production and recovery efficiencies above that obtained by alkaline floods with hydroxides of univalent ions with or without high electrolyte concentration because divalents promote W/O emulsions.

Other Alkaline Flooding Applications

Other mechanisms, which are not discussed here, are more or less related to emulsification and reduced IFT due to in situ generation of soap. One application based on these mechanisms is to inject alkaline solution and gas, simultaneously or alternately, to improve sweep efficiency. As we know, there is a viscous fingering problem for gas injection only. Injection of an alkaline solution in a reservoir with active crude oil will generate O/W and W/O emulsions. The high viscous emulsions and foam formed through gas injection will reduce the viscous fingering problem. In this case, CO_2 cannot be injected because it will neutralize the alkaline solution.

As mentioned earlier, a silicate reacts with calcium chloride to generate precipitates. The precipitates reduce permeability; thus, sweep efficiency is improved, as suggested by Sarem (1974). This process is known as mobility controlled caustic flood (MCCF). If a silicate and a divalent such as calcium ion are mixed near the injection well, precipitates will be formed and a severe plugging will occur near the injection well. This result is not desired, however. We need partial plugging, and plugging should start some distance away from the injection well. In other words, silicate and calcium are not fully mixed near the injection well. To achieve that, silicate slug and calcium chloride are alternately injected, and these slugs are separated by a fresh water buffer slug so that silicate and calcium chloride are gradually mixed. It is understandable that as the alternate slug sizes and fresh water buffer size become smaller, mixing between the slugs becomes easier and quicker; then precipitates will be generated faster. We need to optimize the slug sizes according to the required permeability reduction, distribution of the precipitation, and so on. Similarly, sodium hydroxide and iron chloride can do the same.

Heavy oils are generally produced through thermal recovery methods, such as steam flooding and hot water flooding. One of the problems with these thermal recovery methods is viscous fingering. This problem can be mitigated by the alkaline emulsification of active crude oils. Heavy oils commonly have high content of acidic components. During steam flooding, some oil is left at the bottom of the formation because of the gravity override. Injected alkaline solution will prefer to flow through the less-swept bottom formation and reduce the remaining oil saturation there. Tiab et al. (1982) found that caustic hot water and caustic steam flooding recovered 14.5% more original oil in place than conventional hot water and steam flooding recovery under similar reservoir conditions.

10.6.2 IFT Function in Alkaline Flooding

Alkalis react with naphthenic acid in crude oil to generate soap. The soap, an in situ generated surfactant, reduces the interfacial tension between the alkaline solution and oil. It is intuitive to infer that the main mechanism in alkaline flooding is low IFT.

Castor et al. (1981b) observed that the IFT in the alkaline flooding was on the order of 0.1 mN/m. Their capillary numbers of alkaline floods are presented in Figure 10.18. The capillary number of alkaline floods was about 100 times higher than the capillary numbers in waterfloods. The alkaline flooding results from Castor et al. show that the recovery efficiencies could be better correlated with the stability of emulsions and wettability alteration than with IFT of the systems.

Figure 10.19 shows the reduction in residual oil saturation by alkaline flood versus different acid numbers. These data are calculated from those presented by Ehrlich and Wygal (1977), so are the data in Figures 10.20 through 10.22. The alkali used was 0.1% NaOH. Figure 10.19 shows that those two variables were not correlated.

The reduction in residual oil saturation versus equilibrium IFT is plotted in Figure 10.20, which shows that the S_{or} reduction was not correlated with the equilibrium IFT. Figure 10.21 shows equilibrium IFT versus acid number. Apparently, there was weak correlation between IFT and acid number when acid numbers were low.

Figure 10.22 shows both equilibrium and nonequilibrium caustic/oil IFT for some oils with different acid numbers. For any pair of IFTs, the equilibrium

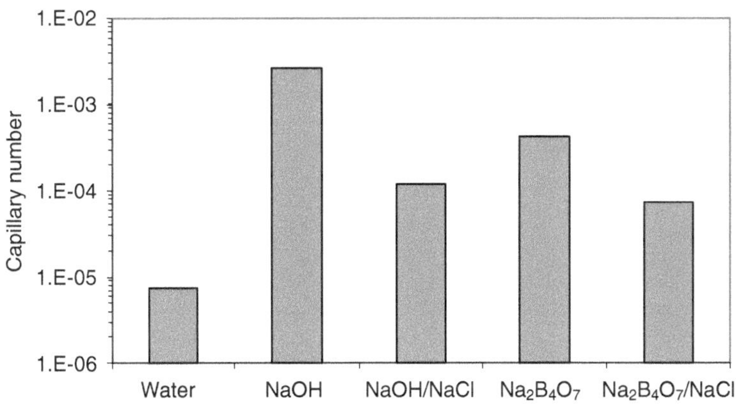

FIGURE 10.18 Capillary numbers in alkaline floods. *Source: Data from Castor et al. (1981b).*

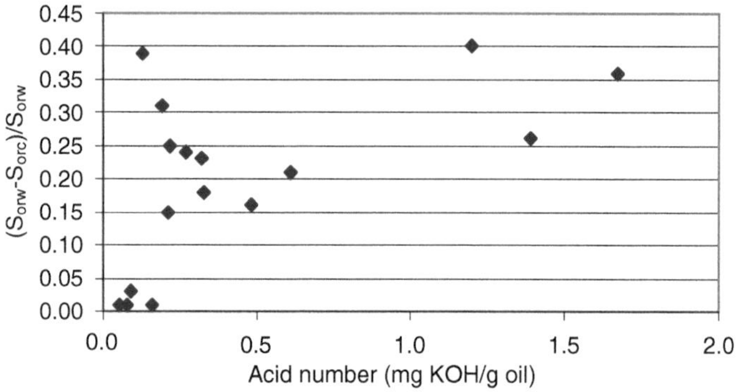

FIGURE 10.19 Reduction in residual oil saturation versus acid number.

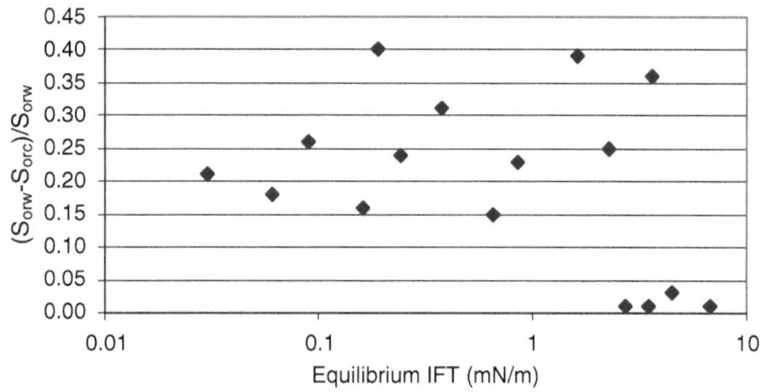

FIGURE 10.20 Reduction in residual oil saturation versus IFT.

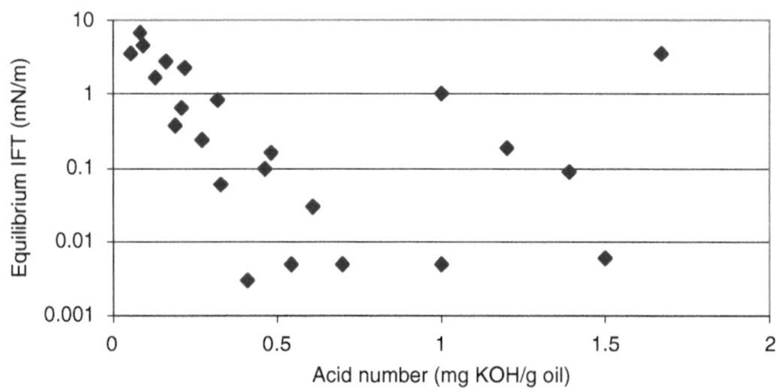

FIGURE 10.21 Equilibrium IFT versus acid number.

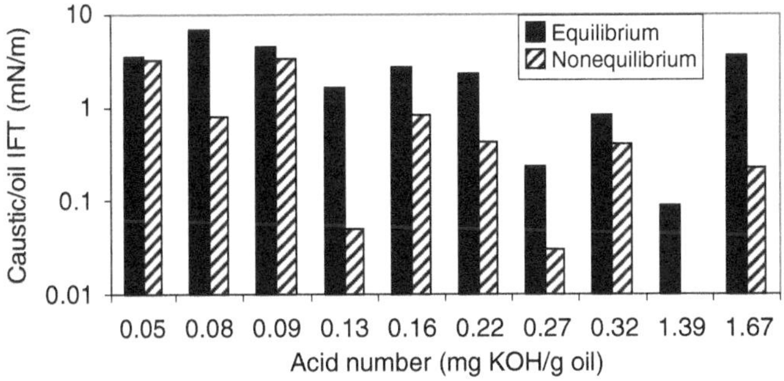

FIGURE 10.22 Equilibrium and nonequilibrium caustic/crude oil IFTs.

IFT was higher than the nonequilibrium one. In other words, IFT increased with time. This phenomenon was noted by McCaffery (1976). The soap is generated initially at the water/oil interface. Then it dissipates into oil and water phases. The dissipation determines the equilibrium time required to reach a stable interfacial tension. The surfactant migration explains that the IFT time-dependent behavior: initially, the tension decreases very fast to a minimum and then increases slowly and stabilizes at some value. It also explains the temperature effect: when the temperature is increased, the IFT increases because of the increased dissipation. The dissipation of soap is speeded up as the concentration of surfactant in the brine is lowered. This will take place if the oil–water bank does not move through the reservoir in a piston-like manner so that the surfactant solution is diluted by existing water in the formation. In other words, the surfactant loss at the interface will be more rapid if fresh brine replaces surfactant solution. Because of this migration of surfactant between oil and water, IFT reduction under reservoir conditions is difficult to predict (Gogarty, 1983a). Some mathematical models have been developed to predict IFT changes with time (England and Berg, 1971; Radke and Somerton, 1978).

From the previous discussions, we can see that ultralow IFT cannot be reached in alkaline flooding. The incremental oil recovery is not correlated with the IFT or crude acid number. The low IFT mechanism may not be the dominant mechanism. However, a reasonably low IFT is required for emulsification to occur, which is another proposed mechanism and summarized in the previous section.

10.7 SIMULATION OF ALKALINE FLOODING

Although alkaline flooding only is not conducted as often as polymer flooding or surfactant flooding, alkaline injection is conducted together with surfactant and polymer injection. Simulation of alkaline flooding is very difficult because of complex chemical reactions. These complex reactions include at least the following:

- Reaction between alkali and acidic components of crude oil to generate soap in situ
- Injected alkali reaction with formation brine (precipitation) and minerals (dissolution and ion exchange)
- Effects of reaction products on other transport phenomena
- Interactions with surfactant and polymer if they are injected

These reactions are discussed in previous sections of this chapter and subsequent chapters. This section first describes a geochemistry program called EQBATCH, which performs batch reaction equilibrium calculations. After introducing EQBATCH, this section briefly presents an alkaline flooding model. The objective of this section is to provide the logic behind the simulation model to help the reader use the model. Finally, this section presents an example to simulate alkaline flooding.

EQBATCH is based on the framework established by Bhuyan (1989), which has been presented elsewhere (e.g., Bhuyan et al., 1990). In EQBATCH, local thermodynamic equilibrium is assumed. It is also assumed that precipitation/dissolution, and cation exchange have a negligible effect on porosity and permeability. Ideal solutions are assumed so that the activity coefficients of the species are equal to unity. As a result, it is possible for activities to be replaced by their respective molar concentrations. For pure solids, activities are considered equal to unity. There are many species and reactions in alkaline flooding.

We use this program to estimate the initial equilibrium state of the reservoir. EQBATCH estimates the initial equilibrium based on the formation and water composition, the acid number of crude oil, and water and oil saturations. The initial equilibrium data from EQBATCH batch calculation are input into a UTCHEM alkaline model. The UTCHEM model then continues simulation of the oil recovery process.

Other uses of EQBATCH include the determination of compatibility between injection water and resident water, equilibrium composition and compatibility of mixing injection water from different sources, and equilibrium composition and the resulting pH of the injection water after the addition of various electrolytes.

10.7.1 Mathematical Formulation of Reactions and Equilibria

Here, we identify N elements from formation and water composition and then define J fluid species, K solid species, I matrix-adsorbed cations, and M micelle-associated cations all made up of N elements. There are then (J + K + I + M) unknown equilibrium concentrations. To determine the equilibrium state of the system, we need (J + K + I + M) number of independent equations. These equations follow (Bhuyan, 1989).

Elemental mass balances provide N equations of the form

$$C_n^t = \sum_{j=1}^{J} h_{nj} C_j + \sum_{k=1}^{K} g_{nk} C_{k(s)} + \sum_{i=1}^{I} f_{ni} \bar{C}_i + \sum_{m=1}^{m} e_{nm} \bar{\bar{C}}_m \quad \text{for } n = 1, \ldots, N. \tag{10.44}$$

where C_n^t is the total concentration of component n, C_j is the concentration of fluid species j, $C_{k(s)}$ is the concentration of solid species k, $\bar{C}_i$ is the concentration of matrix-adsorbed cation i, and $\bar{\bar{C}}_m$ is the concentration of micelle-associated cation m, all in mole/L water. The coefficient before each concentration is the stoichiometric coefficient.

From the J fluid chemical species, we can arbitrarily select N independent species such that the concentrations of the remaining (J – N) fluid species can be expressed in terms of the concentrations of these N independent species through equilibrium relationships of the following form:

$$C_r = K_r^{eq} \prod_{j=1}^{N} C_j^{w_{rj}} \quad \text{for } r = (N+1), \ldots, J. \tag{10.45}$$

For example, $C_{H_2CO_3} = K_{H_2CO_3}^{eq} C_H^2 C_{CO_3}$. In this section, in Eq. 10.45 and subsequent equations, the superscript w denotes the exponent of concentration.

For each solid, there is a solubility product constraint:

$$K_k^{sp} \geq \prod_{j=1}^{N} C_j^{w_{kj}} \quad \text{for } k = 1, \ldots, K. \tag{10.46}$$

The solubility product constants K_k^{sp} are defined in terms of the concentrations C_j of the independent chemical species only. If a solid is not present, the corresponding solubility product constraint is the inequality; if the solid is present, the constraint is the equality. For example, $K_{CaCO_3}^{sp} \geq [Ca^{2+}][CO_3^{2-}]$ for the presence of solid $CaCO_3$.

For I adsorbed cations on the matrix, there are (I – 1) independent exchange equilibrium relationships of the form

$$K_p^{ex} = \prod_{j=1}^{N} C_j^{w_{pj}} \prod_{i=1}^{I} \bar{C}_i^{w_{pi}} \quad \text{for } p = 1, \ldots, (I-1). \tag{10.47}$$

where C_j is the concentration in fluid, and $\bar{C}_i$ is the concentration on matrix in mole/L. w_{pj} and w_{pi} are the exponents for the concentrations C_j and $\bar{C}_i$, respectively. For each of these exponents, it is negative if the species is on the left side of the equilibrium equation; it is positive if the species is on the right side. For example, for

$$Ca^{2+} + 2\overline{Na^+} \xleftrightarrow{K_{Ca-Na}^{ex}} \overline{Ca^{2+}} + 2Na^+,$$

we have

$$K^{ex}_{Ca-Na} = \frac{\left[\overline{Ca^{2+}}\right][Na^+]^2}{\left[\overline{Na^+}\right]^2[Ca^{2+}]} = [Ca^{2+}]^{-1}[Na^+]^2\left[\left[\overline{Na^+}\right]^{-2}\left[\overline{Ca^{2+}}\right]\right].$$

There is one electroneutrality condition given by

$$Q_v = \sum_{i=1}^{I} \overline{z}_i \overline{C}_i, \tag{10.48}$$

where Q_v is the cation exchange capacity (CEC) in mole/L pore volume (PV); $\overline{C}_i$ is the concentration of cation i adsorbed on the matrix in mole/L PV, and $\overline{z}_i$ is the charge of the adsorbed cation i.

Similarly, for M cations associated with surfactant micelles there are (M − 1) cation exchange (on micelle) equilibrium relations of the following form:

$$K^{exm}_q = \prod_{j=1}^{N} C_j^{w_{qj}} \prod_{m=1}^{M} \overline{C}_m^{w_{qm}} \quad \text{for } q = 1, \ldots, (M-1). \tag{10.49}$$

Additionally, the electroneutrality conditions for the micelles as a whole provide one more equation:

$$C_{A^-} + C_{S^-} = \sum_{m=1}^{M} \overline{\overline{z}}_m \overline{\overline{C}}_m. \tag{10.50}$$

In Eq. 10.50, C_{A^-} and C_{S^-} are the in situ generated surfactant and injected surfactant concentrations, respectively, $\overline{\overline{C}}_m$ is the concentration of cation m adsorbed on micelles in mole/L, and $\overline{\overline{z}}_m$ is the charge of the adsorbed cation m.

Now we have N mass balances, (J – N) aqueous reaction equilibrium relations, K solubility–product–constant equations, (I − 1) cation-exchange on matrix equilibrium relations and one electroneutralinity condition, (M − 1) cation-exchange on micelle equilibrium relations and one electroneutralinity condition for the micelles, giving a total of (J + K + I + M) independent equations to solve the same number of concentration unknowns. For the detailed calculation algorithm, see Bhuyan (1989).

10.7.2 Mathematical Formulation of Alkaline Flooding

To formulate an isothermal flow problem, we start with mass balance equations. The following are the material balance equations of N components (Bhuyan et al., 1990):

$$\frac{\partial(\phi C_n^t)}{\partial t}+\frac{\partial}{\partial x}\sum_{j=1}^{n_p}\left[C_{nj}u_{xj}-\phi S_j\left(D_{xxnj}\frac{\partial C_{nj}}{\partial x}+D_{xynj}\frac{\partial C_{nj}}{\partial y}+D_{xznj}\frac{\partial C_{nj}}{\partial z}\right)\right]+$$
$$\frac{\partial}{\partial y}\sum_{j=1}^{n_p}\left[C_{nj}u_{yj}-\phi S_j\left(D_{yynj}\frac{\partial C_{nj}}{\partial y}+D_{yznj}\frac{\partial C_{nj}}{\partial z}+D_{yxnj}\frac{\partial C_{nj}}{\partial x}\right)\right]+$$
$$\frac{\partial}{\partial z}\sum_{j=1}^{n_p}\left[C_{nj}u_{zj}-\phi S_j\left(D_{zznj}\frac{\partial C_{nj}}{\partial z}+D_{zynj}\frac{\partial C_{nj}}{\partial y}+D_{zxnj}\frac{\partial C_{nj}}{\partial x}\right)\right]=r_n,$$
$$n=1,\ldots,N, \tag{10.51}$$

where D_{xx}, D_{yy}, D_{zz}, and so on, are the elements of dispersion tensor; n_p is the number of phases; S_j is the saturation of phase j; u is the Darcy velocity; ϕ is the porosity; and r_n is the source (+) or sink (−) term.

The overall mass-continuity equation is obtained by summarizing the conservation equations over all components:

$$\phi c_t\frac{\partial p}{\partial t}+\vec{\nabla}\cdot\sum_{j=1}^{n_p}\vec{u}_j\sum_{n=1}^{N_v}(1+c_n\Delta p)C_{nj}=q, \tag{10.52}$$

Here, $q=\sum_{n=1}^{N_v}q_n$, N_v is the total number of volume-occupying components, q_n is the specific volumetric rate of injection (+) or production (−) of component n, c_n is the compressibility of volume-occupying component n, c_t is the total compressibility, and p is the pressure.

The solution scheme is as follows. First, we solve Eq. 10.52 for pressure. With pressure determined, we can calculate fluxes. Then we solve Eq. 10.51 for the total concentrations of N components. With this information known, we can use the reaction equilibrium equations as described in the previous section to calculate all the concentrations. When we know the concentrations, we can determine the phase concentrations, phase saturations, and other physical and transport properties required to solve for pressure for the next time level.

Example 10.3 Convert an Acid Number into mmol/mL Water

Because the calculations in EQBATCH are performed on the basis of unit water volume, we need to convert the acid number of 0.81 mg KOH/g oil into mmol/mL water. Other available data are as follows: oil density ρ_o = 0.82 g/mL and oil saturation S_w = 0.383.

Solution

First, convert the acid number into mmol/mL oil:

$$[HA_o]\left[\frac{\text{mmol}}{\text{mL oil}}\right]=[HA_o]\left[\frac{\text{mg KOH}}{\text{g oil}}\right]\left[\frac{\text{mmol}}{(MW)_{KOH}\,\text{mg KOH}}\right]\left[\frac{\rho_o\text{ g oil}}{\text{mL oil}}\right]$$
$$=[HA_o]\left[\frac{\text{mg KOH}}{\text{g oil}}\right]\times\frac{\rho_o}{(MW)_{KOH}}.$$

Continued

Example 10.3 Convert an Acid Number into mmol/mL Water—*Continued*

Then convert $[HA_o]$ in mmol/mL oil into $[HA_o]$ in mmol/mL water. According to the material balance,

$$[HA_o]\left[\frac{\text{mmol}}{\text{mL oil}}\right] \times S_o = [HA_o]\left[\frac{\text{mmol}}{\text{mL water}}\right] \times S_w.$$

we have

$$[HA_o]\left[\frac{\text{mmol}}{\text{mL water}}\right] = [HA_o]\left[\frac{\text{mmol}}{\text{mL oil}}\right] \times \frac{S_o}{S_w}.$$

Then

$$[HA_o]\left[\frac{\text{mmol}}{\text{mL water}}\right] = [HA_o]\left[\frac{\text{mg KOH}}{\text{g oil}}\right] \times \frac{S_o}{S_w} \times \frac{\rho_o}{(MW)_{KOH}}.$$

For this problem, we have

$$[HA_o]\left[\frac{\text{mmol}}{\text{mL water}}\right] = 2\left[\frac{\text{mg KOH}}{\text{g oil}}\right] \times \frac{0.617}{0.383} \times \frac{0.82}{56} = 0.019\left[\frac{\text{mmol}}{\text{mL water}}\right].$$

Similar conversions are required for concentrations of solids, adsorbed ions, and cation exchange capacities of the rock from per-unit-PV to per-unit-water basis.

10.7.3 EQBATCH and UTCHEM

In this section we provide an example to illustrate how to use EQBATCH and UTCHEM to simulate alkaline flooding.

Example 10.4 Use EQBATCH and UTCHEM to Simulate Alkaline Flooding

In this hypothetical case, the objective is to provide detailed procedures to simulate an alkaline flood. The concentrations of formation water (initial water) and injected water, with some calculated concentrations, are shown in Table 10.8. The initial formation water pH is 8.1; the acid number is 0.81 g KOH/g oil; and the initial water and oil saturations are 0.383 and 0.617, respectively. The task is to set up a UTCHEM alkaline simulation model based on these data.

Solution

To set up a UTCHEM alkaline simulation model, we have to use EQBATCH to get initial equilibrium concentrations for the model. In other words, we need to set up the EQBATCH model first. The name of the EQBATCH input file must be EQIN, and the output file is EQOUT. The general procedures follow.

Step 1: Define Elements and Species

The reaction chemistry depends on the chemical composition of the formation and injected chemicals. EQBATCH and UTCHEM set a framework that allows us to specify a suitable chemical description for a given application.

Example 10.4 ***Continued***

TABLE 10.8 Formation and Injection Water Analysis

	Formation Water		Injection Water			Formation Water	Injection Water
Ion	*mg/L*	*meq/mL*	*mg/L*	*meq/mL*	Ion	*mmol/mL*	*meq/mL*
Na^+	2272.6	0.098809	47.0	0.002043			
K^+	214.0	0.005487	11.0	0.0002821	Na^++K^+	0.104296	0.002326
Mg^{2+}	34.5	0.002840	11.5	0.0009498			
Ca^{2+}	57.2	0.002860	67.1	0.0033565	Ca^{2+}+ Mg^{2+}	0.002850	0.004306
Cl^-	2091.0	0.058901	138.5	0.003901	Cl^-	0.058901	0.003901
HCO_3^-	2623.4	0.043006	150.5	0.0024672			
CO_3^{2-}	240.0	0.008000	7.0	0.0002333	CO_3^{2-}+ HCO_3^-	0.047006	0.005168

First, define the elements, independent species, dependent species, solid species, adsorbed cations on matrix, and surfactant-associated cations, based on the compositions shown in Table 10.8. These defined elements and species are listed in Table 10.9. This is a critical step in building an alkaline model. For this case, 6 elements (N = 6), 6 independent species and 8 dependent species with a total of 14 fluid species (J = 14), 2 solid species (K = 2), 3 adsorbed cations on matrix (I = 3), and 2 surfactant-associated cations (M = 2) are defined. Note that the subscripts a and s for $Ca(OH)_2$ and $CaCO_3$ mean in aqueous and solid states, respectively. A^-, HA_o, and HA_w represent petroleum acid anion, petroleum acid in oleic phase, and petroleum acid in aqueous phase, respectively. The last fluid species must be HA_w. In principle, we can arbitrarily select N independent species. Practically, we select the species that are similar to the elements, and they are simple species so that other dependent species can be defined from them with equilibrium constants. Chlorine is a nonreactive species; therefore, it is not selected as an independent species. Of course, it will not appear in any reaction equation.

Step 2: Define Reaction and Equilibrium Equations

After defining the species, write down the relevant reaction and equilibrium equations that follow. The cation exchanges with $\overline{Na^+}$ on matrix and $\overline{\overline{Na^+}}$ in micelle are defined. The following constants and coefficients have been cross-checked from different sources. These data should be typical for most applications. These data are for 25°C, but data at any other reservoir temperature can be estimated or obtained from available databases or software such as Geochemist's Workbench or PHREEQC. For information on the softwares, check the websites *http://wwwbrr.cr.usgs.gov/projects/GWC_coupled/phreeqc/* and *http://www.geology.uiuc.edu/Hydrogeology/hydro_gwb.htm*.

TABLE 10.9 Elements and Reactions Species

Order No.	Elements or Pseudoelements	Independent Fluid Species	Dependent Fluid Species	Solid Species	Adsorbed Cation on Matrix	Surfactant-Associated Cations
1	Calcium	H^+		$CaCO_{3(s)}$	$\overline{H^+}$	$\overline{\overline{Na^+}}$
2	Carbonates	Na^+		$Ca(OH)_{2(s)}$	$\overline{Na^+}$	$\overline{\overline{Ca^{2+}}}$
3	Sodium	Ca^{2+}			$\overline{Ca^{2+}}$	
4	Hydrogen	CO_3^{2-}				
5	Petroleum acid	HA_o				
6	Chlorine	H_2O				
7			$Ca(OH)^+$			
8			$Ca(HCO_3)^+$			
9			A^-			
10			$(OH)^-$			
11			HCO_3^-			
12			H_2CO_3			
13			$CaCO_{3(a)}$			
14			HA_w			

Example 10.4 *Continued*

Aqueous Reaction	Definition	Species	Equilibrium Constant
	K_1^{eq}	H^+	1
	K_2^{eq}	Na^+	1
	K_3^{eq}	Ca^{2+}	1
	K_4^{eq}	CO_3^{2-}	1
	K_5^{eq}	HA_o	1
	K_6^{eq}	H_2O	1
$Ca^{2+} + H_2O \overset{K_7^{eq}}{\leftrightarrow} Ca(OH)^+ + H^+$	$K_7^{eq} = \frac{[Ca(OH)^+][H^+]}{[Ca^{2+}]}$	$Ca(OH)^+$	1.2050E-13
$Ca^{2+} + H^+ + CO_3^{2-} \overset{K_8^{eq}}{\leftrightarrow} Ca(HCO_3)^+$	$K_8^{eq} = \frac{[Ca(HCO_3)^+]}{[Ca^{2+}][CO_3^{2-}][H^+]}$	$Ca(HCO_3)^+$	1.4142E+11
$HA_w + OH^- \overset{K_9^{eq}}{\leftrightarrow} A^- + H_2O$	$K_9^{eq} \equiv K_A = \frac{[A^-][H^+]}{[HA_w]}$	A^-	1.0000E-12
$H_2O \overset{K_{10}^{eq}}{\leftrightarrow} H^+ + OH^-$	$K_{10}^{eq} \equiv K_w = [H^+][OH^-]$	$(OH)^-$	1.0093E-14
$H^+ + CO_3^{2-} \overset{K_{11}^{eq}}{\leftrightarrow} HCO_3^-$	$K_{11}^{eq} = \frac{[HCO_3^-]}{[H^+][CO_3^{2-}]}$	HCO_3^-	2.1380E+10
$2H^+ + CO_3^{2-} \overset{K_{12}^{eq}}{\leftrightarrow} H_2CO_3$	$K_{12}^{eq} = \frac{[H_2CO_3]}{[H^+]^2[CO_3^{2-}]}$	H_2CO_3	3.9811E+16
$Ca^{2+} + CO_3^{2-} \overset{K_{13}^{eq}}{\leftrightarrow} CaCO_{3(a)}$	$K_{13}^{eq} = \frac{[CaCO_{3(a)}]}{[Ca^{2+}][CO_3^{2-}]}$	$CaCO_{3(a)}$	1.5849E+03
$HA_o \overset{K_D}{\leftrightarrow} HA_w$	$K_D = \frac{[HA_w]_{water}}{[HA_o]_{oil}}$	HA_w	1.0000E-04

Dissolution/Precipitation Reaction	Definition	Solubility Product Constant
$CaCO_{3(s)} \overset{K_1^{sp}}{\leftrightarrow} Ca^{2+} + CO_3^{2-}$	$K_1^{sp} = [Ca^{2+}][CO_3^{2-}]$	8.7E-09
$Ca(OH)_2 \overset{K_2^{sp}}{\leftrightarrow} + Ca^{2+} + 2OH^-$	$K_2^{sp} = [Ca^{2+}][H^+]^{-2}$	4.7315E+22

Exchange Reaction on Matrix	Definition	Exchange Constant
$2\overline{Na^+} + Ca^{2+} \overset{K_1^{ex}}{\leftrightarrow} 2Na^+ + \overline{Ca^{2+}}$	$K_1^{ex} = \frac{\left[\overline{Ca^{2+}}\right][Na^+]^2}{\left[\overline{Na^+}\right]^2[Ca^{2+}]}$	2.623E+02
$\overline{H^+} + Na^+ + OH^- \overset{K_2^{ex}}{\leftrightarrow} \overline{Na^+} + H_2O$	$K_2^{ex} = \frac{\left[\overline{H^+}\right][Na^+]}{\left[\overline{Na^+}\right][H^+]}$	1.460E+07

Exchange Reaction on Micelle	Definition	Exchange Constant
$2\overline{\overline{Na^+}} + Ca^{2+} \overset{K_1^{exm}}{\leftrightarrow} 2Na^+ + \overline{\overline{Ca^{2+}}}$	$K_1^{exm} = \frac{\left[\overline{\overline{Ca^{2+}}}\right][Na^+]^2}{\left[\overline{\overline{Na^+}}\right]^2[Ca^{2+}]}$	$K_1^{exm} = \beta_1^{exm}([A^-]+[S^-])$ based on the Hirasaki (1982b) model. $\beta_1^{exm} = 2.5$

Example 10.4 Use EQBATCH and UTCHEM to Simulate Alkaline Flooding—*Continued*

The UTCHEM calculations are performed on the basis of unit water volume. Thus, the unit of the acid composition should be converted. If we define $[HA_o]_w$ as the moles of HA_o associated with oil per liter of water, and $[HA_o]_o$ as the moles of HA_o per liter of oil volume, then

$$[HA_o]_w = \frac{\text{moles of } HA_o}{\text{water volume } (V_w)} = \frac{[HA_o]_o \text{ oil volume } (V_o)}{\text{water volume } (V_w)} = \frac{S_o}{S_w}[HA_o]_o.$$

In the previous table, K_D is defined as

$$K_D \equiv \frac{[HA_w]_{\text{water volume}}}{[HA_o]_{\text{oil volume}}} \equiv \frac{[HA_w]_w}{[HA_o]_o} = \frac{[HA_w]_w}{[HA_o]_w}\frac{S_o}{S_w}.$$

Note that when the value of one of the previous equilibrium constants, exchange constants, or solubility products is taken from other sources, attention should be paid to the definition. For example, the solubility product of $Ca(OH)_2$ is commonly defined as

$$K^{sp}_{Ca(OH)_2} = [Ca^{2+}][OH^-]^2.$$

However, we have defined the solubility product in the preceding as

$$K^{sp}_2 = [Ca^{2+}][H^+]^{-2}.$$

Because OH^- is not defined as one independent species, we have to use the independent species H^+ to define the solubility product.

Because $K_w = [H^+][OH^-]$, we have the relation between the two solubility product constants:

$$K^{sp}_2 = [Ca^{2+}][OH^-]^2(K_w)^{-2} = K^{sp}_{Ca(OH)_2}(K_w)^{-2}.$$

Step 3: List Stoichiometric Coefficients, Exponents, and Charges

List stoichiometric coefficients, exponents, and charges in equations based on the previously identified reaction equations and equilibria. Most of them are listed in an array form. For any two-dimensional (not one-dimensional) array, for example, AR(I,J), the I index is in rows, and the J index is in columns. The orders of elements and fluid species must be the same as those in which their names are listed in the EQBATCH input file, EQIN. However, there are several exceptions in the example presented in Appendix C of the UTCHEM Technical Manual. For the exchange equilibrium constants (Table C.13), exponents (Table C.14), and valence differences on the matrix (Table C.15), the order is Ca-Na, Mg-Na, and H-Na.

Table 10.10a lists the stoichiometric coefficient of the Ith element in the Jth fluid species [AR(I,J) Array]. The elements must be arranged in rows, and the fluid species are arranged in columns. Table 10.10b shows the stoichiometric coefficients of the Ith element in the Jth solid species for the BR array, adsorbed solid cation for the DR array, and surfactant-associated cation for the ER array.

TABLE 10.10a Stoichiometric Coefficient of Ith Element in Jth Fluid Species [AR(I,J) Array]

	H^+	Na^+	Ca^{2+}	CO_3^{2-}	HA_o	H_2O	$Ca(OH)^+$	$Ca(HCO_3)^+$	A^-	OH^-	HCO_3^-	H_2CO_3	$CaCO_{3(a)}$	HA_w
Ca	0	0	1	0	0	0	1	1	0	0	0	0	1	0
CO_3	0	0	0	1	0	0	0	1	0	0	1	1	1	0
Na	0	1	0	0	0	0	0	0	0	0	0	0	0	0
H	1	0	0	0	1	2	1	1	0	1	1	2	0	1
A	0	0	0	0	1	0	0	0	1	0	0	0	0	1

Example 10.4 Use EQBATCH and UTCHEM to Simulate Alkaline Flooding—*Continued*

TABLE 10.10b Stoichiometric Coefficient of Ith Element in Jth Species

	BR Array		DR Array			ER Array	
	$CaCO_{3(s)}$	$Ca(OH)_{2(s)}$	$\overline{H^+}$	$\overline{Na^+}$	$\overline{Ca^{2+}}$	$\overline{\overline{Na^+}}$	$\overline{\overline{Ca^{2+}}}$
Ca	1	1	0	0	1	0	1
CO_3	1	0	0	0	0	0	0
Na	0	0	0	1	0	1	0
H	0	2	1	0	0	0	0
A	0	0	0	0	0	0	0

Table 10.10c shows the exponents of the Jth independent fluid species, adsorbed species, and surfactant-associated species for the BB(I,J) array. The rows (I index) list the independent and dependent fluid species, adsorbed species on matrix, and surfactant-associated species. In the columns (J index), the species listed are basically the same as those in the rows (I index) with the exception that the dependent species are not there. The order in which these species are placed in the table is the same as that in the input file, EQIN. Note that the exponents of H^+ for $Ca(OH)^+$ and OH^- are −1 because we have to substitute $(H^+)^{-1}$ for OH^- to define these two dependent species.

Table 10.10d lists the exponents of the Jth independent species in the Ith solid for EXSLD(I,J). Table 10.10e lists the charge of the fluid species for the CHARGE(I,J) array. Table 10.10f lists the charges of adsorbed species for the SCHARG(I,J) array, and Table 10.10g lists the valence differences between cations involved in exchange for the REDU array. Note that this table shows the adsorbed cations in the reverse order they appear in EQIN. Table 10.10h lists the charges of surfactant-associated species for the CHACAT(I,J) array.

In the EXEX(I,J) array shown in Table 10.10i, the cation exchanges are arranged in the rows (I index), and the species are arranged in the columns (J index). The order of the cation exchanges follows that of the elements listed in the input file, EQIN. The orders of independent species, adsorbed species on matrix, and surfactant-associated species follow those of their respective species listed in the input file, EQIN. Note that the cation exchange is not related to surfactant-associated species, but in Table 10.10i, the species are listed redundantly. Similarly, the three adsorbed species on matrix are not relevant to the surfactant-associated equilibrium Ca-Na in the EXACAT(I,J) array in Table 10.10j, but they also are listed in the table redundantly.

The equilibrium constants for fluid species, the exchange equilibrium constants for adsorbed cations, the solubility product constants for solids, and the

TABLE 10.10c Exponent of Jth Independent Fluid, Adsorbed, and Surfactant-Associated Species [BB(I,J) Array]

	H^+	Na^+	Ca^{2+}	CO_3^{2-}	HA_o	H_2O	$\overline{H^+}$	$\overline{Na^+}$	$\overline{Ca^{2+}}$	$\overline{\overline{Na^{2+}}}$	$\overline{\overline{Ca^{2+}}}$
H^+	1	0	0	0	0	0	0	0	0	0	0
Na^+	0	1	0	0	0	0	0	0	0	0	0
Ca^{2+}	0	0	1	0	0	0	0	0	0	0	0
CO_3^{2-}	0	0	0	1	0	0	0	0	0	0	0
HA_o	0	0	0	0	1	0	0	0	0	0	0
H_2O	0	0	0	0	0	1	0	0	0	0	0
$Ca(OH)^+$	–1	0	1	0	0	0	0	0	0	0	0
$Ca(HCO_3)^+$	1	0	1	1	0	0	0	0	0	0	0
A^-	–1	0	0	0	1	0	0	0	0	0	0
OH^-	–1	0	0	0	0	0	0	0	0	0	0
HCO_3^-	1	0	0	1	0	0	0	0	0	0	0
H_2CO_3	2	0	0	1	0	0	0	0	0	0	0
$CaCO_3$	0	0	1	1	0	0	0	0	0	0	0
HA_w	0	0	0	0	1	0	0	0	0	0	0
$\overline{H^+}$	0	0	0	0	0	0	1	0	0	0	0
$\overline{Na^+}$	0	0	0	0	0	0	0	1	0	0	0
$\overline{Ca^{2+}}$	0	0	0	0	0	0	0	0	1	0	0
$\overline{\overline{Na^+}}$	0	0	0	0	0	0	0	0	0	1	0
$\overline{\overline{Ca^{2+}}}$	0	0	0	0	0	0	0	0	0	0	1

TABLE 10.10d Exponent of Jth Independent Species in Ith Solid [EXSLD(I,J)]

	H^+	Na^+	Ca^{2+}	CO_3^{2-}	HA_o	H_2O
$CaCO_{3(s)}$	0	0	1	1	0	0
$Ca(OH)_{2(s)}$	–2	0	1	0	0	0

TABLE 10.10e Charge of Jth Fluid Species [CHARGE(J) Array]

	H^+	Na^+	Ca^{2+}	CO_3^{2-}	HA_o	H_2O	$Ca(OH)^+$	$Ca(HCO_3)^+$	A^-	OH^-	HCO_3^-	H_2CO_3	$CaCO_{3(a)}$	HA_w
Charge	1	1	2	–2	0	0	1	1	–1	–1	–1	0	0	0

Example 10.4 *Continued*

TABLE 10.10f Charge of Jth Adsorbed Species [SCHARG(J) Array]

	$\overline{H^+}$	$\overline{Na^+}$	$\overline{Ca^{2+}}$
Charge	1	1	2

TABLE 10.10g Valence Differences between Cations Involved in Exchange [REDU(I,J) Array]

	$\overline{Ca^{2+}}$	$\overline{H^+}$
$\overline{Na^+}$	−1	0

TABLE 10.10h Charge of Jth Surfactant-Associated Species [CHACAT(J) Array]

	$\overline{Na^+}$	$\overline{Ca^{2+}}$
Charge	1	2

equilibrium constants for surfactant-associated cations listed beside the reaction and equilibrium equations can be input in EQIN in the same order as they are presented in the proceeding tables. So far we have defined all the constants, charges, and exponents. Next, we need to input initial concentrations.

Step 4: Input Initial Concentrations in EQIN

Some of the initial concentrations required by the EQBATCH input file EQIN are listed in Table 10.11. EQBATCH uses these initial input values and the other input constants discussed in Steps 2 and 3 to regulate their final output values. Some of the EQOUT output will be copied and pasted into the UTCHEM model as input. Because most of the initial concentrations are guessed values, we have to discuss the effect of initial concentrations. Here, we compare the EQIN input and EQOUT output from the two sets of initial concentrations. The objectives of the comparison are (1) to see the difference between the input initial concentrations and the output initial concentrations (ideally, the input and output initial concentration should be very close); (2) to see how significantly the input initial concentrations change the output initial concentrations.

TABLE 10.10i Exponent of Jth Independent Species in Ith Cation Exchanges on Rock [EXEX(I,J) Array]

	6 Independent Fluid Species (N = 6)						3 Adsorbed Species on Matrix (I = 3)			2 Surfactant-Associated Species (M = 2)	
	H^+	Na^+	Ca^{2+}	CO_3^{2-}	HA_o	H_2O	$\overline{H^+}$	$\overline{Na^+}$	$\overline{Ca^{2+}}$	$\overline{\overline{Na^+}}$	$\overline{\overline{Ca^{2+}}}$
Ca-Na (K_1^{ex})	0	2	–1	0	0	0	0	–2	1	0	0
H-Na (K_2^{ex})	–1	1	0	0	0	0	1	–1	0	0	0

TABLE 10.10j Exponent of Jth Independent Species in Ith Cation Exchange on Surfactant Micelles [EXACAT(I,J) Array]

	6 Independent Fluid Species (N = 6)						3 Adsorbed Species on Matrix (I = 3)			2 Surfactant-Associated Species (M = 2)	
	H^+	Na^+	Ca^{2+}	CO_3^{2-}	HA_o	H_2O	$\overline{H^+}$	$\overline{Na^+}$	$\overline{Ca^{2+}}$	$\overline{\overline{Na^+}}$	$\overline{\overline{Ca^{2+}}}$
Ca-Na (K_1^{exm})	0	2	–1	0	0	0	0	0	0	–2	1

Example 10.4 ***Continued***

TABLE 10.11 Input versus Output of Initial Concentrations

C5I, C61, meq/mL W	Cl^-	Ca			
EQIN input 1	0.059				
EQOUT output 1	0.059	2.60E-05			
EQIN input 2	0.059				
EQOUT output 2	0.059	3.04E-03			
CELAQI	**Ca**	**CO_3**	**Na**	**H**	**A**
EQIN input 1 (moles/L W)	2.85E-03	0.047	0.1043	111.154	0.019
EQOUT output 1 (eq/L W)		0.08832	0.1003	111.160	0.019
EQIN input 2 (moles/L W)	2.85E-03	0.047	0.1043	111.154	0.019
EQOUT output 2 (eq/L W)		0.0914	0.1017	111.1567	0.019
CSLDI	**$CaCO_3$**	**$Ca(OH)_2$**			
EQIN input 1 (moles/L W)	1.003	0			
EQIN output 1 (moles/L W)	1.003	0			
EQOUT output 1 (moles/L PV)	0.384	0			
EQIN input 2 (moles/L W)	0.38	0			
EQOUT output 2 (moles/L W)	0.38	0			
EQOUT output 2 (moles/L PV)	0.146	0			
CSORBI	**$\overline{H^+}$**	**$\overline{Na^+}$**	**$\overline{Ca^{2+}}$**		
EQIN input 1 (moles/L W)	0.01	0	0		
EQOUT output 1 (moles/L W)	4.01E-03	3.99E-03	1.20E-07		
EQOUT output 1 (moles/L PV)	1.54E-03	1.53E-03	4.60E-08		
EQIN input 2 (moles/L W)	4.01E-03	3.99E-03	1.20E-07		
EQOUT output 2 (moles/L W)	1.33E-03	6.61E-03	2.96E-05		
EQOUT output 2 (moles/L PV)	5.11E-04	2.53E-03	1.13E-05		

Continued

Example 10.4 Use EQBATCH and UTCHEM to Simulate Alkaline Flooding—*Continued*

TABLE 10.11 Input versus Output of Initital Concentrations—*Continued*

CAQI, moles/L W	H^+	Na^+	Ca^{2+}	CO_3^{2-}	HA_O	H_2O
EQIN input 1	1.20E-06	0.01	1.00E-05	3.09E-09	5.40E-04	55.4999
EQOUT output 1	3.74E-08	0.10	9.60E-06	5.15E-05	1.90E-02	55.5470
EQIN input 2	1.00E-07	0.10	2.85E-03	4.70E-02	1.90E-02	55.5470
EQOUT output 2	7.61E-09	0.10	8.84E-04	2.71E-04	1.90E-02	55.5460

CAQI, moles/L W	$\overline{H^+}$	$\overline{Na^+}$	$\overline{Ca^{2+}}$	$\overline{\overline{Na^+}}$	$\overline{\overline{Ca^{2+}}}$
EQIN input 1	1.00E-06	0.01	1.00E-03	1.00E-06	1.00E-08
EQOUT output 1	4.01E-03	3.99E-03	1.20E-07	5.08E-09	3.12E-28
EQIN input 2	4.01E-03	3.99E-03	1.20E-07	5.08E-09	3.12E-28
EQOUT output 2	5.11E-04	2.53E-03	1.13E-05	2.50E-06	3.33E-18

Note that the correct units of solids, adsorbed cations on the matrix, cation exchange capacity, and surfactant-associated cations in EQBATCH *input* are moles/L water, although their units in the program description of EQBATCH in the technical manual are moles/L PV. This is a typographical error in the manual. Their correct units in the UTCHEM *input* are moles/L PV. In EQBATCH output, the values in both mole/L PV and mole/L PV water are reported. Now we discuss each initial concentration.

- **C5I, initial concentration of chloride ion in equivalents/liter water (eq/L water).** Because chloride is a nonreaction species, the concentration in the formation water (0.059 eq/L water) is the correct initial concentration. Therefore, the EQOUT output is exactly the same as the EQIN input.
- **CELAQI(J), initial concentrations of elemental fluid species.** The concentration of calcium is printed in the same line as chloride in EQOUT, although it is the input in the CELAQI(J) line in EQIN. The unit of CELAQI(J) concentrations in EQIN is mole/L, whereas it is eq/L in EQOUT.

The concentrations from the formation water analysis are good initial concentrations for these elemental fluid species. Thus, the two sets of initial input data are the same. The total amount of the element A (petroleum acid) is the same as that in oil. We showed in Example 10.3 that the acid number of 0.81 g KOH/g rock is converted to 0.019 eq/L water. The input value is equal to the output value for each set of initial concentrations. Most of the hydrogen is in water, which is about 1000 g/L/(18 g/mole)*2 = 111.11 moles/L. Thus, the input and output values should be very close to this value. For a similar reason, the input and output values of sodium are close to each other. The concentrations of

Example 10.4 ***Continued***

carbonate from the two sets are almost the same (0.08832 eq/L or 0.04416 moles/L), which are close to the input value of 0.047 moles/L in formation water. The output calcium concentrations for the first set of data are much lower than the initial input value (2.85E-3). In the second set, we increased the solubility product for $CaCO_3$ up to 2.4E-07, which is much higher than the published value (8.7E-09). By using such a high solubility product, we made the initial calcium concentration close to that in the formation water.

- **CSLDI(J), for J = 1, NSLD, initial solid concentrations.** The number of solids, NSLD, is 2. These two values are generally unknown. The initial concentrations for calcium hydroxide are zero for both the input and output in the two sets of data. For calcium carbonate, the input 1 is 1.003 mole/L water, which is equivalent to 0.383 moles/L pore volume because the initial water saturation in this case is 0.383. Therefore, output 1 is almost the same as input 1. Also, output 2 is the same as input 2.
- **CSORBI(J), for J = 1, NSORB, initial concentrations of adsorbed cations**. The number of adsorbed cations (NSORB) is 3. Table 10.11 shows that the EQOUT output values are quite different from the initial guessed values for the first set of data. For the second set of data, the EQOUT output 1 values are used as EQIN input 2 values. The values in the EQOUT output 2 are several times different from their respective input values, especially for Ca^{2+}. Additional information (e.g., adsorption fraction of each cation) is needed to fine-tune the initial values.
- **CAQI(J), for J = 1, NIND + NSORB + NACAT, initial concentrations for the independent species concentrations, the adsorbed species, and the surfactant-associated species.** The numbers of independent fluid species (NIND), adsorbed species (NSORB), and surfactant-associated species (NACAT) are 6, 3, and 2, respectively. Table 10.11 shows that if the initial guessed values are close to their real values such as H_2O in input 1 and Na^+ and HA_o in input 2, the EQOUT output values are close to the initial guessed values. Otherwise, the output values sometimes could be orders of magnitude different from the input values. In input 2, the initial values are several times and two orders of magnitude higher than their respective input values of Ca^{2+} and CO_3^{2-}. The important parameter in alkaline flood, pH, from output 2 is 7.61E-09, which is 8.11—very close to the initial formation water pH, 8.1.

Note that we are required to input the initial concentrations of adsorbed cations on the matrix twice: once in CSORBI and the other time in CAQI. This is probably an error in the input design of EQBATCH or in the manual. The output values of surfactant-associated cations are quite different from their respective input values. We therefore need extra information to adjust the initial values or their exchange constants to close their gaps.

Because some uncertainties exist in the initial concentrations, we need to look at the effect of initial concentrations on oil recovery. Figure 10.23 compares oil recovery factors from the two alkaline flooding models using the two sets of initial concentrations listed in Table 10.11. This figure shows that the recovery factor

Example 10.4 Use EQBATCH and UTCHEM to Simulate Alkaline Flooding—*Continued*

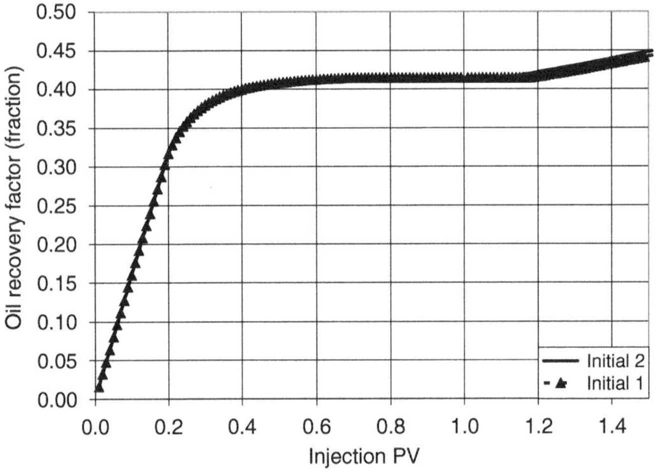

FIGURE 10.23 Effect of initial concentrations on oil recovery factor.

curves from the two models almost overlap each other, showing that the initial guessed concentrations do not affect the results significantly in this case. In general, depending on the problem to solve, we may further adjust the initial concentrations or other parameters such as reaction and equilibrium constants to get more concentrations matched with or closer to their known values. One good parameter to match, as we did in this case, is the water pH value because alkaline flooding is a pH-sensitive process. The other important parameters to adjust are those related to soap generation. Based on our sensitivity tests, those parameters are very sensitive to oil recovery.

Figure 10.24 shows that the oil bank breaks through earlier in the model with the Initial 2 data. The initial formation water pH in Initial 2 is about 8.1 compared to 7.4 in Initial 1. Soap is probably generated earlier in Initial 2 than in Initial 1. Thus, the oil bank is formed earlier and breaks through earlier in Initial 2.

In modeling alkaline floods, we need to input many species concentrations. To make clear their relations, Table 10.12 lists the EQOUT concentrations for the preceding initial output 1 in Table 10.11 and explains their relations.

Step 5: Set Up the UTCHEM Model

After setting the initial concentrations in the input file EQIN, we run EQBATCH. The output file is EQOUT. For the UTCHEM input, copy the lines starting at the line shown here to the end of EQBATCH output file, EQOUT, into the UTCHEM model:

FOLLOWING LINES OF DATA FORMATTED FOR UTCHEM

We do not discuss the details of building a UTCHEM model in this book, but we show how to prepare the concentration data of injection water.

Example 10.4 *Continued*

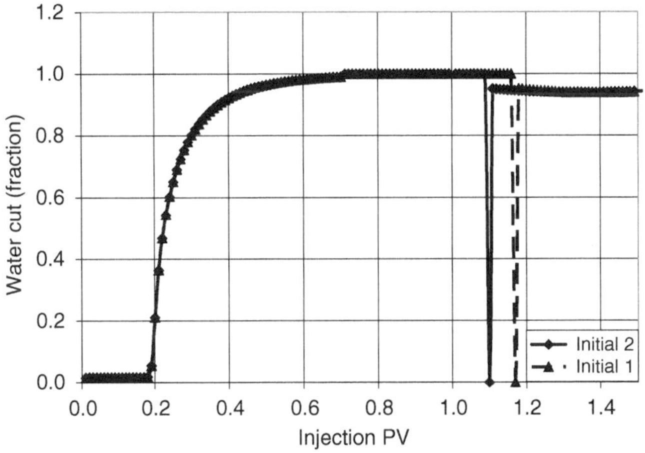

FIGURE 10.24 Effect of initial concentrations on oil recovery factor.

TABLE 10.12 Relations among Species Concentrations in EQOUT

Fluid Species	moles/L W	Solid Species	moles/L W
H^+	3.7415E-08	$CaCO_{3(s)}$	1.0028E+00
Na^+	1.0031E-01	$Ca(OH)_{2(s)}$	0.0000E+00
Ca^{2+}	9.6032E-06		
CO_3^{2-}	5.1546E-05	**Adsorbed Species**	**moles/LW**
HA_o	1.8998E-02	$\overline{H^+}$	4.01E-03
H_2O	5.5547E+01	$\overline{Na^+}$	3.99E-03
$Ca(OH)^+$	3.0929E-11	$\overline{Ca^{2+}}$	1.20E-07
$Ca(HCO_3)^+$	2.6160E-06		
A^-	5.0777E-09	**Surfactant-Associated Species**	**moles/LW**
$(OH)^-$	2.6976E-07	$\overline{\overline{Na^+}}$	5.08E-09
HCO_3^-	4.1233E-02	$\overline{\overline{Ca^{2+}}}$	3.12E-28
H_2CO_3	2.8725E-03		
$CaCO_{3(a)}$	7.8453E-07		
HA_w	1.8998E-06		

Continued

Example 10.4 Use EQBATCH and UTCHEM to Simulate Alkaline Flooding—*Continued*

TABLE 10.12 Relations among Species Concentrations in EQOUT—*Continued*

Fluid Elemental Concentration		
CELAQI	*eq/L water*	
Ca	2.6007E-05	$=([Ca^{2+}]+[Ca(OH)^+]+[Ca(HCO_3)^+]+[CaCO_{3(a)}])\times 2$
CO_3	0.088320248	$=([CO_3^{2-}]+[Ca(HCO_3)^+]+[HCO_3^-]+[H_2CO_3]+[CaCO_{3(a)}])\times 2$
Na	1.0031E-01	$=[Na^+]$
H	1.1116E+02	$=[H^+]+[HA_o]+[H_2O]\times 2+[Ca(OH)^+]+[Ca(HCO_3)^+]+[(OH)^-]+[HCO_3^-]+[H_2CO_3]\times 2+[HA_w]$
A	1.9000E-02	$=[HA_o]+[A^-]+[HA_w]$
Total Elemental Concentration		
	moles/L water	
Calcium	1.00285	$=[Ca^{2+}]/2+[CaCO_{3(s)}]+[Ca(OH)_{2(s)}]+[\overline{Ca^{2+}}]+[\overline{\overline{Ca^{2+}}}]$
Carbonate	1.04700	$=[CO_3]/2+[CaCO_{3(s)}]$
Sodium	0.10430	$=[Na]+[\overline{Na^+}]+[\overline{\overline{Na^+}}]$
Hydrogen	111.16400	$=[H]+[Ca(OH)_{2(s)}]+[\overline{H^+}]$
Acid	0.01900	$=[A]$

In this example, the injection water compositions were shown in Table 10.8. To simplify UTCHEM simulation, we combine some ions into "pseudo-ions": Na^++K^+ to Na^+, Ca^{2+}+Mg^{2+} to Ca^{2+}, and CO_3^{2-}+HCO_3^- to CO_3^{2-}. Table 10.13 shows the injection scheme and detailed components of injection water. Slug 1 is 0.7 PV water injection with 1.5% NaCl added in the injection water (IW), followed by Slug 2 of 0.3 PV alkali injection, and Slug 3 of 0.5 PV water injection with 0.5% NaCl added. In addition to the BATCH input parameters listed previously, the key parameters used in the UTCHEM model are listed in Table 10.14.

Step 6: Performance Analysis

The following paragraphs give the results of the one-dimensional alkaline core flood simulation run using the second initial concentrations (output 2). The alkaline injection is from 0.7 PV to 1.0 PV (Slug 2 in Table 10.13). The profiles of concentrations at 0.9 PV during alkaline injection are presented. Figure 10.25 shows the pH and generated surfactant (soap) concentration profiles.

Example 10.4 ***Continued***

TABLE 10.13 Injection Scheme

	(IW)	Slug 1 (0.7 PV) IW + 1.5% NaCl		Slug 2 (0.3 PV) IW + 1.6% Na_2CO_3		Slug 3 (0.5 PV) IW + 0.5% NaCl	
	meq/ mLW	*1.5% NaCl*	*Total*	*1.6% Na_2CO_3*	*Total*	*0.5% NaCl*	*Total*
Cl^-	0.003901	0.2564	0.260301		0.003901	0.085467	0.089368
Ca^{2+}	0.004306		0.004306		0.004306		0.004306
CO_3^{2-}	0.005168		0.005168	0.3019	0.307068		0.005168
Na^+	0.002326	0.2564	0.258726	0.3019	0.304226	0.085467	0.087793

TABLE 10.14 Key Parameters in the Alkaline Flood Simulation Model

Parameters	UTCHEM Parameter	Parameter Value
Unit system	IUNIT	0
Grid blocks	NX, NY, NZ	80, 1, 1
Grid block size	DX1, DY1, DZ1	0.011, 0.11, 0.11
Components (min.)	W, O, P, S, Cl, Ca, alcohol 1, alcohol 2, CO_3, Na, H^+, Acid	
Time control in PV	ICUMTM, ISTOP	1,1
Flag to output KC profile	IPRFLG(KC)	1 for all
Flag to output concentrations	ICKL	1
Flag to output Pc, S, IFT	ICNM	0
Flag to output effective salinity	ICSE	1
Porosity	PORC1	0.1988
Permeability	PERMXC, PERMYC, PERMZC	236, 236, 118
Initial water saturation	SWI	0.3829
Initial salinities and harness	C50, C60	0.059, 0.0057
Oil concentration at left plait point	C2PLC	0

Continued

Example 10.4 Use EQBATCH and UTCHEM to Simulate Alkaline Flooding—*Continued*

TABLE 10.14 Key Parameters in the Alkaline Flood Simulation Model—*Continued*

Parameters	UTCHEM Parameter	Parameter Value
Oil concentration at right plait point	C2PRC	1
Critical micelle concentration (CMC)	EPSME	0.0001
Flag to input binodal curve	IFGHBN	0
Slope of C_{33maxi} versus alcohol 1	HBNS70, HNBS71, HBNS72	0
C_{33max0}, C_{33max1}, C_{33max2} for alcohol 1	HBNC70, HNBC71, HBNC72	0.04, 0.025, 0.12
Slope of C_{33maxi} versus alcohol 2	HBNS80, HNBS81, HBNS82	0, 0, 0
C_{33max0}, C_{33max1}, C_{33max2} for alcohol 2	HBNC80, HNBC81, HBNC82	0, 0, 0
C_{sel}, C_{seu}	CSEL7, CSEU7	0.2, 0.35
Huh IFT equation	CHUH, AHUH	0.35, 10
Flag for N_c dependency	ITRAP	1
CDC curves	T11, T22, T33	1965, 8000, 364.2
Effective salinity slope for Ca^{2+}	BETA6	0.8
Effective salinity slope for alcohol 1	BETA7	−2
Effective salinity slope for alcohol 2	BETA8	0
Residual saturations at low N_c, $(N_C)_c$	S1RWC, S2RWC, S3RWC	0.382, 0.3803, 0.382
Endpoint relative permeabilities at $(N_C)_c$	P1RWC, P2RWC, P3RWC	0.03, 1.0, 0.03
Relative permeability exponents at $(N_C)_c$	E1WC, E2WC, E3WC	1.12, 1.3, 1.12

Example 10.4 ***Continued***

TABLE 10.14 *Continued*

Parameters	UTCHEM Parameter	Parameter Value
Residual saturations at high N_c, $(N_C)_{max}$	S1RC, S2RC, S3RC	0.0, 0.0, 0.0
Endpoint relative permeabilities at $(N_C)_{max}$	P1RC, P2RC, P3RC	0.5, 1.0, 1.0
Relative permeability exponents, $(N_C)_{max}$	E13C, E23C, E33C	1.1, 1.1, 1.1
Viscosities	VIS1, VIS2	0.995, 24.3
Compositional phase viscosity	ALPHAV(I)	0, 0, 0, 1, 1.7
Densities	DEN1, DEN2, DEN23, DEN3	0.433, 0.364, 0.364, 0.4247
Surfactant adsorption	AD31, AD32, B3D	14, 26, 1000
Cation exchange capacity	QV, XKC, XKS, EQW	0.00306, 0.25, 0.2, 450
Print all species option	IRSPS	2
Surfactant adsorption pH-dependent	IPHAD	1
pH-dependent parameters	PHC, PHT, PHT1, HPHAD	7, 13, 13, 0.25
Production pressure	PWF(M)	14.5
Injection rate	QI(M,L)	0.05

The soap concentration in vol.% per unit water volume basis in this section is converted from moles/L water. The soap concentration profile is parallel to that of the hydrogen ion (pH) except near the injection end of the core. This figure also shows that pH higher than 9.5 is required to generate soap. This high-pH front is at the fractional distance of 0.6.

Figure 10.26 shows the profiles of petroleum acids in water and oil phases, HA_w and HA_o, at 0.9 PV injection. Both of the concentrations are converted to the volume fractions in water phase volume. These two profiles parallel each other. HA_w is almost four orders of magnitude lower than HA_o. Near the injection end, these two concentrations are lower because some acid components are dissociated as soap; thus, the acid components is depleted.

Example 10.4 Use EQBATCH and UTCHEM to Simulate Alkaline Flooding—*Continued*

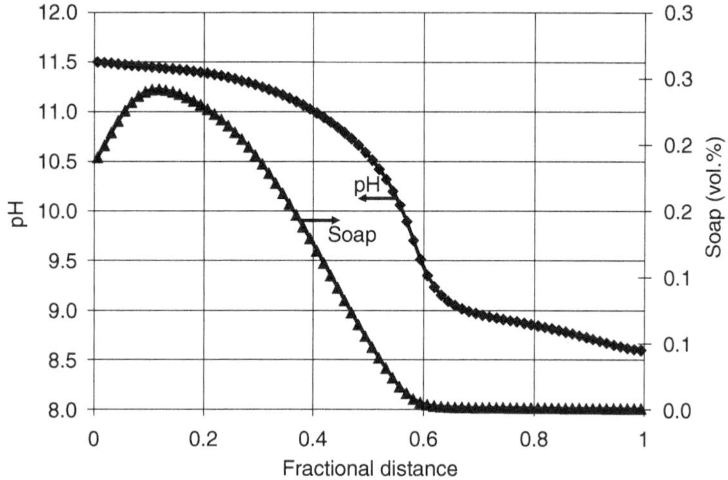

FIGURE 10.25 Graphic representation of pH and soap concentration profiles along fractional distance at 0.9 PV injection.

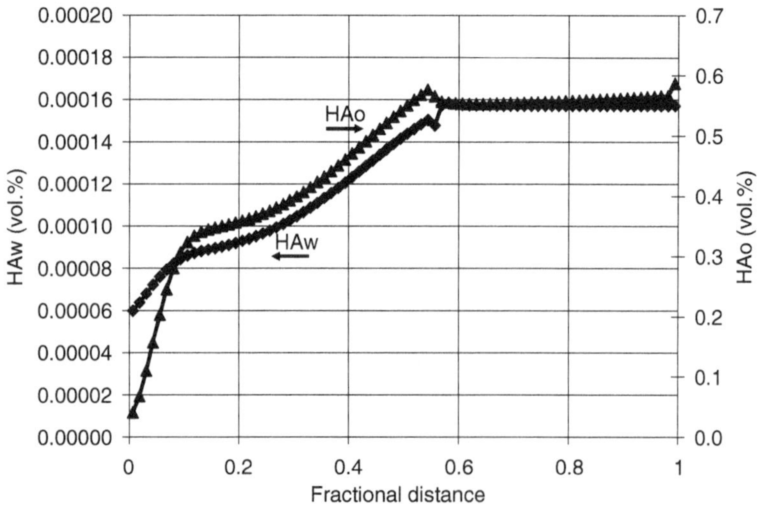

FIGURE 10.26 Profiles of petroleum acids in the water and the oil phases, HA_w and HA_o, at 0.9 PV.

Example 10.4 ***Continued***

Figure 10.27 shows water, oil, and microemulsion phase saturation profiles at 0.9 PV injection. The oil bank ahead of high-pH front is almost invisible (0.02 saturation jump) at the fractional distance of 0.6. The significantly reduced oil saturation region is only near the injection end, although effective salinity is in the type III region along the whole core at this time (see Figure 10.28). Figure

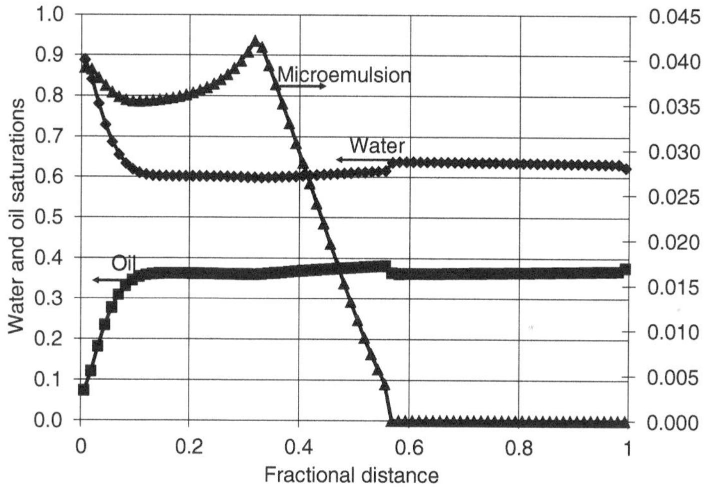

FIGURE 10.27 Saturation profiles (water, oil, and microemulsion) at 0.9 PV injection.

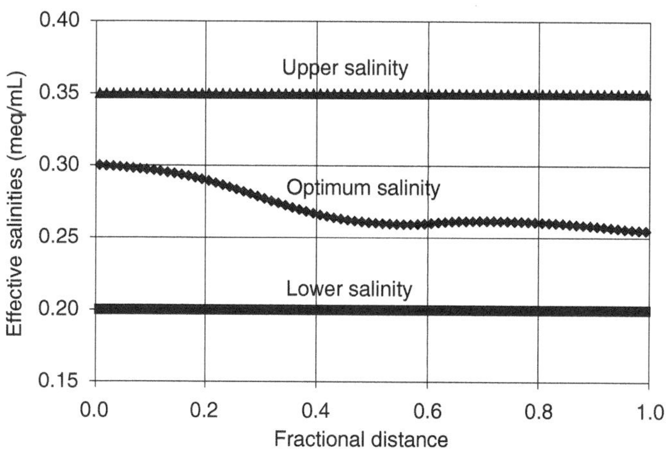

FIGURE 10.28 Effective salinity and effective salinity limits for Type III at 0.9 PV injection.

Example 10.4 Use EQBATCH and UTCHEM to Simulate Alkaline Flooding—*Continued*

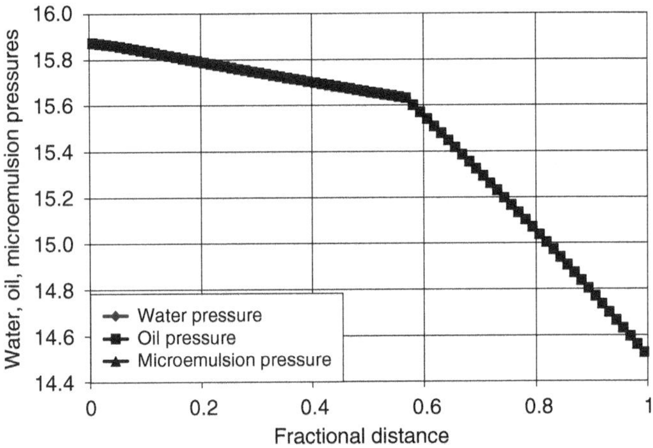

FIGURE 10.29 Phase pressures (water, oil, and microemulsion) at 0.9 PV injection.

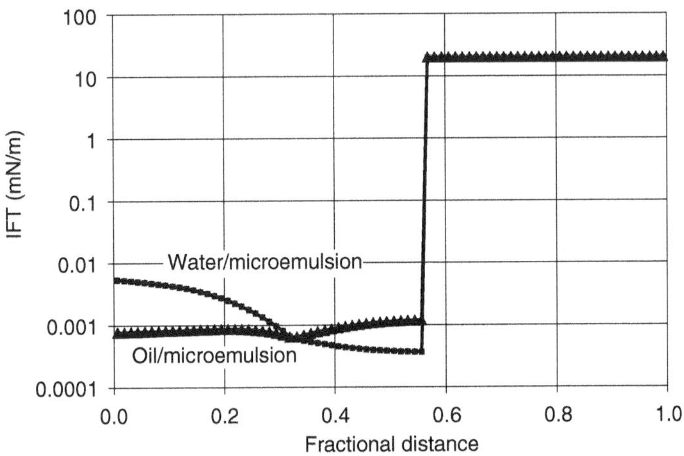

FIGURE 10.30 Profiles of interfacial tensions (water/microemulsion and oil/microemulsion) at 0.9 PV injection.

10.29 shows the pressures profiles at the same time. The pressure gradient behind the high pH is lower than that before. The flow mobility behind the front is higher than that before the front, which moves at an adverse mobility ratio. In this case, a mobility control agent such as polymer is needed.

Figure 10.30 shows the IFT profiles of water/microemulsion and oil/microemulsion. Behind the high-pH front, the oil/microemulsion is in the range of 0.0001 to 0.01 mN/m. The IFT before the front is 20 mN/m.

Example 10.4 ***Continued***

The oil recovery factor is shown in Figure 10.23. From this figure, we can see that the incremental oil recovery factor of alkaline flooding over waterflooding is about 4%. Table 10.13 also serves as an example explaining how to input salinity data into a performance prediction model.

Thus, we have completed all the steps to set up a UTCHEM model. This example is a typical alkaline flood case in terms of compositions in the system. If magnesium is included, we simply add all of the lines related to calcium and modify those lines for magnesium. A case with clay and silica dissolution/precipitation is briefly discussed in the next section.

10.7.4 A Case with Clay and Silica Dissolution/Precipitation Included

The basic framework and procedures to simulate alkaline-related processes are presented in Example 10.4. When a particular case is specified, the details of the reaction chemistry and input data set required must conform to that particular case. This section briefly provides one more case that includes clay and silica dissolution/precipitation based on the description by Bhuyan (1989). This section presents only the elements, species, reactions, and equilibria that are not listed or different from those in Example 10.4. We add this case because clay and silica dissolution or precipitation is a common problem. For more cases, see Mohammadi (2008).

This case is used when clay and silica dissolution/precipitation is important and different sodium silicates are injected. Compared with Example 10.4, additional elements or pseudoelements are aluminum and silicon. Additional independent aqueous species include Al^{3+} and H_4SiO_4. Additional dependent aqueous species are $Al(OH)^{2+}$, $Al(OH)_2^+$, $Al(OH)_4^-$, $H_3SiO_4^-$, $H_2SiO_4^{2-}$, and $HSi_2O_6^{3-}$. Additional solid species are SiO_2 (silica), $Al_2Si_2O_5(OH)_4$ (kaolinite), and $NaAlSiO_2O_6 \cdot H_2O$ (analcite).

Additional Aqueous Reactions	Definition	Species	Equilibrium Constant
$Al^{3+} + H_2O \leftrightarrow Al(OH)^{2+} + H^+$	$K_i^{eq} = \frac{[Al(OH)^{2+}][H^+]}{[Al^{3+}]}$	$Al(OH)^{2+}$	5.9252E-05
$Al^{3+} + 2H_2O \leftrightarrow Al(OH)_2^+ + 2H^+$	$K_i^{eq} = \frac{[Al(OH)_2^+][H^+]^2}{[Al^{3+}]}$	$Al(OH)_2^+$	3.5108E-09
$Al^{3+} + 4H_2O \leftrightarrow Al(OH)_4^- + 4H^+$	$K_i^{eq} = \frac{[Al(OH)_4^-][H^+]^4}{[Al^{3+}]}$	$Al(OH)_4^-$	9.7364E-20
$H_4SiO_4 \leftrightarrow H^+ + H_3SiO_4^-$ (quartz dissolution)	$K_i^{eq} = \frac{[H^+][H_3SiO_4^-]}{[H_4SiO_4]}$	$H_3SiO_4^-$	1.6939E-10

Additional Aqueous Reactions	Definition	Species	Equilibrium Constant
$H_4SiO_4 \leftrightarrow 2H^+ + H_2SiO_4^{2-}$	$K_i^{eq} = \frac{[H^+]^2[H_2SiO_4^{2-}]}{[H_4SiO_4]}$	$H_2SiO_4^{2-}$	4.2658E-22
$2H_4SiO_4 \leftrightarrow 2H_2O + 3H^+ + HSi_2O_6^{3-}$	$K_i^{eq} = \frac{[H^+]^3[HSi_2O_6^{3-}]}{[H_4SiO_4]^2}$	$HSi_2O_6^{3-}$	1.1482E-32
$2H_4SiO_4 \leftrightarrow 3H_2O + 2H^+ + Si_2O_5^{2-}$	$K_i^{eq} = \frac{[H^+]^2[Si_2O_5^{2-}]}{[H_4SiO_4]^2}$	$Si_2O_5^{2-}$	7.2444E-20

As mentioned earlier, several forms of soluble silica exist in the aqueous phase. The preceding reactions are based on Eqs. 10.20 through 10.24. Mohammadi (2008) considered only $Al(OH)_4^-$ and $H_3SiO_4^-$ (Eq. 10.20). Additional dissolution/precipitation reactions include the following:

Formula	Definition	Solubility Product
SiO_2	$K_i^{sp} = [H_4SiO_4]$	1.0000E-04
$Al_2Si_2O_5(OH)_4$	$K_i^{sp} = [H^+]^{-6}[Al^{3+}]^2[H_4SiO_4]^2$	5.6234E-05
$NaAlSiO_2O_6{\cdot}H_2O$	$K_i^{sp} = [H^+]^{-4}[Na][Al^{3+}][H_4SiO_4]^2$	1.9498E-08

For the preceding kaolinite dissolution/precipitation reactions, the congruent dissolution is

$$\begin{aligned} &Al_2Si_2O_5(OH)_4(\text{kaolinite}) + 4OH^- + 3H_2O \\ &\qquad \leftrightarrow 2Al(OH)_4^- + 2H_3SiO_4^-. \end{aligned} \tag{10.53}$$

The incongruent dissolution is

$$\begin{aligned} &Al_2Si_2O_5(OH)_4 + 2Na^+ + 2OH^- + 2H_4SiO_4 \\ &\leftrightarrow 2NaAlSi_2O_6 \cdot H_2O(\text{analcite}) + 5H_2O. \end{aligned} \tag{10.54}$$

The cation exchanges on matrix and associated with micelle are the same as in Example 10.4.

10.8 ALKALINE CONCENTRATION AND SLUG SIZE IN FIELD PROJECTS

Injected alkaline concentration and volume appear to vary depending on the recovery mechanism involved. Concentrations are generally lowest for emulsification mechanisms, from about 0.001 to 0.500 wt.%. Higher concentrations ranging from about 0.5 to 3.0 wt.% or even as high as 15.0 wt.% usually have been required for wettability reversal. Generally, a slug of alkaline solution can be nearly as effective as continuous injection, although the mechanism of emulsification and entrainment will require a sufficient volume to ensure production of the alkaline emulsion. If the volume is small and the alkaline is consumed by rock reaction, the emulsion may become trapped again before it reaches the producing wells. Other mechanisms in which mobility-ratio improvement plays an important role appear to require a slug size no more than about 10 to 30% PV to be effective (Johnson, 1976).

Mayer et al. (1983) summarized alkaline flooding field projects up to 1983. We then analyzed the data of injection concentrations and slug sizes from these projects using the statistical method. Figure 10.31 shows the cumulative percentage versus alkali concentration. This figure shows that the average alkali concentration (at 50% cumulative) is about 0.5 wt.%. For most of the field projects, alkali concentration is less than 1.0 wt.%. Figure 10.32 shows the cumulative percentage versus injected alkali slug with data sources the same as in Figure 10.31. At the 50% cumulative percentage, the injected alkali solution slug is about 15% PV.

Figure 10.33 shows the cumulative percentage versus the total injected alkali, which is presented by the product of concentration (%) and slug size (% PV). At the 50% cumulative percentage, the product is 17. For most of the field projects, the incremental oil recovery factors were 1 to 2%.

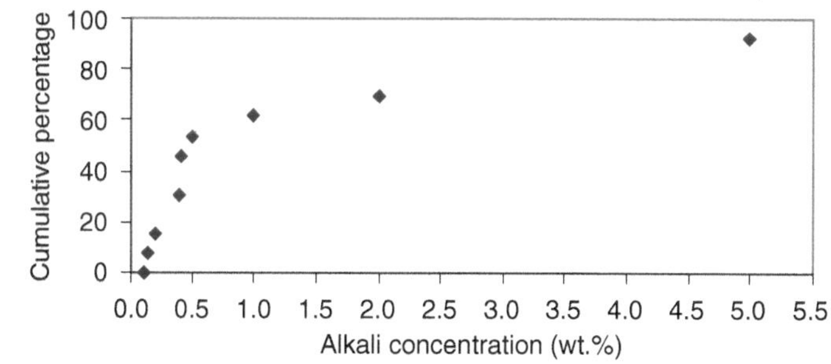

FIGURE 10.31 Cumulative percentage versus alkali concentration for field projects.

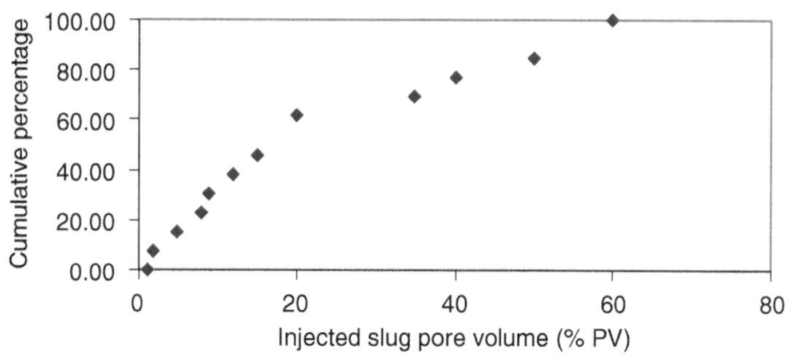

FIGURE 10.32 Cumulative percentage versus injected alkali slug for field projects.

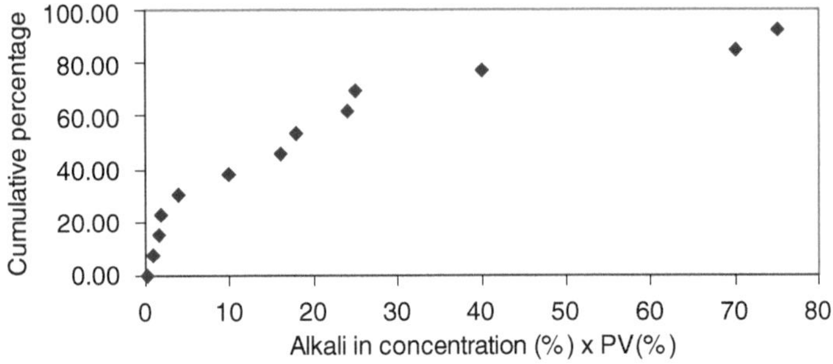

FIGURE 10.33 Cumulative percentage versus total amount of injected alkali.

These low incremental oil recovery factors are consistent with several facts about alkaline flooding: it is difficult to obtain ultralow IFT (a very low alkali concentration is required, which is not practical); the low IFT range is narrow; mobility control is limited, and so on. Considering the synergy among alkaline flooding, surfactant flooding, and/or polymer flooding, it appears that alkaline flooding without adding surfactant or polymer has lost its attraction, at least in light oil reservoirs.

10.9 SURVEILLANCE AND MONITORING IN PILOT TESTING

Because the mechanisms of alkaline flooding are complex and we lack experience in field implementation, we cannot take every aspect of field operation into account. Therefore, we need to make laboratory measurements as much as possible and conduct detailed pilot designs. During pilot testing, a surveillance and monitoring program is extremely important. It will help design future field implementations. More importantly, a go/no go business decision regarding the field scale extension is based on pilot testing performance. Table 10.15 shows a recommended surveillance and monitoring program for an alkaline flooding project.

A pilot pattern should be chosen so that the injected fluid is well controlled within the pattern. Otherwise, the fluid may be "lost" through directional flow channels. Then any interpretation or evaluation of the pilot performance would be difficult. When evaluation wells are drilled, cores should be taken in a closed-loop method so that reservoir conditions are maintained. These cores are used to evaluate alkaline consumption, measure relative permeabilities, and so on. Formation evaluation tests are conducted at evaluation wells. Finally, simulation models (sector models) are built to integrate all the data taken to evaluate the alkaline flooding performance.

10.10 APPLICATION CONDITIONS OF ALKALINE FLOODING

Injection wells should be located within oil zones not in the peripheral aquifer to avoid alkali consumption caused by reaction with divalents. However, if the

TABLE 10.15 Recommended Surveillance and Monitoring Program for an Alkaline Flooding Project

Tests and Data	Frequency
Production Wells	
Liquid production rate	Daily
Water cut	Daily
Formation pressure	Quarterly
Well bottom hole pressure (BHP)	Quarterly
Pressure buildup tests	Quarterly
Production profiles	Quarterly
Produced water analysis	Weekly
Composition analysis	Weekly
Alkali content	Weekly
Tracer content	Weekly
Produced oil composition	Weekly
Injection Wells	
Well injection rate	Daily
Wellhead pressure	Monthly
Well bottom hole pressure (BHP)	Monthly
Injection profiles	Quarterly
Injection water analysis	Quarterly
Composition analysis	Weekly
Alkali content	Daily
Mechanical impurity analysis	Quarterly
Observation Wells	
Wellhead pressure	Daily
Well bottom hole pressure (BHP)	Daily
Sample analysis	Daily

bottom water has a high divalent content, an alkaline solution may be injected to form precipitates so that bottom water coning may be mitigated by precipitates. The reservoir does not have a gas cap.

For reservoirs with oils having high acid numbers, alkaline flooding can be executed at any development stage. However, for reservoirs with oils having low acid numbers, alkaline flooding in an earlier stage performs better. In this case, remaining oil saturation should be higher than 0.4. There is no temperature limitation for alkaline flooding.

Alkaline consumption by chemical reaction and ion exchange is mainly due to the existence of clays. Thus, clay content should not exceed 15 to 25%. Formation permeability should be greater than 100 md.

Oil viscosity should be less than 50 to 100 cP. However, currently alkaline-surfactant injection into very high viscous oils has attracted more and more interest (see Chapter 12 on alkaline-surfactant flooding). Formation water salinity should be less than 20%, and the divalents in the injection water should be less than 0.4%.

In the design of an alkaline flooding project, the following facts may be taken into account:

- Alkaline consumption in the field is higher than that in the laboratory because the contact time of alkalis with rocks in the former is much longer than that in the latter.
- Oil recovery factor in the field is generally lower than that in the laboratory.
- When alkaline flooding is combined with other methods, such as polymer flooding, surfactant flooding, hydrocarbon gas injection, or thermal recovery, much better effects will be obtained.
- Alkaline injection could cause scale problems in the reservoir, wellbore, and surface facility and equipment.
- When an alkaline solution makes contact with oil, stable emulsions may be formed. This will increase the cost to treat produced fluids on the surface.

Chapter 11

Alkaline-Polymer Flooding

11.1 INTRODUCTION

Many field tests have revealed that alkaline flooding is not a simple method but requires careful project design and monitoring techniques. One reason that the results from conventional alkaline flooding have not been encouraging is that low alkaline concentrations required for obtaining low interfacial tension are not capable of propagating alkali because of the consumption by ion exchange and dissolution, and precipitation processes. Another reason is the lack of mobility control. Therefore, the combination of alkaline and polymer floods seems to be a better option.

The theories of alkaline flooding and polymer flooding alone are discussed in their respective chapters. This chapter focuses on the interaction and synergy between alkali and polymer. It also presents a field application.

11.2 INTERACTION BETWEEN ALKALI AND POLYMER

The interaction between alkali and polymer, to be discussed in this section, includes alkaline effect on polymer viscosity, polymer effect on alkaline/oil IFT, and alkaline consumption in alkaline-polymer systems.

11.2.1 Alkaline Effects on Polymer Viscosity

This section discusses alkaline dynamic effect and its concentration effect on polymer viscosity.

Alkaline Dynamic Effect

Alkali and polymer reaction hydrolyzes polymer. Alkali is consumed by the reaction. Thus, the alkaline concentration and pH decrease, as shown in Figure 11.1. This pH decrease phenomenon cannot be ignored. For example, the pH decreases from 10.4 to 9.9 when a 2% Na_2CO_3 and 1000 mg/L HPAM solution is aged for 120 days. Using the systems of polymer and buffer alkalis could be a potential solution (e.g., sodium silicates with SiO_2/Na_2O = 1.6–2.4, Na_2CO_3, $NaHCO_3$, Na_2HPO_4). These systems have medium pH, and pH does not change much in a wide range of alkaline concentrations. The alkaline consumption will

Modern Chemical Enhanced Oil Recovery. DOI: 10.1016/B978-1-85617-745-0.00011-5

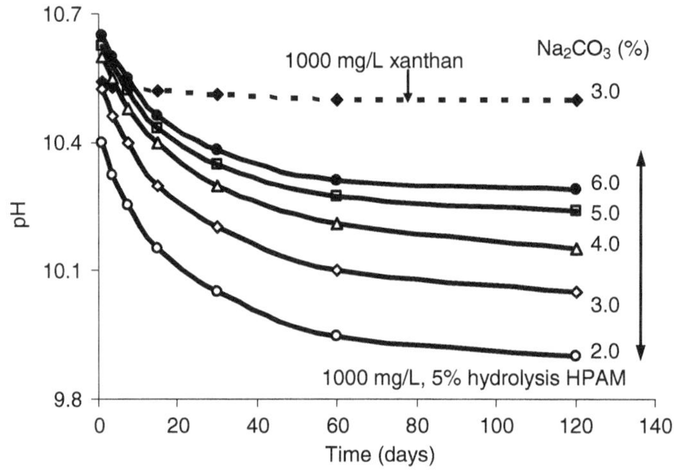

FIGURE 11.1 Graphic of pH changes with aging time: Anaerobic at 60°C, 3216 mg/L TDS. *Source: Sheng et al. (1994).*

be lower because of their low alkaline strength. These systems can also precipitate divalents to protect polymer.

In a polymer solution prepared using distilled water at less than 90°C, when alkali was added, the viscosity decreased initially due to the salt effect; then increased with aging time as hydrolysis was increased, later gradually stabilized. It is well known that polymer viscosity increases with hydrolysis. In saline water, hydrolysis increases in the early time. Then the solution viscosity increases. As the salt effect increases with hydrolysis, the solution viscosity decreases later. When the salinity is low and/or the divalent concentration is low, adding alkali results in flocculation. Then the viscosity of the top clear solution without white flocks becomes lower, as shown by the dotted line in Figures 11.2 and 11.3. In Figure 11.2, white flocks appeared after 30 days; in Figure 11.3, after 60 days. Na_4SiO_4 is more tolerant to divalents than Na_2CO_3 or NaOH. However, if the salinity or divalent concentration is very high, the alkaline effect on the viscosity will not be significant (Sheng et al., 1994). Although polymer viscosity increases with hydrolysis, it will not increase further above 30 to 40% owing to the salt effect. Figure 11.3 shows that xanthan gum viscosity slightly decreased because of its weak reaction. Note that hydrolysis also happens without the addition of alkali, but it is very slow.

Alkaline Concentration Effect

The previous section showed that polymer solution viscosity increases with time initially when alkali is added. In other words, the polymer viscosity at $t > 0$ is higher than its initial viscosity at $t = 0$ for a given polymer concentration. That does not mean adding alkali will increase polymer viscosity. Figure 11.4

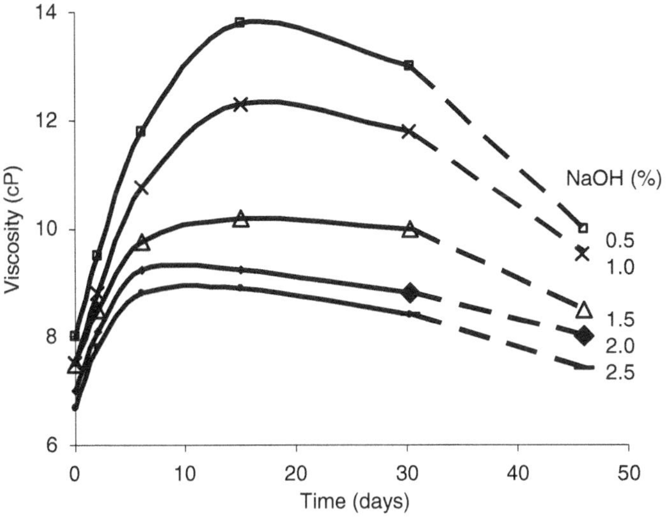

FIGURE 11.2 NaOH–HPAM solution viscosity versus time: 21.5% hydrolysis, 1000 mg/L HPAM, 60°C, 3215 mg/L TDS. *Source: Sheng et al. (1994).*

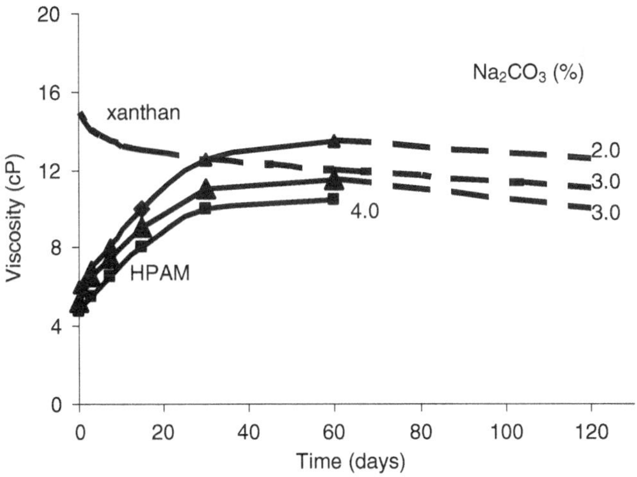

FIGURE 11.3 Na_2CO_3–polymer solution viscosity versus time: 5% hydrolysis, 1000 mg/L HPAM, 60°C, 3215 mg/L TDS. *Source: Sheng et al. (1994).*

shows the HPAM viscosity at different NaOH concentrations and at shear rates. This figure shows that as alkali concentration was increased, the polymer viscosity decreased. This result is due to the salt effect (cation electric shield effect), which reduces the stretch of polymer molecules in the solution. Actually, Figures 11.2 and 11.3 do show that the polymer viscosity was higher at a

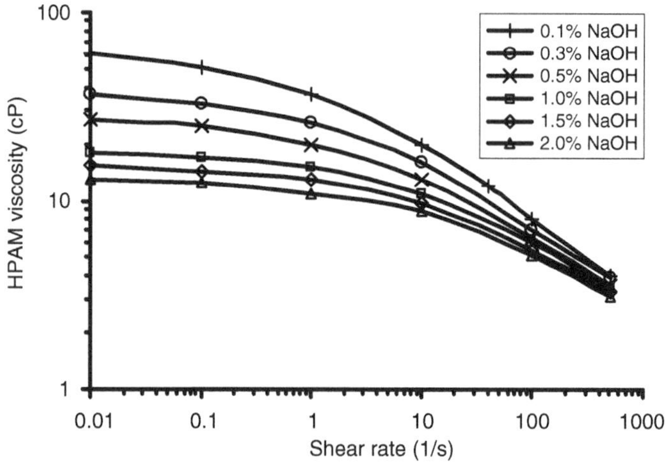

FIGURE 11.4 Alkaline effect on polymer (1000 mg/L 1275A) viscosity. *Source: Kang (2001).*

lower alkaline concentration; these figures show the aging effect (not alkaline concentration) on polymer when alkalis are added.

Alkali has two main effects on polymer viscosity: increased salt effect and increased hydrolysis effect. The former decreases viscosity, whereas the latter increases viscosity. The final viscosity increases or decreases depends on the balance of the two effects. In general, the salt effect is greater. Thus, polymer viscosity decreases with alkaline concentration.

11.2.2 Polymer Effect on Alkaline/Oil IFT

There is no consensus regarding the polymer effect on alkaline/oil IFT. Generally, it is believed that polymer has little effect on the IFT. However, in many polymers, small amount of surfactants are added; thus, polymer may have some contribution to the IFT reduction (Potts and Kuehne, 1988). Sheng et al. (1993) made the following observations (the first three points are supported by the data in Figure 11.5):

1. The addition of polymer could increase or decrease IFT depending on the type of alkali in the system. The Na_2CO_3 + HAPM/crude IFT was lower than that of NaOH + HPAM/crude at the same alkaline concentration.
2. Alkali + HPAM/crude IFT decreased with the time during which the alkaline-polymer solution contacted with crude.
3. Alkali + HPAM/crude IFT decreased with polymer hydrolysis.
4. IFT also depended on alkaline concentration because of alkaline salt effect. There was an optimum alkaline concentration at which IFT was the lowest.
5. Alkali + xanthan/crude IFT was lower than alkali + HPAM/crude IFT, but the former IFT had less reduction with aging time than the latter.

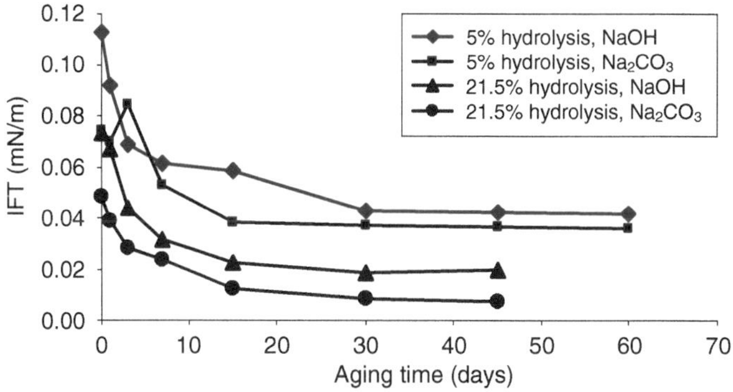

FIGURE 11.5 Effects of polymer hydrolysis, type of alkali, and contact time with crude on IFT, 3315 mg/L TDS water, 60°C, 0.50 mg KOH/g acid number. The NaOH concentration was 1%, whereas the Na_2CO_3 concentration was 3%. *Source: Data from Sheng et al. (1993).*

TABLE 11.1 IFT (mN/m) of Crude Oil-Alkali-Polymer System

Na_2CO_3 (%)	0.6	0.8	1.5	2.0	3.0
Crude + Na_2CO_3	2.68	0.73	0.084	0.037	0.088
Crude + Na_2CO_3 + 0.2% Polymer (3530S)	1.35	0.002	0.0024	0.63	

Source: *Song (1993)*

Table 11.1 shows the IFT between oil and a chemical solution. The conclusions according to the data in the table are mixed. Apparently, when polymer existed in the solution, the IFT was lower. In general, polymer does not affect equilibrium IFT significantly, but it increases the time to reach equilibrium IFT because high viscous solution reduces the rate for surfactant to transfer to the interface. In some cases, HPAM reduces IFT probably because the polymer has both hydrophilic and hydrophobic groups like a surfactant.

11.2.3 Alkaline Consumption in Alkaline-Polymer Systems

Laboratory test results show that alkaline consumption in an alkaline-polymer system is lower than in the alkaline solution itself. The reason is probably that polymer covers some rock surfaces to reduce alkali-rock contact. In an alkaline-polymer system, alkali competes with polymer for positive-charged sites. Thus, polymer adsorption is reduced because the rock surfaces become more negative-charged sites (Krumrine and Falcone, 1987). Mihcakan and van Kirk (1986) observed that alkaline consumption in a radial core is smaller than that in a linear core.

11.3 SYNERGY BETWEEN ALKALI AND POLYMER

In alkaline-polymer flooding, alkaline reaction with crude oil results in soap generation, wettability alteration, and emulsification; and polymer provides the required mobility control. Alkaline-polymer flooding can displace more residual oil than individual alkaline flooding or polymer flooding. The combination of alkaline and polymer flooding can have three variations: (1) alkaline injection followed by polymer injection (A/P), (2) polymer injection followed by alkaline injection (P/A), and (3) alkaline and polymer co-injection (A+P). The recovery factor from the third injection mode is not only higher than the alkaline injection alone or polymer injection alone, but also higher than that from the first or second mode (Sheng et al., 1994).

Table 11.2 shows some experimental data for different scenarios. From this table, we can observe the following:

- Comparing the incremental recovery factors (RFs) of Case 4 versus Case 8, Case 6 versus Case 10, and Case 7 versus Case 11 shows that the injec-

TABLE 11.2 Alkaline-Polymer Flooding Results

Case No.	Injection Sequence	Concentration, %		Injection PV		RF, %	Incremental RF, %
		Alkaline	*Polymer*	*Alkaline*	*Polymer*		
1	W	0.0	0.0	0.0	0.0	75.0	0.0
2	P, W	0.0	0.1	0	0.3	81.8	6.8
3	A, W	0.1	0.0	0.3	0	80.7	5.7
4	A, P, W	0.1	0.1	0.3	0.3	91.8	10.8
5	A, P, W	0.25	0.1	0.3	0.3	96.1	21.1
6	A, P, W	0.5	0.1	0.3	0.3	80.9	5.9
7	A, P, W	1.0	0.1	0.3	0.3	77.5	2.5
8	P, A, W	0.1	0.1	0.3	0.3	94.0	19.0
9	P, A, W	0.25	0.1	0.3	0.3	88.8	13.8
10	P, A, W	0.5	0.1	0.3	0.3	86.8	11.8
11	P, A, W	1.0	0.1	0.3	0.3	80.9	5.9
12	A+P, W	0.1	0.1	0.3		91.9	16.9
13	A+P, W	0.25	0.1	0.3		91.2	16.2
14	A+P, W	0.5	0.1	0.3		90.5	16.5
15	A+P, W	1.0	0.1	0.3		88.2	13.2

tion sequence of polymer followed by alkaline (P/A) was better than alkaline followed by polymer (A/P), with the one exception of Case 5 versus Case 9. The incremental RF from Case 2 for P/W was higher than in Case 3 for A/W.

- Comparing the incremental RFs of Cases 12 through 15 with Cases 4 through 7 or Cases 8 through 11 shows that, overall, the simultaneous injection of alkali and polymer (A+P) was better than the sequential injection of alkali and polymer (A/P) or polymer and alkali (P/A).
- The incremental RF of 16.9% in Case 12 was higher than the sum of Cases 2 and 3 (12.5%), also demonstrating the synergy between alkali and polymer.
- The results show that too high alkaline concentration (e.g., > 0.5) was not beneficial. Apparently, the optimum alkaline concentration for the test cases was 0.25%.

In summary, P/A was better than A/P, and A+P was better than A/P or P/A. These observations were also made by Chen et al. (1999b). Figure 11.6 shows the residual oil recovery factor after waterflooding. The system was as follows: oil viscosity, 180 mPa·s at room temperature; polymer, 5000 mg/L HPAM; and alkali, Na_2SiO_4. This figure shows that A+P was better than any sequential or single injection process. This result was also observed when a biopolymer or less-viscous oil (62 mPa·s) was used. Although in this example A/P was much better than P/A, P/A was better than A/P in the case of 62 mPa·s (Krumrine and Falcone, 1983).

Some Russian and Chinese researchers (Chen et al., 1999b) observed in the laboratory that as pH was increased, polymer adsorption was reduced. However,

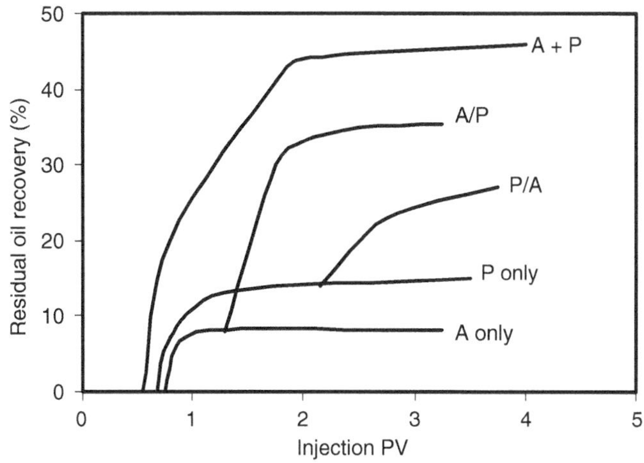

FIGURE 11.6 Comparison of residual oil recovery factors in alkaline flood, polymer flood, and any combination of these two floods. *Source: Krumrine and Falcone (1983).*

this phenomenon is not documented as well as that for surfactant reduction (Labrid, 1991). Meanwhile, polymer adsorption reduces alkali reaction with rocks, as mentioned in Section 11.2.3. It also has been observed in the laboratory that the addition of polymer in alkaline solution can reduce swelling of montmorillonite.

In alkaline-polymer flooding, in addition to the polymer mobility control effect, the precipitation (e.g., $Ca(OH)_2$ and $Mg(OH)_2$) caused by alkali also helps to increase sweep efficiency. Precipitates formed by alkalis may be able to flow through pores without blocking any flow, or reduce both oil and water permeabilities. However, precipitates combined with polymer can effectively reduce water permeability because polymer is in the water phase.

As we know, adding alkali in a polymer solution will reduce the polymer solution's viscosity. We may take advantage of this fact in low-injectivity wells. Initially, the polymer solution with alkali has a low viscosity. As the alkali is consumed by reacting with formation water and rocks, the polymer solution's viscosity will become higher than the initial value. Thus, initially the injectivity and later the sweep efficiency will be improved. Note that the polymer concentration will also be reduced by adsorption. The final effect is determined by the balance between the two effects of the reduced alkaline and polymer concentrations.

The synergy between alkaline and polymer flooding may be summarized as follows:

- Alkaline-polymer can reduce polymer adsorption and alkaline consumption as well.
- Polymer makes the alkaline-polymer solution more viscous to improve sweep efficiency. Thus, polymer "brings" alkaline solution to the oil zone, where the alkaline cannot go without polymer. More oil can be displaced by lowered IFT owing to alkaline-generated soap. In other words, alkaline and polymer work together to improve both sweep efficiency and displacement efficiency.
- The alkaline-polymer environment may decrease biodegradation.
- Alkaline may reduce polymer viscosity in the near wellbore region so that the injectivity is improved.

11.4 FIELD AP APPLICATION EXAMPLE: LIAOHE FIELD

This section presents an alkaline-polymer (AP) pilot test performed in the western part of the Xing-28 block in the Liaohe Oilfield (Zhang et al., 1999).

Reservoir Characterization and Production Status

The Xing-28 block had 2.05 km^2, reservoir thickness of 3 m, and original oil in place (OOIP) of 0.96 million tons. It had an anticline structure and a gas cap of 1.08 km^2 with a thickness of 2.6 m. It also had edge water, and the reservoir depth was 1650 to 1730 m. The formation porosity was 0.276 and permeability was 2063 md. The formation was water-wet sandstone.

In this test, the dead oil had a density of 0.9059 g/cm^3 and viscosity of 24.02 mPa·s at 50°C. The reservoir live oil had a density of 0.8174 g/cm^3 and viscosity of 6.3 mPa·s at the reservoir temperature of 56.6°C. The paraffin content in the oil was 4.14%. The initial reservoir pressure was 17.29 MPa, and the formation water TDS was 3112 mg/L with Ca^{2+} and Mg^{2+} of 14 mg/L.

The oil production was started in September 1971, and the water injection was started in November 1974. By December 1993, the cumulative oil production was 0.437 million tons; that is equivalent to a 45.5% recovery factor. The cumulative water production was 1.55 million m^3. The cumulative water injection was 2.18 million m^3, and the ratio of the cumulative water injection to the cumulative liquid production was above 0.9. In December 1993, when the water cut was 91.9%, 4 producers and 3 injectors were active. The average oil and liquid production rates were 5.6 tons/d and 62.2 tons/d, respectively. The average water injection rate was 70 m^3/d. In December 1994 (just before the AP pilot test), the water cut was 96.2%, and the oil recovery was 46.75%.

The pilot test area was located in a high structure with Well X7 as the center, as shown in Figure 11.7. The target zones were IV_2 and IV_3. Zone IV_2 was in the southern structure, and Zone IV_3 extended from the southern part to the northern part near Wells X19 and X1.

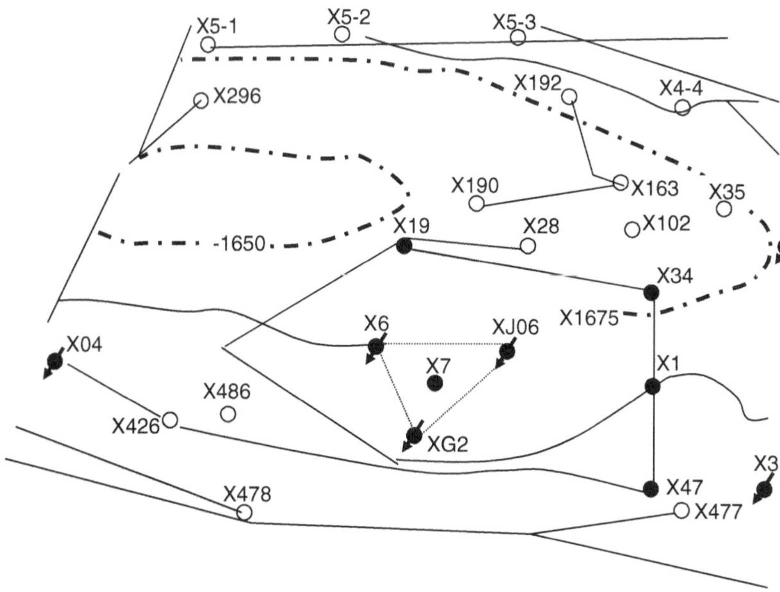

FIGURE 11.7 Well locations in the pilot area of AP flooding.

Residual Oil Saturation Distribution and Waterflood Recovery Factor

To establish a baseline for the alkaline-polymer flooding, the operator used several empirical correlations and reservoir simulation to estimate the waterflood recovery factor, which was 50%. To study residual oil saturation distribution, the operator used several approaches such as pressure coring, C/O logging, core wafer, and waterflood performance analysis. Finally, all data were integrated into a simulation model to output the residual oil saturation distribution. The average residual oil saturation was 0.33. The gas cap shrank and existed only in the north area to Wells X19 and X35. This area was far away from the AP flooding area so that it was not affected by AP.

Selection of Well Pattern

The criteria to select the well pattern were (1) the pattern had higher residual oil saturation zones, (2) the pattern location was away from the gas cap or water zones, and (3) existing wells could be used to reduce or eliminate drilling costs.

Based on the preceding criteria, the selected central well pattern included three injectors (X6, XJ06, XG2) and one observation (production) well (X7). The well spacing (interwell distance) was 160 to 190 m. The four peripheral observation (production) wells, which included X19, X34, X1, and X47, were well connected. The central pilot area covered 0.037 km^2 with a thickness of 7.4 m. The rock had 6.3% carbonate content and 2.5% clay content.

Chemical Formulation

To select alkali, six alkalis—NaOH, Na_2SiO_3, Na_4SiO_4, Na_3PO_4, $NaHCO_3$, and Na_2CO_3—were used to compare IFT reduction, emulsification, alkaline consumption, and alkaline-polymer interaction. The results were as follows:

- *NaOH:* Strong emulsification, low IFT, narrow range of optimum concentrations, high alkaline consumption, and quick polymer hydrolysis so that it was difficult to control.
- *Na_4SiO_4:* Similar to NaOH.
- *Na_2SiO_3:* Strong emulsification, low IFT, medium alkaline consumption, and flocculation occurred after a long reaction with polymer.
- *Na_3PO_4:* Similar to Na_2SiO_3, but a weaker alkali.
- *Na_2CO_3:* Strong emulsification, low equilibrium IFT, wide range of optimum concentrations, high alkaline consumption, slower polymer hydrolysis so that it was easy to control, good supply sources, and low cost. However, the concentration required was higher than NaOH and Na_2SiO_3.
- *$NaHCO_3$:* Weak alkali, limited capability to reduce IFT if used alone.

The final pick was Na_2CO_3.

HPAM and xanthan gum were considered for selection. Xanthan gum was compatible with alkalis and stable in the reservoir, but it had a limited supply source and it was more expensive. Plus, domestic xanthan gums had poorer injectivity. Therefore, the operator decided to use HPAM. After that, 8 HPAM-type products were evaluated: Nanzhong II Xiaoji (China), Nanzhong II84-2.43 (China), AC530 (Japan), AC430 (Japan), 3430S (US), 3530S (US), Tongde #3 (China), and 1175A (UK). The evaluation showed that 3530S and 1175A were the best. The final selection was 1175A based on its lower price.

Core flood tests were used to compare polymer flood only and alkaline-polymer performance. To model in situ oil/water viscosity ratio correctly, the operator mixed the crude oil with kerosene at a ratio of 100:26. Single-, double-, and triple-column tests were conducted. In the single-column tests, polymer flood increased sweep efficiency over waterflood by 5.6 to 9.77%, and AP flood increased by 13.7 to 19.3%. On average, AP outperformed polymer flood by 8.8%. In the double- and triple-column tests, AP recovery factors were about 18 to 20% higher than waterflood recovery factors. Half of the incremental recovery came from the low permeability column.

One natural core was used to compare the performance of waterflood (W), AP flood, and ASP flood. The recovery factors for W, AP, and ASP were 50%, 69.7%, and 86.4%, respectively. These core flood tests were history matched, and the history-matched model was extended to a real field model including alkaline consumption and chemical adsorption mechanisms. A layered heterogeneous model was set up by taking into account the pilot geological characteristics. The predicted performance is shown in Table 11.3. In the table, C_a, C_s, and C_p denote alkaline, surfactant, and polymer concentrations, respectively. After the designed PV of chemical slug was injected, water was injected until almost no oil was produced. The total injection PV for each case is shown in the table as well. The cost is the chemical cost per barrel of incremental oil produced. An exchange rate of 7 Chinese yuan per U.S. dollar was used. From

TABLE 11.3 Performance Comparison of A, AP, and ASP Floods (Simulation Results)

C_a, %	C_s, %	C_p, mg/L	Slug, PV	RF, %	Total PV	Cost, $/bbl oil	Process
0	0	0		44.3			W
0	0	800	0.5	47.1	0.94	7.0	P
2	0	800	0.5	52.9	1.30	7.8	AP
2	0.4	800	0.5	57.6	1.30	11.3	ASP
2	0	800	0.3	51.1	1.07	5.9	AP
2	0.4	800	0.3	54.4	1.08	8.9	ASP

Table 11.3, we can see that the economics of AP flooding were better than that of ASP flooding. Therefore, the AP option was selected for this pilot test.

Optimization of Injection Scheme

The injection scheme was optimized by simulation using the modified UTCHEM 6.0. The final selected formula was 2% NaCO3 + 1000 mg/L 1175A. The simulation results showed that the performance from different injection sequences was similar for the same mass of chemicals. Operation experience shows that polymer viscosity would be reduced by about 50% from the surface to the wellbore by mechanical shear loss, iron effect, and bacteria degradation. Therefore, in the performance prediction, the viscosity was assumed to be 50% of the measured viscosity in the laboratory. Economic analysis was also included in the optimization process to select the best injection scheme.

Implementation and Performance

After all the preceding studies were done, the AP pilot test was implemented from January 1995 to August 1998. The AP flood increased the oil recovery by 1.98% (OOIP) for the whole pilot area and 18.5% (OOIP) for the central well area, respectively. From January 1995 to the time the water cut reached 98.0%, the oil recovery was 3.34% (OOIP) for the whole pilot area, and AP had given an ultimate oil recovery of about 50% (OOIP). However, it was found that the AP flood conducted in this pilot area was not economically attractive owing to larger amount of capital investment and the low oil price at that time. For more details of this performance, see Zhang et al. (1999).

Chapter 12

Alkaline-Surfactant Flooding

12.1 INTRODUCTION

For oil displacement purposes, alkali can be co-injected with any displacing agents except an acid or carbon dioxide. For example, alkaline-polymer (AP), alkaline-surfactant (AS), alkaline-gas, alkaline-steam, alkaline-hot water, and more can be used. This chapter discusses alkaline-surfactant flooding.

Alkali saponifies the naphthenic acids in crude oil to generate sodium naphthenate (soap) in situ. Some may believe the purpose of adding alkali to surfactant flooding is to generate soap so that the amount of injected surfactant can be reduced. Although generating soap is important, the following mechanisms are probably even more important:

- Reduction in surfactant adsorption because of the high pH from the alkali injection
- The synergy between the in situ generated soap and the injected synthetic surfactant

The focus here is on these mechanisms. A high content of naphthenic acids in heavy oils is a good property for soap generation and emulsification. Therefore, this chapter also presents the synergy between alkali and surfactant in heavy oil reservoirs. First, it discusses the phase behavior of the mixed system of soap and surfactant. Then it describes how to build up a UTCHEM phase behavior model and how to use the model to analyze phase behavior. In addition, this chapter investigates a number of parameters related to phase behavior.

12.2 PHASE BEHAVIOR TESTS FOR THE ALKALINE-SURFACTANT PROCESS

Phase behavior tests performed in glass sample tubes (pipettes) for the alkaline-surfactant process include aqueous tests, a salinity scan (alkalinity scan), and an oil scan. The aqueous tests and salinity scan are the same as those for surfactant flooding. For the salinity scan in AS or alkaline-surfactant-polymer (ASP) cases, alkali also works as salt. There are two ways to change salinity. One is to change the salt content while fixing the alkali content; the other is to change the alkali content while fixing the salt content. Therefore, the salinity

Modern Chemical Enhanced Oil Recovery. DOI: 10.1016/B978-1-85617-745-0.00012-7

scan in AS or ASP is sometimes called the alkalinity scan. When conducting the salinity scan, we generally set the water/oil (WOR) ratio equal to 1 or a fixed value. In the oil scan, we repeat the salinity scan by changing WOR only. The oil scan is used to generate an activity map, which will be discussed later in this chapter.

Alkalis can also provide electrolytes, which are required to achieve optimum conditions. However, the relationship between alkaline concentration and the salinity provided is complex. In other words, the salinity that 1 meq/mL of alkali provides may not be 1 meq/mL.

Martin and Oxley (1985) studied the effect of different alkalis on surfactant systems. They showed that the presence of any alkali lowered the optimum salinity of the surfactant system. This phenomenon is caused by two facts: (1) alkali can provide electrolytes; and (2) alkali reacts with crude oil to generate soap, and soap has lower optimum salinity (see the next section). Martin and Oxley found a linear relationship between the optimum salinity and sodium concentration. The addition of any alkali agents results in a decrease in the optimum salinity of the system. However, alkali anions have very little effect on the phase behavior.

12.3 QUANTITATIVE REPRESENTATION OF PHASE BEHAVIOR OF AN ALKALINE-SURFACTANT SYSTEM

Generally, injected alkali content is high enough to react with all the naphthenic acid in crude oil. More soap would likely be generated in a system with higher oil content. Thus, the phase behavior of the two-surfactant system depends on the water/oil ratio (WOR). Figure 12.1 shows that the optimum sodium chloride concentration of an alkaline system is a function of surfactant concentration

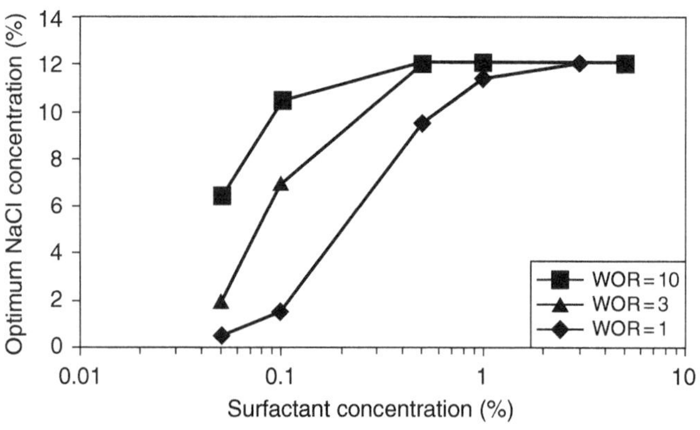

FIGURE 12.1 Optimum salinity of an alkaline-surfactant system as a function of WOR and surfactant concentration. *Source: Zhang et al. (2006).*

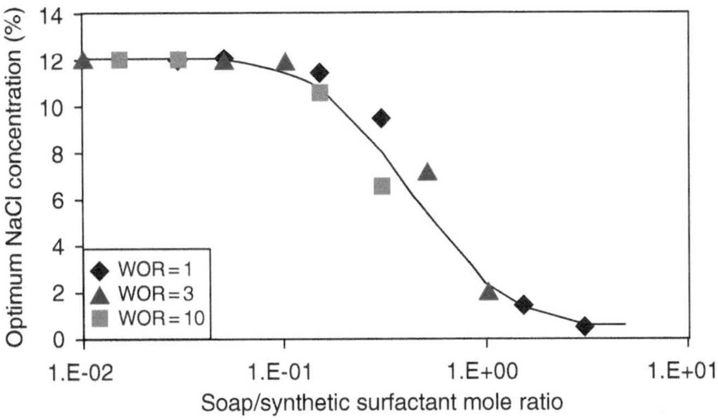

FIGURE 12.2 Optimum salinity of an alkaline-surfactant system as a function of the ratio of soap-to-synthetic surfactant concentration. *Source: Zhang et al. (2006).*

and WOR. When the optimum sodium chloride concentration in Figure 12.1 is plotted as a function of the ratio of soap-to-surfactant concentration, all the data points fall on almost the same single curve, as shown in Figure 12.2. Therefore, the optimum salinity of an alkaline-surfactant system should be a function of soap/surfactant ratio.

Figure 12.2 shows that the optimum salinity increases as the soap/surfactant ratio decreases. During ASP flooding, oil saturation decreases from the downstream (the displacing front) to the upstream. Because soap concentration is proportional to oil saturation, the soap/surfactant ratio would likely decrease. The soap generated in situ is a surfactant different from the injected synthetic surfactant. These two surfactants have different properties. Generally, the injected surfactant is more hydrophilic than the soap. Thus, the optimum salinity of soap is lower than that of the synthetic surfactant. As the soap/surfactant ratio decreases, the optimum salinity would increase. Consequently, the salinity upstream would likely be lower than the optimum salinity, resulting in a local Winsor I environment. Such a microemulsion environment is desirable.

Because soap and injected synthetic surfactant have different properties, a mixing rule is needed to model the properties of the two-surfactant system. Based on experimental data, Salager et al. (1979a) proposed the following logarithmic mixing rule for optimum salinities:

$$\ln\left(C_{se,m}^{opt}\right) = X_1 \ln\left(C_{se,1}^{opt}\right) + X_2 \ln\left(C_{se,2}^{opt}\right), \tag{12.1}$$

Here, $C_{se,m}^{opt}$, $C_{se,1}^{opt}$, and $C_{se,2}^{opt}$ are the optimum salinities of the mixture, surfactant components 1 and 2, respectively. The surfactant mole fractions are X_1 and X_2. Salager et al. also proposed that other characteristic parameters could follow a linear mixing rule.

Puerto and Gale (1977) used a linear mixing rule on optimum salinity to fit their data. In fact, the logarithmic mixing rule has been found to slightly underestimate the optimum salinity, whereas the linear mixing rule has been found to slightly overestimate the optimum salinity. In general, for high electrolyte concentrations, a linear rather than logarithmic mixing rule is best to obtain the mixture optimum salinity, whereas at low electrolyte concentrations, a logarithmic mixing rule should be better (Bourrel and Schechter, 1988). Mohammadi et al. (2008) found that for the optimum solubilization ratios, both logarithmic and linear mixing rules are satisfactory:

$$R_{23,m}^{opt} = X_1 R_{23,1}^{opt} + X_2 R_{23,2}^{opt}, \tag{12.2}$$

or

$$\ln\left(R_{23,m}^{opt}\right) = X_1 \ln\left(R_{23,1}^{opt}\right) + X_2 \ln\left(R_{23,2}^{opt}\right), \tag{12.3}$$

Here, $R_{23,m}^{opt}$, $R_{23,1}^{opt}$, and $R_{23,2}^{opt}$ are the optimum solubilization ratios of the mixture, surfactant components 1 and 2, respectively.

Liu (2007) reported a special characteristic of phase behavior in an alkaline-surfactant system. When the salinity was below the optimum salinity, some materials lighter than the lower phase microemulsion rose to the oil–microemulsion interface over time. As a result, a thin layer of a colloidal dispersion formed with 23 days' settling, but it was not seen at 4 hours' settling, as shown in Figure 12.3. The low density of the dispersed material suggests a higher ratio of oil to brine than in the lower phase. The volume of this dispersion increased with increasing salinity below optimum salinity. Moreover, its volume was significantly greater for the same surfactant concentration at WOR = 1, which contained more soap but less surfactant than the cases at WOR > 1.

This result suggests that the dispersed material had a higher soap-to-surfactant ratio than the lower phase and hence was more lipophilic than the lower

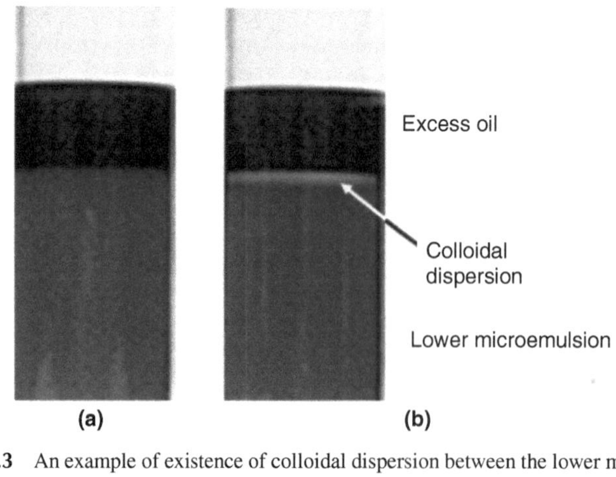

FIGURE 12.3 An example of existence of colloidal dispersion between the lower microemulsion phase and upper extra oil phase: (a) 4 hours' settling and (b) 23 days' settling. *Source: Liu (2007).*

phase being capable of solubilizing more oil but less brine. The dispersed material may be a second microemulsion, which, because it is more lipophilic than the lower phase microemulsion, would have a lower IFT with excess oil. The existence of two microemulsions in equilibrium is possible for mixtures of surfactants very different in structure and in hydrophilic/lipophilic properties, as in the case Liu (2007) reported. Low IFT between the microemulsion phases would contribute to the ease of dispersion of one in the other. Such behavior represents a deviation from the classical Winsor I behavior of a single microemulsion plus excess oil under underoptimum conditions.

The preceding discussions focus on the phase behavior of the middle-phase microemulsion. Actually, when an alkali is added to a surfactant system, a "mixed phase" is formed. It takes some time for a clear middle-phase microemulsion in equilibrium to be formed. The process could take a long time, from several days to weeks. Sometimes, a cloudy middle phase (without a middle-phase microemulsion) is formed between the upper oil phase and lower water phase. The mixed phase has several characteristics (Li et al., 2002):

- The particle size is predominantly in the order of 1 μm or below. The IFT between water and oil is low. When the particle size is about 0.1 μm, the IFT becomes ultralow.
- There are some micelles, micellar aggregates, microemulsions, emulsions, and dispersed liquid crystals in the mixed phase. A proper match between the size of liquid crystals and size of small particles results in ultralow IFT.
- The dispersed liquid crystals can stabilize the mixed phase.
- The energy required to form the mixed phase is lower than that to form an emulsion.
- In the middle phase microemulsion, the particle size is in the order of nanometers (Miller et al., 1977).

Section 12.6 further discusses the behavior of emulsions.

12.4 ACTIVITY MAPS

The purpose of an activity map is to show at what range of concentrations in a system and how a chemical flood will work. For a given reservoir where the temperature, composition of crude oil, and residual oil saturation are fixed, five kinds of variables are under our control: types of alkalis, concentrations of alkalis, types of surfactants, concentrations of surfactants, and salinity. Another important variable that is not under our direct control is the type and amount of petroleum acid that will convert to soap when contacted by the alkalis. As discussed earlier, the amount of soap will determine the concentrations of alkali and surfactant injected. In other words, to generate an activity map, we have to know the amount of soap that can be generated. Because the alkali concentration typically is much greater than that required to convert all the petroleum acids in the oil to soap, the petroleum soap concentration (meq/L) is calculated approximately from the acid number and weight of the oil in the test tube, although such estimation is generally not correct, as we can see from Example 10.4.

Nelson et al. (1984) used an activity map (see Figure 12.4) showing the active region as a function of petroleum soap concentration and salinity when the alkali type, alkali concentration, surfactant type, surfactant concentration, type of oil, and temperature are fixed. In the figure, the shaded areas represent the type III region. Above and below the shaded areas represent the type II and type I regions, respectively. The numbers in the shaded areas are cosurfactant concentrations. Nelson et al. used the term cosurfactant for the injected synthetic surfactant, NEODOL 25-3S. In the horizontal axes, both petroleum soap concentration and volume percent oil in the test tubes are marked. These two variables are proportional. In other words, the soap concentration was calculated from the oil volume present in the test tubes assuming all naphthenic acids were reacted with the alkali. The alkali used was 1.55% $Na_2O{\cdot}SiO_2$. In the

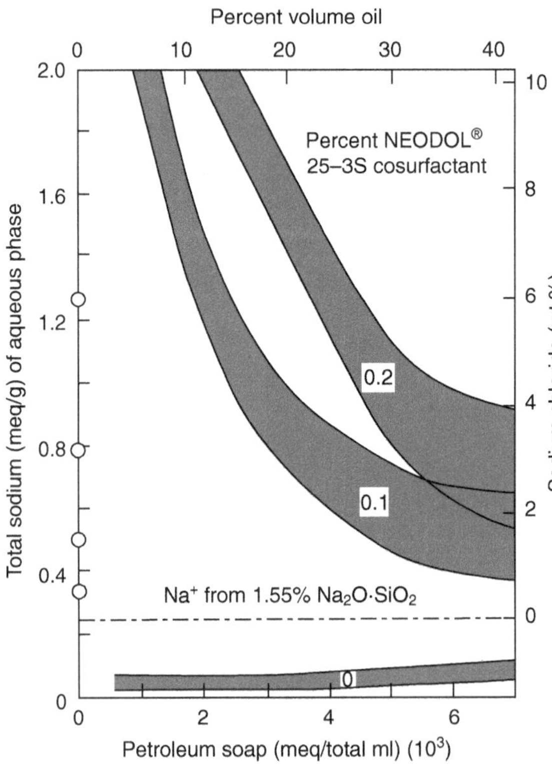

FIGURE 12.4 Activity maps for 1.55% $Na_2O{\cdot}SiO_2$ and 0, 0.1, and 0.2% NEODOL 25-3S with a Gulf Coast crude oil at 30.2°C. *Source: Nelson et al. (1984).*

vertical axes, sodium chloride concentrations are marked to represent the salinity. This figure is discussed in more detail in Section 12.5.

The procedure for constructing an activity map is similar to that used in constructing a phase diagram. Glass sample tubes (pipettes) containing oil, aqueous alkaline, and surfactant solution are equilibrated at the test temperature with periodic shaking. For the oil volume, we generally start with WOR = 1 and then reduce it up to WOR equal to 10 or 20. The chemical concentrations are varied around target concentrations.

As mentioned earlier, we sometimes conduct an alkalinity scan by changing alkali concentration while the salinity is fixed. Then the activity map can be presented by alkali concentration versus oil volume percent or the ratio of oil volume percent to surfactant volume or weight percent, schematically shown in Figure 12.5.

Zhao et al. (2008) used an activity map like the one in Figure 12.5 showing the sodium carbonate concentrations as a function of the ratio of oil concentration in vol.% to surfactant concentration in wt.%. They assumed that this ratio corresponds to the ratio of soap to surfactant because the amount of soap generated by the alkali is proportional to the oil concentration. They used the concentration ratio of oil to surfactant based on the finding shown in Figure 12.2 that the optimum salinity is independent of WOR when the ratio is used. They used oil concentration instead of soap concentration for convenience (without any calculation in the laboratory). As the ratio decreases, the more hydrophilic mixture causes the phase behavior to change from Winsor III to Winsor I.

This is the favorite direction for an ASP flood as the oil saturation becomes lower in the upstream. This type of activity diagram is robust because a type I phase environment is formed at the end of chemical injection, which is desirable, as we discussed in Chapter 8. However, the slope should not be too steep.

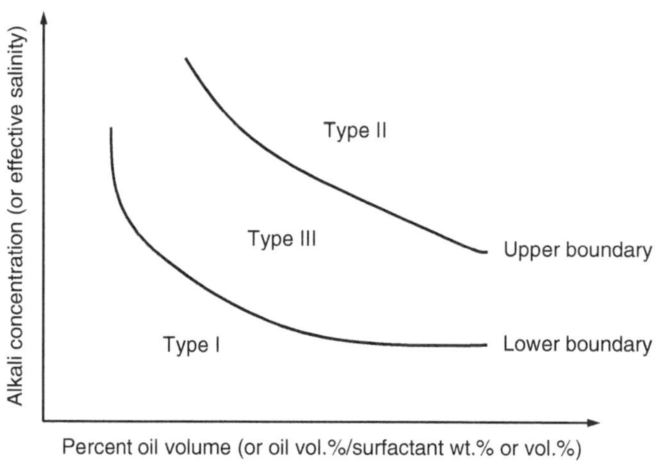

FIGURE 12.5 Variations of activity maps.

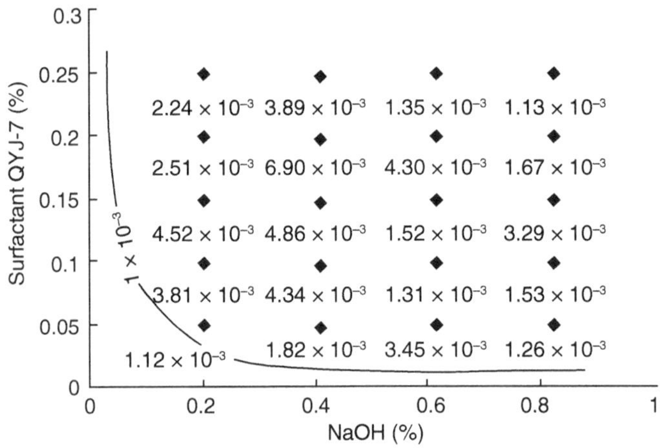

FIGURE 12.6 Equilibrium IFT activity map for QYJ-7 + NaOH system. *Source: Zhou et al. (2009).*

Otherwise, the formula selected is very sensitive to oil content, which is an uncontrollable variable. Note that the representation by Zhao et al. using oil vol.%/surfactant vol.% did not take into account the changes in surfactant concentration as the flood proceeds. In other words, the surfactant concentration was constant, which is true in laboratory test tubes but not in the laboratory corefloods or in the field because of surfactant retention. They also showed that the upper and lower boundaries are straight lines.

Based on Eq. 12.1, optimum salinity follows the logarithmic mixing rule. Mohammadi et al. (2008) replaced the ratio of oil to surfactant concentration shown in Figure 12.5 by soap molar fraction and used the more generally effective salinity in the vertical axis. They did so because they could get these values from UTCHEM simulation models. Based on the logarithmic mixing rule, both axes in such activity maps are in logarithmic scales, and the upper and lower boundaries should be linear.

In describing surfactant phase behavior or activities, Chinese methodology is to use interfacial tension (probably their philosophy is to rely on IFT measurement). Therefore, their activity map is to show the IFT at different surfactant and alkaline concentrations. Figure 12.6 is an example of such an activity map. In this figure, the region of ultralow IFT (10^{-3} mN/m) is marked.

12.5 SYNERGY BETWEEN ALKALI AND SURFACTANT

Many investigators have observed that the lowest interfacial tensions between a crude oil and alkali frequently occur at very low concentrations of alkali (Nelson et al., 1984). Lieu et al. (1982) reported that the concentration range in their cases was in the region of 0.2% sodium hydroxide. Green and Willhite (1998) also reported that the concentration range is in the 0.1 wt.%. The concentrations of this level could not survive where there was a considerable duration of time in the reservoir. Therein lies a dilemma: alkali consumption by a reservoir requires that a concentration of alkali higher than that which produces minimum interfacial tension be used if the alkaline bank is to propagate through the reservoir at an economically acceptable rate. Therefore, we apparently must choose between best oil displacement (lowest interfacial tension) and satisfactory propagation rate.

Nutting (1925) was the first to notice this problem and suggested the use of weaker bases such as sodium carbonates and silicates for improving waterflood performance. This problem can be resolved by applying cosurfactant design concepts used in chemical flooding—in other words, by adding synthetic surfactants in the alkaline solution, as discussed later. This concept was first proposed by Reisberg and Doscher (1956). Low concentrations (~ 0.5%) of a nonionic surfactant mixed with 1 to 2 wt.% sodium hydroxide produced additional oil from their laboratory sand packs and consolidated cores. Later Nelson et al. (1984) further described this concept.

At the concentrations of alkali above that required for minimum interfacial tension, the systems become overoptimum. The excess alkali plays the same role as excess salt. When synthetic surfactants are added, the salinity requirement of alkaline flooding system is increased. NEODOL 25-3S is such a synthetic surfactant used by Nelson et al. (1984). Figure 12.4, shown earlier, is a composite of three activity maps for 0, 0.1, and 0.2% of NEODOL 25-3S as a synthetic surfactant for 1.55% sodium metasilicate with Oil G at 30.2°C. We can see in the figure that without the synthetic surfactant, the active region of this system is below the sodium ion concentration supplied by the alkali. However, with 0.1 and 0.2% of NEODOL 25-3S (60% active) present, the active region is above the sodium ion concentration supplied by the alkali, so additional sodium ions must be added to reach optimum salinity.

The shape of the active region in the presence of the synthetic surfactant, as shown in Figure 12.4, is typical of this type of activity map. If the concentration of synthetic surfactant is constant, moving from the right to the left on the map increases the synthetic surfactant-to-petroleum soap ratio. The salinity requirement for the system therefore is increased. As the concentration of petroleum soap goes to 0, the active region rises to the salinity requirement of the synthetic surfactant—approaching 20% sodium chloride for NEODOL 25-3S. That feature is the principal difference between this type of activity map and a salinity requirement diagram. In a salinity requirement diagram, the ratio of synthetic surfactant to the cosurfactant remains almost unchanged throughout the diagram.

Figure 12.7 shows another example of soap–surfactant synergy. This figure shows IFT between Yates oil and the microemulsion that was formed by 0.2% 4:1 mixture by weight of Neodol 67-7PO sulfate and internal olefin sulfonate 15–18, with water/oil ratio = 3 (Liu et al., 2008). The width of the low IFT region (< 10^{-2} mN/m) is much wider with sodium carbonate added than the

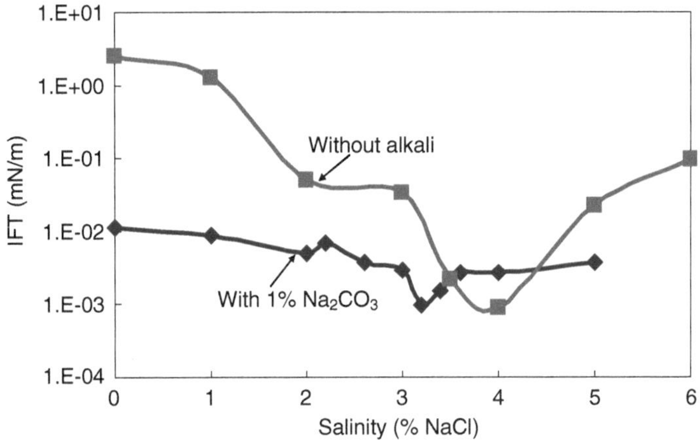

FIGURE 12.7 IFT between Yates oil and the microemulsion. *Source: Liu (2007).*

case without alkali. Martin and Oxley (1985) attributed such behavior to ionization of the carboxylic acid by alkali.

Jackson (2006) tested the effect of sodium carbonate on the phase behavior of surfactants using a crude oil with little or no acid. He observed that the equilibration is more rapid for the sample containing a higher sodium carbonate concentration, and it also shortened the time required for microemulsion to form.

Kang et al. (1997) investigated the IFT of an AS system with Daqing oil. They found that the low IFT was mainly due to carboxylates, whereas the dynamic low IFT resulted from the in situ generated soap adsorption and diffusion at the interface of alkaline solution/crude oil. Kang et al. also investigated the surfactant molecular weight distribution on IFT. Zhang et al. (1998b) investigated IFT in the system of alkali, nonionic surfactant OP-10, sulfonate CY, and oil. The system demonstrated obvious synergy. The synergy more likely affected the early-stage IFT in a low ionic strength condition. In a high ionic strength condition, the IFT was more affected by the added synthetic surfactant. The surfactant concentrations and their ratios determine the value of IFT. Zhang et al. found that the system of single molecular weight surfactant and alkali did not reduce the IFT to an ultralow level, but the system with some distributed molecular weight surfactant did. This result was also attributed to the synergy between different surfactants.

12.6 SYNERGY BETWEEN ALKALI AND SURFACTANT IN HEAVY OIL RECOVERY

When the concentration of surfactant is above a critical micelle concentration (CMC), two phenomena occur: solubilization and emulsification. The former is thermodynamically stable, whereas the latter is thermodynamically unstable. For some systems, emulsions may disappear after a long time. In some surfactant-polymer systems or in surfactant-heavy oil systems, because of high viscosity of polymer solutions or heavy oils, emulsions could be very stable.

Because heavy oils have higher content of acid components, alkali and oil reaction will generate in situ surfactant (soap). It is expected that alkalis would play a more important role in heavy oil recovery and the synergy between alkali and surfactant would be more significant. This section considers the work of Liu et al. (2006b) as an example to illustrate the alkaline-surfactant synergy in heavy oils.

Liu et al. (2006b) conducted bottle tests to emulsify a heavy oil using the alkali Na_2CO_3. They used the surfactant S4, which is alkyl ether surfactant with a molecular weight of 441. The heavy oil viscosity was 1800 mPa·s at 22°C. They first used 0.15 to 1.2% Na_2CO_3 solution and 1 to 2000 ppm S4 solution separately. In these tests, the heavy oil could not be emulsified in either alkaline solution or surfactant solution. However, when they used 50 ppm S4 and 0.15 to 1.2% Na_2CO_3 together, they observed emulsification.

Figure 12.8 shows the IFT curves when only alkali was used. The IFT decreased from 9.5 to 3.5 dyne/cm when the Na_2CO_3 concentration was increased from 0.15 to 1.2 wt.% (see the 40-minute curve). Although the IFT decreased with alkaline concentration, the equilibrium IFT at 1.2% alkaline concentration was still much higher than the ultralow value (e.g., < 10^{-2} dyne/cm). The dynamic reduction in IFT was not significant.

Figure 12.9 shows the IFT curves when only surfactant was used, from 30 to about 3 dyne/cm when the surfactant concentration was increased from 0 to

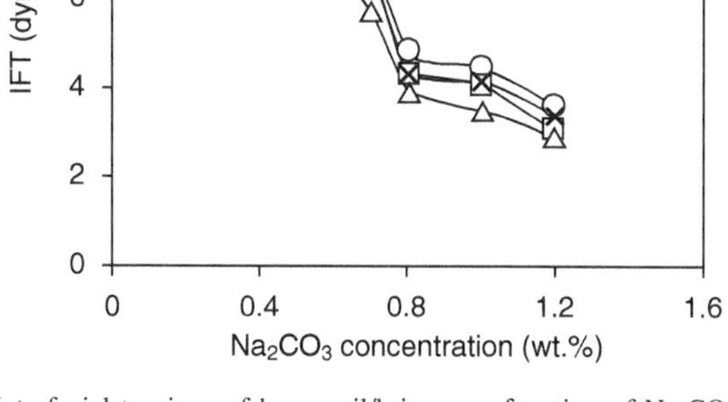

FIGURE 12.8 Interfacial tensions of heavy oil/brine as a function of Na_2CO_3 concentration at different measurement times. *Source: Liu et al. (2006b).*

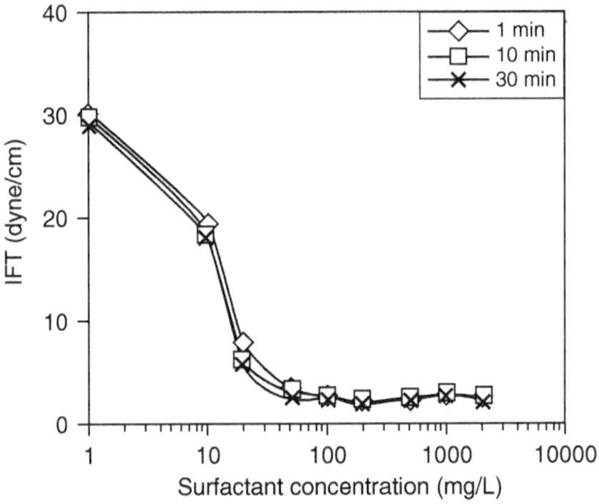

FIGURE 12.9 IFT of heavy oil/surfactant solution at different concentrations and at different measurement times. *Source: Liu et al. (2006b).*

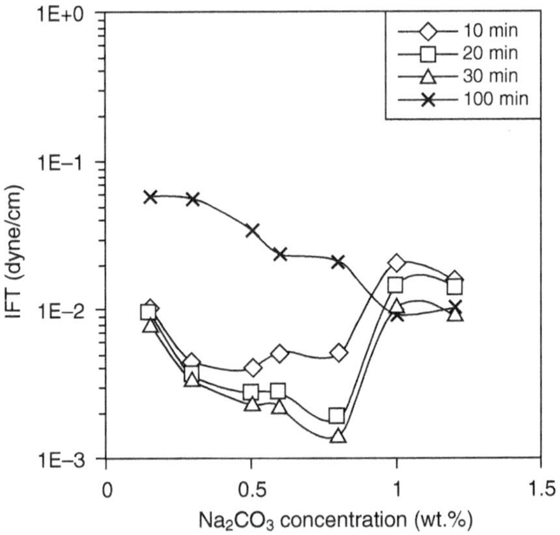

FIGURE 12.10 Interfacial tension of heavy oil/brine as a function of Na_2CO_3 concentration with 50 mg/L surfactant present. *Source: Liu et al. (2006b).*

50 mg/L. The interfacial tension could not be decreased further even though the surfactant concentration was increased to 2000 mg/L. The IFT dynamic effect was negligible.

Figure 12.10 shows IFT versus Na_2CO_3 concentration in the presence of 50 mg/L surfactant S4. The change in IFT was more obvious and lasted longer than that in the alkali-only or the surfactant-only systems. Compared with the results shown in Figures 12.8 and 12.9, the addition of only 50 mg/L surfactant in Figure 12.10 reduced the IFT from the range of several dynes/cm to approximately 5×10^{-3} dyne/cm for a wide range of Na_2CO_3 concentrations. For the samples with Na_2CO_3 concentrations lower than 1.0 wt.%, the dynamic IFT decreased with time in the first 30 minutes and then was stable for some time. The dynamic IFT increased quickly after 80 minutes and gradually stabilized after 100 minutes. The IFTs at 100 minutes were in the range of 0.01 to 0.05 dyne/cm, which are about two orders of magnitude lower than those for the surfactant-only and alkali-only samples.

As discussed previously, ultralow IFT can be obtained owing to the synergy between an alkali and a surfactant. Both low IFT and high surface charge (expressed in ζ-potential or electrophoretic mobility) are the result of the maximum adsorption of the ionic surfactant at the oil/water interface. Therefore, the synergy between alkali and surfactant should result in high ζ-potential as well. Figure 12.11 shows the ζ-potential versus surfactant concentration of systems with and without 0.15 wt.% Na_2CO_3 in brine. For the two systems, the magnitude of ζ-potential first increased rapidly, then decreased, and finally stabilized.

The addition of Na_2CO_3 in the brine can lead to an increase in surface charge in two ways: (1) ionization of the organic acid at oil/water interface; (2) adsorption of hydroxyl ions. Figure 12.11 shows that the ζ-potential for oil in brine was –20 mv and oil in 0.15 wt.% Na_2CO_3 solution in brine was –23 mv at 0% surfactant. The stabilized ζ-potential of the surfactant-only system was about –33 mv. The ζ-potential of the surfactant solution with 0.15 wt.% Na_2CO_3 present stabilized at about –55 mv.

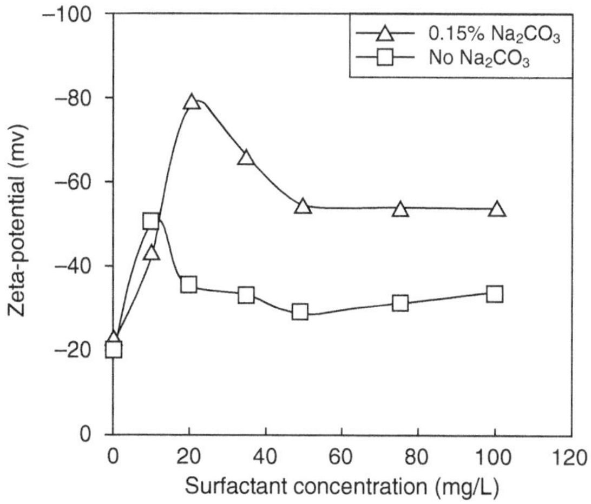

FIGURE 12.11 Zeta-potential of emulsions as a function of surfactant concentration. *Source: Liu et al. (2006b).*

The synergistic enhancement between alkali and the surfactant adsorption at the oil/water interface can be demonstrated by the data that follows: the addition of alkali caused a slight increase in ζ-potential from –20 to –23 mv at 0% surfactant; the addition of the surfactant increased the ζ-potential from –20 to –33 mv (no Na_2CO_3 present); and addition of both the alkali and the surfactant increased the ζ-potential from –20 to –55 mv. The synergistic enhancement of Na_2CO_3 and the surfactant could make the oil droplets much more negatively charged. The high surface charge density at the oil/water interface had a favorable effect in suppressing coalescence. Therefore, the heavy oil was emulsified in the alkaline and surfactant solution. In this case, the average diameter of the emulsion particles was about 15 μm.

Dong et al. (2009) measured IFT between an Alberta oil and chemical solutions. The oil viscosity was 1266 mPa·s at 22°C with an acid number of 1.19 mg KOH/g oil. They found the IFT was about 0.01 dyne/cm at 0.01 wt.% surfactant, and 0.4 wt.% Na_2CO_3 plus 0.2 wt.% NaOH, resulting from the synergy between alkali and surfactant, compared with 0.07 dyne/cm at 0.2 wt.% Na_2CO_3 plus 0.1 wt.% NaOH without surfactant. They also found the combined alkali of Na_2CO_3 and NaOH worked better than Na_2CO_3 only or NaOH only. Tertiary oil recoveries of about 22 to 23% OOIP were obtained for the tests in sand packs using a solution of 0.4 wt.% Na_2CO_3, 0.2 wt.% NaOH, and 0.045 wt.% surfactant. In these tests, the slug sizes were about 1 PV.

12.7 pH EFFECT ON SURFACTANT ADSORPTION

The primary mechanism for the adsorption of anionic surfactants on sandstone and carbonate formation material is the ionic attraction between mineral sites and surfactant anion (Zhang and Somasundaran, 2006). The generation of surface charge on the mineral particles is considered to be either due to preferential dissolution or due to hydrolysis of surface species followed by pH-dependent dissociation of surface hydroxyl groups. For oxides such as silica and alumina, the hydrolysis of surface species followed by pH-dependent dissociation is considered to be a major mechanism:

$$-M(H_2O)^+_{surface} \xleftarrow{H^+} (MOH)_{surface} \xrightarrow{OH^-} MO^-_{surface} + H_2O. \tag{12.4}$$

We can see from the preceding equation that the surface would be positively charged under low pH conditions and negatively charged under high pH conditions (Somasundaran and Hanna, 1977). The pH at which the net charge of the surface is zero is called the point of zero charge (PZC). From the preceding equation, H^+ and OH^- are the potential determining ions, those that determine the surface charge, for oxide minerals. Silica is negatively charged at reservoir conditions and exhibits negligible adsorption of anionic surfactants at high pH (Hirasaki et al., 2008).

In more detail, the reaction for quartz is written as follows:

$$SiO_2(s) + OH^- + H_2O \leftrightarrow H_3SiO_4^-, \tag{12.5}$$

Kaolinite would react according to the following equation:

$$\begin{aligned}[(Al_2O_3), (SiO_2)_2, (H_2O)_2](s) + 4OH^- + 3H_2O \\ \leftrightarrow 2H_3SiO_4^- + 2Al(OH)_4^-,\end{aligned} \tag{12.6}$$

At high pH values, the solid surfaces acquire a negative charge that gives rise to a large repulsion term.

For salt-type minerals such as calcite and apatite, the preferential hydrolysis of the surface species and preferential dissolution of ions have been proposed to be the major controlling mechanisms. Dissolution of ions is often accompanied by reactions with the solution constitutes and possible uptake of the solid. For example, calcite can undergo the following reactions upon contact with water and generate a number of complexes (Somasundaran and Agar, 1967):

$$CaCO_{3(s)} \leftrightarrow CaCO_{3(aq)} \quad K_1 = 10^{-5.09}$$

$$CaCO_{3(aq)} \leftrightarrow Ca^{2+} + CO_3^{2-} \quad K_2 = 10^{-3.25}$$

$$CO_3^{2-} + H_2O \leftrightarrow HCO_3^- + OH^- \quad K_3 = 10^{-3.67}$$

$$HCO_3^- + H_2O \leftrightarrow H_2CO_3 + OH^- \quad K_4 = 10^{-7.65}$$

$$H_2CO_3 \leftrightarrow CO_{2(g)} + H_2O \quad K_5 = 10^{1.47}$$

$$Ca^{2+} + HCO_3^- \leftrightarrow CaHCO_3^+ \quad K_6 = 10^{0.82}$$

$$CaHCO_3^+ \leftrightarrow H^+ + CaCO_{3(aq)} \quad K_7 = 10^{-7.90}$$

$$Ca^{2+} + OH^- \leftrightarrow CaOH^+ \quad K_8 = 10^{1.40}$$

$$CaOH^+ + OH^- \leftrightarrow Ca(OH)_{2(aq)} \quad K_9 = 10^{1.37}$$

$$Ca(OH)_{2(aq)} \leftrightarrow Ca(OH)_{2(s)} \quad K_9 = 10^{2.45}$$

From the preceding equations, we can see that when calcite approaches equilibrium with water at high pH, an excess of negative HCO_3^- and CO_3^{2-} will exist, whereas at low pH an excess of positive Ca^{2+} and $CaHCO_3^-$ and $CaOH^+$ will occur. These ionic species may be produced at the solid/solution interface or may form in solution and subsequently adsorb on the mineral in amounts proportional to their concentration in solution. In either case, the net result will be a positive charge on the surface at low pH and a negative charge at high pH.

Hirasaki and Zhang (2004) found that potential determining ions (CO_3^{2-}) can change the surface charge and reduce the anionic surfactant adsorption on calcite. Carbonate formations and sandstone-cementing material can be calcite or dolomite. These two minerals have an isoelectric point (or PZC) of about pH 9. In these cases, carbonate ion and calcium and magnesium ions are more significant potential determining ions (Hirasaki et al., 2008). In the case of seawater injection into fractured chalk formation, sulfate ions adsorb on the chalk surfaces to alter the surface potential (Austad et al., 2005). In this case, sulfate ion is the potential determining ion.

Hydroxyl ion can change pH so that the zeta potential of the carbonate/brine interface changes from a positive charge to a negative charge (Thompson and Pownall, 1989). Also, sulfate ion could be a potential determining ion, as mentioned previously. However, experimental data from Liu (2007) show that either hydroxyl ion or sulfate ion could not decrease the surfactant adsorption on the dolomite surface, as shown in Figure 12.12.

This figure shows that the adsorption on a dolomite surface with these ions is the same as that without. Although the data in the figure are for dolomite power, Liu et al. (2008) further reported that a series of experiments with the TC blend showed that the adsorption per unit area was nearly the same for calcite and dolomite powers. The TC blend was a 1 : 1 mixture by weight of C_{12} ethoxylated sulfate (3 EO) and iso-C13 propoxylated sulfate (4 PO). For the same TC blend surfactant, Zhang et al. (2006) showed in Figure 12.13 that the adsorption on dolomite sand was reduced by a factor of 10 with sodium carbonate added compared to that without. Their argument was that either hydroxide or sulfate is not a potential determining ion for carbonate surfaces, but carbonate ion is.

Clay minerals that have layered structures consisting of sheets of SiO_4 tetrahedra and sheets of AlO_6 octahedra linked with each other by means of shared oxygen ions are negatively charged under most natural conditions

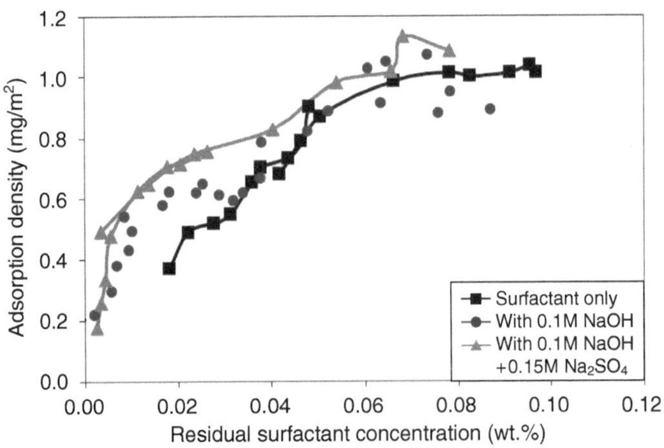

FIGURE 12.12 Adsorption of TC blend on dolomite with hydroxyl ion and sulfate ion. *Source: Liu (2007).*

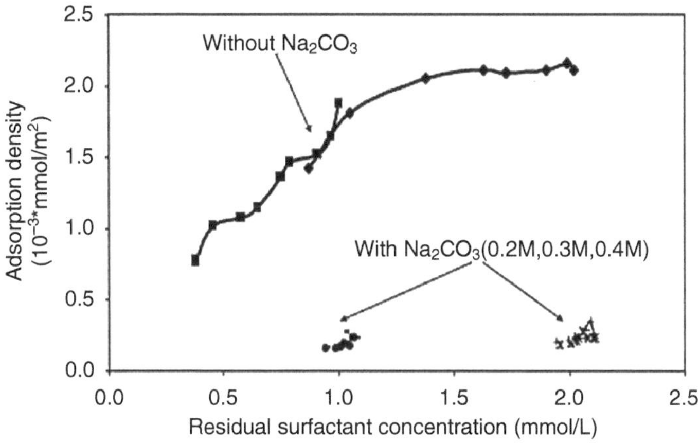

FIGURE 12.13 Static adsorption of blend surfactant on dolomite. *Source: Zhang et al. (2006).*

mainly owing to the substitution, for example, of Al^{3+} for Si^{4+} in the silica tetrahedral. This charge, which is internal to the structure, is not dependent on solution concentrations. The edges of the clay particles will, on the other hand, exhibit pH-dependent charge characteristics due to hydroxylation and ionization of the broken Si–O and Al–O bonds at the edges (Somasundaran and Hanna, 1977). At neutral pH, clays have a negative charge on the faces and a positive charge at the edges. The clay edges are alumina-like and thus are expected to reverse their charge at a pH of about 9 (Hirasaki et al., 2008). The point of zero charge of the clay is thus determined by the algebraic sum of face and edge charges. It is to be noted that at the point of zero charge, both faces and sides could be charged and thus possess adsorptive properties that other minerals might not possess at their points of zero charge (Somasundaran and Hanna, 1977).

Hanna and Somasundaran (1977) conducted tests on Berea sandstone/Mahogany sulfonate and kaolinite/dodecylsulfonate systems to determine the effect of solution pH on adsorption. For the former system at a constant ionic strength of 0.01 M NaCl, the adsorption densities were found to be 0.66 and 0.4 mg/m^2 for the initial pH conditions of 5 and 11, respectively, and the corresponding final pH values were not much different from each other (12.3 and 12.8). The results obtained from the kaolinite/dodecylsulfonate system also showed that the adsorption of sulfonate on kaolinite decreased with increase in pH. These observations are in agreement with what would be expected from the fact that the mineral will become increasingly negatively charged with an increase in pH and thereby possibly retard the adsorption of an anionic surfactant such as sulfonate. Another mechanism for alkaline additives to reduce surfactant retention may be caused by the removal of multivalent ions.

TABLE 12.1 Surfactant Retention in Berea Cores

Alkali	Alkali Concentration (wt.%)	Surfactant Retention (mg/g)
No alkali	0.0	0.68
NaOH	0.38	0.65
Na_2CO_3	0.38	0.26
Na_3PO_4	0.38	0.28
Na_4SiO_4	1.00	0.25
Na_2O, $1.6SiO_2$	0.43	0.20
Na_2O, $3.2SiO_2$	0.38	0.15

Source: *Krumrine and Falcone (1987).*

Table 12.1 shows some results of surfactant retention experiments. In these experiments, a 0.235 PV NaCl solution was injected ahead of the dilute alkaline-surfactant slug. A reduction of retention of 60 to 80% was obtained by adding alkalis to the surfactant solution.

Because a high pH environment can reduce surfactant adsorption and precipitate divalent ions, an alkaline preflush (increasing pH) has been proposed to meet these goals. However, there is no consensus that the preflush will always work, and there are different opinions regarding preflushing to adjust salinity in general. A preflush did not always work, as reported by Pursley et al. (1973) for Loudon field, whereas Rivenq et al. (1985) reported that experimental results confirmed that using an Na_2CO_3 preflush increased the oil recovery rate up to twice its value without preflush, depending on preflush size and quantity of microemulsion injected, and correspondingly reduced surfactant retention. French et al. (1973) supported the idea of preflushing low-salinity water to displace the high-salinity formation water. Reed and Healy (1977) stated that it had not been established that a preflush was a practical way to substantially and sufficiently reduce total salinity.

In UTCHEM, the Langmuir-type equation is used to describe pH-dependent adsorption. To include pH effect, we can use the following empirical equation to modify the input parameter a_3 in Eq. 7.154 (with permeability correction omitted here):

$$a_3 = \begin{cases} a_{31} + a_{32}C_{se} & \text{pH} \le \text{PHC} \\ (a_{31} + a_{32}C_{se})\left(1 - \dfrac{\text{pH} - \text{PHC}}{\text{PHT} - \text{PHC}}\right) & \text{PHC} < \text{pH} < \text{PHT1}. \\ \text{HPHAD} & \text{pH} \ge \text{PHT1} \end{cases} \quad (12.7)$$

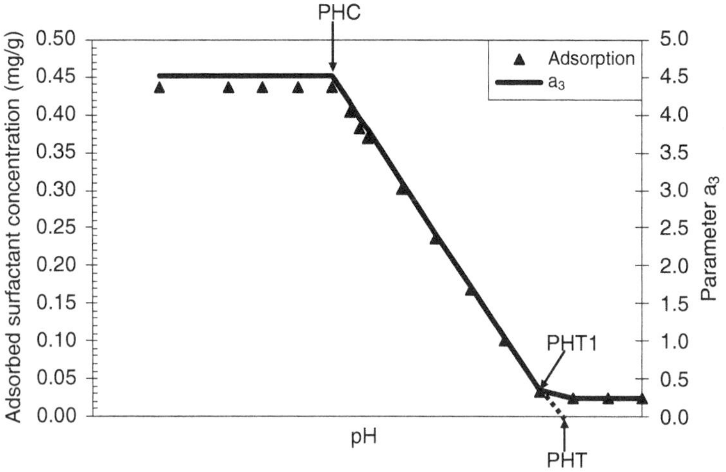

FIGURE 12.14 Graphic representation of a pH-dependent surfactant adsorption model.

In Eq. 12.7, PHC is the critical pH above which surfactant adsorption is pH dependent, PHT is the extrapolated pH value at zero surfactant adsorption, PHT1 is the pH value above which surfactant adsorption is constant, and HPHAD is the UTCHEM parameter for a_3, which makes adsorption equal the constant when pH is above PHT1. Figure 12.14 shows an example of pH-dependent adsorption and pH-dependent a_3. The UTCHEM input parameters—PHC, PHT, PHT1, and HPHAD—are also marked in the figure.

Note that a_3 in the second line of Eq. 12.7 could be smaller than HPHAD when PHT1 and PHT are very close. To avoid that situation, we should change Eq. 12.7 to

$$a_3 = \begin{cases} a_{31} + a_{32}C_{se} & \text{pH} \le \text{PHC} \\ \max\left\{(a_{31} + a_{32}C_{se})\left(1 - \dfrac{\text{pH} - \text{PHC}}{\text{PHT} - \text{PHC}}\right), \text{ HPHAD}\right\} & \text{PHC} < \text{pH} < \text{PHT1} \\ \text{HPHAD} & \text{pH} \ge \text{PHT1}. \end{cases} \quad (12.8)$$

In low alkaline concentrations, as the alkaline concentration is increased, anionic surfactant adsorption is reduced, as discussed previously. However, if the alkaline concentration is high, as the concentration is increased, ionic strength is increased. Then flocculation of surfactant micelles may occur. Also, as ionic strength is increased, the counter ions in the diffusion layer may enter the adsorption layer to reduce the electrostatic repulsion between the anionic surfactant and sand surface. Consequently, surfactant adsorption may be increased with alkaline concentration. Also, cationic surfactant adsorption increases with pH (Yang et al., 2002a).

12.8 RECOVERY MECHANISMS

The recovery mechanisms associated with alkaline-surfactant flooding may be summarized as follows:

- Reduced surfactant adsorption
- In situ soap generation
- Synergy between in situ generated soap and injected surfactant
- Wettability alteration attributable to the injected alkali
- Ability of the alkali to work as a sacrificial agent by reacting with the divalents

12.9 SIMULATION OF PHASE BEHAVIOR OF THE ALKALINE-SURFACTANT SYSTEM

In chemical flooding, the most challenging tasks are the quantification of surfactant phase behavior and alkaline reactions. Simulation of phase behavior of an alkaline-surfactant system that combines these two tasks in a single model may be the most challenging one. This section uses EQBATCH and UTCHEM to investigate several aspects of the phase behavior of alkaline-surfactant systems.

12.9.1 Setup of Alkaline-Surfactant Batch Model

To begin this simulation, we first need to set up an EQBATCH model. The difference between a phase behavior model and a flow model of an alkaline-surfactant system is that the matrix does not exist in the phase behavior test tube; thus, there is no ion exchange on the matrix in the phase behavior model. Therefore, in the phase behavior model, we define 6 elements and 14 fluid species based on Example 10.4 and remove the cation exchanges only on the matrix. In particular, we keep the solid species $Ca(OH)_{2(s)}$ and $CaCO_{3(s)}$. At least one advantage is that we can ensure that there should not be any solid precipitation in the model, or any precipitation should be consistent with the observation in the test tube. The rest of the procedures to set up the EQBATCH model are similar to those in Example 10.4.

Second, we need to set up an UTCHEM batch model. Because Example 7.2 is a UTCHEM batch model for a surfactant pipette test without alkaline reactions, we can simply include the initialization output of EQBATCH in a UTCHEM batch model that is built based on this example. Basically, we combine the EQBATCH model in Example 10.4 and the UTCHEM batch model in Example 7.2 to build a new AS phase behavior batch model.

For this alkaline-surfactant phase behavior batch model, the initialization data from EQBATCH are the same as output 2 of Example 10.4, and the other UTCHEM parameters are the same as those in Tables 7.2 and 7.4 of Example 7.2. To validate this AS batch model, we have to check whether this model

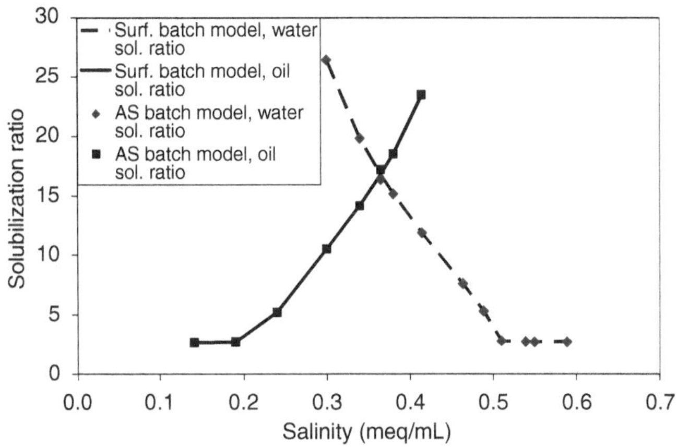

FIGURE 12.15 Comparison of phase behavior data from the surfactant batch model and AS batch model.

could reproduce the phase behavior data of Example 7.2 by simply setting a negligible acid number in the model. Figure 12.15 compares the phase behavior data from the surfactant batch model (Example 7.2) and the AS batch model. The figure shows that the AS model has reproduced the phase behavior data from the surfactant batch model. Therefore, this AS model is validated.

12.9.2 Analysis of Alkaline-Surfactant Phase Behavior

This section uses a set of sample data to investigate alkaline-surfactant phase behavior. The alkaline-surfactant data in Table 12.2 replace the data in the AS batch model used to produce Figure 12.15.

Calculation of Soap-Related Parameters

Parameters, such as soap mole fraction X_{soap}—a fraction of petroleum acid converted to soap, are not the direct output parameters from a UTCHEM model. Before we investigate alkaline-surfactant phase behavior, we need to know how to calculate soap-related parameters. Table 12.3 lists the parameters related to soap and surfactant that are calculated or from the UTCHEM output files. These data are for the base case. This table helps us to understand the relationships of these parameters.

Amount of Soap Generated

We want to see how much acid content in the crude oil is converted into soap, which helps to solubilize oil and water. Figure 12.16 shows the converted fraction of acid into soap at different alkali concentrations. It shows that up to 15 wt.% sodium carbonate, less than half of the acid component, is converted into soap. In practice, alkaline concentration is less than 2%. Then

Table 12.2 Input Data of the Base AS Model

Injection Water	
NaCl, %	0.6
Na_2CO_3, %	1.9
Ca^{2+}, meq/mL	0.001
WOR	1
Acid number, mg KOH/g oil	0.467
Surfactant concentration, vol.%	0.2
Surfactant Phase Behavior Data	
Lower salinity limit, meq/mL	0.55
Upper salinity limit, meq/mL	1.1
Input parameter, C_{33max0}, at zero salinity	0.03
Maximum height of binodal curve at optimum salinity, C_{33max1}	0.015
Input parameter, C_{33max2}, at twice optimum salinity	0.03
Soap Behavior Data	
Lower salinity limit, meq/mL	0.1
Upper salinity limit, meq/mL	0.2

only about a quarter of the acid can be converted into soap according to Figure 12.16.

The next question is: what is the molar fraction of soap in the total surfactant? Figure 12.17 shows the molar fraction of the generated soap in the total moles of surfactants at different alkali concentrations. It shows that up to 15 wt.% sodium carbonate, the generated soap, is less than half of the total moles of surfactants. In a practical alkaline concentration, the generated soap is less than one third of the total moles of surfactants.

Effect of Soap on Solubilization Ratios

When an alkali is injected into a reservoir with acidic crude oil, a fraction of acid components are converted into soap, which helps to solubilize oil and water into the microemulsion phase. Figure 12.18 shows the water and oil solubilization ratios at different effective salinities, based on the two definitions. One definition is the ratio of water or oil volume (V_w or V_o) to the volume of injected synthetic surfactant in the microemulsion phase. The other definition is the ratio of water or oil volume (V_w or V_o) to the total volume of injected

TABLE 12.3 Parameters Related to Surfactant and Soap Calculated from UTCHEM Output Files

	Parameter	Value	Calculation Formula or Data Source
A	B	C	D
1	Soap MW, g/mole	500	Input data
2	Surfactant MW, g/mole	420	Input data
3	Water saturation (S_w), fraction	0.5	Input data
4	Petroleum acid (HA), meq/mL water	8.32E-03	Input data
5	Surfactant vol. fraction (S_3)	9.20E-02	*.SATP
6	Surfactant fraction of PV (C_3)	9.61E-04	*.CONP surf. vol. (C_3)
7	Surfactant + soap, vol. fraction of ME	1.60E-02	*.ALKP surf. conc. in PHASE 3
8	Surfactant + soap, vol. fraction of PV	1.47E-03	*.AlKP TOTAL (INJ. + GEN.)
9	Soap fraction of PV	5.12E-04	= C8–C6
10	Soap (A^-), meq/mL PV	1.02E-03	= C9/C1*1000, soap in ME
11	Petroleum acid (HA), meq/mL PV	4.16E-03	= C4*C3
12	Fraction of converted soap	0.246	= C10/C11
13	Soap molar fraction (X_{soap})	0.309	= (C9/C1)/(C9/C1+C6/C2)
14	Surfactant + soap, vol. fraction of ME	1.60E-02	= C8/C5, should equal C7
15	Soap, vol. fraction of ME	5.56E-03	= C9/C5
16	Surfactant, vol. fraction of ME (C_{33})	1.04E-02	*.COMP_ME

synthetic surfactant and the generated soap in the microemulsion phase. We generally use the former definition for convenience because the volume of soap is unknown without using an AS model like the one presented here. This figure shows that the solubilization ratios in the latter definition are lower than those that are in the former definition. However, the differences are not significant.

Figure 12.19 shows the ratios of water and oil solubilization ratios based on the two definitions. In the figure, $(SR)_{total}$ is the solubilization ratio when

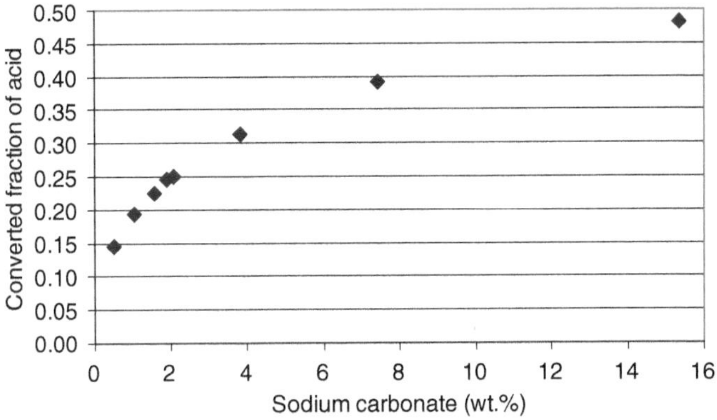

FIGURE 12.16 Converted fraction of acid into soap at different alkali concentrations.

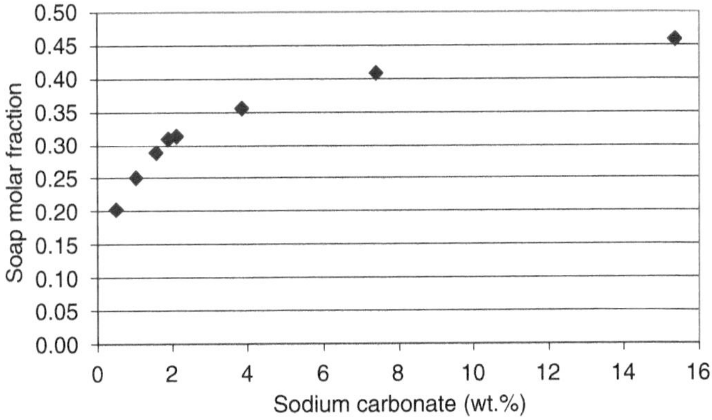

FIGURE 12.17 Molar fraction of soap in the total amount of surfactants at different alkali concentrations.

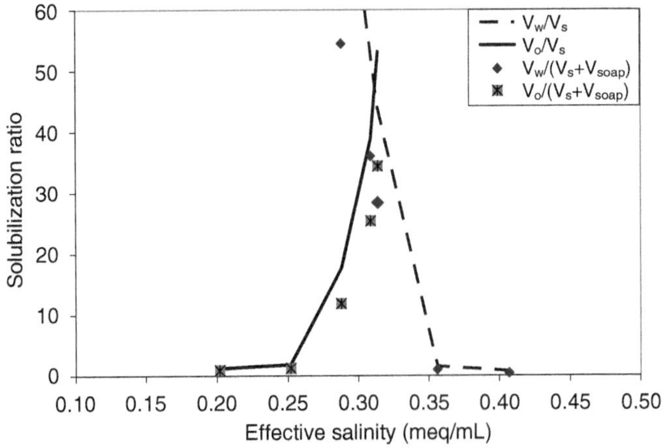

FIGURE 12.18 Water and oil solubilization ratios at different effective salinities.

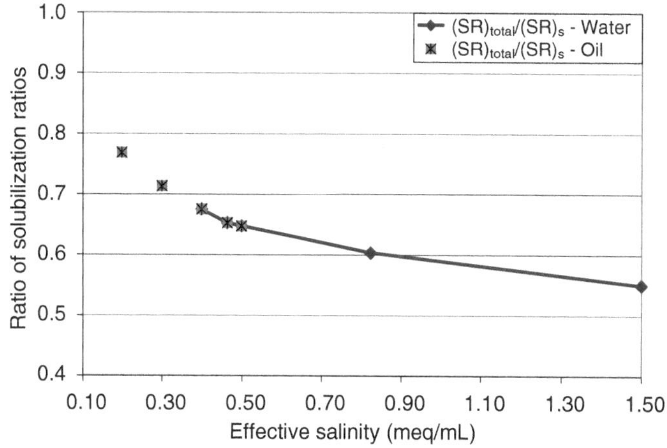

FIGURE 12.19 Ratios of water and oil solubilization ratios based on the two definitions.

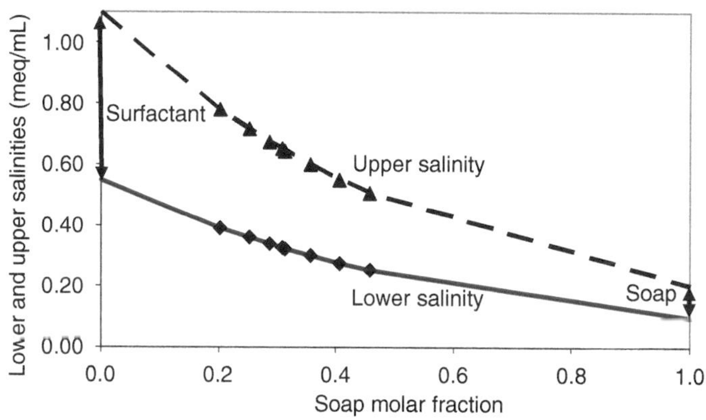

FIGURE 12.20 Lower- and upper-salinity limits of type III at different soap molar fractions.

the total volume of surfactant and soap is used to define the ratio, and $(SR)_s$ is the solubilization ratio when only the surfactant volume is used. This figure shows that the ratios of SR range from 0.55 to 0.77.

Effect of Soap on Salinity Boundaries

Figure 12.20 shows the lower and upper salinity limits of type III at different soap molar fractions from the simulation model. The salinity limits for the soap in this case are 0.1 and 0.2 meq/mL, and the limits for the surfactant are 0.55 and 1.1 meq/mL, respectively. As the soap fraction is increased by adding the alkali, the type III salinity range becomes narrower. This result is not consistent with the experimental observation in Figure 12.7, which shows that the low IFT range becomes wider as the alkali is added. Whether the logarithmic mixing

TABLE 12.4 Effect of Soap Salinity Limits of Type III

Soap		Soap+surfactant									
C_{sel}	C_{seu}	C_{se}	C_{sel}	C_{seu}	Type	A^-/HA	X_{soap}	R_{13s}	R_{23s}	R_{13t}	R_{23t}
0.1	**0.2**	**0.46**	**0.32**	**0.65**	**III**	**0.246**	**0.309**	**55.22**	**38.85**	**36.03**	**25.35**
0.55	1.1	0.46	0.55	1.10	I	0.248	0.303		2.08		1.37

rule could be used in this situation may need more research work. In Figure 12.20 the dot points are the values calculated by hand (not by simulation model) according to the logarithmic mixing rule. These dotted points fall on the lower and upper limit curves from the UTCHEM simulation model, confirming that the logarithmic mixing rule for optimum salinity (Eq. 12.3) is extended to the lower and upper salinity limits in UTCHEM.

In the base case, the lower and upper limits of type III for the generated soap are 0.1 and 0.2 meq/mL, respectively. The limits for the synthetic surfactants are 0.55 and 1.1 meq/mL, respectively. Generally, the optimum salinity and thus the two limits for soap are lower than those for a synthetic surfactant. Let us assume, however, the soap and surfactant have the same limits. The results are compared in Table 12.4. The base case is listed in bold. For this change, the fraction of petroleum acid converted to soap (A^-/HA) and the fraction of soap in the total surfactant, X_{soap}, have minor changes. However, the type III salinity limits for the mixture are quite different from those in the base case. Because of the changes in salinity limits, the new case is in type I microemulsion. The resulting solubility ratios are much lower than those in the base case. This comparison demonstrates that the salinity limits for soap are very sensitive. In the table, R_{i3s} and R_{i3t} (i = 1, 2) denote the solubilization ratios based on surfactant volume only and the total surfactant volume in the microemulsion phase, respectively. These ratios are significantly altered by assuming the soap and synthetic surfactant have the same limits.

Effect of Partition Coefficient and Dissociation Constant

In the base case, the fraction of petroleum acid converted to soap (A/HA) is only 0.246, and the soap molar fraction is 0.309 (see Table 12.4). These values are affected by the partition coefficient K_D between water and oil and the acid dissociation constant K_A. Now let us see how sensitive these two parameters are. The data in Table 12.5 show that K_D is insensitive, whereas K_A is very sensitive. As K_A is increased, more acid is converted to soap. Accordingly, the soap molar fraction in the total surfactant becomes higher. As X_{soap} is increased from the base case, the type III salinity limits are closer to those for the soap, which are lower. Thus, the mixture surfactant system becomes type II. As X_{soap}

TABLE 12.5 Effect of Partition Coefficient and Dissociation Constant

K_D	K_A	A^-/HA	X_{soap}	Type	R_{13s}	R_{23s}	R_{13t}	R_{23t}
1.00E-02	1.00E-12	0.244	0.308	III	55.32	38.38	36.17	25.09
1.00E-03	1.00E-12	0.246	0.309	III	55.23	38.81	36.05	25.33
1.00E-04	**1.00E-12**	**0.246**	**0.309**	**III**	**55.22**	**38.85**	**36.03**	**25.35**
1.00E-05	1.00E-12	0.246	0.309	III	55.22	38.85	36.03	25.35
1.00E-04	1.00E-10	0.953	0.627	II	6.74		2.25	
1.00E-04	1.00E-11	0.723	0.560	II	5.78		2.30	
1.00E-04	**1.00E-12**	**0.246**	**0.309**	**III**	**55.22**	**38.85**	**36.03**	**25.35**
1.00E-04	1.00E-13	0.035	0.058	I		1.13		1.05
1.00E-04	1.00E-14	0.004	0.006	I		0.93		0.92

TABLE 12.6 Effect of Water Saturation

C_{sel}	C_{seu}	S_w	A^-/HA	X_{soap}	Type	R_{13s}	R_{23s}	R_{13t}	R_{23t}
0.24	0.48	0.4	0.336	0.482	III	17.62	149.14	8.36	70.80
0.32	**0.65**	**0.5**	**0.246**	**0.309**	**III**	**55.22**	**38.86**	**36.03**	**25.35**
0.38	0.77	0.6	0.226	0.213	III	68.69	16.86	52.00	12.77
0.48	0.96	0.7	0.011	0.079	I		1.23		1.12
0.52	1.05	0.8	0.068	0.029	I		1.01		0.98

is decreased from the base case, the system becomes type I. As K_A becomes 10^{-10}, almost all the acid is converted to soap.

Effect of Water Saturation (Water/Oil Ratio)

As oil saturation is decreased (water saturation is increased), the acid content in the oil is decreased. Consequently, the soap molar fraction X_{soap} is decreased, as Table 12.6 shows. As X_{soap} is decreased, type III salinity limits are closer to those of surfactant. Thus, the limits are increased, and the optimum salinity is increased as well. The system is changed from type III to type I. This transition from type III to type I is exactly the salinity gradient we need. In practical alkaline-surfactant flooding, water saturation will be increased from the flood front to the upstream, and the microemulsion system will change from type III

TABLE 12.7 Effect of Acid Number (AN, mg KOH/g oil)

C_{sel}	C_{seu}	AN	A^-/HA	X_{soap}	Type	R_{13s}	R_{23s}	R_{13t}	R_{23t}
0.40	0.80	0.234	0.257	0.187	III	76.76	13.19	60.26	10.35
0.32	**0.65**	**0.467**	**0.246**	**0.309**	**III**	**55.22**	**38.86**	**36.03**	**25.35**
0.25	0.50	0.934	0.228	0.458	III	23.73	109.45	11.82	54.51

or type II to type I. When the water saturation is above 0.7, X_{soap} is very low, and when the system becomes Type I, the solubilization ratio becomes very low.

Effect of Acid Number

Table 12.7 shows that as the acid number is increased, X_{soap} is increased, and the system moves closer to a type II system. The fraction of acid converted to soap is decreased as the acid number is increased.

Chapter 13

Alkaline-Surfactant-Polymer Flooding

13.1 INTRODUCTION

Alkaline-surfactant-polymer flooding is the combination of alkaline flooding, surfactant flooding, and polymer flooding. Its displacement mechanisms are consequently the combination of those individual processes. The theories of each individual process and some of their combinations are addressed as follows: Chapter 5 (polymer flooding), Chapter 7 (surfactant flooding), Chapter 9 (surfactant-polymer flooding), Chapter 10 (alkaline flooding), Chapter 11 (alkaline-polymer flooding), and Chapter 12 (alkaline-surfactant flooding). This chapter focuses on the practical issues of the alkaline-surfactant-polymer (ASP) process. Some theories and mechanisms are briefly discussed and summarized here. Eleven pilot tests and field applications were carefully selected so that each example has unique issues for discussion. In addition, this chapter provides more detailed discussion about emulsion, which has become an important subject in chemical flooding.

13.2 SYNERGY OF ALKALI, SURFACTANT, AND POLYMER

Synergy is discussed in previous chapters. Here, we provide extra evidence to demonstrate the synergy in ASP. Core samples were waterflooded to residual oil saturation and then injected with polymer, alkaline-polymer (AP), or ASP. The results, in Table 13.1 (Ball and Surkalo, 1988), show that adding alkali further reduced residual oil saturation by 0.137, compared with polymer flooding. Through the further addition of only 0.1 wt.% surfactant, an additional 0.136 residual oil saturation was reduced. In these samples, ASP was the most efficient approach, demonstrating the synergy of alkali, surfactant, and polymer floods.

13.3 INTERACTIONS OF ASP FLUIDS AND THEIR COMPATIBILITY

This section discusses the effect of alcohol on AS compatibility, and the effects of alkali, surfactant, and polymer in ASP systems. Factors affecting phase separation, IFT, and wettability are discussed as well.

Modern Chemical Enhanced Oil Recovery. DOI: 10.1016/B978-1-85617-745-0.00013-9

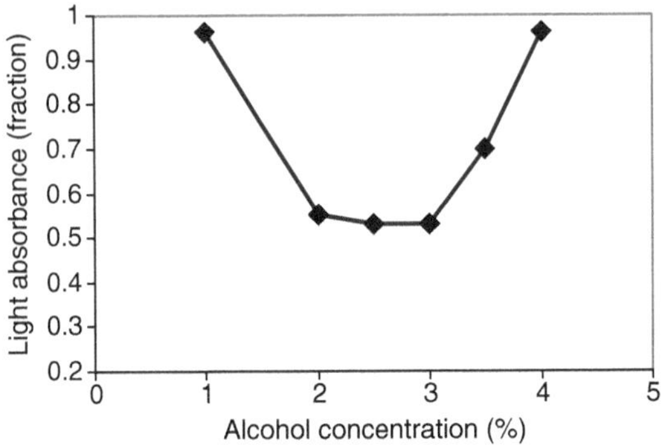

FIGURE 13.1 Alcohol effect on the compatibility of the AS system.

TABLE 13.1 Tertiary Oil Recovery of an Alberta System

Process	S_{or}	ΔS_{or}
Polymer	0.388	
Alkaline-polymer	0.251	0.137
Alkaline-surfactant (0.1 wt.%)-polymer	0.115	0.136

13.3.1 Alcohol Effect on AS Compatibility

An emulsion tends to have a cloudy appearance because phase interfaces scatter the light that passes through them. Light absorbance indicates the compatibility of an alkaline-surfactant system, with a low number for better compatibility. Figure 13.1 shows the alcohol concentration effect on the compatibility of an AS system (1.2%NaOH + 0.6% ORS-41; Kang, 2001). When alkaline concentration was increased from 1 to 2%, the system compatibility was improved, and when the concentration was increased from 2 to 3%, the compatibility was almost unchanged. When the concentration was further increased from 3%, the compatibility became worse. This figure shows a proper range of alcohol concentrations are needed to improve the compatibility. The system becomes less compatible when the alcohol concentration is either too low or too high.

13.3.2 Alkaline and Surfactant Effects in ASP Systems

An accepted principle is that the existence of alkali reduces surfactant adsorption. When alkali concentration is too high, however, due to increased ionic strength, it becomes easier for the opposite-charged ions to enter the adsorption layer from the diffusion layer. The static electric repulsion between the rock surface and surfactant becomes weaker. This makes it easier for surfactant to adsorb on the rock surface, thus resulting in increased adsorption. Chen and Chen (2002) observed that as anionic surfactant concentration was increased, the ASP system viscosity decreased. This result was probably due to the electric shield effect.

13.3.3 Polymer Effect in ASP Systems

Figure 13.2 shows the dynamic IFT for the two systems: (1) 0.2% OP (nonionic) + 0.2% PS (petroleum sulfonate) + 1.1% NaCl (without polymer), and (2) the same as (1) but with 0.1% 3530S polymer. From this figure, we can see that the IFTs for the two systems were almost the same. This figure demonstrates that there was not a strong interaction between the polymer and surfactants. However, polymer increases water viscosity to affect surfactant transport, so dynamic IFT was affected within a short time. Figure 13.2 shows that the dynamically stable IFT with addition of polymer was a little bit higher than that without polymer.

The other observations were reported elsewhere, however. Figure 13.3 shows polymer made the surfactant system emulsification better, and Figure 13.4 shows polymer slightly changed the value of electrophoretic mobility. The addition of polymer into an ASP system does not change IFT but shortens the phase separation time of emulsions. In these examples, when alkali concentration was below 1%, as the concentration was increased, the phase separation time decreased. When alkali concentration was above 1%, the phase separation time increased with the concentration. Thus, polymer apparently reduced the interaction between oil and alkali when alkali concentration was high.

The existence of polymer in an ASP system reduces surfactant adsorption. This result is due to the competition of adsorption sites between polymer and

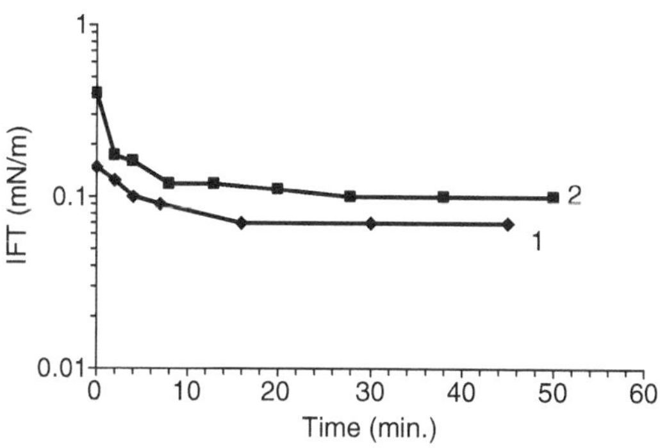

FIGURE 13.2 Effect of polymer on IFT. 1, 0.2% OP + 0.2% PS + 1.1% NaCl; and 2, 0.2% OP + 0.2% PS + 1.1% NaCl + 0.1% 3530S (polymer). *Source: Yu et al. (2002).*

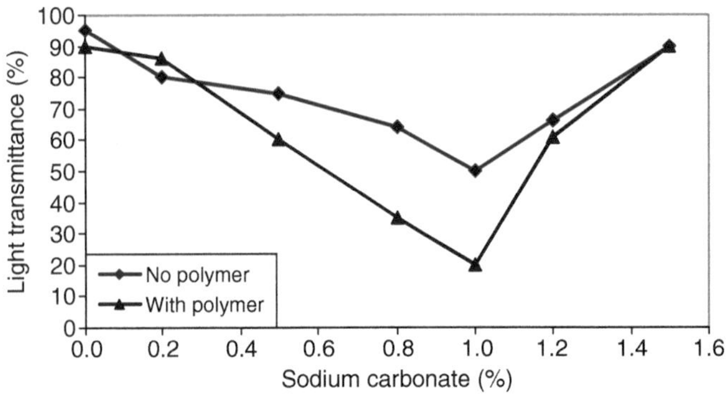

FIGURE 13.3 The effect of polymer (1% 3530S) effect on emulsion (surfactant, 0.2% OP + 0.2% PS). *Source: Yu et al. (2002).*

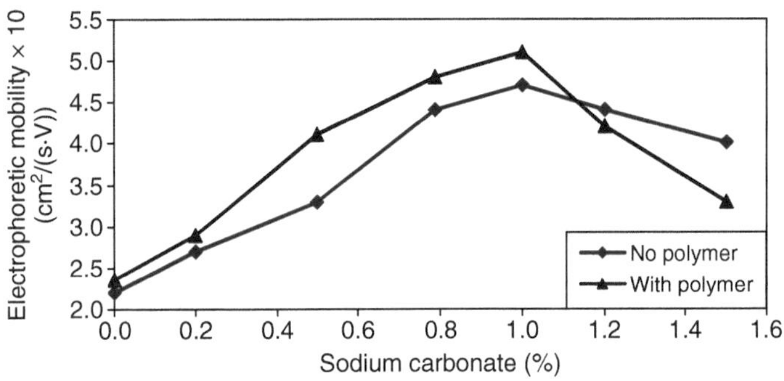

FIGURE 13.4 The effect of polymer (1% 3530S) on electrophoretic mobility (surfactants, 0.2% OP + 0.2% PS). *Source: Yu et al. (2002).*

surfactant. Large polymer molecules may also protect some electrically positive sites from occupation by anionic surfactants, thus reducing surfactant adsorption. Polymer does not significantly affect alkaline consumption, however.

13.3.4 Factors Affecting Phase Separation

The factors that affect phase separation discussed in this section include anion effect, divalent effect, alkaline effect, mixing effect of interstitial flow, and the synergy of mixed surfactants.

Anion Effect

Figure 13.5 shows that anions in solution significantly affected the surfactant-polymer (SP) phase separation for this example. When Na^+ concentration was less than 0.35 mol/L, sodium carbonate, sodium sulfate, or sodium phosphate

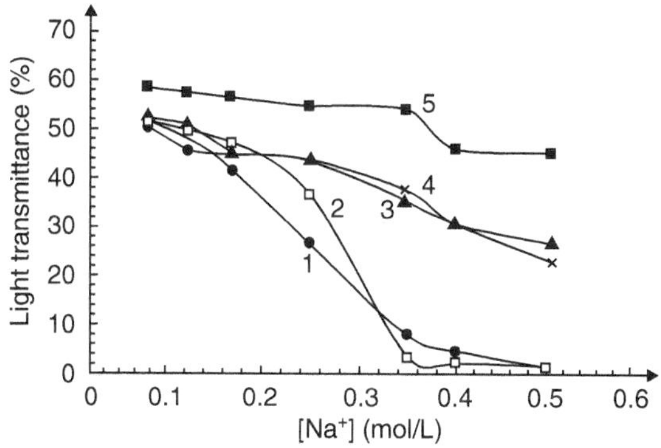

FIGURE 13.5 Anion effect on ASP system stability with Na^+ concentration being the same (800 ppm HPAM + 0.2% JH6A (surfactant)): 1, NaCl; 2, NaOH; 3, Na_2CO_3; 4, Na_2SO_4; and 5, Na_3PO_4. *Source: Yang et al. (2002b).*

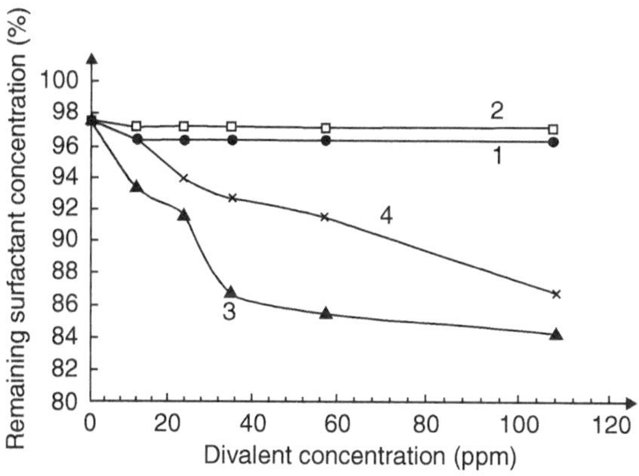

FIGURE 13.6 Surfactant concentration remaining in solution after 35 days at the reservoir temperature (1% Na_2CO_3 + 0.08% HPAM + 0.4% JH6A). 1, Ca^{2+} ASP; 2, Ca^{2+} AS; 3, Mg^{2+} ASP; and 4, Mg^{2+} AS. *Source: Yang et al. (2002b).*

did not significantly affect ASP compatibility. To maintain the compatibility, Na^+ concentration must be lower than 0.25 mol/L for sodium chloride or sodium hydroxide. When Na^+ concentration was higher than 0.25 mol/L, the ASP solution became cloudy and the light transmittance sharply decreased, indicating obvious phase separation. At an equal cation concentration, the order in which phase separation was affected was as follows: monovalent anions > divalent anions > trivalent anions.

Divalent (Ca^{2+} and Mg^{2+}) Effect

Figure 13.6 shows the surfactant concentration remaining in a solution at different divalent concentrations. From this figure, we can see that the Ca^{2+} effect on the AS system was less than that on the ASP system, but the difference was not much. The Mg^{2+} effect on the ASP system was obviously larger than that on the AS system. This result indicates that Mg^{2+} not only associates with the surfactant to yield precipitation in the AS system, but also it causes noncompatibility between surfactant and polymer in the ASP system, in addition to precipitation.

Mixing Effect Due to Interstitial Flow

The compatibility of ASP fluids is not consistent with ultralow IFT. Generally, if the surfactant is more soluble in water, and the ASP fluids are more compatible, having ultralow IFT would be more difficult. If the surfactant is more hydrophobic, and the ASP fluids are less compatible, it would be easier for the IFT to reach ultralow levels. If ASP fluids are less compatible, causing phase separation would be easier. Then phase separation would cause IFT to be increased. Fortunately, Yang et al. (2002b) observed that interstitial flow has a mixing effect so that phase separation can be reduced when ASP fluids flow in the core.

Effect of Alkali

As alkaline concentration is increased in an ASP system, HPAM hydrolysis is increased. Then the COO^- (negative charge) is increased. Thus, the electrostatic repulsion with anionic surfactant is increased, reducing the surfactant's solubility in water. When alkali concentration is further increased, IFT could become higher, which would reduce the system fluid's compatibility.

Synergism of Mixed Surfactants

Some mixtures of surfactants behave better than their individual surfactants alone. This is called positive synergism of mixed surfactants. Although the synergism has been known and applied, quantitative description of the interaction of the mixed surfactants was not developed until the 1980s. According to a regularization theory, we can predict whether there is synergism between two surfactants and at what portions two surfactants can be mixed so that their synergism is optimum. The regularization theory can be applied to only two pure surfactants so far (Yang et al., 2002b).

13.3.5 Factors Affecting IFT

The factors affecting IFT discussed in this section include polymer effect and oil composition effect.

Polymer Effect on IFT

There is no consensus on how polymer affects the IFT between alkaline solution and crude oil. It is commonly believed that the effect is not significant. Some believe that polymer can protect surfactant from being associated with divalents to form complexes, thus making the interfacial tension lower. Others believe that polymer can adsorb in the interface with surfactant to form an adsorption layer of mixtures, and this mixed layer has higher electrical density so that the interfacial tension is further reduced. Kang and Dan (1998) observed that for a given nonionic surfactant ASP system, the HPAM effect on IFT was not significant, and xanthan gum had a positive effect on reducing IFT. For a given anionic surfactant ASP system, the HPAM effect was significant in low concentrations, but not in high concentrations.

Oil Composition Effect on IFT

One important condition for ultralow IFT to occur is surfactant molecules' aggregation at the water/oil interface. Daqing oil has a low acid numbers. Low concentration alkali and oil could not have ultralow IFT, as shown in Figure 13.7. However, as alkali (NaOH) concentration was increased, the alkaline/oil IFT decreased. When NaOH was increased to 0.5%, IFT was about 2 mN/m, indicating that the alkali and oil could generate surface active materials. The surface active molecules aggregated at the alkaline/oil interface. When polymer was added, the alkaline/oil IFT further decreased, indicating polymer could increase interface activity. When the alkaline concentration was above 1%, however, the polymer effect was negligible (Kang et al., 1999).

Adding surfactant ORS-41 in the alkaline/oil system reduced the IFT to an ultralow level. Figure 13.8 shows the ASP/oil IFT at different distillate fractions; here, the IFT increased with the heavy fractions among X1, X2, and X3. Because the oil fraction had lighter components, the reaction with the hydrophobic composition became stronger, and the IFT was lower. However, for the heavy distillate fraction obtained above 360°C, the IFT was lower than those for X2 and X3. This result indicates that the dynamic IFT between ASP/oil is determined by the surface active materials generated from the reaction between the heavy components of crude oil and alkali. When the boiling temperature

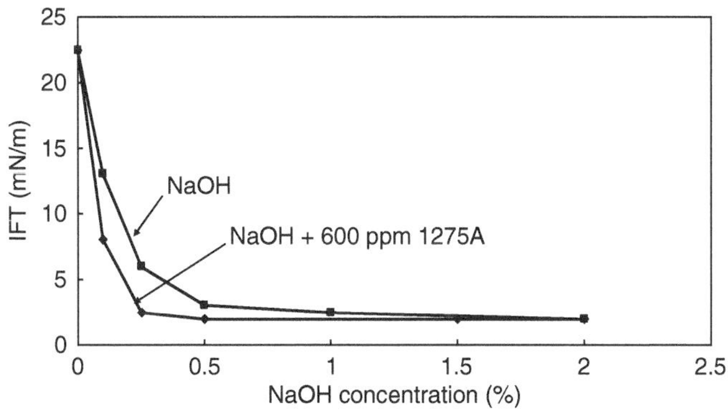

FIGURE 13.7 Alkaline/Daqing oil IFT at 45°C. *Source: Kang et al. (1999).*

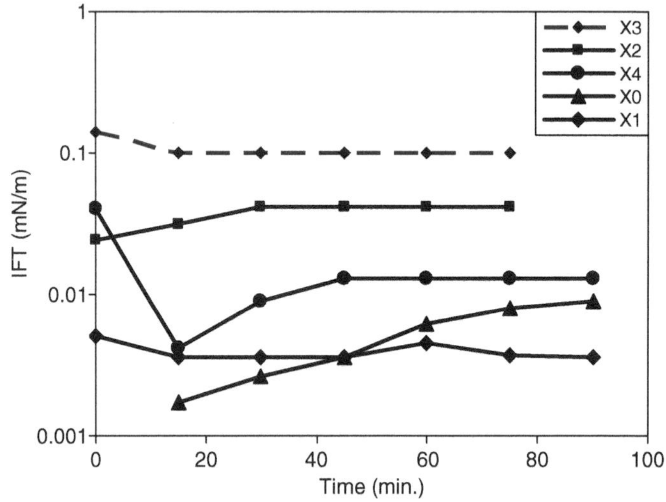

FIGURE 13.8 Dynamic IFT of ASP oil with different distillate fractions. ASP: 0.6% ORS + 1.2% NaOH + 1200 ppm HPAM. In the figure, X0, = crude oil; X1, < 200°C; X2, 200 to 300°C; X3, 300 to 360°C; and X4, > 360°C. *Source: Kang et al. (1999).*

was above 360°C, the oil had more polar components such as asphaltene. Its reaction with surfactant and alkali was increased, so the interfacial activity increased. Because oil had heavier composition (higher molecular weight), molecular polarity and interaction increased, and the formed emulsion became more stable. Buijse et al. (2010) also pointed out that for a heavy oil, even it has a near-zero acid number, it contains a significant fraction of heavier components, such as asphaltenes and resins, which are surface active and can interfere with the surfactant in the oil/brine interfacial layer.

13.3.6 Factors Affecting Wettability

The factors affecting wettability discussed in this section include polymer effect and ASP effect.

Polymer Effect on Wettability

A well-known fact is that alkaline solutions and some surfactant solutions can change wettability toward more water-wet. However, the polymer effect on wettability is not much discussed in the literature. Figure 13.9 shows the changes in contact angles when silica rocks were immersed in an HPAM solution for 24 hours at 60°C. The polymer solution was 1000 ppm 3530S (12.9% hydrolysis, 15.71 million Dalton molecular weight). We can see that the contact angles became larger after polymer contacted the rock, especially for initially water-wet rocks. The reason for this result is that the polymer adsorbs on the rock and reduces the rock's polarity.

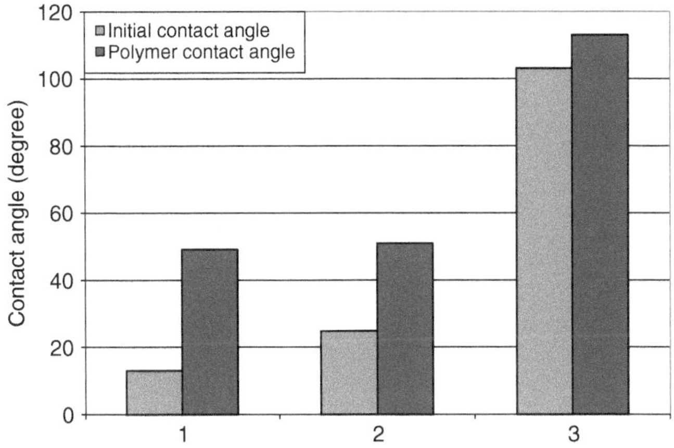

FIGURE 13.9 Polymer effect on silica wettability. *Source: Data from Bi et al. (1997).*

ASP Effect on Wettability

A number of papers (e.g., Standnes and Austad, 2000; Hirasaki and Zhang, 2004; Adibhatla and Mohanty, 2008) report that alkaline and/or surfactant solutions can change rock wettability and favorably change from more oil-wet to water-wet. However, there are also reports that ASP solutions increase the contact angle—for example, an ASP solution of 3530S polymer + ADF-4 surfactant + Na_2CO_3 alkali (Yang et al., 2002b).

13.4 RELATIVE PERMEABILITIES IN ASP

In ASP flooding, alkaline, surfactant, and polymer have different effects on relative permeabilities. Table 13.2 shows our attempt to summarize these effects compared with waterflood. From Table 13.2, we can see that the effect of alkaline flood in terms of emulsification is similar to the polymer effect, whereas its effect in terms of IFT is similar to the surfactant effect. Less rigorously, we may say that only polymer reduces k_{rw}, and only surfactant reduces IFT. In ASP flooding, the viscosity of the aqueous phase that contains the polymer is multiplied by the polymer permeability reduction factor in polymer flooding and the residual permeability reduction factor in postpolymer waterflooding to consider the polymer-reduced k_{rw} effect. Then we can use the k_r curves (water, oil, and microemulsion) from surfactant flooding or alkaline-surfactant flooding experiments without polymer.

Ye and Peng (1995) measured ASP solution/oil relative permeabilities based on the preceding principle. They first conducted a core flood test using an ASP solution and calculated the Darcy viscosity for the solution, which included the polymer permeability reduction factor. Then they conducted ASP/

TABLE 13.2 Effects of Alkaline, Surfactant, and Polymer on k_r

Parameter	Polymer	Alkaline (emulsification)	Surfactant or Alkaline (IFT)
S_{wi}	↑	↑	↓
S_{or}	Slightly ↓	↓	↓
k_{rw}	↓	↓	↑
k_{ro}	Little changed	Slightly ↑	↑

oil relative permeability tests. The Darcy viscosity was used for the ASP solution when calculating k_r curves. They also conducted AS solution/oil relative permeability tests using the cores preflushed with polymer solution. When they calculated AS/oil k_r curves, they divided the k_{rw} by the residual permeability reduction factor. They found that the AS/oil k_r curves matched the ASP/oil k_r curves. Their work demonstrates that for ASP/oil k_r curves, polymer mainly reduces k_{rw} by the permeability reduction factor, and alkaline and surfactant mainly affect k_r curves through the IFT effect. In other tests, Delshad et al. (1987) measured two- and three-phase micellar fluids with and without polymer.

13.5 EMULSIONS IN ASP FLOODING

Emulsion is a system in which the droplets of one liquid are distributed immiscibly in another continuous liquid phase. An emulsion is thermodynamically unstable and thus does not form spontaneously. Energy input through shaking, stirring, homogenizing, or spray processes are needed to form an emulsion. Over time, an emulsion tends to revert to the stable state of the phases comprising the emulsion. However, if an emulsifier (also known as an emulgent) exists or is added in the system, the emulsion can be stable, or the stability is improved.

Stable emulsions can be formed in surfactant, alkaline, and even in water injection. In water injection, stable emulsions can be formed because crude oil has natural emulsifiers such as asphaltene. In surfactant injection, surfactant reduces the water/oil interfacial tension so that stable emulsions can be formed. In alkaline flooding, stable emulsions can be formed because alkali reacts with crude oil to generate in situ surfactant (soap). Although polymer helps to stabilize emulsions, it cannot form emulsions with oils. Here, being stable or unstable is relative. The stability is controlled by creaming, flocculation, and coalescence processes. In the creaming process, a concentration gradient of dispersed droplets is formed under the influence of buoyancy. In the flocculation process, the distances between liquid droplets are significantly reduced owing to their attraction force. In coalescence, smaller droplets contact one another and larger droplets form.

13.5.1 Types of Emulsions

There exist mixed types of emulsions in ASP systems. In phase behavior pipette tests, the emulsions we discuss are microemulsions. However, even in pipettes, other types of emulsions can be present in the early stages. Because of their unstable nature, macroemulsions (or most of them) are phase-separated after a long time of equilibrium. The focus in pipette tests is the microemulsion at equilibrium. According to their droplet sizes, emulsions have macroemulsion, miniemulsion, microemulsion, and micelle. Table 13.3 lists the typical sizes and the shapes of the aggregates present in a common alkaline-surfactant/oil system (Kang, 2001). This section mainly discusses macroemulsions, or simply emulsions.

According to their structures, there are four types of emulsions: W/O, O/W, W/O/W, and O/W/O. Sometimes, W/O/W and/or O/W/O are called multiple types. In waterflooding, most of the produced emulsions are the W/O type, with some O/W. In chemical flooding, the type of emulsion depends on the types of chemicals used and their concentrations, water, and oil properties. It also depends on the water/oil ratio (WOR). Generally, W/O emulsions are formed at low WOR. As the water cut in emulsions increases, the W/O type will be transferred to the O/W type. In other words, when one phase volume is much larger, this phase will be continuous. Therefore, the water/oil ratio can change emulsion type.

Other special rules are (1) when surfactants are easily soluble in one phase, this is a continuous phase; and (2) surfactants made from monovalent metal cations tend to produce O/W emulsion, whereas those made from polyvalent metal cations produce W/O. This is called the oriented wedge theory (Bryan and Kantzas, 2007). Another related theory is the phase volume theory, proposed by Wilhelm Ostwald (winner of the Nobel Prize in chemistry, 1909):

TABLE 13.3 Typical Sizes and Shapes of the Aggregates in an AS System

	Size, μm	Shape	Stability
Micelle	< 0.01	Spherical, cylindrical, dishlike, lamellar	Stable
Microemulsion	0.01–0.1	Spherical	Stable
Miniemulsion	0.1–0.5	Spherical	Unstable
Macroemulsion	0.5–50	Spherical	Unstable

when a bunch of solid spheres are put together, the volume fraction occupied by the spheres is 0.74, and the void space is 0.26. In practice, however, the dispersed phase volume fraction could be higher than 0.74. The reasons are that (1) the liquid droplet sizes are not uniformly the same, so the volume fraction could be higher than 0.74 when they are closely packed; and (2) when the dispersed phase volume fraction is very high, the liquid droplets can deform and have different shapes and sizes. They can be separated by a thin layer of dispersing phase. Other theories on the types of emulsions are the Bancroft rules and coalescence velocity theory (Kang, 2001).

The emulsions formed by water/crude oil or alkali/crude oil are generally the W/O type if the water cut is less than 50% and the O/W type if the water cut is higher than 50%. W/O type emulsions have a small-medium droplet size and a narrow range of size distributions. In contrast, O/W emulsions have a large-medium droplet size and a wide range of size distributions (Kang, 2001), as shown in Figure 13.10. From this figure, we can see that in the W/O type of emulsion, the dispersed water droplets are very small; in the O/W type of emulsion, the dispersed oil droplets are relatively large; and in the multiple type, there are different sizes of dispersed droplets. The W/O type of emulsion is usually seen when low water-cut fluid is produced; W/O, O/W, or even multiple types can be seen when the water cut is higher.

The water cut at which a W/O emulsion is transferred to an O/W emulsion is called the type transferring point or critical water cut. Table 13.4 lists the critical water cuts for several emulsions at which the emulsions were transferred from W/O to O/W. From Table 13.4, we can see that adding surfactant and polymer reduced their critical water cuts below 50%, whereas adding 1.2% alkali did not reduce the water/oil critical water cut. Table 13.4 indicates that under ASP flood conditions (high water saturation), most likely, O/W emulsion will be formed.

Table 13.5 presents the types of emulsions produced at the effluent end in laboratory core floods at different injection pore volumes. This table shows that initially W/O emulsions were produced, followed by W/O and O/W emulsions,

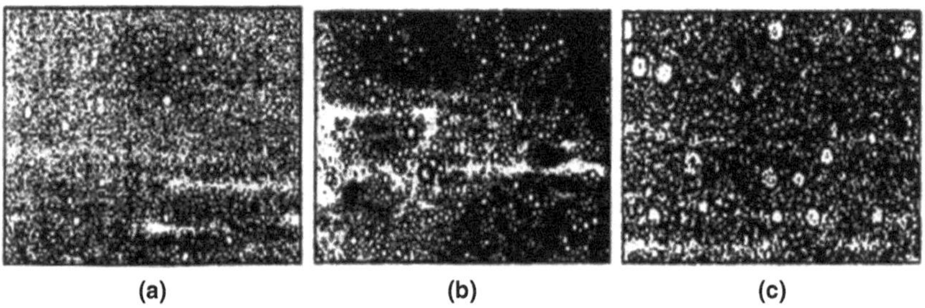

FIGURE 13.10 Three types of produced emulsions: (a) W/O emulsion, (b) O/W emulsion, and (c) multiple emulsion. *Source: Li (2007).*

TABLE 13.4 Critical Water Cuts from W/O Type to O/W Type

Emulsion System	Critical Water Cut
Water, low concentration ASP	50
1.2% NaOH/ or Na_2CO_3/crude oil	50
0.3% surfactant (ORS-41)/crude oil	10
0.12% HPAM/crude oil	20

Source: Kang (2001).

TABLE 13.5 Emulsion Types and Chemical Concentrations at the Effluent

Injection PV	HPAM, mg/L	NaOH, %	ORS-41, %	Emulsion Type
0–0.5	0	0	0	W/O
0.5–0.6	72–970	0.21	0.032	O/W
0.6–0.82	970–1010	0.37	0.086	W/O, O/W
0.82–1	1010–1164	0.37–0.4	0.086–0.12	O/W

Source: Kang (2001).

then O/W emulsions. In the beginning, an oil bank was produced and water cut was low. Under this condition, it was easy for W/O emulsions to form. After the oil bank was produced, water cut started to increase. Gradually, O/W emulsions formed. The final emulsions were of O/W type because of very high water cut. This order for the types of emulsions to appear is similar to what is observed in field tests, except that multiple types of W/O/W or O/W/O were not observed in the laboratory, probably due to the simple core flood conditions compared with field conditions.

Rock wettability has a strong influence over which type of emulsion will form in the reservoir, according to Huang and Yu (2002). In an oil-wet porous medium, forming W/O emulsion is easy. In a continuous heavy oil system, due to high oil viscosity, water droplets collide less frequently than oil droplets in a less viscous water phase. For this reason, W/O is much more common than O/W emulsion in heavy oil systems.

13.5.2 Properties of Emulsions

This section discusses the droplet sizes, viscosity, and stability of emulsions.

Emulsion Droplet Sizes and Their Distribution

The size of emulsion plays an important role in stability. Smaller size means an emulsion is more stable. Emulsion sizes can be from one tenth of a micrometer to tens of micrometers. Figure 13.11 shows the dispersed droplet size (diameter) distribution of produced W/O and O/W emulsions from a well in the ZX block of the Daqing oilfield. The average diameter of dispersed water droplets in the W/O emulsion was 5.9 μm, and the average diameter of dispersed oil droplets in the O/W emulsion was 8 μm. Generally, the average diameter of produced emulsions is less than 10 μm (Li, 2007). In Daqing ASP flooding, the sizes of emulsion particles were on the order of 0.1–10's μm.

Figure 13.12 shows another example of produced emulsion size distribution. The mean size of this example was less than 10 μm. The produced emulsions in most Daqing ASP systems were quite stable (Kang, 2001). If the mean emulsion size becomes larger, the sizes are more widely distributed, and the size distribution moves to the right in the figure with time, the emulsion system is not stable. If the mean emulsion size is smaller, the distribution is dominated by a narrow range of sizes, and the size distribution does not change much with time, the emulsion system is stable.

Viscosity of Produced Emulsions

Emulsion viscosity is higher than either dispersed phase viscosity or continuous phase viscosity. As the dispersed volume fraction is increased, the emulsion viscosity is increased. For a W/O emulsion, the viscosity could be increased from 10s mPa·s to 100s mPa·s. For example, in the Daqing ZX block, the viscosity of a produced emulsion from an ASP well was 195 cP. After dehydration, the oil viscosity was 48 cP, which is still higher than the oil viscosity of 37 cP from a waterflood well.

Emulsions in ASP flooding belong to non-Newtonian fluids. When shear rates are applied, the emulsion viscosity with a higher dispersed phase volume

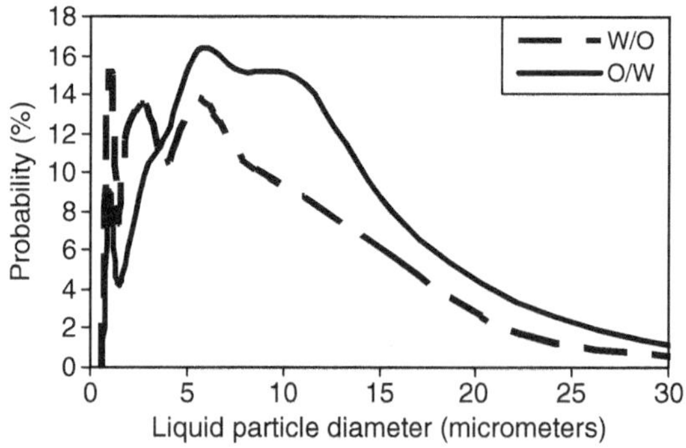

FIGURE 13.11 Dispersed droplet size distribution of produced emulsions. *Source: Data from Li (2007).*

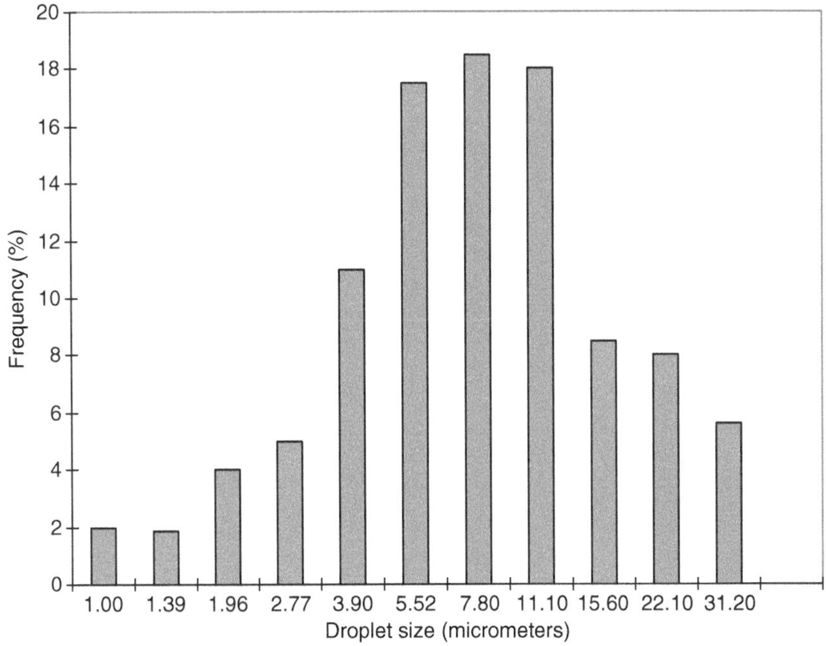

FIGURE 13.12 Example of histogram of size distribution of produced emulsions in ASP flood. *Source: Kang (2001).*

fraction will be decreased more than that with a lower dispersed phase volume fraction (see Figure 13.13). When the water cut was as high as 70%, some water became free water; thus, the viscosity significantly decreased.

Stability of Emulsions

Emulsion stability may be described by the half-life of the emulsion following the concept used for foam stability (Sheng et al., 1997). The half-life corresponds to the time at which the emulsion volume has decayed to half its initial volume. Figure 13.14 shows the half-life times versus droplet volumes for a Daqing-produced fluid. Here, the emulsion was stable for smaller dispersed droplets. The surfactant B-100 with 0.2% was used.

Figure 13.15 shows the half-life times versus surfactant (B-100) concentrations of W/O and O/W emulsions without the presence of alkali (Li, 2007). Here, the W/O emulsion was more stable than the O/W emulsion, and both of the emulsions were more stable at higher surfactant concentrations. There was no alkali in the solution.

Figure 13.16 shows the half-life times versus surfactant (B-100) concentrations of W/O and O/W emulsions in the presence of 0.5% Na_2CO_3. This figure shows that the W/O emulsion was still more stable than the O/W emulsion, and

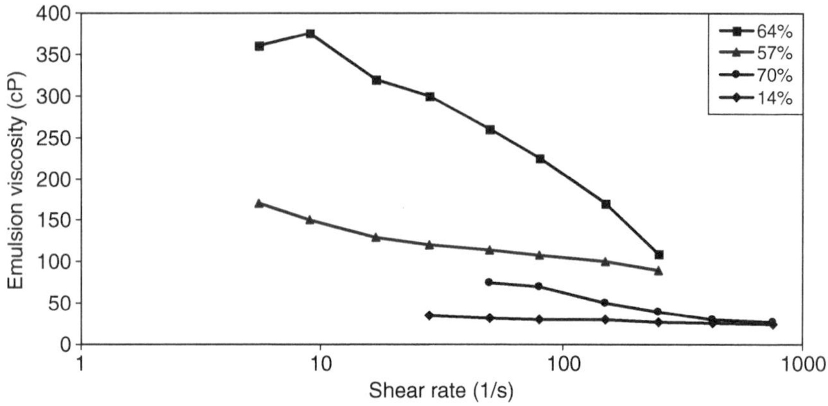

FIGURE 13.13 Example of emulsion viscosity versus shear rate. *Source: Kang (2001).*

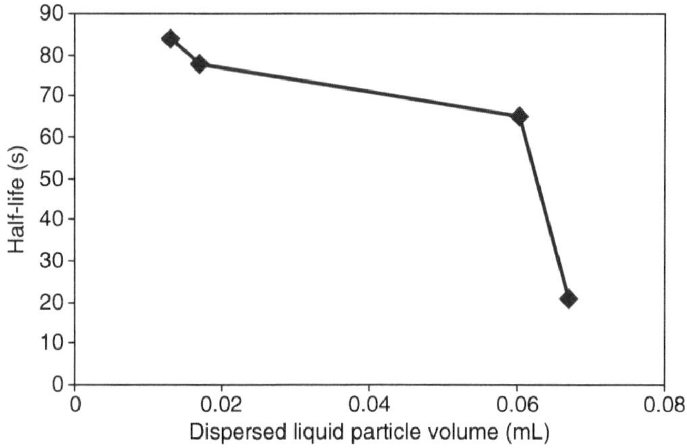

FIGURE 13.14 Emulsion half-life versus dispersed droplet volume. *Source: Data from Li (2007).*

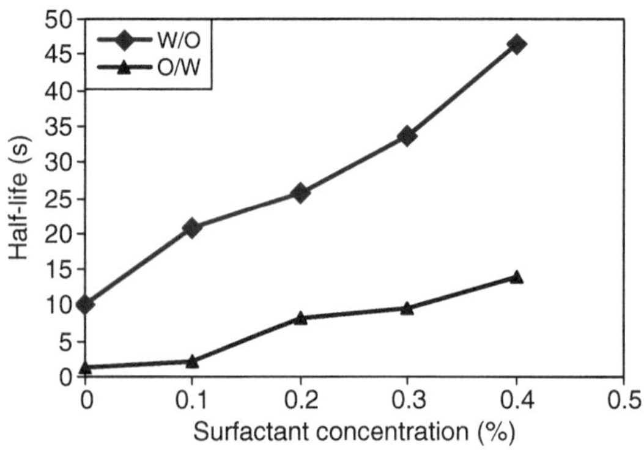

FIGURE 13.15 Emulsion half-life versus surfactant concentration without alkali. *Source: Data from Li (2007).*

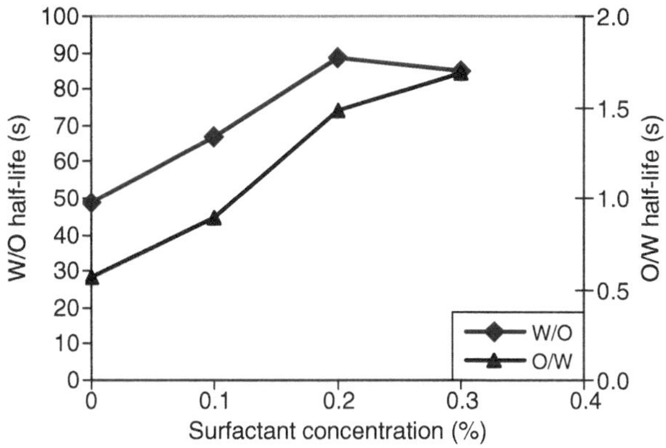

FIGURE 13.16 Emulsion half-life versus surfactant concentration with 0.5% Na_2CO_3. *Source: Data from Li (2007).*

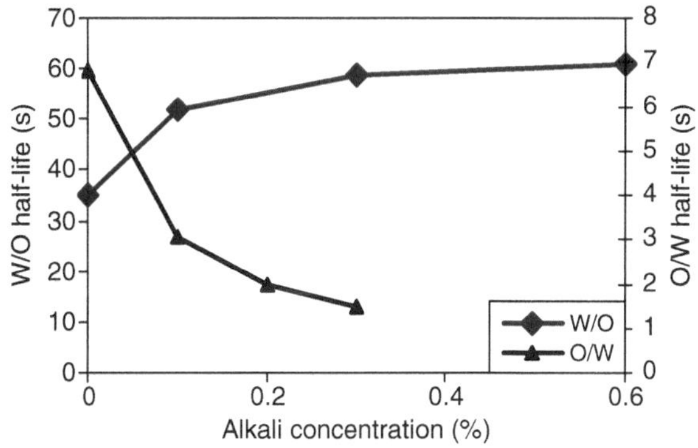

FIGURE 13.17 Emulsion half-life versus alkali concentration. *Source: Data from Li (2007).*

both of the emulsions were more stable at higher surfactant concentrations. However, the half-life times of the O/W emulsion were much shorter than those of the W/O emulsion in the presence of Na_2CO_3. Note that the second vertical axis with a much smaller scale is used to present the half-life of the O/W emulsion.

Figure 13.17 shows the half-life times versus alkali concentrations of W/O and O/W emulsions. Here again, the W/O emulsion was more stable than the O/W emulsion. The W/O emulsion was more stable at high alkali concentrations, whereas the O/W emulsion was less stable. The surfactant in this case was 0.2% B-100.

Figure 13.18 shows the half-life times versus polymer concentrations of W/O and O/W emulsions when 0.6% Na_2CO_3 and 0.2% B-100 surfactant were used. Here, the W/O emulsion was more stable than the O/W emulsion, and both of the emulsions were more stable at higher polymer concentrations.

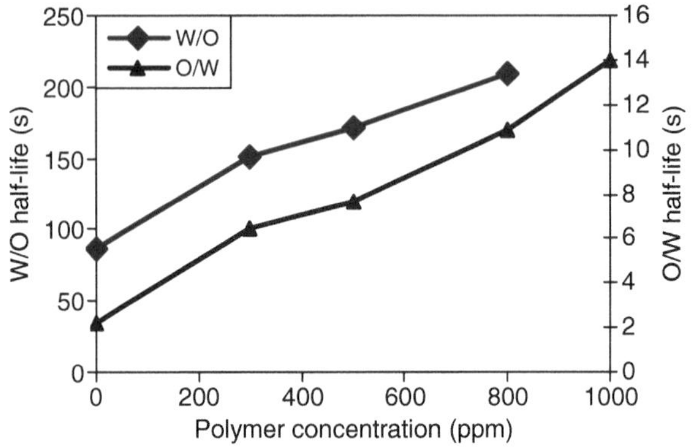

FIGURE 13.18 Emulsion half-life versus polymer concentration. *Source: Data from Li (2007).*

Polymer increases water phase viscosity, which helps to stabilize the O/W emulsion. In this case, the polymer also helped to stabilize the W/O emulsion, indicating that polymer affects the interfaces (films) between the external oil phase and water droplets.

The preceding data show that W/O emulsion was more stable than O/W emulsion in these tests. Demulsification is closely related to emulsion stability. Therefore, the way demulsification is affected by other factors can be partly inferred from the preceding discussions. For a detailed discussion of demulsification, see Li (2007). More factors that affect emulsion stability are discussed next.

13.5.3 Factors Affecting Emulsion Stability

Emulsion stability is affected by temperature, continuous phase viscosity, droplet sizes and their distribution, interfacial tension (IFT), and interfacial film properties. Some of these effects were discussed in the preceding section. This section discusses the effects of viscosity, polymer, IFT, and interfacial film.

Effect of Oil Composition

In W/O emulsions, stability depends on oil film strength. As oil molecular weight becomes larger, the molecular interaction is larger; then the oil film strength is increased. Therefore, emulsion becomes more stable if the oil has more heavy components. High oil viscosity is also beneficial to W/O emulsion stability, similar to foamy oil (gas/oil) stability (Sheng et al., 1997).

IFT Effect

The energy required to create emulsions is

$$E = \Delta A \cdot \sigma, \tag{13.1}$$

where ΔA is the increased interface area, and σ is the IFT. The emulsion will be more stable if the energy is lower. According to this equation, if the IFT is reduced, the energy is lower and the emulsion will be more stable. For example, consider a case in which the kerosene/water IFT is high. If a small amount of surfactant is added to reduce the IFT, the formed emulsion will be stable. However, some emulsions are unstable even though they have very low IFT. Some emulsions with polymeric film are stable even though they have high IFT. Therefore, low IFT is a favorable factor but not a determining factor for emulsion stability (Kang, 2001).

Effect of Interfacial Film

Emulsion stability is determined by the strength of the interfacial film and the way the adsorbed molecules in it are packed. If the adsorbed molecules in the film are closely packed, and it has some strength and viscoelasticity, it is difficult for the emulsified liquid droplets to break the film. In other words, coalescence is difficult. The emulsion is therefore stable. The molecular structure and the properties of the emulsifiers in the film affect the film's properties. The molecules in the film are more closely packed if the emulsifier has straight chains rather than branched chains. The film strength is increased if mixed emulsifiers are used rather than a single one. The reasons are that (1) the molecules in the film are closely packed, (2) mixed liquid crystals are formed between droplets, and (3) molecular complexes are formed in the interface by emulsifier compositions. For example, an oil-soluble surfactant mixed with a water-soluble surfactant works very well to stabilize emulsions (Kang, 2001).

If a proper emulsifier is selected, the interfacial film can be tolerant to local mechanical compression and has some viscoelastic behavior. When the film is damaged, the viscoelasticity can heal the film. The film's viscoelasticity behavior therefore plays an important role in stabilizing emulsions.

Some solid particles can also stabilize emulsions. These solid particles, which have wetting angles with oil and water, are much smaller than liquid droplets so that they can be adsorbed on the interface. When a proper wetting angle is formed, the emulsion can be very stable (Kang, 2001).

Effect of Polymer

Because emulsion stability is affected by continuous phase viscosity, polymer plays an important role in stabilizing emulsions. The effect of polymer in the W/O type of emulsions is different from that in the O/W type. Figure 13.19 shows the polymer (HPAM) effect on a W/O emulsion stability that is represented by the percent of water separated from the emulsion at different times. In this case, the oil was kerosene. As the polymer concentration was increased, the water/oil emulsion stability increased (less percentage of water separated). W/O emulsion stability mainly depends on the strength of the oil film between water droplets. The oil film strength is determined by

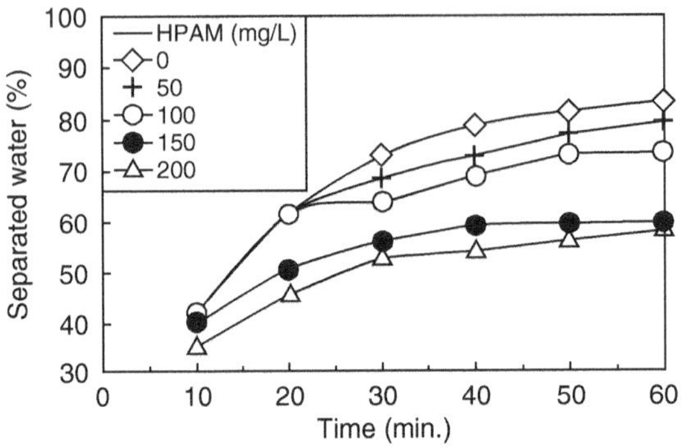

FIGURE 13.19 Polymer effect on W/O emulsion stability. *Source: Kang (2001).*

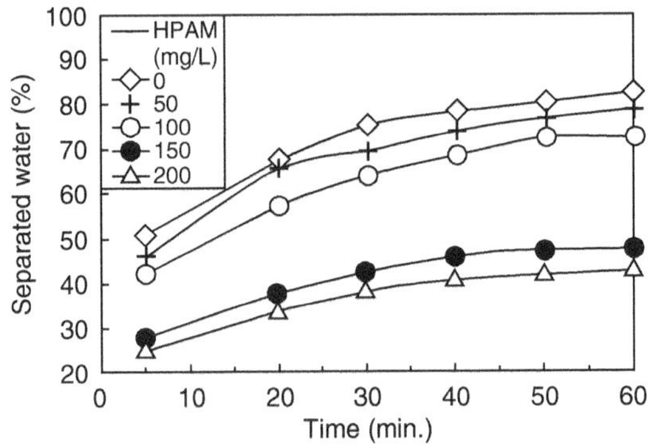

FIGURE 13.20 Polymer effect on O/W emulsion stability. *Source: Kang (2001).*

the oil phase properties. Because the polymer does not exist in the oil phase, oil phase properties such as viscosity are barely changed. Also, HPAM has a minor effect on the water/oil interfacial tension. Therefore, the polymer effect on W/O emulsion stability is not very significant (not sensitive to polymer concentration).

Figure 13.20 shows the polymer (HPAM) effect on O/W emulsion stability. In low polymer concentrations, the stability was not very sensitive to the concentration. When the concentration was above 150 mg/L, the stability was significantly improved. O/W emulsion stability is controlled by the strength of the water film between oil droplets. The existence of polymer in water significantly increases the water film's strength and water phase viscosity. Therefore, HPAM has a significant effect on O/W emulsion stability. It has also been observed that emulsion stability increased with polymer hydrolysis (Kang, 2001).

13.5.4 Effect of Emulsion in Oil Recovery

Chapter 10 on alkaline flooding lists emulsification as one important mechanism in oil recovery. Experiments showed that if the color of produced fluid was dark brown, and the water color was dark yellow, the oil was emulsified. In these experiments, the oil recovery was in the range of 18 to 22%. If water and oil came out of the core alternately, and the water was clear, the oil was not emulsified. In these cases, the oil recovery was in the range of 14 to 16% (Cheng et al., 2001). In other words, emulsification increased the oil recovery factor by about 5%. Many wells in Daqing ASP applications showed that if the produced fluids were more emulsified, the decrease in water cut would be higher.

As shown in the ASP examples later in this chapter, emulsions increase injection pressure and decrease water injection rate and liquid production rate. However, the advantages of emulsions appear to be greater than the disadvantages (Cheng et al., 2001).

13.6 DISPLACEMENT MECHANISMS

In ASP flooding, when surfactant mixes with alkali, crude oil, and formation water, emulsification can occur. The important mechanisms are the alkali emulsification and soap generation owing to its reaction with the crude oil. Surfactant makes emulsions stable through reduced IFT, increases interface (film) strength, and generates charge at the interface. More importantly, added surfactant makes the low IFT salinity range wider because of the synergism with in situ generated soap. Polymer increases water viscosity. In turn, higher external viscosity can reduce the diffusion of droplets, resulting in less probability of coalescence. Thus, emulsion stability is improved. When the number of dispersed droplets increases, emulsion viscosity is increased and stability is improved.

Displacement mechanisms in ASP may be summarized as follows:

- Increased capillary number effect to reduce residual oil saturation because of low to ultralow IFT.
- Improved macroscopic sweep efficiency because of the viscous polymer drive.
- Improved microscopic sweep efficiency and displacement efficiency as a result of polymer viscoelastic property. Oil in the dead ends is pulled out, and the oil films on the pore walls are "peeled" off owing to the high-velocity gradient.
- Emulsification, entrainment, and entrapment of oil droplets because of surfactant and alkaline effects.
- Improved sweep efficiency by emulsions.

13.7 DESIGN OPTIMIZATION OF ASP INJECTION SCHEMES

This section presents two examples to illustrate the effect of chemical concentration gradients on oil recovery. In these examples, the oil had an acid number of 0.3 mg KOH/g oil, and the cores were sandstone with a permeability of 1100 to 1300 md. The ASP formula was 1.2% Na_2CO_3 + 0.2% KPS + 1200 ppm HPAM.

13.7.1 Effect of Alkali and Surfactant Concentration Gradients

The following injection schemes were tested:

- Scheme 1: ASP 0.6 PV
- Scheme 2: alkali concentration gradient: (1.4 → 1.2 → 1.0) %
- Scheme 3: alkali concentration gradient: (1.0 → 1.2 → 1.4) %
- Scheme 4: surfactant concentration gradient: (0.3 → 0.2 → 0.1) %
- Scheme 5: surfactant concentration gradient: (0.1 → 0.2 → 0.3) %

For all these cases, the total amount of each chemical was the same. The core flood results are shown in Figure 13.21. We can see that the incremental oil recovery factors over waterflooding in Schemes 2 and 4 were obviously higher than that in Scheme 1. The alkali and surfactant concentration gradients from high to low can overcome the negative effects at the displacement front caused by dilution, alkali consumption, and surfactant adsorption.

13.7.2 Reduced Surfactant Injection

Because alkali can contact crude oil to reduce IFT, we can inject a preflush slug of AP before an ASP slug and reduce the amount of surfactant in the ASP slug. For this reason, the three schemes listed next were tested.

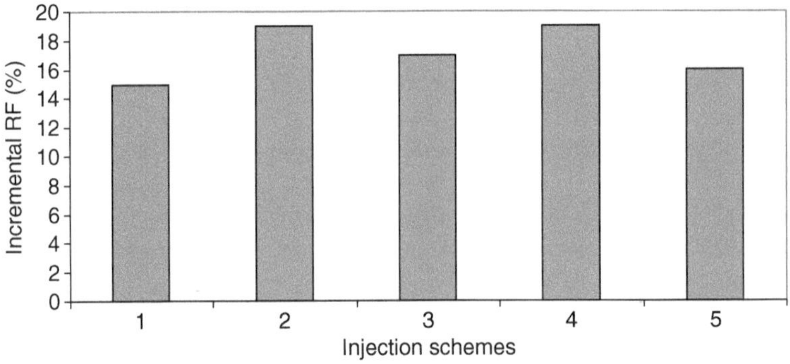

FIGURE 13.21 Effect of alkali and surfactant concentration gradients on oil recovery. *Source: Yang et al. (2002b).*

- Scheme 1: A single ASP slug of 0.6 PV
- Scheme 2: 0.3 PV AP + 0.3 PV ASP
- Scheme 3: 0.4 PV AP + 0.2 PV ASP

In these three schemes, the amount of alkali or polymer was the same. Only the injected surfactant was gradually reduced. In these schemes, it was observed that the incremental oil recovery factors over waterflooding were almost the same. Because less surfactant was injected in Schemes 2 and 3, Schemes 2 and 3 were economically more attractive than Scheme 1.

13.8 AMOUNTS OF CHEMICALS INJECTED IN FIELD ASP PROJECTS

From the 1990s to early 2000s, almost all field pilot tests and field applications were conducted in China. Table 13.6 summarizes the amounts of chemicals injected in the ASP projects from 1992 to 2004; field names and references are provided in Table 13.7. For most of the ASP projects, polymer was injected before and after ASP slugs for conformance control and mobility control. The injection concentrations of alkali, surfactant, and polymer in the ASP slugs are shown in Figures 13.22 through 13.24, respectively. The average injection concentrations of alkali, surfactant, and polymer were 1.28%, 0.28%, and 0.15%, respectively. For the micellar flood or micellar-polymer flood projects in 1980s, the injected surfactant concentration was more than a few percent. In these alkali-surfactant-polymer projects, the surfactant concentrations were one order of magnitude lower. Several factors contributed to the surfactant deduction. Although some of these factors are not new, they have been validated in the field: (1) alkalis can reduce surfactant adsorption significantly; (2) alkalis react with crude oils to generate in situ surfactant (soap); (3) modern surfactants have been improved; and (4) the synergistic effect of alkaline, surfactant, and polymer results in less surfactant required to recover significantly incremental oil.

Corresponding to those chemical concentrations, the injection pore volumes are shown in Figure 13.25. The average injected pore volume was 41.8% PV. Injection schemes may be optimized, for example, using graded polymer concentrations (Claridge, 1978; Ligthelm, 1989). However, our simulation sensitivity study showed that the incremental oil recovery mainly depended on the amounts of chemicals injected.

Figures 13.26 through 13.28 (see pages 527 and 528) show the total amounts of alkali, surfactant, and polymer injected, respectively. Their respective averages were 53.44, 11.53, and 8.08. Their unit is the product of injection volume in PV% and the concentration in %. These field data may serve as a reference or guide for other field projects.

Chang et al. (2006) listed the following chemical costs (US $/lb): alkaline agents, 0.12; surfactants, 2.2; and HPAM polymer, 1.03. When these chemical

TABLE 13.6 Amounts of Chemicals Injected in ASP Projects

	Preslug		ASP Slug				Post-slug		Total Injection, PV(%) × Concentration(%)		
Case No.	*PV*, %	*P*, %	*PV*, %	*A*, %	*S*, %	*P*, %	*PV*, %	*P*, %	*A*	*S*	*P*
1			32.0	1.25	0.30	0.12	28.3	0.06	40.00	9.60	5.54
2			37.0	1.31	0.32	0.14	30.8	0.11	48.47	11.84	8.57
3	3.75	0.15	45.1	1.20	0.26	0.22			54.12	11.73	10.44
4			48.0	1.20	0.27	0.18	20.0	0.08	57.60	12.96	10.24
5			55.2	1.20	0.30	0.12			66.24	16.56	6.62
6	3.75	0.15	45.4	1.02	0.18	0.14	20.0	0.08	46.31	8.17	8.56
7	4.00	0.14	50.0	1.54	0.21	0.18	20.0	0.14	77.00	10.50	12.36
8	5.00	0.10	40.0	1.50	0.38	0.09	10.0	0.05	60.00	15.20	4.60
9	9.70	0.20	30.9	1.20	0.30	0.17	5.0	0.15	37.08	9.27	8.07
10			34.0	1.40	0.30	0.13	15.0	0.10	47.60	10.2	5.92
Avg.	5.24	0.15	41.8	1.28	0.28	0.15	18.6	0.10	53.44	11.53	8.08

TABLE 13.7 References for Field Cases Listed in Table 13.6

Case No.	Field	References
1	Sa-Zhong-Xi, Daqing	Wang et al. (1997c); Gao et al. (1996); Li et al. (1999b)
2	Xing-5, Daqing	Wang et al. (1997c); Han (2001); Wang et al. (2006b)
3	Xing-2-Xi, Daqing	Wang et al. (1998a)
4	Bei-1-Xi, Daqing	Wang et al. (1999b)
5	Bei-2-Dong, Daqing	Wang et al. (2001a); Zhao et al. (2005a)
6	Xing-2-Zhong, Daqing	Li et al. (2003); Wang et al. (2006b)
7	Bei-2-Dong, Daqing	Wan et al. (2006)
8	Gudong, Shengli	Song et al. (1995); Wang et al. (1997b); Qu et al. (1998)
9	Gudao, Shengli	Yang et al. (2002c); Cao et al. (2002); Chang et al. (2006)
10	Er-Zhong-Bei, Karamy	Gu et al. (1998); Delshad et al. (1998); Qiao et al. (2000); Han (2001); Chang et al. (2006)

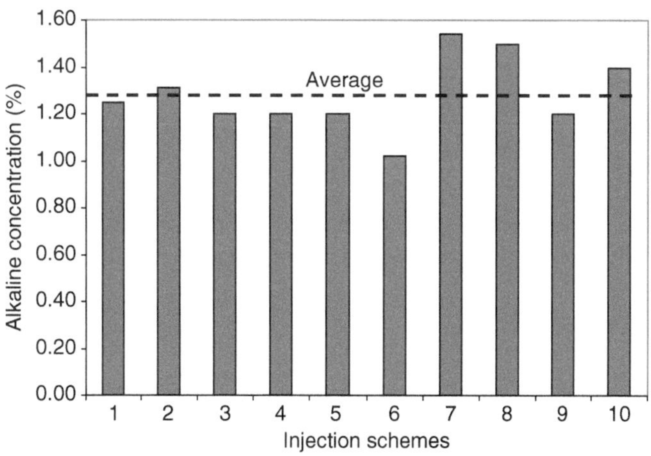

FIGURE 13.22 Injected alkaline concentrations in the field ASP projects.

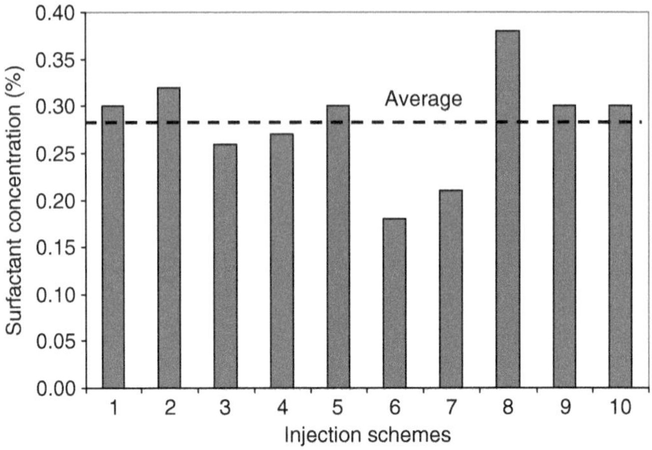

FIGURE 13.23 Injected surfactant concentrations in the field ASP projects.

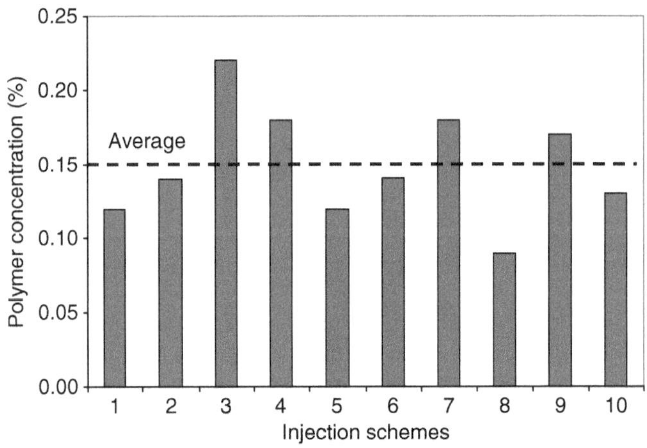

FIGURE 13.24 Injected polymer concentrations in the field ASP projects.

prices were used, the chemical cost ratio of alkali, surfactant, and polymer was 1:4:1.3. Surfactant cost more than half of the total chemical cost.

13.9 VERTICAL LIFT METHODS IN ASP FLOODING

In Daqing ASP projects, three lift methods were used: sucker rod pump, screw pump, and electric pump. Table 13.8 later in this chapter lists the average work lives of screw pumps in chemical flooding before any repair was needed (Wang et al., 2006b). According to this table, the work life during ASP flooding was

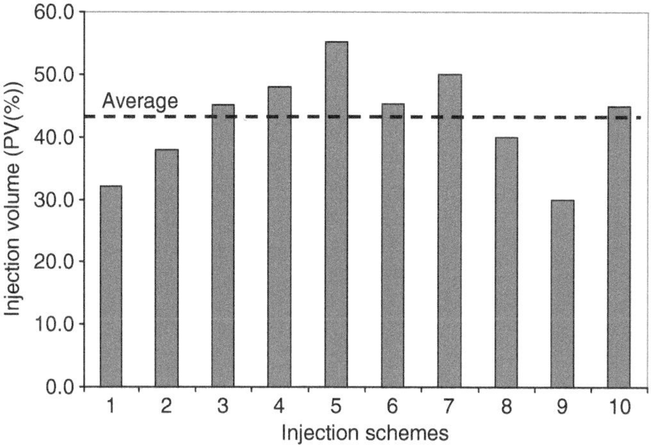

FIGURE 13.25 Injected pore volumes in the field ASP projects.

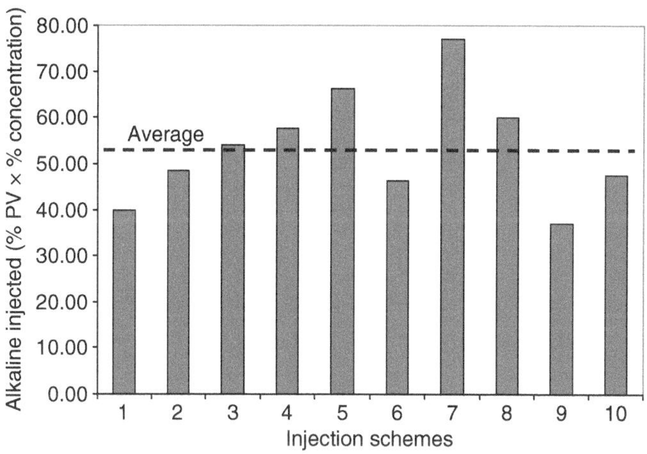

FIGURE 13.26 Alkaline injected in the field ASP projects.

significantly decreased compared with those for polymer flooding and waterflooding. The screw rods were broken because of the scale deposited in ASP flooding. For some wells, the work lives of screw rods were only 30 days. However, their work lives were still longer than those of sucker rod pumps. Improvement has been made for screw pumps in recent years (Li et al., 2005b).

13.10 PROBLEMS ASSOCIATED WITH ASP

This section discusses the problems associated with ASP, including chromatographic separation, precipitation, scaling, formation damage, and the problems caused by produced emulsion.

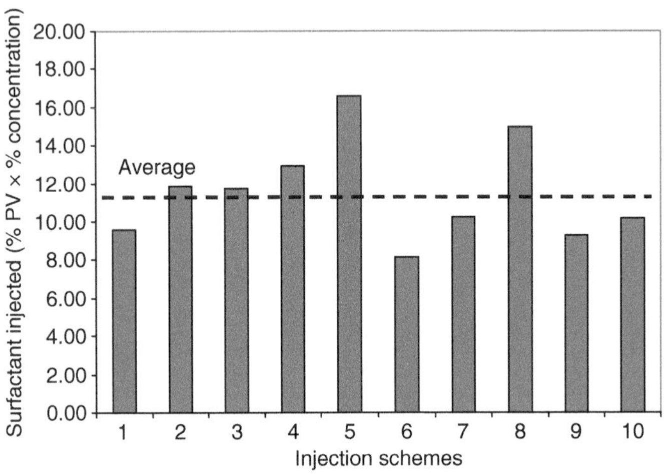

FIGURE 13.27 Surfactant injected in the field ASP projects.

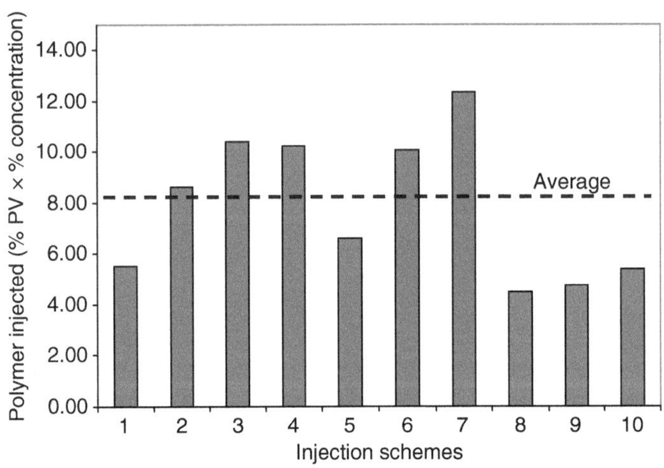

FIGURE 13.28 Polymer injected in the field ASP projects.

13.10.1 Chromatographic Separation of Alkali, Surfactant, and Polymer

Figure 13.29 shows the effluent concentration histories of an ASP slug injection. The vertical axis illustrates the relative concentration of polymer, alkali, and surfactant—the effluent concentrations relative to its respective injection concentrations. The horizontal axis illustrates the injection pore volume. This figure shows different effluent histories for the polymer, alkali, and surfactant.

TABLE 13.8 Average Work Lives of Screw Pumps in Waterflooding and Chemical Flooding

Process	Average Work Life
Waterflooding	618 days
Polymer flooding	375 days
ASP flooding	97 days

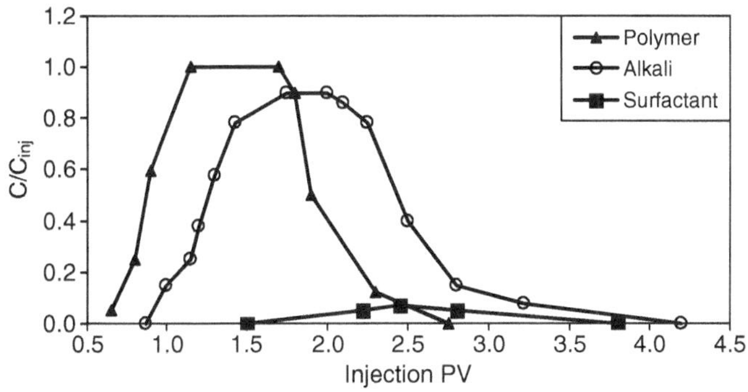

FIGURE 13.29 Effluent concentration histories of polymer, alkali, and surfactant. *Source: Huang and Yu (2002).*

We can see that, first, breakthrough times were different. In this case, polymer broke through first, then alkali, and finally surfactant. Second, each maximum relative concentration depended on its retention or consumption in the pore medium. The maximum relative polymer concentration was 1, the maximum relative alkali concentration was 0.9, and the maximum relative surfactant concentration was 0.09 in this case. Third, their concentration ratios in the system were constantly changing. In other words, the chemical injection concentrations were not proportionally decreased.

As shown in Sections 2.4 and 3.3.6, the higher the retention, the slower the transport velocity will be. Generally, surfactant retention is higher than alkaline, and alkaline retention is higher than polymer. Therefore, the order of breakthrough is: polymer, alkali, and then surfactant. The difference in breakthrough time between the alkali and the polymer is not much, but the surfactant breaks through far behind the alkali. Daqing experimental data showed that the chromatographic separation between surfactant and polymer, or alkali, is more significant in water-wet conditions than in oil-wet, while the separation between alkali and polymer is similar in both wet conditions. As the permeability becomes higher, the separation becomes smaller (Li, 2007).

13.10.2 Chromatographic Separation of Surfactant Compositions

A good surfactant like ORS-41 often has different fractions of carbon chains, and benzene ring could attach at different locations in the carbon chain. An injected solution having different compositions may have chromatographic separation issues when flowing through a porous medium. Following is an example of chromatographic separation.

A two-meter long oil column (9.8 mm inner diameter) was used to investigate the chromatographic separation of surfactant components (Yang et al., 2002a). The sand column porosity was 30 to 40%, the temperature was 68°C, the flow velocity was 0.3 mL/min, and 0.5 PV of 0.3% SDBS surfactant mixture was injected. Figure 13.30 shows the composition concentrations, c/c_o, at the effluent end, where c is the effluent concentration and c_o is the inlet concentration. This figure shows that the shorter chain surfactant C_{10} flowed out first, but the surfactant with benzene ring replacing the second carbon chain flowed more slowly. C_{11} was slower than C_{10}, and C_{12} and C_{13} were the slowest.

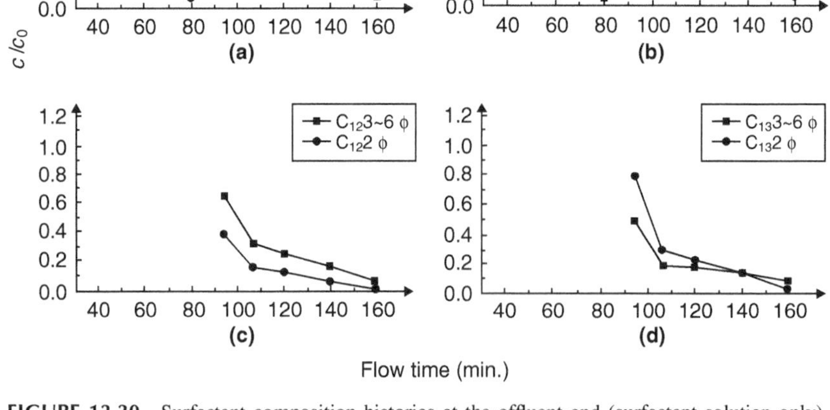

FIGURE 13.30 Surfactant composition histories at the effluent end (surfactant solution only). *Source: Yang et al. (2002a).*

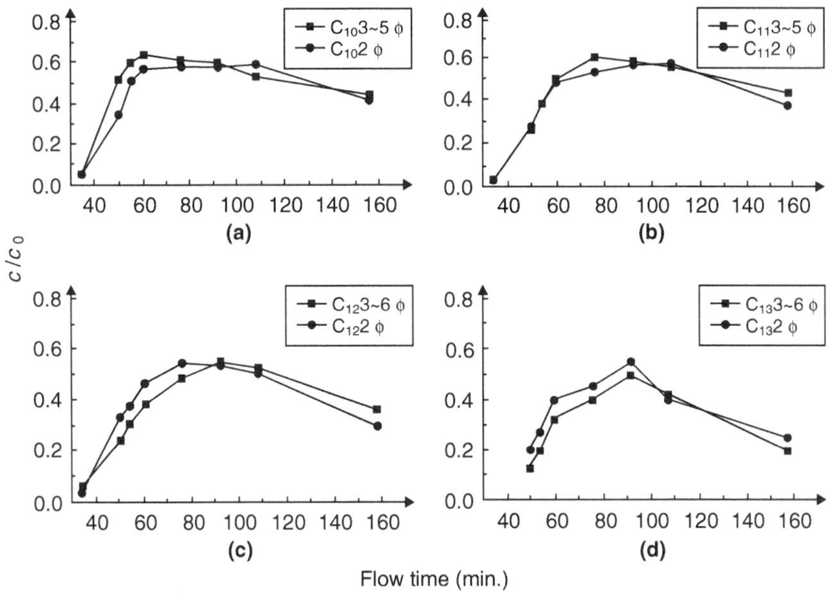

FIGURE 13.31 Surfactant composition histories at the effluent end (surfactant and alkali). *Source: Yang et al. (2002a).*

Figure 13.31 shows the surfactant composition histories at the effluent end when 1.5% sodium carbonate was added. We can see that all the surfactants with different carbon chains and those with the second carbon chain replaced by the benzene ring flowed out earlier than without the alkali (refer to Figure 13.30). C_{10}, C_{11}, and C_{12} and the ones with their second carbon replaced by benzene ring flowed almost at the same velocity. The flow velocities of C_{13} and the one with the second carbon replacement changed. This figure shows that the alkali changed the rock surface properties and thus changed the chromatographic separation.

Figure 13.32 shows the surfactant composition histories at the effluent end when 0.1% HPAM was added. All the surfactants flowed more slowly. Figure 13.33 shows the synergistic effect of alkali and polymer that reduced chromatographic separation significantly.

Surfactant structure affects chromatographic separation. For the structures of alkyl benzene sulfonate shown in Figure 13.34, the location of benzene ring in the straight carbon chain generally has the following effects from a to e (Yang et al., 2002a): (1) solubility is increased, (2) polarity is increased, (3) hydrophilicity is increased, (4) adsorption is decreased, and (5) chromatographic separation is decreased.

For the structures shown in Figure 13.35, the length of carbon chain has the following effects from a to d (Yang et al., 2002a): (1) cohensive action is

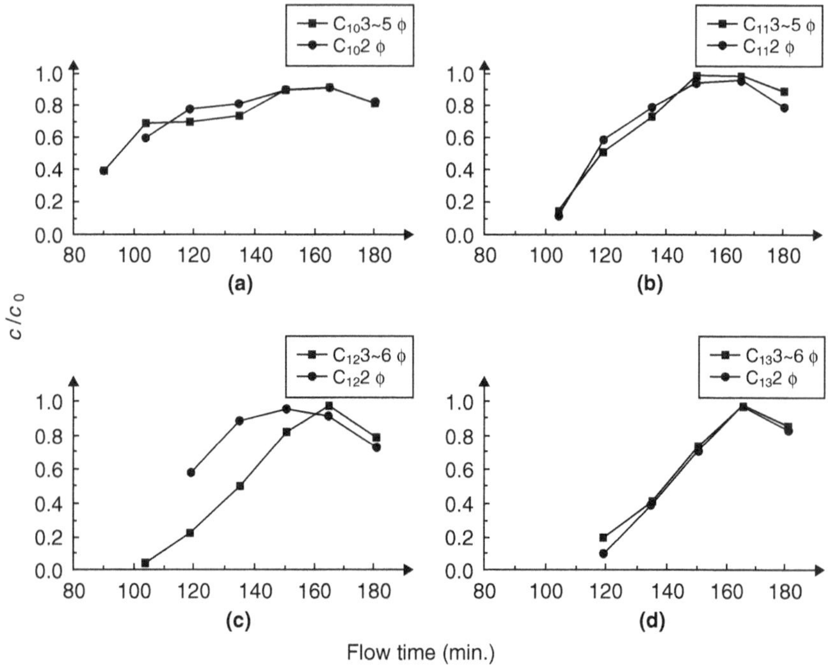

FIGURE 13.32 Surfactant composition histories at the effluent end (surfactant and polymer). *Source: Yang et al. (2002a).*

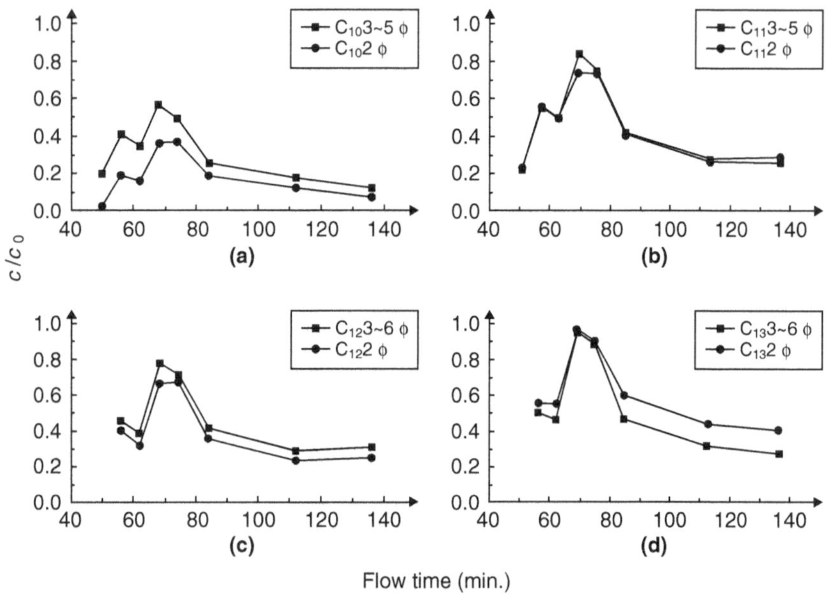

FIGURE 13.33 Surfactant composition histories at the effluent end (surfactant, alkali, and polymer). *Source: Yang et al. (2002a).*

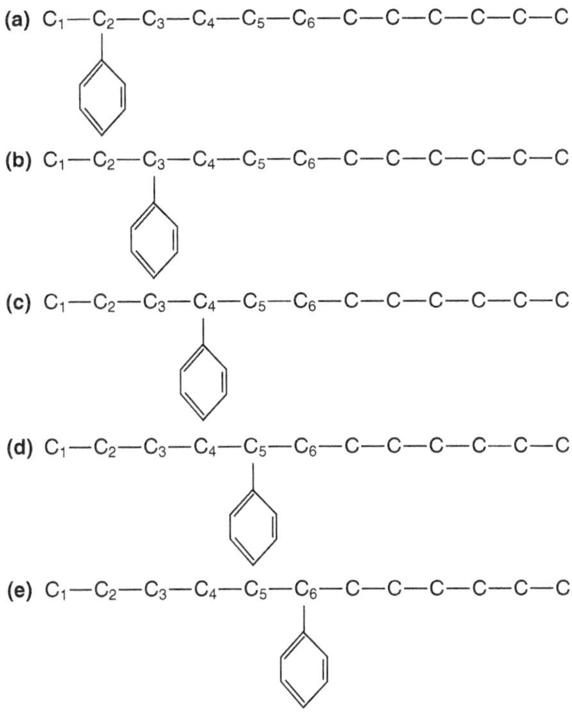

FIGURE 13.34 Different benzene ring locations of alkyl benzene sulfonate.

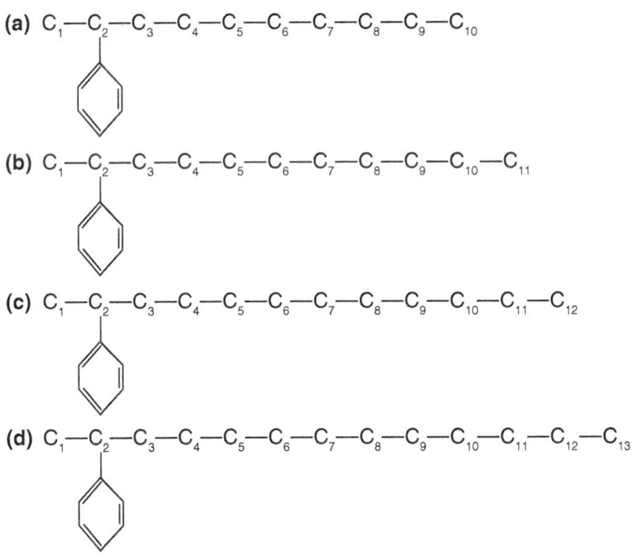

FIGURE 13.35 Different carbon chain lengths of alkyl benzene sulfonate.

increased, (2) hydrophilicity is decreased, (3) adsorption is increased, and (4) chromatographic separation is increased.

13.10.3 Precipitation and Scale Problems

In the process of alkaline flooding, alkali reacts with reservoir rock, resulting in the dissolution of some rock materials. The flooding liquid carries the dissolved materials to the production wells. The blending of produced liquid coming from different layers, decrease of pressure, loss of the dissolved gas, and decrease of temperature result in precipitation and its deposition on the tubing, surface pipelines, pumps, and so on. The deposition can cause severe pipe plugging, tubing plugging, and thus breaking of beam pump rods. Frequent operation failures of the production wells were observed in Daqing chemical flooding projects. In addition to the dissolved rock materials, reactions of alkalis with divalents such as calcium and magnesium lead to the formation of precipitates. The divalents could be from mixing with resident brines and the ion exchange process.

Although formation water compositions could be different in different fields or even in different layers or blocks within the same field, the main compositions include cations (Na^+, K^+, Ca^{2+}, Mg^{2+}, Fe^{2+}, F^{3+}) and anions (Cl^-, $HCO3^-$, SO_4^{2-}, CO_3^{2-}). NaCl and Na_2CO_3 are the main compositions (more than 90%) in formation water. The divalents are mainly Ca^{2+} and Mg^{2+} in a range of 20 to 2000 mg/L. Mg^{2+} is generally much lower than Ca^{2+} in formation water (but higher in seawater). The concentrations of other cations are low, less than 1 mg/L. Therefore, the dominating ions that cause scaling are Ca^{2+} and CO_3^{2-}.

When an alkaline solution is injected into a formation, the concentrations of OH^-, CO_3^{2-}, and SiO_3^{2-} are increased. The increase in OH^- is from the injected alkaline solution. The increase in CO_3^{2-} is from HCO_3^- because high OH^- makes the reservoir an alkaline environment and converts HCO_3^- into CO_3^{2-}. SiO_3^{2-} is a result of the reaction between the injected alkali and formation minerals. Strong alkalis dissolve the clay minerals and damage the rock surface electric charge and microstructures, which finally results in dispersion and migration of the clay minerals. If seawater is injected, SO_4^{2-} is increased. Ca^{2+} and Mg^{2+} are from the formation water, cation exchange, and reactions between the injected solution and rock minerals. Sometimes, there is Al^{3+}. Several inorganic scales can be formed as follows:

Carbonate scales:

$$Ca^{2+} + CO_3^{2-} = CaCO_3 \downarrow$$

$$Mg^{2+} + CO_3^{2-} = MgCO_3 \downarrow$$

Hydroxyl scales:

$$Ca^{2+} + 2OH^- = Ca(OH)_2 \downarrow$$

$$Mg^{2+} + 2OH^- = Mg(OH)_2 \downarrow$$

Silicate scales:

$$Ca^{2+} + SiO_3^{2-} = CaSiO_3 \downarrow$$

$$Mg^{2+} + SiO_3^{2-} = MgSiO_3 \downarrow$$

$$Fe^{2+} + SiO_3^{2-} = FeSiO_3 \downarrow$$

Sulfate scales:

$$Ca^{2+} + SO_4^{2-} = CaSO_4 \downarrow$$

$$Mg^{2+} + SO_4^{2-} = MgSO_4 \downarrow$$

Silicic scales:

$$2H^{+} + SiO_3^{2-} = H_2SiO_3 \downarrow$$

$$4H^{+} + SiO_4^{4-} = H_4SiO_4 \downarrow$$

Here is an example. An ASP flood was initiated in the N-1DX block in Daqing on March 15, 1997. A pump malfunction occurred with a scale thickness of 1 cm over the wellbore by September 23, 1998. By the end of 1999, serious scale problems were observed in surface facilities, such as the oil–water separation system and the produced water disposal system. Scale usually occurred on pipe walls and in valves, pump heads, vane wheels, and flow meters. In another case, some scale samples from Well Z1-35 looked white, and the scale thickness was 0.5 to 1.0 cm. The scale in the delivery pipes of Z1-35 reached about 0.3 cm (Wang et al., 2004b).

13.10.4 Formation Damage

Alkaline solutions erode formation rocks and clays. In some cases, the permeability could be increased owing to the erosion. In most cases, permeability is reduced because eroded rocks and clays migrate and block pore throats. The permeability reduction in low permeability rocks is higher than that in high permeability rocks, as shown in Figure 13.36. The reason is that the pore throats

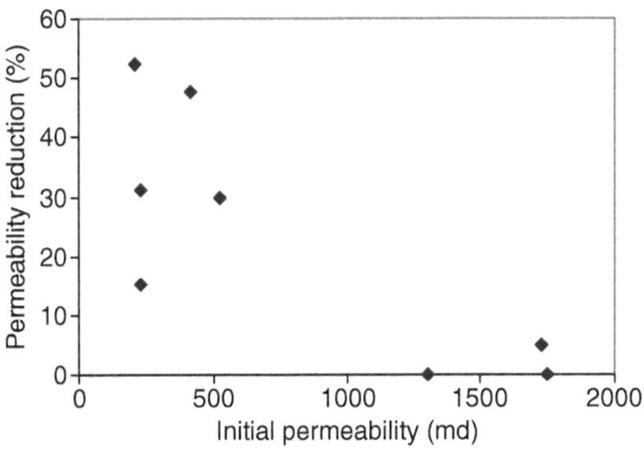

FIGURE 13.36 Permeability reduction after alkaline injection. *Source: Data from Li (2007).*

in low permeability rocks are smaller and are easier to be blocked. Here, 1.2% NaOH was used in the core floods.

Permeability reduction in alkaline-surfactant is similar to that in alkaline flooding. In such situations, permeability reduction could also be caused by scales and precipitates that are formed through reactions of alkalis and surfactants with rock minerals. In ASP flooding, the formation damage is less than that caused by alkaline flooding alone. The permeability can be recovered in some degree during the post-ASP waterflooding.

13.10.5 Produced Emulsions

Although emulsion can increase sweep efficiency in a reservoir, it also can cause difficulties in transportation and oil/water separation. In a Shengli ASP pilot test started in 1992, it was difficult to separate water from oil even though the weak alkali Na_2CO_3 was injected. In addition, because it was difficult to treat produced water, the produced water had to be re-injected into the reservoir.

Chinese operators have experienced severe scaling and emulsion problems in their surface facilities caused by strong alkalis such as NaOH. In addition, high-concentration alkalis reduce SP solution viscosity and viscoelasticity. To avoid the preceding problems, China is proposing: (1) alkali-free SP flooding and (2) dynamic IFT of 10^{-2} mN/m, not 10^{-3} mN/m. Interfacial tension of 10^{-2} mN/m may be low enough to reach a high incremental oil recovery factor. The main reason is that it is quite difficult to reach 10^{-3} mN/m IFT. To achieve such low IFT, sometimes you need to use high surfactant and alkaline concentrations. Thus, the chemical cost will go up. A high alkaline concentration will cause more problems, such as emulsion, scaling, precipitation, formation damage, and so on. A high alkaline concentration will also reduce polymer viscosity. As a result, more polymer will be needed to reach a required viscosity.

13.11 ASP EXAMPLES OF FIELD PILOTS AND APPLICATIONS

This section presents eleven field pilot and application examples. These examples were carefully selected so that each one can provide unique observations. Using these examples we summarize the methods used, lessons learned, and the observations made.

13.11.1 Daqing ASP Pilot Test in Sa-Zhong-Xi (S-ZX)

This section describes the first ASP pilot test in Daqing. Although several laboratory and simulation studies were conducted, only a brief summary is presented here. For more details, see Gao et al. (1996), Wang et al. (1997c), Li et al. (1999b), and Wang et al. (2006b).

Reservoir and Fluid Description

Figure 13.37 shows the S-ZX pilot test location and locations of other ASP pilot tests and commercial applications in the Daqing oilfield. The target layer, SII_{1-3}, was composed mainly of braided channel sand formed by Lamadian-Saxi fluvial system. Some of the reservoir and fluid data are shown in Table 13.9.

The pilot test consisted of 4 five-spot patterns, schematically shown in Figure 13.38. The test involved 4 injectors (I01, I02, I03, and I04), 9 producers

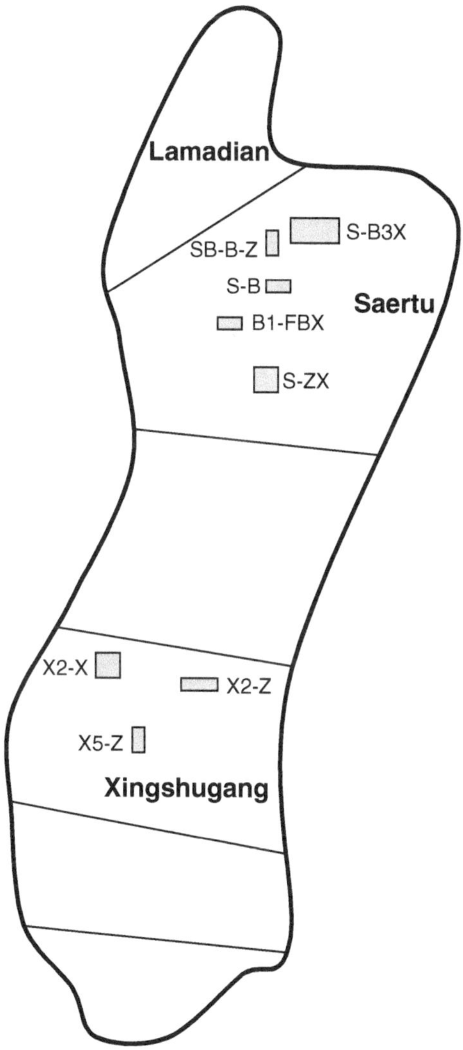

FIGURE 13.37 ASP pilot locations in the Daqing oilfield. *Source: Chang et al. (2006).*

TABLE 13.9 S-ZX Reservoir and Fluid Data

Pilot area, km^2	0.09
Formation depth (subsea), m	814
Permeability, md	509–1426
Porosity, %	26
Total PV, m^3	203,300
Permeability variation coefficient	0.5–0.8
Effective thickness, m	8.6
OOIP, tons	117,300
Formation temperature,°C	45
Formation water TDS, mg/L	6800
Fresh water salinity, mg/L	600
Oil acid number, mg KOH/g	0.01
Oil viscosity at the reservoir temperature, mPa·s	11.5

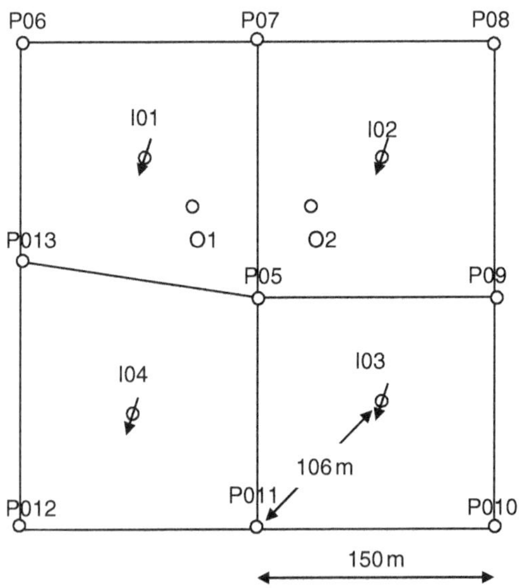

FIGURE 13.38 S-ZX ASP pilot well patterns.

(P05, P06, P07, P08, P09, P010, P011, P012, and P013), 1 sampling well (O1), and 1 observation well (O2).

Laboratory Study

The chemical formula in this pilot test was 1.25% Na_2CO_3 + 0.3% B-100 + 1200 ppm 1275A blended in fresh water. Table 13.10 shows the IFT values at different alkaline and surfactant concentrations. The table also shows that ultralow IFT of 10^{-3} mN/m was reached within a large range of concentrations, particularly near the designed injection concentrations. The chemical adsorption or consumption for alkali, surfactant, and polymer were 1.065, 0.455, 0.169 (mg/mL PV), respectively. The residual resistance factor in this test was 2.0695. The injection scheme was 0.32 PV ASP solution, then 600 mg/L polymer buffer solution, followed by water until the water cut reached 98%.

Linear and radial core flood tests were conducted to determine the polymer concentration for mobility control requirement. Figure 13.39 shows Brookfield (UL adapter) viscosity properties for the Alcoflood 1275A polymer in injection water and in an alkaline-surfactant solution. Note that the AS dramatically decreased the viscosity, and a higher polymer concentration was required to provide the same viscosity.

Pilot Performance

To establish the residual oil saturation as a reference for evaluating ASP performance, waterflood was re-initiated in the pilot area in June 1993, ended in

TABLE 13.10 IFT Values at Different Alkaline and Surfactant Concentrations (mN/m)

	Surfactant Concentration, wt.%				
Alkaline, wt.%	*0.0*	*0.05*	*0.11*	*0.2*	*0.5*
0.00	22.9	1.072	2	1.8	1.6
0.50	16.9	0.155	0.019	0.075	0.008
0.75	14.6	0.005	0.007	0.007	0.005
1.00	13.7	0.002	0.005	0.005	0.003
1.25	13.0	0.0001	0.0018	0.0018	0.0017
1.50	11.3	0.001	0.0008	0.0008	0.0004
1.75	10.9	0.001	0.0008	0.0008	0.0004
2.00	9.9	0.0055	0.0015	0.0015	0.006

Source: Wang et al. (2006b).

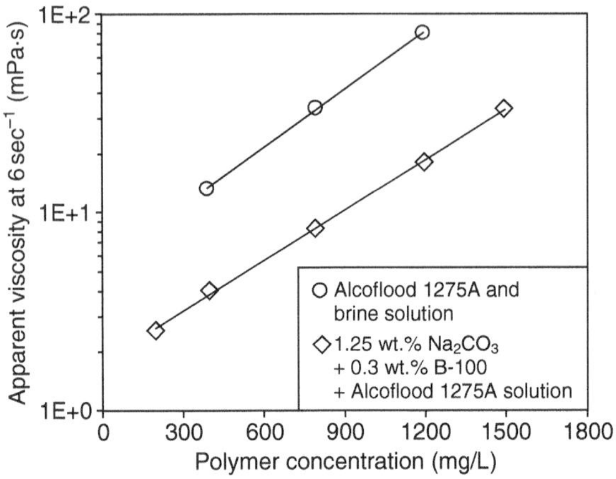

FIGURE 13.39 Viscosity versus polymer concentration. *Source: Gao et al. (1996).*

September 23, 1994, followed by 0.327 PV ASP flood, and 0.273 PV polymer drive and water drive. The ASP solution viscosity was 16 mPa·s. During water preflush, the oil recovery before ASP was 31.63% from the $SII_{1\text{-}3}$ layer. The response to ASP injection was observed in November 1994 (after 0.0693 PV of injection). The average water cut in the entire pilot area decreased from 82.7% to a low of 59.7%, and the daily oil production increased from 37 m^3/d to a peak of 91.5 m^3/d. The water injectivity decreased from 1.75 m^3/(m·d·MPa), stabilized at about 1.42 m^3/(m·d·MPa), and then dropped to 1.19 m^3/(m·d·MPa). In general, after an ASP slug is injected, flow resistance increases, and water injectivity decreases. The simulation prediction showed about 20% incremental oil recovery factor over waterflood. The early performance matched very well with the simulation prediction. In this pilot test, the simulator used was GCOMP.

Alkali, surfactant, and polymer concentrations in the produced fluid at Well P05, as well as the oil cut, are shown in Figure 13.40. The peak oil cut was just ahead of the produced peak chemical concentrations, indicating an oil bank was formed ahead of the chemical slug. This figure shows that chromatographic separation was not a serious issue. No difficulties were encountered with oil-water separation.

13.11.2 Daqing ASP Pilot in Xing-5-Zhong (X5-Z)

This section presents an example showing some emulsion problems.

Reservoir and Fluid Description

The pilot location is marked X5-Z in Figure 13.37. The target layers were PI_2^2 and PI_3^3. Some of the reservoir and fluid data for this test are shown in Table 13.11. This pilot test had one pattern of 1 injector and 4 producers.

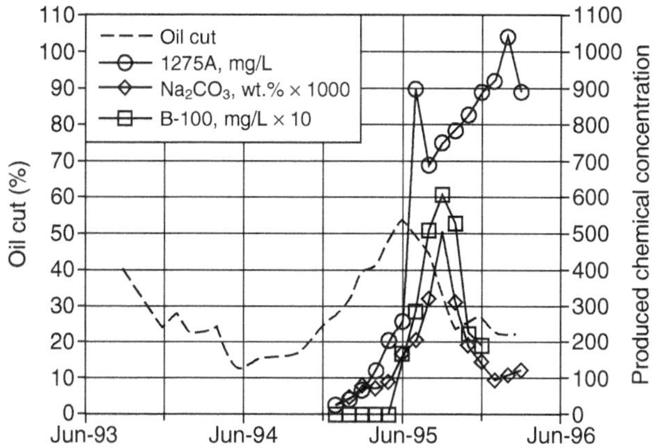

FIGURE 13.40 P05 (center well) oil cut and produced chemical concentrations in water. *Source: Gao et al. (1996).*

TABLE 13.11 X5-Z Reservoir and Fluid Data

Pilot area, km^2	0.04
Permeability, md	589
Porosity, %	26
Total PV, m^3	68,000
Permeability variation coefficient	0.63
Effective thickness, m	6.8
Total thickness, m	8.4
OOIP, tons	37,000
Formation temperature,°C	45
Oil viscosity at the reservoir temperature, mPa·s	6.1

Chemical Formulation

The chemical formula used for the main 0.37 PV ASP slug was 1.31% NaOH + 0.32% ORS-41 + 0.14% 1275A. The IFT activity map in Figure 13.41 shows the IFT values at different alkaline and surfactant concentrations. For the IFT value of 7.8×10^{-3}, for example, –3 is shown on the top and 7.8 is shown in the bottom near the point in the figure. Here, ultralow IFT of 10^{-3} mN/m was reached within a large range of concentrations. The injection scheme was 0.37 PV ASP solution followed by graded polymer drive.

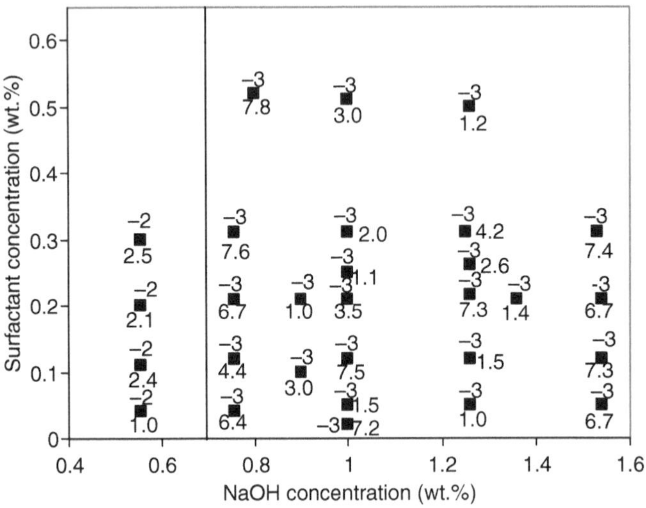

FIGURE 13.41 IFT activity map. *Source: Wang et al. (2006b).*

Pilot Performance and Emulsion Problem

ASP injection for this pilot test was started in January 1995, and the response (about 0.08 PV) was observed on March 30 of the same year. The water cut in the pilot area decreased from 96.6 to 80.7%. An emulsion problem was obvious in this pilot test. From one well sample, the emulsion viscosity was 40 mPa·s, which is about twice the viscosity of unemulsified fluid. However, it was observed that emulsions improved sweep efficiency. Table 13.12 reports the observations from the different injection phases, and Table 13.13 lists the oil concentrations in water after a 30-minute settlement. In this case, the demulsifier was SP169. For more details on this pilot test, see Wang et al. (1997c) and Wang et al. (2006b).

13.11.3 ASP Pilot in Xing-Er-Xi (X2-X)

In the ASP pilot test described in this section, the water cut in the test area was already 98% before the ASP flood, and some well water cut reached 100%. Optimization of injection schemes was conducted. In this case, a scaling problem was experienced in the pilot area. For more details, see Wang et al. (1998a).

Reservoir and Fluid Description

The location of this ASP pilot test is marked X2-X in Figure 13.37. The target layers were $PI3_3$, and there were 9 producers and 4 injectors drilled in five-spot patterns in the pilot area, as shown in Figure 13.42. The distance between

TABLE 13.12 Emulsions at Different Injection Phases

Phase	Time Interval	Injection PV	Emulsion Type	Oil Production, %	Observations
1	February–May 1995	0–0.209		14.1	No emulsion
2	June–November 1995	0.209–0.516	W/O	57	400 ppm P, low S, no A, f_w decreased
3	November 1995–March 1996	0.516–0.76	W/O, O/W, a few multiple	22.5	f_w increased
4	April 1996–	0.76–	O/W	6.4	Low A, low P, high S, f_w > 90%

TABLE 13.13 Oil Concentration in Water after 30-Minute Settlement

Emulsion Type	Demulsifier Concentration, ppm	Oil Concentration, ppm
W/O	15	5700
W/O, O/W, multiple	50	7000
O/W	30	5500

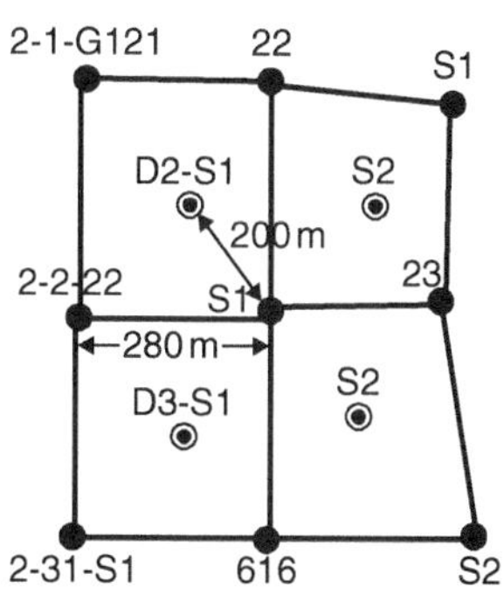

FIGURE 13.42 Well patterns in the ASP pilot test at Xing-er-xi (X2-X).

injector and producer was 200 m. Some of the reservoir and fluid data are shown in Table 13.14.

Optimization of Injection Schemes

Experimental data show that for the same amount of chemicals, graded chemical concentrations (high to low) could increase the recovery factor by 4.1%. In this pilot test, the incremental oil recovery factor increased with the main ASP slug size up to 0.35 PV. After that, a further increase in the main slug size did not increase the oil recovery factor. Similarly, the second ASP slug should be 0.1 PV maximum. For the same amount of polymer injected, the effects of placing polymer before and after the ASP slug were investigated. It was found that adding a preflush polymer drive could increase the recovery factor by 2%. The post polymer slug should be 0.2 PV. The final injection scheme was as follows:

- Preflush polymer: 0.0375 PV, 1500 mg/L
- Main ASP slug: 0.35 PV, 1.2% NaOH + 0.3% S + 2300 mg/L P
- Second ASP slug: 0.1 PV, 1.2% NaOH + 0.1% S + 1800 mg/L P
- Post-flush graded polymer drive: 0.2 PV, 1000, 700, 500 mg/L

Pilot Performance

After 0.097 PV ASP injection, the response was observed at Well X-2-1-22 (marked 22 in the well patterns). From the 9 producers, the oil rate increased from 24 to 134 t/d, and the water cut decreased from 98 to 80.2%. However, Well X-2-2-S1 (S1 in the well patterns) stopped pumping after 0.3678 PV

TABLE 13.14 X2-X Reservoir and Fluid Data

ASP area, km^2	0.3
Formation depth (subsea), m	800–1200
Permeability, md	675
Total PV, m^3	435,000
Permeability variation coefficient	0.65
Effective thickness, m	5.8
Total thickness, m	7.0
OOIP, tons	240,100
Formation oil viscosity, mPa·s	7.1

injection (in February 1998) owing to a scale problem. White material was deposited on the rod and inner tubing walls. However, the well worked after workover. The pumping was stopped again in April and June 1998. Afterwards, the rod pump was changed to a screw pump, which worked better. It was observed that scaling reduced liquid production rate. Fracturing therefore had to be carried out for some seriously damaged wells. Polymer, alkali, and surfactant reached the center well (S1) after 0.264, 0.34, and 0.345 PV injection, respectively. Although emulsion was also observed, it helped to improve sweep efficiency.

13.11.4 Daqing's Largest ASP Application (X2-Z)

This section describes the largest Daqing ASP project; it is probably the largest ASP project in the world so far. Compared with other small pilot tests, the formation connectivity in this test was an important factor affecting ASP performance. In this case, an emulsion problem was noticed.

Reservoir and Fluid Description

The pilot location is marked X2-Z in Figure 13.37. The target layers were $PI2^1$ to $PI3^3$, and there were 27 producers and 17 injectors drilled in regular five-spot patterns in the pilot area, as shown in Figure 13.43. The distance between injector and producer was 250 m. Some of the reservoir and fluid data for this pilot test are shown in Table 13.15 (Li et al., 2003; Wang et al., 2006b).

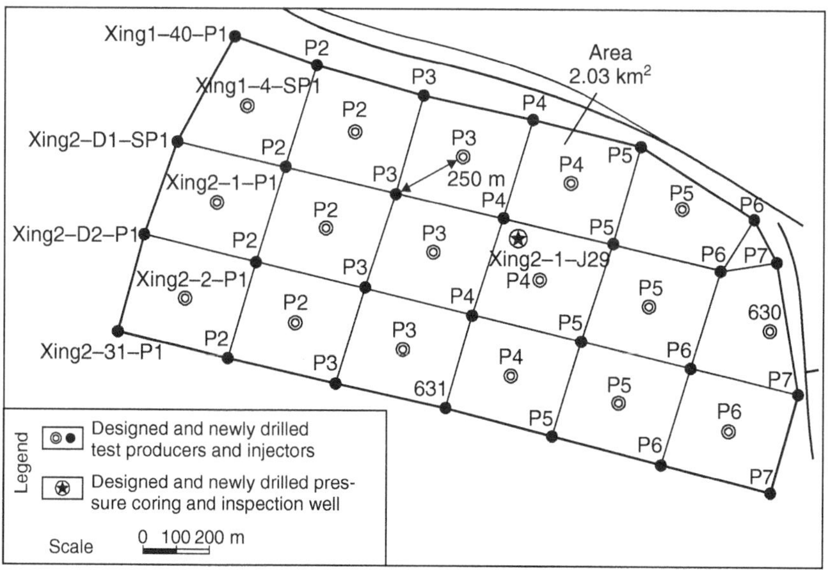

FIGURE 13.43 Well locations in the X2-Z ASP project.

TABLE 13.15 X2-Z Reservoir and Fluid Data

ASP area, km^2	2.03
Formation depth (subsea), m	947–1029
Permeability, md	69–673
Average porosity, %	25
Total PV, m^3	3,654,000
Permeability variation coefficient	0.57
Effective thickness, m	7.2
Total thickness, m	10.1
OOIP, tons	2,017,000
Formation temperature,°C	45
Formation water TDS, mg/L	4144

Designed Injection Scheme

The surfactant in this test was alkyl benzene sulfonate made in China. Based on experimental work, 0.025 to 0.3% S + 0.4 to 1.0% A could make IFT reach 10^{-3} mN/m. As a result, the following injection scheme was designed:

- Preflush polymer slug: 0.0375 PV × 1400 mg/L, viscosity 40 mPa·s.
- ASP main slug: 0.35 PV, 1% NaOH + 0.2% S + 1650 mg/L P, viscosity 40 mPa·s.
- Second ASP slug: 0.1 PV, 1% NaOH + 0.1% S + 1000 mg/L P, viscosity 35 mPa·s.
- Polymer drive slugs: 0.1 PV × 1000 mg/L, viscosity 30 mPa·s; 0.1 PV × 630 mg/L, viscosity 15 mPa·s.
- Water drive: Water drive until f_w in the central area reaches 98%.

Field Injection Scheme

Waterflooding was started in October 1998 and ended in March 2000 with 0.2002 PV injection. Then preflush polymer flood was started in April 2000 and ended in April 2001 (0.128 PV injection). Throughout the testing, the average polymer concentration was 1538 mg/L with viscosity of 40.9 mPa·s. An injection of the main ASP slug was started on May 1, 2001. By November 2004, 0.354 PV was injected. The average injection concentrations of alkali, surfactant, and polymer were 1.02%, 0.201%, and 1407 mg/L, respectively. The wellhead sample viscosity was 30.2 mPa·s, and the IFT between the ASP system and crude oil was below 10^{-2}. The second ASP slug was started on December 1, 2004.

Field Performance

Figure 13.44 shows the injection rate and pressure versus injection PV in the whole ASP injection area. We can see that as polymer was injected, the injection rate decreased and injection pressure increased. In the following ASP injection, the injection rate and pressure gradually stabilized. Although the fluid production rate *decreased*, the oil rate increased and water cut decreased after polymer injection and ASP injection, as shown in Figure 13.45.

Table 13.16 lists the breakthrough times of polymer, surfactant, and alkali at some producers. For all these wells, polymer was seen first, then surfactant,

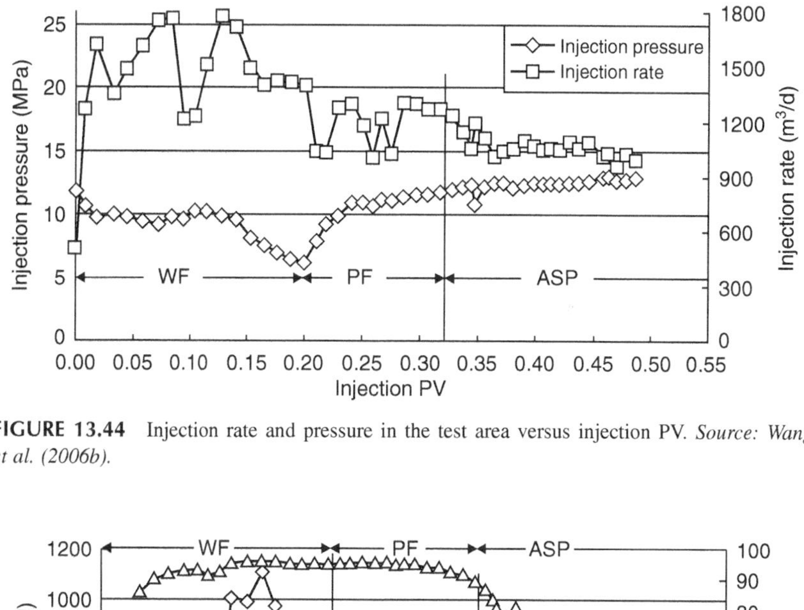

FIGURE 13.44 Injection rate and pressure in the test area versus injection PV. *Source: Wang et al. (2006b).*

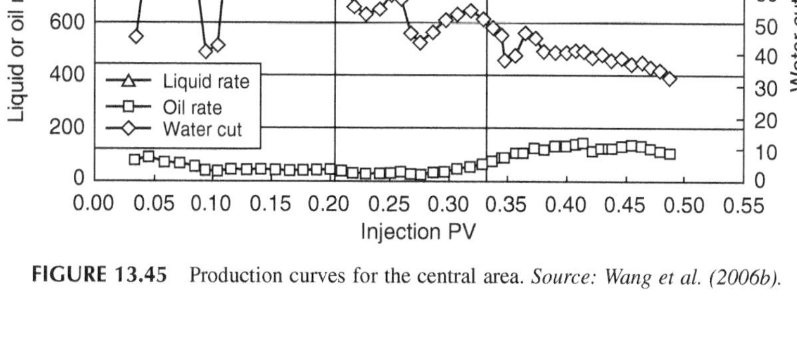

FIGURE 13.45 Production curves for the central area. *Source: Wang et al. (2006b).*

TABLE 13.16 Chemical Breakthrough Time, PV

Well	Polymer	Surfactant	Alkali
Xing2-D2-P4	0.164	0.168	0.193
Xing2-D2-P5	0.164	0.168	0.176
Xing2-D1-SP1	0.161	0.175	0.193
Xing2-1-J29	0.164	0.168	0.184
Average	0.163	0.17	0.187

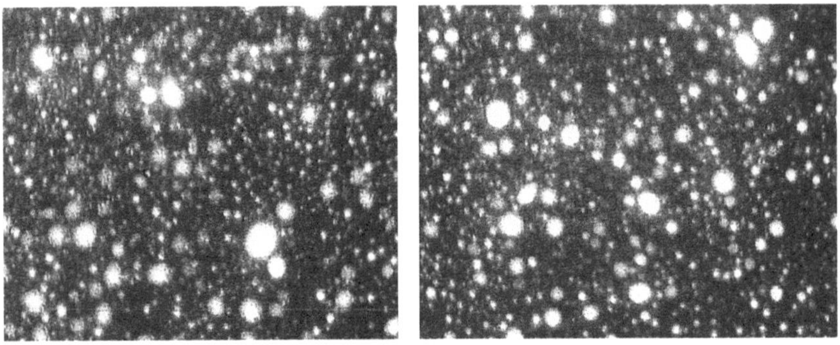

FIGURE 13.46 Emulsion picture of the sample taken from Xing2-D2-P3. *Source: Wang et al. (2006b).*

followed by alkali. In this case, emulsion was a problem. Figure 13.46 shows a picture of the W/O type emulsion from Well Xing2-D2-P3 on September 17, 2003. The viscosity was 185 mPa·s. Another sample was taken on November 20, 2003. No water was separated after 129 day settlement at room temperature, indicating the emulsion was stable.

Lessons Learned

Figure 13.47 shows the well recovery factor during ASP flood versus formation connectivity for some wells. A connectivity equal to 100% indicates the formation between injector and producer was well connected; 0% was not connected. The figure shows that the oil recovered from a production well during ASP flood was directly proportional to the formation connectivity between the producer and neighboring injectors.

In a large-scale ASP application, the production performance and ASP responses from different layers or wells are quite different owing to the vertical and areal heterogeneities. This phenomenon is more obvious in this large-scale application compared with other single-well pattern pilots or small-scale pilots.

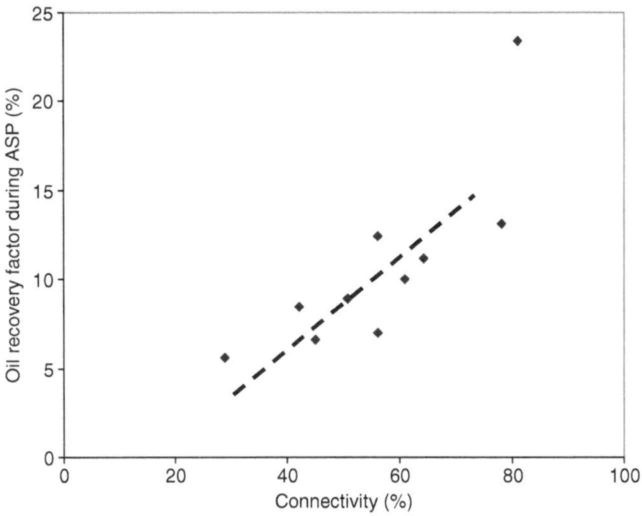

FIGURE 13.47 Oil recovery factor during ASP versus formation connectivity.

13.11.5 Daqing ASPF Pilot Test

ASPF (alkaline-surfactant-polymer-foam) flood was developed based on foam and ASP flooding (Wang et al., 2001a; Zhao et al., 2005a; Li et al., 2008). In this type of flooding, nitrogen or natural gas can be used. The synergy between foam and ASP is expected. The term LIFTF, for Low IFT foam, has also been used in the literature to describe this process (Zhao et al., 2005a). A similar process called alkaline-surfactant-gas (ASG), with gas replacing polymer in ASP, was also proposed (Srivastava et al., 2009). This section presents the first ASPF pilot tests published in literature.

Reservoir and Fluid Description

For this test, the target layer was $SIII^{3\text{-}7}$, a lacustrine deltaic sandstone deposition. In a pilot area, 10 producers and 6 injectors were drilled in five-spot patterns, as shown in Figure 13.48. The distance between injector and producer was 200 m. The pilot area had been waterflooded for 20 years and then flooded by water-alternate-gas (WAG) for another 5 years. Before the ASPF flooding, the recovery factor was 49.9%. Some of the reservoir and fluid data for this test are shown in Table 13.17.

Laboratory Study

Laboratory experiments were conducted to study the foaming stability and optimum chemical concentrations (Wang et al., 2001a). Table 13.18 shows the half-lives of different foaming systems. From this table, we can see that

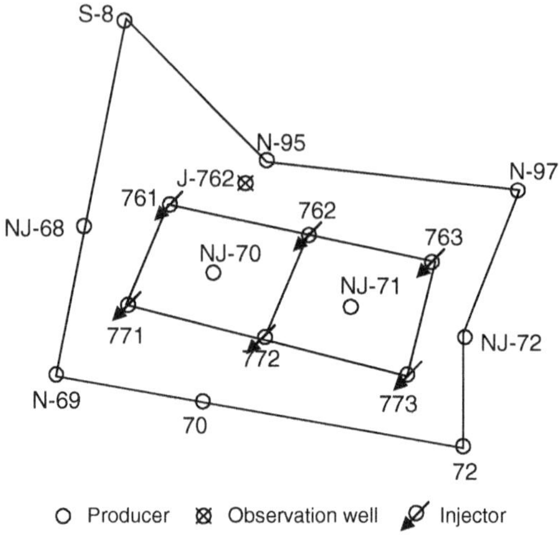

FIGURE 13.48 ASPF pilot well pattern.

TABLE 13.17 ASPF Pilot Reservoir and Fluid Data

ASP area, km^2	0.39
Permeability variation coefficient	0.755
Effective thickness, m	6.3
Total thickness, m	8.7
OOIP, tons	358,200

TABLE 13.18 Stability of Different Foam Systems

Foaming Agents	Initial Foam Volume, mL	Half-life, min.
Alky benzene sulfonate (A)	92	20
Nonionic (N)	17	17
N:A = 1:3	97	25
1.2% NaOH + 0.3% S + 0.12% P	94	35

ASP-generated foam was the most stable. In particular, Wang et al. (2001a) stated that polymer increased foam stability.

Salinity was found to decrease foam stability. The surfactant concentrations in which foaming ability increased with concentration were 0 to 0.5%. The optimum polymer molecular weight for foaming ability was around 17 million. Core flood tests showed that ASPF incremental oil recovery factor over ASP was above 10% because the ASPF sweep efficiency was higher than the ASP efficiency.

Injection Schemes

Based on laboratory study, the following scheme was proposed:

1. Preflush slug: 0.02 PV, 0.3% S (ORS-41) + 1.2% NaOH + 0.12% P. The polymer molecular weight was 15 million.
2. Main ASPF slug: 0.55 PV, gas volume:liquid volume = 1 : 1 (reservoir), alternating in 7 days. The ASP formula was the same as that in the preflush slug. Natural gas was used.
3. Second ASPF slug: 0.3 PV, gas volume:liquid volume = 1 : 1 (reservoir), alternating in 7 days. ASP formula: 0.1% S (ORS-41) + 1.2% NaOH + 0.12% P.
4. Post-flush: 0.2 PV graded polymer drive.

In the actual main ASPF, the liquid and gas volumes injected were 0.368 PV and 0.14 PV, respectively. In the second ASPF flood, the liquid and gas volumes injected were 0.184 PV and 0.047 PV, respectively. Overall, the ratio of injected gas volume to liquid volume was 0.34 : 1, instead of 1 : 1 as designed. The reasons are provided in the following section on pilot performance.

Pilot Performance

After the WAG flood and before the ASPF flood, the pilot area was flooded again by water. From January 1, 1997 to February 24, 1997, 0.067 PV of water was injected. By the end of waterflood, the water cut was 94.2%, and the recovery factor was 50.58%. From February 25, 1997, the ASPF flood was injected until May 2002. By that time, gas injection was stopped because the injection pressure was too high for injection pumps. Because the reservoir pressure was already high, to be able to drill neighboring wells, gas injection had to be terminated. Therefore, the injected gas/liquid ratio was 0.34 : 1, which was much lower than the designed 1:1.

The ASPF pilot was executed according to the designed program (Zhao et al., 2005a). However, according to Wang et al. (2001a), the preflush slug 1 and the post-flush slug 4 in the designed program were not executed. Following are some of observations from this case:

- The peak production rate in the pilot area increased from 27 t/d to 84 t/d. The water cut decreased from 94.2 to 83.9%. For the ASP projects near this pilot area, low water cut was maintained about 1 year. Most of the wells in this pilot had a longer period of low water cut.

- After ASPF flooding, the chloride ion concentration of the effluent fluid at all the producers increased significantly, as shown in Figure 13.49. We can see that the volumetric sweep efficiency was increased because the formation water salinity was much higher.
- Emulsification was more severe. The emulsion viscosity of a sample was 300 mPa·s, which is higher than the emulsion viscosity from a similar ASP fluid (100 to 150 mPa·s).
- The increase in injection pressure in this ASPF pilot was higher than that in water, polymer, or ASP flood in the neighboring areas, as shown in Table 13.19 (Wang et al., 2001a). The high pressure made it difficult to inject gas

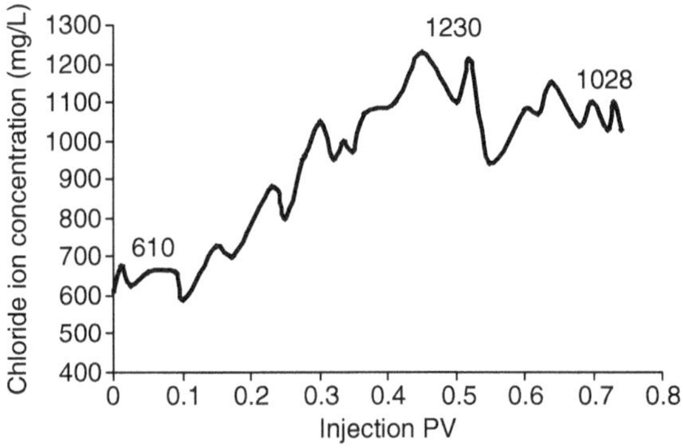

FIGURE 13.49 Chloride ion concentration in the produced water from a central well. *Source: Wang et al. (2001a).*

TABLE 13.19 Injection Pressure in Different Pilots

Pilot	Injection Pressure, MPa		
	Before	*After*	*Increase, %*
ASPF	7.0	14.0	100
ASP in S-ZX	3.6	5.9	63.0
ASP in Bei-1-X	9.2	12.5	35.9
ASP in SB-B2-Z	7.2	9.3	29.2
ASP in X2-X	8.1	11.0	35.8
WAG	7.4	9.2	24.3

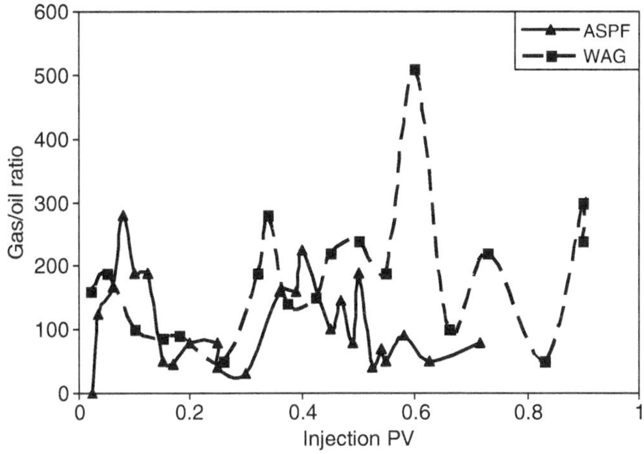

FIGURE 13.50 Comparison of GOR during WAG and ASPF flooding. *Source: Wang et al. (2001a).*

continuously. In fact, that was the reason for the actual injection gas/liquid ratio lower than the designed value.

- The gas/oil ratio (GOR) during the ASPF period was lower than that during the WAG period in the same flood area (see Figure 13.50), indicating less gas fingering and channeling. From Figure 13.50 we can see that at the beginning, the GOR of ASPF flooding was high; then it got markedly lower, about equal to the original GOR of the formation. The reason might be that this area previously was WAG flooded and free gas accumulated at the local high areas of the pilot. When ASPF flooding, because the volumetric sweep efficiency increased, the free gas in the local high areas were first displaced and pushed toward the producers, causing the GOR to be high at the beginning of the ASPF flood and then to decrease.

13.11.6 Shengli Gudong ASP Pilot

This section describes the first ASP pilot in China, started in 1992; the Shengli Gudong pilot test is earlier than the first Daqing ASP pilot test in S-ZX. It was operated by the Shengli Petroleum Administration Bureau (Shengli Oilfield), a subsidiary of SINOPEC, China Petroleum & Chemical Corporation.

Reservoir and Fluid Description

This pilot location was in the Gudong field in the northeast of Shangdong Province, The target layer was $Ng5_{2+3}$ of the tertiary Guantao member. This unconsolidated sandstone reservoir, formed by fluvial sedimentation, was characterized by severe heterogeneity and high fluid viscosity. In this test, 9 producers, 4 injectors, and 2 observation wells were drilled in five-spot patterns, as shown in Figure 13.51. Here, the well spacing was 50 m. Well spacing is defined as the distance between an injection well and its adjacent production wells in China. Some of the reservoir and fluid data are shown in Table 13.20 (Song et al., 1995; Wang et al., 1997b; Qu et al., 1998).

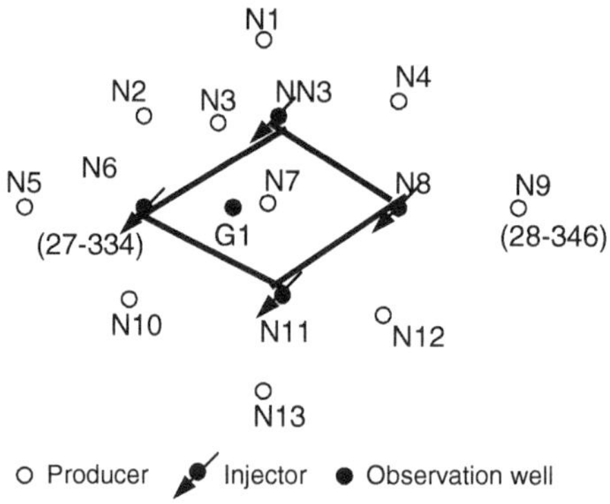

FIGURE 13.51 Well patterns in the pilot area.

TABLE 13.20 Gudong Pilot Reservoir and Fluid Data

ASP area, km^2	0.031
Formation depth (subsea), m	1272
Permeability, md	2560
Average porosity, %	35
Total PV, m^3	119,227
OOIP, tons	77,952
Permeability variation coefficient	0.33
Effective thickness, m	11
Total thickness, m	12.5
Formation temperature,°C	68
Formation water TDS, mg/L	3259
Formation water divalents, mg/L	19–92
Injection water salinity, mg/L	737
Oil viscosity in place, mPa·s	41.3
Oil acid number, mg KOH/g	3.11

Injection Scheme

Based on laboratory study, the following scheme was designed:

1. Preflush slug: 0.05 PV, 0.1% 3530S polymer
2. AS slug: 0.05 PV, 1.5% Na_2CO_3 + 0.2% OP-10 + 0.2% CY1
3. Main ASP slug: 0.35 PV, 1.5% Na_2CO_3 + 0.2% OP-10 + 0.2% CY1 + 0.1% 3530S
4. Post-flush: 0.1 PV 0.05% 3530S

Pilot Performance

In this test, the field was waterflooded from 1987 to 1992. A total 5.06 PV of water was injected before the ASP pilot test, with 98.4% water cut and 54.4% oil recovery factor. The ASP injection was started August 1, 1992, and ended June 24, 1993. A total of 0.592 PV ASP was injected. Basically, the designed injection scheme was executed, except in the main ASP slug, Na_2CO_3 concentration was reduced to 1.1%, and at a later stage, CY was replaced by OP because of the short supply. In addition, 0.045 PV more than the designed ASP slug was injected. The viscosities of injected fluids at wellhead at the four injection stages were 16.5, 0.47, 7.86, and 6.16 mPa·s, respectively. After July 1993, the water was reinjected. By July 1994, the water cut returned to the level before the ASP pilot, and the chemical concentrations in produced water were very low. Then the test was stopped. When 0.49 PV of fluid was injected, the water cut in the pilot area reached a minimum of 74.2%. The final incremental recovery was 13.4 % OOIP.

Injectivity and Injection Rate

The sand in this test was unconsolidated. It would be interesting to see how the injectivity and injection rate changed. It was observed that when high viscous polymer in the preflush was injected, the injectivity decreased by 27%, and the injection pressure increased by 1.07 MPa. The injection rate was almost unchanged. After that, the well injectivity increased rapidly along with the decrease in the viscosity of injected fluids.

Chromatographic Separation

In most of the Daqing ASP tests, polymer broke through first, followed by alkali and surfactant. However, this pilot test showed that alkali broke through first, followed by polymer, and later by surfactants, as shown in Figure 13.52. Their breakthrough concentrations are shown in Figure 13.53 (Qu et al., 1998). The reason for this outcome is unknown. The consumptions of alkaline, polymer, and surfactants (CY-1 and OP-1O) were 62.1%, 69.7%, and (99.5% and 99.4%) of the injected, respectively. The consumptions in this test were higher than those in laboratory experiments (Song et al., 1995).

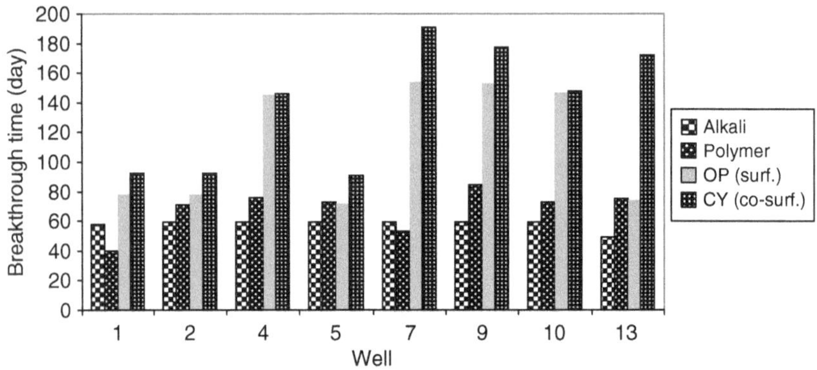

FIGURE 13.52 Chemical breakthrough times at different wells.

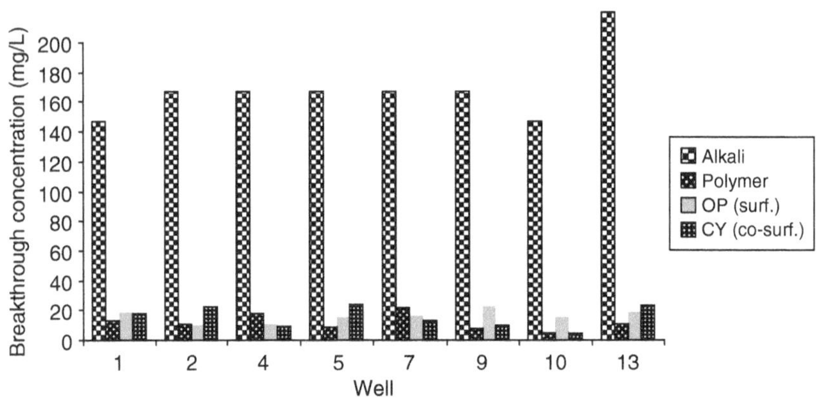

FIGURE 13.53 Chemical breakthrough concentrations at different wells.

13.11.7 Shengli Gudao ASP Pilot

The ASP pilot test described in this section was in a relatively viscous oilfield (with the in-place oil viscosity being 70 mPa·s).

Reservoir and Fluid Description

The ASP pilot was located in the Gudao field, Shengli. The target layer, Guan 4^2–4^4, was characterized by severe heterogeneity and high fluid viscosity. There were 13 producers and 6 injectors drilled in 6 rectangular inverted five-spot patterns, as shown in Figure 13.54. The central producers were 5-142 and 6-121. In this test, the well spacing between an injector and an adjacent producer was 200 m. Some of the reservoir and fluid data are shown in Table 13.21 (Cao et al., 2002; Chang et al., 2006).

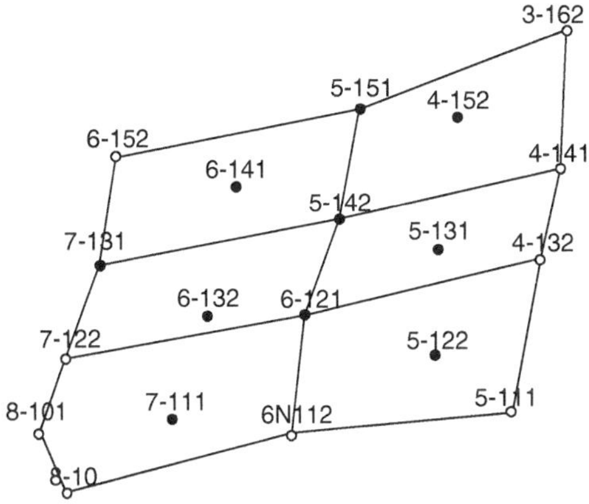

FIGURE 13.54 Well patterns in the pilot area.

TABLE 13.21 Gudao Pilot Reservoir and Fluid Data

ASP area, km^2	0.61
Formation depth (subsea), m	1190–1310
Permeability, md	1520
Average porosity, %	32
Total PV, m^3	3,160,000
OOIP, tons	1,972,000
Effective thickness, m	16.2
Formation temperature,°C	69
Formation water TDS, mg/L	6864
Formation water divalents, mg/L	143
Oil viscosity in place, mPa·s	70
Oil acid number, mg KOH/g	1.7

Injection Scheme

The injection scheme for the Shengli Gudao pilot test was designed based on laboratory studies and simulation optimization. The actual injection scheme, which follows, was similar to the designed:

- Preflush slug: 0.097 PV, 2000 mg/L 3530S polymer
- ASP slug: 0.309 PV, 1.2% Na_2CO_3, 0.3% surfactant (mixture of anionic BES and lignosulfonate PS), and 0.17% polymer
- Post-flush: 0.05 PV, 0.15% polymer

Pilot Performance

The polymer preflush was started in May 1997, followed by an ASP slug in May 1998. The test ended in November 2001. Waterflooding followed the pilot test. After 0.03 PV ASP slug injection, producers started to respond to the chemical injection. The water cut decreased by 10.2% from 94.7 to 84.5%, and the production rate increased from 82 to 194 t/d. The estimated incremental oil recovery from this test was 15.5%.

13.11.8 Karamay ASP Pilot

The ASP pilot was located in the Karamay field, Xinjiang Province, China. For this pilot, extensive laboratory chemical screening and UTCHEM simulations, including alkaline reactions, were conducted.

Reservoir and Fluid Description

In the pilot test in the Er-Zhong-Bei block in the Karamay field, the target layers were S_7^3 and S_7^4. Nine producers, 4 injectors, and 2 observation wells were drilled in the 4 inverted five-spot patterns, as shown in Figure 13.55. Outside the pilot zone, there were 6 pressure equilibrium wells: 9-2B, 10-2A, 10-3, 9-4, 323, and T7-7. Wells 9-3 and 8-3A were used for observation. The well spacing from injector to producer was 31.2 to 60.45 m, and the spacing between two injectors was about 70 m. Some of the reservoir and fluid data are shown in Table 13.22 (Gu et al., 1998; Delshad et al., 1998; Qiao et al., 2000).

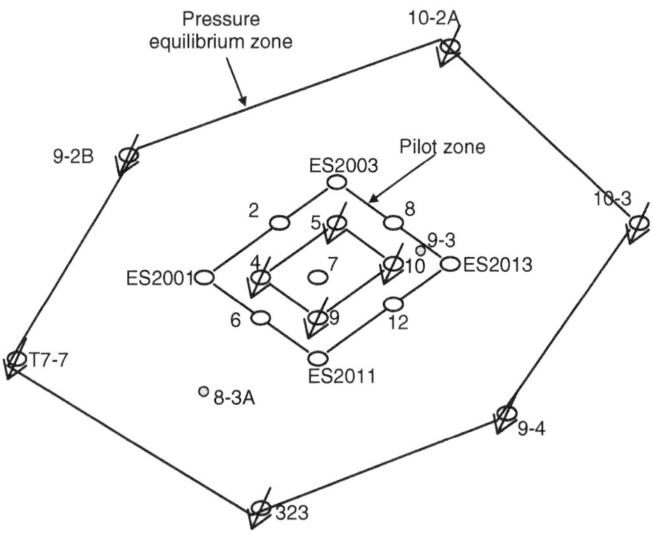

FIGURE 13.55 Well patterns in the Karamay pilot area.

TABLE 13.22 Karamay ASP Pilot Reservoir and Fluid Data

Formation depth (subsea), m	599–678
Average permeability, md	157
Average porosity, %	18.1
Effective thickness, m	15–22
Formation temperature,°C	23
Formation water TDS, mg/L	~700
Formation water divalents, mg/L	~70
Formation water pH	6
Injection water salinity, mg/L	387
Oil viscosity in place, mPa·s	8.82, 17.2*
Oil acid number, mg KOH/g	0.35–1.5

**Reported in Delshad et al. (1998)*

Laboratory Studies

To select alkali, static alkaline consumption and mineral dissolution studies were conducted using field sand. The results showed that Na_2CO_3 consumption was almost unchanged as Na_2CO_3 concentration increased. The consumption was about 6 mg/g sand when Na_2CO_3 concentration reached 0.8%. In contrast, NaOH consumption increased significantly with its concentration; NaOH consumption reached 40 mg/g sand at 0.8% NaOH concentration. Silica dissolution was insensitive to Na_2CO_3 concentration and at a constant value of 0.25 mg SiO_2/g sand, but silica dissolution increased with NaOH. The silicon dissolution was 0.63 mg SiO_2/g sand, which is 2.5 times the dissolution with Na_2CO_3. The minimum IFT in the order of 10^{-3} mN/m corresponded to 0.1% NaOH concentration and 0.8% Na_2CO_3 concentration (Qiao et al., 2000). Based on these results, Na_2CO_3 was selected. Many surfactants, including some surfactants made locally, were tested. Finally, KPS-1, made locally, was selected owing to its low cost and good phase behavior.

Figure 13.56 shows the IFT at different alkaline and surfactant concentrations. It shows that the ultralow IFT depended on the alkaline and surfactant concentrations; 1.4% Na_2CO_3 concentration should be chosen for 0.3% KPS-1 for the low IFT. Figure 13.57 shows that the polymer (3530S) concentrations affected IFT, with higher concentrations corresponding to lower IFT. For the mobility control purpose, a higher concentration of polymer was needed. These

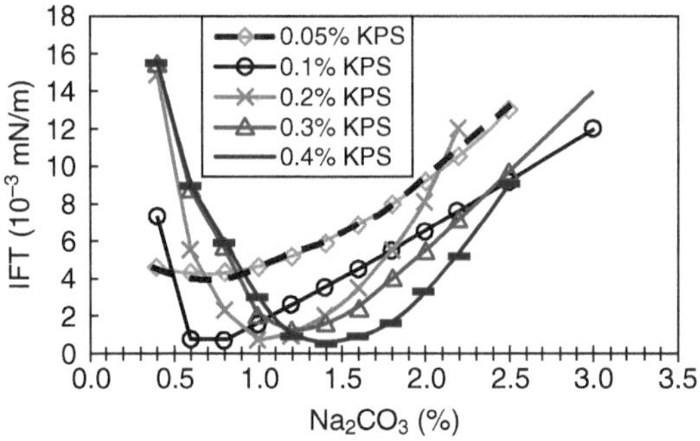

FIGURE 13.56 Effect of AS (Na_2CO_3-KPS) on IFT. *Source: Qiao et al. (2000).*

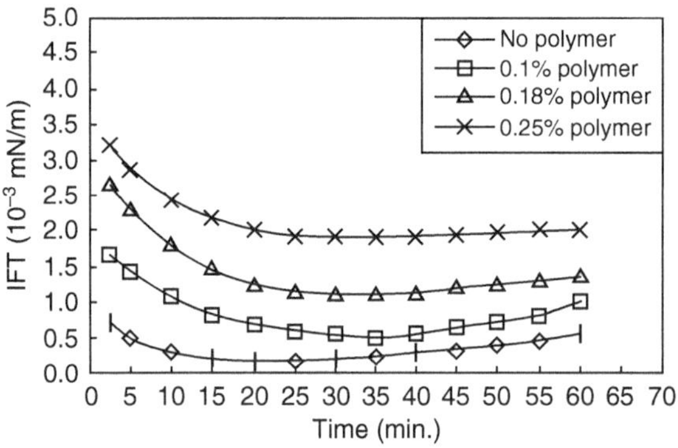

FIGURE 13.57 Polymer (3530S) concentration effect on IFT. *Source: Qiao et al. (2000).*

two factors must be balanced; therefore, the polymer concentrations of 0.13% in the ASP slug and 0.1% in the polymer drive, respectively, were selected.

A chemical slug will be diluted by formation water or by the slugs pre-injected and followed. The dilution will affect the IFT and phase behaviors of the injected chemical slugs. Therefore, a dilution study was conducted to ensure the selected chemical slug was robust to dilution.

Four injection schemes—AS/ASP, AP/ASP, AS/P, and ASP/P—were evaluated to optimize the process. All the injection schemes except AS/P (AS followed by P) showed over 25% incremental oil recovery factor in test conditions. The chemical cost for the ASP/P scheme (ASP followed by P) was much lower than the other cases. Therefore, ASP/P was selected.

Simulation Studies

A black oil model was built based on matching 35 years of waterflooding history. This model was used to define the injection and production rates, limits of well injection, and flowing pressure so that the injection and production were equilibrated in the pressure equilibrium zone. The model was also used to define the injection and production rates in the pilot zone to ensure the injected chemicals were confined in the pilot zone. This model was used to successfully predict the preflush history.

A UTCHEM chemical flood model was built based on history-matching core flood experiments. The chemical parameters were calibrated through matching experiments. The calibrated parameters were used to build a field sector model to simulate the ASP pilot test. In the model, detailed alkaline reactions were considered. It was probably the first time such an ASP model was applied in a real field scale. The model was used to optimize injection schemes. For example, the model showed that when the post-flush slug volume was in the range of 0.15 to 0.35 PV, the pilot performance was insensitive to the slug size. Therefore, 0.15 PV of post-flush slug was selected for the injection scheme.

The UTCHEM model was updated with the ASP flood history (see Figure 13.58). The updated model was then used to optimize or modify the injection scheme. For example, the chemical formulations were adjusted in January 1997 based on the results from the simulation model, which followed up the ASP flood history, together with the monitoring results of injection and production fluids. The surfactant concentration in the ASP slug was changed from the designed 0.3 to 0.35% and Na_2CO_3 from 1.4 to 1.2%.

Injection Scheme

Based on the chemical screening and core flood tests in the laboratory and a numerical simulation study, the following scheme was designed:

- Preflush slug: 0.4 PV, 1.5% NaCl brine
- ASP slug: 0.34 PV, 1.4% Na_2CO_3 + 0.1% $Na_5P_3O_{10}$ + 0.3% S (KPS-1) + 0.13% 3530S
- Post-flush: 0.15 PV, 0.1% P (3530S) + 0.4% NaCl

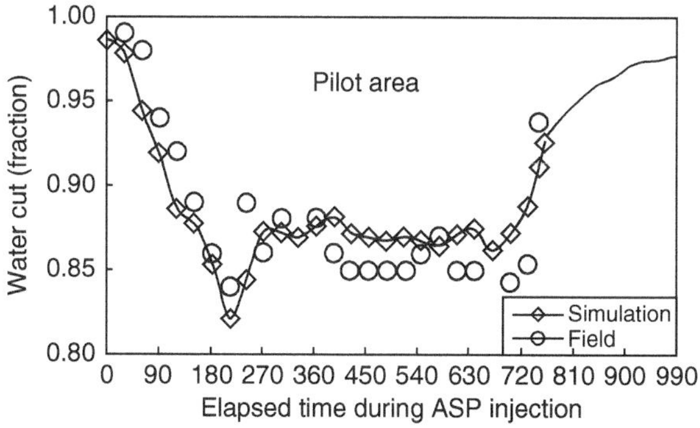

FIGURE 13.58 Water cut history match plot. *Source: Qiao et al. (2000).*

NaCl brine preflush was used to condition the reservoir, and $Na_5P_3O_{10}$ was used as a divalent chelating agent.

Pilot Performance

The NaCl brine preflush slug was injected from August 18, 1995, to July 21, 1996, after about 35 years of waterflooding. This process lasted 338 days, including the waiting period for the injection facilities. The ASP injection, which was started on July 22, 1996, and ended on June 12, 1997, lasted 319 days. In the 174-day period from June 13, 1997, to December 4, 1997, the polymer slug was injected. A water drive, which was conducted after the chemical injection processes, took about another half year, and the pilot test ended in the middle of 1998.

At the end of preflush, the oil recovery factor in the pilot area was 49.65% with 99% water cut, and 50.14% oil recovery factor with 99% water cut in the central well. An increased recovery was observed after 0.043 PV of ASP slug had been injected. At 0.202 PV injection, the oil rate reached 1.27 t/d, which is 6.35 times that before the ASP injection, and the water cut decreased from 99 to 84%. The oil recovery factor was increased by 24% OOIP for the whole pilot and 25% for the central well. Severe emulsions in produced fluids were observed, and difficulties were encountered in breaking the emulsions (Chang et al., 2006).

13.11.9 Jilin Honggang ASP Pilots

The ASP pilot test described in this section was conducted in the Honggang field in Jilin Province, China. Three well patterns were used in this pilot test: Well 7-3, Well 8-4, and Well 125. Figure 13.59 shows the schematic for the

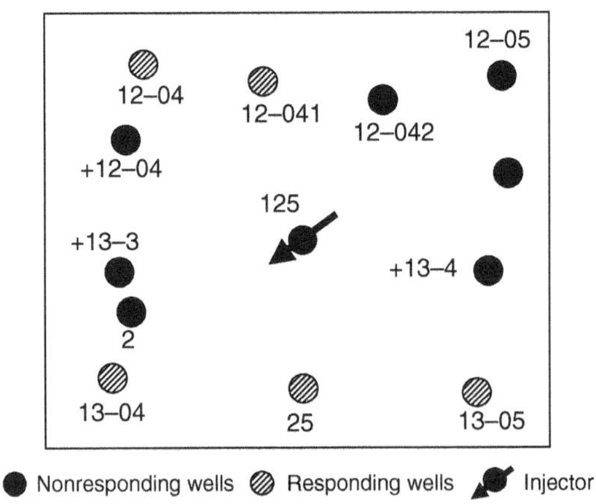

FIGURE 13.59 Well location in the well 125 pattern.

TABLE 13.23 125 Pattern Reservoir and Fluid Data

ASP area, km^2	0.36
Permeability, md	163
Total PV, m^3	388,230
OOIP, tons	450,000
Permeability variation coefficient	0.425
Formation temperature,°C	55
Oil viscosity in place, mPa·s	12.9
Oil acid number, mg KOH/g	Low

Well 125 pattern, which was an inverted 13-spot pattern. The well spacing from an injector to a producer was 200 m. Some of the reservoir and fluid data are shown in Table 13.23 (Zhang et al., 2001). The other two patterns were similar to the 125 pattern.

The injection scheme for this test was 0.25% A1 + 0.5% A2 + 0.06% S + 0.15% P. The pilot was started in October 1997 and was stopped in June 1999 because the casing for the injector 125 was broken. The test was resumed in October 1999, and a chemical injection was ended in June 2000. Only the wells on the southern and northern sides of the injector 125 responded to the chemical injection (water cut reduced and oil rate increased), not the wells on the eastern and western sides, because the main water injection stream was oriented in the eastern–western direction so that the wells in this direction were well flushed by waterflood before the ASP injection. The ASP injection improved sweep efficiency on the southern and northern sides. Thus, the wells on these sides responded to the ASP injection. The injected surfactant concentration was low (0.06%), and the crude oil had a low acid number. The main mechanism in this pilot was probably the sweep efficiency improvement by polymer injection.

Table 13.24 shows the produced oil analysis results from one of the producers (Zhang et al., 2001). From this table, we can see that the heavy nonhydrocarbon components in the produced crude oil increased and wax melting point became higher. The reason is that these heavy components are adsorbed initially on rock surfaces and are desorbed when the ASP solution contacts them. This phenomenon was also observed in Gudong ASP pilot (Song et al., 1994).

13.11.10 Zhongyuan Huzhuangji ASP Pilot

The Zhongyuan Huzhuangji ASP pilot was conducted in a high-temperature and high-salinity field by the Zhongyuan Oilfield, a subsidiary of SINOPEC,

TABLE 13.24 Produced Oil Sample Analysis

Sampling Date	Asphaltene, %	Resin, %	Wax, %	Wax Melting Point,°C	Viscosity, mPa·s
Before ASP	2.9	14.2	18.2		63.3
April 1998	0.1	23.72	19.96	36.5–37.5	
November 1998	0.11	26.97	16.76	40–41	50.16
February 1999	0.14	26.01	19.42	48.5–49.5	78.45

TABLE 13.25 Huzhuangji Pilot Reservoir and Fluid Data

ASP area, m^2	410
Formation depth (subsea), m	2000
OOIP, tons	737,000
Porosity, %	19.6
Permeability, md	129–277
Formation temperature,°C	86
Formation water TDS, mg/L	170,000
Formation water Ca^{2+}, mg/L	4799
Formation water Mg^{2+}, mg/L	830

China Petroleum and Chemical Corporation. The polymer in this pilot test was xanthan gum.

Reservoir and Fluid Description

The target layer of the pilot test performed in the Huzhuangji field, Henan Province, was S3 Zhong1-2 in Hu 5-15 block. There were 5 producers and 5 injectors. Some of the reservoir and fluid data from this test are shown in Table 13.25 (Jiang et al., 2003b).

Injection Scheme

The injection scheme for this pilot test follows.

- Preflush: 2920 m^3, 1% KCl to reduce formation salinity
- Polymer slug: 2920 m^3, 0.15% xanthan gum

- Main ASP slug: 49,640 m^3, 0.5% mixed alkalis (Na_2CO_3 + $NaHCO_3$ at 1 : 1 ratio) + 1.5% natural mixed carboxylates + 0.08% alkyl sulfobetaines (DSB) + 0.08% xanthan. The mixed alkalis were injected through only 1 of 5 injectors
- Polymer drive: 2920 m^3, 0.1% xanthan gum

Natural mixed carboxylates could be resistant to 380 mg/L divalents. If combined with DSB, they could be resistant to 5000 mg/L divalents. These surfactants were made in China. Ryles (1983) raised the question about whether xanthan can be used with alkali in such a high temperature, however.

Pilot Performance

The pilot test was started on January 24, 2000, and ended on March 20, 2001. The liquid rate in the pilot decreased from 377.5 to 219.8 t/d. The oil rate increased from 10.5 to 20.1 t/d, and the water cut decreased from 97.2 to 90.9%. By December 31, 2001, the incremental oil production was 6208 m^3. By this time, the benefit had already lasted 20 months. Some wells still benefited from the chemical injection after that time. However, one ton of incremental oil costs 40 kg alkalis, 126 L of surfactants, and 7.6 kg of polymer!

13.11.11 Yumen Laojunmiao A/S/P Pilot

The Yumen Laojunmiao A/S/P pilot test was not the same as the ASP pilots presented previously where alkali, surfactant, and polymer were injected simultaneously in the same slug. In this pilot test, alkali, surfactant, and polymer were injected sequentially. This pilot, conducted in a reservoir of initially high salinity, was operated by Yumen Oilfield, a subsidiary of PetroChina.

Reservoir and Fluid Description

In this pilot test performed in the Laojunmiao field, Gangsu Province, the target layer was L_1. This test involved 1 injector (H184) and 4 producers (406, 930, F185, and G185), forming an inverted 5-spot pattern, as shown in Figure 13.60. The average well spacing between an injector and a producer was 70 m. The clay content was 11.6%, which was composed of montomorillonite. Some of the reservoir and fluid data for this oil-wet reservoir are shown in Table 13.26 (Wang et al., 1999a).

Actual Injection Scheme

The actual injection scheme for this pilot test follows.

- KCl preflush: 3427 m^3 (equivalent 0.24 PV), 4% KCl
- KCl and alkaline preflush: 3971 m^3 (equivalent to 0.28 PV), 2% KCl + 4% Na_2CO_3
- Micellar/polymer flooding: 4807 m^3 (equivalent to 0.33 PV), 5% YPS-3A (local surfactant) + 3% n-butanol (cosolvent) + 0.4% K_2CO_3 that is followed

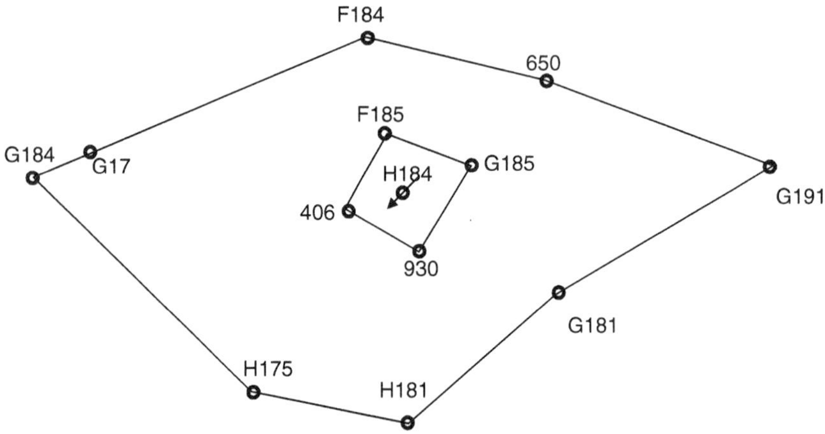

FIGURE 13.60 Liaojunmiao A/S/P pilot well pattern.

TABLE 13.26 Laojummiao Pilot Reservoir and Fluid Data

A/S/P area, m^2	9247
Formation depth (subsea), m	500
Pore volume, m^3	14,250
OOIP, tons	8200
Effective thickness, m	6.42
Porosity, %	24
Permeability, md	374
Formation temperature,°C	28
Formation water TDS, mg/L	63,700
Formation water Ca^{2+} and Mg^{2+}, mg/L	4335
Formation water TDS before test, mg/L	7996
Formation water Ca^{2+} and Mg^{2+}, mg/L	566
Oil viscosity original in place, mPa·s	4.17
Oil viscosity before pilot test, mPa·s	7.14
Crude oil acid number, mg KOH/g oil	0.19

by AD-27 (SNF product, 9 million MW, 17% hydrolysis) with graded concentrations

Pilot Execution and Performance Evaluation

This subsection presents the alkaline preflush and micellar-polymer flooding in this pilot test. The economic data is also presented.

Alkaline Preflush

Considering the high clay content, KCl and alkaline preflush were injected to reduce the divalent content. First, a 4% KCl solution was injected in March 1994, followed by a 2% KCl and 4% Na_2CO_3 injection, which ended in May 1995. The injection rate was 30 m^3/d. The total produced Ca^{2+} and Mg^{2+} from the three wells—406, F185, and 930—was 5772 kg. The total amount of divalents in the preflushed zone was 10,125 kg; 57% of the divalents had been flushed out. pH increased from 6.2 before the preflush to 8.0. The injection pressure increased from 4.7 MPa before the flush to 5.0 MPa at the end of 4% KCl injection to 5.6 MPa at the end of alkaline injection. The pressure increase was thought to be caused by divalent precipitation.

Micellar-Polymer Flooding

The molecular weight of the surfactant YPS-3A used in this test was 430. The IFT with oil was 0.052 mN/m at 7000 mg/L salinity. During this injection period, no response was observed from the producers. The reason is probably that the IFT was so high the oil films on oil-wet rock surfaces could not be displaced. Another reason for the noneffectiveness is that YPS-3A was so lipophilic that most of the injected surfactant partitioned in the oil phase. Because the producer 930 was converted from an injector, low residual oil saturation existed around the well. Thus, no oil was produced during the pilot test period.

At the end of polymer drive—cumulative 0.33 PV injection (4807 m^3), including surfactant injection—the producers started to respond. The water cut decreased from 99% before the test to 87%, and the oil rate increased from 0.2 to 1.9 t/d at the peak rate. However, the response lasted only 90 days. In this pilot, the viscosities of micellar solution and polymer solution were 15.7 mPa·s and 14 mPa·s, respectively, about twice the oil viscosity. A good mobility ratio might be the main reason for the positive response.

Economics

The incremental oil recovery from this pilot test was 1.82%. According to the oil price at that time, the ratio of investment to return was 10 : 1. The chemical cost was 67.6% of the total cost. The pilot test itself was not economical. From this pilot test, it was concluded that the surfactant YPS-3A was not feasible for the application in this field.

References

Abrams, A., 1975. The influence of fluid viscosity, interfacial tension, and flow velocity on residual oil saturation left by waterflood. Society of Petroleum Engineers Journal (SPEJ) (October), 437–447.

Adamson, A.W., 1997. Physical Chemistry of Surfaces. Interscience Publishers.

Adibhatla, B., Mohanty, K.K., 2008. Oil recovery from fractured carbonates by surfactant-aided gravity drainage: laboratory experiments and mechanistic simulations. SPE Reservoir Engineering and Evaluation (SPEREE) (February), 119–130.

Adibhatla, B., Sun, X., Mohanty, K.K., 2005. Numerical studies of oil production from initially oil-wet fracture blocks by surfactant brine imbibitions. Paper SPE 97687 presented at the SPE International Improved Oil Recovery Conference in Asia Pacific, Kuala Lumpur, 5–6 December.

Akstinat, M.H., 1980. Polymers for enhanced oil recovery in reservoirs of extremely high salinities and high temperatures. Paper SPE 8979 presented at the SPE Fifth International Symposium on Oilfield and Geothermal Chemistry, Stanford, 28–30 May.

Akstinat, M.H., 1981. Surfactants for EOR process in high-salinity systems: product selection and evaluation. In: Fayers, F.J. (Ed.), Enhanced Oil Recovery. Elsevier, pp. 43–62.

Al-Anazi, H.A., Sharma, M.M., 2002. Use of pH sensitive polymer for conformance control. Paper SPE 73782 presented at SPE International Symposium Formation Damage Control, Lafayette, Louisiana, 20–21 February.

Al-Azri, N., Jamal, E., Murshidi, A., Al-Mahrouqi, A., Al-Busaicdi, I., Kazzaz, A., AI-Ajmi, W., Ramalingam, S., De Kruijf. A., Al-Kharusi, B., Brooks, D., 2010. Polymer injection in heavy oil reservoir under strong bottom water drive. Paper SPE 129177 presented at the SPE EOR Conference at Oil & Gas West Asia, Muscat, 11–13 April.

Al-Bahar, M.A., Merrill, R., Peake, W., Jumaa, M., Oskui, R., 2004. Evaluation of IOR potential within Kuwait. Paper SPE 88716 presented at the Abu Dhabi International Conference and Exhibition, Abu Dhabi, 10–13 October.

Amaefule, J.O., Handy, L.L., 1982. The effect of interfacial tension on relative oil and water permeabilities of consolidated porous media. SPEJ (June), 371–381.

Anderson, G.A., Delshad, M., King, C.B., Mohammadi, H., Pope, G.A., 2006. Optimization of chemical flooding in a mixed-wet dolomite reservoir. Paper SPE 100082 presented at the SPE/DOE Symposium on Improved Oil Recovery, Tulsa, 22–26 April.

Anderson, W.G., 1987. Wettability literature survey part 5: the effects of wettability on relative permeability. Journal of Petroleum Technology (JPT) 39 (11), 1453–1468.

Appelo, C.A., Postma, D., 2007. Geochemistry, Groundwater and Pollution, second ed. A.A. Balkema Publishers.

Arya, A., 1986. Dispersion and Reservoir Heterogeneity. Ph.D. dissertation, University of Texas at Austin.

Arya, A., Hewett, T.A., Larson, R.G., Lake, L.W., 1988. Dispersion and reservoir heterogeneity. SPE Reservoir Engineering (SPERE) (February), 139–148.

Atkins, P.W., de Paula, J., 2006. Physical Chemistry, eigth ed. W.H. Freeman.

Austad, T., Matre, B., Milter, J., Sævereid, A., Øyno, L., 1997. Chemical flooding of oil reservoirs 8. Spontaneous oil expulsion from oil- and water-wet low permeable chalk material by imbibition of aqueous surfactant solutions. Colloids Surface A: Physicochemical and Engineering Aspects 137 (1–3), 117–129.

Austad, T., Strand, S., Høgnesen, E.J., Zhang, P., 2005. Seawater as IOR fluid in fractured chalk. Paper SPE 93000 presented at SPE International Symposium on Oilfield Chemistry, Woodlands, Texas, 2–4 February.

Austad, T., Strand, S., Madland, M.V., Puntervold, T., Korsnes, R.I., 2008. Seawater in chalk: an EOR and compaction fluid. SPEREE (August), 648–654.

Ayirala, S., Uehara-Nagamine, E., Matzakos, A., Chin, R., Doe, P., van Den Hoek, P., 2010. A designer water process for offshore low salinity and polymer flooding applications. Paper SPE 129926 presented at the SPE Improved Oil Recovery Symposium, Tulsa, 24–28 April.

Babadagli, T., Al-Bemani, A., Boukadi, F., Al-Maamri, R., 2002. A laboratory feasibility study of dilute surfactant injection for the Yibal Field, Oman. Paper SPE 78352 presented at the SPE 13th European Petroleum Conference, Aberdeen, 29–31 October.

Bakhitov, G.G., Polishchuk, A.M., Ogandzhanyants, V.G., 1980. Experimental investigation into the influence of polymer additives in water on the relative permeabilities of porous media. Fluid Dynamics Izv. Akad. Nauk, Moscow, 611–615.

Bai, B.-J., Li, L.-X., Liu, Y.-Z., Liu, H., Wang, Z.-G., You, C.-M., 2007. Preformed particle gel for conformance control: Factors affecting its properties and applications. SPEREE 10 (4), 415–422.

Baijal, S.K., 1981. Polymer molecules interact with porous matrix. Oil and Gas J. (March), 106–107.

Ball, J., Surkalo, H., 1988. New variations of chemical EOR and their economic potential: The ASP process (alkaline-surfactant-polymer). Proceedings of the Improved Oil Recovery Conference, Abilene, September (DOE/BC/14286–1, March 1989), 229–250.

Barakat, Y., Fortney, L.N., Schechter, R.S., Wade, W.W., Yiv, S.H., Graciaa, A., 1983. Criteria for structuring surfactants to maximize solubilization of oil and water. Journal of Colloid and Interface Science 92, 561–574.

Baviere, M., Wade, W.H., Schechter, R.S., 1981. The effect of salt, alcohol and surfactant on optimum middle phase composition. In: Shah, D.O. (Ed.), Surface Phenomena in Enhanced Oil Recovery. Plenum, pp. 117–135.

Bear, J., 1972. Dynamics of Fluids in Porous Media. Elsevier.

Berger, P.D., Lee, C.H., 2006. Improved ASP process using organic alkali. Paper SPE 99581 presented at the 2006 SPE/DOE Symposium on Improved Oil Recovery, Tulsa, 22–26 April.

Bernard, G.G., 1967. Effect of floodwater salinity on recovery of oil from cores containing clays. Paper SPE 1725 presented at the 38th Annual California Regional Meeting, Los Angeles, 26–27 October.

Bhuyan, D., 1989. Development of an Alkaline/Surfactant/Polymer Flood Compositional Reservoir Simulator. Ph.D. dissertation, University of Texas at Austin.

Bhuyan, D., Pope, G.A., Lake, L.W., 1990. Mathematical modeling of high-pH chemical flooding. SPERE (May), 213–220.

Bi, Z.-C., Qian, Y.-Y., Xu, F., Yu, J.Y., 1997. Formation of decane oil film and wettability on silica surfaces. Oilfield Chemistry 14 (4), 345–348.

Bird, R.B., Armstrong, R.C., Hassager, O., 1987. Dynamics of Polymeric Liquids. Vol. 1, Fluid Mechanics, second ed. Wiley.

Bird, R.B., Stewart, W.E., Lightfoot, E.N., 1960. Transport Phenomena. Wiley.

Bock, J., Valin Jr., P.L., Pace, S.J., et al., 1988. Hydrophobically associating polymers. In: Stahl, G.A., Schulz, D.N. (Eds.), Water-Soluble Polymers for Petroleum Recovery. Plenum, pp. 147–160.

Bondor, P.L., Hirasaki, G.J., Tham, M.J., 1972. Mathematical simulation of polymer flooding in complex reservoirs. SPEJ (October), 369–382.

Boom, W., Wit, K., Schulte, A.M., Oedai, S., Zeelenberg, J.P.W., Maas, J.G., 1995. Experimental evidence for improved condensate mobility at near-wellbore flow conditions. Paper SPE 30766 presented at the SPE Annual Technical Conference and Exhibition, Dallas, 22–25 October.

Boom, W., Wit, K., Zeelenberg, J.P.W., Weeda, H.C., Maas, J.G., 1996. On the use of model experiments for assessing improved under near-wellbore flow conditions. Paper SPE 36714 presented at the SPE Annual Technical Conference and Exhibition, Denver, 6–9 October.

Boston, W.G., Brandner, C.F., Foster, W.R., 1969. Recovery of oil by waterflooding from an argillaceous, oil-containing subterranean formation. US Patent 3,740,956, 7 October.

Bourdarot, G., Sardin, M., Putz, A., 1984. Chateaurenard field test recovery mechanisms and interpretation. Paper SPE 12685 presented at the SPE Enhanced Oil Recovery Symposium, Tulsa, 15–18 April.

Bourrel, M., Schechter, R.S., 1988. Microemulsions and related systems, formulation, solvency, and physical properties. Marcel Dekker, Inc.

Bragg, J.R., Gale, W.W., McElhannon Jr., W.A., Davenport, O.W., Petrichuk, M.D., Ashcraft, T.L., 1982. Loudon surfactant flood pilot test. Paper SPE 10862 presented at the SPE/DOE Enhanced Oil Recovery Symposium, Tulsa, 4–7 April.

Breeuwsma, A., Wösten, J.H.M., Vleeshouwer, J.J., Van Slobbe, A.M., Bouma, J., 1986. Derivation of land qualities to assess environmental problems from soil survey. Soil Science Society of America Journal (Soil Sci. Soc. Am. J.) 50, 186–190.

Brigham, W.E., 1974. Mixing equations in short laboratory cores. SPEJ (February), 91–99.

Brooks, A.D., de Zwart, A.H., Bychkov, A., Al-Azri, N., Hern, C.Y., Al-Ajmi, W., Mukmin, M., 2010. Evaluation of EOR techniques for medium-heavy oil reservoirs with a strong bottom aquifer in South Oman. Paper SPE 129149 presented at the SPE EOR Conference at Oil & Gas West Asia, Muscat, 11–13 April.

Brooks, R.H., Corey, A.T., 1966. Properties of porous media affecting fluid flow. Journal of the Irrigation and Drainage Division (J. Irrig. Drain. Div.) 6, 61.

Brownell, L.E., Katz, D.L., 1947. Flow of fluids through porous media, part II, simultaneous flow of two homogeneous phases. Chemical Engineering Process 43, 601–612.

Bruggenwert, M.G.M., Kamphorst, A., 1982. Survey of experimental information on cation exchange in soil systems. In: Bolt, G.H. (Ed.), Soil Chemistry B. Physico-Chemical Methods. Elsevier, pp. 141–203.

Bryan, J., Kantzas, A., 2007. Enhanced heavy-oil recovery by alkali–surfactant flooding. Paper SPE 110738 presented at the SPE Annual Technical Conference and Exhibition, Anaheim, 11–14 November.

Buell, R.S., Kazemi, H., Poettmann, F.H., 1990. Analyzing injectivity of polymer solutions with the Hall plot. SPERE (February), 41–46.

Buckley, J.S., Liu, Y., Monsterleet, S., 1998. Mechanisms of wetting alteration by crude oils. SPE Journal (SPE J.) (March), 54–61.

Buckley, S.E., Leverett, M.C., 1942. Mechanism of fluid displacements in sands. Trans. AIME 146, 107–116.

Buijse, M.A., Prelicz, R.M., Barnes, J.R., Cosmo, C., 2010. Application of internal olefin sulfonates and other surfactants to EOR. Part 2: the design and execution of an ASP field test. Paper SPE 129769 presented at the SPE Improved Oil Recovery Symposium, Tulsa, 24–28 April.

Bunge, A.L., Radke, C.J., 1982. Migration of alkaline pulses in reservoir sands. SPEJ (December), 998–1012.

Burk, J.H., 1987. Comparison of sodium carbonate, sodium hydroxide, and sodium orthosilicate for EOR. SPERE (February), 9–16.

Camilleri, D., 1983. Micellar/Polymer Flooding Experiments and Comparison with an Improved 1-D Simulator. M.S. thesis, University of Texas at Austin, May.

Camilleri, D., Fil, A., Pope, G.A., Rouse, B.A., Sepehrnoori, K., 1987. Improvements in physical-property models used in micellar/polymer flooding. SPERE (November), 433–440.

Campbell, T.C., 1981. The role of alkaline chemicals in oil displacement mechanisms. In: Shah, D.O. (Ed.), Surface Phenomena in Enhanced Oil Recovery. Plenum, pp. 293–306.

Campbell, T.C., 1982. The role of alkaline chemicals in the recovery of low-gravity crude oils. JPT (November), 2510–2516.

Cannella, W.J., Huh, C., Seright, R.S., 1988. Prediction of xanthan rheology in porous media. Paper SPE 18089 presented at the 63rd Annual Technical Conference and Exhibition, Houston, 2–5 October.

Cao, X.-L., Sun, H.-Q., Jiang, Y.-B., Zhang, X.-S., Guo, L.-L., 2002. ASP pilot test in the Western Guodao Field. Oilfield Chemistry 19 (4), 350–353.

Carreau, P.J., 1972. Rheological Equations from Molecular Network Theories. Trans. Soc. Rheol. 16 (1), 99–127 (Ph.D. dissertation, University of Wisconsin, Madison, 1968).

Castelletto, V., Hamley, I.W., Xue., W., Sommer, C., Pederson, J.S., Olmsted, P.D., 2004. Rheological and structural characterization of hydrophobically modified polyacrylamide solutions in the semidilute Regime. Macromolecules 37, 1492–1501.

Castor, T.P., Edwards, J.B., Passman, F.J., 1981a. Response of mobility control agents to shear, electrochemical, and biological stress. In: Shah, D.O. (Ed.), Surface Phenomena in Enhanced Oil Recovery. Plenum, pp. 773–820.

Castor, T.P., Somerton, W.H., Kelly, J.F., 1981b. Recovery mechanisms of alkaline flooding. In: Shah, D.O. (Ed.), Surface Phenomena in Enhanced Oil Recovery. Plenum, pp. 249–291.

Cayias, J.L., Schechter, R.S., Wade, W.H., 1977. The utilization of petroleum sulfonates for producing low interfacial tensions between hydrocarbons and water. Journal of Colloid and Interface Science 59 (1), 31–38.

Chan, K.S., Shah, D.O., 1979. The effect of surfactant partitioning on the phase behavior and phase inversion of the middle phase microemulsions. Paper SPE 7869 presented at the SPE International Symposium on Oilfield and Geothermal Chemistry, Houston, 22–24 January.

Chan, K.S., Shah, D.O., 1981. The physico-chemical conditions necessary to produce ultralow interfacial tension at the oil/brine interface. In: Shah, D.O. (Ed.), Surface Phenomena in Enhanced Oil Recovery. Plenum, pp. 53–72.

Chang, H.L., 1976. Oil recovery by alkaline waterflooding, Canadian Patent No. 1037863, 24 February.

Chang, H.L., Zhang, Z.-Q., Wang, Q.-M., Xu, Z.-S., Guo, Z.-D., Sun, H.-Q., Cao, X.-L., Xiao, Q., 2006. Advances in polymer flooding and alkaline/surfactant/polymer processes as developed and applied in the People's Republic of China. JPT (February), 84–89.

Chang, K.T., Cheung, S., Frampton, H., Neal, P., Ng, R., 2007. Update on an in-depth conformance control system for improving waterflood efficiency. Paper presented at the 6th International Conference on Production Optimization–Reservoir Conformance, Profile Control and Water and Gas Shut-off, Houston, 6–7 November.

Chang, M.-S.M., Wasan, D.T., 1980. Emulsion characteristics associated with an alkaline water flooding process. Paper SPE 9001 presented at the SPE Oilfield and Geothermal Chemistry Symposium, Stanford, 28–30 May.

Chatzis, I., Morrow, N.R., 1984. Correlation of capillary number relationships for sandstone. SPEJ 24 (5), 555–562.

Chauveteau, G., 1981. Molecular interpretation of several different properties of flow of coiled polymer solutions through porous media in oil recovery conditions. Paper SPE 10060 presented at the SPE 56th Annual Technical Conference and Exhibition, San Antonio, 5–7 October.

Chauvetean, G., 1982. Rodlike polymer solution flow through fine pores: Influence of pore size on rheological behavior. J. Rheology 26 (2), 111–142.

Chauveteau, G., Combe, J., Han, D., 1988. Preparation of two polymer pilot tests in Daqing Oil Field. Paper SPE 17632 presented at the International Meeting on Petroleum Engineering, Tianjin, 1–4 November.

Chauveteau, G., et al., 2001. New size-controlled microgels for oil production. Paper SPE 64988 presented at the SPE International Symposium on Oilfield Chemistry, Houston, 13–16 February.

Chauveteau, G., Kohler, N., 1974. Polymer flooding: The essential elements for laboratory evaluation. Paper SPE 4745 presented at the SPE Improved Oil Recovery Symposium, Tulsa, 22–24 April.

Chauveteau, G., Kohler, N., 1984. The influence of microgels in xanthan polysaccharide solutions on their flow through porous media. SPEJ (June), 361–368.

Chauveteau, G., Tabary, R., Blin, N., Renard, M., Rousseau, D., Faber, R., 2004. Disproportionate permeability reduction by soft preformed microgels. Paper SPE 89390 presented at the 14th SPE/DOE Improved Oil Recovery Symposium, Tulsa, 17–21 April.

Chauveteau, G., Tabary, R., Le Bon, C., Renard, M., Feng, Y., Omari, A., 2003. In-depth permeability control by adsorption of soft size-controlled microgels. Paper SPE 82228 presented at the SPE European Formation Damage Conference, Hague, 13–14 May.

Chauveteau, G., Taraby, R., Renard, M., Omari, A., 2000. Method for preparing microgels of controlled size. US Patent 6579909 B1, 17 June 2003.

Chen, H., Zhang, S.H., Chu, Y.B., Yang, H.X., Liu, F.L., 2004. Development and application of hydrophobically associating polymer gel in high-temperature and high-salinity reservoirs for profile modification. Oilfield Chemistry 21 (4), 343–346.

Chen, T.-L., Tang, J.-X., Peng, K.-Z., Xie, F., 1998. HPAM rheology in subsurface flow. In: Chemical Flooding Symposium–Research Results during the Eighth Five-Year Period (1991–1995), Vol. I. Petroleum Industry Press, pp. 112–119.

Chen, T.-P., Pu, C.-S., 2006. Study on ultralow interfacial tension chemical flooding in low permeability reservoirs. Journal of Xi'an Shiyou University 21 (3), 30–33.

Chen, T.-R., Chen, Z., 2002. Multiphase physicochemical flow in porous media and measurement of relative permeability curves. In: Yu, J.-Y., Song, W.-C., Li, Z.-P., et al. (Eds.), Fundamentals and Advances in Combined Chemical Flooding, Chapter 16. China Petrochemical Press.

Chen, X.-Z., 2005. A portable polymer injection skid used in an offshore platform–application case. In: Yan, C.-Z., Li, Y. (Eds.), Tertiary Oil Recovery Symposium. Petroleum Industry Press, pp. 58–68.

Chen, Y.-M., Jiao, L.-M., Li, Z.-P., 1999a. Synergy among petroleum sulfonate polar compositions. ACTA PETROLEI SINICA 20 (2), 73–77.

Chen, Z.-Y., Sun, W., Yang, H.-J., 1999b. Mechanistic study of alkaline–polymer flooding in Yangsanmu Field. Journal of Northwest University (Natural Science Edition) 29 (3), 237–240.

Cheng, K.H., 1986. Chemical consumption during alkaline flooding: A comparative evaluation. Paper SPE 14944 presented at the SPE/DOE Symposium on Enhanced Oil Recovery, Tulsa, 20–23 April.

Cheng, J.-C., Liao, G.-Z., Yang, Z.-Y., Li, Q., Yao, Y.-M., Xu, D.-P., 2001. Overview of Daqing ASP pilots. Petroleum Geology & Oilfield Development in Daqing (P.G.O.D.D.) 20 (2), 46–49.

Cheng, R.-X., Zheng, W.-N., 1988. Study of crude oil active components in alkaline flooding. Oilfield Chemistry 5, 8.

Childs, E.C., 1969. The Physcial Basics of Soil Water Phenomena. Wiley and Sons.

Choi, S.K., Ermel, Y.M., Bryant, S.L., Huh, C., Sharma, M.M., 2006. Transport of a pH-sensitive polymer in porous media for novel mobility control applications. Paper SPE 99656 presented at the SPE/DOE Symposium on Improved Oil Recovery, Tulsa, 22–26 April.

Christopher, R.H., Middleman, S., 1965. Industrial and Engineering Chemistry Fundamentals (Ind. Eng. Chem. Fund.) 4, 422–426.

Cinar, Y., Marquez, S., Orr Jr., F.M., 2007. Effect of IFT variation and wettability on three-phase relative permeability. SPEREE (June), 211–220.

Claridge, E.L., 1978. A method for designing graded viscosity banks. SPEJ (October), 315–324; Trans. AIME, 265.

Collins, R.E., 1961. Flow of Fluids through Porous Materials. Petroleum Publishing Co., Tulsa.

Cooke, C.E., Williams, R.E., Kolodzie, P.A., 1974. Oil recovery by alkaline waterflooding, JPT (February), 1365–1374.

Corlay, P., Lemouzy, P., Eschard, R., Zhang, L.-R., 1992. Fully integrated reservoir study and numerical forecast simulations of two-polymer pilots in Daqing Field. Paper SPE 22364 presented at the International Meeting on Petroleum Engineering, Beijing, 24–27 March.

Coste, J.P., Liu, Y., Bai, B.-J., LI, Y., Shen, P., Wang, Z., Zhu, G., 2000. In-depth fluid diversion by pre-gelled particles, laboratory study and pilot testing. Paper SPE 59362 presented at the SPE/DOE Improved Oil Recovery Symposium, Tulsa, 3–5 April.

Craig, F.F., 1971. The Reservoir Engineering Aspects of Waterflooding, Society of Petroleum Engineers, Dallas.

Craig Jr., F.F., Geffen, T.M., Morse, R.A., 1955. Oil recovery performance of pattern gas or water injection operations from model tests. Trans. AIME 204, 7–15.

Crocker, M.E., Donaldson, E.C., Marchin, L.M., 1983. Comparison and analysis of reservoir rocks and related clays. Paper SPE 11973 presented at the SPE Annual Technical Conference and Exhibition, San Francisco, 5–8 October.

Cui, C.X., 2004. Pilot test of crosslinked AMPS with low concentrations for profile modification in Mazhai Field. Journal of Jianghan Petroleum Institute 26 (2), 118–119.

Culberson, C.H., Pytkowicz, R.M., Atlas, E.L., 1975. Hydrogen ion exchange on amorphous silica in sea water. Marine Chemistry (Mar. Chem.) 3, 43–54.

Davies, J.T., 1957. A quantitative kinetic theory of emulsion type, I. physical chemistry of the emulsifying agent, gas/liquid and liquid/liquid interface. Proceedings of the International Congress of Surface Activity, 426–438.

Davison, P., Mentzer, E., 1980. Polymer flooding in North Sea reservoirs. Paper SPE 9300 presented at the SPE 55th Annual Fall Conference, Dallas, 21–24 September; SPEJ (June 1982), 353–362.

Delamaide, E., Corlay, P., Wang, D.-M., 1994. Daqing Oil Field: the success of two pilots initiates first extension of polymer injection in a Giant Oil Field. Paper SPE/DOE 27819 presented at the Improved Oil Recovery Symposium, Tulsa, 17–20 April.

Delshad, M., Delshad, M., Bhuyan, D., Pope, G.A., Lake, L.W., 1986. Effect of capillary number on the residual saturation of a three-phase micellar solution. Paper SPE14911 presented at the SPE/DOE fifth Symposium on Enhanced Oil Recovery, Tulsa, 20–23 April.

Delshad, M., Delshad, M., Pope, G.A., Lake, L.W., 1987. Two- and three-phase relative permeabilities of micellar fluids. SPEFE (September), 327–337.

Delshad, M., Han, W., Pope, G.A., Sepehrnoori, K., Wu, W., Yang, R., Zhao, L., 1998. Alkaline/surfactant/polymer flood predictions for the Karamay Oil Field, paper SPE 39610 presented at the SPE/DOE Improved Oil Recovery Symposium, Tulsa, 19–22 April.

Delshad, M., Kim, D.H., Magbagbeola, O.A., Huh, C., Pope, G.A., Tarahhom, F., 2008. Mechanistic interpretation and utilization of viscoelastic behavior of polymer solutions for improved polymer-flood efficiency. Paper SPE 113620 presented at the SPE/DOE Improved Oil Recovery Symposium, Tulsa, 19–23 April.

Delshad, M., MacAllister, D.J., Pope, G.A., Rouse, B.A., 1985. Multiphase dispersion and relative permeability experiments. SPEJ (August), 524–534.

Delshad, M., Pope, G.A., Sepehrnoori, K., 2004. A framework to design and optimize chemical flooding processes. DOE report 835937, November.

deZabala, E.F., Vislocky, J.M., Rubin, E., Radke, C.J., 1982. A chemical theory for linear alkaline flooding. SPEJ (April), 245–258.

Dickson, J.L., Leahy-Dios, A., Wylie, P.L., 2010. Development of improved hydrocarbon recovery screening methodologies. Paper SPE 129768 presented at the SPE Improved Oil Recovery Symposium, Tulsa, 24–28 April.

DiMarzio, E.A., Guttman, C.M., 1970. Separation by flow. Macromolecules 3, 131.

Dombrowski, H.S., Brownell, L.E., 1954. Residual equilibrium saturation of porous media. Industrial and Engineering Chemistry (Ind. Eng. Chem.) 46, 1207.

Dominguez, J.G., Willhite, G.P., 1977. Retention and flow characteristics of polymer solutions in porous media. SPEJ (April), 111–121.

Dong, H.-Z. Fang, S.-F., Wang, D.-M., Wang, J.-Y., Liu, Z., Hong, W.-H., 2008. Review of practical experience & management by polymer flooding at Daqing. Paper SPE 114342-MS presented at the SPE/DOE Symposium on Improved Oil Recovery, 20–23 April, Tulsa.

Dong, M., Ma, S., Liu, Q., 2009. Enhanced heavy oil recovery through interfacial instability: A study of chemical flooding for Brintnell heavy oil. Fuel 88, 1049–1056.

Dou, Z.-L., Zhao, W., Chen, B.-P., 1999. Polymer flooding pilot test and performance analysis in Layer 5^{2+3} of the Western Block 7, Gudong field. Oil and Gas Recovery Technology 6 (1), 12–16.

Dria, M.A., Schechter, R.S., Lake, L.W., 1988. An analysis of reservoir chemical treatments. SPEPE (February), 52–62.

Dunning, H.N., Moore, J.W., Denekas, M.O., 1953. Interfacial activities and porphyrin contents of petroleum extracts. Ind. Eng. Chem. 45 (August), 1759–1765.

Durst, F., Haas, R., Interthal, W., 1982. Laminar and turbulent flows of dilute polymer solutions: A physical model, Rheologica Acta 21 (4–5), 572–577.

Dwarakanath, V., 1997. Characterization and Remediation of Aquifers Contaminated by Nonaqueous Phase Liquids Using Partitioning Tracers and Surfactants. Ph.D. dissertation, The University of Texas at Austin.

Dwarakanath, V., Chaturvedi, T., Jackson, A.C., Malik, T., Siregar, A., Zhao, P., 2008. Using co-solvents to provide gradients and improve oil recovery during chemical flooding in a light oil reservoir. Paper SPE 113965 presented at the SPE/DOE Symposium on Improved Oil Recovery, Tulsa, 20–23 April.

Dyes, A.B., Caudle, B.H., Erickson, R.A., 1954. Oil production after breakthrough—as influenced by Mobility Ratio. Trans. AIME 201, 81–86.

Dykstra, H., Parsons, R.L., 1950. The prediction of oil recovery by water flood. Secondary Recovery of Oil in the United States, second ed. API, Washington, DC, pp. 160–174.

Ehrlich, R., Hasiba, H.H., Raimondi, P., 1974. Alkaline waterflooding for wettability alternation evaluating a potential field application, JPT (September), 1335–1352.

Ehrlich, R., Wygal, R.J., 1977. Interaction of crude oil and rock properties with the recovery of oil by caustic waterflooding. SPEJ (August), 263–279.

Eisenzimmer, A., Desmarquest, J.-P., 1981. Selection of composition for an optimal micellar system in reservoir conditions. In: Shah, D.O. (Ed.), Surface Phenomena in Enhanced Oil Recovery. Plenum, pp. 105–116.

England, D.C., Berg, J.C., 1971. Transfer of surface-active agents across a liquid-liquid interface. American Institute of Chemical Engineers J. (AIChE J.) (March), 313–322.

Fan, T., Buckley, J.S., 2006. Acid number measurements revisited. Paper SPE 99884 presented at the SPE/DOE Symposium on Improved Oil Recovery, Tulsa, 22–26 April.

Fan, Z.-H., Kong, B.-L., She, Y.-M., Tang, J.-L., 2004. Polymer flooding test in the heterogeneous Xia-er-men field of large channels. Petroleum Exploration and Development 31 (1), 110–112.

Feng, Y., Tabary, R., Renard, M., Le Bon, C., Omari, A., Chauveteau, G., 2003. Characteristics of microgels designed for water shutoff and profile control. Paper SPE 80203 presented at the SPE International Symposium Oilfield Chemistry, Houston, 5–7 February.

Fielding Jr., R.C., Gibbons, D.H., Legrand, F.P., 1994. In-depth drive fluid diversion using an evolution of colloidal dispersion gels and new bulk gels: An operational case history of North Rainbow Ranch Unit. Paper SPE 27773 presented at the SPE/DOE Improved Oil Recovery Symposium, Tulsa, 17–20 April.

Flaaten, A.K., Nguyen, Q.P., Zhang, J.-Y., Mohammadi, H., Pope, G.A., 2008. Chemical flooding without the need for soft water. Paper SPE 116754 presented at the SPE Annual Technical Conference and Exhibition, Denver, 21–24 September.

Fleming, P.D. III, Sitton, D.M., Hessert, J.E., Vinatieri, J.E., Boneau, D.F., 1978. Phase properties of oil-recovery systems containing petroleum sulfonates. Paper SPE 7576 presented at the SPE 53rd Annual Technical Conference and Exhibition, Houston, 1–3 October.

Fletcher, A.J.P., Flew, S., Forsdyke, I.N., Morgan, J.C., Rogers, C., Suttles, D., 1992. Deep diverting gels for very cost-effective waterflood control. Journal of Petroleum Science and Engineering 7, 33–43.

Flory, P.J., 1953. Principles of Polymer Chemistry. Cornell University Press.

Flournoy, K.H., Carlin, J.T., Cardenas, R.L., 1977. Surfactant oil recovery process. US Patent No. Re. 29,219.

Foster, W.R., 1973. A low tension waterflooding process employing a petroleum sulfonate, inorganic salts, and a biopolymer. JPT 25 (2), 205–210.

Frampton, H., Morgan, J.C., Cheung, S.K., Munson, L., Chang, K.T., Williams, D., 2004. Development of a novel waterflood conformance control system. Paper SPE 89391 presented at the SPE/DOE 14th Symposium on Improved Oil Recovery, Tulsa, 17–21 April.

French, M.S., Keys, G.W., Stegemeier, G.L., Ueber, R.C., Abrams, A., Hill, H.J., 1973. Field test of an aqueous surfactant system for oil recovery, Benton Field, Illinois. JPT 25 (2), 195–204.

Fu, M.R., 1997. Evaluation of AM/DMDAAC/Na-AMPS in profile modification. Drilling and Production Techniques 20 (5), 60–62.

Fulcher, R.A., Ertekin, T., Stahl, C.D., 1985. Effect of capillary number and its constituents on two-phase relative permeability curves. JPT (February), 249–260.

Gaines, G.L., Thomas, H.C., 1953. Adsorption studies on clay minerals. II. A formulation of the thermodynamics of exchange adsorption. Journal Chemical Physics 21, 714–718.

Gale, W.W., Sandvik, E.I., 1973. Tertiary surfactant flooding: Petroleum sulfonate composition efficacy studies. SPEJ (August), 191–199.

Gao, S.-T., Li, H.-B., Yang, Z.-Y., Pitts, M.J., Surkalo, H., Kon, W., 1996. Alkaline/surfactant/polymer pilot performance of the West Central Saertu, Daqing Oil Field. SPERE (August), 181–188.

Gapon, E.N., 1933. Theory of exchange adsorption. Journal of General Chemistry (USSR) 3, 667–669 (Chemical Abstracts 28, 4516, 1934).

Garnes, J.M., Mathisen, A.M., Scheie, A., Skauge, A., 1990. Capillary number relations for some North Sea reservoir sandstones. Paper SPE 20264 presented at the SPE/DOE 7th Symposium on Enhanced Oil Recovery, Tulsa, 22–25 April.

Gelhar, L.W., 1997. Perspectives on field-scale application of stochastic subsurface hydrology. In: Dagan, G., Neuman, S.P. (Eds.), Subsurface Flow and Transport: A Stochastic Approach. Cambridge University Press, pp. 157–176.

Geochemist's Workbench®, 2009. *http://www.geology.uiuc.edu/Hydrogeology/hydro_gwb.htm.*

Gerbacia, W., McMillen, T.J., 1982. Oil-recovery surfactant formulation development using surface design experiments. SPE J. (April), 237–244.

Giordano, R.M., Salter, S.J., 1984. Comparison of simulation and experiments for compositionally well-defined corefloods. Paper SPE 12697 presented at the 4th Joint SPE/DOE Symposium on Enhanced Oil Recovery, Tulsa, 15–18 April.

Glinsmann, G.R., 1978. Aqueous surfactant systems for in situ multiphase microemulsion formation. US Patent 4,125,156, 14 November.

Glover, C.J., Puerto, M.C., Maerker, J.M., Sandvik, E.L., 1979. Surfactant phase behavior and retention in porous media. SPEJ (June), 183–193.

Gogarty, W.B., 1967. Rheological properties of pseudoplastic fluids in porous media. SPEJ (June), 149–160.

Gogarty, W.B., 1969. Process for the simultaneous displacement of petroleum and water in formations. US Patent No. 3,443,636, 13 May.

Gogarty, W.B., 1976. Status of surfactant or micellar methods. JPT 28 (1), 93–102.

Gogarty, W.B., 1983a. Enhanced oil recovery through the use of chemicals–part 1. JPT 35 (9), 1581–1590.

Gogarty, W.B., 1983b. Enhanced oil recovery through the use of chemicals–part 2. JPT 35 (10), 1767–1775.

Gogarty, W.B., Meabon, H.P., Milton, H.W., 1970. Mobility control design for miscible-type waterfloods using micellar solutions. JPT (February), 141–147.

Graciaa, A., Fortney, L.N., Schechter, R.S., Wade, W.W., Yiv, S., 1982. Criteria for structuring surfactants to maximize solubilization of oil and water: part 1–commercial nonionics. SPEJ (October), 743–749.

Gramain, P., Myard, P., 1981. Adsorption studies of polyacrylamides in porous media. Journal of Colloid and Interface Science (J. Coll. Int. Sci.), 84 (1), 114–126.

Green, D.W., Willhite, G.P., 1998. Enhanced Oil Recovery. Society of Petroleum Engineers, Dallas.

Griffin, W.C., 1949. Classification of surface-active agents by "HLB". J. Society of Cosmetic Chemists 1, 311.

Griffin, W.C., 1954. Calculation of HLB values of non-ionic surfactants. J. Society of Cosmetic Chemists 5, 259.

Grim, R.E., 1939. Properties of clays. Recent Marine Sediments. AAPG, Tulsa, pp. 466–495.

Gu, H.-J., Yang, R.-Q., Guo, S.G., Guan, W.-D., Yue, X.-J., Pan, Q.-Y., 1998. Study on reservoir engineering: ASP flooding pilot test in Karamay Oilfield. Paper SPE 50918 presented at the International Oil and Gas Conference and Exhibition, Beijing, 2–6 November.

Guan, S.-H., Yu, P., Guo, J.-L., Su, F., Song, R.-Q., 2002. Analysis of the modification of the polymer flooding program in the Gutao Zhong-2-Nan block. Testing and Production Technology 23 (4), 20–22.

Gupta, S.P., 1984. Compositional effects on displacement mechanisms of the micellar fluid injected in the Sloss field test. SPEJ 24 (1), 38–48.

Gupta, S.P., Trushenski, S.P., 1979. Micellar flooding–compositional effects on oil displacement. SPEJ (April), 116–128.

Hagoort, J., Leijnse, A., van Poelgeest, F., 1976. Steam-strip drive: A potential tertiary recovery process. JPT (December), 1409–1419.

Hall, H.N., 1963. How to analyze waterflood injection well performance? World Oil (October), 128–130.

Han, D., 2001. Surfactant Flooding: Principles and Applications. Petroleum Industry Press.

Han, J., Kong, B.L., Lu, X.H., 2006a. Study on changes of hydrolysis degree and solution viscosity of HPAM at different reservoir temperatures. Oilfield Chemistry 23 (3), 235–238, 255.

Han, M., Xiang, W.-T., Zhang, J., Jiang, W., Sun, F.-J., 2006b. Application of EOR technology by means of polymer flooding in Bohai Oil Fields. Paper SPE 104432 presented at the International Oil & Gas Conference and Exhibition, Beijing, 5–7 December.

Han, X., Wang, W., Xu, Y., 1995. The viscoelastic behavior of HPAM solutions in porous media and its effects on displacement efficiency. Paper SPE 30013 presented at the International Petroleum Meeting, Beijing, 14–17 November.

Hand, D.B., 1939. Dineric distribution: I. The distribution of a consolute liquid between two immiscible liquids. J. Physics and Chem. 34, 1961–2000.

Hanna, H.S., Somasundaran, P., 1977. Physico-chemical aspects of adsorption at solid/liquid interfaces, II: Mahogany sulfonate/Berea sandstone, kaolinite. In: Shah, D.O., Schechter, R.S. (Eds.), Improved Oil Recovery by Surfactant and Polymer Flooding. Academic Press, pp. 253–274.

Harmsen, G.J., 1971. Oil recovery by hot-water and steam injection. Paper PD9 presented at the Eighth World Petroleum Congress, Moscow, 13–16 June.

He, H.-P, Fan, S.-M., Mao, Z.-Y., Zhong, J., Zhang, S.-M., Zhang, C.-P., 2006. Effect of surfactant injection on waterflooding performance in a low permeability reservoir. Henan Petroleum 20 (4), 37–38.

He, M., Chen, W., 1998. High Molecular Physics. Fudan University Publishing.

Healy, R.N., Reed, R.L., 1977a. Immiscible microemulsion flooding. SPEJ (April), 129–139.

Healy, R.N., Reed, R.L., 1977b. Some physical-chemical aspects of microemulsion flooding: a review. In: Shah, D.O., Schechter, R.S. (Eds.), Improved Oil recovery by Surfactant and Polymer Flooding. Academic Press, pp. 383–437.

Healy, R.N., Reed, R.L., Carpenter Jr., C.W., 1975. A laboratory study of microemulsion flooding. SPEJ (Feb.), 87–103; Trans. AIME, 259.

Healy, R.N., Reed, R.L., Stenmark, D.G., 1976. Multiphase microemulsion systems. SPEJ (June), 147–160; Trans. AIME, 261.

Henderson, G.D., Danesh, A., Tehrani, D.H., Al-Kharusi, B., 2000. The relative significance of positive coupling and inertial effects on gas condensate relative permeabilities at high velocity. Paper SPE 62933 presented at the SPE Annual Technical Conference and Exhibition, Dallas, 1–4 October.

Henderson, G.D., Danesh, A., Tehrani, D.H., Al-Shaidi, S., Peden, J.M., 1998. Measurement and correlation of gas condensate relative permeability by the steady-state method. SPEREE (April), 134–140.

Hill, H.J., Lake, L.W., 1978. Cation exchange in chemical flooding: Part 3—experimental. SPEJ 18, 445–456.

Hill, H.J., Reisberg, J., Helfferich, F.G., Lake, L.W., Pope, G.A., 1978. Ion exchange controlled chemically aided waterflood process. US Patent 4,074,755, 21 February.

Hirasaki, G.J., 1981. Application of the theory of multicomponent, multiphase displacement to three-component, two-phase surfactant flooding. SPEJ (April), 191–204.

Hirasaki, G.J., 1982a. Interpretation of the change in optimal salinity with overall surfactant concentration. SPEJ (December), 971–982.

Hirasaki, G.J., 1982b. Ion exchange with clays in the presence of surfactant. SPEJ 22 (2), 181–192.

Hirasaki, G.J., Miller, C.A., Pope, G.A., 2006. Surfactant based enhanced oil recovery and foam mobility control. 3rd Annual & Final Technical Report for DOE project (DE-FC26-03NT15406, July.

Hirasaki, G.J., Miller, C.A., Puerto, M., 2008. Recent advances in surfactant EOR. Paper SPE 115386 presented at the SPE Annual Technical Conference and Exhibition, Denver, Colorado. 21–24 September. The revised version presented at the International Petroleum Technology Conference, Kuala Lumpur, 3–5 December.

Hirasaki, G.J., Pope, G.A., 1974. Analysis of factors influencing mobility and adsorption in the flow of polymer solution through porous media. SPEJ (August), 337–346.

Hirasaki, G.J., van Domselaar, H.R., Nelson, R.C., 1983. Evaluation of the salinity gradient concept in surfactant flooding. SPEJ 23 (3), 486–500.

Hirasaki, G.J., Zhang, D.L., 2004. Surface chemistry of oil recovery from fractured, oil-wet, carbonate formations. SPE J. (June), 151–162.

Holm, L.W., Robertson, S.D., 1981. Improved micellar/polymer flooding with high-pH chemicals. JPT 33 (1), 161–172.

Hou, T.J., Zhao, R.B., Mo, B., Wu, Y.S., Zhang, J.L., Liu, S.Q., Xie, L.X., Zhang, H.E., 2003. Determination of AMPS copolymer parameters used in simulation. Petroleum Exploration and Development 30 (4), 100–101.

Hou, W.-G., 1993. Surfactant applications in enhanced oil recovery. In: Wang, Y.-F. (Ed.), Surfactants and Their Application in Oil and Gas Fields. Petroleum Industry Press, pp. 258–289.

Hsieh, W.C., Shah, D.O., 1977. The effect of chain length of oil and alcohol as well as surfactant to alcohol ratio on the solubilization, phase behavior and interfacial tension of oil/brine/surfactant/alcohol systems. Paper SPE 6594 presented at the SPE International Oilfield and Geothermal Chemistry Symposium, San Diego, 27–29 June.

Huang, P.-C., Li, D.-M., Xing, H.-B., Li, H.-S., Li, C.-J., Liu, S.-F., 1998a. Prediction of HPAM permeability reduction factor. In: Gang, Q.-L., et al. (Eds.), Chemical Flooding Symposium–Research Results during the Eighth Five-Year Period (1991–1995), vol. I. Petroleum Industry Press, pp. 240–245.

Huang, S., Liu, J.-H., Xie, X.-Q., 1998b. Improved dissolution and maturation unit and its effect on polymer solubility. In: Gang, Q.-L., et al. (Eds.), Chemical Flooding Symposium–Research Results during the Eighth Five-Year Period (1991–1995), vol. I. Petroleum Industry Press, pp. 409–413.

Huang, S., Xie, X.-Q., Liu, J.-H., Zhang, X.-B., 1998c. Surface injection flow system to control oxygen content. In: Gang, Q.-L., et al. (Eds.), Chemical Flooding Symposium–Research Results during the Eighth Five-Year Period (1991–1995), vol. I. Petroleum Industry Press, pp. 404–408.

Huang, Y.-D., et al., 1987. Dynamic interfacial tension of acid oil/alkaline solution. Petroleum Exploration and Development 5, 81–88.

Huang, Y.-Z., Yu, D.-S., 2002. Mechanisms and principles of physicochemical flow. In: Yu, J.-Y., Song, W.-C., Li, Z.-P., et al. (Eds.), Fundamentals and Advances in Combined Chemical Flooding, Chapter 15. China Petrochemical Press.

Hughes, D.S., Teeuw, D., Cottrell, C.W., Tollas, J.M., 1990. Appraisal of the use of polymer injection to suppress aquifer influx and to improve volumetric sweep in a viscous oil reservoir. SPERE (February), 33–40.

Huh, C., 1979. Interfacial tension and solubilizing ability of a microemulsion phase that coexists with oil and brine. J. Coll. Int. Sci. 71, 408–428.

Huh, C., Choi, S.K., Sharma, M.M., 2005. A rheological model for pH-sensitive ionic polymer solutions for optimal mobility-control applications. Paper SPE 96914 presented at the SPE Annual Technical Conference and Exhibition, Dallas, 9–12 October.

Huh, C., Pope, G.A., 2008. Residual oil saturation from polymer floods: Laboratory measurements and theoretical interpretation, paper SPE 113417 presented at the SPE/DOE Improved Oil Recovery Symposium, Tulsa, 19–23 April.

Hurd, B.G., 1969. Method of desorbing surfactant and reusing it in flooding water. US Patent No. 3,474,864, 28 October.

Isaacs, E.E., Iyory, J., Green, M.K., 1994. Steam-foams for heavy oil and bitumen recovery. In: Schramm, L.L. (Ed.), Foams: Fundamentals and Applications in the Petroleum Industry. American Chemical Society, Washington, DC, pp. 235–258.

Israelachvili, J., Drummond, C., 1998. Fundamental studies of oil–surface–water interactions and its relationship to wettability. Paper presented at the International Symposium on Evaluation of Reservoir Wettability and its Effect on Oil Recovery, Trondheim, 22–24 June.

Jackson, A.C., 2006. Experimental Study of the Benefits of Sodium Carbonate on Surfactants for Enhanced Oil Recovery. M.S. thesis, University of Texas at Austin.

Jadhunandan, P., Morrow, N.R., 1991. Spontaneous imbibition of water by crude oil/brine/rock systems. In Situ 15 (4), 319–345.

Jadhunandan, P., Morrow, N.R., 1995. Effect of wettability on waterflood recovery for crude oil/Brine/rock systems. SPERE 10 (1), 40–46.

James, D.F., McLaren, D., 1975. Journal of Fluid Mechanics 70, 733.

Jennings Jr., H.Y., Johnson Jr., C.E., McAuliffe, C.D., 1974. A caustic waterflooding process for heavy Oils. JPT (December), 1344–1352.

Jennings, R.R., Rogers, J.H., West, T.J., 1971. Factors influencing mobility control by polymer solutions. JPT (March), 391–401; Trans. AIME, 251.

Jiang, G.Y., Ren, K., Xu, C.M., Lin, M.Q., Luo, W.Q., 2003a. Synthesis and solution properties of hydrophobically association polymers. Journal of the University of Petroleum (China) 27 (6), 80–82.

Jiang, J.-L., Guo, D.-F., Li, X.-F., Zhao, P.-W., Wang, X.-J., Wu, X.-L., Li, G.-Q., 2003b. Pilot field trial of natural mixed carboxylates/xanthan flood at Well H5-15 block in Huzhuangji Oil Field. Oilfield Chemistry 20 (1), 58–60.

Jiang, Y., Sun, B.-J., Li, Y.-A., Cui, X.-Y., Gao, C.-T., 2003c. Measures and results of the polymer flooding in the Zhong-2-Nan block in the Gudong field. Oil and Gas Surface Engineering 22 (4), 22.

Jiang, Y.-B., Dan, L.-T., Liao, H.-Y., Zhou, Z.-M., Sun, Y.-H., 2001. Learning from the polymer injection in a multilayer sandstone reservoir of viscous oil in the Gudong field. In: Wang, D.-P. (Ed.), Symposium on Oil and Gas Recovery Technology. Petroleum Industry Press.

Jin, M., 1995. A Study of Nonaqueous Phase Liquid Characterization and Surfactant Remediation. Ph.D. dissertation, University of Texas at Austin.

Johnson Jr., C.E., 1976. Status of caustic and emulsion methods. JPT (January), 85–92.

Johnson, J.R., Dehghani, K., Hawkins, D.B., 1988. Caustic consumption by kaolinite and quartz and their mixtures at temperature up to 120°C. Paper SPE 17410 presented at the SPE California Regional Meeting, Long Beach, 23–25 March.

Jones, W.M., 1980. Polymer additives in reservoir flooding for oil recovery: shear thinning or shear thickening? Journal of Physics D: Applied Physics (J. Phys. D: Appl. Phys.) 13 (5), L87–L88.

Jørgenvåg, A.H., Sagli, J.R., 2008. The EOR research program. Paper OTC 19407 presented at the Offshore Technology Conference, Houston, 5–8 May.

Kaczmar, B.U., 1980. Report SFB 80/E/174. University of Karlsruhe, FRG.

Kalpakci, B., Arf, T.G., Barker, J.W., Krupa, A.S., Morgan, J.C., Neira, R.D., 1990. Paper SPE 20220 presented at the SPE/DOE Enhanced Oil Recovery Symposium, Tulsa, 22–25 April.

Kamath, J., Meyer, R.F., Nakagawa, F,M., 2001. Understanding waterflood residual oil saturation of four carbonate rock types. SPE paper 71505 presented at the 2001 SPE Annual Technical Conference and Exhibition, New Orleans, 30 September–3 October.

Kang, W.-L., 2001. Study of Chemical Interactions and Drive Mechanisms in Daqing ASP Flooding. Petroleum Industry Press.

Kang, W.-L., Dan, X.-L., 1998. IFT in nonionic surfactant ASP systems. P.G.O.D.D. 17 (2), 32–33, 4.

Kang, W.-L, Hu, J.-B., Song, W.-L., 1997. Investigation of AS system/Daqing oil low dynamic IFT. Chemistry and Adhension 4, 202.

Kang, W.-L., Li, J.-G., Dan, X.-L., Yang, R.-M., Qi, Y.-B., 1999. The interfacial tension between Daqing crude oil fractions and ASP or AS solution and the stability of their emulsion. Oilfield Chemistry 16 (4), 345–347.

Kang, W.-L., Qi, B.-Y., Dan, X.-L., Xi, H.-Q., Wang, H.-L., 1998. Increase the interfacial activity of alkaline-surfactant-oil system by adding complex agents. J. Daqing Petroleum Institute 22 (1), 17–18.

Khilar, K., Fogler, H., 1984. The existence of a critical salt concentration for particle release. J. Coll. Int. Sci. 101 (1), 214–224.

Kia, S.F., Fogler, H.S., Reed, M.G., Vaidya, R.N., 1987. Effect of salt composition on clay release in Berea sandstone. SPEPE (November), 277–283.

Klenk, I.D., Grathwohl, P., 2002. Transverse vertical dispersion in groundwater and the capillary fringe. Journal of Contaminant Hydrology 58 (1–2), 111–128.

Knickerbocker, B.M., Pesheck, C.V., Davis, H.T., Scriven, L.E., 1982. Patterns of three-liquid phase behavior illustrated by alcohol-hydrocarbon-water-salt mixtures. The Journal of Physical Chemistry B (J. Phys. Chem. B) 86 (3), 393–400.

Knoll, S.K., Prud'homme, R.K., 1987. Interpretation of dynamic oscillatory measurements for characterization of well completion fluids. Paper SPE 16283 presented at the SPE International Symposium on Oilfield Chemistry, San Antonio, 4–6 February.

Kolodziej, E.J., 1987. Mechanisms of microgel formation in Xanthan biopolymer solutions. Paper SPE 16730 presented at the SPE Annual Technical Conference and Exhibition, Dallas, 27–30 September.

Kong, B.-L., 1996. Aging effect at high temperature on HPAM solution behavior. Oil and Gas Recovery Technology 3 (4), 7–12.

Kong, B.-L., Tang, J.-X., Chang, R.-Z., Bai, H.-X., 2005. Pilot test and dynamic performance of microgel injection in the Shuanghe field. In: Yan, C.Z., Li, Y. (Eds.), Tertiary Oil Recovery Symposium. Petroleum Industry Press, pp. 197–202.

Koval, E.J., 1963. A Method for predicting the performance of unstable miscible displacement in heterogeneous media. SPEJ 3 (2), 145–154.

Kremesec, V.J., Raterman, K.T., Taggart, D.L., 1988. Laboratory evaluation of a crude-oil-sulfonate/nonionic-cosurfactant micellar fluid. SPERE 3 (3), 801–808.

Krumrine, P.H., Falcone, J.S., 1983. Surfactant, polymer, and alkali interactions in chemical flooding processes. Paper SPE 11778 presented at the SPE Oilfield and Geothermal Chemistry Symposium, Denver, 1–3 June.

Krumrine, P.H., Falcone, J.S., 1987. Beyond alkaline flooding: Design of complete chemical systems. Paper SPE 16280 presented at the SPE International Symposium on Oilfield Chemistry, San Antonio, 4–6 February.

Krumrine, P.H., Falcone, J.S., Campbell, T.C., 1982. Surfactant flooding 2: the effect of alkaline additives on permeability and sweep efficiency. SPEJ (December), 983–992.

Labrid, J., 1991. The use of alkaline agents in enhanced oil recovery processes. In: Bavière, M. (Ed.), Basic Concepts in Enhanced Oil Recovery Processes. Elsevier Science, pp. 123–155.

Lager, A., Webb, K.J., Black, C.J., 2006. Low salinity oil recovery—an experimental investigation. Paper SCA 2006-36 presented at the International Symposium of the Society of Core Analysts, Trondheim.

Lakatos, I., Lakatos-Szabo, J., 1980. Acta Chimica Academiae Scientiarum Hungaricae (Acta Chim. Acad. Sci. Hung.) 105, 57–72.

Lakatos., I., Lakatos-Szabo, J., Toth, J., 1979. Factors influencing polyacrylamide adsorption in porous media and their effect on flow behavior, Paper presented at the Symposium on Surface Phenomena in EOR, Stockholm, August, Plenum.

Lake, L.W., 1989. Enhanced Oil Recovery. Prentice-Hall.

Lallemand-Barrès, A., Peaudecerf, P., 1978. Recherche des relations entre le valeur de la dispersivité macroscopique d'un milieu aquifere, ses autres caractéristiques et les conditions de mesure. *Bull.*, BRGM 2e Series, Sec. III, No. 4.

Larson, R.G., 1979. The influence of phase behavior on surfactant flooding. SPEJ (December), 411–422.

Larson, R.G., Davis, H.T., Scriven, L.E., 1982. Elementary mechanisms of oil recovery by chemical methods. JPT 34 (2), 243–258.

Lefebvre du Prey, E.J., 1973. Factors Affecting Liquid-Liquid Relative Permeabilities of a Consolidated Porous Media. SPEJ 13 (1), 39–47.

Lei, Z.-X., Chen, Y.-M., Chen, Y.-W., Yan, J., Jin, X.-P., 2006. Preliminary results of the pilot test of deep profile control using inverse polymer emulsion. Oilfield Chemistry 23 (1), 81–84.

Leverett, M.C., 1941. Capillary behavior in porous solids. Trans. AIME 142, 152–169.

Levitt, D.B., Jackson, A.C., Heinson, C., Britton, L.N. Malik, T., Dwarakanath, V., Pope, G.A., 2006. Identification and evaluation of high performance EOR surfactants. Paper SPE 100089 presented at the SPE/DOE Symposium on Improved Oil Recovery, Tulsa, 22–26 April.

Li, G.-Z., Shen, Q., Zheng, L.-Q., Mao, H.-Z., 1999a. Natural mixed carboxylates as EOR surfactants. Oilfield Chemistry 16 (1), 57–59, 63.

Li, H.-B., 2007. Advances in Alkaline-Surfactant-Polymer Flooding and Pilot Tests. Science Press.

Li, H.-F., Liao, G.Z., Han, P.-H., Yang, Z.Y., Wu, X.-L., Chen, G.-Y., Xu, D.-P., Jin, P.Q., 2003. Alkaline/surfactant/polymer (ASP) commercial flooding test in the central Xing2 area of Daqing Oilfield. Paper SPE 84896 presented at the International Improved Oil Recovery Conference in Asia Pacific, Kuala Lumpur, 20–21 October.

Li, H.-P., Li, H.-F., Yang, Z.-Y., Ye, Z.-G., Cheng, J.-C., 1999b. ASP pilot test in Daqing Sa-Zhong-Xi Region. Oil and Gas Recovery Technology 6 (2), 15–19.

Li, H.-P., Zhao, H.T., Chen, H., Zhao, P.C., 2006a. Field application of hydrophobically-associating-polymer-based gel for profile modification in Zhongyuan oil fields. Oilfield Chemistry 23 (1), 54–58.

Li, H.-P., Zhao, H.T., Zhao, P.C., Chen, H., 2006b. Laboratory study of hydrophobically-associating-polymer-based gel as a profiling agent in high-temperature and high-salinity Zhongyuan oil fields. Oilfield Chemistry 23 (1), 50–53.

Li, J.-L., Liu, D.-S., Li, T.-D., Jiang, Z.-B., 2005a. Pump problems caused by the scale in ASP and the solutions. In: Yan, C.Z., Li, Y. (Eds.), Tertiary Oil Recovery Symposium. Petroleum Industry Press, pp. 275–278.

Li, J.-G., Wang, X.-J., Zhou, W.-F., Zhang, L.-M., 2005a. Application of aluminum citrate dispersion gel in a Daqing field. In: Yan, C.-Z., Li, Y. (Eds.), Tertiary Oil Recovery Symposium. Petroleum Industry Press, pp. 332–338.

Li, J.-L., Liu, D.-S., Li, T.-D., Jiang, Z.-B., 2005b. Pump problems caused by the scale in ASP and the solutions. In: Yan, C.-Z., Li, Y. (Eds.), Tertiary Oil Recovery Symposium, Petroleum Industry Press, pp. 275–278.

Li, L.-C., Zheng, J.-M., Ma, J.-L., Li, Q., 2005c. Profile modification in a viscous oil reservoir under edge water drive. In: Yan, C.-Z., Li, Y. (Eds.), Tertiary Oil Recovery Symposium. Petroleum Industry Press, pp. 211–220.

Li., R.F., Yan, W., Liu, S., Hirasaki, G.J., Miller, C.A., 2008. Foam mobility control for surfactant EOR. Paper SPE 113910 presented at the SPE/DOE Symposium on Improved Oil Recovery, Tulsa, 20–23 April.

Li, X.-J., Wang, Z.-D., Zong, D.-Q., 2005d. Technology development and application of polymer mixing and produced liquid treatment in Daqing. In: Yan, C.Z., Li, Y. (Eds.), Tertiary Oil Recovery Symposium. Petroleum Industry Press, pp. 48–53.

Li, Y.-X., Shu, Z., Zhang, L., Zhang, X.-J., 2004. Effect of injection pressure on polymer flooding performance in the II4–5 layer of the northern block in the Shuanghe field. J. Jianghan Petroleum Institute 26 (supplemental issue), 257–258.

Li, Z.-P., Zheng, H.-X., He, Z.-S., Cheng, Y.M., 2002. Compatibility of oil-displacing agents. In: Yu, J.-Y., Song, W.-C., Li, Z.-P., et al. (Eds.), Fundamentals and Advances in Combined Chemical Flooding, Chapter 13, China Petrochemical Press.

Li, Z.-Q., 2004a. Preliminary test results of KYPAM, a comb-shaped salinity-resistant polymer, in Shengli Field. Oilfield Chemistry 21 (2), 165–167.

Li, Z.-Q., 2004b. A commercial scale polymer flooding test in Gudao Zhong-1 block of high water cut. Petroleum Exploration and Development 31 (2), 119–121.

Li, Z.-Q., 2004c. Polymer flooding test progress in Shengtuo block 1 of Shengli Field. Petroleum Processing and Petrochemicals 35 (10), 56–59.

Liang, B., Dai, H., Huang, R.H., 1997. Viscosity properties of aqueous AM/DMAM/AMPS copolymer solutions. Oilfield Chemistry 14 (4), 357–360.

Liang, J.-T., Sun, H., Seright, R.S., 1995. Why do gels reduce water permeability more than oil permeability? SPERE 10 (4), (November), 282–286.

Liao, G.-Z., 1998. Comprehensive Study of Daqing Oilfield Alkaline-Surfactant-Polymer Flooding. Post-doctoral report, Daqing.

Lieu, V.T., Miller, S.G., Staphanos, S.J., 1982. Long-term consumption of caustic and silicate solutions by petroleum reservoir sands. Soluble Silicates, Am. Chem. Soc. Symposium Series 194, 215–226.

Ligthelm, D.J., 1989. Reservoir engineering approach to viscosity grading in polymer drives. Journal of Petroleum Science and Engineering (J. Pet. Sci. Eng.) 2, 159–166.

Lim, T., Uhl, J.T., Prud'homme, R.K., 1986. The interpretation of screen-factor measurements. SPERE 1 (3), 272–276.

Lin, J., Mao, Y., Yang, M.-L., Kan, S.-H., 2005. Performance analysis of polymer flooding in the Feiyantan field. Fault-Block Oil & Gas Fields 12 (5), 41–43.

Lipatov, Y.S., Sergeeva, L.M., 1974. Adsorption of Polymers. Wiley.

Liu, H., 1995. Status and prospect of polymer flooding research in Daqing Field. Petroleum Exploration and Development 22 (2), 64–67.

Liu, H., Wang, Y., Liu, Y.Z., 2006a. Techniques of polymer solution mixing, transport and injection, and oil production. In: Shen, P.-P., Liu, Y.-Z., Liu, H.-R. (Eds.), Enhanced Oil Recovery—Polymer Flooding. Petroleum Industry Press, pp. 157–181.

Liu, J.-H., Huang, S., Xie, X.-Q., Peng, S., 1998. Polymer shear loss in surface facilities at high temperature. In: Gang, Q.-L., et al. (Eds.), Chemical Flooding Symposium—Research Results during the Eighth Five-Year Period (1991–1995), Vol. I. Petroleum Industry Press, pp. 398–403.

Liu, J.-J., Feng, D., Ma, W.-D., Yan, Z.-Q., Fu, J., 2002. Polymer flooding pilot in Shengtuo Block 1 Sa2 1–3 group. Oil and Gas Surface Engineering 21 (2), 45–47.

Liu, L., Wang, L.-J., Sui, X.-G., Song, K.-P., 2005a. Performance evaluation and analysis of the typical polymer flooding in Daqing Sa-Zhong fields. In: Yan, C.Z., Li, Y. (Eds.), Tertiary Oil Recovery Symposium. Petroleum Industry Press, pp. 16–22.

Liu, L.-W., 2003. Properties of the salinity-resistant polymer KYPAM. Fault Block Oil & Gas Field 10 (4), 62–63.

Liu, P., Hu, X.-P., Yan, Z.-J., 2005b. LPS polymer flooding pilot test in the Ke-shang layer of Block 3–2, Karamay Field. In: Yan, C.-Z., Li, Y. (Eds.), Tertiary Oil Recovery Symposium. Petroleum Industry Press, pp. 28–35.

Liu, S., 2007. Alkaline Surfactant Polymer Enhanced Oil Recovery Processes. Ph.D. dissertation, Rice University, Houston.

Liu, Q, Dong, M.-Z., Asghari, K., Tu, Y., 2007. Wettability alteration by magnesium ion binding in heavy oil/brine/chemical/sand systems—analysis of electrostatic forces. J. Pet. Sci. Eng. 59 (1–2), 147–156.

Liu, Q., Dong, M.-Z., Yue, X., Hou, J., 2006b. Synergy between alkali and surfactant in emulsification of heavy oil in brine. Colloids and Surfaces A: Physicochemical and Engineering Aspects 273 (1–3), 219–228.

Liu, Q., Dong, M., Zhou, W., Ayub, M., Zhang, Y.P., Huang, S., 2004. Improved oil recovery by adsorption–desorption in chemical flooding. J. Pet. Sci. Eng. 43 (1–2), 75–86.

Liu, S., Li, R.F., Miller, C., Hirasaki, G.J., 2008a. ASP processes: Wide range of conditions for good recovery. Paper SPE 113936 presented at the SPE/DOE Improved Oil Recovery Symposium, Tulsa, 19–23 April.

Liu, S., Zhang, D.L., Yan, W., Puerto, M., Hirasaki, G.A., Millar, C.A., 2008b. Favorable attributes of alkaline-surfactant-polymer flooding. SPEJ (March), 5–16.

Liu, Y.Z., Bai, B., Shuler, P.J., 2006c. Application and development of chemical-based conformance control treatments in China oil fields. Paper SPE 99641 presented at the SPE/DOE Symposium on Improved Oil Recovery, Tulsa, 22–26 April.

Lötsch, T., 1988. Untersachungen zum retentionsverhalten von polymeren beim viskosen fluten. Polymer Flooding. In: Littman, W. (Ed.), Developments in Petroleum Science. Elsevier Science Publishers City, p. 24.

Lu, R., Chen, L.D., 1996. The Application of 2-acrylamide-2-methyl-propane-sulfonic acid. Speciality Chemicals 4, 43–46.

Lu, Y., 1994. A Study of Residual Oil Saturation in Heterogeneous Sandstone Experiments. M.S. thesis, University of Texas at Austin.

Luo, G.M., Lin, R.S., 2003. Synthesis and evaluation of copolymer PL-1. P.G.O.D.D. 22 (6), 63–64.

Luo, J.-H., Bu, R.-Y., Wang, P.-M., Bai, F.-L., Zhang, Y., Yang, J.-B., Liu, Y.-Z., 2002. Properties of KYPAM, a salinity-resistant polymer used in EOR. Oilfield Chemistry 19 (1), 64–67.

Luo, J.-H., Liu, Y.-Z., Zhu, P., 2006. Polymer solution properties and displacement mechanisms. In: Shen, P.-P., Liu, Y.-Z., Liu, H.-R. (Eds.), Enhanced Oil Recovery–Polymer Flooding. Petroleum Industry Press, pp. 1–72.

Luo, J.H., Cheng, G.Y., 1993. Synthesis of salinity-resistant and high-temperature-resistant RTS series of polymers and their properties. Oilfield Chemistry 10 (4), 331–335.

Luo, P.Y., Li, H.B., Zheng, Y., Liao, G.Z., Yang, Z.Y., 2001. Evaluation of a hydrophobically associative polymer in Daqing ASP flooding. P.G.O.D.D. 20 (6), 1–4.

Lyman, W.J., Reehl, W.F., Rosenblatt, D.H., 1990. Handbook of Chemical Property Estimation Methods. American Chemical Society, Washington, DC.

Ma, Q.-K., Wu, X.-C., Zeng, J.-C., 2005. Research status of the flow behavior of movable polymer gels. In: Yan, C.Z., Li, Y. (Eds.), Tertiary Oil Recovery Symposium. Petroleum Industry Press, pp. 183–189.

Mack, J.C., Smith, J.E., 1994. In-depth colloidal dispersion gels improve oil recovery efficiency. Paper SPE/DOE 27780 presented at the SPE/DOE Improved Oil Recovery Symposium, Tulsa, 17–20 April.

MacWilliams, D.C., Rogers, J.H., West, T.J., 1973. Water soluble polymers in petroleum recovery. In: Bikales, N.M. (Ed.), Water Soluble Polymers. Plenum Press, pp. 105–126.

Maerker, J.M., 1975. Shear degradation of partially hydrolyzed polyacrylamide solutions. SPEJ (August), 311–321; Trans. AIME 259.

Maerker, J.M., 1976. Mechanical degradation of partially hydrolyzed polyacrylamide solutions in unconsolidated porous media. SPEJ (August), 172–174.

Manrique, E.J., Muci, V.E., Gurfinkel, M.E., 2007. EOR field experiences in carbonate reservoirs in the United States. SPEREE (December), 667–686.

Marshall, R.J., Mentzner, A.B., 1964. Ind. Eng. Chem. Fund. 6, 393–400.

Martin, F.D., Hatch, M.J., Shepitka, J.S., Ward, J.S., 1983. Improved water-soluble polymers for enhanced recovery of oil. Paper SPE 11786 presented at the SPE Oilfield and Geothermal Chemistry Symposium, Denver, 1–3 June.

Martin, F.D., Oxley, J.C., 1985. Effect of various chemicals on phase behavior of surfactant/brine/oil mixtures. Paper SPE 13575 presented at the International Symposium on Oilfield and Geothermal Chemistry, Phoenix, 9–11 April.

Martin, J.C., 1959. The effect of clay on the displacement of heavy oil by water. Paper SPE 1411-G presented at the 3rd Annual Meeting, Caracas, 14–16 October.

Masalmeh, S.K., 2002. The effect of wettability on saturation functions and impact on carbonate reservoirs in the Middle East. Paper SPE 78515 presented at Abu Dhabi International Petroleum Exhibition and Conference, Abu Dhabi, 13–16 October.

Masuda, Y., Tang, K., Miyazawa, M., Tanaka, S., 1992. 1D simulation of polymer flooding including the viscoelastic effect of polymer solution. SPEREE (May), 247–252.

Mayer, E.H., Berg, R.L., Carmichael, J.D., Weinbrandt, R.M., 1983. Alkaline injection for enhanced oil recovery–a status report. JPT (January), 209–221.

McCaffery, F.G., 1976. Interfacial tensions and aging behavior of some crude oils against caustic solution. J. Can. Pet. Tech. (July–September).

McCormick, C.L., Nonaka, T., Johnson, C.B., 1988. Polymer. 29, 731.

McGuire, P.L., Chatham, J.R., Paskvan, F.K., Sommer, D.M., Carini, F.H., 2005. Low salinity oil recovery: an exciting new EOR opportunity for Alaska's North Slope. Paper SPE 93903 presented at the Western Regional Meeting, Irvine, 30 March–1 April.

McNeil, J.D., 1980. Electrical Conductivity of Soils and Rocks. Geonics Ltd., Missisauga.

Meter, D.M., Bird, R.B., 1964. Tube flow of non-Newtonian polymer solutions, Parts I and II—Laminar Flow and Rheological Models. AIChE J. (November), 878–881, 1143–1150.

Meyers, K.O., Salter, S.J., 1981. The effect of oil/brine ratio on surfactant adsorption from microemulsion. SPEJ (August), 500–512.

Michaeli, I., 1978. Colston Pap. 29 (Ions Macromol. Biol. Syst.), 121–131.

Michels, A.M., Djojosoeparto, R.S., Haas, H., Mattern, R.B., van der Weg, P.B., Schulte, W.M., 1996. Enhanced waterflooding design with dilute surfactant concentrations for North Sea conditions. SPERE (August), 189–195.

Mihcakan, I.M., Van Kirk, C.W., 1986. Blending alkaline and polymer solutions together into a single slug improves EOR. Paper presented at the SPE Rocky Mountain Regional Meeting, Billings, 19–21 May.

Miller, C.A., 1975. Stability of moving surfaces in fluid systems with heat and mass transport; III. Stability of Displacement Fronts in Porous Media. AIChE J. (May), 474–479.

Miller, C.A., Reinan, H., Benton, W.J., Tomlinson, F., 1977. A mechanism for ultralow interfacial tension in systems containing microemulsion—theoretical consideration and experiments with ultracentrifuge. J. Colloid Interface Sci. 61, 554.

Mohammadi, H., 2008. Mechanistic Modeling, Design and Optimization of Alkaline/Surfactant/Polymer Flood. Ph.D. dissertation, University of Texas at Austin.

Mohammadi, H., Delshad, M., Pope, G., 2008. Mechanistic modeling of alkaline/surfactant/polymer floods. Paper SPE 110212 presented at SPE/DOE Improved Oil Recovery Symposium, Tulsa, 19–23 April.

Mohanty, K.K., Salter, S.J., 1983. Multiphase flow in porous media: III. Oil mobilization, transverse dispersion and wettability. Paper SPE 12127 presented at the 58th Annual Conference of the SPE, San Francisco, 5–8 October.

Mohnot, S.M., Bae, J.H., 1989. A study of mineral–alkali reactions—Part II. SPERE (August), 381–390.

Mohnot, S.M., Bae, J.H., Foley, W.L., 1987. A study of mineral–alkali reactions. SPERE (November), 653–663.

Montaron, B., 2008. Carbonate evolution. Oil & Gas Middle East (August), 26–31.

Moore, T.F., Slobod, R.C., 1955. Displacement of oil by water–effect of wettability, rate, and viscosity on recovery. Paper SPE 502-G presented at the SPE Annual Fall Meeting, New Orleans, 2–5 October.

Moradi-Araghi, A., Doe, P.H., 1984. Hydrolysis and precipitation of polyacrylamides in hard brines at elevated temperatures. Paper SPE 13033 presented at the SPE Annual Technical Conference and Exhibition, Houston, 16–19 September.

Morel, D., Vert, M., Jouenne, S., Nahas, E., 2008. Polymer injection in deep offshore field: the Dalia Angola case. Paper SPE 116672 presented at the SPE Annual Technical Conference and Exhibition, Denver, 21–24 September.

Morgan, J.C., Schechter, R.S., Wade, W.H., 1977. Recent advances in the study of low interfacial tensions. In: Shah, D.O., Schechter, R.S. (Eds.), Improved Oil Recovery by Surfactant and Polymer Flooding. Academic Press, pp. 101–118.

Morrow, N.R., Chatzis, I., Lim, H.T., 1983. Relative permeabilities at reduced residual saturations. Paper 83–34–28 presented at the Canadian Petroleum Society Annual Technical Meeting, Banff, 10–13 May.

Morrow, N.R., Chatzis, I., Taber, J.J., 1988. Entrapment and mobilization of residual oil in bead packs. SPERE 3 (3), 927–934.

Morrow, N.R., Lim, H.T., Ward, J.S., 1986. Effect of crude-oil-induced wettability changes on oil recovery. SPEFE (February), 89–103.

Morrow, N.R., Songkran, B., 1981. Effect of viscous and buoyancy forces on nonwetting phase trapping in porous media. In: Shah, D.O. (Ed.), Surface Phenomena in Enhanced Oil Recovery. Plenum Press, pp. 387–411.

Moritis, G., 2000. EOR weathers low oil prices. Oil & Gas J. (Special issue), 20 March, 44.

Mungan, N., 1966a. Certain wettability effects in laboratory waterfloods. JPT (February), 247–252.

Mungan, N., 1966b. Interfacial effects in immiscible liquid-liquid displacement in porous medium. SPEJ (September), 247–253.

Mungan, N., 1969. Rheology and adsorption of aqueous polymer solutions. J. Can. Pet. Tech. 8 (April–June), 45.

Murtada, H., Marx, C., 1982. Evaluation of the low tension flood process for high-salinity reservoirs–laboratory investigation under reservoir conditions. SPEJ 22 (6), 831–846.

Naiki, M., 1979. Numerical Simulation of Polymer Flooding Including the Effects of Salinity. Ph.D. dissertation, University of Texas at Austin.

Nelson, R.C., 1981. Further studies on phase relationships in chemical flooding. In: Shah, D.O. (Ed.), Surface Phenomena in Enhanced Oil Recovery. Plenum Press, pp. 73–104.

Nelson, R.C., 1982. The salinity-requirement diagram—a useful tool in chemical flooding research and development. SPEJ (April), 259–270.

Nelson, R.C., 1983. The effect of live crude on phase behavior and oil-recovery efficiency of surfactant flooding systems. SPEJ (June), 501–510.

Nelson, R.C., Lawson, J.B., Thigpen, D.R., Stegemeier, G.L., 1984. Cosurfactant-enhanced alkaline flooding. Paper SPE 12672 presented at the SPE/DOE Fourth Symposium on Enhanced Oil Recovery, Tulsa, 15–18 April.

Nelson, R.C., Pope, G.A., 1978. Phase relationships in chemical flooding. SPEJ (October), 325–338; Trans. AIME, 265.

Nie, X.-B., Wang, Y.-Q., Guo, H.-J., Dong, H.-P., 2005. Application of weak gel technology in a low permeability sandstone reservoir. In: Yan, C.Z., Li, Y. (Eds.), Tertiary Oil Recovery Symposium. Petroleum Industry Press, pp. 190–196.

Niu, J.-G., Chen, P., Shao, Z.-B., Wang, D.-M., Sun, G., Li, Y., 2006. Research and development of polymer enhanced oil recovery. In: Cao, H.-Q. (Ed.), Research and Development of Enhanced Oil Recovery in Daqing. Petroleum Industry Press, pp. 227–325.

Novosad, J., 1982. Surfactant retention in Berea sandstone—effects of phase behavior and temperature. SPEJ (December), 962–970.

Novosad, Z., et al., 1981. Comparison of oil recovery potential of sodium orthosilicate and sodium hydroxide for the Wainwright reservoir, Alberta. Petroleum Recovery Institute Report No. 81-10 (July).

Nutting, P.G., 1925. Chemical problems in the water driving of petroleum from oil sands. Ind. Eng. Chem. 17, 1035–1036.

O'Leary, W.B., Boivin, J.W., Dasinger, B.L., Beck, D., Goldman, I.M., Wernau, W.C., 1985. Biocide evaluation against sessile xanthan polymer-degrading bacteria. Paper SPE 13588 presented at the SPE International Symposium on Oilfield and Geothermal Chemistry, Phoenix, 9–11 April.

Olsen, D.K., Hicks, M.D., Hurd, B.G., Sinnokrot, A.A., Sweigart, C.N., 1990. Design of a novel flooding system for an oil-wet Central Texas carbonate reservoir. Paper SPE 20224 presented at the SPE/DOE Seventh Symposium on Enhanced Oil Recovery, Tulsa, 22–25 April.

Ottewill, R.H., 1984. Introduction. In: Tadros, T.F. (Ed.), Surfactants. Academic Press, pp. 1–18.

Ou-Yang, J., Sun, G.-H., Wang, G.-J., Zhu, Z.-Y., Wang, F., 2004. Study of rheological behavior and oil displacement efficiency of the temperature- and salt-resistant copolymer, TS-45. Oilfield Chemistry 21 (4), 327, 330–332.

Pang, Z.-W., Li, J.-L., Liu, H.-B., Li, Y., 1998a. Polymer viscosity loss in injection and production processes. In: Gang, Q.-L., et al. (Eds.), Chemical Flooding Symposium–Research Results during the Eighth Five-Year Period (1991–1995), Vol. I. Petroleum Industry Press, pp. 385–394.

Pang, Z.-W., Li, J.-L., Shi, S.-K., Li, Y., Liu, H.-B., Chen, J.-S., Zhang, F.-L., Zhang, Z.-Y., 1998b. Effect of polymer molecular weight on residual permeability reduction factor. In: Gang, Q.-L., et al. (Eds.), Chemical Flooding Symposium–Research Results during the Eighth Five-Year Period (1991–1995), Vol. I. Petroleum Industry Press, pp. 138–149.

Parlar, M., Yortsos, Y.C., 1987. Percolation theory of steam/water relative permeability. Paper SPE 16969 presented at the SPE 62nd Annual Technical Conference and Exhibition, Dallas, 27–30 September.

Pasquarelli, C.H., Wasan, D.T., 1979. The effect of film-forming materials on the dynamic interfacial properties of crude oil–aqueous system. Paper presented at the Third International Conference on Surface and Colloid Science, Stockholm, 20–25 August.

Pennell, K.D., Abriola, L.M., Weber, W.J., 1993. Surfactant enhanced solubilization of residual dodecane in soil columns: 1. Experimental Investigation. Environmental Science and Technology 27 (12), 2332–2340.

Pennell, K.D., Pope, G.A., Abriola, L.M., 1996. Influence of viscous and buoyancy forces on the mobilization of residual tetrachloroethylene during surfactant flushing. Environmental Science and Technology 30 (4), 1328–1335.

Perkins, T.K., Johnston, O.C., 1963. A review of diffusion and dispersion in porous media. SPEJ (March), 70–81; Trans. AIME, 228.

PHREEQC, 2008. *http://wwwbrr.cr.usgs.gov/projects/GWC_coupled/phreeqc/.*

Pickens, J.F., Grisak, G.E., 1981. Scale-dependent dispersion in a stratified granular aquifer. Water Resources Research 17 (4), 1191–1211.

Potts, D.E., Kuehne, D.L., 1988. Strategy for alkaline polymer flood design with Berea and reservoir-rock corefloods. SPERE 3 (4), 1143–1152.

Pope, G.A., 1980. The application of fractional flow theory to enhanced oil recovery. SPEJ (June), 191–205.

Pope, G.A., Baviere, M., 1991. Basic concepts in enhanced oil recovery processes. In: Baviere, M. (Ed.), Critical Reports on Applied Chemistry, Vol. 33. Elsevier Science, p. 101.

Pope, G.A., Lake, L.W., Helfferich, F.G., 1978. Cation exchange in chemical flooding: part 1—basic theory without dispersion. SPEJ 18, 418–434.

Pope, G.A., Nelson, R.C., 1978. A chemical flooding compositional simulator. SPEJ 18, 339–354.

Pope, G.A., Tsaur, K., Schechter, R.S., Wang, B., 1982. The effect of several polymers on the phase behavior of micellar fluids. SPEJ (December), 816–830.

Pope, G.A., Wang, B., Tsaur, K., 1979. A sensitivity study of micellar/polymer flooding. JPT (December), 357–368.

Pritchett, J., Frampton, H., Brinkman, J., Cheung, S., Morgan, J., Chang, K.T., Williams, D., Goodgame, J., 2003. Field application of a new in-depth waterflood conformance improvement tool. Paper SPE 84897 presented at the SPE International Oil Recovery Conference in Asian Pacific, Kuala Lumpur, 20–21 October.

Prats, M., 1982. Thermal Recovery. Society of Petroleum Engineers, Dallas.

Prud'Homme, R.K., Uhl, J.T., Poinsatte, J.P., Halverson, F., 1983. Rheological monitoring of the formation of polyacrylamide/Cr^{+3} gels. SPEJ (October), 804–808.

Puerto, M.C., Gale, W.W., 1977. Estimation of optimal salinity and solubilization parameters for alkylorthoxylene sulfonate mixtures. SPEJ 17 (3), 193–200.

Pursley, S.A., Graham, H.L., 1975. Borregos surfactant pilot test. JPT (June), 695–700.

Pursley, S.A., Healy, R.N., Sandvik, E.I., 1973. A field test of surfactant flooding, Loudon, Illinois. JPT (July), 793–802.

Pusch, G., Lotsch, T., Muller, T., 1987. Investigation of the oil displacing efficiency of suitable polymer products in porous media, aspects of recovery mechanisms during polymer flooding. DGMK—Report, German Society Petrol. Sci. Coal Chem., 295–296, Hamburg.

Pyziak, D., Smith, D., 2007. Update on anton irish conformance efforts. Paper presented at the 6th International Conference on Production Optimization–Reservoir Conformance–Profile Control–Water and Gas Shutoff, Houston, 6–7 November.

Qi, L.-Q., Feng, H.-Q., 1998. Optimum amount of polymer injection in polymer flooding. In: Gang, Q.-L., et al. (Eds.), Chemical Flooding Symposium–Research Results during the Eighth Five-Year Period (1991–1995), Vol. I. Petroleum Industry Press, pp. 277–279.

Qiao, Q., Gu, H.-J., Li, D.-W., Dong, L., 2000. The pilot test of ASP combination flooding in the Karamay oil field, paper SPE 64726 presented at the International Oil and Gas Conference and Exhibition in China, Beijing, 7–10 November.

Qu, Z.-J., Zhang, Y.-G., Zhang, X.-S., Dai, J.-L., 1998. A successful ASP flooding pilot in the Gudong oil field, paper SPE/DOE 39613 presented at Improved Oil Recovery Symposium, Tulsa, 19–22 April.

Radke, C.J., Somerton, W.H., 1978. Enhanced recovery with mobility and reactive tension agents. In Enhanced Oil and Gas Recovery and Improved Drilling Methods, Vol. 1A–Oil, B-2, Fourth Annual U.S. DOE Symposium, 29–31 October.

Reed, R.L., Carpenter Jr., C.W., 1982. Shear-stabilized emulsion flooding process. US patent No. 4,337,159, 29 June.

Reed, R.L., Healy, R.N., 1977. Some physico-chemical aspects of microemulsion flooding: a review. In: Shah, D.O., Schechter, R.S. (Eds.), Improved Oil recovery by Surfactant and Polymer Flooding. Academic Press, pp. 383–437.

Reed, R.L., Healy, R.N., 1979. Contact angles for equilibrated microemulsion systems. Paper SPE 8262 presented at the SPE Annual Technical Conference and Exhibition, Las Vegas, 23–26 September.

Reisberg, J., Doscher, T.M., 1956. Interfacial phenomena in crude oil–water systems. Production Monthly (November), 43–50.

Rivenq, R., Sardin, M., Schweich, D., Putz, A., 1985. Sodium carbonate preflush theoretical analysis and application to Chateaurenard field test. Paper SPE 14294 presented at the SPE Annual Technical Conference and Exhibition, Las Vegas, 22–26 September.

Roland, C.M., Archer, L.A., Mott, P.H., Sanchez-Reyes, J., 2004. Determining Rouse relaxation times from the dynamic modulus of entangled polymers. Journal of Rheology (J. Rheol.) 48 (2), 395–403.

Rosen, M.J., 1978. Surfactants and Interfacial Phenomena. Wiley.

Rousseau, D., Chauveteau, G., Renard, M., Tabary, R., Zaitoun, A., Mallo, P., Braun, O., Omari, A., 2005. Rheology and transport in porous media of new water shutoff/conformance control microgels. Paper SPE 93254 presented at the SPE International Symposium on Oilfield Chemistry, Woodlands, 2–4 February.

Rudin, J., Wasan, D.T., 1992a. Mechanisms for lowering of interfacial tension in alkali/acidic oil systems 1. Experimental studies. Colloid and Surfaces 68, 67–79.

Rudin, J., Wasan, D.T., 1992b. Mechanisms for lowering of interfacial tension in alkali/acidic oil systems: effect of added surfactant. Industrial & Engineering Chemistry Research (Ind. Eng. Chem. Res.) 31 (8), 1899–1906.

Rudzinski, W.E., Oehlers, L., Zhang, Y., 2002. Tandem mass spectrometric characterization of commercial naphthenic acids and a Maya crude oil. Energy Fuels 16 (5), 1178–1185.

Ryles, R.G., 1983. Elevated temperature testing of motility-control reagents. Paper SPE 12008 presented at the SPE Annual Technical Conference and Exhibition, San Francisco, 5–8 October.

Ryles, R.G., 1988. Chemical stability limits of water-soluble polymers used in oil recovery. SPERE (Feb.), 23–34.

Sadowski, T.J., Bird, R.B., 1965. Non-Newtonian flow through porous media. I. theoretical. Transaction of the Society of Rheology (Trans. Soc. Rheol.) 9 (2), 243–250.

Salager, J.L., Bourrel, M., Schechter, R.S., Wade, W.H., 1979a. Mixing rules for optimum phase-behavior formulations of surfactant/oil/water systems. SPEJ (October), 271–278.

Salager, J.L., Morgan, J.C., Schechter, R.S., Wade, W.H., Vasquez, E., 1979b. Optimum formulation of surfactant/water/oil systems for minimum interfacial tension or phase behavior. SPEJ 23 (April), 107–115.

Salinas, J.L., Miller, C.A., Yoo, K.H.K., Puerto, M., Hirasaki, G.A., 2009. Viscometer for opaque, sealed microemulsion samples. Paper presented at the SPE International Symposium on Oilfield Chemistry, Woodlands, 20–22 April.

Salter, S.J., Mohanty, K.K., 1982. Multiphase flow in porous media: I. Macroscopic observations and modeling. Paper SPE 11017 presented at the SPE Annual Technical Conference and Exhibition, New Orleans, 26–29 September.

Sarem, A.M., 1974. Secondary and tertiary recovery of oil by MCCF (mobility-controlled caustic flooding) process. Paper SPE 4901 presented at the SPE–AIME 44th Annual California Regional Meeting, San Francisco, 4–5 April.

Savins, J.G., 1969. Non-Newtonian flow through porous media. Ind. Eng. Chem. 61 (10), 18–47.

Schneider, F.N., Owens, W.W., 1982. Steady state measurements of relative permeability of polymer/oil systems. SPEJ (February), 79–86.

Schwartzenbach, R.P., Gschwend, P.M., Imboden, D.M., 1993. Environmental Organic Chemistry. Wiley.

Seccombe, J.C., Lager, A., Webb, K., Jerauld, G., Fueg, E., 2008. Improving waterflood recovery: LoSal™ EOR field evaluation. Paper SPE 113480 presented at the SPE/DOE Symposium on Improved Oil Recovery, Tulsa, 20–23 April.

Seethepalli, A., Adibhatla, B., Mohanty, K.K., 2004. Physicochemical interactions during surfactant flooding of fractured carbonate reservoirs. SPE J. 9 (4), 411–418.

Seifert, W.K., 1975. Carboxylic Acids in Petroleum Sediments. Progress in Chemistry of Natural Products, Springer-Verlag.

Selamat, S., Teletzke, G.F., Patel, P.D., Darman, N., Suhaimi, M.A., 2008. EOR: The new frontier in the Malay basin development. Paper IPTC 12805 presented at the International Petroleum Technology Conference, Kuala Lumpur, 3–5 December.

Seright, R.S., 2004. Conformance improvement using gels. Annual Technical Progress Report (U.S. DOE Report DOE/BC/15316–6), U.S. DOE Contract DE-FC26-01BC15316.

Seright, R.S., 2006. Discussion of SPE 89175. JPT (May), 80.

Seright, R.S., Adamski, R.P., Roffall, J.C., Liauh, W.W., 1983. Rheology and mechanical degradation of EOR polymers. Paper presented at the SPE/British Society of Rheology Conference on Rheology in Crude Oil Production, Imperial College, 13–15 April.

Seright, R.S., Liang, J., 1994. A survey of field applications of gel treatments for water shut-off. Paper SPE 26991 presented at III Latin American/Caribbean Petroleum Engineering Conference, Buenos Aires, 27–29 April.

Seright, R.S., Maerker, J.M., Holzwarth, G., 1981. Polymer Preprints (Polym. Prepr.) 22, 30–33.

Shah, B., 1978. An Experimental Study of Inaccessible Pore Volume as a Function of Polymer Concentration during Flow Though Porous Media. M.S. thesis, University of Kansas, Lawrence.

Shao, Z.-B., Chen, P., Wang, D.-M., Wu, L.-J., 2005. Polymer flooding performance in Daqing commercial scale applications and the learning. In: Yan, C.Z., Li, Y. (Eds.), Tertiary Oil Recovery Symposium. Petroleum Industry Press, pp. 1–8.

Sharma, M.M., Filoco, P.R., 2000. Effect of brine salinity and crude-oil properties on oil recovery and residual saturations. SPE J. 5 (3), 293–300.

Sharma, M.M., Jang, L.K., Yen, T.F., 1989. Transient interfacial tension behavior of crude-oil/caustic interfaces. SPERE (May), 228–236.

Shen, M.-D., Chen, Z., 1996. Chemical behavior of minerals in alkaline solutions. In: Liu, S.-G., Lu, Z.-Y. (Eds.), Advance in Petroleum Geology. Xichun Science and Technology Press, pp. 144–148.

Sheng, D.-C., Yang, P.-H., Liu, Y.-L., 1993. Alkali-polymer interaction and alkali+polymer/crude IFT. Oilfield Chemistry 10 (1), 46–50, 61.

Sheng, D.-C., Yang, P.-H., Liu, Y.-L., 1994. Effect of alkali-polymer interaction on the solution properties. Petroleum Exploration and Development 21 (2), 81–85.

Sheng, J.J., Maini, B.B., Hayes, R.E., Tortike, W.S., 1997. Experimental study of foamy oil stability. J. Can. Pet. Tech. 36 (4), 31–37.

Sherborne, J.E., Sarem, A.M., Sandiford, B.B., 1967. Flooding oil containing formations with solutions of polymer in water. Paper presented at the Seventh World Petroleum Congress, Mexico City, 2–8 April.

Shuler, P.J., Kuehne, D.L., Lerner, R.M., 1989. Improving chemical flood efficiency with micellar/alkaline/polymer process. JPT (January), 80–88.

Slobod, R.L., Caudle, B.H., 1952. X-Ray shadowgraph studies of areal sweepout efficiencies. Trans. AIME 195, 265–270.

Smith, F.W., 1970. The behavior of partially hydrolyzed polyacrylamide solutions in porous media. JPT 22, 148–156.

Smith, J.E., Liu, H., Guo, Z.-D., 2000. Laboratory studies of in-depth colloidal dispersion gel technology for Daqing Oil Field. Paper SPE 62610 presented at the SPE/AAPG Western Regional Meeting, Long Beach, 19–22 June.

Somasundaran, P., Agar, G.E., 1967. The zero point of charge of calcite. J. Colloid Interface Sci. 24, 433–440.

Somasundaran, P., Ananthapadmanabhan, K., Celik, M.S., Manev, E.D., 1984a. A thermodynamic model of redissolution of calcium sulfonate precipitates in NaCl solutions. SPEJ (December), 667–676.

Somasundaran, P., Celik, M.S., Goyal, A., Manev, E.D., 1984b. The role of surfactant precipitation and redissolution in the adsorption of sulfonate on minerals. SPEJ (April), 233–239.

Somasundaran, P., Hanna, H.S., 1977. Physico-chemical aspects of adsorption at solid/liquid interfaces, I. Basic principles. In: Shah, D.O., Schechter, R.S. (Eds.), Improved Oil Recovery by Surfactant and Polymer Flooding. Academic Press, pp. 205–251.

Somerton, W.H., Radke, C.J., 1983. Role of clays in the enhanced recovery of petroleum from some California sands. JPT (March), 643–654.

Song, J.-P., 1993. A study of the interfacial tension between a heavy oil and an alkaline solution with or without polymer. Petroleum Exploration and Development 20 (4), 81–86.

Song, W.-C., Yang, C.-Z., Han, D.-K., Qu, Z.-J., Wang, B.Y. Jia, W.-L., 1995. Alkaline-surfactant-polymer combination flooding for improving recovery of the oil with high acid value, paper SPE 29905 presented at the International Meeting on Petroleum Engineering, Beijing, 14–17 November.

Song, W.-C., Zhang, Y.-G., Wang, Y.-Q., Zhao, J.-D., 1994. Gudong ASP pilot performance. Oil and Gas Recovery Technology 1 (2), 8–12.

Sorbie, K.S., 1991. Polymer-Improved Oil Recovery. CRC Press.

Sorbie, K.S., Collins, I.R., 2010. A proposed pore-scale mechanism for how low salinity waterflooding works. Paper SPE 129833 presented at the SPE Improved Oil Recovery Symposium, Tulsa, 24–28 April.

Southwick, J.G., 1985. Solubility of crystalline silica in alkaline solutions: implications for alkaline flooding. SPEJ (December), 857–864.

Srivastava, M., Zhang, J.Y., Nguyen, Q.P., Pope, G.A., 2009. A systematic study of alkaline-surfactant-gas injection as an EOR technique. Paper SPE 124752 presented at the SPE Annual Technical Conference and Exhibition, New Orleans, 4–7 October.

Standard Oil., 1951. Fluid distributions in porous systems. Technical film produced by Standard Oil and Gas Co., Tulsa.

Standnes, D.C., Austad, T., 2000. Wettability alteration in chalk 2. Mechanism for wettability alteration from oil-wet to water-wet using surfactants. J. Pet. Sci. Eng. 28, 123–143.

Stegemeier, G.L., 1977. Mechanisms of entrapment and mobilization of oil in porous media. In: Shah, D.O., Schechter, R.S. (Eds.), Improved Oil Recovery by Surfactant and Polymer Flooding. Academic Press, pp. 55–91.

Stoneberger, M.W., Claridge, E.L., 1988. Graded-viscosity-bank design with pseudoplastic fluids. SPERE (November), 1221–1232.

Stosur, G.J., Hite, J.R., Carnahan, N.F., Miller, K., 2003. The alphabet soup of IOR, EOR and AOR: effective communication requires a definition of terms. Paper SPE 84908 presented at the SPE International Improved Oil Recovery Conference in Asia Pacific, Kuala Lumpur, 20–21 October.

Strand, S., Høgnesen, E.J., Austad, T., 2006. Wettability alteration of carbonates—effects of potential determining ions (Ca^{2+} and SO_4^{2-}) and temperature. Colloids and Surfaces A: Physicochemical and Engineering Aspects 275 (2006), 1–10.

Stumm, W., Morgan, J.J., 1970. Aquatic Chemistry. Wiley–Interscience.

Subkow, P., 1942. Process for the removal of bitumen from bituminous deposits, US Patent No. 2,288,857, 7 July.

Sun, A.-J., Zhu, L.-J., Wu, Z.-L., Lin, M.-Q., 2005. Viscoelasticity of low concentration HPAM/AlCit cross-linked system—mathematical characterization. In: Yan, C.-Z., Li, Y. (Eds.), Tertiary Oil Recovery Symposium. Petroleum Industry Press, pp. 324–331.

Sun, H.-Q., 2005. Formula selection and pilot test of a surfactant-polymer flooding. In: Yan, C.-Z., Li, Y. (Eds.), Tertiary Oil Recovery Symposium. Petroleum Industry Press, pp. 116–122.

Sun, H.-Q., Wang, T., Xiao, J.-H., Chen, H., 2006. A new micro-polymer ball for deep profile control. Petroleum Geology and Recovery Efficiency 13 (4), 77–79.

Sydansk, R.D., Al-Dhafeeri, A.M., Xiong, Y., Seright, R.S., 2004. Polymer gels formulated with a combination of high- and low-molecular-weight polymers provide improved performance for water-shutoff treatments of fractured production wells. SPE Production & Facilities 19 (4), 229–236.

Sydansk, R.D., Xiong, Y., Al-Dhafeeri, A.M., Schrader, R.J., Seright, R.S., 2005. Characterization of partially formed polymer gels for application to fractured production wells for water-shutoff purposes. SPE Production & Facilities 20 (3), 240–249.

Szabo, M.T., 1975. Some aspects of polymer retention in porous media using a ^{14}C-tagged hydrolyzed polyacrylamide. SPEJ (August), 323–337.

Szabo, M.T., 1979. An evaluation of water-soluble polymers for secondary oil recovery. parts I and II. JPT (May), 553–570.

Taber, J.J., 1969. Dynamic and static forces required to remove a discontinuous oil phase from porous media containing both oil and water. SPEJ 9 (1), 2–12.

Taber, J.J., Kirby, J.C., Schroeder, F.U., 1973. Studies on the displacement of residual oil: viscosity and permeability effects. AIChE Symp. Series 69, 53.

Taber, J.J., Martin, F.D., 1983. Technical screening guides for the enhanced recovery of oil. Paper SPE 12069 presented at the SPE Annual Technical Conference and Exhibition, San Francisco, 5–8 October.

Taber, J.J., Martin, F.D., Seright, R.S., 1997a. EOR screening criteria revisited–part 1: introduction to screening criteria and enhanced recovery field projects. SPEREE (August), 189–198.

Taber, J.J., Martin, F.D., Seright, R.S., 1997b. EOR screening criteria revisited–part 2: applications and impact of oil prices. SPEREE (August), 199–205.

Tan, K.H., 1982. Principles of Soil Chemistry. Marcel Dekker.

Tan, Z.-L., 1998. HAPM thermal stability under reservoir conditions. In: Gang, Q.-L., et al. (Eds.), Chemical Flooding Symposium–Research Results during the Eighth Five-Year Period (1991–1995), Vol. I. Petroleum Industry Press, pp. 107–111.

Tang, G.-Q., Firoozabadi, A., 2002. Relative permeability modification in gas/liquid systems through wettability alteration to intermediate gas wetting. SPEREE 5 (6), 427–436.

Tang, G.-Q., Morrow, N.R., 1997. Salinity, temperature, oil composition and oil recovery by waterflooding. SPERE (November), 269–276.

Tang, G.-Q., Morrow, N.R., 1999. Influence of brine composition and fines migration on crude/oil/rock interactions and oil recovery. J. Pet. Sci. Eng. 24, 99–111.

Tang, J.-X, Cheng, T.-L., He, J.-S., Peng, K.-Z., 1998. Relative permeabilities in polymer flooding. In: Gang, Q.-L., et al. (Eds.), Chemical Flooding Symposium—Research Results during the Eighth Five-Year Period (1991–1995), Vol. I. Petroleum Industry Press, pp. 120–124.

Teeuw, D., Hesselink, F.T., 1980. Power law flow and hydrodynamic behavior of biopolymer solutions in porous media. Paper SPE 8982 presented at the 5th SPE International Symposium on Oilfield and Geothermal Chemistry, Stanford, 28–30 May.

Tham, M.K., Lorenz, P.B., 1981. The EACN of crude oil: variations with cosurfactant and water–oil–ratio. Paper presented at the European Symposium on Enhanced Oil Recovery, Bournemouth, 21–23 September.

Thomas, S., 2008. Enhanced oil recovery—an overview. Oil & Gas Science and Technology—Rev. IFP 63 (1), 9–19.

Thompson, D.W., Pownall, P.G., 1989. Surface electrical properties of calcite. J. Coll. Int. Sci. 131 (1), 74–82.

Tiab, D., Okoye, C.U., Osman, M.M., 1982. Caustic steam flooding. JPT (August), 1817–1827.

Todd, M.R., Dietrich, J.K., Goldburg, A., Larson, R.G., 1978. Numerical simulation of competing chemical flood designs. Paper SPE 7077 presented at the SPE Symposium on Improved Methods of Oil Recovery, Tulsa, 16–17 April.

Trantham, J.C., Patterson, H.L., Boneau, D.F., 1978. The North Burbank unit, Tract 97 surfactant/polymer pilot—operation and control. JPT (July), 1068–1074.

Trushenski, S.P., 1977. Micellar flooding: sulfonate-polymer interaction. In: Shah, D.O., Schechter, R.S. (Eds.), Improved Oil Recovery by Surfactant and Polymer Flooding. Academic Press, pp. 555–575.

Trushenski, S.P., Dauben, D.L., Parrish, D.R., 1974. Micellar flooding fluid propagation, interaction, and mobility. SPEJ (December), 633–642; Trans. AIME, 257.

UTCHEM-9.0., 2000. Technical Documentation for UTCHEM-9.0, A Three-Dimensional Chemical Flood Simulator, Austin, July.

Valdya, R.N., Fogler, H.S., 1992. Fines migration and formation damage: influence of pH and ion exchange. SPEPE (November), 325–330.

Van Horn, L.E., 1981. El Dorado micellar-polymer demonstration project, sixth annual report for the period September 1979–August 1980. Report DOE/ET/130070–63, U.S. DOE, Washington, DC.

Vanselow, A.P., 1932. Equilibria of the base-exchange reactions of bentonites, permutites, soil colloids and zeolites. Soil Science 33, 95–113.

Veizen, J.F.G.V., Leeriooljer, K., 1992. Impairment of a water injection well by suspended solids: testing and prediction. Paper SPE 23822 presented at the SPE international symposium on formation damage control, Lafayette, 26–27 February.

Vela, S., Peaceman, D.W., Sandvik, E.I., 1976. Evaluation of polymer flooding in a layered reservoir with crossflow, retention and degradation. SPEJ (April), 82–96.

Volpert, E., Selb, J., Candau, F., 1998. Associating behavior of polyacrylamide hydrophobically modified with dihexylacrylamide. Polymer 39, 1025–1033.

Wagner, O.R., Leach, R.O., 1959. Improving oil displacement efficiency by wettability adjustment. Trans. AIME 216, 65–72.

Wan, X.-D., Huang, M., Yu, H.-Y., Zhang, Y.-Z., Chen, W.-D., Gao, E.S., 2006. Pilot test of weak alkaline system ASP flooding in secondary layer with small well spacing. Paper SPE 104416 presented at the International Oil & Gas Conference and Exhibition, Beijing, 5–7 December.

Wang, B.-Y., Zhang, J., Jia, Z.-S., 1997a. Study of commercial scale application of polymer flooding in Daqing Oil Field. Petroleum Exploration and Development 24 (6), 51–55.

Wang, C.-L., 2006. Performance analysis of polymer flooding in Gudong Block 6. Neigang Science and Technology 2, 139.

Wang, C.-L., Wang, B.-Y., Cao, X.-L., Li, H.-C., 1997b. Application and design of alkaline-surfactant-polymer system to close well spacing pilot Gudong oilfield. Paper SPE 38321 presented at the SPE Western Regional Meeting, Long Beach, 25–27 June.

Wang, D.-C., Yang, T.-R., Du, T.-P., Fang, B.-F., Yang, C.-Z., 1999a. Micellar/polymer flooding pilot test in the H184 well pattern in the Laojunmiao field. Petroleum Exploration and Development 26 (1), 47–49.

Wang, D.-M., 2001. Development of new tertiary recovery theories and technologies to sustain Daqing oil production. P.G.O.D.D. 20 (3), 1–7.

Wang, D.-M., Cheng, J.-C., Li, Q., Li, L.-Z., Zhao, C.-J., Hong, J.-C., 1999b. An alkaline biosurfactant polymer flooding pilots in Daqing Oil Field. Paper SPE 57304 presented at the SPE Asia Pacific Improved Oil Recovery Conference, Kuala Lumpur, 25–26 May.

Wang, D.-M., Cheng, J.-C., Wu, J.-Z., Wang, F.-L., Li, H.-B., Gong, X.-H., 1998a. An alkaline/surfactant/polymer field test in a reservoir with a long-term 100% water cut. Paper SPE 49018 presented at the SPE Annual Technical Conference and Exhibition, New Orleans, 27–30 September.

Wang, D.M., Cheng, J.C., Yang, Q.Y., Gong, W.C., Li, Q., Chen, F.M., 2000. Viscous-elastic polymer can increase microscale displacement efficiency in cores, paper SPE 63227 presented at the SPE Annual Technical Conference and Exhibition, Dallas, 1–4 October.

Wang, D.-M., Cheng, J.C., Yang, Z.-Y., Li, Q., Wu, W.-X., Yu, H.-Y., 2001a. Successful field test of the first ultra-low interfacial tension foam flood. Paper SPE 72147 presented the SPE Asia Pacific Improved Oil Recovery Conference, Kuala Lumpur, 6–9 October.

Wang, D.-M., Dong, H.-Z., Lv, C.-S., Fu, X.-F., Nie, J., 2009. Review of practical experience by polymer flooding at Daqing. SPEREE 23 (3), 470–476.

Wang, D.-M., Han, P., Shao, Z., Chen, J., Seright, R.S., 2006a. Sweep improvement options for the Daqing Oil Field. Paper SPE 99441 presented at the SPE/DOE Symposium on Improved Oil Recovery, Tulsa, 22–26 April.

Wang, D.-M., Jiang, Y., Wang, Y., Gong, X., Wang, G., 2004a. Viscous-elastic polymer fluids rheology and its effect upon production equipment. SPE Production & Facilities (November), 209–216.

Wang, D.-M., Liu, H., Niu, J.-G., Chen, F.M., 1998b. Application results and understanding of several problems of industrial scale polymer flooding in Daqing Oil Field. Paper SPE 50928 presented at the SPE International Oil and Gas Conference and Exhibition, Beijing, 2–6 November.

Wang, D.-M., Wang, G., Wu, W., Xia, H.F., Yin, H.J., 2007. The influence of viscoelasticity on displacement efficiency—from Micro- to macroscale. Paper SPE 109016 presented at the SPE Annual Technical Conference and Exhibition, Anaheim, 11–14 November.

Wang, D.-M., Xia, H.-F., Liu, Z.-C., Yang, Q.-Y., 2001b. Study of the mechanism of polymer solution with viscoelastic behavior increasing microscopic oil displacement efficiency and the forming of steady "Oil Thread" flow channels. Paper SPE 68723 prepared at the SPE Asia Pacific Oil and Gas Conference and Exhibition, Jakarta, 17–19 April.

Wang, D.M., Zhang, Z.-H., Cheng, J.-C., Yang, J.-C., Gao, S.-T., Li, L., 1997c. Pilot tests of alkaline/surfactant/polymer flooding in Daqing Oil Field. SPERE (November), 229–233.

Wang, F.H.L., 1993. Effects of reservoir anaerobic, reducing conditions on surfactant retention in chemical flooding. SPERE 8 (2), 108–116.

Wang, H.-Z., Liao, G.-Z., Song, J., 2006b. Combined chemical flooding technologies. In: Shen, P.P. (Ed.), Technological Developments in Enhanced Oil Recovery. Petroleum Industry Press, pp. 126–188.

Wang, J., Wang, D.-M., Sui, X.G., Zeng, H.-M., Bai, W.-G., 2006c. Combining small well spacing with polymer flooding to improve oil recovery of marginal reservoirs. Paper SPE 96946 presented at the SPE/DOE Symposium on Improved Oil Recovery, Tulsa, 22–26 April.

Wang, J., Zheng, Y., Feng, Y.J., Luo, P.Y., 1999c. Evaluation of a new hydrophobically associative polymer. Oilfield Chemistry 16 (2), 149–152.

Wang, K.-L., Liao, G.-Z., Yang, Z.-Y., Li, J.-L., 2001c. Experimental study of the effect of displacing fluid viscosity in polymer and ASP flooding. Oilfield Chemistry 18 (4), 354–357.

Wang, M., 1995. Laboratory Investigation of Factors Affecting Residual Oil Saturation by Polymer Flooding. Ph.D. dissertation, University of Texas at Austin.

Wang, T., Sun, H.-Q., Xiao, J.-H., Chen, H., Wang, W.-M., Song, D.-F., 2005. Polymer microball pilot test in the well 1–14 pattern in the eastern block of the Gudao field. Journal of Oil and Gas Technology 27 (6), 779–781.

Wang, Y., Liu, J., Liu, B., Liu, Y., Wang, H., Chen, G., 2004b. Why does scale form in ASP flood? How to prevent from it?—a case study of the technology and application of scaling mechanism and inhibition in ASP flood pilot area of N–1DX block in Daqing. Paper SPE 87469 presented at the 6th International Symposium on Oilfield Scale, Aberdeen, 26–27 May.

Wang, Y., Wang, D.-M., Sun, Z., Han, H., Gong, M., Wang, G., 2004c. Technology for controlling eccentric wear of sucker rods and tubing in pumping wells lifting fluids containing polymer. Paper SPE 89927 presented at the SPE Annual Technical Conference and Exhibition, Houston, 26–29 September.

Wang, Y.-P., Luo, J.-H., Bu, R.Y., Wang, P.M., Bai, F.-L., Liu, Y.Z., 2003a. Analysis of polymers used in tertiary recovery which are temperature- and salinity-resistant. Chemical Industry and Engineering Progress 22 (3), 271–274.

Wang, Y.-P., Luo, J.H., Bu, R.Y., Wang, P.M., Liu, Y.Z., 2003b. Field test of KYPAM, a comb-shaped and salinity-resistant polymer. Chemical Industry and Engineering Progress 22 (5), 509–511.

Webb, K.J., Black, C.J.J., Al-Jeel, H., 2004. Low salinity oil recovery—log-inject-log. Paper SPE 89379 presented at the SPE/DOE 14th Symposium on Improved Oil Recovery, Tulsa, 17–21 April.

Webb, K,J., Black, C.J.J., Edmonds, I.J., 2005. Low salinity oil recovery: the role of reservoir condition corefloods. Paper presented at the 13th European Symposium on Improved Oil Recovery, Budapest, 25–27 April.

Welge, H.J., 1952. A simplified method for computing oil recovery by gas or water drive. Trans. AIME 195, 91–98.

Wellington, S.L., 1980. Biopolymer solution viscosity stabilization–polymer degradation and antioxidant use. SPEJ (December), 901–912.

White, J.L., Goddard, J.E., Phillips, H.M., 1973. Use of polymers to control water production in oil wells. JPT (February), 143–150.

Whittington, L.E., Naae, D.G., 1992. Conformance control gels–formation by contact with brine. J. Pet. Sci. Eng. 7, 45–51.

Willhite, G.P., 1986. Waterflooding, Textbook series. SPE, Richardson, Texas.

Willhite, G.P., Dominguez, J.G., 1977. Mechanisms of polymer retention in porous media. In: Shah, D.O., Schechter, R.S. (Eds.), Improved Oil Recovery by Surfactant and Polymer Flooding. Academic Press, pp. 511–554.

Willhite, G.P., Uhl, J.T., 1986. Correlation of the mobility of biopolymer with polymer concentration and rock properties in sandstone. Proceedings of the ACS Division of PMSE Fall Meeting, Anaheim (September) 55, 577.

Winsor, P.A., 1948. Hydrotropy, solubilization and related emulsification processes, part I. Trans. Faraday Soc. 44, 376–398.

Winsor, P.A., 1954. Solvent Properties of Amphiphilic Compounds. Butterworth Scientific Publications.

Wreath, D.G., 1989. A Study of Polymer Flooding and Residual Oil Saturation. M.S. thesis, University of Texas at Austin.

Wreath, D.G., Pope, G.A., Sepehrnoori, K.S., 1990. Dependence of polymer apparent viscosity on the permeable media and flow conditions. In Situ 14 (3), 263–284.

Wu, W., 1996. Optimum Design of Field-Scale Chemical Flooding Using Reservoir Simulation. Ph.D. dissertation, University of Texas at Austin.

Wu, X.-C., Zeng, Q.-Q., Wang, B.-X., Tian, K.-Z., 2005. Pilot test of movable gel in the Menggulin field. In: Yan, C.-Z., Li, Y. (Eds.), Tertiary Oil Recovery Symposium. Petroleum Industry Press, pp. 339–347.

Wu, X.-L., Zong, L.-P., Chen, J., Wang, H.-F., Yang, Y., Shan, C.-L., Geng, J., 2008. Application of surfactants with narrow equivalent weight distribution and desirable structure to Daqing

ASP flooding. Paper SPE 114345 presented at the SPE/DOE Symposium on Improved Oil Recovery, Tulsa, 20–23 April.

Wyatt, K, Pitts, M., Surkalo, H., 2008. Economics of field proven chemical flooding technologies. Paper SPE 113126 presented at SPE/DOE Improved Oil Recovery Symposium, Tulsa, 19–23 April.

Xia, H.F., Wang, D.M., Liu, Z.C., Yang, Q.Y., 2001. Mechanisms of viscous polymer solutions to increase microscopic displacement efficiency. ACTA PETROLEI SINICA 22 (4), 60–65.

Xia, H.F., Wang, D.M., Wang, G., Ma, W.G., Liu, J., 2008. Mechanism of the effect of micro-forces on residual oil in chemical flooding. Paper SPE 114335 presented at the SPE/DOE Improved Oil Recovery Symposium, Tulsa, 19–23 April.

Xiang, W.-T., Han, M., Zhang, J., Jiang, W., Sun, F.-J., 2005. Polymer pilot test in an offshore viscous oil field. In: Yan, C.-Z., Li, Y. (Eds.), Tertiary Oil Recovery Symposium. Petroleum Industry Press, pp. 23–27.

Xie, F., Ma, Y., Deng, Z.-Z., Liu, W.-H., Lü, X.-H., 2001. Polymer flooding test in the heterogeneous Xia-er-men oilfield of large channels. Petroleum Exploration and Development 28 (4), 86–88, 98.

Yan, J.-Xi., Liu, H., Ni, F.-T., 2006. Examples of field polymer flooding tests. In: Shen, P.-P., Liu, Y.-Z., Liu, H.-R. (Eds.), Enhanced Oil Recovery–Polymer Flooding. Petroleum Industry Press, pp. 191–228.

Yan, Y.-L., Zheng, H.-M., Liu, B., Jiang, F.-J., 2005. Alternate polymer injection through the same well to improve injection profile in high permeability contrast layers. In: Yan, C.-Z., Li, Y. (Eds.), Tertiary Oil Recovery Symposium. Petroleum Industry Press, pp. 42–47.

Yang, A-P., 1990. Stochastic Heterogeneity and Dispersion. Ph.D. dissertation, University of Texas at Austin.

Yang, B., Feng, G.-B., 2001. Pilot test of polymer flooding in the Wangchang oilfield in Jianghan. Oil and Gas Surface Engineering 20 (3), 1–3.

Yang, C.-Z., Yan, H.-K., Li, G.-Z., Cui, G.-Z., Yuan, H., 2002a. Adsorption mechanisms and chromatographic separation. In: Yu, J.-Y., Song, W.-C., Li, Z.-P., et al. (Eds.), Fundamentals and Advances in Combined Chemical Flooding, Chapter 6. China Petrochemical Press.

Yang, F.-L., Wang, D.-M., Wang, G., Sui, X.-G., Liu, W.-J., Kan, C.-L., 2006. Study on high-concentration polymer flooding to further enhance oil recovery. Paper SPE 101202 presented at the SPE Annual Technical Conference and Exhibition, San Antonio, 24–27 September.

Yang, H.-J., Ni, F.-T., 1998. Experimental study of the mechanism of blocking polymer channeling. In: Gang, Q.-L., et al. (Eds.), Chemical Flooding Symposium–Research Results during the Eighth Five-Year Period (1991–1995), Vol. I. Petroleum Industry Press, pp. 331–335.

Yang, S.H., Treiber, L.E., 1985. Chemical stability of polyacrylamide under simulated field conditions. Paper SPE 14232 presented at the SPE Annual Technical Conference and Exhibition, Las Vegas, 22–26 September.

Yang, H.-W., Zhang, Z.-C., Feng, Q.-Y., Xu, B.-Y., Liu, M., Li, C.-Z., Zhang, J.-Z., 1996. Field cases of polymer flooding. In: Zhang, Z.-H. (Ed.), Pilot Tests of Polymer Flooding. Petroleum Industry Press, pp. 196–242.

Yang, L., Me, S.-C., 2006. Research and testing of combinations of chemical enhanced oil recovery. In: Cao, H.-Q. (Ed.), Research and Development of Enhanced Oil Recovery in Daqing. Petroleum Industry Press, pp. 326–351.

Yang, P.-H., et al., 1992. Oil recovery mechanisms of alkaline/polymer flooding. Paper presented at the Second and Tertiary Recovery Symposium, Chengdu.

Yang, P.-H., Wang, H.-Z., Li, Z.P., Yan, H.-K., 2002b. Synergism in combined chemical systems. In: Yu, J.-Y., Song, W.-C., Li, Z.-P., et al. (Eds.), Fundamentals and Advances in Combined Chemical Flooding, Chapter 11. China Petrochemical Press.

Yang, S.-Y., 2000. Mechanisms of Wettability for Crude Oil/Brine/Mica System. Ph.D. dissertation, Rice University, Houston.

Yang, Z.-J, Zhu, Y., Ma, X.-Y., Li, X.-H., Fu, J., Jiang, P., Xu, Q.-Y., 2002c. ASP Pilot test in Guodao Field. J. Jianghan petroleum Institute 24 (1), 62–64.

Yang, Z.-Y., Yao, S.-C., 2006. Research and testing of combined chemical enhanced oil recovery, In: Cao, H.-C. (Ed.), Research and Practices of Enhanced Oil Recovery in Daqing Oil Field. Petroleum Industry Press, pp. 326–351.

Yao, H.-X., Zheng, J., Li, L.-C., Zheng, J.-P., 2005. A feasibility study of crosslinked polymer injection into a bottom and edge aquifer drive reservoir. In: Yan, C.-Z., Li, Y. (Eds.), Tertiary Oil Recovery Symposium. Petroleum Industry Press, pp. 36–41.

Ye, Z.-P., Peng, K.-Z., 1995. Methods to measure relative permeability curves in ASP systems. J. Southwestern Petroleum Institute 17 (3), 85–91.

Yi, G.C., Gu, Y.-Q., Gu, Y.-T., Zheng, T.-C., 1999. Performance analysis of polymer flooding in Gudong Block 8. Oil and Gas Recovery Technology 6 (4), 13–18.

Yildiz, H.O., Morrow, N.R., 1996. Effect of brine composition on recovery of Moutray crude oil by waterflooding. J. Pet. Sci. Eng. 14, 159–168.

Yin, H.J., Wang, D.M., Zhang, H., 2006. Study of flow behaviors of viscoelastic polymer solution in Micropore with dead end. Paper SPE 101950 presented at the SPE Annual Technical Conference and Exhibition, San Antonio, 24–27 September.

Yu, J.-Y., Tang, J.-A., Zhang, L., Li, Z.-Q., Ma, H.-D., 2002. Properties of interface and oil-water interfacial film in combined chemical systems. In: Yu, J.-Y., Song, W.-C., Li, Z.-P., et al. (Eds.), Fundamentals and Advances in Combined Chemical Flooding, Chapter 4. China Petrochemical Press.

Zaitoun, A., Kohler, N., 1987. The role of adsorption in polymer propagation through reservoir rock. Paper SPE 16274 presented at the SPE International Symposium on Oilfield Chemistry, San Antonio, 4–6 February.

Zaitoun, A., Kohler, N., 1988. Two phase flow through porous media: Effect of adsorbed polymer layer. Paper SPE 18085 presented at the SPE 63rd Annual Technical Conference, Houston, 2–5 October.

Zaitoun, A., Potie, B., 1983. Limiting conditions for the use of hydrolyzed polyacrylamides in brines containing divalent ions. Paper SPE 11785 presented at the SPE Oilfield and Geothermal Chemistry Symposium, Denver, 1–3 June.

Zaitoun, A., Tabary, R., Rousseau, D., Pichery, T., Nouyoux, S., Mallo, P., Braun, O., 2007. Using microgels to shutoff water in gas storage wells. Paper SPE106042 presented at the SPE International Symposium on Oilfield Chemistry, Houston, 28 February–2 March.

Zerpa, L.E., Queipoa, N.V., Pintosa, T.S., Salagerb, J.L., 2005. An optimization methodology of alkaline-surfactant-polymer flooding processes using field scale numerical simulation and multiple surrogates. J. Pet. Sci. Eng. 47, 197–208.

Zhang, A.-M., Cao, X.-R., Li, X.-L., Jiang, Y.-B., 2004. Program optimization of the surfactant-polymer pilot test in Shengli oil zone. J. Xinjiang Petroleum Institute 16 (3), 40–43.

Zhang, D.L., Liu, S., Puerto, M., Miller, C.A., Hirasaki, G.J., 2006. Wettability alteration and spontaneous imbibition in oil-wet carbonate formations, J. Pet. Sci. Eng. 52, 213–226.

Zhang, J.-C., 1998. Learning in several aspects of polymer flooding in Daqing. In: Gang, Q.-L., et al. (Eds.), Chemical Flooding Symposium–Research Results during the Eighth Five-Year Period (1991–1995), Vol. I. Petroleum Industry Press, pp. 60–70.

Zhang, J.-F, Wang, K.-L., He, F.-Y., Zhang, F.-L., 1999. Ultimate evaluation of the alkali/polymer combination flooding pilot test in XingLongTai Oil Field. Paper SPE 57291 presented at the SPE Asia Pacific Improved Oil Recovery Conference, Kuala Lumpur, 25–26 October.

Zhang, J-.Y., Yang, P.-H., 1998. HPAM molecular weight compatibility with rock permeability. In: Gang, Q.-L., et al. (Eds.), Chemical Flooding Symposium–Research Results during the Eighth Five-Year Period (1991–1995), Vol. I. Petroleum Industry Press, pp. 150–154.

Zhang, L., Zhao, S., Liu, L., Yu., J.-Y., 1998a. Separation, analysis and interfacial activity of acidic components in Gudong crude oil. Oilfield Chemistry 15 (4), 344–347.

Zhang, L., Zhao, S., Liu, L., Yu., J.-Y., 1998b. A study of synergy between the interfacial tension between alkaline-surfactant solution and Gudong crude oil. Oilfield Chemistry 15 (4), 348–353.

Zhang, P., Tweheyo, M.T., Austad, T., 2007a. Wettability alteration and improved oil recovery by spontaneous imbibition of seawater into chalk: Impact of the potential determining ions Ca^{2+}, Mg^{2+}, and SO_4^{2+}. Colloids and Surfaces A: Physicochemical Engineering Aspects 301, 199–208.

Zhang, R., Dan, D.-S., Peng, L., Cai, J.-J., Hu, Y.-T., 2003. Investigation of the factors causing the difference in the effectiveness of polymer flooding in Xia-er-men field. Henan Petroleum 17 (supplemental), 40–42.

Zhang, R., Somasundaran, P., 2006. Advances in adsorption of surfactants and their mixtures at solid/solution interfaces. Advance of Colloid and Interface Science (Adv. Colloid Interface Sci.) 123–126, 213–229.

Zhang, Y., Morrow, N.R., 2006. Comparison of secondary and tertiary recovery with change in injection brine composition for crude oil/sandstone combinations. Paper SPE 99757 presented at SPE/DOE Symposium on Improved Oil Recovery, Tulsa, 22–26 April.

Zhang, Y., Xie, X., Morrow, N.R., 2007b. Waterflood performance by injection of brine with different salinity for reservoir cores. Paper SPE 109849 presented at the 2007 SPE Annual Technical Conference and Exhibition, Anaheim, 11–14 November.

Zhang, Z.-L., Yang, Y.-M., Hong, L., Peng, T., Xuan, M.-R., 2001. ASP pilot test in the Honggang field. J. Southwest Petroleum Institute 23 (3), 47–49, 57.

Zhao, C.-J., Ma, C.-J., Yang, Z.-Y., Yao, S.-C., 2005a. Pilots of ultra-low-IFT foaming flooding. Petroleum Exploration and Development 32 (1), 127–130.

Zhao, F.-L., 1991. Chemistry in Oil Production. University of Petroleum, China.

Zhao, P., Jackson, A.C., Britton, C., Kim, D.H., Britton, L.N., Levitt, D.B., Pope, G.A., 2008. Development of high-performance surfactants for difficult oils. Paper SPE 113432 presented at SPE/DOE Improved Oil Recovery Symposium, Tulsa, 19–23 April.

Zhao, R.B., Yue, X.A., Zhang, H.F., 2005b. Flow behavior of 2-Acrylamide-2-Methyl Propane-Sulfonic acid copolymer solution in porous medium. ACTA PETROLEI SINICA 26 (2), 85–87, 91.

Zhao, S., Zhang, L., Luo, L., Yu, J.Y., 2002. Synergy between displacement agents and the active components of crude oils. In: Yu, J.-Y., Song, W.-C., Li, Z.-P., et al. (Eds.), Fundamentals and Advances in Combined Chemical Flooding, Chapter 2. China Petrochemical Press.

Zhao, T.H., Pan, G.F., Guo, S.F., Hu, X.Q., 2006. Synthesis and evaluation of AMPS/AM. Journal of Southwest Petroleum Institute 28 (4), 82–84.

Zhou, H., Huang, R.H., 1997. Study of the solution properties of water-soluble hydrophobically associating Acrylamide/n-Octylacrylate copolymers. Oilfield Chemistry 14 (3), 252–256.

Zhou, S.W., Han, M., Xiang, W.T., Zhang, J., Zhou, W., Jiang, W., 2006. Application of polymer flooding technology in a Bohai Offshore Field. China Offshore Oil and Gas 18 (6), 386–412.

Zhou, Y.-P., Liu, Q.-C., Liu, B.-L., Zhao, L.-H., Pang, S.-B., 2009. Laboratory investigation of "2+3" recovery mechanisms to increase oil recovery in light oil reservoirs. Chemical Engineer 165 (6), 58–62.

Zhu, L., Zheng, Y.-X., 1998. Preliminary analysis of the polymer viscosity loss in an injection system. In: Gang, Q.-L., et al. (Eds.), Chemical Flooding Symposium—Research Results during the Eighth Five-Year Period (1991–1995), Vol. I. Petroleum Industry Press, pp. 395–397.

Ziegler, V.M., 1988. Laboratory investigation of high temperature surfactant flooding. SPERE 3 (2), 586–596.

Index

A

Note: Page numbers followed by f indicate figures, by t indicate tables, and by b indicate examples.

Modern Chemical Enhanced Oil Recovery. DOI: 10.1016/B978-1-85617-745-0.00016-4

B

C

R

T

X

Y

Z

Kolibri 360 GmbH
Ferdinand-Jühlke-Straße 7
99095 Erfurt, Deutschland
produktsicherheit@kolibri360.de